Carlos S. Kubrusly

# Elements of Operator Theory

Birkhäuser
Boston • Basel • Berlin

Carlos S. Kubrusly
Catholic University
R. Marques de S. Vicente 225
22453-900, Rio de Janeiro, Brazil
e-mail:carlos@ele.puc-rio.br

**Library of Congress Cataloging-in-Publication Data**

Kubrusly, Carlos S., 1947-
   Elements of operator theory / Carlos S. Kubrusly
      p. cm.
   Includes bibliographical references and index.
   ISBN 0-8176-4174-2  (acid-free paper) – ISBN 3-7643-4174-2 (acid-free paper)
      1. Operator theory.  I. Title.
   QA329.K79 2001
   515'.724–dc21

                                                               2001018439

AMS Subject Classifications: 47-XX, 47-01, 47A05, 47A10, 47A12, 47A15, 47A75, 47B10, 47B15, 47B20, 47B37, 47C05, 47L05, 46-XX, 46-01,46A22, 46A30, 46B10, 46B15, 46B20, 46B45, 46B50, 46C05,46C07, 46C15, 54-XX, 54-01, 54A20, 54B05, 54B10, 54B15, 54C05, 54C20, 54E35, 54E45, 54E50, 54E52, 15-XX, 15-01, 15A03, 15A04, 03Exx, 03E10, 03E20

Printed on acid-free paper.
©2001 Birkhäuser Boston *Birkhäuser* 

ISBN 0-8176-4174-2    SPIN 10754156
ISBN 3-7643-4174-2

Reformatted from author's files in LATEX2ε by TEXniques, Inc., Cambridge, MA
Printed and bound by Hamilton Printing, Rensselaer, NY
Printed in the United States of America

9 8 7 6 5 4 3 2 1

*To the memory of my father*

*The truth, he thought, has never been of any real value to any human being — it is a symbol for mathematicians and philosophers to pursue. In human relations kindness and lies are worth a thousand truths. He involved himself in what he always knew was a vain struggle to retain the lies.*

Graham Greene

# Preface

"Elements" in the title of this book has its standard meaning, namely, basic principles and elementary theory. The main focus is operator theory, and the topics range from sets to the spectral theorem. Chapter 1 (*Set-Theoretic Structures*) introduces the reader to ordering, lattices and cardinality. Linear spaces are presented in Chapter 2 (*Algebraic Structures*). Metric (and topological) spaces in Chapter 3 (*Topological Structures*). The purpose of Chapter 4 (*Banach Spaces*) is to put algebra and topology to work together. Continuity plays a central role in the theory of topological spaces, and linear transformation plays a central role in the theory of linear spaces. When algebraic and topological structures are compatibly laid on the same underlying set, leading to the notion of topological vector spaces, then we may consider the concept of continuous linear transformations. By an operator we mean a continuous linear transformation of a normed space into itself. Chapter 5 (*Hilbert Spaces*) is central. There a geometric structure is properly added to the algebraic and topological structures. The spectral theorem is a cornerstone in the theory of operators on Hilbert spaces. It gives a full statement on the nature and structure of normal operators, and is considered in Chapter 6 (*The Spectral Theorem*).

The book is addressed to graduate students, both in mathematics or in one of the sciences, and also to working mathematicians getting into operator theory and scientists willing to apply operator theory to their own subject. In the former case it actually is a first course. In the latter case it may serve as a basic reference on the so-called elementary theory of single operator theory. Its primary intention is to introduce operator theory to a new generation of students and provide the necessary background for it. Technically, the prerequisite for this book is some mathematical maturity that a first-year graduate student in mathematics, engineering or in one of the formal sciences is supposed to have already acquired. The book is largely self-contained. Of course, a formal introduction to analysis will be helpful, as well as an introductory course on functions of a complex variable. Measure and integration are not required up to the very last section of the last chapter. Each section of each chapter has a short and concise (sometimes a compound) title. They were selected in such a way that, when put together in the contents, give a brief outline of the book to the right audience.

The focus of this book is on concepts and ideas as an alternative to the computational approach. The proofs avoid computation whenever possible or convenient. Instead, I try to unfold the structural properties behind the statements of theorems, stressing mathematical ideas rather than long calculations. Tedious and ugly (all right, "ugly" is subjective) calculations were avoided where a more conceptual way to explain the stream of ideas was possible. Clearly, this is not new. In any event, every single proof in this book was specially tailored to meet this requirement but they (at least the majority of them) are standard proofs, perhaps with a touch of what may reflect some of the author's minor idiosyncrasies. In writing this book I kept my mind focused on the reader. Sometimes I am talking to my students and sometimes to my colleagues (they surely will identify in each case to whom I am talking). For my students the objective is to teach mathematics (ideas, structures and problems).

There are 300 problems throughout the book, many of them with multiple parts. These problems, at the end of each chapter, comprise complements and extensions of the theory, further examples and counterexamples, or auxiliary results that may be useful in the sequel. They are an integral part of the main text, which makes them different from traditional classroom exercises. Many of these problems are accompanied by hints, which may be a single word or a sketch, sometimes long, of a proof. The idea behind providing these long and detailed hints is that just talking to students is not enough. One has to motivate them too. In my view, motivation (in this context) is to reveal the beauty of pure mathematics, and to challenge students with a real chance to reconstruct a proof for a theorem that is "new" to them. Such a real chance can be offered by a suitable, sometimes rather detailed, hint.

At the end of each chapter, just before the problems, the reader will find a list of suggested readings that contains only books. Some of them had a strong influence in preparing this book, and many of them are suggested as a second or third reading. The reference section comprises a list of all those books and just a few research papers (82 books and 11 papers), all of them quoted in the text. Research papers are only mentioned to complement occasional historical remarks so that the few articles cited there are, in fact, classical breakthroughs. For a glance at current research in operator theory the reader is referred to recent research monographs suggested in Chapters 5 and 6.

I started writing this book after lecturing on its subject at Catholic University of Rio de Janeiro for over 20 years. In general, the material is covered in two one-semester beginning graduate courses, where the audience comprises mathematics, engineering, economics and physics students. Quite often senior undergraduate students joined the courses. The dividing line between these two one-semester courses depends a bit on the pace of lectures but is usually somewhere at the beginning of Chapter 5. Questions asked by generations of students and colleagues have been collected. When the collection was big enough some former students, as well as current students, insisted upon a new book but urged that it should not be a mere collection of lecture notes and exercises bound together. I hope not to disappoint them so much.

At this point, where a preface is coming to an end, one has the duty and pleasure to acknowledge the participation of those people who somehow effectively contributed in connection with writing the book. Certainly, the students in those courses were a big help and a source of motivation. Some friends among students and colleagues have collaborated by discussing the subject of this book for a long time many on occasions. They are: Gilberto O. Corrêa, Oswaldo L. V. Costa, Giselle M. S. Ferreira, Marcelo D. Fragoso, Ricardo S. Kubrusly, Abilio P. Lucena, Helios Malebranche, Carlos E. Pedreira, Denise O. Pinto, Marcos A. da Silveira, and Paulo César M. Vieira. Special thanks are due to my friend and colleague Augusto C. Gadelha Vieira who read part of the manuscript and made many valuable suggestions. I am also grateful to Ruth F. Curtain who, back in the early seventies, introduced me to functional analysis. I wish to thank Catholic University of Rio de Janeiro for providing the release time that made this project possible. Let me also thank the staff of Birkhäuser Boston and Elizabeth Loew of TeXniques for their ever efficient and friendly partnership. Finally, it is just fair to mention that this project was supported in part by CNPq (Brazilian National Research Council) and FAPERJ (Rio de Janeiro State Research Council).

Carlos S. Kubrusly
Rio de Janeiro
November, 2000

# Contents

# 1
# Set-Theoretic Structures

The purpose of this chapter is to present a very brief review of some basic set-theoretic concepts that will be needed in the sequel. By basic concepts we mean standard notation and terminology, and a few essential results that will be required in later chapters. We assume the reader is familiar with the notion of *set* and *elements* (or *members*, or *points*) of a set, as well as with the basic set operations. It is convenient to reserve certain symbols for certain sets, especially for the basic *number systems*. The set of all *nonnegative integers* will be denoted by $\mathbb{N}_0$, the set of all *positive integers* (i.e., the set of all *natural numbers*) by $\mathbb{N}$, and the set of all *integers* by $\mathbb{Z}$. The set of all *rational numbers* will be denoted by $\mathbb{Q}$, the set of all *real numbers* (or the *real line*) by $\mathbb{R}$, and the set of all *complex numbers* by $\mathbb{C}$.

## 1.1 Background

We shall also assume that the reader is familiar with the basic rules of elementary (classical) logic, but acquaintance with formal logic is not necessary. The foundations of mathematics will not be reviewed in this book. However, before starting our brief review of set-theoretic concepts, we shall introduce some preliminary notation, terminology and logical principles as a background for our discourse.

If a *predicate* $P(\ )$ is meaningful for a subject $x$, then $P(x)$ (or simply $P$) will denote a *proposition*. The terms *statement* and *assertion* will be used as synonyms for proposition. A statement on statements is sometimes called a *formula* (or a *secondary proposition*). Statements may be *true* or *false* (not true). A *tautology* is a formula that is true regardless of the truth of the statements in it. A *contradiction* is a formula that is false regardless of the truth of the statements in it. The symbol $\Rightarrow$ denotes *implies* and the formula $P \Rightarrow Q$ (whose logical definition is "either $P$ is false or $Q$ is true") means "the statement $P$ implies the statement $Q$". That is, "*if $P$ is true, then $Q$ is true*", or "$P$ is a *sufficient condition* for $Q$". We shall also use the symbol $\nRightarrow$ for the denial of $\Rightarrow$, so that $\nRightarrow$ denotes *does not imply* and the formula $P \nRightarrow Q$ means "the statement $P$ does not imply the statement $Q$". Accordingly, let $\not P$ stand for the denial of $P$ (read: *not $P$*). If $P$ is a statement, then $\not P$ is its *contradictory*.

Let us first recall one of the basic rules of deduction called *modus ponens*: "if a statement $P$ is true and if $P$ implies $Q$, then the statement $Q$ is true" — "anything implied by a true statement is true". Symbolically, $\{P \text{ true and } P \Rightarrow Q\} \Longrightarrow \{Q \text{ true}\}$. A *direct proof* is essentially a chain of modus ponens. For instance, if $P$ is true, then the string of implications $P \Rightarrow Q \Rightarrow R$ ensures that $R$ is true. Indeed, if we can establish that $P$ holds, and also that $P$ implies $Q$, then (modus ponens) $Q$ holds. Moreover, if we can also establish that $Q$ implies $R$, then (modus ponens again) $R$ holds.

However, modus ponens alone is not enough to ensure that such a reasoning may be extended to an arbitrary (endless) string of implications. In certain cases the *Principle of Mathematical Induction* provides an alternative reasoning. Let $\mathbb{N}$ be the set of all natural numbers. A set $S$ of natural numbers is called *inductive* if $n + 1$ is an element of $S$ whenever $n$ is. The Principle of Mathematical Induction states that "if 1 is an element of an inductive set $S$, then $S = \mathbb{N}$". This leads to a second scheme of proof, called *proof by induction*. For instance, for each natural number $n$ let $P_n$ be a proposition. If $P_1$ holds true and if $P_n \Rightarrow P_{n+1}$ for each $n$, then $P_n$ holds true for every natural number $n$. The scheme of proof by induction works for $\mathbb{N}$ replaced with $\mathbb{N}_0$. There is nothing magical about the number 1 as far as a proof by induction is concerned. All that is needed is a "beginning" and the notion of "induction". Example: Let $i$ be an arbitrary integer and let $\mathbb{Z}_i$ be the set made up of all integers greater than or equal to $i$. For each integer $k$ in $\mathbb{Z}_i$ let $P_k$ be a proposition. If $P_i$ holds true and if $P_k \Rightarrow P_{k+1}$ for each $k$, then $P_k$ holds true for every integer $k$ in $\mathbb{Z}_i$ (particular cases: $\mathbb{Z}_0 = \mathbb{N}_0$ and $\mathbb{Z}_1 = \mathbb{N}$).

"If a statement leads to a contradiction, then this statement is false." This is the rule of a *proof by contradiction — reductio ad absurdum*. It relies on the *Principle of Contradiction*, which states that "$P$ and $\not P$ are impossible". In other words, the Principle of Contradiction says that the formula "$P$ and $\not P$" is a contradiction. But this alone does not ensure that any of $P$ or $\not P$ must hold. The *Law of the Excluded Middle* (or *Law of the Excluded Third — tertium non datur*) does: "either $P$ or $\not P$ hold". That is, the Law of the Excluded Middle simply says that the formula "$P$ or $\not P$" is a tautology. Therefore, the formula $\not Q \Rightarrow \not P$ means "$P$ holds *only if* $Q$ holds", or "$Q$ is a *necessary condition* for $P$". If $P \Rightarrow Q$ and $Q \Rightarrow P$, then we shall write $P \Leftrightarrow Q$ which means "$P$ *if and only if* $Q$", or "$P$ is a *necessary and sufficient condition* for $Q$", or "$P$ and $Q$ are *equivalent*" (and vice versa). Indeed, the formulas $P \Rightarrow Q$ and $\not Q \Rightarrow \not P$ are equivalent: $\{P \Rightarrow Q\} \Longleftrightarrow \{\not Q \Rightarrow \not P\}$. Such an equivalence is the basic idea behind a *contrapositive proof*: "to verify that a proposition $P$ implies a proposition $Q$ prove, instead, that the denial of $Q$ implies the denial of $P$".

We conclude this introductory section by pointing out another usual but slightly different meaning for the term "proposition." We shall often say "prove the following proposition" instead of "prove that the following proposition holds true". Here the term *proposition* is being used as a synonym for *theorem* (a true statement for which we demand a proof of its truth), and not as a synonym for an assertion or statement (that may be either true or false). A *conjecture* is a statement that has not been proved yet — it may turn out to be either true or false once a proof of its truth or falsehood is supplied. If a conjecture is proved to be true, then it becomes a theorem. Note that there is no "false theorem" — if it is false, it is not a theorem. Another synonym for theorem is *lemma*. There is no logical difference among the terms "theorem", "lemma" and "proposition" but it is usual to endow them with a psychological hierarchy. Generally, a theorem is supposed to bear a greater importance (which is subjective) and a lemma is often viewed as an intermediate theorem (which may be very important indeed) that will be applied to prove a further theorem. Propositions are sometimes placed a step below, either as an isolated theorem or as an auxiliary result. A *corollary* is, of course, a theorem that comes out as a consequence of a previously proved theorem (i.e., whose proof is mainly based on an application of that previous theorem). Unlike "conjecture", "proposition", "lemma", "theorem" and "corollary", the term *axiom* (or *postulate*) is applied to a fundamental statement (or assumption, or hypothesis) upon which a theory (i.e., a set of theorems) is built. Clearly, a set of axioms (or, more appropriately, a system of axioms) should be *consistent* (i.e., they should not lead to a contradiction), and they are said to be *independent* if none of them is a theorem (i.e., if none of them can be proved by the remaining axioms).

## 1.2   Sets and Relations

If $x$ is an element of a set $X$, then we shall write $x \in X$ (meaning that $x$ *belongs* to $X$, or $x$ is *contained* in $X$). Otherwise (i.e., if $x$ is not an element of $X$), $x \notin X$. We also write $A \subseteq B$ to mean that a set $A$ is a *subset* of a set $B$ ($A \subseteq B \iff \{x \in A \Rightarrow x \in B\}$). In such a case $A$ is said to be *included* in $B$. The *empty set*, which is a subset of every set, will be denoted by $\varnothing$. Two sets $A$ and $B$ are *equal* (notation: $A = B$) if $A \subseteq B$ and $B \subseteq A$. If $A$ is a subset of $B$ but not equal to $B$, then we say that $A$ is a *proper subset* of $B$ and write $A \subset B$. In such a case $A$ is said to be *properly included* in $B$. A *nontrivial subset* of a set $X$ is a nonempty proper subset of it. If $P(\ )$ is a predicate that is meaningful for every element $x$ of a set $X$ (so that $P(x)$ is a proposition for each $x$ in $X$), then $\{x \in X : \ P(x)\}$ will denote the subset of $X$ consisting of all those elements $x$ of $X$ for which the proposition $P(x)$ is true. The *complement* of a subset $A$ of a set $X$, denoted by $X \backslash A$, is the subset $\{x \in X : \ x \notin A\}$. If $A$ and $B$ are sets, the *difference* between $A$ and $B$, or the *relative complement* of $B$ in $A$, is the set

$$A \backslash B \ = \ \big\{x \in A : \ x \notin B\big\}.$$

We shall also use the standard notations $\cup$ and $\cap$ for *union* and *intersection*, respectively ($x \in A \cup B \iff \{x \in A$ or $x \in B\}$ and $x \in A \cap B \iff \{x \in A$ and $x \in B\}$). The sets $A$ and $B$ are *disjoint* if $A \cap B = \varnothing$ (i.e., if they have an empty intersection). The *symmetric difference* (or *Boolean sum*) of two sets $A$ and $B$ is the set

$$A \triangledown B \ = \ (A \backslash B) \cup (B \backslash A) \ = \ (A \cup B) \backslash (A \cap B).$$

The terms *class*, *family* and *collection* (as their related terms prefixed with "sub") will be used as synonyms for set (usually applied for sets of sets, but not necessarily) without imposing any hierarchy among them. If $\mathcal{X}$ is a collection of subsets of a given set $X$, then $\bigcup \mathcal{X}$ will denote the union of all sets in $\mathcal{X}$. Similarly, $\bigcap \mathcal{X}$ will denote the intersection of all sets in $\mathcal{X}$ (alternative notation: $\bigcup_{A \in \mathcal{X}} A$ and $\bigcap_{A \in \mathcal{X}} A$). An important statement about complements that exhibits the duality between union and intersection is the so-called *De Morgan laws*:

$$X \backslash \left( \bigcup_{A \in \mathcal{X}} A \right) = \bigcap_{A \in \mathcal{X}} (X \backslash A) \quad \text{and} \quad X \backslash \left( \bigcap_{A \in \mathcal{X}} A \right) = \bigcup_{A \in \mathcal{X}} (X \backslash A).$$

The *power set* of any set $X$, denoted by $\wp(X)$, is the collection of all subsets of $X$. Note that $\bigcup \wp(X) = X \in \wp(X)$ and $\bigcap \wp(X) = \varnothing \in \wp(X)$.

A *singleton* in a set $X$ is a subset of $X$ containing one and only one point of $X$ (notation: $\{x\} \subseteq X$ is a singleton on $x \in X$). A *pair* (or a *doubleton*) is a set containing just two points, say $\{x, y\}$, where $x$ is an element of a set $X$ and $y$ is an element of a set $Y$. A pair of points $x \in X$ and $y \in Y$ is an *ordered pair*, denoted by $(x, y)$, if $x$ is regarded as the first member of the pair and $y$ is regarded as the second. The *Cartesian product* of two sets $X$ and $Y$, denoted by $X \times Y$, is the set of all ordered pairs $(x, y)$ where $x \in X$ and $y \in Y$. A *relation* $R$ between two sets $X$ and $Y$ is any subset of the Cartesian product $X \times Y$. If $R$ is a relation between $X$ and $Y$, and if $(x, y)$ is a pair in $R \subseteq X \times Y$, then we say that $x$ is *related* to $y$ under $R$ (or $x$ and $y$ are related by $R$), and write $x R y$ (instead of $(x, y) \in R$). Tautologically, for any ordered pair $(x, y) \in X \times Y$, either $(x, y) \in R$ or $(x, y) \notin R$ (i.e., either $x R y$ or $x \not{R} y$). A relation between a set $X$ and itself is called a *relation on $X$*. If $X$ and $Y$ are sets and if $R$ is a relation between $X$ and $Y$, then the *graph* of the relation $R$ is the subset of $X \times Y$

$$G_R \ = \ \big\{(x, y) \in X \times Y : \ x R y\big\}.$$

A relation $R$ clearly coincides with its graph $G_R$.

## 1.3   Functions

Let $x$ be an arbitrary element of a set $X$ and let $y$ and $z$ be arbitrary elements of a set $Y$. A relation $F$ between the sets $X$ and $Y$ is a *function* if $x F y$ and $x F z$ imply $y = z$. In other words, a relation $F$ between a set $X$ and a set $Y$ is called a *function from $X$ to $Y$* (or a *mapping of $X$ into $Y$*) if for each $x \in X$ there exists a *unique* $y \in Y$ such that $x F y$. The terms *map* and *transformation* are often used as synonyms for function and mapping. (Sometimes the terms *correspondence* and *operator* are also used but we shall keep them for special kinds of functions.) It is usual to write $F : X \to Y$ to indicate that $F$ is a mapping of $X$ into $Y$, and $y = F(x)$ (or $y = Fx$) instead of $x F y$. If $y = F(x)$, we say that $F$ *maps $x$ to $y$*, so that $F(x) \in Y$ is the *value* of the function $F$ at $x \in X$. Equivalently, $F(x)$, which is a point in $Y$, is the *image* of the point $x$ in $X$ under $F$. It is also customary to use the abbreviation "the function $X \to Y$ defined by $x \mapsto F(x)$" for a function from $X$ to $Y$ that assigns to each $x$ in $X$ the value $F(x)$ in $Y$. A *$Y$-valued function on $X$* is precisely a function from $X$ to $Y$. If $Y$ is a subset of the set $\mathbb{C}$, $\mathbb{R}$ or $\mathbb{Z}$, then *complex-valued function, real-valued function* or *integer-valued function*, respectively, are usual terminologies. An $X$-valued function on $X$ (i.e., a function $F : X \to X$ from $X$ to itself) is referred to as a *function on $X$*. The collection of all functions from a set $X$ to a set $Y$ will be denoted by $Y^X$. Indeed, $Y^X \subseteq \wp(X \times Y)$.

Consider a function $F : X \to Y$. The set $X$ is called the *domain* of $F$ and the set $Y$ is called the *codomain* of $F$. If $A$ is a subset of $X$, then the *image of $A$ under $F$*, denoted by $F(A)$, is the subset of $Y$ consisting of all points $y$ of $Y$ such that $y = F(x)$ for some $x \in A$:

$$F(A) = \big\{ y \in Y : \; y = F(x) \text{ for some } x \in A \subseteq X \big\}.$$

On the other hand, if $B$ is a subset of $Y$, then the *inverse image of $B$ under $F$* (or the *pre-image of $B$ under $F$*), denoted by $F^{-1}(B)$, is the subset of $X$ made up of all points $x$ in $X$ such that $F(x)$ lies in $B$:

$$F^{-1}(B) = \big\{ x \in X : \; F(x) \in B \subseteq Y \big\}.$$

The *range* of $F$, denoted by $\mathcal{R}(F)$, is the image of $X$ under $F$. Thus

$$\mathcal{R}(F) = F(X) = \big\{ y \in Y : \; y = F(x) \text{ for some } x \in X \big\}.$$

If $\mathcal{R}(F)$ is a singleton, then $F$ is said to be a *constant* function. If the range of $F$ coincides with the codomain (i.e., if $F(X) = Y$), then $F$ is a *surjective* function. In this case $F$ is said to map *$X$ onto $Y$*. The function $F$ is *injective* (or $F$ is a *one-to-one* mapping) if its domain $X$ does not contain two elements with the same image. In other words, let $x$ and $x'$ be arbitrary elements of $X$. A function $F : X \to Y$ is injective if $F(x) = F(x')$ implies $x = x'$. A *one-to-one correspondence* between a set $X$ and a set $Y$ is a one-to-one mapping of $X$ onto $Y$. That is, a surjective and injective function (also called a *bijective* function).

If $A$ is an arbitrary subset of $X$ and $F$ is a mapping of $X$ into $Y$, then the function $G: A \to Y$ such that $G(x) = F(x)$ for each $x \in A$ is *the restriction of $F$ to $A$*. Conversely, if $G: A \to Y$ is the restriction of $F: X \to Y$ to some subset $A$ of $X$, then $F$ is *an extension of $G$ over $X$*. It is usual to write $G = F|_A$. Note that $\mathcal{R}(F|_A) = F(A)$.

Let $A$ be a subset of $X$ and consider a function $F: A \to X$. An element $x$ of $A$ is a *fixed point* of $F$ (or *$F$ leaves $x$ fixed*) if $F(x) = x$. The function $J: A \to X$ defined by $J(x) = x$ for every $x \in A$ is the *inclusion map* (or the *embedding*, or the *injection*) of $A$ into $X$. In other words, the inclusion map of $A$ into $X$ is the function $J: A \to X$ that leaves each point of $A$ fixed. The inclusion map of $X$ into $X$ is called the *identity map on $X$* and denoted by $I$, or by $I_X$ when necessary (i.e., the identity on $X$ is the function $I: X \to X$ such that $I(x) = x$ for every $x \in X$). Thus the inclusion map of a subset of $X$ is the restriction to that subset of the identity map on $X$. Now consider a function on $X$; that is, a mapping $F: X \to X$ of $X$ into itself. A subset of $X$, say $A$, is *invariant for $F$* (or *invariant under $F$*, or *$F$-invariant*) if $F(A) \subseteq A$. In this case the restriction of $F$ to $A$, $F|_A: A \to X$, has its range included in $A$: $\mathcal{R}(F|_A) = F(A) \subseteq A \subseteq X$. Therefore we shall often think of the restriction of $F: X \to X$ to an invariant subset $A \subseteq X$ as a mapping of $A$ into itself: $F|_A: A \to A$. It is in this sense that the inclusion map of a subset of $X$ can be thought of as the identity map on that subset: they differ only in that one has a larger codomain than the other.

Let $F: X \to Y$ be a function from a set $X$ to a set $Y$, and let $G: Y \to Z$ be a function from the set $Y$ to a set $Z$. Since the range of $F$ is included in the domain of $G$, $\mathcal{R}(F) \subseteq Y$, consider the restriction of $G$ to the range of $F$, $G|_{\mathcal{R}(F)}: \mathcal{R}(F) \to Z$. The *composition* of $G$ and $F$, denoted by $G \circ F$ (or simply by $GF$), is the function from $X$ to $Z$ defined by $(G \circ F)(x) = G|_{\mathcal{R}(F)}(F(x)) = G(F(x))$ for every $x \in X$. It is usual to say that the diagram

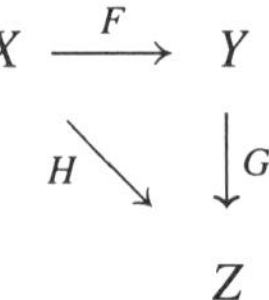

$$X \xrightarrow{\;F\;} Y$$
$$H \searrow \quad \downarrow G$$
$$Z$$

*commutes* if $H = G \circ F$. Although the above diagram is said to be *commutative* whenever $H$ is the composition of $G$ and $F$, the composition itself is not a commutative operation even when such a commutation makes sense. For instance, if $X = Y = Z$ and $F$ is a constant function on $X$, say $F(x) = a \in X$ for every $x \in X$, then $G \circ F$ and $F \circ G$ are constant functions on $X$ as well: $(G \circ F)(x) = G(a)$ and $(F \circ G)(x) = a$ for every $x \in X$. However $G \circ F$ and $F \circ G$ need not be the same (unless $a$ is a fixed point of $G$). Composition may not be commutative but it is always associative. If $F$ maps $X$ into $Y$, $G$ maps $Y$ into $Z$ and $K$ maps $Z$ into $W$, then we can consider the compositions $K \circ (G \circ F): X \to W$ and $(K \circ G) \circ F: X \to W$. It is readily verified that $K \circ (G \circ F) = (K \circ G) \circ F$. For this reason we may and shall drop the parentheses. In other words, the diagram

$$X \xrightarrow{\ F\ } Y$$

commutes (i.e., $H = G \circ F$, $L = K \circ G$, and $K \circ H = L \circ F$). If $F$ is a function
on set $X$, then the composition of $F: X \to X$ with itself, $F \circ F$, is denoted by $F^2$.
Likewise, for any positive integer $n \in \mathbb{N}$, $F^n$ denotes the composition of $F$ with
itself $n$ times, $F \circ \cdots \circ F: X \to X$, which is called the *n*th *power* of $F$. A function
$F: X \to X$ is *idempotent* if $F^2 = F$ (and hence $F^n = F$ for every $n \in \mathbb{N}$). It is
easy to show that the range of an idempotent function is precisely the set of all its
fixed points. In fact, $F = F^2$ if and only if $\mathcal{R}(F) = \{x \in X: \ F(x) = x\}$.

Suppose $F: X \to Y$ is an injective function. Thus, for an arbitrary element of
$\mathcal{R}(F)$, say $y$, there exists a *unique* element of $X$, say $x_y$, such that $y = F(x_y)$. This
defines a function from $\mathcal{R}(F)$ to $X$, $F^{-1}: \mathcal{R}(F) \to X$, such that $x_y = F^{-1}(y)$.
Hence $y = F(F^{-1}(y))$. On the other hand, if $x$ is an arbitrary element of $X$,
then $F(x)$ lies in $\mathcal{R}(F)$ so that $F(x) = F(F^{-1}(F(x)))$. Since $F$ is injective, $x = F^{-1}(F(x))$. Conclusion: For every injective function $F: X \to Y$ there exists a
(unique) function $F^{-1}: \mathcal{R}(F) \to X$ such that $F^{-1}F: X \to X$ is the identity on
$X$ (and $FF^{-1}: \mathcal{R}(F) \to \mathcal{R}(F)$ is the identity on $\mathcal{R}(F)$). $F^{-1}$ is called the *inverse
of $F$ on $\mathcal{R}(F)$: an injective function has an inverse on its range.* If $F$ is also surjective, then $F^{-1}: Y \to X$ is called the *inverse* of $F$. Thus an injective and surjective
function is also called an *invertible* function (in addition to its other names).

## 1.4   Equivalence Relations

Let $x$, $y$ and $z$ be arbitrary elements of a set $X$. A relation $R$ on $X$ is

> *reflexive* if $x R x$ for every $x \in X$,
>
> *transitive* if $x R y$ and $y R z$ imply $x R z$, and
>
> *symmetric* if $x R y$ implies $y R x$.

An *equivalence relation* on a set $X$ is a relation $\sim$ on $X$ that is reflexive, transitive
and symmetric. If $\sim$ is an equivalence relation on a set $X$, then the *equivalence
class* of an arbitrary element $x$ of $X$ (with respect to $\sim$) is the set

$$[x] = \{x' \in X: \ x' \sim x\}.$$

Given an equivalence relation $\sim$ on a set $X$, the *quotient space of $X$ modulo* $\sim$,
denoted by $X/\sim$, is the collection

$$X/\sim \ = \{[x] \subseteq X: \ x \in X\}$$

of the equivalence classes (with respect to $\sim$) of every $x \in X$. For each $x$ in $X$ set $\pi(x) = [x]$ in $X/\sim$. This defines a surjective map $\pi : X \to X/\sim$ which is called the *natural mapping* of $X$ onto $X/\sim$.

Let $\mathcal{X}$ be any collection of nonempty subsets of a set $X$. $\mathcal{X}$ *covers* $X$ (or $\mathcal{X}$ is a *covering* of $X$) if $X = \bigcup \mathcal{X}$ (i.e., if every point in $X$ belongs to some set in $\mathcal{X}$). $\mathcal{X}$ is *disjoint* if the sets in $\mathcal{X}$ are pairwise disjoint (i.e., $A \cap B = \varnothing$ for every pair of distinct sets $A$ and $B$ in $\mathcal{X}$). A *partition* of a set $X$ is a disjoint covering of $X$

Let $\approx$ be a an equivalence relation on a set $X$, and let $X/\approx$ be the quotient space of $X$ modulo $\approx$. It is clear that

$$X/\approx \ \text{ is a partition of } X.$$

Conversely, let $\mathcal{X}$ be any partition of a set $X$ and define a relation $\sim/\mathcal{X}$ on $X$ as follows: for every $x, x'$ in $X$, $x$ is related to $x'$ under $\sim/\mathcal{X}$ (i.e., $x \sim/\mathcal{X} x'$) if $x$ and $x'$ belong to the same set in $\mathcal{X}$. In fact,

$$\sim/\mathcal{X} \ \text{ is an equivalence relation on } X,$$

which is called the *equivalence relation induced* by a partition $\mathcal{X}$. It is readily verified that the quotient space of $X$ modulo the equivalence relation induced by the partition $\mathcal{X}$ coincides with $\mathcal{X}$ itself, just as the equivalence relation induced by the quotient space of $X$ modulo the equivalence relation $\approx$ on $X$ coincides with $\approx$. Symbolically,

$$X/(\sim/\mathcal{X}) = \mathcal{X} \quad \text{and} \quad \sim/(X/\approx) = \approx.$$

Thus an equivalence relation $\approx$ on $X$ induces a partition $X/\approx$ of $X$, which in turn induces back an equivalence relation $\sim/(X/\approx)$ on $X$ that coincides with $\approx$. On the other hand, a partition $\mathcal{X}$ of $X$ induces an equivalence relation $\sim/\mathcal{X}$ on $X$, which in turn induces back a partition $X/(\sim/\mathcal{X})$ of $X$ that coincides with $\mathcal{X}$. Conclusion: *The collection of all equivalence relations on a set $X$ is in a one-to-one correspondence with the collection of all partitions of $X$.*

## 1.5  Ordering

Let $x$ and $y$ be arbitrary elements of a set $X$. A relation $R$ on $X$ is

$$\textit{antisymmetric} \ \text{ if } x R y \text{ and } y R x \text{ imply } x = y.$$

A relation $\leq$ on a nonempty set $X$ is a *partial ordering* of $X$ if it is reflexive, transitive and antisymmetric. If $\leq$ is a partial ordering on a set $X$, the notation $x < y$ means $x \leq y$ and $x \neq y$. Moreover, $y > x$ and $y \geq x$ are just another way to write $x < y$ and $x \leq y$, respectively. Thus a *partially ordered set* is a pair $(X, \leq)$ where $X$ is a nonempty set and $\leq$ is a partial ordering of $X$ (i.e., a nonempty set equipped with a partial ordering on it). Warning: It may happen that $x \nleq y$ and $y \nleq x$ for some

$(x, y) \in X \times X$. Let $(X, \leq)$ be a partially ordered set, and let $A$ be a subset of $X$. Note that $(A, \leq)$ is a partially ordered set as well. An element $x \in X$ is an *upper bound* for $A$ if $y \leq x$ for every $y \in A$. Similarly, an element $x \in X$ is a *lower bound* for $A$ if $x \leq y$ for every $y \in A$. A subset $A$ of $X$ is *bounded above* in $X$ if it has an upper bound in $X$, and *bounded below* in $X$ if it has a lower bound in $X$. A *bounded* subset of $X$ is one that is bounded both above and below.

If a subset $A$ of a partially ordered set $X$ is bounded above in $X$ and if some upper bound of $A$ belongs to $A$, then this (*unique*) element of $A$ is the *maximum* of $A$ (or the *greatest* or *biggest* element of $A$), denoted by max $A$. Similarly, if $A$ is bounded below in $X$ and if some lower bound of $A$ belongs to $A$, then this (*unique*) element of $A$ is the *minimum* of $A$ (or the *least* or *smallest* element of $A$), denoted by min $A$. An element $x \in A$ is *maximal* in $A$ if there is no element $y \in A$ such that $x < y$ (equivalently, if $x \not< y$ for every $y \in A$). Similarly, an element $x \in A$ is *minimal* in $A$ if there is no element $y \in A$ such that $y < x$ (equivalently, if $y \not< x$ for every $y \in A$). Note that $x \not< y$ (or $y \not< x$) does not mean that $y \leq x$ (or $x \leq y$) so that the concepts of *a* maximal (or *a* minimal) element in $A$ and that of *the* maximum (or *the* minimum) element of $A$ do not coincide.

**Example 1A.** A collection of many (e.g., two) pairwise disjoint nonempty subsets of a nonempty set, equipped with the partial ordering defined by the inclusion relation $\subseteq$, has no maximum, no minimum, and every element in it is both maximal and minimal. On the other hand, the collection of all infinite subsets of an infinite set, whose complements are also infinite, has no maximal element in the inclusion ordering $\subseteq$. (The notion of infinite sets will be introduced later in Section 1.8 — for instance, the set all even natural numbers is an infinite subset of $\mathbb{N}$ which has an infinite complement).

Let $A$ be a subset of a partially ordered set $X$. Let $U_A \subseteq X$ be the set of all upper bounds of $A$, and let $V_A \subseteq X$ be the set of all lower bounds of $A$. If $U_A$ is nonempty and has a minimum element, say $u = \min U_A$, then $u \in U_A$ is called the *supremum* (or the *least upper bound*) of $A$ (notation: $u = \sup A$). Similarly, if $V_A$ is nonempty and has a maximum, say $v = \max V_A$, then $v \in V_A$ is called the *infimum* (or the *greatest lower bound*) of $A$ (notation: $v = \inf A$). A bounded set may not have a supremum or an infimum. However, if a set $A$ has a maximum (or a minimum), then $\sup A = \max A$ (or $\inf A = \min A$). Moreover, if a set $A$ has a supremum (or an infimum) *in* $A$, then $\sup A = \max A$ (or $\inf A = \min A$). If a pair $\{x, y\}$ of elements of a partially ordered set $X$ has a supremum or an infimum in $X$, then we shall use the following notation: $x \vee y = \sup\{x, y\}$ and $x \wedge y = \inf\{x, y\}$.

Let $F : X \to Y$ be a function from a set $X$ to a partially ordered set $Y$. Thus the range of $F$, $F(X) \subseteq Y$, is a partially ordered set. An *upper bound* for $F$ is an upper bound for $F(X)$, and $F$ is *bounded above* if it has an upper bound. Similarly, a *lower bound* for $F$ is a *lower bound* for $F(X)$, and $F$ is *bounded below* if it has lower bound. If a function $F$ is bounded both above and below, then it is said to be *bounded*. The *supremum* of $F$, $\sup_{x \in X} F(x)$, and the *infimum* of $F$, $\inf_{x \in X} F(x)$, are defined by $\sup_{x \in X} F(x) = \sup F(X)$ and $\inf_{x \in X} F(x) = \inf F(X)$. Now suppose $X$ also is partially ordered and take an arbitrary pair of points $x_1, x_2$ in $X$. $F$ is an *increasing* function if $x_1 \leq x_2$ in $X$ implies $F(x_1) \leq F(x_2)$ in $Y$, and *strictly increasing* if $x_1 < x_2$ in $X$ implies $F(x_1) < F(x_2)$ in $Y$. (For notational simplicity we are using the same symbol $\leq$ to denote both the partial ordering of $X$ and the partial ordering of $Y$). Note that $F$ *is strictly increasing if and only if it is increasing and injective.* In a similar way we can define *decreasing* and *strictly decreasing* functions between partially ordered sets. If a function is either decreasing or increasing, then it is said to be *monotone*.

## 1.6    Lattices

Let $X$ be a partially ordered set. If every pair $\{x, y\}$ of elements of $X$ is bounded above, then $X$ is a *directed set* (or the set $X$ is said to be *directed upward*). If every pair $\{x, y\}$ is bounded below, then $X$ is said to be *directed downward*. $X$ is a *lattice* if every pair of elements of $X$ has a supremum and an infimum in $X$ (i.e., if there exits a unique $u \in X$ and a unique $v \in X$ such that $u = x \vee y$ and $v = x \wedge y$ for every pair $x \in X$ and $y \in X$). A nonempty subset $A$ of a lattice $X$ that contains $x \vee y$ and $x \wedge y$ for every $x$ and $y$ in $A$ is a *sublattice* of $X$ (and hence a lattice itself). Every lattice is directed both upward and downward. If every bounded subset of $X$ has a supremum and an infimum, then $X$ is a *boundedly complete lattice*. If every subset of $X$ has a supremum and an infimum, then $X$ is a *complete lattice*. The following chain of implications

$$\textit{complete lattice} \Rightarrow \textit{boundedly complete lattice} \Rightarrow \textit{lattice} \Rightarrow \textit{directed set}$$

is clear enough, and none of them can be reversed. If $X$ is a complete lattice, then $X$ has a supremum and an infimum in $X$, which actually are the maximum and the minimum of $X$, respectively. Since $\min X \in X$ and $\max X \in X$, this shows that a complete lattice in fact is nonempty (even if this had not been assumed when we defined a partially ordered set). Likewise, the empty set $\varnothing$ of a complete lattice $X$ has a supremum and an infimum. Since every element of $X$ is both an upper and a lower bound for $\varnothing$, it follows that $U_\varnothing = V_\varnothing = X$. Hence $\sup \varnothing = \min X$ and $\inf \varnothing = \max X$.

**Example 1B.** The power set $\wp(X)$ of a set $X$ is a complete lattice in the inclusion ordering $\subseteq$, where $A \vee B = A \cup B$ and $A \wedge B = A \cap B$ for every pair $\{A, B\}$ of subsets of $X$. In this case, $\sup \varnothing = \min \wp(X) = \varnothing$ and $\inf \varnothing = \max \wp(X) = X$.

**Example 1C.** The real line $\mathbb{R}$ with its natural ordering $\leq$ is a boundedly complete lattice but not a complete lattice (and so is its sublattice $\mathbb{Z}$ of all integers). The set $A = \{x \in \mathbb{R}: \ 0 \leq x \leq 1\}$, as a sublattice of $(\mathbb{R}, \leq)$, is a complete lattice where $\sup \varnothing = \min A = 0$ and $\inf \varnothing = \max A = 1$. The set of all rational numbers $\mathbb{Q}$ is a sublattice of $\mathbb{R}$ (in the natural ordering $\leq$) but not a boundedly complete lattice —

e.g., the set $\{x \in \mathbb{Q}: \ x^2 \le 2\}$ is bounded in $\mathbb{Q}$ but has no infimum and no supremum in $\mathbb{Q}$.

**Example 1D.** The notion of connectedness needs topology and we shall define it in due course. However, if the reader is already familiar with it, then he can appreciate now a rather simple example of a directed set that is not a lattice (connectedness will be defined in Chapter 3). The subcollection of $\wp(\mathbb{R}^2)$ made up of all connected subsets of the Euclidean plane $\mathbb{R}^2$ is a directed set in the inclusion ordering $\subseteq$ (both upward and downward) but not a lattice.

**Lemma 1.1.** (Banach–Tarski). *An increasing function on a complete lattice has a fixed point.*

*Proof.* Let $(X, \le)$ be a partially ordered set, consider a function $F: X \to X$, and set $A = \{x \in X: \ F(x) \le x\}$. Suppose $X$ is a complete lattice. Then $X$ has a supremum in $X$ ($\sup X = \max X$). Since $\max X \in X$, it follows that $F(\max X) \in X$ so that $F(\max X) \le \max X$. Conclusion: $A$ is nonempty. Take $x \in A$ arbitrary and let $a$ be the infimum of $A$ ($a = \inf A \in X$). If $F$ is increasing, then $F(a) \le F(x) \le x$ because $a \le x$ and $x \in A$. Therefore $F(a)$ is a lower bound for $A$, and hence $F(a) \le a$. Thus $a \in A$. On the other hand, since $F(x) \le x$ and since $F$ is increasing, $F(F(x)) \le F(x)$. Thus $F(x) \in A$ so that $F(A) \subseteq A$, and hence $F(a) \in A$ (for $a \in A$), which implies that $a = \inf A \le F(a)$. Therefore $a \le F(a) \le a$. Thus (antisymmetry) $F(a) = a$. $\qquad\square$

The next theorem is an extremely important result that plays a central role in Section 1.8. Its proof is based on the previous lemma.

**Theorem 1.2.** (Cantor–Bernstein). *If there exist an injective mapping of $X$ into $Y$ and an injective mapping of $Y$ into $X$, then there exists a one-to-one correspondence between the sets $X$ and $Y$.*

*Proof.* First note that the theorem statement can be translated into the following problem. Given an injective function from $X$ to $Y$ and also an injective function from $Y$ to $X$, construct a bijective function from $X$ to $Y$. Thus consider two functions $F: X \to Y$ and $G: Y \to X$. Let $\wp(X)$ be the power set of $X$. For each $A \in \wp(X)$ set

$$\Phi(A) \ = \ X \backslash G(Y \backslash F(A)).$$

It is readily verified that $\Phi: \wp(X) \to \wp(X)$ is an increasing function with respect to the inclusion ordering of $\wp(X)$. Therefore, by the Banach–Tarski Lemma, it has a fixed point in the complete lattice $\wp(X)$. That is, there exists $A_0 \in \wp(X)$ such that $\Phi(A_0) = A_0$. Hence $A_0 = X \backslash G(Y \backslash F(A_0))$ so that

$$X \backslash A_0 \ = \ G(Y \backslash F(A_0)).$$

Thus $X \setminus A_0$ is included in the range of $G$. If $F \colon X \to Y$ and $G \colon Y \to X$ are both injective, then it is easy to show that the function $H \colon X \to Y$, defined by

$$
H(x) = \begin{cases} F(x), & x \in A_0 A, \\ G^{-1}(x), & x \in X \setminus A_0, \end{cases}
$$

is injective and surjective. $\qquad\qquad\qquad\qquad\qquad\qquad\qquad\qquad\qquad\qquad\square$

If $X$ is a partially ordered set such that for every pair $x$, $y$ of elements of $X$ either $x \le y$ or $y \le x$, then $X$ is *simply ordered* (synonyms: *linearly ordered, totally ordered*). A simply ordered set is also called a *chain*. Note that, in this particular case, the concepts of maximal element and maximum element (as well as minimal element and minimum element) coincide. It is clear that every simply ordered set is a lattice. For instance, any subset of the real line $\mathbb{R}$ (e.g., $\mathbb{R}$ itself or $\mathbb{Z}$) is simply ordered.

**Example 1E.** Let $\le$ be a simple ordering on a set $X$ and recall that $x < y$ means $x \le y$ and $x \ne y$. This defines a transitive relation $<$ on $X$ that satisfies the *trichotomy law*: for every pair $\{x, y\}$ in $X$ exactly one of the three statements $x < y$, $x = y$, or $y < x$ is true. Conversely, if $<$ is a transitive relation on a set $X$ that satisfies the trichotomy law, and if a relation $\le$ on $X$ is defined by setting $x \le y$ whenever either $x < y$ or $x = y$, then $\le$ is a simple ordering on $X$. Thus, according to the above notation, $\le$ is a simple ordering on a set $X$ if and only if $<$ is a transitive relation on $X$ that satisfies the trichotomy law.

If $X$ is a partially ordered set such that every nonempty subset of it has a minimum, then $X$ is said to be *well-ordered*. Every well-ordered set is simply ordered. Example: Any subset of the set $\mathbb{N}_0$ of all nonnegative integers is well-ordered.

## 1.7   Indexing

Let $F$ be a function from a set $X$ to a set $Y$. Another way to look at the range of $F$ is: for each $x \in X$ set $y_x = F(x) \in Y$ and note that $F(X) = \{y_x \in Y \colon x \in X\}$, which can also be written as $\{y_x\}_{x \in X}$. Thus the domain $X$ can be thought of as an *index set*, the range $\{y_x\}_{x \in X}$ as a family of elements of $Y$ indexed by an index set $X$ (an *indexed family*), and the function $F \colon X \to Y$ as an *indexing*. An indexed family $\{y_x\}_{x \in X}$ may contain elements $y_a$ and $y_b$, for $a$ and $b$ in $X$, such that $y_a = y_b$. If $\{y_x\}_{x \in X}$ has the property that $y_a \ne y_b$ whenever $a \ne b$, then it is said to be an indexed family of *distinct* elements. Observe that $\{y_x\}_{x \in X}$ is a family of distinct elements if and only if the function $F \colon X \to Y$ (i.e., the indexing process) is injective. The identity mapping on an arbitrary set $X$ can be viewed as an indexing of $X$, the *self-indexing* of $X$. Thus any set $X$ can be thought of as an indexed family (the range of the self-indexing of itself).

A mapping of the set $\mathbb{N}$ (or $\mathbb{N}_0$, but not $\mathbb{Z}$) into a set $Y$ is called a *sequence* (or an *infinite sequence*). Notations: $\{y_n\}_{n\in\mathbb{N}}$, $\{y_n\}_{n\geq 1}$, $\{y_n\}_{n=1}^{\infty}$ or simply $\{y_n\}$. Thus a $Y$-*valued sequence* (or a *sequence of elements in $Y$*, or even a *sequence in $Y$*) is precisely a function from $\mathbb{N}$ to $Y$, which is commonly thought of as an indexed family (indexed by $\mathbb{N}$) where the indexing process (i.e., the function itself) is often omitted. The elements $y_n$ of $\{y_n\}$ are sometimes referred to as the *entries* of the sequence $\{y_n\}$. If $Y$ is a subset of the set $\mathbb{C}$, $\mathbb{R}$ or $\mathbb{Z}$, then *complex-valued sequence, real-valued sequence* or *integer-valued sequence*, respectively, are usual terminologies.

Let $\{X_\gamma\}_{\gamma\in\Gamma}$ be an indexed family of sets. The *Cartesian product* of $\{X_\gamma\}_{\gamma\in\Gamma}$, denoted by $\prod_{\gamma\in\Gamma}X_\gamma$, is the set consisting of all indexed families $\{x_\gamma\}_{\gamma\in\Gamma}$ such that $x_\gamma \in X_\gamma$ for every $\gamma \in \Gamma$. In particular, if $X_\gamma = X$ for all $\gamma \in \Gamma$, where $X$ is a fixed set, then $\prod_{\gamma\in\Gamma}X_\gamma$ is precisely the collection of all functions from $\Gamma$ to $X$. That is,

$$\prod_{\gamma\in\Gamma} X = X^{\Gamma}.$$

Recall: $X^{\Gamma}$ denotes the collection of all functions from a set $\Gamma$ to a set $X$. Suppose $\Gamma = \mathbb{I}_n$, where $\mathbb{I}_n = \{i \in \mathbb{N}: i \leq n\}$ for some $n \in \mathbb{N}$ ($\mathbb{I}_n$ is called an *initial segment* of $\mathbb{N}$). The Cartesian product of $\{X_i\}_{i\in\mathbb{I}_n}$ (or $\{X_i\}_{i=1}^{n}$), denoted by $\prod_{i\in\mathbb{I}_n}X_i$ or $\prod_{i=1}^{n}X_i$, is the set $X_1\times\cdots\times X_n$ of all *ordered n-tuples* $(x_1,\dots,x_n)$ with $x_i \in X_i$ for every $i \in \mathbb{I}_n$. Moreover, if $X_i = X$ for all $i \in \mathbb{I}_n$, then $\prod_{i\in\mathbb{I}_n}X$ is the Cartesian product of $n$ copies of $X$ which is denoted by $X^n$ (instead of $X^{\mathbb{I}_n}$). The n-tuples $(x_1,\dots,x_n)$ in $X^n$ are also called *finite sequences* (as functions from an initial segment of $\mathbb{N}$ into $X$). Accordingly, $\prod_{n\in\mathbb{N}}X$ is referred to as the Cartesian product of countably infinite copies of $X$, which coincides with $X^{\mathbb{N}}$: the set of all $X$-valued (infinite) sequences.

A remarkably useful way of defining an infinite sequence is given by the *Principle of Recursive Definition* which says that, *if $F$ is a function from a nonempty set $X$ into itself, and if $x$ is an arbitrary element of $X$, then there exists a unique $X$-valued sequence $\{x_n\}_{n\in\mathbb{N}}$ such that $x_1 = x$ and $x_{n+1} = F(x_n)$ for every $n \in \mathbb{N}$.* The existence of such a unique sequence is intuitively clear, and it can be easily proved by induction (i.e., by using the Principle of Mathematical Induction). A slight generalization reads as follows. *For each $n \in \mathbb{N}$ let $G_n$ be a mapping of $X^n$ into $X$, and let $x$ be an arbitrary element of $X$. Then there exists a unique $X$-valued sequence $\{x_n\}_{n\in\mathbb{N}}$ such that $x_1 = x$ and $x_{n+1} = G_n(x_1,\dots,x_n)$ for every $n \in \mathbb{N}$.*

Since sequences are functions of $\mathbb{N}$ (or of $\mathbb{N}_0$) to a set $X$, the terms associated with the notion of boundedness clearly apply to sequences in a partially ordered set $X$. In particular, if $X$ is a partially ordered set, and if $\{x_n\}$ is an $X$-valued sequence, then $\sup_n x_n$ and $\inf_n x_n$ are defined as the supremum and infimum, respectively, of the partially ordered indexed family $\{x_n\}$. Since $\mathbb{N}$ and $\mathbb{N}_0$ (with their natural ordering) are partially ordered sets (well-ordered, really), the terms associated with the property of being monotone also apply to sequences in a partially ordered set $X$. Let $\{z_n\}_{n\in\mathbb{N}}$ be a sequence in a set $Z$, and let $\{n_k\}_{k\in\mathbb{N}}$ be a strictly increasing sequence of positive integers (i.e., a strictly increasing sequence in $\mathbb{N}$). If we think of

$\{n_k\}$ and $\{z_n\}$ as functions, then the range of the former is a subset of the domain of the latter (i.e., the indexed family $\{n_k\}_{k\in\mathbb{N}}$ is a subset of $\mathbb{N}$). Thus we may consider the composition of $\{z_n\}$ with $\{n_k\}$, say $\{z_{n_k}\}$, which is again a function of $\mathbb{N}$ to $Z$; that is, $\{z_{n_k}\}$ is a sequence in $Z$. Moreover, since $\{n_k\}$ is strictly increasing, to each element of the indexed family $\{z_{n_k}\}_{k\in\mathbb{N}}$ there corresponds a unique element of the indexed family $\{z_n\}_{n\in\mathbb{N}}$. In this case the $Z$-valued sequence $\{z_{n_k}\}$ is called a *subsequence* of $\{z_n\}$.

A sequence is a function whose domain is either $\mathbb{N}$ or $\mathbb{N}_0$, but a similar concept could be likewise defined for a function on a well-ordered domain. Even in this case, a function with domain $\mathbb{Z}$ (equipped with its natural ordering) would not be a sequence. Now recall the string of (nonreversible) implications

$$\textit{well-ordered} \;\Rightarrow\; \textit{simply ordered} \;\Rightarrow\; \textit{lattice} \;\Rightarrow\; \textit{directed set}.$$

This might suggest an extension of the concept of sequence by allowing functions whose domains are directed sets. A *net* in a set $X$ is a family of elements of $X$ indexed by a directed set $\Gamma$. In other words, if $\Gamma$ is a directed set and $X$ is an arbitrary set, then an indexed family $\{x_\gamma\}_{\gamma\in\Gamma}$ of elements of $X$ indexed by $\Gamma$ is called a net in $X$ indexed by $\Gamma$. Examples: Every $X$-valued sequence $\{x_n\}$ is a net in $X$. In fact, sequences are prototypes of nets. Every $X$-valued function on $\mathbb{Z}$ (notations: $\{x_k\}_{k\in\mathbb{Z}}$, $\{x_k\}_{k=-\infty}^{\infty}$ or $\{x_k;\; k = 0, \pm 1, \pm 2, \dots\}$) is a net (sometimes called *double sequences* or *bisequences*, although these nets are not sequences themselves).

## 1.8   Cardinality

Two sets, say $X$ and $Y$, are said to be *equivalent* (denoted by $X \leftrightarrow Y$) if there exists a one-to-one correspondence between them. Clearly (see Problems 1.8 and 1.9), $X \leftrightarrow X$ (reflexivity), $X \leftrightarrow Y$ if and only if $Y \leftrightarrow X$ (symmetry), and $X \leftrightarrow Z$ whenever $X \leftrightarrow Y$ and $Y \leftrightarrow Z$ for some set $Z$ (transitivity). Thus, if there exists a set upon which $\leftrightarrow$ is a relation, then it is an equivalence relation. For instance, if the notion of equivalent sets is restricted to subsets of a given set $X$, then $\leftrightarrow$ is an equivalence relation on the power set $\wp(X)$. If $C = \{x_\gamma\}_{\gamma\in\Gamma}$ is an indexed family of *distinct* elements of a set $X$ indexed by a set $\Gamma$ (so that $x_\alpha \neq x_\beta$ for every $\alpha \neq \beta$ in $\Gamma$), then $C \leftrightarrow \Gamma$ (the very indexing process sets a one-to-one correspondence between $\Gamma$ and $C$).

Let $\mathbb{N}$ be the set of all natural numbers and, for each $n \in \mathbb{N}$, consider the initial segment

$$\mathbb{I}_n = \{i \in \mathbb{N}\colon\; i \leq n\}.$$

A set $X$ is *finite* if it is either empty or equivalent to $\mathbb{I}_n$ for some $n \in \mathbb{N}$. A set is *infinite* if it is not finite. If $X$ is finite and $Y$ is equivalent to $X$, then $Y$ is finite. Therefore, if $X$ is infinite and $Y$ is equivalent to $X$, then $Y$ is infinite. It is easy to show by induction that, for each $n \in \mathbb{N}$, $\mathbb{I}_n$ has no proper subset equivalent to it. Thus (see Problem 1.12), every finite set has no proper subset equivalent to it. That is, *if a set has a proper equivalent subset, then it is infinite*. Moreover, such a subset must be infinite too (since it is equivalent to an infinite set).

**Example 1F.** $\mathbb{N}$ *is infinite*. Indeed, it is easy to show that $\mathbb{N}_0$ is equivalent to $\mathbb{N}$ (the function $F: \mathbb{N}_0 \to \mathbb{N}$ such that $F(n) = n + 1$ for every $n \in \mathbb{N}_0$ will do the job). Thus $\mathbb{N}_0$ is infinite, because $\mathbb{N}$ is a proper subset of $\mathbb{N}_0$ which is equivalent to it, and so is $\mathbb{N}$.

To verify the converse (i.e., to show that *every infinite set has a proper equivalent subset*) we apply the Axiom of Choice.

**Axiom of Choice.** *If $\{X_\gamma\}_{\gamma \in \Gamma}$ is an indexed family of nonempty sets indexed by a nonempty index set $\Gamma$, then there exists an indexed family $\{x_\gamma\}_{\gamma \in \Gamma}$ such that $x_\gamma \in X_\gamma$ for each $\gamma \in \Gamma$.*

**Theorem 1.3.** *A set is infinite if and only if it has a proper equivalent subset.*

*Proof.* We have already seen that every set with a proper equivalent subset is infinite. To prove the converse take an arbitrary element $x_0$ from an infinite set $X_0$, and an arbitrary $k$ from $\mathbb{N}_0$. The Principle of Mathematical Induction allows us to construct, for each $k \in \mathbb{N}_0$, a finite family $\{X_n\}_{n=0}^{k+1}$ of infinite sets as follows. Set $X_1 = X_0 \backslash \{x_0\}$ and, for every nonnegative integer $n \leq k$, let $X_{n+1}$ be recursively defined by the formula

$$X_{n+1} = X_n \backslash \{x_n\},$$

where $\{x_n\}_{n=0}^{k}$ is a finite set of pairwise distinct elements, each $x_n$ being an arbitrary element taken from each $X_n$. However, if we consider the (infinite) indexed family $\{X_n\}_{n \in \mathbb{N}_0} = \bigcup_{k \in \mathbb{N}_0} \{X_n\}_{n=0}^{k+1}$, then we need the Axiom of Choice to ensure the existence of the indexed family $\{x_n\}_{n \in \mathbb{N}_0} = \bigcup_{k \in \mathbb{N}_0} \{x_n\}_{n=0}^{k}$ where each $x_n$ is arbitrarily taken from each $X_n$. Now set $A_0 = \{x_n\}_{n \in \mathbb{N}_0} \subseteq X_0$, $A = \{x_n\}_{n \in \mathbb{N}} \subset A_0$, and

$$X = A \cup (X_0 \backslash A_0) \subset A_0 \cup (X_0 \backslash A_0) = X_0.$$

Note that $A_0 \leftrightarrow \mathbb{N}_0$ and $A \leftrightarrow \mathbb{N}$ (since the elements of $A_0$ are distinct). Thus $A_0 \leftrightarrow A$ (because $\mathbb{N}_0 \leftrightarrow \mathbb{N}$), and hence $X_0 \leftrightarrow X$ (see Problem 1.20). Conclusion: Any infinite set $X_0$ has a proper equivalent subset (i.e., there exists a proper subset $X$ of $X_0$ such that $X_0 \leftrightarrow X$). $\qquad\qquad\square$

If $X$ is a finite set, so that it is equivalent to an initial segment $\mathbb{I}_n$ for some natural number $n$, then we say that its *cardinality* (or its *cardinal number*) is $n$. Thus the cardinality of a finite set $X$ is just the number of elements of $X$ (where, in this case, "numbering" means "indexing" as a finite set may be naturally indexed by an index set $\mathbb{I}_n$). We shall use the symbol # for cardinality. Thus $\#\mathbb{I}_n = n$, and so $\#X = n$ whenever $X \leftrightarrow \mathbb{I}_n$. For infinite sets the concept of cardinal number is a bit more complicated. We shall not define a cardinal number for an infinite set as we did for finite sets (which "number" should it be?) but define the following concept instead.

Two sets $X$ and $Y$ are said to have the *same cardinality* if they are equivalent. Thus, to each set $X$ we shall assign a symbol $\#X$, called the *cardinal number* of $X$ (or the *cardinality* of $X$) according to the following rule: $\#X = \#Y \iff X \leftrightarrow Y$ — two sets have the same cardinality if and only if they are equivalent; otherwise (i.e., if they are not equivalent) we shall write $\#X \neq \#Y$. We say that the cardinality of a set $X$ is *less than or equal to* the cardinality of a set $Y$ (notation: $\#X \leq \#Y$) if there exists an injective mapping of $X$ into $Y$ (i.e., if there exists a subset $Y'$ of $Y$ such that $\#X = \#Y'$). Equivalently, $\#X \leq \#Y$ if there exists a surjective mapping of $Y$ onto $X$ (see Problem 1.6). If $\#X \leq \#Y$ and $\#X \neq \#Y$, then we shall write $\#X < \#Y$.

**Theorem 1.4.** (Cantor). $\#X < \#\wp(X)$ *for every set* $X$.

*Proof.* Consider the function $F: X \to \wp(X)$ defined by $F(x) = \{x\}$ for every $x \in X$, which is clearly injective. Thus $\#X \leq \#\wp(X)$. Hence $\#X < \#\wp(X)$ if and only if $\#X \neq \#\wp(X)$. Suppose $\#X = \#\wp(X)$ so that there exists a surjective function $G: X \to \wp(X)$. Consider the set $A = \{x \in X : x \notin G(x)\}$ in $\wp(X)$ and take $a \in X$ such that $G(a) = A$ (recall: G is surjective). If $a \in A$, then $a \notin G(a)$ and hence $a \notin A$, which is a contradiction. Conclusion 1: $a \notin A$. On the other hand, if $a \notin A$, then $a \in G(a)$ and hence $a \in A$, which is another contradiction. Conclusion 2: $a \in A$. Therefore $\{a \notin A \text{ and } a \in A\}$, which is impossible (i.e., which also is a contradiction). Final conclusion: $\#X \neq \#\wp(X)$.  $\square$

Let $A$ be a subset of a set $X$. The *characteristic function* of the set $A$ is the map $\chi_A : X \to \{0, 1\}$ such that

$$\chi_A(x) = \begin{cases} 1, & x \in A, \\ 0, & x \in X \setminus A. \end{cases}$$

It is clear that the correspondence between the subsets of $X$ and their characteristic functions is one-to-one. Hence $\#\wp(X)$ coincides with the cardinality of the collection of all characteristic functions of subsets of $X$. More generally, let $2^X$ denote the collection of all maps of a set $X$ into the set $\{0, 1\}$ (i.e., set $2^X = \{0, 1\}^X$).

**Theorem 1.5.** $\#\wp(X) = \#2^X$ *for every set* $X$.

*Proof.* Let $F$ be a function from the collection $2^X$ to the power set $\wp(X)$ that assigns to each map $\varphi: X \to \{0, 1\}$ the inverse image of the singleton $\{1\}$ under $\varphi$. That is, consider the function $F: 2^X \to \wp(X)$ defined by $F(\varphi) = \varphi^{-1}(\{1\})$ for every $\varphi$ in $2^X$.

*Claim* 1. F is surjective.

*Proof.* If $A$ is a subset of $X$, then the characteristic function of $A$, $\chi_A$ in $2^X$, is such that $\chi_A^{-1}(\{1\}) = A$. Thus $F(\chi_A) = A$ for every $A \in \wp(X)$. Therefore, $\wp(X) = \bigcup_{A \in \wp(X)} F(\chi_A) \subseteq \bigcup_{\varphi \in 2^X} F(\varphi) = \mathcal{R}(F).$  $\square$

*Claim* 2. F is injective.

*Proof.*  Take $\varphi, \psi \in 2^X$. If $\varphi \neq \psi$, then $\varphi(x) \neq \psi(x)$ for some $x \in X$. Thus $\varphi(x) = 0$ and $\psi(x) = 1$ (or vice versa), so that $x \notin \varphi^{-1}(\{1\})$ and $x \in \psi^{-1}(\{1\})$. Hence $\varphi^{-1}(\{1\}) \neq \psi^{-1}(\{1\})$ so that $F(\varphi) \neq F(\psi)$. $\square$

Conclusion: $F$ is a one-to-one correspondence between $2^X$ and $\wp(X)$.     $\square$

Although the next theorem may come as no surprise, it is all but trivial. This actually is a rewriting, in terms of the concept of cardinality, of the rather important Theorem 1.2

**Theorem 1.6.**  (Cantor–Bernstein). *Let $X$ and $Y$ be any sets. If $\#X \leq \#Y$ and $\#Y \leq \#X$, then $\#X = \#Y$.*

The Cantor–Bernstein Theorem exhibits an antisymmetry property. Note that reflexivity and transitivity are readily verified (see Problem 1.22) which, together with Theorem 1.6, lead to a partial ordering property. Behind the antisymmetry property of Theorem 1.6 there is in fact a simple ordering property. To establish it (Theorem 1.7 below) we shall rely on the following axiom.

**Zorn's Lemma.**  *Let $X$ be a partially ordered set. If every simply ordered subset of $X$ (i.e., if each chain in $X$) has an upper bound in $X$, then $X$ has a maximal element.*

The label "Zorn's Lemma" is inappropriate but has already been consecrated in the literature. It should read "Zorn's Axiom" instead, for it really is an axiom equivalent to the Axiom of Choice.

**Theorem 1.7.**  *For any two sets $X$ and $Y$, either $\#X \leq \#Y$ or $\#Y \leq \#X$.*

*Proof.*  Consider two sets $X$ and $Y$. Let $\mathcal{I}$ be the collection of all injective functions from subsets of $X$ to $Y$. That is,

$$\mathcal{I} = \left\{ F \in Y^A : \ A \subseteq X \text{ and } F \text{ is injective} \right\}.$$

Recall that (see Problem 1.17), as a subset of $\mathcal{F} = \bigcup_{A \in \wp(X)} Y^A$, $\mathcal{I}$ is a partially ordered set in the extension ordering, and every simply ordered subset of it has a supremum in $\mathcal{I}$. Thus, according to Zorn's Lemma, $\mathcal{I}$ contains a maximal function. Let $F_0 : A_0 \to Y$ be a maximal function of $\mathcal{I}$, where $A_0 \subseteq X$ and $\#F_0(A_0) = \#A_0$ (since $F_0$ is injective). Suppose $A_0 \neq X$ and $F_0(A_0) \neq Y$. Take $x_0 \in X \backslash A_0$ and $y_0 \in Y \backslash F_0(A_0)$, and consider the function $F_1 : A_0 \cup \{x_0\} \to Y$ defined by

$$F_1(x) = \begin{cases} F_0(x) \in F_0(A_0), & x \in A_0, \\ y_0 \in Y \setminus F_0(A_0), & x = x_0 \in X \setminus A_0, \end{cases}$$

which is injective (because $F_0$ is injective and $y_0 \notin F_0(A_0)$). Since $F_1 \in \mathcal{I}$ and $F_0 = F_1|_{A_0}$, it follows that $F_0 \leq F_1$, which contradicts the fact that $F_0$ is a maximal of $\mathcal{I}$ (for $F_0 \neq F_1$). Hence, either $A_0 = X$ or $F_0(A_0) = Y$. If $A_0 = X$, then

$F_0 \colon X \to Y$ is injective and so $\#X \leq \#Y$. If $F_0(A_0) = Y$, then $\#Y = \#F_0(A_0) = \#A_0 \leq \#X$ (for $A_0 \subseteq X$ — see Problem 1.21(a)).    $\square$

We have already seen that $\mathbb{N} \leftrightarrow \mathbb{N}_0$. Thus $\mathbb{N}$ and $\mathbb{N}_0$ have the same cardinality. It is usual to assign a special symbol (*aleph naught*) to such a cardinal number: $\#\mathbb{N} = \#\mathbb{N}_0 = \aleph_0$. We have also seen (cf. proof of Theorem 1.3) that, if $X$ is an infinite set, then there exists a subset of it, say $A$, which is equivalent to $\mathbb{N}$. Thus $\#\mathbb{N} = \#A \leq \#X$, and hence $\aleph_0$ *is the smallest infinite cardinal number* in the sense that $\aleph_0 \leq \#X$ for every infinite set $X$ (see Problems 1.21(a) and 1.22). A set $X$ such that $\#X = \aleph_0$ is said to be *countably infinite* (or *denumerable*). Therefore, *every infinite set has a countably infinite subset*. A set that is either finite or countably infinite (i.e., a set $X$ such that $\#X \leq \aleph_0$) is said to be *countable*; otherwise it is said to be *uncountable* (or *uncountably infinite*, or *nondenumerable*).

**Proposition 1.8.** $\#(X \times X) = \#X$ *for every countably infinite set* $X$.

*Proof.* Suppose $\#X = \#\mathbb{N}$. According to Problems 1.26, 1.23(b) and 1.25(a) we get $\#X \leq \#(X \times X) \leq \#(\mathbb{N} \times \mathbb{N}) = \#\mathbb{N} = \#X$. Hence the identity $\#X = \#(X \times X)$ follows by the Cantor–Bernstein Theorem.    $\square$

Note that $\#X \leq \#(X \times X)$ for any set $X$ (see Problem 1.26). Moreover, it is easy to show that $\#X < \#(X \times X)$ whenever $X$ is a finite nonempty set. Thus, if a nonempty set $X$ is such that $\#X = \#(X \times X)$, then it is an infinite set. The previous proposition ensured the converse for countably infinite sets. The next theorem (which is another application of Zorn's Lemma) ensures the converse for every infinite set. Therefore, the identity $\#X = \#(X \times X)$ actually characterizes the infinite sets (of any cardinality) among the nonempty sets.

**Theorem 1.9.** *If* $X$ *is an infinite set, then* $\#X = \#(X \times X)$.

*Proof.* First we verify the following auxiliary result.

*Claim* 0. Let $C$, $D$ and $E$ be nonempty sets. If $\#(E \times E) = \#E$, then $\#(C \cup D) \leq \#E$ whenever $\#C \leq \#E$ and $\#D \leq \#E$.

*Proof.* The claimed result is a straightforward application of Problems 1.26, 1.23(b) and 1.22: $\#(C \cup D) \leq \#(C \times D) \leq \#(E \times E) = \#E$.    $\square$

Now, back to the theorem statement. Let $X$ be a set and let $\mathcal{J}$ be the collection of all injective functions from subsets of $X$ to $X \times X$ such that the range of each function in $\mathcal{J}$ coincides with the Cartesian product of its domain with itself. That is,

$$\mathcal{J} = \left\{ F \in (X \times X)^A \colon \ A \subseteq X, \ F \text{ is injective and } F(A) = A \times A \right\}.$$

Note that $\mathcal{J}$ is nonempty (at least the empty function is there). From now on suppose $X$ is infinite. Thus $X$ has a countably infinite subset, and so Proposition 1.8 ensures the existence of a function in $\mathcal{J}$ with infinite (at least countably infinite) domain.

Recall that $\mathcal{J}$ is a partially ordered set in the extension ordering, and that every chain in $\mathcal{J}$ (i.e., every simply ordered subset of $\mathcal{J}$) has an injective supremum (see Problem 1.17).

*Claim* 1. Such a supremum in fact lies in $\mathcal{J}$.

*Proof.* Let $\{F_\gamma\}$ be an arbitrary chain in $\mathcal{J}$, and let $\mathcal{D}(F_\gamma)$ and $\mathcal{R}(F_\gamma)$ denote the domain and range of $F_\gamma$, respectively. Thus each $F_\gamma$ is an injective function from $A_\gamma$ to $X \times X$, with $A_\gamma = \mathcal{D}(F_\gamma) \subseteq X$ and $F_\gamma(A_\gamma) = A_\gamma \times A_\gamma$. Now let $\bigvee_\gamma F_\gamma : \bigcup_\gamma A_\gamma \to X \times X$ be the supremum of $\{F_\gamma\}$, and note that (see Problem 1.17) $(\bigvee_\gamma F_\gamma)(\bigcup_\gamma A_\gamma) = \mathcal{R}(\bigvee_\gamma F_\gamma) = \bigcup_\gamma \mathcal{R}(F\gamma) = \bigcup_\gamma (A_\gamma \times A_\gamma)$. Clearly, $\bigcup_\gamma(A_\gamma \times A_\gamma) \subseteq (\bigcup_\gamma A_\gamma) \times (\bigcup_\gamma A_\gamma)$. On the other hand, if $\{F_\lambda, F_\mu\}$ is an arbitrary pair from $\{F_\gamma\}$, then $F_\lambda \leq F_\mu$ (or vice versa), so that $A_\lambda \times A_\mu \subseteq A_\mu \times A_\mu$ (for $A_\lambda \subseteq A_\mu$). Thus $(\bigcup_\gamma A_\gamma) \times (\bigcup_\gamma A_\gamma) \subseteq \bigcup_\gamma(A_\gamma \times A_\gamma)$. Therefore $(\bigvee_\gamma F_\gamma)(\bigcup_\gamma A_\gamma) = (\bigcup_\gamma A_\gamma) \times (\bigcup_\gamma A_\gamma)$, and hence $\bigvee_\gamma F_\gamma \in \mathcal{J}$. $\square$

Conclusion: Every chain in $\mathcal{J}$ has an upper bound (a supremum, actually) in $\mathcal{J}$. Thus, according to Zorn's Lemma, $\mathcal{J}$ contains a maximal element. Let $F_0 : A_0 \to X \times X$ be a maximal function of $\mathcal{J}$, where $\varnothing \neq A_0 \subseteq X$ and $F_0(A_0) = A_0 \times A_0$. Since $F_0$ is injective,

$$\#(A_0 \times A_0) = \#A_0,$$

which implies that the nonempty set $A_0$ is an infinite set.

*Claim* 2. If $\#A_0 < \#X$, then $\#A_0 < \#(X \backslash A_0)$.

*Proof.* If $\#(X \backslash A_0) \leq \#A_0$, then (cf. Problems 1.26 and 1.23(b)) $\#X = \#(A_0 \cup X \backslash A_0) \leq \#(A_0 \times (X \backslash A_0)) \leq \#(A_0 \times A_0) = \#A_0$. Conclusion: $\#(X \backslash A_0) \leq \#A_0$ implies $\#X \leq \#A_0$ (see Problem 1.22). Equivalently, $\#A_0 < \#X$ implies $\#A_0 < \#(X \backslash A_0)$ by Theorem 1.7. $\square$

Note that $\#A_0 \leq \#X$ (for $A_0 \subseteq X$). Suppose $\#A_0 < \#X$. In this case $\#A_0 < \#(X \backslash A_0)$ by Claim 2. Thus there exists a proper subset of $X \backslash A_0$, say $A_1$, such that $\#A_0 = \#A_1$. Therefore (Problem 1.23(b)) $\#(A_i \times A_j) \leq \#(A_0 \times A_0) = \#A_0$ for all combinations of $i$, $j$ in $\{0, 1\}$, and hence $\#[(A_0 \times A_1) \cup (A_1 \times A_0) \cup (A_1 \times A_1)] \leq \#A_0$ according to Claim 0. Since the reverse inequality is trivially verified, it follows by Theorem 1.6 that $\#[(A_0 \times A_1) \cup (A_1 \times A_0) \cup (A_1 \times A_1)] = \#A_0$. Set

$$A = A_0 \cup A_1$$

and observe that $(A \times A) \backslash (A_0 \times A_0) = (A_0 \times A_1) \cup (A_1 \times A_0) \cup (A_1 \times A_1)$ because $A_0$ and $A_1$ are disjoint. Thus

$$\#[(A \times A) \backslash (A_0 \times A_0)] = \#A_0 = \#A_1,$$

which ensures the existence of an injective function $F_1 : A_1 \to X \times X$ such that $F_1(A_1) = (A \times A) \backslash (A_0 \times A_0)$. Now consider a function $F$ from $A$ to $X \times X$ defined as follows.

$$F(x) = \begin{cases} F_0(x) \in A_0 \times A_0, & x \in A_0, \\ F_1(x) \in (A \times A) \setminus (A_0 \times A_0), & x \in A_1 = A \setminus A_0. \end{cases}$$

$F : A \to X \times X$ is injective (because $F_0$ and $F_1$ are injective functions with disjoint ranges) and $F(A) = A \times A$ (for $F_0(A_0) \cup F_1(A_1) = A \times A$). Since $F \in \mathcal{J}$ and $F_0 = F|_{A_0}$, it follows that $F_0 \leq F$, which contradicts the fact that $F_0$ is a maximal of $\mathcal{J}$ (for $F_0 \neq F$). Therefore

$$\#A_0 = \#X,$$

and hence (cf. Problem 1.23(b)) $\#(A_0 \times A_0) = \#(X \times X)$. Conclusion: $\#X = \#A_0 = \#(A_0 \times A_0) = \#(X \times X)$. $\qquad\square$

Theorem 1.9 is a natural extension of Proposition 1.8 which in turn generalizes Problem 1.25(a). Another important and particularly useful result in this line is given by the next theorem and its corollary.

**Theorem 1.10.** *Let $X$ be a set and consider an indexed family of sets $\{X_\gamma\}_{\gamma \in \Gamma}$. If $\#X_\gamma \leq \#X$ for all $\gamma \in \Gamma$, then*

$$\#\left( \bigcup_{\gamma \in \Gamma} X_\gamma \right) \leq \#(\Gamma \times X).$$

*Proof.* Take an indexed family of sets $\{X_\gamma\}_{\gamma \in \Gamma}$ and suppose there exists a set $X$ such that $\#X_\gamma \leq \#X$ for all $\gamma \in \Gamma$. Thus for each $\gamma \in \Gamma$ there exists a surjective mapping of $X$ onto $X_\gamma$, say, $F_\gamma : X \to X_\gamma$. Now consider a function $G : \Gamma \times X \to \bigcup_{\gamma \in \Gamma} X_\gamma$ defined by $G(\gamma, x) = F_\gamma(x)$ for every $\gamma \in \Gamma$ and every $x \in X$. Take an arbitrary $y \in \bigcup_{\gamma \in \Gamma} X_\gamma$ so that $y \in X_\gamma$ for some $\gamma \in \Gamma$. Since $F_\gamma : X \to X_\gamma$ is surjective, there exists $x \in X$ such that $y = F_\gamma(x)$. Thus $y = G(\gamma, x)$; that is, $y \in \mathcal{R}(G)$. Hence $G : \Gamma \times X \to \bigcup_{\gamma \in \Gamma} X_\gamma$ is surjective, and so $\#(\bigcup_{\gamma \in \Gamma} X_\gamma) \leq \#(\Gamma \times X)$. $\qquad\square$

**Corollary 1.11.** *A countable union of countable sets is countable.*

*Proof.* Consider a countable family of countable sets $\{X_n\}_{n \in N}$, where $\#N \leq \#\mathbb{N}$ and $\#X_n \leq \#\mathbb{N}$ for all $n \in N$. Theorem 1.10 ensures that $\#(\bigcup_{n \in N} X_n) \leq \#(N \times \mathbb{N})$. However, $\#(N \times \mathbb{N}) \leq \#(\mathbb{N} \times \mathbb{N}) = \#\mathbb{N}$. Thus

$$\#\left( \bigcup_{n \in N} X_n \right) \leq \#\mathbb{N}.$$

(See Problems 1.23(b), 1.25(a) and 1.22.) $\qquad\square$

## 1.9   Remarks

We assume the reader is familiar with the definition of an *interval* of the real line $\mathbb{R}$. An interval of $\mathbb{R}$ is *nondegenerate* if it does not collapse to a singleton. It is easy to show that the cardinality of the real line $\mathbb{R}$ is the same as the cardinality of any nondegenerate interval of $\mathbb{R}$. A typical example: The function $F\colon [0,1] \to [-1,1]$ given by $F(x) = 2x - 1$ for every $x \in [0,1]$ is injective and surjective, and so is the function $G\colon (-1,1) \to \mathbb{R}$ defined by $G(x) = (1 - |x|)^{-1}x$ for every $x \in (-1,1)$. Thus $\#[0,1] = \#[-1,1]$ and $\#(-1,1) = \#\mathbb{R}$. Since $(-1,1) \subset [-1,1] \subset \mathbb{R}$, it follows that $\#\mathbb{R} = \#(-1,1) \leq \#[-1,1] \leq \#\mathbb{R}$, and hence (Cantor–Bernstein Theorem) $\#[-1,1] = \#\mathbb{R}$. Therefore,

$$\#[0,1] \;=\; \#(-1,1) \;=\; \#[-1,1] \;=\; \#\mathbb{R}.$$

Moreover, we can also prove that $\#\mathbb{R} = \#2^{\mathbb{N}}$. Indeed, consider the function $F\colon 2^{\mathbb{N}} \to [0,1]$ that assigns to each sequence $\{\alpha_n\}$ in $2^{\mathbb{N}}$ (i.e., to each sequence $\{\alpha_n\}_{n\in\mathbb{N}}$ with values either 0 or 1) a real number in $[0,1]$, in ternary expansion, as follows.

$$F(\{\alpha_n\}) \;=\; 0.(2\alpha_1)(2\alpha_2)\,\ldots$$

for every $\{\alpha_n\} \in 2^{\mathbb{N}}$. It can be shown that $F$ is injective.

Reason: Every real number $x \in [0,1]$ can be written as $\sum_{n=1}^{\infty}\beta_n\, p^{-n}$ for a given positive integer $p$ greater than one (i.e., for a given *base p*). In this case $0.\beta_1\beta_2\,\ldots$ is a *p-ary* expansion of $x$, where $\{\beta_n\}_{n\in\mathbb{N}}$ is a sequence of nonnegative integers ranging from 0 to $p - 1$. That is, $\{\beta_n\}_{n\in\mathbb{N}}$ is a sequence of *digits* with respect to the base $p$ — e.g., if $p = 2, 3$, or 10, then $0.\beta_1\beta_2\,\ldots$ is a *binary*, *ternary* or *denary* (i.e., *decimal*) expansion of $x$, respectively. A $p$-ary expansion (with respect to a base $p$) is not unique — e.g., $0.499\,\ldots$ and $0.500\,\ldots$ are decimal expansions for $x = \frac{1}{2}$. However, if two $p$-ary expansions of $x$ differ, then the absolute difference between the first digits in which they differ is equal to 1. Thus, if we take a $p$-ary expansion whose digits are either 0 or 2 (as we did), then it is *unique*. This can only be done for a $p$-ary expansion with respect to a base $p \geq 3$, so that a ternary expansion is enough.

Thus $\#2^{\mathbb{N}} \leq \#[0,1]$. On the other hand, let $G\colon 2^{\mathbb{N}} \to [0,1]$ be the function that assigns to each sequence $\{\alpha_n\}$ in $2^{\mathbb{N}}$ a real number in $[0,1]$, in binary expansion, as follows.

$$G(\{\alpha_n\}) \;=\; 0.\alpha_1\alpha_2\,\ldots$$

for every $\{\alpha_n\} \in 2^{\mathbb{N}}$. It can also be shown that $G$ is surjective.

Reason: Every real number $x \in [0,1]$ can be written as $\sum_{n=1}^{\infty}\alpha_n\,2^{-n}$ so that $0.\alpha_1\alpha_2\,\ldots$ is a binary expansion of it for some sequence $\{\alpha_n\} \in 2^{\mathbb{N}}$.

Thus $\#[0,1] \leq \#2^{\mathbb{N}}$. Therefore $\#[0,1] = \#2^{\mathbb{N}}$ by the Cantor–Bernstein Theorem. Hence

$$\#\mathbb{R} \;=\; \#2^{\mathbb{N}}$$

(for $\#[0, 1] = \#\mathbb{R}$). Using Theorems 1.4 and 1.5 we may conclude that

$$\#\mathbb{N} \; < \; \#\mathbb{R}.$$

Such a fundamental result can also be derived by the celebrated Cantor's *diagonal procedure* as follows. Clearly, $\#\mathbb{N} \leq \#\mathbb{R}$ (since $\mathbb{N} \subset \mathbb{R}$). Suppose $\#\mathbb{N} = \#\mathbb{R}$. This implies that $\#\mathbb{N} = \#[0, 1]$ (for $\#[0, 1] = \#\mathbb{R}$). Thus the interval $[0, 1]$ can be indexed by $\mathbb{N}$ so that $[0, 1] = \{x_n\}_{n\in\mathbb{N}}$. Write each $x_n$ in decimal expansion:

$$x_n \; = \; 0.\alpha_{n_1}\alpha_{n_2} \cdots$$

where each $\alpha_{n_k}$ ($k \in \mathbb{N}$) is a nonnegative integer between 0 and 9. Now consider the point $x \in [0, 1]$ with the following decimal expansion.

$$x \; = \; 0.\alpha_1\alpha_2 \cdots$$

where, again, each $\alpha_n$ ($n \in \mathbb{N}$) is a nonnegative integer between 0 and 9 but $\alpha_1 \neq \alpha_{1_1}$, $\alpha_2 \neq \alpha_{2_2}$, and so on. That is, $\alpha_n \neq \alpha_{n_n}$ for each $n \in \mathbb{N}$ (e.g., take $\alpha_n$ diametrically opposite to $\alpha_{n_n}$ so that, for each $n \in \mathbb{N}$, $\alpha_n = \alpha_{n_n} + 5$ if $0 \leq \alpha_{n_n} \leq 4$ or $\alpha_n = \alpha_{n_n} - 5$ if $5 \leq \alpha_{n_n} \leq 9$). Thus $x \neq x_n$ for every $n \in \mathbb{N}$. Hence $x \notin [0, 1]$, which is a contradiction. Therefore $\#\mathbb{N} \neq \#\mathbb{R}$. Equivalently (since $\#\mathbb{N} \leq \#\mathbb{R}$), $\#\mathbb{N} < \#\mathbb{R}$.

We have denoted $\#\mathbb{N}$ by $\aleph_0$. Let us now denote $\#2^{\mathbb{N}}$ by $2^{\aleph_0}$ so that

$$\#\mathbb{N} \; = \; \aleph_0 \; < \; 2^{\aleph_0} \; = \; \#\mathbb{R}.$$

Cantor conjectured in 1878 that there is no cardinal number between $\aleph_0$ and $2^{\aleph_0}$. The conjecture is called the Continuum Hypothesis.

**Continuum Hypothesis**. *There is no set whose cardinality is greater than $\#\mathbb{N}$ and smaller than $\#\mathbb{R}$.*

The Generalized Continuum Hypothesis is the conjecture that naturally generalizes the Continuum Hypothesis.

**Generalized Continuum Hypothesis**. *For any infinite set $X$, there is no cardinal number between $\#X$ and $\#2^X$.*

There are several different axiomatic set theories, each based on a somewhat different axiom system. The most popular is probably the axiom system *ZFC*. It comprises the axiom system *ZF* ("Z" for Zermelo and "F" for Fraenkel) plus the Axiom of Choice. The Axiom of Choice actually is a genuine axiom to be added to *ZF*. Indeed, Gödel proved in 1939 that the Axiom of Choice is consistent with *ZF*, and Cohen proved in 1963 that the Axiom of Choice is independent of *ZF*. The situation of the Continuum Hypothesis with respect to *ZFC* is somewhat similar to that of the Axiom of Choice with respect to *ZF*, although the Continuum Hypothesis itself is not as primitive as the Axiom of Choice (even if the Axiom of Choice might be regarded as not primitive enough). Gödel proved in 1939 that the Generalized Continuum Hypothesis is consistent with *ZFC*, and Cohen proved in 1963 that the denial of the Continuum Hypothesis also is consistent with *ZFC*. Thus both the Continuum Hypothesis and the Generalized Continuum Hypothesis are consistent with *ZFC* and also independent of *ZFC*: neither of them can be proved or disproved on the basis of *ZFC* alone (i.e., they are *undecidable* statements in *ZFC*). The Generalized Continuum Hypothesis in fact is stronger than the Axiom of Choice: Sierpinski showed in 1947 that the Generalized Continuum Hypothesis implies the Axiom of Choice.

We have already observed that the Axiom of Choice and Zorn's Lemma are equivalent. There is a myriad of axioms equivalent to the Axiom of Choice. Let us mention just two of them.

**Hausdorff Maximal Principle**. *Every partially ordered set contains a maximal chain* (i.e., *a maximal simply ordered subset*).

**Zermelo Well-Ordering Principle**. *Every set may be well-ordered.*

In particular, the set $\mathbb{R}$ of all reals may be well-ordered. This is a pure existence result, not exhibiting (or constructing or even defining) a well-ordering of $\mathbb{R}$. In fact, Feferman showed in 1965 that no *defined* partial ordering can be proved in *ZFC* to well-order the set $\mathbb{R}$.

If $X$ and $Y$ are any sets, properly well-ordered, and if there exists a one-to-one order-preserving correspondence between them (i.e., an injective and surjective mapping $\Phi\colon X \to Y$ such that $x_1 < x_2$ in $(X, \leq)$ if and only if $\Phi(x_1) < \Phi(x_2)$ in $(Y, \leq)$), then $X$ and $Y$ are said to have the same *ordinal number*. If two well-ordered sets have the same ordinal number, then they have the same cardinal number. However, unlike the notion of cardinal number, the notion of ordinal number depends on the well-orderings that well-order the sets.

**Proposition 1.12.** *There is an uncountable set $X$, well-ordered by a relation $\leq$ on it, with the following properties. $X$ has a greatest element $\Omega$, and the set $\{x \in X\colon\ x < z\}$ is countable for every $z$ in $X \setminus \{\Omega\}$.*

*Proof.* Let $Y$ be an uncountable set. By the Well-Ordering Principle, there exists a well-ordering of $Y$. Take $\zeta$ not in $Y$, set $Z = Y \cup \{\zeta\}$, and extend the well-ordering of $Y$ to $Z$ by setting $y < \zeta$ for every $y \in Y$. Consider the set $A = \{\alpha \in Z\colon\ \{z \in Z\colon\ z < \alpha\}$ is an uncountable set$\}$. $A$ is nonempty ($\zeta \in A$ because $Y = \{z \in Z\colon\ z < \zeta\}$ is uncountable), and hence it has a minimum (for $\varnothing \neq A \subseteq Z$ and $Z$ is well-ordered). Set $\Omega = \min A$ so that $X = \{z \in Z\colon\ z \leq \Omega\}$ is the required set. $\qquad\square$

Moreover, it can be shown that such a well-ordered set $X$ is unique in the sense that, if $Y$ is any well-ordered set with the same properties of $X$, then there exists a one-to-one order-preserving correspondence between $X$ and $Y$ (i.e., then $X$ and $Y$ have the same ordinal number). The greatest (or the last) element $\Omega$ in $X$ is called the *least* or *first uncountable ordinal*, and the elements $x$ of $X$ such that $x < \Omega$

are called *countable ordinals*. The greatest elements of the finite subsets of $X$ are called *finite ordinals*. If $\omega$ is the *first infinite ordinal* (i.e., the least nonfinite ordinal), then the set $\{x \in X : x < \omega\}$ of all finite ordinals and the set of all natural numbers $\mathbb{N}$ (equipped with its natural ordering) have the same ordinal number. It is usual to assign the symbol $\omega$ as the ordinal number of any well-ordered set that is in a one-to-one order-preserving correspondence with $\mathbb{N}$.

## Suggested Reading

Binmore [1]

Brown and Pearcy [2]

Crossley et al. [1]

Dugundji [1]

Fraenkel, Bar-Hillel and Levy [1]

Halmos [3]

Kelley [1]

Kolmogorov and Fomin [1]

Moore [1]

Royden [1]

Simmons [1]

Suppes [1]

Vaught [1]

Wilder [1]

# Problems

**Problem 1.1.** Let $A$, $B$ and $C$ be arbitrary sets. Prove that

(a)  $(A \backslash B) \cup (B \backslash A) = (A \cup B) \backslash (A \cap B)$    (symmetric difference);

(b)  $(A \triangledown B) \cup (B \triangledown C) = (A \cup B \cup C) \backslash (A \cap B \cap C)$;

(c)  $X \backslash (A \cup B) = (X \backslash A) \cap (X \backslash B)$  and  $X \backslash (A \cap B) = (X \backslash A) \cup (X \backslash B)$ whenever $A$ and $B$ are subsets of $X$    (De Morgan laws).

**Problem 1.2.** Consider a function $F : X \to Y$ from a set $X$ to a set $Y$. Let $A$, $A_1$ and $A_2$ be arbitrary subsets of $X$, and let $B$, $B_1$ and $B_2$ be arbitrary subsets of $Y$. Verify the following propositions.

(a)  $F(X \backslash A) = F(X) \backslash F(A)$.

(b)  $F^{-1}(Y \backslash B) = X \backslash F^{-1}(B)$.

(c)  $A_1 \subseteq A_2 \implies F(A_1) \subseteq F(A_2)$.

(d)  $B_1 \subseteq B_2 \implies F^{-1}(B_1) \subseteq F^{-1}(B_2)$.

(e)  $F(A_1 \cup A_2) = F(A_1) \cup F(A_2)$.

(f)  $F(A_1 \cap A_2) \subseteq F(A_1) \cap F(A_2)$.

(g)  $F^{-1}(B_1 \cup B_2) = F^{-1}(B_1) \cup F^{-1}(B_2)$.

(h)  $F^{-1}(B_1 \cap B_2) = F^{-1}(B_1) \cap F^{-1}(B_2)$.

(i)  $A \subseteq F^{-1}(F(A))$.

(j)  $F(F^{-1}(B)) \subseteq B$.

**Problem 1.3.** Consider the setup of Problem 1.2. Show that

(a)  $F$ is injective if and only if the inverse image under $F$ of each singleton in $\mathcal{R}(F)$ is a singleton in $X$;

(b)  $F$ is injective if and only if $F(A_1 \cap A_2) = F(A_1) \cap F(A_2)$ for every $A_1$, $A_2 \subseteq X$;

(c)  $F$ is injective if and only if the images of disjoint sets in $X$ are disjoint in $Y$;

(d)  $F$ is injective if and only if $A = F^{-1}(F(A))$ for every $A \subseteq X$;

(e)  $F$ is surjective if and only if the inverse image under $F$ of each nonempty subset of $Y$ is a nonempty subset of $X$;

(f)  $F$ is surjective if and only if $F(F^{-1}(B)) = B$ for every $B \subseteq Y$.

**Problem 1.4.** Verify that a function $F: X \to X$ is idempotent if and only if the range of $F$ coincides with the set of all fixed points of $F$.

**Problem 1.5.** A function $L: Y \to X$ is said to be a *left inverse* of a function $F: X \to Y$ if $LF = I_X$, the identity on $X$. $F$ is injective if and only if it has a left inverse. The restriction of a left inverse of $F$ to the range of $F$ is unique and coincides with the inverse of $F$ on $\mathcal{R}(F)$. (Recall: an injective function has an inverse on its range.) Prove.

**Problem 1.6.** A function $R: Y \to X$ is said to be a *right inverse* of a function $F: X \to Y$ if $FR = I_Y$, the identity on $Y$. Show that $F$ is surjective if and only if it has a right inverse. Note that any right inverse of $F$ is injective (for it has a left inverse). Similarly, any left inverse of $F$ is surjective (for it has a right inverse). Conclusion: There exists an injective mapping of $X$ into $Y$ if and only if there exists a surjective mapping of $Y$ onto $X$.

**Problem 1.7.** A function $F: X \to Y$ is injective and surjective if and only if there exists a function $G: Y \to X$ such that $GF = I_X$ (the identity on $X$) and $FG = I_Y$ (the identity on $Y$). Prove the above proposition and show that, in this case, such a function $G$ is unique and coincides with the inverse of $F$.

**Problem 1.8.** If $F: X \to Y$ is an invertible function, then so is its inverse $F^{-1}:$ $Y \to X$ and $(F^{-1})^{-1} = F$. Moreover, if $A \subseteq X$, then $F|_A: A \to F(A)$ is invertible and $(F|_A)^{-1} = F^{-1}|_{F(A)}$. Prove.

**Problem 1.9.** Verify the following propositions.

(a)  The composition of two injective functions is an injective function.

(b)  The composition of two surjective functions is a surjective function.

(c)  The composition of two invertible functions is an invertible function.

Note: When we speak of the composition $G \circ F$ of two functions $F$ and $G$ it is assumed that the domain of $G$ includes the range of $F$.

**Problem 1.10.** Let $F: X \to Y$ and $G: Y \to Z$ be invertible mappings. Show that $(G \circ F)^{-1} = F^{-1} \circ G^{-1}$.

**Problem 1.11.** A function $F: X \to X$ is an *involution* if $F^2 = I$. Verify that an involution is precisely an invertible function on $X$ that coincides with its inverse: $F = F^{-1}$. Show that the composition of two involutions is again an involution if and only if they commute.

**Problem 1.12.** Let $F: X \to Y$ be a one-to-one mapping of a set $X$ onto a set $Y$. Let $G: X \to A$ be a one-to-one mapping of $X$ onto a subset $A$ of $X$. Prove: If $A$ is a proper subset of $X$, then $F(A)$ is a proper subset of $Y$.

*Hint*: Consider the commutative diagram

$$
\begin{array}{ccc}
X & \xleftarrow{\;F^{-1}\;} & Y \\
{\scriptstyle G}\big\downarrow & & \big\downarrow{\scriptstyle H} \\
A & \xrightarrow{\;F|_A\;} & F(A).
\end{array}
$$

**Problem 1.13.** Let $X$ be a set with more than one element, and consider the following relations $R_1$, $R_2$, $R_3$, $R_4$ and $R_5$ on the power set $\wp(X)$ of $X$. For every pair $\{A, B\}$ of subsets of $X$

$A\, R_1\, B$  if  $A \triangledown B = \varnothing$    (i.e., if $A = B$),

$A\, R_2\, B$  if  $A \triangledown B$  is  finite,

$A\, R_3\, B$  if  $A \subseteq B$  or  $B \subseteq A$    (i.e., if $A \backslash B = \varnothing$ or $B \backslash A = \varnothing$),

$A\, R_4\, B$  if  $A \subseteq B$    (i.e., if $A \backslash B = \varnothing$).

Show that the table below properly classifies these relations according to reflexivity, transitivity, symmetry and antisymmetry.

|        | Reflexive | Transitive | Symmetric | Antisymmetric |
|--------|-----------|------------|-----------|---------------|
| $R_1$  | ✓         | ✓          | ✓         | ✓             |
| $R_2$  | ✓         | ✓          | ✓         |               |
| $R_3$  |           | ✓          | ✓         |               |
| $R_4$  | ✓         |            | ✓         |               |
| $R_5$  | ✓         | ✓          |           | ✓             |

*Hint*: To verify that $R_2$ is transitive use Problem 1.1(b) and recall that the union of two sets is finite if and only if each of them is finite.

**Problem 1.14.** Consider the functions $\Phi: \wp(X) \to \wp(X)$ and $H: X \to Y$ defined in the proof of Theorem 1.2. Show that $\Phi$ is an increasing function with respect to the inclusion ordering of $\wp(X)$, and that $H$ is injective and surjective.

**Problem 1.15.** Let $Y^X$ denote the collection of all functions from a set $X$ to a set $Y$. Suppose $Y$ is partially ordered and let $\leq$ be a partial ordering of $Y$. Now consider a relation on $Y^X$, also denoted by $\leq$ and defined as follows. For any pair $\{F, G\}$ of functions in $Y^X$,

$$F \leq G \quad \text{if} \quad F(x) \leq G(x) \ \text{ for every } \ x \in X.$$

(a) Show that $\leq$ is a partial ordering of $Y^X$.

(b) Prove: If $(Y, \leq)$ is a lattice, then $(Y^X, \leq)$ is a lattice.

> *Hint*: Suppose $(Y, \leq)$ is a lattice. Take $F$ and $G$ from $Y^X$ and let $U$ and $V$ be functions in $Y^X$ defined as follows. $U(x) = F(x) \vee G(x)$ and $V(x) = F(x) \wedge G(x)$ for every $x \in X$. Show that $F \vee G = U$ and $F \wedge G = V$.

**Problem 1.16.** Set $Y = \{0, 1\}$ and let $\chi_A: X \to \{0, 1\}$ be the characteristic function of an arbitrary subset $A$ of a set $X$. Thus, for every $A \in \wp(X)$, $\chi_A \in Y^X = 2^X$. Let $A$ and $B$ be subsets of $X$, and consider the partial ordering of $2^X$ introduced in Problem 1.15. Prove the following propositions.

(a) $\chi_A \leq \chi_B \iff A \subseteq B$.

(b) $\chi_A \vee \chi_B = \chi_{A \cup B}$.

(c) $\chi_A \wedge \chi_B = \chi_{A \cap B} = \chi_A \chi_B$.

(d)  $A \cap B = \varnothing \iff \chi_{A \cap B} = 0 \iff \chi_{A \cup B} = \chi_A + \chi_B.$

**Problem 1.17.** Let $\mathcal{F}$ be the collection of all functions from subsets of a set $X$ to a set $Y$. That is,

$$\mathcal{F} = \{F \in Y^A : A \subseteq X\} = \bigcup_{A \in \wp(X)} Y^A.$$

The unique function whose domain is the empty set $\varnothing$ is called the *empty function* in $\mathcal{F}$. Consider the following relation $\leq$ on $\mathcal{F}$. For any pair $\{F, G\}$ of functions in $\mathcal{F}$, $F \leq G$ if $F$ is a restriction of $G$ (equivalently, if $G$ is an extension of $F$). That is, $F \leq G$ if and only if $F: A \to Y$, $G: B \to Y$, $A \subseteq B \subseteq X$, and $F = G|_A$ (i.e., $F(x) = G(x)$ for every $x \in A$).

  (a)  Show that the relation $\leq$ on $\mathcal{F}$ is a partial ordering (called the *extension ordering*).

A function $V : C \to Y$ in $\mathcal{F}$ is a lower bound for a pair of functions $F : A \to Y$ and $G: B \to Y$ in $\mathcal{F}$ if $C \subseteq B, C \subseteq A$ and $V = F|_C = G|_C$. A function $U : D \to Y$ in $\mathcal{F}$ is an upper bound for the pair $\{F, G\}$ if $A \subseteq D$, $B \subseteq D$, $U|_A = F$ and $U|_B = G$.

  (b)  Show that every pair of functions $F$ and $G$ in $\mathcal{F}$ has an infimum $F \wedge G$ on $\mathcal{F}$. (In particular, if the domain $A$ of $F$ and the domain $B$ of $G$ are disjoint, then $F \wedge G$ is the empty function — which function is $F \wedge G$ if $A$ and $B$ are not disjoint but $F(A)$ and $G(B)$ are disjoint?)

Let $\{F_\gamma\}$ be an indexed family of functions in $\mathcal{F}$. For each $F_\gamma$ let $\mathcal{D}(F_\gamma)$ and $\mathcal{R}(F_\gamma)$ denote the domain and range of $F_\gamma$, respectively. Prove the following propositions.

  (c)  If $\{F_\gamma\}$ is simply ordered (i.e., if $\{F_\gamma\}$ is a chain in $\mathcal{F}$), then it has a supremum $\bigvee_\gamma F_\gamma$ in $\mathcal{F}$, whose domain and range are $\mathcal{D}(\bigvee_\gamma F_\gamma) = \bigcup_\gamma \mathcal{D}(F_\gamma)$ and $\mathcal{R}(\bigvee_\gamma F_\gamma) = \bigcup_\gamma \mathcal{R}(F_\gamma)$. Moreover, if each $F_\gamma$ is injective, then so is $\bigvee_\gamma F_\gamma$.

  (d)  If the domains $\{\mathcal{D}(F_\gamma)\}$ are pairwise disjoint, then the family $\{F_\gamma\}$ has a supremum $\bigvee_\gamma F_\gamma : \bigcup_\gamma \mathcal{D}(F_\gamma) \to \bigcup_\gamma \mathcal{R}(F_\gamma)$ in $\mathcal{F}$. Moreover, if the ranges $\{\mathcal{R}(F_\gamma)\}$ are also pairwise disjoint, and if each $F_\gamma$ is injective, then so is $\bigvee_\gamma F_\gamma$.

**Problem 1.18.** Let $\{X_n\}_{n \in \mathbb{N}}$ be a sequence of sets. Set $Y_1 = X_1$ and

$$Y_{n+1} = \left( \bigcup_{k=1}^{n+1} X_k \right) \backslash \left( \bigcup_{k=1}^{n} X_k \right)$$

for each $n \in \mathbb{N}$. Show by induction that

$$\bigcup_{k=1}^{n} Y_k = \bigcup_{k=1}^{n} X_k$$

for every $n \in \mathbb{N}$. (*Hint*: $A \cup (B \backslash A) = A \cup B$.) Verify that $Y_n \subseteq X_n$ for each $n \in \mathbb{N}$, and $Y_m \cap Y_n = \varnothing$ for every pair of distinct natural numbers $m$ and $n$. Moreover, show that

$$\bigcup_{n=1}^{\infty} Y_n = \bigcup_{n=1}^{\infty} X_n.$$

Such a disjoint sequence $\{Y_n\}_{n \in \mathbb{N}}$ is referred to as the *disjointification* of $\{X_n\}_{n \in \mathbb{N}}$.

**Problem 1.19.** Let $\mathscr{P}(X)$ be the power set of a set $X$ and consider the inclusion ordering of $\mathscr{P}(X)$. Let $\{X_n\}_{n=1}^{\infty}$ be an arbitrary $\mathscr{P}(X)$-valued sequence (i.e., a sequence of subsets of a given set $X$). Recall that, for each $n \in \mathbb{N}$, $\{X_k\}_{k=n}^{\infty}$ is an indexed family of subsets of $X$, and hence a subset of the complete lattice $\mathscr{P}(X)$. Let $\inf_{n \leq k} X_k$ and $\sup_{n \leq k} X_k$ denote $\inf\{X_k\}_{k=n}^{\infty}$ and $\sup\{X_k\}_{k=n}^{\infty}$, respectively; and set

$$Y_n = \inf_{n \leq k} X_k = \bigcap_{k=n}^{\infty} X_k \quad \text{and} \quad Z_n = \sup_{n \leq k} X_k = \bigcup_{k=n}^{\infty} X_k$$

so that $\{Y_n\}_{n=1}^{\infty}$ and $\{Z_n\}_{n=1}^{\infty}$ are $\mathscr{P}(X)$-valued sequences as well.

(a) Verify that $\{Y_n\}_{n=1}^{\infty}$ is an increasing sequence and $\{Z_n\}_{n=1}^{\infty}$ is a decreasing sequence.

The union $\bigcup_{n=1}^{\infty} Y_n$ and the intersection $\bigcap_{n=1}^{\infty} Z_n$, which are elements of $\mathscr{P}(X)$, are usually denoted by

$$\bigcup_{n-1}^{\infty} Y_n = \liminf_{n} X_n \quad \text{and} \quad \bigcap_{n=1}^{\infty} Z_n = \limsup_{n} X_n,$$

and called *limit inferior* and *limit superior* of $\{X_n\}_{n=1}^{\infty}$, respectively.

(b) Show that

$$\liminf_{n} X_n \subseteq \limsup_{n} X_n.$$

If $\liminf_{n} X_n = \limsup_{n} X_n$, then we say that the sequence $\{X_n\}_{n=1}^{\infty}$ *converges* to the *limit*

$$\lim_{n} X_n = \liminf_{n} X_n = \limsup_{n} X_n.$$

Prove the following propositions.

(c) If $\{X_n\}_{n=1}^{\infty}$ is an increasing sequence, then $Y_n = X_n$ for each $n \in \mathbb{N}$ and $Z_n = \bigcup_{m=1}^{\infty} X_m = \sup_{m} X_m$ for every $n \in \mathbb{N}$, so that

$$\liminf_{n} X_n = \limsup_{n} X_n = \sup_{n} X_n.$$

(d) If $\{X_n\}_{n=1}^{\infty}$ is itself decreasing, then $Y_n = \bigcap_{m=1}^{\infty} X_m = \inf_m X_m$ for every $n \in \mathbb{N}$ and $Z_n = X_n$ for each $n \in \mathbb{N}$, so that

$$\liminf_n X_n = \limsup_n X_n = \inf_n X_n.$$

Therefore, an increasing sequence of sets converges to its union $(\sup_n X_n = \bigcup_n X_n)$ and, dually, a decreasing sequence of sets converges to its intersection $(\inf_n X_n = \bigcap_n X_n)$; so that every monotone sequence of sets converges. Thus, since $\{Y_n\}_{n=1}^{\infty}$ and $\{Z_n\}_{n=1}^{\infty}$ are always increasing and decreasing, respectively, they do converge: $\{Y_n\}_{n=1}^{\infty}$ converges to its union $\bigcup_{n=1}^{\infty} Y_n = \bigcup_{n=1}^{\infty} \bigcap_{k=n}^{\infty} X_k$, and $\{Z_n\}_{n=1}^{\infty}$ converges to its intersection $\bigcap_{n=1}^{\infty} Z_n = \bigcap_{n=1}^{\infty} \bigcup_{k=n}^{\infty} X_k$.

(e) Verify the following identities.

$$\liminf_n X_n = \lim_n \left( \inf_{n \le k} X_k \right) \quad \text{and} \quad \limsup_n X_n = \lim_n \left( \sup_{n \le k} X_k \right).$$

(f) Now show that

$$X \setminus \limsup_n X_n = \liminf_n (X \setminus X_n) \quad \text{and} \quad X \setminus \liminf_n X_n = \limsup_n (X \setminus X_n).$$

Thus a sequence $\{X_n\}_{n=1}^{\infty}$ converges if and only if the sequence of its complements $\{X \setminus X_n\}_{n=1}^{\infty}$ converges and, in this case,

$$\lim_n (X \setminus X_n) = X \setminus \lim_n X_n.$$

**Problem 1.20.** Let $A$ and $B$ be subsets of the sets $X$ and $Y$, respectively. Show that

$$A \leftrightarrow B \quad \text{and} \quad X \setminus A \leftrightarrow Y \setminus B \quad \text{imply} \quad X \leftrightarrow Y.$$

(Warning: $X \leftrightarrow Y$ and $A \leftrightarrow B$ do not imply $X \setminus A \leftrightarrow Y \setminus B$).

**Problem 1.21.** Let $A$ and $B$ be any sets. Prove that

(a) $\qquad A \subseteq B \quad \text{implies} \quad \#A \le \#B \quad$ (hint: inclusion map),

(b) $\qquad A \subseteq B,\ B$ is finite, and $\#A = \#B \quad \text{imply} \quad A = B$;

and show that assertion (b) does not hold if $B$ is infinite.

**Problem 1.22.** For any sets $A$, $B$ and $C$, verify that

$$\#A \le \#B \le \#C \quad \text{implies} \quad \#A \le \#C.$$

Moreover, if $\#A < \#B \le \#C$ or $\#A \le \#B < \#C$, then $\#A < \#C$. (*Hint*: Cantor–Bernstein Theorem.)

**Problem 1.23.** Suppose the sets $A$, $B$, $C$ and $D$ are such that

$$\#A \le \#C \quad \text{and} \quad \#B \le \#D.$$

Prove the following propositions.

(a)  $\#(A \cup B) \le \#(C \cup D)$  whenever  $C \cap D = \varnothing$.

(b)  $\#(A \times B) \le \#(C \times D)$.

Moreover, if $\#A = \#C$ and $\#B = \#D$, then the cardinalities in (b) coincide too.

**Problem 1.24.** Let $Y^X$ denote the collection of all mappings of a set $X$ into a set $Y$. Show that, if $Y$ has more than one element, then

$$\#\wp(X) \le \#Y^X \quad \text{for every set } X.$$

(In fact, $\#\wp(X) = \#2^X = \#Y^X$ if $X$ is infinite and $2 \le \#Y \le \#X$).

**Problem 1.25.** Let $\mathbb{N}$, $\mathbb{Z}$ and $\mathbb{Q}$ have their standard meanings and set $\aleph_0 = \#\mathbb{N}$ as usual. Verify the following identities.

(a)  $\#(\mathbb{N} \times \mathbb{N}) = \aleph_0$.

*Hint*: Consider the function $F : \mathbb{N} \times \mathbb{N} \to \mathbb{N}$ defined by the formula $F(m, n) = \frac{(m+n-1)(m+n-2)}{2} + m$ for every $m, n \in \mathbb{N}$. $F$ is injective and surjective. The array

<pre>
n ↑
      5  11  ·   ·    ·    ·
      4   7  12  ·    ·    ·
      3   4   8  13   ·    ·
      2   2   5   9  14    ·
      1   1   3   6  10   15
                              m
          1   2   3   4   5   →
</pre>

may be suggestive.

(b)  $\#\mathbb{Z} = \aleph_0$.

*Hint*: Let $\mathbb{N}_e$ denote the set of all nonnegative even integers (including zero), and let $\mathbb{N}_o$ denote the set of all positive odd integers. Recall that $\#\mathbb{N}_0 = \#\mathbb{N}$, and note that $\#\mathbb{N}_0 = \#\mathbb{N}_e$ and $\#\mathbb{N} = \#\mathbb{N}_o$. (Reason: the functions $F : \mathbb{N}_0 \to \mathbb{N}_e$ and $G : \mathbb{N} \to \mathbb{N}_o$, given by $F(n) = 2n$ for every $n \in \mathbb{N}_0$ and $G(n) = 2n - 1$ for every $n \in \mathbb{N}$, are injective and surjective). Set $\mathbb{N}_- = \{k \in \mathbb{Z} : -k \in \mathbb{N}\}$ so that $\#\mathbb{N}_- = \#\mathbb{N}_o$. Use Problem 1.23(a) to show that $\#\mathbb{Z} \le \#\mathbb{N}$.

(c)  $\#\mathbb{Q} = \aleph_0$.

*Hint*: The function $F : \mathbb{Z} \times \mathbb{N} \to \mathbb{Q}$ defined by $F(k, n) = \frac{k}{n}$ for every $k \in \mathbb{Z}$ and every $n \in \mathbb{N}$ is surjective. Use Problem 1.23(b).

**Problem 1.26.** The purpose of this problem is to prove that, if $X$ and $Y$ are nonempty sets, then
$$\#(X \cup Y) \leq \#(X \times Y).$$

(a)  First verify that the above assertion holds whenever $X$ and $Y$ are both (nonempty) finite sets.

Consider the relations $\sim_X$ and $\sim_Y$ on the Cartesian product $X \times Y$ defined as follows. For every pair of pairs $(x_1, y_1)$ and $(x_2, y_2)$ in $X \times Y$,
$$(x_1, y_1) \sim_X (x_2, y_2) \quad \Longleftrightarrow \quad x_1 = x_2,$$
$$(x_1, y_1) \sim_Y (x_2, y_2) \quad \Longleftrightarrow \quad y_1 = y_2.$$

(b)  Show that $\sim_X$ and $\sim_Y$ are both equivalence relations on $X \times Y$.

Now take $x \in X$ and $y \in Y$ arbitrary and consider the equivalence classes, $[x] \subseteq X \times Y$ and $[y] \subseteq X \times Y$, of the point $(x, y) \in X \times Y$ with respect to $\sim_X$ and $\sim_Y$, respectively.

(c)  Show that  $[x] \leftrightarrow Y$  and  $[y] \leftrightarrow X$.

*Hint*:

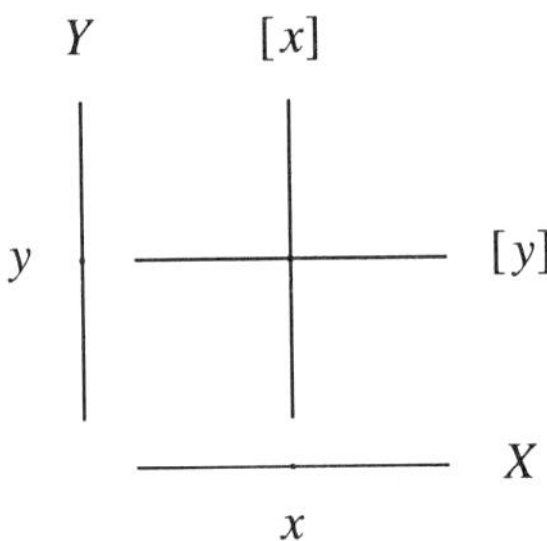


Next suppose one of the sets $X$ or $Y$, say $X$, is infinite and consider the singleton $\{(x, y)\}$ on $(x, y) \in X \times Y$.

(d)  Verify that  $\#X = \#([y] \setminus \{(x, y)\})$  and  $\#Y = \#[x]$.

(e)  Finally, apply the results of Problems 1.21(a), 1.22 and 1.23(a) to conclude that $\#(X \cup Y) \leq \#(X \times Y)$.

**Problem 1.27.** Prove the following propositions.

(a)  The union of two sets is countable if and only if each of them is countable.

   *Hint*: Problems 1.22, 1.23, 1.25 and 1.26.

(b)  Let $X$ be an infinite set, and let $A$ and $B$ be arbitrary subsets of $X$. The relation $\sim$ on $\wp(X)$ defined by

$$A \sim B \quad \text{if} \quad A \triangledown B \text{ is countable}$$

is an equivalence relation — compare with Problem 1.13.

**Problem 1.28.**  (a) Let $E$ be an infinite set. Suppose $A$ and $B$ are sets such that

$$\#A \le \#E \quad \text{and} \quad \#B \le \#E.$$

The following propositions, which are naturally linked to Problems 1.23 and 1.26, are in fact corollaries of Theorem 1.9. Prove them.

$$\#(A \cup B) \le \#E \quad \text{and} \quad \#(A \times B) \le \#E.$$

(b)  Now use Problem 1.23(b) and Theorem 1.9 to show by induction that, if $X$ is an infinite set, then

$$\#X^n = \#X \text{ for every } n \in \mathbb{N}.$$

**Problem 1.29.**  Let $\mathcal{I}$ be the collection of all injective functions from subsets of a set $X$ to itself. That is, set

$$\mathcal{I} = \left\{ F \in X^A: \quad A \subseteq X \text{ and } F \text{ is injective} \right\}.$$

We know from Problem 1.17 that, as a subset of $\bigcup_{A \in \wp(X)} X^A$, $\mathcal{I}$ is partially ordered in the extension ordering. Let $\mathcal{J}$ be the collection of all those functions $F$ in $\mathcal{I}$ for which the range $\mathcal{R}(F)$ is disjoint with the domain $\mathcal{D}(F)$:

$$\mathcal{J} = \left\{ F \in \mathcal{I}: \quad \mathcal{R}(F) \subseteq X \backslash \mathcal{D}(F) \right\}.$$

Problem 1.17 also tells us that every chain $\{F_\gamma\}$ in $\mathcal{J}$ has a supremum $\bigvee_\gamma F_\gamma$ in $\mathcal{I}$, and also that $\mathcal{D}\left(\bigvee_\gamma F_\gamma\right) = \bigcup_\gamma \mathcal{D}(F_\gamma)$ and $\mathcal{R}\left(\bigvee_\gamma F_\gamma\right) = \bigcup_\gamma \mathcal{R}(F_\gamma)$.

(a)  Show that $\bigvee_\gamma F_\gamma$ in fact lies in $\mathcal{J}$.

   *Hint*: Take $F_\lambda$ and $F_\mu$ arbitrary from $\{F_\gamma\} \subseteq \mathcal{J}$ so that $F_\lambda \le F_\mu$ (or vice versa). Note that $\mathcal{R}(F_\lambda) \cap \mathcal{D}(F_\mu) \subseteq \mathcal{R}(F_\mu) \cap \mathcal{D}(F_\mu) = \varnothing$, and conclude that $\bigcup_\gamma \mathcal{R}(F_\gamma) \cap \bigcup_\gamma \mathcal{D}(F_\gamma)$ is empty too.

Therefore, every chain in $\mathcal{J}$ has an upper bound (a supremum, actually) in $\mathcal{J}$. Thus, according to Zorn's Lemma, $\mathcal{J}$ contains a maximal function. Let $F_0$ be a maximal function of $\mathcal{J}$ and let $A_0$ be the domain of $F_0$, so that $F_0(A_0) \subseteq X \backslash A_0$. Suppose

$$\#A_0 < \#(X \backslash A_0)$$

and show that, if $X$ is an infinite set, then

(b)  there exist two distinct points, say $x_0$ and $x_1$, in $(X \setminus A_0) \setminus F_0(A_0)$.

*Hint*: Under the above assumption $\#F_0(A_0) < \#(X \setminus A_0)$ (why?) and $X \setminus A_0$ is infinite — recall: the union of finite sets is finite.

Now set $A_1 = A_0 \cup \{x_0\}$ and consider the function $F_1 : A_1 \to X$ defined by the formula

$$F_1(x) = \begin{cases} F_0(x) \in F_0(A_0), & x \in A_0, \\ x_1 \in X \setminus F_0(A_0), & x = x_0 \in X \setminus A_0. \end{cases}$$

(c)  Show that $F_1 \in \mathcal{J}$.

Since $F_0 = F_1|_{A_0}$ it follows that $F_0 \leq F_1$, which contradicts the fact that $F_0$ is a maximal of $\mathcal{J}$ (for $F_0 \neq F_1$). Therefore (by Theorem 1.7),

$$\#(X \setminus A_0) \leq \#A_0.$$

Next verify that $\#A_0 = \#F_0(A_0) \leq \#(X \setminus A_0)$ and conclude (by the Cantor–Bernstein Theorem) that

$$\#A_0 = \#(X \setminus A_0).$$

Finally, by using Problem 1.28(a), show that

$$\#A_0 = \#X.$$

Outcome: *If $X$ is an infinite set, then there exists a subset of it, say $A_0$, such that*

$$\#A_0 = \#X = \#(X \setminus A_0).$$

**Problem 1.30.**  Let $A$ be an arbitrary set and write $A' = A \times \{A\}$.

(a)  Show that $\#A' = \#A$.

*Hint*: The function that assigns to each $a \in A$ the ordered pair $(a, A)$ in the Cartesian product $A \times \{A\}$ is a one-to-one mapping of $A$ onto $A'$.

Clearly, $A \cap A' = \varnothing$. Moreover, if $B$ is any set such that $B \neq A$, then $A' \cap B' = \varnothing$ where $B' = B \times \{B\}$. Conclusion 1: If $A$ and $B$ are any sets, then there exist sets $C$ and $D$ such that

$$\#A = \#C, \quad \#B = \#D \quad \text{and} \quad C \cap D = \varnothing.$$

Now suppose $C_1, C_2, D_1$ and $D_2$ are sets with the following properties. $C_1 \cap D_1 = \varnothing$, $C_2 \cap D_2 = \varnothing$, $\#C_1 = \#C_2$ and $\#D_1 = \#D_2$.

(b)  Show that $\#(C_1 \cup D_1) = \#(C_2 \cup D_2)$.

Conclusion 2: $\#(C \cup D)$ is independent of the particular pair of sets $\{C, D\}$ employed in Conclusion 1. We are now prepared to define the sum of cardinal numbers. If $A$ and $B$ are sets, then

$$\#A + \#B \ = \ \#(C \cup D)$$

for any pair of sets $\{C, D\}$ such that

$$\#A = \#C, \quad \#B = \#D \quad \text{and} \quad C \cap D = \varnothing.$$

In particular, if $A \cap B = \varnothing$, then $\#A + \#B = \#(A \cup B)$.

(c) Use Problem 1.29 to show that $\#X + \#X = \#X$ for every infinite set $X$.

The definition of product of cardinal numbers is much simpler:

$$\#A \cdot \#B \ = \ \#(A \times B)$$

for any sets $A$ and $B$. According to Theorem 1.9, $\#X \cdot \#X = \#X$ for every infinite set $X$.

(d) Prove: If $X$ and $Y$ are two sets, at least one of which is infinite, then
$$\#X + \#Y = \#X \cdot \#Y = \max\{\#X, \#Y\}.$$

*Hint*: If $\#Y \leq \#X$, then $\#X \leq \#X + \#Y \leq \#X + \#X$ and $\#X \leq \#X \cdot \#Y \leq \#X \cdot \#X$.

# 2
# Algebraic Structures

The main algebraic structure that will be involved with the subject of this book is that of a "linear space" (or "vector space"). A linear space is a set endowed with an extra structure in addition to its set-theoretic structure (i.e., an extra structure that goes beyond the notions of functions and ordering, for instance). Roughly speaking, linear spaces are sets where two operations, called "addition" and "scalar multiplication", are properly defined so that we can refer to the "sum" of two points in a linear space, as well as to the "product" of a point in it by a "scalar". Although the reader is supposed to have already had some contact with linear algebra and, in particular, with "finite-dimensional vector spaces", we shall proceed from the very beginning. Our approach avoids the parochially "finite-dimensional" constructions (whenever this is possible), and focuses either on general results that do not depend on the "dimensionality" of the linear space, or on abstract "infinite-dimensional" linear spaces.

## 2.1   Linear Spaces

A *binary operation* on a set $X$ is a mapping of $X \times X$ into $X$. If $F$ is a function from $X \times X$ to $X$, then we generally write $z = F(x, y)$ to indicate that $z$ in $X$ is the value of $F$ at the point $(x, y)$ in $X \times X$. However, to emphasize the rule of the binary operation (the outcome of a binary operation on two points of $X$ is again a point of $X$), it is convenient (and usual) to adopt a different notation. Moreover, in order to emphasize the abstract character of a binary operation, it is also common to use a noncommittal symbol to denote it. Thus, if $\star$ is a binary operation on $X$ (so that $\star : X \times X \to X$), then we shall write $z = x \star y$ instead of $z = \star(x, y)$ to indicate that $z$ in $X$ is the value of $\star$ at the point $(x, y)$ in $X \times X$. If a binary operation $\star$ on $X$ has the property that

$$x \star (y \star z) = (x \star y) \star z$$

for every $x$, $y$ and $z$ in $X$, then it is said to be *associative*. In this case we shall drop the parentheses and write $x \star y \star z$. If there exists an element $e$ in $X$ such that

$$x \star e = e \star x = x$$

for every $x \in X$, then $e$ is said to be the *neutral element* (or the *identity element*) with respect to the binary operation $\star$ on $X$. It is easy to show that, if a binary operation $\star$ has a neutral element $e$, then $e$ is *unique*. If an associative binary operation $\star$ on $X$ has a neutral element $e$ in $X$, and if for some $x \in X$ there exists $x^{-1} \in X$ such that

$$x \star x^{-1} = x^{-1} \star x = e,$$

then $x^{-1}$ is called the *inverse* of $x$ with respect to $\star$. It is also easy to show that, if the inverse of $x$ exists with respect to an associative binary operation $\star$, then it is *unique*. A *group* is a nonempty set $X$ on which is defined a binary operation $\star$ such that

(a)  $\star$ is associative,

(b)  $\star$ has a neutral element in $X$, and

(c)  every $x$ in $X$ has an inverse in $X$ with respect to $\star$.

If a binary operation $\star$ on $X$ has the property that

$$x \star y = y \star x$$

for every $x$ and $y$ in $X$, then it is said to be *commutative*. If $X$ is a group with respect to a binary operation $\star$, and if

(d)  $\star$ is commutative,

then $X$ is said to be an *Abelian* (or *commutative*) *group*.

**Example 2A.**  The collection of all injective mappings of a nonempty set $X$ onto itself (i.e., the collection of all invertible mappings on $X$) is a non-Abelian group with respect to the composition operation $\circ$. The neutral element (or the identity element) of such a group is, of course, the identity map on $X$.

An *additive Abelian group* is an Abelian group $X$ for which the underlying binary operation is interpreted as an *addition* and denoted by $+$ (instead of $\star$). In this case the element $x + y$ (which lies in $X$ for every $x$ and $y$ in $X$) is called the *sum* of

$x$ and $y$. The (unique) neutral element with respect to addition is denoted by 0 (instead of $e$) and called *zero*. The (unique) inverse of $x$ under addition is denoted by $-x$ (instead of $x^{-1}$) and called the *negative* of $x$. Thus $x + 0 = 0 + x = x$ and $x + (-x) = (-x) + x = 0$ for every $x \in X$. Moreover, the operation of *subtraction* is defined by $x - y = x + (-y)$ and $x - y$ is called the *difference* between $x$ and $y$.

If $\diamond : X \times X \to X$ is another binary operation on $X$, and if

$$x \diamond (y \star z) = (x \diamond y) \star (x \diamond z) \quad \text{and} \quad (y \star z) \diamond x = (y \diamond x) \star (z \diamond x)$$

for every $x$, $y$ and $z$ in $X$, then $\diamond$ is said to be *distributive* with respect to $\star$. The above properties are the so-called *distributive laws*. A *ring* is an additive Abelian group $X$ with a second binary operation on $X$, called *multiplication* and denoted by $\cdot$, such that

(e)  the multiplication operation is associative and

(f)  distributive with respect to the addition operation.

In this case the element $x \cdot y$ (which lies in $X$ for every $x$ and $y$ in $X$) is called the *product* of $x$ and $y$ (alternative notation: $xy$ instead of $x \cdot y$). A *commutative ring* is a ring for which

(g)  the multiplication operation is commutative.

A *ring with identity* is a ring $X$ such that

(h)  the multiplication operation has a neutral element in $X$.

In this case such a (unique) neutral element in $X$ with respect to the multiplication operation is denoted by 1 (so that $x \cdot 1 = 1 \cdot x = x$ for every $x \in X$) and called the *identity*.

**Example 2B.**  The power set $\wp(X)$ of a nonempty set $X$ is a commutative ring with identity if addition is interpreted as symmetric difference (or Boolean sum) and multiplication as intersection (i.e., $A + B = A \triangledown B$ and $A \cdot B = A \cap B$ for all subsets $A$ and $B$ of $X$). Here the neutral element under addition (i.e., the zero) is the empty set $\varnothing$, and the neutral element under multiplication (i.e., the identity) is $X$ itself.

A ring with identity is *nontrivial* if it has another element besides the identity. (The set $\{0\}$ with the operations $0 + 0 = 0 \cdot 0 = 0$ is the trivial ring whose only element is the identity). If a ring with identity is nontrivial, then the neutral element under addition and the neutral element under multiplication never coincide (i.e., $0 \neq 1$). In fact, $x \cdot 0 = 0 \cdot x = 0$ for every $x$ in $X$ whenever $X$ is a ring (with or without identity). Incidentally (or not) this also shows that, in a nontrivial ring with identity, zero has no inverse with respect to the multiplication operation (i.e., there is no $x$ in $X$ such that $0 \cdot x = x \cdot 0 = 1$). A ring $X$ with identity is called a *division ring* if

(i) each nonzero $x$ in $X$ has an inverse in $X$ with respect to the multiplication operation.

That is, if $x \neq 0$ in $X$, then there exists a (unique) $x^{-1} \in X$ such that $x \cdot x^{-1} = x^{-1} \cdot x = 1$.

**Example 2C.** Let the addition and multiplication operations have their ordinary ("numerical") meanings. The set of all natural numbers $\mathbb{N}$ is not a group under addition; neither is the set of all nonnegative integers $\mathbb{N}_0$. However, the set of all integers $\mathbb{Z}$ is a commutative ring with identity, but not a division ring. The sets $\mathbb{Q}$, $\mathbb{R}$ and $\mathbb{C}$ (of all rational, real and complex numbers, respectively), when equipped with their respective operations of addition and multiplication, are all commutative division rings. These are infinite commutative division rings but there are finite commutative division rings (e.g., if we declare that $1 + 1 = 0 + 0 = 0$, $0 + 1 = 1 + 0 = 1$, $0 \cdot 0 = 0 \cdot 1 = 1 \cdot 0 = 0$, and $1 \cdot 1 = 1$, then $\{0, 1\}$ is a commutative division ring).

Roughly speaking, commutative division rings are the *number systems* of mathematics, and so they deserve a name of their own. A *field* is a nontrivial commutative division ring. The elements of a field are usually called *scalars*. We shall be particularly concerned with the fields $\mathbb{R}$ and $\mathbb{C}$ (the *real field* and the *complex field*). An arbitrary field will be denoted by $\mathbb{F}$. Summing up: A field $\mathbb{F}$ is a set with more than one element (at least 0 and 1 are distinct elements of it) equipped with two binary operations, called addition and multiplication, that satisfy all the properties (a) to (i) — clearly, with $\star$ replaced by $+$ in properties (a) to (d).

**Definition 2.1.** A *linear space* (or *vector space*) over a field $\mathbb{F}$ is a nonempty set $\mathcal{X}$ (whose elements are called *vectors*) satisfying the following axioms.

*Vector Addition.* $\mathcal{X}$ is an additive Abelian group under a binary operation $\dotplus$ called *vector addition.*

*Scalar Multiplication.* There is given a mapping of $\mathbb{F} \times \mathcal{X}$ into $\mathcal{X}$ that assigns to each scalar $\alpha$ in $\mathbb{F}$ and each vector $x$ in $\mathcal{X}$ a vector $\alpha x$ in $\mathcal{X}$. Such a mapping defines an operation, called *scalar multiplication*, with the following properties. For all scalars $\alpha$ and $\beta$ in $\mathbb{F}$, and all vectors $x$ and $y$ in $\mathcal{X}$,

$$\begin{aligned}
1x &= x, \\
\alpha(\beta x) &= (\alpha \cdot \beta)x, \\
\alpha(x \dotplus y) &= \alpha x \dotplus \alpha y, \\
(\alpha + \beta)x &= \alpha x \dotplus \beta x.
\end{aligned}$$

Some remarks on notation and terminology. The *underlying set* of a linear space is the nonempty set upon which the linear space is built. We shall use the same notation $\mathcal{X}$ for both the linear space and its underlying set, even though the underlying set alone has no algebraic structure of its own. A set $\mathcal{X}$ needs a binary operation on it, a field, and another operation involving such a field with $\mathcal{X}$ to acquire the necessary algebraic structure that will grant it the status of a linear space. The scalar 1 in the above definition stands, of course, for the identity in the field $\mathbb{F}$ with respect to the multiplication $\cdot$ in $\mathbb{F}$, and $+$ denotes the addition in $\mathbb{F}$. Observe that $+$ (addition in the field $\mathbb{F}$) and $\dotplus$ (addition in the group $\mathcal{X}$) are different binary operations. However, once the difference has been pointed out, we shall from now on use the same symbol $+$ to denote both of them. Moreover, we shall drop the dot also from the multiplication notation in $\mathbb{F}$, and write $\alpha\beta$ instead of $\alpha \cdot \beta$. The neutral element under the vector addition in $\mathcal{X}$ (i.e., the vector zero) is referred to as the *origin* of the linear space $\mathcal{X}$. Again, we shall use the same symbol 0 to denote both the origin in $\mathcal{X}$ and the scalar zero in $\mathbb{F}$. A linear space over $\mathbb{R}$ is called a *real linear space* and a linear space over $\mathbb{C}$ is called a *complex linear space*.

**Example 2D.** $\mathbb{R}$ itself is a linear space over $\mathbb{R}$. That is, the plain set $\mathbb{R}$ when equipped with the ordinary binary operations of addition and multiplication becomes a field, also denoted by $\mathbb{R}$. If vector addition is identified with scalar addition, then it becomes a real linear space, denoted again by $\mathbb{R}$. More generally, for each $n \in \mathbb{N}$, let $\mathbb{F}^n$ denote the Cartesian product of n copies of a field $\mathbb{F}$ (i.e., the set of all ordered n-tuples of scalars in $\mathbb{F}$). Now define vector addition and scalar multiplication coordinatewise, as usual:

$$x + y = (\xi_1 + \upsilon_1, \dots, \xi_n + \upsilon_n) \quad \text{and} \quad \alpha x = (\alpha\xi_1, \dots, \alpha\xi_n)$$

for every $x = (\xi_1, \dots, \xi_n)$ and $y = (\upsilon_1, \dots, \upsilon_n)$ in $\mathbb{F}^n$ and every $\alpha$ in $\mathbb{F}$. This makes $\mathbb{F}^n$ into a linear space over $\mathbb{F}$. In particular, $\mathbb{R}^n$ (the Cartesian product of $n$ copies of $\mathbb{R}$) and $\mathbb{C}^n$ (the Cartesian product of $n$ copies of $\mathbb{C}$) become real and complex linear spaces, respectively, whenever vector addition and scalar multiplication are defined coordinatewise. However, if we restrict scalar multiplication to real multiplication only, then $\mathbb{C}^n$ can also be made into a real linear space.

**Example 2E.** Let $S$ be a nonempty set and let $\mathbb{F}$ be an arbitrary field. Consider the set

$$\mathcal{X} = \mathbb{F}^S$$

of all functions from $S$ to $\mathbb{F}$ (i.e., the set of all *scalar-valued functions* on $S$, where "scalar-valued" stands for "$\mathbb{F}$-valued"). Let vector addition and scalar multiplication be defined pointwise. That is, if $x$ and $y$ are functions in $\mathcal{X}$ and $\alpha$ is a scalar in $\mathbb{F}$, then $x + y$ and $\alpha x$ are functions in $\mathcal{X}$ defined by

$$(x + y)(s) = x(s) + y(s) \quad \text{and} \quad (\alpha x)(s) = \alpha(x(s))$$

for every $s \in S$. Now it is easy to show that $\mathcal{X}$, when equipped with these two operations, in fact is a linear space over $\mathbb{F}$. Particular cases: $\mathbb{F}^{\mathbb{N}}$ (the set of all *scalar-valued sequences*) and $\mathbb{F}^{[0,1]}$ (the set of all scalar-valued functions on the interval

[0, 1]) are linear spaces over $\mathbb{F}$, whenever vector addition and scalar multiplication are defined pointwise. Note that the linear space $\mathbb{F}^n$ in the previous example also is a particular case of the present example, where the coordinatewise operations are identified with the pointwise operations (recall: $\mathbb{F}^n = \mathbb{F}^{\mathbb{I}_n}$ and $\mathbb{I}_n = \{i \in \mathbb{N} : i \le n\}$).

**Example 2F.** What was the role played by the field $\mathbb{F}$ in the previous example? Answer: Vector addition and scalar multiplication in $\mathbb{F}^S$ were defined pointwise by using addition and multiplication in $\mathbb{F}$. This suggests the following generalization of Example 2E. Let $S$ be a nonempty set, and let $\mathcal{Y}$ be an arbitrary linear space (over a field $\mathbb{F}$). Consider the set

$$\mathcal{X} = \mathcal{Y}^S$$

of all functions from $S$ to $\mathcal{Y}$ (i.e., the set of all $\mathcal{Y}$-valued functions on $S$). Let vector addition and scalar multiplication in $\mathcal{X}$ be defined pointwise by using vector addition and scalar multiplication in $\mathcal{Y}$. That is, if $f$ and $g$ are functions in $\mathcal{X}$ (so that $f(s)$ and $g(s)$ are elements of $\mathcal{Y}$ for each $s \in S$) and $\alpha$ is a scalar in $\mathbb{F}$, then $f + g$ and $\alpha f$ are functions in $\mathcal{X}$ defined by

$$(f + g)(s) = f(s) + g(s) \quad \text{and} \quad (\alpha f)(s) = \alpha(f(s))$$

for every $s \in S$. As before, it is easily verified that $\mathcal{X}$, when equipped with these operations, becomes a linear space over the same field $\mathbb{F}$. The origin of $\mathcal{X}$ is the *null function* $0 \colon S \to \mathcal{Y}$ (which is defined by $0(s) = 0$ for all $s \in S$). Examples 2D and 2E can be thought of as particular cases of this one.

**Example 2G.** Let $\mathcal{X}$ be a linear space over $\mathbb{F}$, and let $x$, $x'$, $y$ and $y'$ be arbitrary vectors in $\mathcal{X}$. An equivalence relation $\sim$ on $\mathcal{X}$ is *compatible with vector addition* if

$$x' \sim x \ \text{ and } \ y' \sim y \quad \text{imply} \quad x' + y' \sim x + y.$$

Similarly, it is said to be *compatible with scalar multiplication* if, for $x$ and $x'$ in $\mathcal{X}$ and $\alpha$ in $\mathbb{F}$,

$$x' \sim x \quad \text{implies} \quad \alpha x' \sim \alpha x.$$

If an equivalence relation $\sim$ on a linear space $\mathcal{X}$ is compatible with both vector addition and scalar multiplication, then we shall say that $\sim$ is a *linear equivalence relation*. Now consider $\mathcal{X}/\!\sim$, the quotient space of $\mathcal{X}$ modulo $\sim$ (i.e., the collection of all equivalence classes $[x]$ with respect to $\sim$), and suppose the equivalence relation $\sim$ on $\mathcal{X}$ is linear. In this case a binary operation $+$ on $\mathcal{X}/\!\sim$ can be defined by setting

$$[x] + [y] = [x + y]$$

for every $[x]$ and $[y]$ in $\mathcal{X}/\!\sim$. Indeed, since $\sim$ is compatible with vector addition, it follows that $[x + y]$ does not depend on which particular members $x$ and $y$ of the equivalence classes $[x]$ and $[y]$ were taken. Thus the operation $+$ actually is a function from $(\mathcal{X}/\sim \, \times \, \mathcal{X}/\sim)$ to $\mathcal{X}/\sim$. This defines vector addition in $\mathcal{X}/\sim$. Scalar multiplication in $\mathcal{X}/\sim$ can be similarly defined by setting

$$\alpha[x] \, = \, [\alpha x]$$

for every $[x]$ in $\mathcal{X}/\sim$ and every $\alpha$ in $\mathbb{F}$. Therefore, if $\sim$ is a linear equivalence relation on a linear space $\mathcal{X}$ over a field $\mathbb{F}$, then $\mathcal{X}/\sim$ becomes a linear space over $\mathbb{F}$ whenever vector addition and scalar multiplication in $\mathcal{X}/\sim$ are defined this way.

It is clear by the definition of a linear space $\mathcal{X}$ that $x + y + z$ is a well-defined vector in $\mathcal{X}$ whenever $x$, $y$ and $z$ are vectors in $\mathcal{X}$. Similarly, if $\{x_i\}_{i=1}^n$ is a *finite* set of vectors in $\mathcal{X}$, then the sum $x_1 + \cdots + x_n$, denoted by $\sum_{i=1}^n x_i$, is again a vector in $\mathcal{X}$. (The notion of *infinite sums* needs *topology* and we shall consider them in Chapters 4 and 5.)

## 2.2   Linear Manifolds

A *linear manifold* of a linear space $\mathcal{X}$ over $\mathbb{F}$ is a nonempty subset $\mathcal{M}$ of $\mathcal{X}$ with the following properties.

$$x + y \in \mathcal{M} \quad \text{and} \quad \alpha x \in \mathcal{M}$$

for every pair of vectors $x$, $y$ in $\mathcal{M}$ and every scalar $\alpha$ in $\mathbb{F}$. It is readily verified that a linear manifold $\mathcal{M}$ of a linear space $\mathcal{X}$ over a field $\mathbb{F}$ is itself a linear space over the same field $\mathbb{F}$. The origin $0$ of $\mathcal{X}$ is the origin of every linear manifold $\mathcal{M}$ of $\mathcal{X}$. The *zero linear manifold* is $\{0\}$, consisting of the single vector $0$. If a linear manifold $\mathcal{M}$ is a proper subset of $\mathcal{X}$, then it is said to be a *proper linear manifold*. A *nontrivial linear manifold* $\mathcal{M}$ of a linear space $\mathcal{X}$ is a nonzero proper linear manifold of it ($\{0\} \neq \mathcal{M} \neq \mathcal{X}$).

**Example 2H.** Let $\mathcal{M}$ be a linear manifold of a linear space $\mathcal{X}$ and consider a relation $\sim$ on $\mathcal{X}$ defined as follows. If $x$ and $x'$ are vectors in $\mathcal{X}$, then

$$x' \sim x \quad \text{if} \quad x' - x \in \mathcal{M}.$$

That is, $x' \sim x$ if $x'$ is *congruent to $x$ modulo $\mathcal{M}$* — notation: $x' \equiv x \pmod{\mathcal{M}}$. Since $\mathcal{M}$ is a linear manifold of $\mathcal{X}$, the relation $\sim$ in fact is an equivalence relation on $\mathcal{X}$ (reason: $0 \in \mathcal{M}$ — reflexivity, $x' - x'' = (x' - x) + (x - x'') \in \mathcal{M}$ whenever $x' - x$ and $x - x''$ lie in $\mathcal{M}$ — transitivity, and $x - x' \in \mathcal{M}$ whenever $x' - x \in \mathcal{M}$ — symmetry). The equivalence class (with respect to $\sim$)

$$[x] \, = \, \big\{x' \in \mathcal{X}\colon \; x' \sim x\big\} \, = \, \big\{x' \in \mathcal{X}\colon \; x' = x + z \text{ for some } z \in \mathcal{M}\big\}$$

of a vector $x$ in $\mathcal{X}$ is called the *coset of $x$ modulo $\mathcal{M}$* — notation: $[x] = x + \mathcal{M}$. The set of all cosets $[x]$ modulo $\mathcal{M}$ for every $x \in \mathcal{X}$ (i.e., the collection of all equivalence classes $[x]$ with respect to the equivalence relation $\sim$ for every $x$ in $\mathcal{X}$) is precisely the quotient space $\mathcal{X}/\sim$ of $\mathcal{X}$ modulo $\sim$. Following the terminology introduced in Example 2G, $\sim$ is a *linear equivalence relation* on the linear space $\mathcal{X}$. Indeed, if $x' - x \in \mathcal{M}$ and $y' - y \in \mathcal{M}$, then $(x' + y') - (x + y) = (x' - x) + (y' - y) \in \mathcal{M}$ and $\alpha x' - \alpha x = \alpha(x' - x) \in \mathcal{M}$ for every scalar $\alpha$, so that $x' \sim x$ and $y' \sim y$ imply $x' + y' \sim x + y$ and $\alpha x' \sim \alpha x$. Therefore, with vector addition and scalar multiplication defined by

$$[x] + [y] = [x + y] \quad \text{and} \quad \alpha[x] = [\alpha x],$$

$\mathcal{X}/\sim$ is made into a linear space. This is usually denoted by $\mathcal{X}/\mathcal{M}$ (instead of $\mathcal{X}/\sim$) and called the *quotient space of $\mathcal{X}$ modulo $\mathcal{M}$*. The origin of $\mathcal{X}/\mathcal{M}$ is, of course, $[0] = \mathcal{M}$.

If $\mathcal{M}$ and $\mathcal{N}$ are linear manifolds of a linear space $\mathcal{X}$, then the *sum* of $\mathcal{M}$ and $\mathcal{N}$, denoted by $\mathcal{M} + \mathcal{N}$, is the subset of $\mathcal{X}$ made up of all sums $x + y$ where $x$ is a vector in $\mathcal{M}$ and $y$ is a vector in $\mathcal{N}$:

$$\mathcal{M} + \mathcal{N} = \{z \in \mathcal{X}: \ z = x + y, \ x \in \mathcal{M} \text{ and } y \in \mathcal{N}\}.$$

It is trivially verified that $\mathcal{M} + \mathcal{N}$ is a linear manifold of $\mathcal{X}$. If $\{\mathcal{M}_i\}_{i=1}^n$ is a finite family of linear manifolds of a linear space $\mathcal{X}$, then the sum $\sum_{i=1}^n \mathcal{M}_i$ is the linear manifold $\mathcal{M}_1 + \cdots + \mathcal{M}_n$ of $\mathcal{X}$ consisting of all sums $\sum_{i=1}^n x_i$ where each vector $x_i$ lies in $\mathcal{M}_i$. More generally, if $\{\mathcal{M}_\gamma\}_{\gamma \in \Gamma}$ is an arbitrary indexed family of linear manifolds of a linear space $\mathcal{X}$, then the *sum* $\sum_{\gamma \in \Gamma} \mathcal{M}_\gamma$ is defined as the set of all sums $\sum_{\gamma \in \Gamma} x_\gamma$ with $x_\gamma \in \mathcal{M}_\gamma$ for each index $\gamma$ and $x_\gamma = 0$ except for some finite set of indices (i.e., $\sum_{\gamma \in \Gamma} \mathcal{M}_\gamma$ is the set made up of *all finite sums* with each summand being a vector in one of the linear manifolds $\mathcal{M}_\gamma$). Clearly, $\sum_{\gamma \in \Gamma} \mathcal{M}_\gamma$ is itself a linear manifold of $\mathcal{X}$, and $\mathcal{M}_\alpha \subseteq \sum_{\gamma \in \Gamma} \mathcal{M}_\gamma$ for every $\mathcal{M}_\alpha \in \{\mathcal{M}_\gamma\}_{\gamma \in \Gamma}$.

A linear manifold of a linear space $\mathcal{X}$ is never empty: the origin of $\mathcal{X}$ is always there. Note that the intersection $\mathcal{M} \cap \mathcal{N}$ of two linear manifolds $\mathcal{M}$ and $\mathcal{N}$ of a linear space $\mathcal{X}$ is itself a linear manifold of $\mathcal{X}$. In fact, if $\{\mathcal{M}_\gamma\}_{\gamma \in \Gamma}$ is an arbitrary collection of linear manifolds of a linear space $\mathcal{X}$, then the intersection $\bigcap_{\gamma \in \Gamma} \mathcal{M}_\gamma$ is again a linear manifold of $\mathcal{X}$. Moreover, $\bigcap_{\gamma \in \Gamma} \mathcal{M}_\gamma \subseteq \mathcal{M}_\alpha$ for every $\mathcal{M}_\alpha \in \{\mathcal{M}_\gamma\}_{\gamma \in \Gamma}$.

Now consider the collection $\mathcal{L}at(\mathcal{X})$ of all linear manifolds of a linear space $\mathcal{X}$. Since $\mathcal{L}at(\mathcal{X})$ is a subcollection of the power set $\wp(\mathcal{X})$, it follows that $\mathcal{L}at(\mathcal{X})$ is partially ordered in the inclusion ordering. If $\{\mathcal{M}_\gamma\}_{\gamma \in \Gamma}$ is any subcollection of $\mathcal{L}at(\mathcal{X})$, then $\sum_{\gamma \in \Gamma} \mathcal{M}_\gamma$ in $\mathcal{L}at(\mathcal{X})$ is an upper bound for $\{\mathcal{M}_\gamma\}_{\gamma \in \Gamma}$ and $\bigcap_{\gamma \in \Gamma} \mathcal{M}_\gamma$ in $\mathcal{L}at(\mathcal{X})$ is a lower bound for $\{\mathcal{M}_\gamma\}_{\gamma \in \Gamma}$. If $\mathcal{U}$ in $\mathcal{L}at(\mathcal{X})$ is an upper bound for $\{\mathcal{M}_\gamma\}_{\gamma \in \Gamma}$ (i.e., if $\mathcal{M}_\gamma \subseteq \mathcal{U}$ for all $\gamma \in \Gamma$), then $\sum_{\gamma \in \Gamma} \mathcal{M}_\gamma \subseteq \mathcal{U}$. Therefore

$$\sum_{\gamma \in \Gamma} \mathcal{M}_\gamma = \sup\{\mathcal{M}_\gamma\}_{\gamma \in \Gamma}.$$

Similarly, if $\mathcal{V}$ in $\mathit{Lat}(\mathcal{X})$ is a lower bound for $\{\mathcal{M}_\gamma\}_{\gamma\in\Gamma}$ (i.e., if $\mathcal{V} \subseteq \mathcal{M}_\gamma$ for all $\gamma \in \Gamma$), then $\mathcal{V} \subseteq \bigcap_{\gamma\in\Gamma}\mathcal{M}_\gamma$. Thus

$$\bigcap_{\gamma\in\Gamma} \mathcal{M}_\gamma \;=\; \inf\{\mathcal{M}_\gamma\}_{\gamma\in\Gamma}.$$

Conclusion: $\mathit{Lat}(\mathcal{X})$ is a complete lattice. *The collection of all linear manifolds of a linear space is a complete lattice in the inclusion ordering.* If $\{\mathcal{M}, \mathcal{N}\}$ is a pair of elements of $\mathit{Lat}(\mathcal{X})$, then $\mathcal{M} \vee \mathcal{N} = \mathcal{M} + \mathcal{N}$ and $\mathcal{M} \wedge \mathcal{N} = \mathcal{M} \cap \mathcal{N}$.

Let $A$ be an arbitrary subset of a linear space $\mathcal{X}$, and consider the subcollection (a sublattice, actually) $\mathcal{L}_A$ of the complete lattice $\mathit{Lat}(\mathcal{X})$,

$$\mathcal{L}_A \;=\; \bigl\{\mathcal{M} \in \mathit{Lat}(\mathcal{X}): \; A \subseteq \mathcal{M}\bigr\},$$

consisting of all linear manifolds of $\mathcal{X}$ that include $A$. Set

$$\operatorname{span} A \;=\; \inf \mathcal{L}_A \;=\; \bigcap \mathcal{L}_A,$$

which is called the (linear) *span* of $A$. Since $A \subseteq \bigcap \mathcal{L}_A$ (for $A \subseteq \mathcal{M}$ for every $\mathcal{M} \in \mathcal{L}_A$), it follows that $\inf \mathcal{L}_A = \min \mathcal{L}_A$ so that $\operatorname{span} A \in \mathcal{L}_A$. Thus span $A$ *is the smallest linear manifold of $\mathcal{X}$ that includes $A$*, which coincides with *the intersection of all linear manifolds of $\mathcal{X}$ that includes $A$*. It is readily verified that $\operatorname{span} \varnothing = \{0\}$, $\operatorname{span} \mathcal{M} = \mathcal{M}$ for every $\mathcal{M} \in \mathit{Lat}(\mathcal{X})$, and $A \subseteq \operatorname{span} A = \operatorname{span}(\operatorname{span} A)$ for every $A \in \wp(\mathcal{X})$. Moreover, if $A$ and $B$ are subsets of $\mathcal{X}$, then

$$A \subseteq B \quad \text{implies} \quad \operatorname{span} A \subseteq \operatorname{span} B.$$

If $\mathcal{M}$ and $\mathcal{N}$ are linear manifolds of a linear space $\mathcal{X}$, then it is clear that $\mathcal{M} \cup \mathcal{N} \subseteq \mathcal{M} + \mathcal{N}$. Moreover, if $\mathcal{K}$ is a linear manifold of $\mathcal{X}$ such that $\mathcal{M} \cup \mathcal{N} \subseteq \mathcal{K}$, then $x + y \in \mathcal{K}$ for every $x \in \mathcal{M}$ and every $y \in \mathcal{N}$, and hence $\mathcal{M} + \mathcal{N} \subseteq \mathcal{K}$. Thus $\mathcal{M} + \mathcal{N}$ is the smallest linear manifold of $\mathcal{X}$ that includes $\mathcal{M} \cup \mathcal{N}$, which means that

$$\mathcal{M} + \mathcal{N} \;=\; \operatorname{span}(\mathcal{M} \cup \mathcal{N}).$$

More generally, let $\{\mathcal{M}_\gamma\}_{\gamma\in\Gamma}$ be an arbitrary subcollection of $\mathit{Lat}(\mathcal{X})$, and suppose $\mathcal{K} \in \mathit{Lat}(\mathcal{X})$ is such that $\bigcup_{\gamma\in\Gamma}\mathcal{M}_\gamma \subseteq \mathcal{K}$. Then every (finite) sum $\sum_{\gamma\in\Gamma} x_\gamma$ with each $x_\gamma$ in $\mathcal{M}_\gamma$ is a vector in $\mathcal{K}$. Thus $\sum_{\gamma\in\Gamma}\mathcal{M}_\gamma \subseteq \mathcal{K}$. Since $\bigcup_{\gamma\in\Gamma}\mathcal{M}_\gamma \subseteq \sum_{\gamma\in\Gamma}\mathcal{M}_\gamma$, it follows that $\sum_{\gamma\in\Gamma}\mathcal{M}_\gamma$ is the smallest element of $\mathit{Lat}(\mathcal{X})$ that includes $\bigcup_{\gamma\in\Gamma}\mathcal{M}_\gamma$. Equivalently,

$$\sum_{\gamma\in\Gamma} \mathcal{M}_\gamma \;=\; \operatorname{span}\left(\bigcup_{\gamma\in\Gamma} \mathcal{M}_\gamma\right).$$

## 2.3   Linear Independence

Let $A$ be a nonempty subset of a linear space $\mathcal{X}$. A vector $x \in \mathcal{X}$ is a *linear combination* of vectors in $A$ if there exist a *finite* set $\{x_i\}_{i=1}^n$ of vectors in $A$ and a *finite* family of scalars $\{\alpha_i\}_{i=1}^n$ such that

$$x = \sum_{i=1}^n \alpha_i x_i.$$

Warning: A linear combination is, by definition, *finite*. That is, a linear combination of vectors in a set $A$ is a weighted sum of a finite subset of vectors in $A$, weighted by a finite family of scalars, no matter whether $A$ is a finite or an infinite set. Since $\mathcal{X}$ is a linear space, any linear combination of vectors in $A$ is a vector in $\mathcal{X}$.

**Proposition 2.2.** *The set of all linear combinations of vectors in a nonempty subset $A$ of a linear space $\mathcal{X}$ is a linear manifold of $\mathcal{X}$ that coincides with* span $A$.

*Proof.* Let $A$ be an arbitrary subset of a linear space $\mathcal{X}$, consider the collection $\mathcal{L}_A$ of all linear manifolds of $\mathcal{X}$ that include $A$, and recall that

$$\text{span } A \ = \ \min \mathcal{L}_A.$$

Suppose $A$ is nonempty and let $\langle A \rangle$ denote the set of all linear combinations of vectors in $A$. It is clear that $A \subseteq \langle A \rangle$ (every vector in $A$ is a trivial linear combination of vectors in $A$), and that $\langle A \rangle$ is a linear manifold of $\mathcal{X}$ (if $x, y \in \langle A \rangle$, then $x + y$ and $\alpha x$ lie in $\langle A \rangle$). Thus

$$\langle A \rangle \in \mathcal{L}_A.$$

Moreover, if $\mathcal{M}$ is an arbitrary linear manifold of $\mathcal{X}$, and if $x \in \mathcal{X}$ is a linear combination of vectors in $\mathcal{M}$, then $x \in \mathcal{M}$ (because $\mathcal{M}$ is itself a linear space). Thus $\langle \mathcal{M} \rangle \subseteq \mathcal{M}$. Since $\mathcal{M} \subseteq \langle \mathcal{M} \rangle$, it follows that $\langle \mathcal{M} \rangle = \mathcal{M}$ for every linear manifold $\mathcal{M}$ of $\mathcal{X}$. Furthermore, if $\mathcal{M} \in \mathcal{L}_A$, then $A \subseteq \mathcal{M}$ and hence $\langle A \rangle \subseteq \langle \mathcal{M} \rangle$ (reason: $\langle A \rangle \subseteq \langle B \rangle$ whenever $A$ and $B$ are nonempty subsets of $\mathcal{X}$ such that $A \subseteq B$). Therefore,

$$\mathcal{M} \in \mathcal{L}_A \quad \text{implies} \quad \langle A \rangle \subseteq \mathcal{M}.$$

Conclusion: $\langle A \rangle$ is the smallest element of $\mathcal{L}_A$. That is,

$$\langle A \rangle \ = \ \text{span } A. \qquad\qquad \square$$

Following the notation introduced in the proof of Proposition 2.2, $\langle A \rangle = \text{span } A$ whenever $A \neq \varnothing$. Set $\langle \varnothing \rangle = \text{span } \varnothing$ so that $\langle \varnothing \rangle = \{0\}$, and hence $\langle A \rangle$ is well-defined for every subset $A$ of $\mathcal{X}$. We shall use one and the same notation, viz. span $A$, for both of them: the set of all linear combinations of vectors in $A$ and the (linear) span of $A$. For this reason span $A$ is also referred to as the *linear manifold generated* (or *spanned*) by $A$. If a linear manifold $\mathcal{M}$ of $\mathcal{X}$ (which may be $\mathcal{X}$ itself) is such that span $A = \mathcal{M}$ for some subset $A$ of $\mathcal{X}$, then we say that $A$ *spans* $\mathcal{M}$.

A subset $A$ of a linear space $\mathcal{X}$ is said to be *linearly independent* if each vector $x$ in $A$ is not a linear combination of vectors in $A\backslash\{x\}$. Equivalently, $A$ is linearly independent if $x \notin \text{span}\,(A\backslash\{x\})$ for every $x \in A$. If a set $A$ is not linearly independent, then it is said to be *linearly dependent*. Note that the empty set $\varnothing$ of a linear space $\mathcal{X}$ is linearly independent (there is no vector in $\varnothing$ that is a linear combination of vectors in $\varnothing$). Any singleton $\{x\}$ of $\mathcal{X}$ such that $x \neq 0$ is linearly independent. Indeed, span $(\{x\}\backslash\{x\}) = \text{span}\,\varnothing = \{0\}$ so that $x \notin \text{span}\,(\{x\}\backslash\{x\})$ if $x \neq 0$. However, $0 \in \text{span}\,(\{0\}\backslash\{0\}) = \{0\}$, and hence the singleton $\{0\}$ is not linearly independent. In fact, every subset of $\mathcal{X}$ that contains the origin of $\mathcal{X}$ is not linearly independent (reason: if $0 \in A \subseteq \mathcal{X}$ and $A$ has another vector besides the origin, say $x \neq 0$, then $0 = 0x$). Thus, if a vector $x$ is an element of a linearly independent subset of a linear space $\mathcal{X}$, then $x \neq 0$.

**Proposition 2.3.** *Let $A$ be a nonempty subset of a linear space $\mathcal{X}$. The following assertions are pairwise equivalent.*

(a) *$A$ is linearly independent.*

(b) *Each nonzero vector in* span $A$ *has a unique representation as a linear combination of vectors in $A$.*

(c) *Every finite subset of $A$ is linearly independent.*

(d) *There is no proper subset of $A$ whose span coincides with* span $A$.

*Proof.* The statement (b) can be rewritten as follows.

(b$'$) For every nonzero vector $x \in \text{span}\,A$ there exist a unique finite family of scalars $\{\alpha_i\}_{i=1}^{n}$ and a unique finite subset $\{a_i\}_{i=1}^{n}$ of $A$ such that $x = \sum_{i=1}^{n}\alpha_i\,a_i$.

*Proof of* (a)$\Rightarrow$(b). Suppose $A \neq \varnothing$ is linearly independent and take an arbitrary nonzero $x \in \text{span}\,A$. Consider two representations of $x$ as a linear combination of vectors in $A$:

$$x = \sum_{i=1}^{n}\beta_i\,b_i = \sum_{i=1}^{m}\gamma_i\,c_i,$$

where each $b_i$ and each $c_i$ are vectors in $A$ (and hence nonzero because $A$ is linearly independent). Since $x \neq 0$ we may assume that the scalars $\beta_i$ and $\gamma_i$ are all nonzero. Set $B = \{b_i\}_{i=1}^{n}$ and $C = \{c_i\}_{i=1}^{m}$, both finite nonempty subsets of $A$. Take an arbitrary $b \in B$ and note that $b$ is a linear combination of vectors in $(B\backslash\{b\}) \cup C$. However, since $b \in A$ and $A$ is linearly independent, it follows that $b$ is not a linear combination of any subset of $A\backslash\{b\}$. Thus $b \in C$. Similarly, take an arbitrary $c \in C$ and conclude that $c \in B$ by using the same argument. Hence

$B \subseteq C \subseteq B$. That is, $B = C$. Therefore $x = \sum_{i=1}^{n} \beta_i b_i = \sum_{i=1}^{n} \gamma_i b_i$, which implies that $\sum_{i=1}^{n} (\beta_i - \gamma_i) b_i = 0$. Since each $b_i$ is not a linear combination of vectors in $B \backslash \{b_i\}$, it follows that $\beta_i = \gamma_i$ for every $i$. Summing up: The two representations of $x$ coincide.

*Proof of* (b)$\Rightarrow$(a). If $A$ is nonempty and every nonzero vector $x$ in span $A$ has a unique representation as a linear combination of vectors in $A$, then the unique representation of an arbitrary $a$ in $A$ as a linear combination of vectors in $A$ is $a$ itself (recall: $A \subseteq$ span $A$). Therefore, every $a \in A$ is not a linear combination of vectors in $A \backslash \{a\}$, which means that $A$ is linearly independent.

*Proof of* (a)$\Leftrightarrow$(c). If $A$ is linearly independent, then every subset of it clearly is linearly independent. If $A$ is not linearly independent, then either $A = \{0\}$ or there exists $x \in A$ that is a linear combination of vectors, say $\{x_i\}_{i=1}^{n}$ for some $n \in \mathbb{N}$, in $A \backslash \{x\} \neq \varnothing$. In the former case $A$ is itself a finite subset of $A$ that is not linearly independent. In the latter case $\{x_i\}_{i=1}^{n} \cup \{x\}$ is a finite subset of $A$ that is not linearly independent. Conclusion: If every finite subset of $A$ is linearly independent, then $A$ is itself linearly independent.

*Proof of* (a)$\Rightarrow$(d). Recalling that $B \subseteq A$ implies span $B \subseteq$ span $A$, the statement (d) can be rewritten as follows.

(d$'$)  $B \subset A$  implies  span $B \subset$ span $A$.

Suppose $A$ is nonempty and linearly independent. Let $B$ be an arbitrary proper subset of $A$. If $B = \varnothing$, then (d$'$) holds trivially ($\varnothing \neq A \neq \{0\}$ so that span $\varnothing \subset$ span $A$). Thus suppose $B \neq \varnothing$ and take any $x \in A \backslash B$. If $x \in$ span $B$, then $x$ is a linear combination of vectors in $B$. This implies that $B \cup \{x\}$ is a subset of $A$ that is not linearly independent, and hence $A$ itself is not linearly independent, which is a contradiction. Therefore, $x \notin$ span $B$ for every $x \in A \backslash B$ whenever $\varnothing \neq B \subset A$. Since $x \in$ span $A$ (because $x \in A$), and since span $B \subseteq$ span $A$ (for $B \subset A$), it follows that span $B \subset$ span $A$ so that (d$'$) holds true.

*Proof of* (d)$\Rightarrow$(a). If $A$ is not linearly independent, then either $A = \{0\}$ or there exists $x \in A$ that is a linear combination of vectors in $A \backslash \{x\}$. In the former case the only proper subset of $A = \{0\}$ is $B = \varnothing$, and span $B = \{0\} =$ span $A$. In the latter case $B = A \backslash \{x\}$ is a proper subset of $A$ such that span $B =$ span $A$ (reason: span $B \subseteq$ span $A$ because $B \subseteq A$, and span $A \subseteq$ span $B$ because every vector in $A$ is a linear combination of vectors in $A \backslash \{x\}$). Therefore, (d$'$) implies (a).    $\square$

## 2.4   Hamel Basis

A linearly independent subset of a linear space $\mathcal{X}$ that spans $\mathcal{X}$ is called a *Hamel basis* (or a *linear basis*) for $\mathcal{X}$. In other words, a subset $B$ of a linear space $\mathcal{X}$ is a

Hamel basis for $\mathcal{X}$ if

$$\text{(i)} \quad B \text{ is linearly independent, and}$$

$$\text{(ii)} \quad \text{span } B = \mathcal{X}.$$

Let $B = \{x_\gamma\}_{\gamma \in \Gamma}$ be an indexed Hamel basis for a linear space $\mathcal{X}$. If $x$ is a nonzero vector in $\mathcal{X}$, then Proposition 2.3 ensures the existence of a unique (similarly indexed) family of scalars $\{\alpha_\gamma\}_{\gamma \in \Gamma}$ (which may depend on $x$) such that $\alpha_\gamma = 0$ for all but a finite set of indices $\gamma$, and $x = \sum_{\gamma \in \Gamma} \alpha_\gamma x_\gamma$. The weighted sum $\sum_{\gamma \in \Gamma} \alpha_\gamma x_\gamma$ (i.e., the unique representation of $x$ as a linear combination of vectors in $B$, or the unique (linear) representation of $x$ in terms of $B$) is called the *expansion* of $x$ on $B$, and the coefficients of it (i.e., the unique indexed family of scalars $\{\alpha_\gamma\}_{\gamma \in \Gamma}$) are called the *coordinates* of $x$ with respect to the indexed basis $B$. If $x = 0$, then its unique expansion on $B$ is the trivial one whose coefficients are all null.

Since $\varnothing$ is linearly independent, and since span $\varnothing = \{0\}$, it follows that the empty set $\varnothing$ is a Hamel basis for the zero linear space $\{0\}$. Now suppose $\mathcal{X}$ is a nonzero linear space. Every singleton $\{x\}$ in $\mathcal{X}$ such that $x \neq 0$ is linearly independent. Thus every nonzero linear space has many linearly independent subsets. If a linearly independent subset $A$ of $\mathcal{X}$ is not already a Hamel basis for $\mathcal{X}$, then we can construct a larger linearly independent subset of $\mathcal{X}$.

**Proposition 2.4.** *If $A$ is a linearly independent subset of a linear space $\mathcal{X}$, and if there exists $x \in \mathcal{X} \backslash \text{span } A$, then $A \cup \{x\}$ is a linearly independent subset of $\mathcal{X}$.*

*Proof.* Suppose there exists a vector $x$ in $\mathcal{X} \backslash \text{span } A$. Note that $x \neq 0$, and hence $\mathcal{X} \neq \{0\}$. If $A = \varnothing$, then the result is trivially verified ($\{x\} = \varnothing \cup \{x\}$ is linearly independent). Thus suppose $A$ is nonempty and set $C = A \cup \{x\} \subset \mathcal{X}$. Since $x \notin \text{span } A$, it follows that $x \notin \text{span } (C \backslash \{x\})$. Take an arbitrary $a \in A$. Suppose $a \in A$ is a linear combination of vectors in $C \backslash \{a\}$. Clearly, $a \neq \alpha x$ for every scalar $\alpha$ (for $x \notin \text{span } A$ and $a \neq 0$ because $A$ is linearly independent) so that

$$a = \alpha_0 x + \sum_{i=1}^{n} \alpha_i a_i,$$

where each $a_i$ is a vector in $A \backslash \{a\}$ and each $\alpha_i$ is a nonzero scalar (recall: $0 \neq a \neq \sum_{i=1}^{n} \alpha_i a_i$ for $A$ is linearly independent). Thus $x$ is a linear combination of vectors in $A$, which contradicts the assumption that $x \notin \text{span } A$. Therefore, every $a \in A$ is not a linear combination of vectors in $C \backslash \{a\}$. Conclusion: Every $c \in C$ is not a linear combination of vectors in $C \backslash \{c\}$, which means that $C$ is linearly independent. $\qquad\square$

Can we proceed this way, enlarging linearly independent subsets of $\mathcal{X}$ in order to form a chain of linearly independent subsets, so that an "ultimate" linearly independent subset becomes a Hamel basis for $\mathcal{X}$? Yes, we can; and it seems reasonable that the Axiom of Choice (or any statement equivalent to it as, for instance, Zorn's

Lemma) might be called into play. In fact, every linearly independent subset of any linear space $\mathcal{X}$ is included in some Hamel basis for $\mathcal{X}$, so that every linear space has a large supply of Hamel bases.

**Theorem 2.5.** *If $A$ is a linearly independent subset of a linear space $\mathcal{X}$, then there exists a Hamel basis $B$ for $\mathcal{X}$ such that $A \subseteq B$.*

*Proof.* Let $\mathcal{X}$ be a linear space and suppose $A$ is a linearly independent subset of $\mathcal{X}$. Set

$$\mathcal{I}_A = \{B \in \wp(\mathcal{X}):\quad B \text{ is linearly independent and } A \subseteq B\},$$

the collection of all linearly independent subsets of $\mathcal{X}$ that include $A$. Recall that, as a nonempty subcollection ($A \in \mathcal{I}_A$) of the power set $\wp(\mathcal{X})$, $\mathcal{I}_A$ is partially ordered in the inclusion ordering.

*Claim* 1. $\mathcal{I}_A$ has a maximal element.

*Proof.* If $\mathcal{X} = \{0\}$, then $A = \varnothing$ and $\mathcal{I}_A = \{A\} = \{\varnothing\} \neq \varnothing$, so that the claimed result is trivially verified. Thus suppose $\mathcal{X} \neq \{0\}$. In this case, the nonempty collection $\mathcal{I}_A$ contains a nonempty set (e.g., if $A = \varnothing$, then every nonzero singleton in $\mathcal{X}$ belongs to $\mathcal{I}_A$; if $A \neq \varnothing$, then $A \in \mathcal{I}_A$). Now consider an arbitrary chain $\mathcal{C}$ in $\mathcal{I}_A$ containing a nonempty set. Recall that $\bigcup \mathcal{C}$ denotes the union of all sets in $\mathcal{C}$. Take an arbitrary finite nonempty subset of $\bigcup \mathcal{C}$, say, a set $D \subseteq \bigcup \mathcal{C}$ such that $\#D = n$ for some $n \in \mathbb{N}$. Each element of $D$ belongs to a set in $\mathcal{C}$ (for $D \subseteq \bigcup \mathcal{C}$). Since $\mathcal{C}$ is a chain, we can arrange the elements of $D$ as follows. $D = \{x_i\}_{i=1}^n$ such that $x_i \in C_i \in \mathcal{C}$ for each index $i$, where $C_1 \subseteq \cdots \subseteq C_n$. Thus $D \subseteq C_n$. Since $C_n$ is linearly independent (because $C_n \in \mathcal{C} \subseteq \mathcal{I}_A$), it follows that $D$ is linearly independent. Conclusion: Every finite subset of $\bigcup \mathcal{C}$ is linearly independent. Therefore $\bigcup \mathcal{C}$ is linearly independent by Proposition 2.3. Moreover, since $A \subseteq C$ for all $C \in \mathcal{C}$ (for $\mathcal{C} \subseteq \mathcal{I}_A$), it also follows that $A \subseteq \bigcup \mathcal{C}$. Hence $\bigcup \mathcal{C} \in \mathcal{I}_A$. Since $\bigcup \mathcal{C}$ clearly is an upper bound for $\mathcal{C}$, we may conclude: Every chain in $\mathcal{I}_A$ has an upper bound in $\mathcal{I}_A$. Thus $\mathcal{I}_A$ has a maximal element by Zorn's Lemma. $\square$

*Claim* 2. Take $B \in \mathcal{I}_A$. $B$ is maximal in $\mathcal{I}_A$ if and only if $B$ is a Hamel basis for $X$.

*Proof.* Again, if $\mathcal{X} = \{0\}$, then $B = A = \varnothing$ is the only (and so a maximal) element in $\mathcal{I}_A$ and span $B = \mathcal{X}$, so that the claimed result holds trivially. Thus suppose $\mathcal{X} \neq \{0\}$, which implies that $\mathcal{I}_A$ contains nonempty sets, and take an arbitrary $B$ in $\mathcal{I}_A$. If span $B \neq \mathcal{X}$ (i.e., if span $B \subset \mathcal{X}$), then take $x \in \mathcal{X} \backslash \text{span } B$ so that $B \cup \{x\} \in \mathcal{I}_A$ (i.e., $B \cup \{x\}$ is linearly independent by Proposition 2.4, and $A \subset B \cup \{x\}$ because $A \subseteq B$). Hence $B$ is not maximal in $\mathcal{I}_A$. Therefore, if $B$ is maximal in $\mathcal{I}_A$, then span $B = \mathcal{X}$. On the other hand, if span $B = \mathcal{X}$, then $B \neq \varnothing$ (for $\mathcal{X} \neq \{0\}$) and every vector in $\mathcal{X}$ is a linear combination of vectors in $B$. Thus $B \cup \{x\}$ is not linearly independent for every $x \in \mathcal{X} \backslash B$. This implies that there is no $B' \in \mathcal{I}_A$ such that $B \subset B'$, which means that $B$ is maximal in $\mathcal{I}_A$. Conclusion: If $B \in \mathcal{I}_A$, then $B$

is maximal in $\mathcal{I}_A$ if and only if span $B = \mathcal{X}$. According to the definition of Hamel basis, $B$ in $\mathcal{I}_A$ is such that span $B = \mathcal{X}$ if and only if $B$ is a Hamel basis for $\mathcal{X}$. $\square$

Claims 1 and 2 ensure that, for each linearly independent subset $A$ of $\mathcal{X}$, there exists a Hamel basis $B$ for $\mathcal{X}$ such that $A \subseteq B$. $\square$

Since the empty set is a subset of every set (in particular, of every linear space), and since the empty set is linearly independent, it follows that the preceding theorem holds for $A = \varnothing$. In this case $\mathcal{I}_\varnothing$ is simply the collection of all linearly independent subsets of the linear space $\mathcal{X}$, and the theorem statement just says that *every linear space has a Hamel basis*. Moreover, Claim 2 says that *a Hamel basis for a linear space is precisely a maximal linearly independent subset of it* (i.e., a Hamel basis is a maximal element of $\mathcal{I}_\varnothing$).

The idea behind the previous theorem was that of enlarging a linearly independent subset of $\mathcal{X}$ to get a Hamel basis for $\mathcal{X}$. Another way of facing the same problem (i.e., another way to obtain a Hamel basis for linear space $\mathcal{X}$) is to begin with a set that spans $\mathcal{X}$ and then to weed out from it a linearly independent subset that also spans $\mathcal{X}$.

**Theorem 2.6.** *If a subset $A$ of a linear space $\mathcal{X}$ spans $\mathcal{X}$, then there exists a Hamel basis $B$ for $\mathcal{X}$ such that $B \subseteq A$.*

*Proof.* Let $A$ be a subset of a linear space $\mathcal{X}$ such that span $A = \mathcal{X}$, and consider the collection $\mathcal{I}'_A$ of all linearly independent subsets of $A$:

$$\mathcal{I}'_A = \big\{ B \in \wp(\mathcal{X}) \colon \ B \text{ is linearly independent and } B \subseteq A \big\}.$$

If $\mathcal{X} = \{0\}$, then either $A = \varnothing$ or $A = \{0\}$. In any case $\mathcal{I}'_A = \{\varnothing\}$ trivially has a maximal element. If $\mathcal{X} \neq \{0\}$, then $A$ has a nonzero vector (for span $A = \mathcal{X}$) and every nonzero singleton $\{x\} \subseteq A$ is an element of $\mathcal{I}'_A$. Thus, proceeding exactly as in the proof of Theorem 2.5 (Claim 1), we can show that $\mathcal{I}'_A$ has a maximal element. Let $A_0$ be a maximal element of $\mathcal{I}'_A$. If $A$ is linearly independent, then we are done (i.e., $A$ is itself a Hamel basis for $\mathcal{X}$ since span $A = \mathcal{X}$). Thus suppose $A$ is not linearly independent so that $A_0$ is a proper subset of $A$. Take an arbitrary $a \in A \backslash A_0$ and consider the set $A_0 \cup \{a\} \subseteq A$, which is not linearly independent because $A_0$ is maximal in $\mathcal{I}'_A$. Since $A_0$ is linearly independent, it follows that $a$ is a linear combination of vectors in $A_0$. Thus $A \backslash A_0 \subseteq$ span $A_0$, and hence $A = A_0 \cup (A \backslash A_0) \subseteq$ span $A_0$. Therefore span $A \subseteq$ span (span $A_0$) = span $A_0 \subseteq$ span $A$, which implies that span $A_0 =$ span $A = \mathcal{X}$. Conclusion: $A_0$ is a Hamel basis for $\mathcal{X}$. $\square$

Since $\mathcal{X}$ trivially spans $\mathcal{X}$, the above theorem holds for $A = \mathcal{X}$. In this case $\mathcal{I}'_\mathcal{X}$ is precisely the collection of all linearly independent subsets of $\mathcal{X}$ (i.e., $\mathcal{I}'_\mathcal{X} = \mathcal{I}_\varnothing$), and the theorem statement again says that every linear space has a Hamel basis.

An ever-present purpose in mathematics is a quest for hidden invariants. The concept of Hamel basis supplies a fundamental invariant for a linear space, namely, the cardinality of all Hamel bases for $\mathcal{X}$.

**Theorem 2.7.** *Every Hamel basis for a given linear space $\mathcal{X}$ has the same cardinality.*

*Proof.* If $\mathcal{X} = \{0\}$, then the result holds trivially. Suppose $\mathcal{X} \neq \{0\}$ and let $B$ and $C$ be arbitrary Hamel bases for $\mathcal{X}$ (so that they are nonempty and do not contain the origin). Proposition 2.3 ensures that for every nonzero vector $x \in \mathcal{X}$ there exists a unique finite subset of the Hamel basis $C$, say $C_x$, such that $x$ is a linear combination of all vectors in $C_x \subseteq C$. Now take an arbitrary $c \in C$ and consider the unique representation of it as a linear combination of vectors in the Hamel basis $B$. Thus $c$ is a linear combination of all vectors in $\{b\} \cup B'$ for some (nonzero) $b \in B$ and some finite subset $B'$ of $B$. Hence $c = \beta b + d$, where $\beta$ is a nonzero scalar and $d$ is a vector in $\mathcal{X}$ different from $c$ (for $\beta b \neq 0$). If $d = 0$, then $c = \beta b$ so that $C_b = \{c\}$, and hence $c \in C_b$ trivially. Suppose $d \neq 0$. Recalling again that $C$ also is a Hamel basis for $\mathcal{X}$, consider the unique representation of the nonzero vector $d$ as a linear combination of vectors in $C$, so that $\beta b = c - d \neq 0$ is a linear combination of vectors in $C$. Thus $b$ is itself a linear combination of all vectors in $\{c\} \cup C'$ for some subset $C'$ of $C$. Since such a representation is unique, it follows that $\{c\} \cup C' = C_b$. Therefore $c \in C_b$. Summing up: For every $c \in C$ there exists $b \in B$ such that $c \in C_b$. Thus

$$C \subseteq \bigcup_{b \in B} C_b.$$

Now we shall split the proof into two parts. One dealing with the case of finite Hamel bases, and the other with infinite Hamel bases.

*Claim* 0. Let $\mathcal{X}$ be a linear space. If a subset $E$ of $\mathcal{X}$ with exactly $n$ elements spans $\mathcal{X}$, then every subset of $\mathcal{X}$ with more than $n$ elements is not linearly independent.

*Proof.* Assume the linear space $\mathcal{X}$ is nonzero (i.e., $\mathcal{X} \neq \{0\}$) to avoid trivialities. Take an integer $n \in \mathbb{N}$ and let $E = \{e_i\}_{i=1}^{n}$ be a subset (with $n$ distinct elements) of $\mathcal{X}$ such that span $E = \mathcal{X}$. Now take an arbitrary subset of $\mathcal{X}$ with $n + 1$ elements, say $D = \{d_i\}_{i=1}^{n+1}$. Suppose $D$ is linearly independent. Next consider the set

$$S_1 \,=\, \{d_1\} \cup E$$

which clearly spans $\mathcal{X}$ (because $E$ already does it). Since span $E = \mathcal{X}$, it follows that $d_1$ is a linear combination of vectors in $E$. Moreover, $d_1 \neq 0$ because $D$ is linearly independent. Thus $d_1 = \sum_{i=1}^{n} \alpha_i e_i$ where at least one, say $\alpha_k$, of the scalars $\{\alpha_i\}_{i=1}^{n}$ is nonzero. Therefore, if we delete $e_k$ from $S_1$, then the resulting set

$$S_1' \,=\, S_1 \backslash \{e_k\} \,=\, \{d_1\} \cup E \backslash \{e_k\}$$

still spans $\mathcal{X}$. That is, in forming this new set $S_1'$ that spans $\mathcal{X}$ we have traded off one vector in $D$ for one vector in $E$. Rename the elements of $S_1'$ by setting $s_i = e_i$ for each $i \neq k$ and $s_k = d_1$, so that $S_1' = \{s_i\}_{i=1}^{n}$. Since $D$ has at least two elements, set

$$S_2 \,=\, \{d_2\} \cup S_1' \,=\, \{d_1, d_2\} \cup E \backslash \{e_k\}$$

which again spans $\mathcal{X}$ (for $S_1'$ spans $\mathcal{X}$). Since span $S_1' = \mathcal{X}$, it follows that $d_2$ is a linear combination of vectors in $S_1'$, say $d_2 = \sum_{i=1}^{n} \beta_i s_i$ for some family of scalars $\{\beta_i\}_{i=1}^{n}$. Moreover, $0 \neq d_2 \neq \beta_k s_k = \beta_k d_1$ because $D$ is linearly independent. Thus there exists at least one nonzero scalar in $\{\beta_i\}_{i=1}^{n}$ different from $\beta_k$, say $\beta_j$. Therefore, if we delete $s_j$ from $S_2$ (recall: $s_j = e_j \neq e_k$), then the resulting set

$$S_2' = S_2 \backslash \{e_j\} = \{d_1, d_2\} \cup E \backslash \{e_k, e_j\}$$

still spans $\mathcal{X}$. Continuing this way we eventually get down to the set

$$S_n' = \{d_i\}_{i=1}^{n} \cup E \backslash \{e_i\}_{i=1}^{n} = D \backslash \{d_{n+1}\}$$

which once again spans $\mathcal{X}$. Thus $d_{n+1}$ is a linear combination of vectors in $D \backslash \{d_{n+1}\}$, which contradicts the assumption that $D$ is linearly independent. Conclusion: Every subset of $\mathcal{X}$ with $n+1$ elements is not linearly independent. Recalling that every subset of a linearly independent set is again linearly independent, it follows that every subset of $\mathcal{X}$ with more than $n$ elements is not linearly independent. $\square$

*Claim* 1. If $B$ is finite, then $\#C = \#B$.

*Proof.* Recall that $C_b$ is finite for every $b$ in $B$. If $B$ is finite, then $\bigcup_{b \in B} C_b$ is a finite union of finite sets. Hence any subset of it is finite. In particular, $C$ is finite. Since $C$ is linearly independent, it follows by Claim 0 that $\#C \leq \#B$. Dually (swap the Hamel bases $B$ and $C$), $\#B \leq \#C$. Hence $\#C = \#B$. $\square$

*Claim* 2. If $B$ is infinite, then $\#C = \#B$.

*Proof.* Since $B$ is infinite, and since $C_b$ is finite for every $b$ in $B$, it follows that $\#C_b \leq \#B$ for all $b$ in $B$. Thus, according to Theorems 1.10 and 1.9,

$$\#\left(\bigcup_{b \in B} C_b\right) \leq \#(B \times B) = \#B$$

because $B$ is infinite. Therefore $\#C \leq \#B$ (recall that $C \subseteq \bigcup_{b \in B} C_b$ and use Problems 1.21(a) and 1.22). Moreover, Claim 1 says that $B$ is finite whenever $C$ is finite. Thus $C$ must be infinite because $B$ is infinite. Since $C$ is infinite we may reverse the argument (swapping again the Hamel bases $B$ and $C$) and get $\#B \leq \#C$. Hence $\#C = \#B$ by the Cantor–Bernstein Theorem (Theorem 1.6). $\square$

Claims 1 and 2 ensure that, if $B$ and $C$ are Hamel bases for a linear Space $\mathcal{X}$, then $B$ and $C$ have the same cardinal number.                                                                $\square$

Such an invariant (i.e., the cardinality of any Hamel basis) is called the *dimension* (or the *linear dimension*) of the linear space $\mathcal{X}$, denoted by dim $\mathcal{X}$. Thus dim $\mathcal{X} = \#B$ for any Hamel basis $B$ for $\mathcal{X}$. If the dimension of $\mathcal{X}$ is finite (equivalently, if any Hamel basis for $\mathcal{X}$ is a finite set) then we say that $\mathcal{X}$ is a *finite-dimensional linear space*. Otherwise (i.e., if any Hamel basis for $\mathcal{X}$ is an infinite set) we say that $\mathcal{X}$ is an *infinite-dimensional linear space*.

**Example 2I.** The *Kronecker delta* (or *Kronecker function*) is the mapping in $2^{\mathbb{Z} \times \mathbb{Z}}$ (i.e., the function from $\mathbb{Z} \times \mathbb{Z}$ to $\{0, 1\}$) defined by

$$
\delta_{ij} = \begin{cases} 1, & i = j, \\ 0, & i \neq j, \end{cases}
$$

for all integers $i, j$. Now consider the linear space $\mathbb{F}^n$ (for an arbitrary positive integer $n$, over an arbitrary field $\mathbb{F}$ — see Example 2D). The subset $B = \{e_i\}_{i=1}^{n}$ of $\mathbb{F}^n$ consisting of the $n$-tuples $e_i = (\delta_{i1}, \ldots, \delta_{in})$, with 1 at the $i$th position and zeros elsewhere, constitute a Hamel basis for $\mathbb{F}^n$. This is called the *canonical basis* (or the *natural basis*) for $\mathbb{F}^n$. Thus $\dim \mathbb{F}^n = n$. As we shall see later, $\mathbb{F}^n$ in fact is a prototype for every finite-dimensional linear space (of dimension $n$) over a field $\mathbb{F}$.

**Example 2J.** Let $\mathbb{F}^{\mathbb{N}}$ be the linear space (over a field $\mathbb{F}$) of all scalar-valued sequences (see Example 2E), and let $\mathcal{X}$ be the subset of $\mathbb{F}^{\mathbb{N}}$ defined as follows. $x = \{\xi_k\}_{k \in \mathbb{N}}$ belongs to $\mathcal{X}$ if and only if $\xi_k = 0$ except for some finite set of indices $k$ in $\mathbb{N}$. That is, $\mathcal{X}$ is the set consisting of *all $\mathbb{F}$-valued sequences with a finite number of nonzero entries*, which clearly is a linear manifold of $\mathbb{F}^{\mathbb{N}}$, and hence a linear space itself over $\mathbb{F}$. For each integer $i \in \mathbb{N}$ let $e_i$ be an $\mathbb{F}$-valued sequence with just one nonzero entry (equal to 1) at the $i$th position; that is, $e_i = \{\delta_{ik}\}_{k \in \mathbb{N}} \in \mathcal{X}$ for every $i \in \mathbb{N}$. Now set $B = \{e_i\}_{i \in \mathbb{N}} \subset \mathcal{X}$. It is readily verified that $B$ is linearly independent and that span $B = \mathcal{X}$ (every vector in $\mathcal{X}$ is a linear combination of vectors in $B$). Thus $B$ is a Hamel basis for $\mathcal{X}$. Since $B$ is countably infinite, $\mathcal{X}$ is an infinite-dimensional linear space with $\dim \mathcal{X} = \aleph_0$. Therefore (see Problem 2.6(b)), $\mathbb{F}^{\mathbb{N}}$ is an infinite-dimensional linear space. Note that $B$ is not a Hamel basis for $\mathbb{F}^{\mathbb{N}}$ (reason: span $B = \mathcal{X}$ and $\mathcal{X}$ is properly included in $\mathbb{F}^{\mathbb{N}}$). The next example shows that $\aleph_0 < \dim \mathbb{F}^{\mathbb{N}}$ whenever $\mathbb{F} = \mathbb{Q}$, $\mathbb{F} = \mathbb{R}$, or $\mathbb{F} = \mathbb{C}$.

**Example 2K.** Let $\mathbb{C}^{\mathbb{N}}$ be the complex linear space of all complex-valued sequences. For each real number $t \in (0, 1)$ consider the real sequence $x_t = \{t^{k-1}\}_{k \in \mathbb{N}} = \{t^k\}_{k \in \mathbb{N}_0} = (1, t, t^2, \ldots) \in \mathbb{C}^{\mathbb{N}}$ whose entries are the nonnegative powers of $t$. Set $A = \{x_t\}_{t \in (0,1)} \subseteq \mathbb{C}^{\mathbb{N}}$. We claim that $A$ is linearly independent. A bit of elementary real analysis (rather than pure algebra) supplies a very simple proof as follows. Suppose $A$ is not linearly independent. Then there exists $s \in (0, 1)$ such that $x_s$ is a linear combination of vectors in $A \backslash \{s\}$. That is, $x_s = \sum_{i=1}^{n} \alpha_i x_{t_i}$ for some $n \in \mathbb{N}$, where $\{\alpha_i\}_{i=1}^{n}$ is a family of nonzero complex numbers and $\{x_{t_i}\}_{i=1}^{n}$ is a (finite) subset of $A$ such that $x_{t_i} \neq x_s$ for every $i = 1, \ldots, n$. Hence $n > 1$ (reason: if $n = 1$, then $x_s = \alpha_1 x_{t_1}$ so that $s^k = \alpha_1 t_1^k$ for every $k \in \mathbb{N}_0$, which implies that $x_s = x_{t_1}$). As the set $\{t_i\}_{i=1}^{n}$ consists of distinct points from $(0, 1)$, suppose it is decreasingly ordered (reorder it if necessary) so that $t_i < t_1$ for each $i = 2, \ldots, n$. Since $s^k = \sum_{i=1}^{n} \alpha_i t_i^k$, it follows that $(s/t_1)^k = \alpha_1 + \sum_{i=2}^{n} \alpha_i (t_i/t_1)^k$ for every $k \in \mathbb{N}_0$. However $\lim_k \sum_{i=2}^{n} \alpha_i (t_i/t_1)^k = 0$, because each $t_i/t_1$ lies in $(0, 1)$, and hence $\lim_k (s/t_1)^k = \alpha_1$. Thus $\alpha_1 = 0$ (recall: $x_s \neq x_{t_1}$ so that $s \neq t_1$) which is a contradiction. Conclusion: $A$ is linearly independent. Therefore, according to

Theorem 2.5, there exists a Hamel basis $B$ for $\mathbb{C}^{\mathbb{N}}$ including $A$. Since $A \subseteq B$ and $\#A = \#(0, 1) = 2^{\aleph_0}$, it follows that $2^{\aleph_0} \leq \#B$. However, $\#\mathbb{C} = \#\mathbb{R} = 2^{\aleph_0} \leq \#B = \dim \mathbb{C}^{\mathbb{N}}$, so that $\#\mathbb{C}^{\mathbb{N}} = \dim \mathbb{C}^{\mathbb{N}}$ (see Problem 2.8). Conclusion: $\mathbb{C}^{\mathbb{N}}$ is an infinite-dimensional linear space such that

$$2^{\aleph_0} \leq \dim \mathbb{C}^{\mathbb{N}} = \#\mathbb{C}^{\mathbb{N}}.$$

Note that the whole argument applies for $\mathbb{C}$ replaced by $\mathbb{R}$, so that

$$2^{\aleph_0} \leq \dim \mathbb{R}^{\mathbb{N}} = \#\mathbb{R}^{\mathbb{N}};$$

but it does not apply to the rational field $\mathbb{Q}$ (the interval $(0, 1)$ is not a subset of $\mathbb{Q}$, and hence the set $A$ is not included in $\mathbb{Q}^{\mathbb{N}}$). However, the final conclusion does hold for the linear space $\mathbb{Q}^{\mathbb{N}}$. Indeed, if $\mathbb{F}$ is an arbitrary infinite field, then $2^{\aleph_0} = \#2^{\mathbb{N}} \leq \#\mathbb{F}^{\mathbb{N}} = \max\{\#\mathbb{F}, \dim \mathbb{F}^{\mathbb{N}}\}$ according to Problems 1.24 and 2.8. Therefore, since $\#\mathbb{Q} = \aleph_0 < 2^{\aleph_0}$ (Problem 1.25(c)), it follows that

$$2^{\aleph_0} \leq \dim \mathbb{Q}^{\mathbb{N}} = \#\mathbb{Q}^{\mathbb{N}}.$$

## 2.5 Linear Transformations

A mapping $L \colon \mathcal{X} \to \mathcal{Y}$ of a linear space $\mathcal{X}$ over a field $\mathbb{F}$ into a linear space $\mathcal{Y}$ over the same field $\mathbb{F}$ is *homogeneous* if

$$L(\alpha x) = \alpha L x$$

for every vector $x \in \mathcal{X}$ and every scalar $\alpha \in \mathbb{F}$. The scalar multiplication on the left-hand side is an operation on $\mathcal{X}$ and that on the right-hand side is an operation on $\mathcal{Y}$ (so that the linear spaces $\mathcal{X}$ and $\mathcal{Y}$ must indeed be defined over the same field $\mathbb{F}$). $L$ is *additive* if

$$L(x_1 + x_2) = L(x_1) + L(x_2)$$

for all vectors $x_1, x_2$ in $\mathcal{X}$. Again, the vector addition on the left-hand side is an operation on $\mathcal{X}$ while the one on the right-hand side is an operation on $\mathcal{Y}$. If $\mathcal{X}$ and $\mathcal{Y}$ are linear spaces over the same scalar field, and if $L$ is a homogeneous and additive mapping of $\mathcal{X}$ into $\mathcal{Y}$, then $L$ is a *linear transformation*: a linear transformation is a homogeneous and additive mapping between linear spaces over the same scalar field. When we say that $L \colon \mathcal{X} \to \mathcal{Y}$ is a linear transformation, it is implicitly assumed that $\mathcal{X}$ and $\mathcal{Y}$ are linear spaces over the same field $\mathbb{F}$. If $\mathcal{X} = \mathcal{Y}$ and $L \colon \mathcal{X} \to \mathcal{X}$ is a linear transformation, then we refer to $L$ as a *linear transformation on* $\mathcal{X}$. Trivial example: The identity $I \colon \mathcal{X} \to \mathcal{X}$ (such that $I(x) = x$ for every $x \in \mathcal{X}$) is a linear transformation on $\mathcal{X}$. Recall that a field $\mathbb{F}$ can be made into a linear space over $\mathbb{F}$ itself (see Example 2D). If $\mathcal{X}$ is a linear space over $\mathbb{F}$, then a linear transformation $f \colon \mathcal{X} \to \mathbb{F}$ is called a *linear functional*: a linear functional is a scalar-valued linear transformation (i.e., a linear transformation of a linear space $\mathcal{X}$ into its scalar field).

If $y \in \mathcal{Y}$ is the value of a linear transformation $L: \mathcal{X} \to \mathcal{Y}$ at $x \in \mathcal{X}$, then we shall often write $y = Lx$ (instead of $y = L(x)$). Since $\mathcal{Y}$ is a linear space, it has an origin. The *null space* (or *kernel*) of a linear transformation $L: \mathcal{X} \to \mathcal{Y}$ is the subset

$$\mathcal{N}(L) = \{x \in \mathcal{X}: \ Lx = 0\} = L^{-1}(\{0\})$$

of $\mathcal{X}$ consisting of all vectors in $\mathcal{X}$ mapped into the origin of $\mathcal{Y}$ by $L$. Since $\mathcal{X}$ also is a linear space, it has an origin too. The origin of $\mathcal{X}$ is always in $\mathcal{N}(L)$ (i.e., $L0 = 0$ for every linear transformation $L$). The *null transformation* (denoted by $O$) is the mapping $O: \mathcal{X} \to \mathcal{Y}$ such that $Ox = 0$ for every $x \in \mathcal{X}$, which certainly is a linear transformation. In fact, if $L: \mathcal{X} \to \mathcal{Y}$ is a linear transformation, then $L = O$ if and only if $\mathcal{N}(L) = \mathcal{X}$. Equivalently, $L = O$ if and only if $\mathcal{R}(L) = \{0\}$. The null space, $\mathcal{N}(L) = L^{-1}(\{0\})$, of any linear transformation $L: \mathcal{X} \to \mathcal{Y}$ is a linear manifold of $\mathcal{X}$, and the range of $L$, $\mathcal{R}(L) = L(\mathcal{X})$, is a linear manifold of $\mathcal{Y}$ (see Problem 2.10). These are indeed particular cases of Problem 2.11: *The linear image of a linear manifold is a linear manifold*, and *the inverse image of a linear manifold under a linear transformation is again a linear manifold.*

The theorem below supplies an elegant and useful, although very simple, necessary and sufficient condition that a linear transformation be injective.

**Theorem 2.8.** *A linear transformation $L$ is injective if and only if $\mathcal{N}(L) = \{0\}$.*

*Proof.* let $\mathcal{X}$ and $\mathcal{Y}$ be linear spaces over the same scalar field, and consider a linear transformation $L: \mathcal{X} \to \mathcal{Y}$. If $L$ is injective, then $L^{-1}(L(\{0\})) = \{0\}$ (see Problem 1.3(d)). But $L(\{0\}) = \{0\}$ (for $L0 = 0$) so that $L^{-1}(\{0\}) = \{0\}$, which means $\mathcal{N}(L) = \{0\}$. On the other hand, suppose $\mathcal{N}(L) = \{0\}$. Take $x_1$ and $x_2$ arbitrary in $\mathcal{X}$, and note that $Lx_1 - Lx_2 = L(x_1 - x_2)$ since $L$ is linear. Thus, if $Lx_1 = Lx_2$, then $L(x_1 - x_2) = 0$ and hence $x_1 = x_2$ (i.e., $x_1 - x_2 = 0$ because $\mathcal{N}(L) = \{0\}$). Therefore $L$ is injective.    $\square$

The collection $\mathcal{Y}^X$ of all mappings of a set $X$ into a linear space $\mathcal{Y}$ over a field $\mathbb{F}$ is itself a linear space over $\mathbb{F}$ (see Example 2F). Now suppose $\mathcal{X}$ is a linear space (over the same field $\mathbb{F}$), and let $\mathcal{L}[\mathcal{X}, \mathcal{Y}]$ denote the collection of all linear transformations of $\mathcal{X}$ into $\mathcal{Y}$. Since $\mathcal{L}[\mathcal{X}, \mathcal{Y}]$ is a linear manifold of $\mathcal{Y}^X$ (Problem 2.13), it follows that $\mathcal{L}[\mathcal{X}, \mathcal{Y}]$ is a linear space over the same field $\mathbb{F}$. Set $\mathcal{L}[\mathcal{X}] = \mathcal{L}[\mathcal{X}, \mathcal{X}]$ for short, so that $\mathcal{L}[\mathcal{X}] \subset \mathcal{X}^{\mathcal{X}}$ is the linear space of all linear transformations on $\mathcal{X}$. The linear space $\mathcal{L}[\mathcal{X}, \mathbb{F}]$ of all linear functionals defined on a linear space $\mathcal{X}$, which is a linear manifold of the linear space $\mathbb{F}^{\mathcal{X}}$ (see Example 2E), is called the *algebraic dual* (or *algebraic conjugate*) of $\mathcal{X}$ and denoted by $\mathcal{X}'$. (Dual spaces will be considered in Chapter 4.)

Let $\mathcal{X}$ and $\mathcal{Y}$ be linear spaces over the same scalar field, and let $L|_{\mathcal{M}}: \mathcal{M} \to \mathcal{Y}$ be the restriction of a linear transformation $L: \mathcal{X} \to \mathcal{Y}$ to a linear manifold $\mathcal{M}$ of $\mathcal{X}$. Since $\mathcal{M}$ is a linear space, it is readily verified that $L|_{\mathcal{M}}$ is a linear transformation. Briefly: The restriction of a linear transformation to a linear manifold is again a linear transformation (Problem 2.14). The next result ensures the converse: If $L \in$

$\mathcal{L}[\mathcal{M}, \mathcal{Y}]$ and $\mathcal{M}$ is a linear manifold of $\mathcal{X}$, then there exists $T \in \mathcal{L}[\mathcal{X}, \mathcal{Y}]$ such that $L = T|_{\mathcal{M}}$, which is called a *linear extension* of $L$ over $\mathcal{X}$.

**Theorem 2.9.** *Let $\mathcal{X}$ and $\mathcal{Y}$ be linear spaces over the same field $\mathbb{F}$, and let $\mathcal{M}$ be a linear manifold of $\mathcal{X}$. If $L\colon \mathcal{M} \to \mathcal{Y}$ is a linear transformation, then there exists a linear extension $T\colon \mathcal{X} \to \mathcal{Y}$ of $L$ defined on the whole space $\mathcal{X}$.*

*Proof.* Set

$$\mathcal{K} = \big\{ K \in \mathcal{L}[\mathcal{N}, \mathcal{Y}] :\ \mathcal{N} \in \mathcal{L}at(\mathcal{X}),\ \mathcal{M} \subseteq \mathcal{N} \text{ and } L = K|_{\mathcal{M}} \big\},$$

the collection of all linear transformations from linear manifolds of $\mathcal{X}$ to $\mathcal{Y}$ that are extensions of $L$. Note that $\mathcal{K}$ is nonempty (at least $L$ is there). Moreover, as a subcollection of $\mathcal{F} = \bigcup_{A \in \mathcal{P}(\mathcal{X})} \mathcal{Y}^A$, $\mathcal{K}$ is partially ordered in the extension ordering (see Problem 1.17). Problem 1.17 also tells us that every chain $\{K_\gamma\}$ in $\mathcal{K}$ has a supremum $\bigvee_\gamma K_\gamma$ in $\mathcal{F}$ with domain $\mathcal{D}\big(\bigvee_\gamma K_\gamma\big) = \bigcup_\gamma \mathcal{D}(K_\gamma)$ and range $\mathcal{R}\big(\bigvee_\gamma K_\gamma\big) = \bigcup_\gamma \mathcal{R}(K_\gamma)$. Since $\mathcal{D}(K_\gamma) \in \mathcal{L}at(\mathcal{X})$ (each $K_\gamma$ is a linear transformation defined on a linear manifold of $\mathcal{X}$), and since $\mathcal{L}at(\mathcal{X})$ is a complete lattice, it follows that $\mathcal{D}\big(\bigvee_\gamma K_\gamma\big)$ is a linear manifold of $\mathcal{X}$ (i.e., $\bigcup_\gamma \mathcal{D}(K_\gamma) \in \mathcal{L}at(\mathcal{X})$). Similarly, $\mathcal{R}\big(\bigvee_\gamma K_\gamma\big)$ is a linear manifold of $\mathcal{Y}$.

*Claim.* The supremum $\bigvee_\gamma K_\gamma$ lies in $\mathcal{K}$.

*Proof.* Take $u$ and $v$ arbitrary in $\mathcal{D}\big(\bigvee_\gamma K_\gamma\big)$, so that $u \in \mathcal{D}(K_\lambda)$ for some $K_\lambda$ in $\{K_\gamma\}$ and $v \in \mathcal{D}(K_\mu)$ for some $K_\mu$ in $\{K_\gamma\}$. Since $\{K_\gamma\}$ is a chain, it follows that $K_\lambda \leq K_\mu$ (or vice versa), so that $\mathcal{D}(K_\lambda) \subseteq \mathcal{D}(K_\mu)$. Thus $\alpha u + \beta v \in \mathcal{D}(K_\mu)$ and hence $K_\mu(\alpha u + \beta v) = \alpha K_\mu u + \beta K_\mu v$ for every $\alpha, \beta \in \mathbb{F}$ (recall: each $K_\gamma$ is linear). However $\big(\bigvee_\gamma K_\gamma\big)|_{\mathcal{D}(K_\mu)} = K_\mu$, which implies that $\big(\bigvee_\gamma K_\gamma\big)(\alpha u + \beta v) = \alpha\big(\bigvee_\gamma K_\gamma\big)u + \beta\big(\bigvee_\gamma K_\gamma\big)v$. That is, $\bigvee_\gamma K_\gamma :\ \mathcal{D}\big(\bigvee_\gamma K_\gamma\big) \to \mathcal{Y}$ is linear. Moreover, since each $K_\gamma$ is such that $K_\gamma|_{\mathcal{M}} = L$, and since $\{K_\gamma\}$ is a chain, it follows that $\big(\bigvee_\gamma K_\gamma\big)|_{\mathcal{M}} = L$. Conclusion: $\bigvee_\gamma K_\gamma \in \mathcal{K}$. $\square$

Therefore, every chain in $\mathcal{K}$ has a supremum (and so an upper bound) in $\mathcal{K}$. Thus, according to Zorn's Lemma, $\mathcal{K}$ contains a maximal element, say $K_0\colon \mathcal{N}_0 \to \mathcal{Y}$. We shall show that $\mathcal{N}_0 = \mathcal{X}$, and hence $K_0$ is a linear extension of $L$ over $\mathcal{X}$. The proof goes by contradiction. Suppose $\mathcal{N}_0 \neq \mathcal{X}$. Take $x_1 \in \mathcal{X}\backslash\mathcal{N}_0$ (so that $x_1 \neq 0$ because $\mathcal{N}_0$ is a linear manifold of $\mathcal{X}$) and consider the sum of $\mathcal{N}_0$ and the one-dimensional linear manifold of $\mathcal{X}$ spanned by $\{x_1\}$,

$$\mathcal{N}_1 = \mathcal{N}_0 + \text{span}\,\{x_1\},$$

which is a linear manifold of $\mathcal{X}$ properly including $\mathcal{M}$ (because $\mathcal{M} \subseteq \mathcal{N}_0 \subset \mathcal{N}_1$). Since $\mathcal{N}_0 \cap \text{span}\,\{x_1\} = \{0\}$, it follows that every $x$ in $\mathcal{N}_1$ has a unique representation as a sum of a vector in $\mathcal{N}_0$ and a vector in span $\{x_1\}$. That is, for each $x \in \mathcal{N}_1$ there exists a unique pair $(x_0, \alpha)$ in $\mathcal{N}_0 \times \mathbb{F}$ such that $x = x_0 + \alpha x_1$. (Indeed, if $x = x_0 + \alpha x_1 = x_0' + \alpha' x_1$, then $x_0 - x_0' = (\alpha' - \alpha)x_1 \in \mathcal{N}_0 \cap \text{span}\,\{x_1\} = \{0\}$

so that $x_0' = x_0$ and $\alpha' = \alpha$ — recall: $x_1 \neq 0$.) Take an arbitrary $y$ in $\mathcal{Y}$ (for instance, $y = 0$) and consider the mapping $K_1 \colon \mathcal{N}_1 \to \mathcal{Y}$ defined by

$$K_1 x = K_0 x_0 + \alpha y$$

for every $x \in \mathcal{N}_1$. Observe that $K_1$ is linear (it inherits the linearity of $K_0$) and $K_0 = K_1|_{\mathcal{N}_0}$ (so that $K_0 \leq K_1$). Since $\mathcal{M} \subseteq \mathcal{N}_0 \subset \mathcal{N}_1$, it follows that $L = K_0|_{\mathcal{M}} = K_1|_{\mathcal{M}}$. Thus $K_1 \in \mathcal{K}$, which contradicts the fact that $K_0$ is maximal in $\mathcal{K}$ (for $K_0 \neq K_1$). Therefore $\mathcal{N}_0 = \mathcal{X}$. $\qquad\qquad\square$

Let $\mathcal{X}$ and $\mathcal{Y}$ be nonzero linear spaces over the same field. Take $x \neq 0$ in $\mathcal{X}$ and $y \neq 0$ in $\mathcal{Y}$, set $\mathcal{M} = \operatorname{span}\{x\}$ in $\mathcal{L}at(\mathcal{X})$, and let $L \colon \mathcal{M} \to \mathcal{Y}$ be defined by $Lu = \alpha y$ for every $u = \alpha x \in \mathcal{M}$. Clearly, $L$ is linear and $L \neq O$. Thus Theorem 2.9 ensures that, *if $\mathcal{X}$ and $\mathcal{Y}$ are nonzero linear spaces over the same field, then there exist many $T \neq O$ in $\mathcal{L}[\mathcal{X}, \mathcal{Y}]$* (at least as many as one-dimensional linear manifolds in $\mathcal{L}at(\mathcal{X})$).

## 2.6   Isomorphisms

Two exemplars of a mathematical structure are indistinguishable, in the context of the theory in which that structure is embedded, if there exists a one-to-one correspondence between them that preserves such a structure. This is a central concept in mathematics. From the point of view of the linear space theory, two linear spaces are essentially the same if there exists a one-to-one correspondence between them that preserves all the linear relations — they may differ on the set-theoretic nature of their elements but, as far as the linear space (algebraic) structure is concerned, they are indistinguishable. In other words, two linear spaces $\mathcal{X}$ and $\mathcal{Y}$ over the same scalar field are regarded as essentially the same linear space if there exists a one-to-one correspondence between them that preserves vector addition and scalar multiplication. That is, if there exists at least one invertible linear transformation from $\mathcal{X}$ to $\mathcal{Y}$ whose inverse from $\mathcal{Y}$ to $\mathcal{X}$ also is linear. The theorem below shows that the inverse of an invertible linear transformation is always linear.

**Theorem 2.10.** *Let $\mathcal{X}$ and $\mathcal{Y}$ be linear spaces over $\mathbb{F}$. If $L \colon \mathcal{X} \to \mathcal{Y}$ is an invertible linear transformation, then its inverse $L^{-1} \colon \mathcal{Y} \to \mathcal{X}$ is a linear transformation.*

*Proof.* Recall that, by definition, a function is invertible if it is injective and surjective. Take $y_1$ and $y_2$ arbitrary in $\mathcal{Y}$ so that there exist $x_1$ and $x_2$ in $\mathcal{X}$ such that $y_1 = Lx_1$ and $y_2 = Lx_2$ (for $\mathcal{Y} = \mathcal{R}(L)$ — i.e., $L$ is surjective). Since $L$ is injective (i.e., $L^{-1}L$ is the identity on $\mathcal{X}$ — see Problems 1.5 and 1.7) and additive, it follows that

$$\begin{aligned}
L^{-1}(y_1 + y_2) &= L^{-1}(Lx_1 + Lx_2) = L^{-1}L(x_1 + x_2) = x_1 + x_2 \\
&= L^{-1}Lx_1 + L^{-1}Lx_2 = L^{-1}y_1 + L^{-1}y_2,
\end{aligned}$$

and hence $L^{-1}$ is additive. Similarly, since $L$ is injective and homogeneous,

$$L^{-1}(\alpha y) = L^{-1}(\alpha L x) = L^{-1}L(\alpha x) = \alpha x = \alpha L^{-1}L x = \alpha L^{-1}y$$

for every $y \in \mathcal{Y} = \mathcal{R}(L)$ and every $\alpha \in \mathbb{F}$, which implies that $L^{-1}$ is homogeneous. Thus $L^{-1}$ is a linear transformation. $\qquad\qquad\qquad\qquad\qquad\square$

An *isomorphism* between linear spaces (over the same scalar field) is an injective and surjective linear transformation. Equivalently, an invertible linear transformation. Two linear spaces $\mathcal{X}$ and $\mathcal{Y}$ over the same field $\mathbb{F}$ are *isomorphic* if there exists an isomorphism (i.e., a linear one-to-one correspondence) of $\mathcal{X}$ onto $\mathcal{Y}$. Thus, according to Theorem 2.8, a linear transformation $L \colon \mathcal{X} \to \mathcal{Y}$ of a linear space $\mathcal{X}$ into a linear space $\mathcal{Y}$ is an isomorphism if and only if $\mathcal{N}(L) = \{0\}$ and $\mathcal{R}(L) = \mathcal{Y}$. In particular, if $\mathcal{N}(L) = \{0\}$, then $\mathcal{X}$ and the range of $L$ ($\mathcal{R}(L) = L(\mathcal{X})$) are isomorphic linear spaces.

We noticed in Example 2I that $\mathbb{F}^n$ is a "prototype" for every n-dimensional linear space over $\mathbb{F}$. What this really means is that every n-dimensional linear space over a field $\mathbb{F}$ is *isomorphic* to $\mathbb{F}^n$, and hence two n-dimensional linear spaces over the same scalar field are isomorphic. In fact, such an isomorphism between linear spaces with the same dimension holds in general, either for finite or infinite dimensional linear spaces. We shall prove this below (Theorem 2.12) but first we need the following auxiliary result.

**Proposition 2.11.** *Let $\mathcal{X}$ and $\mathcal{Y}$ be linear spaces over the same field, and let $B$ be a Hamel basis for $\mathcal{X}$. For each mapping $F \colon B \to \mathcal{Y}$ there exists a unique linear transformation $T \colon \mathcal{X} \to \mathcal{Y}$ such that $T|_B = F$.*

*Proof.* If $B = \{x_\gamma\}_{\gamma \in \Gamma}$ is a Hamel basis for $\mathcal{X}$, indexed by an index set $\Gamma$ (recall: any set can be thought of as an indexed set), then every vector $x$ in $\mathcal{X}$ has a unique expansion on $B$, viz.,

$$x = \sum_{\gamma \in \Gamma} \alpha_\gamma x_\gamma,$$

where $\{\alpha_\gamma\}_{\gamma \in \Gamma}$ is a similarly indexed family of scalars with $\alpha_\gamma = 0$ for all but a finite set of indices $\gamma$ (the coordinates of $x$ with respect to the indexed basis $B$). Now set

$$T x = \sum_{\gamma \in \Gamma} \alpha_\gamma F(x_\gamma)$$

for every $x \in \mathcal{X}$. This defines a mapping $T \colon \mathcal{X} \to \mathcal{Y}$ of $\mathcal{X}$ into $\mathcal{Y}$ which is homogeneous, additive, and equals $F$ when restricted to $B$. That is, $T$ is a linear transformation such that $T|_B = F$. Moreover, if $L \colon \mathcal{X} \to \mathcal{Y}$ is a linear transformation of $\mathcal{X}$ into $\mathcal{Y}$ such that $L|_B = F$, then $L = T$. Indeed, for every $x \in \mathcal{X}$,

$$Lx = L\Big(\sum_{\gamma \in \Gamma} \alpha_\gamma x_\gamma\Big) = \sum_{\gamma \in \Gamma} \alpha_\gamma F(x_\gamma) = T\Big(\sum_{\gamma \in \Gamma} \alpha_\gamma x_\gamma\Big) = T x. \qquad \square$$

**Theorem 2.12.** *Let $\mathcal{X}$ and $\mathcal{Y}$ be linear spaces over the same scalar field. $\mathcal{X}$ and $\mathcal{Y}$ are isomorphic if and only if* $\dim \mathcal{X} = \dim \mathcal{Y}$.

*Proof.* (a) Let $L: \mathcal{X} \to \mathcal{Y}$ be an isomorphism of $\mathcal{X}$ onto $\mathcal{Y}$, and let $B_X$ be a Hamel basis for $\mathcal{X}$. Set $B_Y = L(B_X)$, a subset of $\mathcal{Y}$.

*Claim* 1.  $B_Y$ is linearly independent.

*Proof.* Recall that $L$ is an injective and surjective linear transformation. If $B_Y$ is not linearly independent, then there exists $y \in B_Y$, which is a linear combination of vectors in $B_Y \backslash \{y\}$, say $y = \sum_{i=1}^{n} \alpha_i y_i$ where each $y_i$ is a vector in $B_Y \backslash \{y\}$. Thus $x = L^{-1} y$ in $B_X = L^{-1}(B_Y)$ is a linear combination of vectors in $B_X \backslash \{x\}$. (Indeed, $x = \sum_{i=1}^{n} \alpha_i x_i$ where each $x_i = L^{-1} y_i$ is a vector in $B_X = L^{-1}(B_Y)$ different from $x = L^{-1} y$ — recall: each $y_i$ is a vector in $B_Y$ different from $y$, and $L$ is injective.) But this contradicts the fact that $B_X$ is linearly independent. Conclusion: $B_Y$ is linearly independent. $\square$

*Claim* 2.  $B_Y$ spans $\mathcal{Y}$.

*Proof.* Take $y \in \mathcal{Y}$ arbitrary so that $y = Lx$ for some $x \in \mathcal{X}$ (because $L$ is surjective). Since $\mathrm{span}\, B_X = \mathcal{X}$, it follows that $x$ is a linear combination of vectors in $B_X$. Hence $y = Lx$ is a linear combination of vectors in $B_Y = L(B_X)$ (since $L$ is linear) so that $\mathrm{span}\, B_Y = \mathcal{Y}$. $\square$

Therefore, $B_Y$ is a Hamel basis for $\mathcal{Y}$. Moreover, $\#B_Y = \#B_X$ because $L$ sets a one-to-one correspondence between $B_X$ and $B_Y$. (In fact, the restriction $L|_{B_X}: B_X \to B_Y$ is injective and surjective, since $L$ is injective and $B_Y = L(B_X)$ by definition.) Thus $\dim \mathcal{X} = \dim \mathcal{Y}$.

(b) Let $B_X$ and $B_Y$ be Hamel bases for $\mathcal{X}$ and $\mathcal{Y}$, respectively. If $\dim \mathcal{X} = \dim \mathcal{Y}$, then $\#B_Y = \#B_X$, which means that there exists a one-to-one mapping $F: B_X \to B_Y$ of $B_X$ onto $B_Y$. Let $T: \mathcal{X} \to \mathcal{Y}$ be the unique linear transformation such that $T|_{B_X} = F$ (see Proposition 2.11), and hence $T(B_X) = F(B_X) = B_Y$.

*Claim* 3.  $T$ is injective.

*Proof.* If $\mathcal{X} = \{0\}$, then the result holds trivially. Thus suppose $\mathcal{X} \neq \{0\}$. Take any nonzero vector $x$ in $\mathcal{X}$ and consider its (unique) representation as a linear combination of vectors in $B_X$. Therefore, $Tx$ has a representation as a linear combination of vectors in $B_Y = T(B_X)$ because $T$ is linear. Since $B_Y$ is linearly independent, it follows that $Tx \neq 0$. That is, $\mathcal{N}(T) = \{0\}$ which means, by Theorem 2.8, that $T$ is injective. $\square$

*Claim* 4.  $T$ is surjective.

*Proof.* Take any vector $y \in \mathcal{Y}$ and consider its expansion on $B_Y$, say $y = \sum_{i=1}^{n} \alpha_i y_i$ with each $y_i$ in $B_Y$. Thus $y = \sum_{i=1}^{n} \alpha_i T(x_i)$ with each $x_i$ in $B_X$ because $B_Y = T(B_X)$. Since $T$ is linear, it follows that $y = T\left(\sum_{i=1}^{n} \alpha_i x_i\right)$, where $\sum_{i=1}^{n} \alpha_i x_i$ is a vector in $\mathcal{X}$ (since $\mathcal{X}$ is a linear space). Hence $y \in \mathcal{R}(T)$. $\square$

Therefore, $T\colon \mathcal{X} \to \mathcal{Y}$ is an isomorphism of $\mathcal{X}$ onto $\mathcal{Y}$. $\quad\square$

**Example 2L.** Let $\mathcal{X}$ and $\mathcal{Y}$ be *finite-dimensional* linear spaces over the same field $\mathbb{F}$, with $\dim \mathcal{X} = n$ and $\dim \mathcal{Y} = m$. Let $B_X = \{x_j\}_{j=1}^n$ and $B_Y = \{y_i\}_{i=1}^m$ be Hamel bases for $\mathcal{X}$ and $\mathcal{Y}$, respectively. Take an arbitrary vector $x$ in $\mathcal{X}$ and consider its unique expansion on $B_X$:

$$x = \sum_{j=1}^n \xi_j x_j,$$

where the family of scalars $\{\xi_j\}_{j=1}^n$ consists of the coordinates of $x$ with respect to $B_X$. Now let $A\colon \mathcal{X} \to \mathcal{Y}$ be any linear transformation so that

$$Ax = \sum_{j=1}^n \xi_j A x_j.$$

Each $Ax_j$ is a vector in $\mathcal{Y}$. Consider its unique expansion on $B_Y$:

$$Ax_j = \sum_{i=1}^m \alpha_{ij} y_i$$

where, for each $j$, $\{\alpha_{ij}\}_{i=1}^m$ is a family of scalars — the coordinates of each $Ax_j$ with respect to $B_Y$. Set $y = Ax$ in $\mathcal{Y}$ and consider the unique expansion of $y$ on $B_Y$:

$$y = \sum_{i=1}^m \upsilon_i y_i.$$

Again, $\{\upsilon_i\}_{i=1}^m$ is a family of scalars consisting of the coordinates of $y$ with respect to $B_Y$. Thus the identity $y = Ax$ can be written as

$$\sum_{i=1}^m \upsilon_i y_i = \sum_{i=1}^m \left( \sum_{j-1}^n \xi_j \alpha_{ij} \right) y_j.$$

Since the expansion of $y$ on $B_Y$ is unique, it follows that

$$\upsilon_i = \sum_{j=1}^n \alpha_{ij} \xi_j$$

for every $i = 1, \ldots, m$. This gives an expression for each coordinate of $Ax$ as a function of the coordinates of $x$. In terms of standard matrix notation, and according to the ordinary matrix operations, the matrix equation

$$\begin{pmatrix} \upsilon_1 \\ \vdots \\ \upsilon_m \end{pmatrix} = \begin{pmatrix} \alpha_{11} & \cdots & \alpha_{1n} \\ \vdots & & \vdots \\ \alpha_{m1} & \cdots & \alpha_{mn} \end{pmatrix} \begin{pmatrix} \xi_1 \\ \vdots \\ \xi_n \end{pmatrix}$$

represents the identity $y = Ax$ (the vector $y$ is the value of the linear transformation $A$ at the point $x$), and the $m \times n$ array of scalars

$$[A] = \begin{pmatrix} \alpha_{11} & \cdots & \alpha_{1n} \\ \vdots & & \vdots \\ \alpha_{m1} & \cdots & \alpha_{mn} \end{pmatrix}$$

is the *matrix* that represents the linear transformation $A \colon \mathcal{X} \to \mathcal{Y}$ with respect to the bases $B_X$ and $B_Y$. The matrix $[A]$ of a linear transformation $A$ depends on the bases $B_X$ and $B_Y$. If we change the bases, then the matrix that represents the linear transformation may change as well. Different matrices representing the same linear transformation are simply different representations of it with respect to different bases. However, if we fix the bases $B_X$ and $B_Y$, then the representation $[A]$ of $A$ is unique. But uniqueness is not all. It is easy to show that

(a) the set $\mathbb{F}_{m \times n}$ of all $m \times n$ matrices with entries in $\mathbb{F}$ is a linear space over $\mathbb{F}$ when equipped with the ordinary (entrywise) operations of matrix addition and scalar multiplication.

Moreover, for fixed bases $B_X$ and $B_Y$,

(b) $\mathbb{F}_{m \times n}$ is isomorphic to $\mathcal{L}[\mathcal{X}, \mathcal{Y}]$.

If we fix the bases $B_X$ and $B_Y$, then the relation between $\mathcal{L}[\mathcal{X}, \mathcal{Y}]$ and $\mathbb{F}_{m \times n}$ defined by "$[A]$ represents $A$ with respect to $B_X$ and $B_Y$" in fact is a function from $\mathcal{L}[\mathcal{X}, \mathcal{Y}]$ to $\mathbb{F}_{m \times n}$. It is readily verified that such a function, say $\Phi \colon \mathcal{L}[\mathcal{X}, \mathcal{Y}] \to \mathbb{F}_{m \times n}$, is homogeneous, additive, injective and surjective. In other words, $\Phi$ is an isomorphism. For this reason we may and shall identify a linear transformation $A \in \mathcal{L}[\mathbb{F}^n, \mathbb{F}^m]$ with its matrix $[A] \in \mathbb{F}_{m \times n}$ relative to the *canonical* bases for $\mathbb{F}^n$ and $\mathbb{F}^m$ (which were introduced in Example 2I).

**Example 2M.** Let $\mathbb{F}$ denote either the real field or the complex field. Take an arbitrary nonnegative integer $n$, and let $\mathcal{P}_n[0, 1]$ be the collection of all polynomials in the variable $t \in [0, 1]$ with coefficients in $\mathbb{F}$ of degree no greater than $n$. That is,

$$\mathcal{P}_n[0, 1] = \left\{ p \in \mathbb{F}^{[0,1]} \colon \ p(t) = \textstyle\sum_{i=0}^{n} \alpha_i t^i, \, t \in [0, 1], \text{ with each } \alpha_i \text{ in } \mathbb{F} \right\}.$$

Recall: The degree of a nonzero polynomial $p$ is $m$ if $p(t) = \sum_{i=0}^{m} \alpha_i t^i$ with $\alpha_m \neq 0$ (e.g., the degree of a constant polynomial is zero), and the degree of the zero polynomial is undefined (thus not greater than any $n \in \mathbb{N}_0$). It is readily verified that $\mathcal{P}_n[0, 1]$ is a linear manifold of the linear space $\mathbb{F}^{[0,1]}$ (see Example 2E), and hence a linear space over $\mathbb{F}$. Now consider the mapping $L \colon \mathbb{F}^{n+1} \to \mathcal{P}_n[0, 1]$ defined as follows. For each $x = (\xi_0, \ldots, \xi_n) \in \mathbb{F}^{n+1}$ let $p = Lx$ in $\mathcal{P}_n[0, 1]$ be given by

$$p(t) = \sum_{i=0}^{n} \xi_i t^i$$

for every $t \in [0, 1]$. It is easy to show that $L$ is a linear transformation. Moreover, $\mathcal{N}(L) = \{0\}$ (i.e., if $p(t) = \sum_{i=0}^{n} \xi_i t^i = 0$ for every $t \in [0, 1]$, then $x = (\xi_0, \ldots, \xi_n) = 0$ — a nonzero polynomial has only a finite number of zeros) so that $L$ is injective (see Theorem 2.8). Furthermore, every polynomial $p$ in $\mathcal{P}_n[0, 1]$ is of the form $p(t) = \sum_{i=0}^{n} \xi_i t^i$ for some $x = (\xi_0, \ldots, \xi_n)$ in $\mathbb{F}^{n+1}$, which means that $\mathcal{P}_n[0, 1] \subseteq \mathcal{R}(L)$. Hence $\mathcal{P}_n[0, 1] = \mathcal{R}(L)$; that is, $L$ is also surjective. Therefore, the linear transformation $L$ is an isomorphism between the linear spaces $\mathbb{F}^{n+1}$ and $\mathcal{P}_n[0, 1]$. Thus, since $\dim \mathbb{F}^{n+1} = n + 1$ (see Example 2I), it follows by Theorem 2.12 that

$$\dim \mathcal{P}_n[0, 1] = n + 1.$$

Next consider the collection $\mathcal{P}[0, 1]$ of all polynomials in the variable $t \in [0, 1]$ with coefficients in $\mathbb{F}$ of any degree:

$$\mathcal{P}[0, 1] = \bigcup_{n \in \mathbb{N}_0} \mathcal{P}_n[0, 1].$$

Note that $\mathcal{P}[0, 1]$ contains the zero polynomial together with every polynomial of finite degree. It is again readily verified that, as a linear manifold of $\mathbb{F}^{[0,1]}$, $\mathcal{P}[0, 1]$ is itself a linear space over $\mathbb{F}$. The functions $p_i : [0, 1] \to \mathbb{F}$, defined by $p_i(t) = t^i$ for every $t \in [0, 1]$, clearly belong to $\mathcal{P}[0, 1]$ for each $i \in \mathbb{N}_0$. Consider the set $B = \{p_i\}_{i \in \mathbb{N}_0} \subset \mathcal{P}[0, 1]$. Since any polynomial in $\mathcal{P}[0, 1]$ is, by definition, a (finite) linear combination of vectors in $B$, it follows that $\mathcal{P}[0, 1] \subseteq \operatorname{span} B$. Hence $B$ spans $\mathcal{P}[0, 1]$ (i.e., $\operatorname{span} B = \mathcal{P}[0, 1]$). We claim that $B$ is also linearly independent. Indeed, suppose $B$ is not linearly independent. Then there exists in $B$ a linear combination $p_k$ of vectors in $B \backslash \{p_k\}$. That is, $p_k = \sum_{j=1}^{m} \alpha_j p_{i_j}$ for some integer $m \in \mathbb{N}$, where $\{\alpha_j\}_{j=1}^{m}$ is a family of nonzero scalars and $\{p_{i_j}\}_{j=1}^{m}$ is a finite subset of $B$ such that $p_{i_j} \neq p_k$ (i.e., $i_j \neq k$) for every $j = 1, \ldots, m$. Thus $p = p_k - \sum_{j=1}^{m} \alpha_j p_{i_j}$ is the origin of $\mathcal{P}[0, 1]$, which means that

$$p(t) = t^k - \sum_{j=1}^{m} \alpha_j t^{i_j} = 0$$

for all $t \in [0, 1]$. But this is a contradiction because $p$ is a polynomial of degree equal to $\max \left\{ \{k\} \cup \{i_j\}_{j=1}^{m} \right\} \geq 1$. Conclusion: $B$ is linearly independent. Therefore, the set $B = \{p_i\}_{i \in \mathbb{N}_0}$ is a Hamel basis for $\mathcal{P}[0, 1]$, and hence

$$\dim \mathcal{P}[0, 1] = \aleph_0$$

(for $\#B = \#\mathbb{N}_0 = \aleph_0$). Thus $\mathcal{P}[0, 1]$ is isomorphic to the linear space $\mathcal{X}$ of all $\mathbb{F}$-valued sequences with a finite number of nonzero entries (which was introduced in Example 2J).

## 2.7   Isomorphic Equivalence

Two linear spaces over the same scalar field are regarded as essentially the same linear space if they are isomorphic. Let $\mathcal{X}$, $\mathcal{Y}$ and $\mathcal{Z}$ be linear spaces over the same field $\mathbb{F}$. It is clear that $\mathcal{X}$ is isomorphic to itself (reflexivity), and $\mathcal{Y}$ is isomorphic to $\mathcal{X}$ whenever $\mathcal{X}$ is isomorphic to $\mathcal{Y}$ (symmetry). Moreover, since the composition of two isomorphisms is again an isomorphism (see Problems 1.9(c) and 2.15) it follows that, if $\mathcal{X}$ is isomorphic to $\mathcal{Y}$ and $\mathcal{Y}$ is isomorphic to $\mathcal{Z}$, then $\mathcal{X}$ is isomorphic to $\mathcal{Z}$ (transitivity). Thus, if the notion of isomorphic linear spaces is restricted to a given set (for instance, to the collection of all linear manifolds $\mathcal{Lat}(\mathcal{X})$ of a linear space $\mathcal{X}$), then it is an equivalence relation on that set. We shall now define an equivalence between linear transformations. Recall that $GF\colon \mathcal{X} \to \mathcal{Z}$ denotes the composition of a mapping $G\colon \mathcal{Y} \to \mathcal{Z}$ and a mapping $F\colon \mathcal{X} \to \mathcal{Y}$.

**Definition 2.13.** Let $\mathcal{X}$, $\widetilde{\mathcal{X}}$, $\mathcal{Y}$ and $\widetilde{\mathcal{Y}}$ be linear spaces over the same scalar field, where $\mathcal{X}$ is isomorphic to $\widetilde{\mathcal{X}}$ and $\mathcal{Y}$ is isomorphic to $\widetilde{\mathcal{Y}}$. Two linear transformations $T\colon \mathcal{X} \to \mathcal{Y}$ and $L\colon \widetilde{\mathcal{X}} \to \widetilde{\mathcal{Y}}$ are *isomorphically equivalent* if there exist isomorphisms $X\colon \mathcal{X} \to \widetilde{\mathcal{X}}$ and $Y\colon \mathcal{Y} \to \widetilde{\mathcal{Y}}$ such that

$$YT = LX.$$

That is, $T = Y^{-1}LX$ (or equivalently, $L = YTX^{-1}$) which means that the diagram

$$
\begin{array}{ccc}
\mathcal{X} & \xrightarrow{\ T\ } & \mathcal{Y} \\[2pt]
X\downarrow & & \uparrow Y^{-1} \\[2pt]
\widetilde{\mathcal{X}} & \xrightarrow{\ L\ } & \widetilde{\mathcal{Y}}
\end{array}
$$

commutes. Warning: If $\mathcal{X}$ is isomorphic to $\widetilde{\mathcal{X}}$ and $\mathcal{Y}$ is isomorphic to $\widetilde{\mathcal{Y}}$, then there exists an uncountable supply of isomorphisms between $\mathcal{X}$ and $\widetilde{\mathcal{X}}$ and between $\mathcal{Y}$ and $\widetilde{\mathcal{Y}}$. If we take arbitrary linear transformations $T\colon \mathcal{X} \to \mathcal{Y}$ and $L\colon \widetilde{\mathcal{X}} \to \widetilde{\mathcal{Y}}$, it may happen that the above diagram does not commute (i.e., it may happen that $YT \neq LX$) for all isomorphisms of $\mathcal{X}$ onto $\widetilde{\mathcal{X}}$ and all isomorphisms of $\mathcal{Y}$ onto $\widetilde{\mathcal{Y}}$. In this case $T$ and $L$ are not isomorphically equivalent. However, if there exists at least one pair of isomorphisms $X$ and $Y$ for which $YT = LX$, then $T$ and $L$ are isomorphically equivalent.

It is readily verified that isomorphic equivalence deserves its name. In fact, every $T \in \mathcal{L}[\mathcal{X}, \mathcal{Y}]$ is isomorphically equivalent to itself (reflexivity), and $L \in \mathcal{L}[\widetilde{\mathcal{X}}, \widetilde{\mathcal{Y}}]$ is isomorphically equivalent to $T \in \mathcal{L}[\mathcal{X}, \mathcal{Y}]$ whenever $T$ is isomorphically equivalent to $L$ (symmetry). Moreover, if $T \in \mathcal{L}[\mathcal{X}, \mathcal{Y}]$ is isomorphically equivalent to $L \in \mathcal{L}[\widetilde{\mathcal{X}}, \widetilde{\mathcal{Y}}]$ and $L$ is isomorphically equivalent to $K \in \mathcal{L}[\widehat{\mathcal{X}}, \widehat{\mathcal{Y}}]$ (so that $\mathcal{X}$, $\widetilde{\mathcal{X}}$ and $\widehat{\mathcal{X}}$ are isomorphic linear spaces, as well as $\mathcal{Y}$, $\widetilde{\mathcal{Y}}$ and $\widehat{\mathcal{Y}}$), then it is easy to show that $T$ is isomorphically equivalent to $K$ (transitivity). Indeed, if $\mathcal{X} = \widetilde{\mathcal{X}}$ and $\mathcal{Y} = \widetilde{\mathcal{Y}}$, and if we restrict the concept of isomorphic equivalence to the set $\mathcal{L}[\mathcal{X}, \mathcal{Y}]$ of all linear transformations of $\mathcal{X}$ into $\mathcal{Y}$, then isomorphic equivalence actually is an equivalence relation on $\mathcal{L}[\mathcal{X}, \mathcal{Y}]$.

An important particular case is obtained when $\mathcal{X} = \mathcal{Y}$ and $\widetilde{\mathcal{X}} = \widetilde{\mathcal{Y}}$ so that $T$ lies in $\mathcal{L}[\mathcal{X}]$ and $L$ lies in $\mathcal{L}[\widetilde{\mathcal{X}}]$. Let $\mathcal{X}$ and $\widetilde{\mathcal{X}}$ be isomorphic linear spaces. Two linear transformations $T \colon \mathcal{X} \to \mathcal{X}$ and $L \colon \widetilde{\mathcal{X}} \to \widetilde{\mathcal{X}}$ are *similar* if there exists an isomorphism $W \colon \mathcal{X} \to \widetilde{\mathcal{X}}$ such that

$$WT = LW.$$

Equivalently, if there exists an isomorphism $W$ such the diagram

$$
\begin{array}{ccc}
\mathcal{X} & \xrightarrow{\ T\ } & \mathcal{X} \\
{\scriptstyle W}\big\downarrow & & \big\downarrow{\scriptstyle W} \\
\widetilde{\mathcal{X}} & \xrightarrow{\ L\ } & \widetilde{\mathcal{X}}
\end{array}
$$

commutes. It should be noticed now that the concept of similarity will be redefined later in Chapter 4 where the linear spaces are endowed with an additional (topological) structure. Such a redefinition will assume that all linear transformations involved in the definition of similarity are "continuous" (including the inverse of $W$).

**Example 2N.**  Consider the setup of Example 2L, where $\mathcal{X}$ and $\mathcal{Y}$ are *finite-dimensional* linear spaces over the same field $\mathbb{F}$. Let $X \colon \mathcal{X} \to \mathbb{F}^n$ and $Y \colon \mathcal{Y} \to \mathbb{F}^m$ be two mappings defined by

$$Xx = (\xi_1, \dots, \xi_n) \quad \text{and} \quad Yy = (\upsilon_1, \dots, \upsilon_m)$$

for every $x \in \mathcal{X}$ and every $y \in \mathcal{Y}$, where $\{\xi_j\}_{j=1}^n$ and $\{\upsilon_i\}_{i=1}^m$ consist of the coordinates of $x$ and $y$ with respect to the bases $B_X$ and $B_Y$, respectively. It is readily verified that $X$ and $Y$ are both isomorphisms (for fixed bases $B_X$ and $B_Y$). Let $\mathbb{F}_{n \times 1}$ denote the linear space (over the field $\mathbb{F}$) of all $n \times 1$ matrices (or, if you like, the linear space of all "column $n$-vectors" with entries in $\mathbb{F}$ — Example 2L). Now consider the map $W_n \colon \mathbb{F}^n \to \mathbb{F}_{n \times 1}$ that assigns to each $n$-tuple $(\xi_1, \dots, \xi_n)$ in $\mathbb{F}^n$ the $n \times 1$ matrix

$$
\begin{pmatrix} \xi_1 \\ \vdots \\ \xi_n \end{pmatrix} = W_n(\xi_1, \dots, \xi_n)
$$

in $\mathbb{F}_{n \times 1}$ whose entries are the (similarly ordered) coordinates of the ordered $n$-tuple with respect to the *canonical* basis for $\mathbb{F}^n$. It is easy to show that $W_n$ is an isomorphism between $\mathbb{F}^n$ and $\mathbb{F}_{n \times 1}$. This is called the *natural isomorphism* of $\mathbb{F}^n$ onto $\mathbb{F}_{n \times 1}$. Note that any $m \times n$ matrix (with entries in $\mathbb{F}$) can be viewed as a linear transformation from $\mathbb{F}_{n \times 1}$ to $\mathbb{F}_{m \times 1}$: the action of an $m \times n$ matrix $[\alpha_{ij}] \in \mathbb{F}_{m \times n}$ on an $n \times 1$ matrix

$[\xi_j] \in \mathbb{F}_{n \times 1}$ is simply the matrix product

$$[\alpha_{ij}][\xi_j] = \begin{pmatrix} \alpha_{11} & \cdots & \alpha_{1n} \\ \vdots & & \vdots \\ \alpha_{m1} & \cdots & \alpha_{mn} \end{pmatrix} \begin{pmatrix} \xi_1 \\ \vdots \\ \xi_n \end{pmatrix},$$

which is an $m \times 1$ matrix in $\mathbb{F}_{m \times 1}$. According to Example 2L let $[A] \in \mathbb{F}_{m \times n}$ be the unique matrix representing the linear transformation $A \in \mathcal{L}[\mathcal{X}, \mathcal{Y}]$ with respect to the bases $B_X$ and $B_Y$. Now, if this matrix is viewed as a linear transformation of $\mathbb{F}_{n \times 1}$ into $\mathbb{F}_{m \times 1}$, then the diagram

$$
\begin{array}{ccc}
\mathcal{X} & \stackrel{A}{\longrightarrow} & \mathcal{Y} \\
X \downarrow & & \uparrow Y^{-1} \\
\mathbb{F}^n & & \mathbb{F}^m \\
W_n \downarrow & & \uparrow W_m^{-1} \\
\mathbb{F}_{n \times 1} & \stackrel{[A]}{\longrightarrow} & \mathbb{F}_{m \times 1}
\end{array}
$$

commutes. This shows that the linear transformation $A : \mathcal{X} \to \mathcal{Y}$ is isomorphically equivalent to its matrix $[A]$ with respect to the bases $B_X$ and $B_Y$ when this matrix is viewed as a linear transformation $[A] : \mathbb{F}_{n \times 1} \to \mathbb{F}_{m \times 1}$. That is,

$$(W_m Y)A = [A](W_n X).$$

## 2.8   Direct Sum

Let $\{\mathcal{X}_i\}_{i=1}^n$ be a finite indexed family of linear spaces over the same field $\mathbb{F}$ (but not necessarily linear manifolds of the same linear space). The *direct sum* of $\{\mathcal{X}_i\}_{i=1}^n$, denoted by $\bigoplus_{i=1}^n \mathcal{X}_i$, is the set of all ordered $n$-tuples $(x_1, \ldots, x_n)$, with each $x_i$ in $\mathcal{X}_i$, where vector addition and scalar multiplication are defined as follows.

$$(x_1, \ldots, x_n) \oplus (y_1, \ldots, y_n) = (x_1 + y_1, \ldots, x_n + y_n),$$

$$\alpha(x_1, \ldots, x_n) = (\alpha x_1, \ldots, \alpha x_n)$$

for every $(x_1, \ldots, x_n)$ and $(y_1, \ldots, y_n)$ in $\bigoplus_{i=1}^n \mathcal{X}_i$ and every $\alpha$ in $\mathbb{F}$. It is easy to verify that the direct sum $\bigoplus_{i=1}^n \mathcal{X}_i$ of the linear spaces $\{\mathcal{X}_i\}_{i=1}^n$ is a linear space over $\mathbb{F}$ when vector addition (denoted by $\oplus$) and scalar multiplication are defined as above. The underlying set of the linear space $\bigoplus_{i=1}^n \mathcal{X}_i$ is the Cartesian product $\prod_{i=1}^n \mathcal{X}_i$ of the underlying sets of each linear space $\mathcal{X}_i$. The origin of $\bigoplus_{i=1}^n \mathcal{X}_i$ is the ordered $n$-tuple $(0_1, \ldots, 0_n)$ consisting of the origins of each $\mathcal{X}_i$.

If $\mathcal{M}$ and $\mathcal{N}$ are linear manifolds of a linear space $\mathcal{X}$, then we may consider both their ordinary sum $M + \mathcal{N}$ (defined as in Section 2.2) and their direct sum $\mathcal{M} \oplus \mathcal{N}$. These are different linear spaces over the same field. There is however a *natural mapping* $\Phi \colon \mathcal{M} \oplus \mathcal{N} \to \mathcal{M} + \mathcal{N}$, defined by

$$\Phi((x_1, x_2)) = x_1 + x_2,$$

which assigns to each pair $(x_1, x_2)$ in $\mathcal{M} \oplus \mathcal{N}$ their sum in $\mathcal{M} + \mathcal{N} \subseteq \mathcal{X}$. It is readily verified that $\Phi$ is a surjective linear transformation of the linear space $\mathcal{M} \oplus \mathcal{N}$ onto the linear space $\mathcal{M} + \mathcal{N}$, but $\Phi$ is not always injective. We shall establish below a necessary and sufficient condition that $\Phi$ be injective, viz., $\mathcal{M} \cap \mathcal{N} = \{0\}$. In such a case the mapping $\Phi$ is an isomorphism (called the *natural isomorphism*) of $\mathcal{M} \oplus \mathcal{N}$ onto $\mathcal{M} + \mathcal{N}$, so that the direct sum $\mathcal{M} \oplus \mathcal{N}$ and the ordinary sum $\mathcal{M} + \mathcal{N}$ become isomorphic linear spaces.

**Theorem 2.14.** *Let $\mathcal{M}$ and $\mathcal{N}$ be linear manifolds of a linear space $\mathcal{X}$. The following assertions are pairwise equivalent.*

(a)  $\mathcal{M} \cap \mathcal{N} = \{0\}$.

(b)  *For each $x$ in $\mathcal{M} + \mathcal{N}$ there exists a unique $u$ in $\mathcal{M}$ and a unique $v$ in $\mathcal{N}$ such that $x = u + v$.*

(c)  *The natural mapping $\Phi \colon \mathcal{M} \oplus \mathcal{N} \to \mathcal{M} + \mathcal{N}$ is an isomorphism.*

*Proof.* Take an arbitrary $x$ in $\mathcal{M} + \mathcal{N}$. If $x = u_1 + v_1 = u_2 + v_2$, with $u_1, u_2$ in $\mathcal{M}$ and $v_1, v_2$ in $\mathcal{N}$, then $u_1 - u_2 = v_1 - v_2 \in \mathcal{M} \cap \mathcal{N}$ (for $u_1 - u_2 \in \mathcal{M}$ and $v_1 - v_2 \in \mathcal{N}$). Thus $\mathcal{M} \cap \mathcal{N} = \{0\}$ implies that $u_1 - u_2$ and $v_1 = v_2$, and hence (a)$\Rightarrow$(b). On the other hand, if $\mathcal{M} \cap \mathcal{N} \neq \{0\}$, then there exists a nonzero vector $w$ in $\mathcal{M} \cap \mathcal{N}$. Take any nonzero vector $x$ in $\mathcal{M} + \mathcal{N}$ so that $x = u + v$ with $u$ in $\mathcal{M}$ and $v$ in $\mathcal{N}$. Thus $x = (u + w) + (v - w)$, where $u + w$ is in $\mathcal{M}$ and $v - w$ is in $\mathcal{N}$. Since $w \neq 0$, it follows that $u + w \neq u$, and hence the representation of $x$ as a sum $u + v$ with $u$ in $\mathcal{M}$ and $v$ in $\mathcal{N}$ is not unique. Therefore, if (a) does not hold, then (b) does not hold. Equivalently, (b)$\Rightarrow$(a). Finally, recall that the natural mapping $\Phi$ is linear and surjective. Since $\Phi$ is injective if and only if (b) holds (by its very definition), it follows that (b)$\Leftrightarrow$(c). $\qquad\square$

Two linear manifolds $\mathcal{M}$ and $\mathcal{N}$ of a linear space $\mathcal{X}$ are said to be *disjoint* (or *algebraically disjoint*) if $\mathcal{M} \cap \mathcal{N} = \{0\}$. (Note that, as linear manifolds of a linear space $\mathcal{X}$, $\mathcal{M}$ and $\mathcal{M}$ can never be "disjoint" in the set-theoretical sense — the origin of $\mathcal{X}$ always belongs to both of them.) Therefore, if $\mathcal{M}$ and $\mathcal{N}$ are disjoint linear manifolds of a linear space $\mathcal{X}$, then we may and shall identify their ordinary sum $\mathcal{M} + \mathcal{N}$ with their direct sum $\mathcal{M} \oplus \mathcal{N}$. Such an identification is carried out by the natural isomorphism $\Phi \colon \mathcal{M} \oplus \mathcal{N} \to \mathcal{M} + \mathcal{N}$ (Theorem 2.14). When we identify $\mathcal{M} \oplus \mathcal{N}$ with $\mathcal{M} + \mathcal{N}$, which is a linear manifold of $\mathcal{X}$, we are automatically identifying the pairs $(u, 0)$ and $(0, v)$ in $\mathcal{M} \oplus \mathcal{N}$ with $u$ in $\mathcal{M}$ and with $v$ in $\mathcal{N}$, respectively. More generally, we shall be identifying the direct sums $\mathcal{M} \oplus \{0\}$ and $\{0\} \oplus \mathcal{N}$ with $\mathcal{M}$ and $\mathcal{N}$, respectively. For instance, if $x \in \mathcal{M} \oplus \mathcal{N}$ and $\mathcal{M} \cap \mathcal{N} = \{0\}$, then Theorem 2.14 ensures that there exists a unique $u$ in $\mathcal{M}$ and a unique $v$ in $\mathcal{N}$ such that $x = (u, v)$. Hence $x = (u, 0) \oplus (0, v)$ where $(u, 0) \in \mathcal{M} \oplus \{0\}$ and $(0, v) \in \{0\} \oplus \mathcal{N}$ (recall: $\mathcal{M} \oplus \{0\}$ and $\{0\} \oplus \mathcal{N}$ are both linear manifolds of $\mathcal{M} \oplus \mathcal{N}$). Now identify $(u, 0)$ with $\Phi((u, 0)) = u$ and $(0, v)$ with $\Phi((0, v)) = v$, and write $x = u \oplus v$ where $u \in \mathcal{M}$ and $v \in \mathcal{N}$ (instead of $x = (u, 0) \oplus (0, v) = \Phi^{-1}(u) \oplus \Phi^{-1}(v)$). Outcome: If $\mathcal{M}$ and $\mathcal{N}$ are disjoint linear manifolds of a linear space $\mathcal{X}$, then every $x$ in $\mathcal{M} \oplus \mathcal{N}$ has a *unique decomposition with respect to $\mathcal{M}$ and $\mathcal{N}$*, denoted by $x = u \oplus v$, which is referred to as the *direct sum* of $u$ in $\mathcal{M}$ and $v$ in $\mathcal{N}$. It should be noticed that $u \oplus v$ is just another notation for $(u, v)$ that reminds us of the algebraic structure of the linear space $\mathcal{M} \oplus \mathcal{N}$. What really is being added in $\mathcal{M} \oplus \mathcal{N}$ is $(u, 0) \oplus (0, v)$.

If $\mathcal{M}$ and $\mathcal{N}$ are disjoint linear manifolds of a linear space $\mathcal{X}$, and if their (ordinary) sum is $\mathcal{X}$, then we say that $\mathcal{M}$ and $\mathcal{N}$ are *algebraic complements* of each other. In other words, two linear manifolds $\mathcal{M}$ and $\mathcal{N}$ of a linear space $\mathcal{X}$ form a pair of algebraic complements in $\mathcal{X}$ if

$$\mathcal{X} = \mathcal{M} + \mathcal{N} \quad \text{and} \quad \mathcal{M} \cap \mathcal{N} = \{0\}.$$

Accordingly, this can be written as

$$\mathcal{X} = \mathcal{M} \oplus \mathcal{N} \quad \text{and} \quad \mathcal{M} \cap \mathcal{N} = \{0\}$$

once we have identified the direct sum $\mathcal{M} \oplus \mathcal{N}$ with its isomorphic image $\Phi(\mathcal{M} \oplus \mathcal{N}) = \mathcal{M} + \mathcal{N} = \mathcal{X}$ through the natural isomorphism $\Phi$.

**Proposition 2.15.** *Let $\mathcal{M}$ and $\mathcal{N}$ be linear manifolds of a linear space $\mathcal{X}$, and let $B_M$ and $B_N$ be Hamel bases for $\mathcal{M}$ and $\mathcal{N}$, respectively.*

(a) *$\mathcal{M} \cap \mathcal{N} = \{0\}$ if and only if $B_M \cap B_N = \varnothing$ and $B_M \cup B_N$ is linearly independent.*

(b) *$\mathcal{M} + \mathcal{N} = \mathcal{X}$ and $B_M \cup B_N$ is linearly independent if and only if $B_M \cup B_N$ is a Hamel basis for $\mathcal{X}$.*

*In particular, if $B_M \cup B_N \subseteq B$, where $B$ is a Hamel basis for $\mathcal{X}$, then*

(a′) *$\mathcal{M} \cap \mathcal{N} = \{0\}$ if and only if $B_M \cap B_N = \varnothing$,*

(b′) *$\mathcal{M} + \mathcal{N} = \mathcal{X}$ if and only if $B_M \cup B_N = B$.*

*Proof.* (a)  Recall that

$$\{0\} \subseteq \text{span}\,(B_M \cap B_N) \subseteq \text{span}\,(\mathcal{M} \cap \mathcal{N}) = \mathcal{M} \cap \mathcal{N}.$$

Thus $\mathcal{M} \cap \mathcal{N} = \{0\}$ implies $\text{span}\,(B_M \cap B_N) = \{0\}$, which implies $B_M \cap B_N = \varnothing$ (for $0 \notin B_M \cup B_N$). Moreover, if $\mathcal{M} \cap \mathcal{N} = \{0\}$, then the union of the linearly independent sets $B_M$ and $B_N$ is again linearly independent (see Problem 2.3). On the other hand, recall that

$$\{0\} \subseteq M \cap N = \operatorname{span} B_M \cap \operatorname{span} B_N = \operatorname{span}(B_M \cap B_N)$$

when $(B_M \cup B_N)$ is linearly independent (see Problem 2.4). Thus $B_M \cap B_N = \varnothing$ implies $\operatorname{span}(B_M \cap B_N) = \{0\}$, and hence $M \cap N = \{0\}$.

(b)  Next recall that

$$\operatorname{span}(B_M \cup B_N) = \operatorname{span}(M \cup N) = M + N \subseteq X$$

whenever $B_M$ and $B_N$ are Hamel bases for $M$ and $N$, respectively. If $B_M \cup B_N$ is a Hamel basis for $X$, then $B_M \cup B_N$ is linearly independent and $X = \operatorname{span}(B_M \cup B_N)$ so that $M + N = X$. On the other hand, if $M + N = X$, then $\operatorname{span}(B_M \cup B_N) = X$. Thus, according to Theorem 2.6, there exists a Hamel basis $B'$ for $X$ such that $B' \subseteq B_M \cup B_N$. If $B_M \cup B_N$ is linearly independent, then Theorem 2.5 ensures that there exists a Hamel basis $B$ for $X$ such that $B_M \cup B_N \subseteq B$. Therefore $B' \subseteq B$. But a Hamel basis is maximal (see Claim 2 in the proof of Theorem 2.5) so that $B' = B$. Hence $B_M \cup B_N = B$.    $\square$

**Theorem 2.16.** *Every linear manifold has an algebraic complement.*

*Proof.* Let $M$ be a linear manifold of a linear space $X$, let $B_M$ be a Hamel basis for $M$, and let $B$ be a Hamel basis for $X$ such that $B_M \subseteq B$ (see Theorem 2.5). Set $B_N = B \backslash B_M$ (which, as a subset of a linearly independent set $B$, is linearly independent itself) and $N = \operatorname{span} B_N$ (a linear manifold of $X$). Thus $B_M$ and $B_N$ are Hamel bases for $M$ and $N$, respectively, both included in the Hamel basis $B$ for $X$. Since $B_M \cap B_N = \varnothing$ and $B_M \cup B_N = B$, it follows by Proposition 2.15 that $N$ is an algebraic complement of $M$.    $\square$

**Lemma 2.17.** *Let $M$ be a linear manifold of a linear space $X$. Every algebraic complement of $M$ is isomorphic to the quotient space $X / M$.*

*Proof.* Let $M$ be a linear manifold of a linear space $X$ over a field $\mathbb{F}$, and let $X / M$ be the quotient space of $X$ modulo $M$, which is again a linear space over $\mathbb{F}$ (see Example 2H). Let $\pi$ be the natural mapping of $X$ onto $X / M$ defined in Section 1.4. That is, for each $x \in X$ set $\pi(x) = [x] = x + M \in X / M$. Example 2H shows that $\pi : X \to X / M$ in fact is a linear transformation. Let $K$ be a linear manifold of $X$ and consider the restriction of $\pi$ to $K$, $\pi|_K : K \to X / M$, which is again a linear transformation.

*Claim.* If $K$ is an algebraic complement of $M$, then $\pi|_K$ is invertible.

*Proof.* Take an arbitrary $[x]$ in $X / M$ so that $[x] = x + M$ for some $x$ in $X$. Since $X = M + K$, it follows that $x = u + v$ with $u$ in $M$ and $v$ in $K$. Thus $[x] = [v]$ (reason: $[x] = [u + v] = [u] + [v] = [v]$ because $[u] = u + M = M = [0]$ for every $u$ in $\mathcal{M}$). Hence $[x] = \pi|_{\mathcal{K}}(v)$ and so $[x] \in \mathcal{R}(\pi|_{\mathcal{K}})$. Conclusion: $\pi|_{\mathcal{K}}$ is surjective. Moreover, if $\pi|_{\mathcal{K}}(v) = [0]$ for some $v \in \mathcal{K}$, then $v + \mathcal{M} = [v] = [0] = \mathcal{M}$ and hence $v \in \mathcal{M}$. Since $\mathcal{M} \cap \mathcal{K} = \{0\}$, it follows that $v = 0$. That is, the null space of $\pi|_{\mathcal{K}}$ is the singleton $\{0\}$, $\mathcal{N}(\pi|_{\mathcal{K}}) = \{v \in \mathcal{K}: \ \pi|_{\mathcal{K}}(v) = [0]\} = \{0\}$, which means that $\pi|_{\mathcal{K}}$ is injective (see Theorem 2.8). $\square$

Therefore, $\pi|_{\mathcal{K}}$ is an isomorphism of $\mathcal{K}$ onto $\mathcal{X}/\mathcal{M}$ whenever $\mathcal{K}$ is an algebraic complement of $\mathcal{M}$. $\qquad\square$

**Theorem 2.18.** *Let $\mathcal{M}$ be a linear manifold of a linear space $\mathcal{X}$. Every algebraic complement of $\mathcal{M}$ has the same dimension.*

*Proof.* According to Theorem 2.12 the above statement can be rewritten as follows. If $\mathcal{N}$ and $\mathcal{K}$ are algebraic complements of $\mathcal{M}$, then $\mathcal{K}$ and $\mathcal{N}$ are isomorphic. But this is a straightforward consequence of the previous lemma: $\mathcal{N}$ and $\mathcal{K}$ are both isomorphic to $\mathcal{X}/\mathcal{M}$, and hence isomorphic to each other. $\qquad\square$

The dimension of an algebraic complement of $\mathcal{M}$ is therefore a property of $\mathcal{M}$ (i.e., it is an invariant for $\mathcal{M}$). We refer to this invariant as the *codimension* of $\mathcal{M}$: the codimension of a linear manifold $\mathcal{M}$, denoted by codim $\mathcal{M}$, is the (constant) dimension of any algebraic complement of $\mathcal{M}$.

## 2.9   Projections

A *projection* is an idempotent linear transformation of a linear space into itself. Thus, if $\mathcal{X}$ is a linear space, then $P \in \mathcal{L}[\mathcal{X}]$ is a projection if and only if $P = P^2$. Briefly, projections are the idempotent elements of $\mathcal{L}[\mathcal{X}]$. Clearly, the null transformation $O$ and the identity $I$, both in $\mathcal{L}[\mathcal{X}]$, are projections. A *nontrivial projection* in $\mathcal{L}[\mathcal{X}]$ is a projection $P$ such that $O \neq P \neq I$. It is easy to verify that, if $P$ is a projection, then so is $I - P$. Moreover, the null spaces and ranges of $P$ and $I - P$ are related as follows (see Problem 1.4).

$$\mathcal{R}(P) = \mathcal{N}(I - P) \quad \text{and} \quad \mathcal{N}(P) = \mathcal{R}(I - P).$$

Projections are singularly useful linear transformations. One of their main properties is that the range and the null space of a projection form a pair of algebraic complements.

**Theorem 2.19.** *If $P \in \mathcal{L}[\mathcal{X}]$ is a projection, then $\mathcal{R}(P)$ and $\mathcal{N}(P)$ are algebraic complements of each other.*

*Proof.* Let $\mathcal{X}$ be a linear space, and let $P: \mathcal{X} \to \mathcal{X}$ be a projection. Recall that both the range $\mathcal{R}(P)$ and the null space $\mathcal{N}(P)$ are linear manifolds of $\mathcal{X}$ (since $P$ is linear). Since $P$ is idempotent, it follows that

$$\mathcal{R}(P) = \{x \in \mathcal{X}: \ Px = x\}$$

(the range of an idempotent mapping is the set of all its fixed points — Problem 1.4). If $x \in \mathcal{R}(P) \cap \mathcal{N}(P)$, then $x = Px = 0$, and hence

$$\mathcal{R}(P) \cap \mathcal{N}(P) = \{0\}.$$

Moreover, for an arbitrary vector $x$ in $\mathcal{X}$, write $x = Px + (x - Px)$. Since $P(x - Px) = Px - P^2x = 0$ (recall: $P$ is linear and idempotent), it follows that $(x - Px) \in \mathcal{N}(P)$. Hence $x = u + v$ with $u = Px$ in $\mathcal{R}(P)$ and $v = (x - Px)$ in $\mathcal{N}(P)$. Therefore

$$\mathcal{X} = \mathcal{R}(P) + \mathcal{N}(P). \qquad \square$$

On the other hand, for any pair of algebraic complements there exists a unique projection whose range and null space coincide with them.

**Theorem 2.20.** *Let $\mathcal{M}$ and $\mathcal{N}$ be linear manifolds of a linear space $\mathcal{X}$. If $\mathcal{M}$ and $\mathcal{N}$ are algebraic complements of each other, then there exists a unique projection $P: \mathcal{X} \to \mathcal{X}$ such that $\mathcal{R}(P) = \mathcal{M}$ and $\mathcal{N}(P) = \mathcal{N}$.*

*Proof.* Suppose $\mathcal{M}$ and $\mathcal{N}$ are algebraic complements in a linear space $\mathcal{X}$ so that

$$\mathcal{M} + \mathcal{N} = \mathcal{X} \quad \text{and} \quad \mathcal{M} \cap \mathcal{N} = \{0\}.$$

According to Theorem 2.14, for each $x \in \mathcal{X}$ there exists a unique $u \in \mathcal{M}$ and a unique $v \in \mathcal{N}$ such that $x = u + v$. Let $P: \mathcal{X} \to \mathcal{X}$ be the function that assigns to each $x$ in $\mathcal{X}$ its unique summand $u$ in $\mathcal{M}$ (i.e., $Px = u$). It is easy to verify that $P$ is linear. Moreover, for each vector $x$ in $\mathcal{X}$, $P^2x = P(Px) = Pu = u = Px$ (reason: $u$ is itself its unique summand in $\mathcal{M}$), so that $P$ is idempotent. By the very definition of $P$ we get $\mathcal{R}(P) = \mathcal{M}$ and $\mathcal{N}(P) = \mathcal{N}$. Conclusion: $P: \mathcal{X} \to \mathcal{X}$ is a projection with $\mathcal{R}(P) = \mathcal{M}$ and $\mathcal{N}(P) = \mathcal{N}$. Now let $P': \mathcal{X} \to \mathcal{X}$ be any projection with $\mathcal{R}(P') = \mathcal{M}$ and $\mathcal{N}(P') = \mathcal{N}$. Take an arbitrary $x \in \mathcal{X}$ and consider again its unique representation as $x = u + v$ with $u \in \mathcal{M} = \mathcal{R}(P')$ and $v \in \mathcal{N} = \mathcal{N}(P')$. Since $P'$ is linear and idempotent, it follows that $P'x = P'u + P'v = u = Px$. Therefore $P' = P$. $\qquad \square$

Remark: An immediate corollary of Theorems 2.16 and 2.20 says that any linear manifold of a linear space is the range of some projection. That is, *if $\mathcal{M}$ is a linear manifold of a linear space $\mathcal{X}$, then there exists a projection $P: \mathcal{X} \to \mathcal{X}$ such that $\mathcal{R}(P) = \mathcal{M}$.*

If $\mathcal{M}$ and $\mathcal{N}$ are algebraic complements in a linear space $\mathcal{X}$, then the unique projection $P$ in $\mathcal{L}[\mathcal{X}]$ with range $\mathcal{R}(P) = \mathcal{M}$ and null space $\mathcal{N}(P) = \mathcal{N}$ is called *the projection on $\mathcal{M}$ along $\mathcal{N}$*. If $P$ is the projection on $\mathcal{M}$ along $\mathcal{N}$, then the projection on $\mathcal{N}$ along $\mathcal{M}$ is precisely the projection $E = I - P$ in $\mathcal{L}[\mathcal{X}]$. Note that $EP = PE = O$.

**Proposition 2.21.** *Let $\mathcal{M}$ and $\mathcal{N}$ be linear manifolds of a linear space $\mathcal{X}$. If $\mathcal{M}$ and $\mathcal{N}$ are algebraic complements of each other, then the unique decomposition of each $x$ in $\mathcal{X} = \mathcal{M} \oplus \mathcal{N}$ as a direct sum*

$$x = u \oplus v$$

*of $u$ in $\mathcal{M}$ and $v$ in $\mathcal{N}$ is such that*

$$u = Px \quad \text{and} \quad v = (I - P)x$$

*where $P : \mathcal{X} \to \mathcal{X}$ is the unique projection on $\mathcal{M}$ along $\mathcal{N}$.*

*Proof.* Take an arbitrary $x$ in $\mathcal{X}$ and consider its unique decomposition $x = u \oplus v$ in $\mathcal{X} = \mathcal{M} \oplus \mathcal{N}$. Note that the identification of $\mathcal{M} \oplus \mathcal{N}$ with $\mathcal{M} + \mathcal{N} = \mathcal{X}$ is implicitly assumed in the proposition statement. Now write $x = (u, v)$ and set $Px = (u, 0)$. The very same argument used in the proof of Theorem 2.20 can be applied here to verify that this actually defines a unique projection $P : \mathcal{M} \oplus \mathcal{N} \to \mathcal{M} \oplus \mathcal{N}$ such that $\mathcal{R}(P) = \mathcal{M} \oplus \{0\}$ and $\mathcal{N}(P) = \{0\} \oplus \mathcal{N}$. Finally, identify $\mathcal{M} \oplus \{0\}$ and $\{0\} \oplus \mathcal{N}$ with $\mathcal{M}$ and $\mathcal{N}$ (and hence $(u, 0)$ and $(0, v)$ with $u$ and $v$), respectively.    $\square$

According to Theorem 2.16, every linear space $\mathcal{X}$ can be represented as the sum $\mathcal{X} = \mathcal{M} + \mathcal{N}$ of a pair $\{\mathcal{M}, \mathcal{N}\}$ of algebraic complements in $\mathcal{X}$. If $\mathcal{M} \oplus \mathcal{N}$ is identified with $\mathcal{M} + \mathcal{N}$, then this means that *every linear space $\mathcal{X}$ has a decomposition $\mathcal{X} = \mathcal{M} \oplus \mathcal{N}$ as a direct sum of disjoint linear manifolds of $\mathcal{X}$.*

**Proposition 2.22.** *Let $\mathcal{X}$ be a linear space and consider its decomposition*

$$\mathcal{X} = \mathcal{M} \oplus \mathcal{N}$$

*as a direct sum of disjoint linear manifolds $\mathcal{M}$ and $\mathcal{N}$ of $\mathcal{X}$. Let $P : \mathcal{X} \to \mathcal{X}$ be the projection on $\mathcal{M}$ along $\mathcal{N}$, and let $E = I - P$ be the projection on $\mathcal{N}$ along $\mathcal{M}$. Every linear transformation $L : \mathcal{X} \to \mathcal{X}$ can be written as a $2 \times 2$ matrix with linear transformation entries*

$$L = \begin{pmatrix} A & B \\ C & D \end{pmatrix},$$

*where $A = PL|_{\mathcal{M}} : \mathcal{M} \to \mathcal{M}$, $B = PL|_{\mathcal{N}} : \mathcal{N} \to \mathcal{M}$, $C = EL|_{\mathcal{M}} : \mathcal{M} \to \mathcal{N}$ and $D = EL|_{\mathcal{N}} : \mathcal{N} \to \mathcal{N}$.*

*Proof.* Let $\mathcal{M}$ and $\mathcal{N}$ be linear manifolds of a linear space $\mathcal{X}$. Suppose $\mathcal{M}$ and $\mathcal{N}$ are algebraic complements of each other and consider the decomposition $\mathcal{X} = \mathcal{M} \oplus \mathcal{N}$. Let $L$ be a linear transformation on $\mathcal{M} \oplus \mathcal{N}$ so that $L \in \mathcal{L}[\mathcal{X}]$. Take an arbitrary $x \in \mathcal{X}$ and consider its unique decomposition $x = u \oplus v$ in $\mathcal{X} = \mathcal{M} \oplus \mathcal{N}$ with $u$ in $\mathcal{M}$ and $v$ in $\mathcal{N}$. Now write $x = (u, v)$ so that $Lx = L(u, v) = L((u, 0) \oplus (0, v)) = L(u, 0) \oplus L(0, v) = L|_{\mathcal{M} \oplus \{0\}}(u, 0) \oplus L|_{\{0\} \oplus \mathcal{N}}(0, v)$. Identifying $\mathcal{M} \oplus \{0\}$ and

$\{0\} \oplus \mathcal{N}$ with $\mathcal{M}$ and $\mathcal{N}$ (and hence $(u, 0)$ and $(0, v)$ with $u$ and $v$), respectively, it follows that

$$L x = L|_{\mathcal{M}} u \oplus L|_{\mathcal{N}} v,$$

where $L|_{\mathcal{M}} u$ and $L|_{\mathcal{N}} v$ lie in $\mathcal{X} = \mathcal{M} \oplus \mathcal{N}$. Proposition 2.21 says that we may write

$$
\begin{aligned}
L|_{\mathcal{M}} u &= P L|_{\mathcal{M}} u \oplus E L|_{\mathcal{M}} u, \\
L|_{\mathcal{N}} v &= P L|_{\mathcal{N}} v \oplus E L|_{\mathcal{N}} v,
\end{aligned}
$$

where $P$ is the unique projection on $\mathcal{M}$ along $\mathcal{N}$ and $E = I - P$. Therefore

$$L x = (P L|_{\mathcal{M}} u + P L|_{\mathcal{N}} v) \oplus (E L|_{\mathcal{M}} u + E L|_{\mathcal{N}} v),$$

where $P L|_{\mathcal{M}} u + P L|_{\mathcal{N}} v$ is in $\mathcal{M}$ and $E L|_{\mathcal{M}} u + E L|_{\mathcal{N}} v$ is in $\mathcal{N}$. Since the ranges of $P L|_{\mathcal{M}}$ and $P L|_{\mathcal{N}}$ are included in $\mathcal{R}(P) = \mathcal{M}$, we may think of them as linear transformations into $\mathcal{M}$. Similarly, $E L|_{\mathcal{M}}$ and $E L|_{\mathcal{N}}$ can be thought of as linear transformations into $\mathcal{N}$. Thus set $A = P L|_{\mathcal{M}} \in \mathcal{L}[\mathcal{M}]$, $B = P L|_{\mathcal{N}} \in \mathcal{L}[\mathcal{N}, \mathcal{M}]$, $C = E L|_{\mathcal{M}} \in \mathcal{L}[\mathcal{M}, \mathcal{N}]$ and $D = E L|_{\mathcal{N}} \in \mathcal{L}[\mathcal{N}]$ so that

$$L x = (Au + Bv, \ Cu + Dv) \in \mathcal{M} \oplus \mathcal{N}$$

for every $x = (u, v) \in \mathcal{M} \oplus \mathcal{N}$. In terms of standard matrix notation, the vector $L x$ in $\mathcal{M} \oplus \mathcal{N}$ can be viewed as a $2 \times 1$ matrix with the first entry in $\mathcal{M}$ and the other in $\mathcal{N}$, namely $\left( \begin{smallmatrix} Au+Bv \\ Cu+Dv \end{smallmatrix} \right)$. This is precisely the action of the $2 \times 2$ matrix with linear transformation entries, $\left( \begin{smallmatrix} A & B \\ C & D \end{smallmatrix} \right)$, on the $2 \times 1$ matrix with entries in $\mathcal{M}$ and $\mathcal{N}$ representing $x$, namely $\left( \begin{smallmatrix} u \\ v \end{smallmatrix} \right)$. Thus $L x = \left( \begin{smallmatrix} A & B \\ C & D \end{smallmatrix} \right) \left( \begin{smallmatrix} u \\ v \end{smallmatrix} \right)$, and hence we write $L = \left( \begin{smallmatrix} A & B \\ C & D \end{smallmatrix} \right)$.

$\square$

**Example 2O.** Consider the setup of Proposition 2.22. Note that the projection on $\mathcal{M}$ along $\mathcal{N}$ can be written as

$$P = \begin{pmatrix} I & O \\ O & O \end{pmatrix}$$

with respect to the decomposition $\mathcal{X} = \mathcal{M} \oplus \mathcal{N}$, where $I$ denotes the identity on $\mathcal{M}$. Thus $LP = \left( \begin{smallmatrix} A & O \\ C & O \end{smallmatrix} \right)$ and $PLP = \left( \begin{smallmatrix} A & O \\ O & O \end{smallmatrix} \right)$, so that $LP = PLP$ if and only if $C = O$. Now note that $\mathcal{M}$ is $L$-invariant (i.e., $L(\mathcal{M}) \subseteq \mathcal{M}$) if and only if $P L|_{\mathcal{M}} = L|_{\mathcal{M}}$ or, equivalently, if and only if $E L|_{\mathcal{M}} = O$ (recall: $E = I - P$). Therefore,

$$L(\mathcal{M}) \subseteq \mathcal{M} \quad \Longleftrightarrow \quad A = L|_{\mathcal{M}} \quad \Longleftrightarrow \quad C = O \quad \Longleftrightarrow \quad LP = PLP.$$

Conclusion 1: The following assertions are pairwise equivalent.

(a)  $\mathcal{M}$ is $L$-invariant.

(b)  $L = \left( \begin{smallmatrix} L|_{\mathcal{M}} & B \\ O & D \end{smallmatrix} \right)$.

(c)  $LP = PLP$.

Similarly, if we apply the same argument to $\mathcal{N}$, then

$$L(\mathcal{N}) \subseteq \mathcal{N} \quad \Longleftrightarrow \quad D = L|_{\mathcal{N}} \quad \Longleftrightarrow \quad B = O \quad \Longleftrightarrow \quad PL = PLP.$$

Conclusion 2: The following assertions are pairwise equivalent as well.

(a$'$)  $\mathcal{M}$ and $\mathcal{N}$ are both $L$-invariant.

(b$'$)  $L = \begin{pmatrix} L|_{\mathcal{M}} & O \\ O & L|_{\mathcal{N}} \end{pmatrix}$.

(c$'$)  $L$ and $P$ commute   (i.e., $PL = LP$).


Let $\mathcal{M}$ and $\mathcal{N}$ be algebraic complements in a linear space $\mathcal{X}$. If a linear transformation $L$ in $\mathcal{L}[\mathcal{X}]$ is represented as $L = \begin{pmatrix} A & O \\ O & D \end{pmatrix}$ in terms of the decomposition $\mathcal{X} = \mathcal{M} \oplus \mathcal{N}$ (as in (a$'$) above), where $A \in \mathcal{L}[\mathcal{M}]$ and $D \in \mathcal{L}[\mathcal{N}]$, then it is usual to write $L = A \oplus D$. For instance, the projection on $\mathcal{M}$ along $\mathcal{N}$, which is represented as $P = \begin{pmatrix} I & O \\ O & O \end{pmatrix}$ with respect to the same decomposition $\mathcal{X} = \mathcal{M} \oplus \mathcal{N}$, is usually written as $P = I \oplus O$. These are examples of the following concept.

Let $\{\mathcal{X}_i\}_{i=1}^n$ be a finite indexed family of linear spaces over the same scalar field and consider their direct sum $\bigoplus_{i=1}^n \mathcal{X}_i$. Now let $\{L_i\}_{i=1}^n$ be a (similarly indexed) family of linear transformations such that $L_i \in \mathcal{L}[\mathcal{X}_i]$ for every index $i$. The *direct sum* of $\{L_i\}_{i=1}^n$, denoted by $\bigoplus_{i=1}^n L_i$, is the mapping of $\bigoplus_{i=1}^n \mathcal{X}_i$ into itself defined by

$$\bigoplus_{i=1}^n L_i(x_1, \dots, x_n) = (L_1 x_1, \dots, L_n x_n)$$

for every $(x_1, \dots, x_n)$ in $\bigoplus_{i=1}^n \mathcal{X}_i$. It is readily verified that $\bigoplus_{i=1}^n L_i$ is linear (i.e., $\bigoplus_{i=1}^n L_i \in \mathcal{L}[\bigoplus_{i=1}^n \mathcal{X}_i]$) and also that, for every index $i$,

$$\left( \bigoplus_{i=1}^n L_i \right)\Big|_{\mathcal{X}_i} = L_i.$$

Observe that the above identity actually is a short notation for the following assertion: "if $0_i$ is the origin of each $\mathcal{X}_i$ and $O_i$ is the unique (linear) transformation of $\{0_i\}$ onto itself, then each linear manifold $\{0_1\} \oplus \cdots \oplus \{0_{i-1}\} \oplus \mathcal{X}_i \oplus \{0_{i+1}\} \oplus \cdots \oplus \{0_n\}$ of $\bigoplus_{i=1}^n \mathcal{X}_i$ is invariant for $\bigoplus_{i=1}^n L_i$ and the restriction of $\bigoplus_{i=1}^n L_i$ to that invariant linear manifold is the direct sum $O_1 \oplus \cdots \oplus O_{i-1} \oplus L_i \oplus O_{i+1} \oplus \cdots \oplus O_n$". Of course, we shall always use the short notation. Conversely, if $L \in \mathcal{L}[\bigoplus_{i=1}^n \mathcal{X}_i]$ is such that $L|_{\mathcal{X}_i} \in \mathcal{L}[\mathcal{X}_i]$ for every index $i$, then $L$ is the direct sum of $\{L|_{\mathcal{X}_i}\}_{i=1}^n$. That is, if each $\mathcal{X}_i$ in $\bigoplus_{i=1}^n \mathcal{X}_i$ is invariant for $L \in \mathcal{L}[\bigoplus_{i=1}^n \mathcal{X}_i]$, then

$$L = \bigoplus_{i=1}^n L|_{\mathcal{X}_i}.$$

Summing up: Set $\mathcal{X} = \bigoplus_{i=1}^{n} \mathcal{X}_i$ and consider linear transformations $L_i$ in $\mathcal{L}[\mathcal{X}_i]$ for each $i$ and $L$ in $\mathcal{L}[\mathcal{X}]$.

$$L = \bigoplus_{i=1}^{n} L_i \quad \text{if and only if} \quad L_i = L|_{\mathcal{X}_i}$$

for every index $i$ (so that each $\mathcal{X}_i$, viewed as a linear manifold of the linear space $\bigoplus_{i=1}^{n} \mathcal{X}_i$, is invariant for $L$). The linear transformations $\{L_i\}$ are referred to as the *direct summands* of $L$.

In particular, if we consider the decomposition $\mathcal{X} = \mathcal{M} \oplus \mathcal{N}$ of a linear space $\mathcal{X}$ into the direct sum of a pair of algebraic complements $\mathcal{M}$ and $\mathcal{N}$ in $\mathcal{X}$, and if we take linear transformations $L \in \mathcal{L}[\mathcal{X}]$, $A \in \mathcal{L}[\mathcal{M}]$ and $D \in \mathcal{L}[\mathcal{N}]$, then

$$L = \begin{pmatrix} A & O \\ O & D \end{pmatrix} = A \oplus D \quad \text{if and only if} \quad A = L|_{\mathcal{M}} \text{ and } D = L|_{\mathcal{N}}$$

(so that $\mathcal{M}$ and $\mathcal{N}$ are both $L$-invariant), where $A$ and $D$ are the direct summands of $L$ with respect to the decomposition $\mathcal{X} = \mathcal{M} \oplus \mathcal{N}$.

## Suggested Reading

Brown and Pearcy [2]          Naylor and Sell [1]
Halmos [2]                    Roman [1]
Herstein [1]                  Simmons [1]
MacLane and Birkhoff [1]      Taylor and Lay [1]

# Problems

**Problem 2.1.** Let $\mathcal{X}$ be a linear space over a field $\mathbb{F}$. Take arbitrary $\alpha$ and $\beta$ in $\mathbb{F}$ and arbitrary $x$, $y$ and $z$ in $\mathcal{X}$. Verify the following propositions.

(a) $(-\alpha)x = -(\alpha x)$.

(b) $0x = 0 = \alpha 0$.

(c) $\alpha x = 0 \implies \alpha = 0 \text{ or } x = 0$.

(d) $x + y = x + z \implies y = z$.

(e) $\alpha x = \alpha y \implies x = y \text{ if } \alpha \neq 0$.

(f) $\alpha x = \beta x \implies \alpha = \beta \text{ if } x \neq 0$.

**Problem 2.2.** Let $\mathcal{X}$ be a real or complex linear space. A subset $C$ of $\mathcal{X}$ is *convex* if $\alpha x + (1 - \alpha)y \in C$ whenever $x, y \in C$ and $0 \le \alpha \le 1$. A vector $x \in \mathcal{X}$ is a *convex linear combination* of vectors in $\mathcal{X}$ if there exists a finite set $\{x_i\}_{i=1}^{n}$ of vectors in $\mathcal{X}$ and a finite family of nonnegative scalars $\{\alpha_i\}_{i=1}^{n}$ such that $x = \sum_{i=1}^{n} \alpha_i x_i$ and $\sum_{i=1}^{n} \alpha_i = 1$. If $A$ is a subset of $\mathcal{X}$, then the intersection of all convex sets containing $A$ is called the *convex hull* of $A$, denoted by $co(A)$.

(a)  Show that the intersection of an arbitrary nonempty collection of convex sets is convex.

(b)  Show that $co(A)$ is the smallest (in the inclusion ordering) convex set that contains $A$.

(c)  Show that $C$ is convex if and only if every convex linear combination of vectors in $C$ belongs to $C$.

   *Hint*: To verify that *every convex linear combination of vectors in a convex set $C$ belongs to $C$* proceed as follows. Note that the italicized result holds for any convex linear combination of two vectors in $C$ (by the definition of a convex set). Suppose it holds for every convex linear combination of $n$ vectors in $C$, for some $n \in \mathbb{N}$. This implies that $\alpha \sum_{i=1}^{n} \alpha^{-1} \alpha_i x_i + \alpha_{n+1} x_{n+1}$ lies in $C$ whenever $\{x_i\}_{i=1}^{n+1} \subset C$ and $\sum_{i=1}^{n+1} \alpha_i = 1$ with $0 < \alpha_{n+1}$, where $\alpha = \sum_{i=1}^{n} \alpha_i$ (reason: $\sum_{i=1}^{n} \alpha^{-1} \alpha_i x_i \in C$). Now conclude the proof by induction.

(d)  Show that $co(A)$ coincides with the set of all convex linear combinations of vectors in $A$.

   *Hint*: Let $clc(A)$ denote the set of all convex linear combinations of vectors in $A$. Verify that $clc(A)$ is a convex set. Now use (b) and (c) to show that $co(A) \subseteq clc(A) \subseteq clc(co(A)) = co(A)$.


**Problem 2.3.** Let $\mathcal{M}$ and $\mathcal{N}$ be linear manifolds of a linear space $\mathcal{X}$, and let $A$ and $B$ be linearly independent subsets of $\mathcal{M}$ and $\mathcal{N}$, respectively. If $\mathcal{M} \cap \mathcal{N} = \{0\}$, then $A \cup B$ is linearly independent. (*Hint*: If $a \in A$ is a linear combination of vectors in $A \cup B$, then $a = b + a'$ for some $a' \in \mathcal{M}$ and some $b \in \mathcal{N}$.)


**Problem 2.4.** Let $A$ be a linearly independent subset of a linear space $\mathcal{X}$. If $B$ and $C$ are subsets of $A$, then

$$\text{span}\,(B \cap C) \;=\; \text{span}\,B \cap \text{span}\,C.$$

*Hint*: Show that $\text{span}\,B \cap \text{span}\,C \subseteq \text{span}\,(B \cup C)$ by Proposition 2.3.


**Problem 2.5.** If a subset $A$ of a linear space $\mathcal{X}$ spans $\mathcal{X}$, then the cardinality of every linearly independent subset of $\mathcal{X}$ is less than or equal to the cardinality of $A$.

*Hint*: Suppose $A$ is a subset of a linear space $\mathcal{X}$ that spans $\mathcal{X}$. Let $B$ be an arbitrary Hamel basis for $\mathcal{X}$, and let $C$ be an arbitrary linearly independent subset of $\mathcal{X}$. Show that $\#C \leq \#B \leq \#A$. (Apply Theorems 2.5, 2.6 and 2.7 — see Problems 1.21(a) and 1.22). Note that this generalizes Claim 0 in the proof of Theorem 2.7 for subsets of arbitrary cardinality.

**Problem 2.6.** Let $\mathcal{X}$ be a linear space, and let $\mathcal{M}$ be a linear manifold of $\mathcal{X}$. Verify the following propositions.

(a) $\dim \mathcal{M} = 0$ if and only if $\mathcal{M} = \{0\}$.

(b) $\dim \mathcal{M} \leq \dim \mathcal{X}$.  (*Hint*: Problem 2.5.)

**Problem 2.7.** If $\mathcal{M}$ is a proper linear manifold of a *finite-dimensional* linear space $\mathcal{X}$, then $\dim \mathcal{M} < \dim \mathcal{X}$. Prove the above statement and show that it does not hold for infinite-dimensional linear spaces (e.g., show that $\dim \mathcal{X} = \dim \mathcal{X}_0$ where $\mathcal{X}$ is the linear space of Example 2J and $\mathcal{X}_0 = \{x = (\xi_1, \xi_2, \xi_3, \dots ,) \in \mathcal{X} : \xi_1 = 0\}$).

**Problem 2.8.** Let $\mathcal{X}$ be a nonzero linear space over an *infinite* field, and let $B$ be a Hamel basis for $\mathcal{X}$. Recall that every nonzero vector $x$ in $\mathcal{X}$ has a unique representation in terms of $B$. That is, for each $x \neq 0$ in $\mathcal{X}$ there exists a unique nonempty finite subset $B_x$ of $B$ and a unique finite family of nonzero scalars $\{\alpha_b\}_{b \in B_x} \subset \mathbb{F}$ such that

$$x = \sum_{b \in B_x} \alpha_b b.$$

For each positive integer $n \in \mathbb{N}$, let $X_n$ be the set of all *nonzero* vectors in $\mathcal{X}$ whose representations as a (finite) linear combination of vectors in $B$ have exactly $n$ (nonzero) summands. That is, for each $n \in \mathbb{N}$, set

$$X_n = \{x \in \mathcal{X} : \#B_x = n\}.$$

(a) Prove that $\#X_n = \#(\mathbb{F} \times B)$ for all $n \in \mathbb{N}$.

   *Hint*: Show that $\#X_n = \#(\mathbb{F}^n \times B)$ and recall: if $\mathbb{F}$ is an infinite set, then $\#\mathbb{F}^n = \#\mathbb{F}$ (Problems 1.23 and 1.28).

(b) Apply Theorem 1.10 to show that $\#\left(\bigcup_{n \in \mathbb{N}} X_n\right) \leq \#(\mathbb{F} \times B)$.

(c) Verify that $\{X_n\}_{n \in \mathbb{N}}$ is a partition of $\mathcal{X} \backslash \{0\}$.

Thus conclude from (b) and (c) (see Problem 1.28(a)) that

$$\#\mathcal{X} = \#(\mathbb{F} \times B) = \max\{\#\mathbb{F}, \dim \mathcal{X}\}.$$

**Problem 2.9.** Prove the following proposition, which is known as the *Principle of Superposition*. A mapping $L\colon \mathcal{X} \to \mathcal{Y}$, where $\mathcal{X}$ and $\mathcal{Y}$ are linear spaces over the same scalar field, is a linear transformation if and only if

$$L\left(\sum_{i=1}^{n} \alpha_i x_i\right) = \sum_{i=1}^{n} \alpha_i L x_i$$

for all *finite* sets $\{x_i\}_{i=1}^{n}$ of vectors in $\mathcal{X}$ and all *finite* families of scalars $\{\alpha_i\}_{i=1}^{n}$.

**Problem 2.10.** Let $L\colon \mathcal{X} \to \mathcal{Y}$ be a linear transformation. Show that the null space $\mathcal{N}(L)$ and the range $\mathcal{R}(L)$ of $L$ are linear manifolds (of the linear spaces $\mathcal{X}$ and $\mathcal{Y}$, respectively).

**Problem 2.11.** Let $L\colon \mathcal{X} \to \mathcal{Y}$ be a linear transformation of a linear space $\mathcal{X}$ into a linear space $\mathcal{Y}$. Prove the following propositions.

(a) If $\mathcal{M}$ is a linear manifold of $\mathcal{X}$, then $L(\mathcal{M})$ is a linear manifold of $\mathcal{Y}$ (i.e., the linear image of a linear manifold is a linear manifold).

(b) If $\mathcal{N}$ is a linear manifold of $\mathcal{Y}$, then $L^{-1}(\mathcal{N})$ is a linear manifold of $\mathcal{X}$ (i.e., the inverse image of a linear manifold under a linear transformation is again a linear manifold).

**Problem 2.12.** Let $\mathcal{X}$ and $\mathcal{Y}$ be linear spaces, and let $L\colon \mathcal{X} \to \mathcal{Y}$ be a linear transformation. Show that the following assertions are equivalent.

(a) $A \subseteq \mathcal{X}$ is a linear manifold whenever $L(A) \subseteq \mathcal{Y}$ is a linear manifold.

(b) $\mathcal{N}(L) = \{0\}$.

*Hint*: Give a direct proof for (b)$\Rightarrow$(a) by using Problems 1.3(d) and 2.11(b). Give a contrapositive proof for (a)$\Rightarrow$(b) — recall: if $x$ is a nonzero vector in $\mathcal{X}$, then $\{x\}$ is not a linear manifold of $\mathcal{X}$.

**Problem 2.13.** Show that the set $\mathcal{L}[\mathcal{X}, \mathcal{Y}]$ of all linear transformations of a linear space $\mathcal{X}$ into a linear space $\mathcal{Y}$ is itself a linear space (over the same common field of $\mathcal{X}$ and $\mathcal{Y}$) when vector addition and scalar multiplication in $\mathcal{L}[\mathcal{X}, \mathcal{Y}]$ are defined pointwise as in Example 2F.

**Problem 2.14.** The restriction $L|_{\mathcal{M}}\colon \mathcal{M} \to \mathcal{Y}$ of a linear transformation $L\colon \mathcal{X} \to \mathcal{Y}$ to a linear manifold $\mathcal{M}$ of $\mathcal{X}$ is itself a linear transformation.

**Problem 2.15.** Prove that a composition of two linear transformations is again a linear transformation.

**Problem 2.16.** Let $L\colon \mathcal{X} \to \mathcal{Y}$ be a linear transformation. It is trivially verified that, if $L$ is surjective, then $\dim \mathcal{R}(L) = \dim \mathcal{Y}$. Now verify that, if $L$ is injective, then $\dim \mathcal{R}(L) = \dim \mathcal{X}$.

**Problem 2.17.** Let $L\colon \mathcal{X} \to \mathcal{Y}$ be a linear transformation of a linear space $\mathcal{X}$ into a linear space $\mathcal{Y}$. The dimension of the range of $L$ is the *rank* of $L$, and the dimension of the null space of $L$ is the *nullity* of $L$. Show that rank and nullity are related as follows.

$$\dim \mathcal{N}(L) + \dim \mathcal{R}(L) = \dim \mathcal{X}.$$

(*Hint*: Addition of cardinal numbers was defined in Problem 1.30.)

**Problem 2.18.** If $\dim \mathcal{R}(L)$ is finite, then $L$ is called a *finite-dimensional* (or a *finite-rank*) *linear transformation*. Clearly, if $\mathcal{Y}$ is a finite-dimensional linear space, then every $L \in \mathcal{L}[\mathcal{X}, \mathcal{Y}]$ is finite-dimensional. Verify that, if $\mathcal{X}$ is a finite-dimensional linear space, then every $L \in \mathcal{L}[\mathcal{X}, \mathcal{Y}]$ is finite-dimensional. Moreover, if $L\colon \mathcal{X} \to \mathcal{Y}$ is a finite-dimensional linear transformation (so that $\mathcal{R}(L)$ is a finite-dimensional linear manifold of $\mathcal{Y}$), then show that

(a)   $L$ is injective if and only if $\dim \mathcal{R}(L) = \dim \mathcal{X}$,

(b)   $L$ is surjective if and only if $\dim \mathcal{R}(L) = \dim \mathcal{Y}$.

**Problem 2.19.** let $\mathcal{X}$ be a linear space over a field $\mathbb{F}$, and let $\mathcal{X}^{\mathbb{N}_0}$ be the linear space (over the same field $\mathbb{F}$) of all $\mathcal{X}$-valued sequences $\{x_n\}_{n \in \mathbb{N}_0}$. Suppose $A$ is a linear transformation of $\mathcal{X}$ into itself. Take an arbitrary sequence $u = \{u_n\}_{n \in \mathbb{N}_0}$ in $\mathcal{X}^{\mathbb{N}_0}$ and consider the (unique) sequence $x = \{x_n\}_{n \in \mathbb{N}_0}$ in $\mathcal{X}^{\mathbb{N}_0}$ which is recursively defined as follows. Set $x_0 = u_0$ and, for each $n \in \mathbb{N}_0$, let

$$x_{n+1} = A x_n + u_{n+1}.$$

Prove by induction that

$$x_n = \sum_{i=0}^{n} A^{n-i} u_i$$

for every $n \in \mathbb{N}_0$, where $A^0$ is (by definition) the identity $I$ in $\mathcal{L}[\mathcal{X}]$. Let $L\colon \mathcal{X}^{\mathbb{N}_0} \to \mathcal{X}^{\mathbb{N}_0}$ be the map that assigns to each sequence $u$ in $\mathcal{X}^{\mathbb{N}_0}$ this unique sequence $x$ in $\mathcal{X}^{\mathbb{N}_0}$, so that

$$x = Lu.$$

Show that $L$ is a linear transformation of $\mathcal{X}^{\mathbb{N}_0}$ into itself. The recursive equation (or the *difference equation*) $x_{n+1} = A x_n + u_{n+1}$ is called a *discrete linear dynamical system* because $L$ is *linear*. Its unique solution is given by $x = Lu$ (i.e., $x_n = \sum_{i=0}^{n} A^{n-i} u_i$ for every $n \in \mathbb{N}_0$).

**Problem 2.20.** Let $\mathbb{F}$ denote either the real or complex field, and let $\mathcal{X}$ and $\mathcal{Y}$ be linear spaces over $\mathbb{F}$. For any polynomial $p$ (in one variable in $\mathbb{F}$, with coefficients in $\mathbb{F}$, and of any finite order $n$) set

$$p(L) = \sum_{i=0}^{n} \alpha_i L^i$$

in $\mathcal{L}[\mathcal{X}]$ for every $L \in \mathcal{L}[\mathcal{X}]$, where the coefficients $\{\alpha_i\}_{i=0}^{n}$ lie in $\mathbb{F}$ (note: $L^0 = I$). Take $L \in \mathcal{L}[\mathcal{X}]$, $K \in \mathcal{L}[\mathcal{Y}]$ and $M \in \mathcal{L}[\mathcal{X}, \mathcal{Y}]$ arbitrary, and prove the following implication.

(a) If $ML = KM$, then $Mp(L) = p(K)M$ for any polynomial $p$.

Thus conclude: $p(L)$ *is similar to* $p(K)$ *whenever* $L$ *is similar to* $K$. A linear transformation $L$ in $\mathcal{L}[\mathcal{X}]$ is called *nilpotent* if $L^n = O$ for some integer $n \in \mathbb{N}$, and *algebraic* if $p(L) = O$ for some polynomial $p$. It is clear that every nilpotent linear transformation is algebraic. Prove the following propositions.

(b) A linear transformation is similar to an algebraic (nilpotent) linear transformation if and only if it is itself algebraic (nilpotent).

(c) Sum and composition of nilpotent linear transformations are not necessarily nilpotent.

*Hint*: The matrices $T = \left(\begin{smallmatrix} 0 & 1 \\ 0 & 0 \end{smallmatrix}\right)$ and $L = \left(\begin{smallmatrix} 0 & 0 \\ 1 & 0 \end{smallmatrix}\right)$ in $\mathcal{L}[\mathbb{C}^2]$ are both nilpotent. $L + T$ is an involution. $LT$ and $TL$ are idempotent.

**Problem 2.21.** Let $\mathbb{F}$ denote either the real or complex field, and let $\mathcal{X}$ be a linear space over $\mathbb{F}$. A subset $K$ of $\mathcal{X}$ is a *cone* (with vertex at the origin) if $\alpha x \in K$ whenever $x \in K$ and $\alpha \geq 0$. Recall the definition of a convex set in Problem 2.2 and verify the following assertions.

(a) Every linear manifold is a convex cone.

(b) The union of nonzero disjoint linear manifolds is a nonconvex cone.

Let $S$ be a nonempty set and consider the linear space $\mathbb{F}^S$. Show that

(c) $\{x \in \mathbb{F}^S : x(s) \geq 0 \text{ for all } s \in S\}$ is a convex cone in $\mathbb{F}^S$.

**Problem 2.22.** Show that the implication (a)$\Rightarrow$(b) in Theorem 2.14 does not generalize to three linear manifolds, say $\mathcal{M}$, $\mathcal{N}$ and $\mathcal{R}$, if we simply assume that they are pairwise disjoint. (*Hint*: $\mathbb{R}^3$.)

**Problem 2.23.** Let $\{\mathcal{M}_i\}_{i=1}^{n}$ be a finite collection of linear manifolds of a linear space $\mathcal{X}$. Show that the following assertions are equivalent.

(a) $\mathcal{M}_i \cap \sum_{j=1, j \neq i}^{n} \mathcal{M}_j = \{0\}$ for every $i = 1, \ldots, n$.

(b) For each $x$ in $\sum_{i=1}^{n} \mathcal{M}_i$ there exists a unique $n$-tuple $(x_1, \ldots, x_n)$ in $\prod_{i=1}^{n} \mathcal{M}_i$ such that $x = \sum_{i=1}^{n} x_i$.

*Hint*: (a)$\Rightarrow$(b) for $n = 2$ by Theorem 2.14. Let $n > 2$ and suppose (a)$\Rightarrow$(b) for every $2 \leq m < n$. Show that, if (a) holds true for $m + 1$, then (b) holds true for $m + 1$. Now conclude the proof of (a)$\Rightarrow$(b) by induction in $n$. Next show that (b)$\Rightarrow$(a) by Theorem 2.14.

**Problem 2.24.** Let $\{\mathcal{M}_i\}_{i=1}^{n}$ be a finite collection of linear manifolds of a linear space $\mathcal{X}$, and let $B_i$ be a Hamel basis for each $\mathcal{M}_i$. If $\mathcal{M}_i \cap \sum_{j=1, j \neq i}^{n} \mathcal{M}_j = \{0\}$ for every $i = 1, \ldots, n$, then $\bigcup_{i=1}^{n} B_i$ is a Hamel basis for $\sum_{i=1}^{n} \mathcal{M}_i$. Prove.

*Hint*: The result holds for $n = 2$ by Proposition 2.15. Use the hint of the previous problem.

**Problem 2.25.** Let $\mathcal{M}$ and $\mathcal{N}$ be linear manifolds of a linear space.

(a) If $\mathcal{M}$ and $\mathcal{N}$ are disjoint, then

$$\dim(\mathcal{M} \oplus \mathcal{N}) = \dim(\mathcal{M} + \mathcal{N}) = \dim \mathcal{M} + \dim \mathcal{N}.$$

    *Hint*: Problem 1.30, Theorem 2.14 and Proposition 2.15.

(b) If $\mathcal{M}$ and $\mathcal{N}$ are finite-dimensional, then

$$\dim(\mathcal{M} + \mathcal{N}) = \dim \mathcal{M} + \dim \mathcal{N} - \dim(\mathcal{M} \cap \mathcal{N}).$$

**Problem 2.26.** Let $\mathcal{M}$ be a proper linear manifold of a linear space $\mathcal{X}$ so that $\mathcal{M} \in \mathcal{L}at(\mathcal{X}) \backslash \{\mathcal{X}\}$. Consider the inclusion ordering of $\mathcal{L}at(\mathcal{X})$ and show that

$$\mathcal{M} \text{ is maximal in } \mathcal{L}at(\mathcal{X}) \backslash \{\mathcal{X}\} \iff \text{codim } \mathcal{M} = 1.$$

**Problem 2.27.** Let $\varphi$ be a nonzero linear functional on a linear space $\mathcal{X}$ (i.e., let $\varphi$ be a nonzero element of $\mathcal{X}'$, the algebraic dual of $\mathcal{X}$). Prove that

(a) $\mathcal{N}(\varphi)$ is maximal in $\mathcal{L}at(\mathcal{X}) \backslash \{\mathcal{X}\}$.

That is, the null space of any nonzero linear functional in $\mathcal{X}'$ is a maximal proper linear manifold of $\mathcal{X}$. Conversely, if $\mathcal{M}$ is a maximal linear manifold in $\mathcal{L}at(\mathcal{X}) \backslash \{\mathcal{X}\}$, then there exists a nonzero $\varphi$ in $\mathcal{X}'$ such that $\mathcal{M} = \mathcal{N}(\varphi)$. In other words, prove the following assertion.

(b) Every maximal element of $\mathcal{L}at(\mathcal{X}) \backslash \{\mathcal{X}\}$ is the null space of some nonzero $\varphi$ in $\mathcal{X}'$.

**Problem 2.28.** Let $\mathcal{X}$ be a linear space over a field $\mathbb{F}$. The set

$$H_{\varphi,\alpha} = \{x \in \mathcal{X} : \varphi(x) = \alpha\},$$

determined by a *nonzero* $\varphi$ in $\mathcal{X}'$ and a scalar $\alpha$ in $\mathbb{F}$, is called a *hyperplane* in $\mathcal{X}$. It is clear that $H_{\varphi,0}$ coincides with $\mathcal{N}(\varphi)$, but $H_{\varphi,\alpha}$ is not a linear manifold of $\mathcal{X}$ if $\alpha$ is a nonzero scalar. A *linear variety* is a translation of a proper linear manifold. That is, a linear variety $V$ is a subset of $\mathcal{X}$ that coincides with the coset of $x$ modulo $\mathcal{M}$,

$$V = \mathcal{M} + x = \{y \in \mathcal{X} : y = z + x \text{ for some } z \in \mathcal{M}\},$$

for some $x \in \mathcal{X}$ and some $\mathcal{M} \in \mathcal{L}at(\mathcal{X})\backslash\{\mathcal{X}\}$. If $\mathcal{M}$ is maximal in $\mathcal{L}at(\mathcal{X})\backslash\{\mathcal{X}\}$, then $\mathcal{M} + x$ is called a *maximal linear variety*. Show that a hyperplane is precisely a maximal linear variety.

**Problem 2.29.** Let $\mathcal{X}$ be a linear space over a field $\mathbb{F}$, and let $P$ and $E$ be projections in $\mathcal{L}[\mathcal{X}]$. Suppose $E \neq O$, and let $\alpha$ be an arbitrary *nonzero* scalar in $\mathbb{F}$. Prove the following proposition.

(a)  $P + \alpha E$ is a projection if and only if $PE + EP = (1 - \alpha)E$.

Moreover, if $P + \alpha E$ is a projection, then show that

(b)  $E$ and $P$ commute (i.e., $EP = PE$) and $EP$ is a projection,

(c)  $EP = O$ if and only if $\alpha = 1$ and $EP \neq O$ if and only if $\alpha = -1$.

Therefore,

(d)  if $P + \alpha E$ is a projection, then $\alpha = 1$ or $\alpha = -1$.

Finally conclude that

$$P + E \text{ is a projection if and only if } EP = PE = O,$$

$$P - E \text{ is a projection if and only if } EP = PE = E.$$

**Problem 2.30.** An *algebra* (or a *linear algebra*) is a linear space $\mathcal{A}$ that is also a ring with respect to a second binary operation on $\mathcal{A}$ called *product* (notation: $xy \in \mathcal{A}$ is the product of $x \in \mathcal{A}$ and $y \in \mathcal{A}$). The product is related to scalar multiplication by the property

$$\alpha(xy) = (\alpha x)y = x(\alpha y)$$

for every $x, y \in \mathcal{A}$ and every scalar $\alpha$. We shall refer to a *real* or *complex algebra* if $\mathcal{A}$ is a real or complex linear space. Recall that this new binary operation on $\mathcal{A}$ (i.e., the product in the ring $\mathcal{A}$) is associative,

$$x(yz) = (xy)z,$$

and distributive with respect to vector addition,

$$x(y + z) = xy + xz \quad \text{and} \quad (y + z)x = yx + zx,$$

for every $x$, $y$ and $z$ in $\mathcal{A}$. If $\mathcal{A}$ possesses a neutral element 1 under the product operation (i.e., if there exists $1 \in \mathcal{A}$ such that $x1 = 1x = x$ for every $x \in \mathcal{A}$), then $\mathcal{A}$ is said to be an *algebra with identity* (or a *unital algebra*). Such a neutral element 1 is called the *identity* (or *unit*) of $\mathcal{A}$. If $\mathcal{A}$ is an algebra with identity, and if $x \in \mathcal{A}$ has an *inverse* (denoted by $x^{-1}$) with respect to the product operation (i.e., if there exists $x^{-1} \in \mathcal{A}$ such that $xx^{-1} = x^{-1}x = 1$), then $x$ is an *invertible* element of $\mathcal{A}$. Recall that the identity is unique if it exists, and so is the inverse of an invertible element of $\mathcal{A}$. If the product operation is commutative, then $\mathcal{A}$ is said to be a *commutative algebra*.

(a) Let $\mathcal{X}$ be a linear space of dimension greater than one. Show that $\mathcal{L}[\mathcal{X}]$ is a noncommutative algebra with identity when the product in $\mathcal{L}[\mathcal{X}]$ is interpreted as composition (i.e., $LT = L \circ T$ for every $L, T \in \mathcal{L}[\mathcal{X}]$). The identity $I$ in $\mathcal{L}[\mathcal{X}]$ is precisely the neutral element under the product operation. $L$ is an invertible of $\mathcal{L}[\mathcal{X}]$ if and only if $L$ is injective and surjective.

A *subalgebra* of $\mathcal{A}$ is a linear manifold $\mathcal{M}$ of $\mathcal{A}$ (when $\mathcal{A}$ is viewed as a linear space) which is an algebra in its own right with respect to the product operation of $\mathcal{A}$ (i.e., $uv \in \mathcal{M}$ whenever $u \in \mathcal{M}$ and $v \in \mathcal{M}$). A subalgebra $\mathcal{M}$ of $\mathcal{A}$ is a *left ideal* of $\mathcal{A}$ if $ux \in \mathcal{M}$ whenever $u \in \mathcal{M}$ and $x \in \mathcal{A}$. A *right ideal* of $\mathcal{A}$ is a subalgebra $\mathcal{M}$ of $\mathcal{A}$ such that $xu \in \mathcal{M}$ whenever $x \in \mathcal{A}$ and $u \in \mathcal{M}$. An *ideal* (or a *two-sided ideal* or a *bilateral ideal*) of $\mathcal{A}$ is a subalgebra $\mathcal{I}$ of $\mathcal{A}$ that is both a left ideal and a right ideal.

(b) Let $\mathcal{X}$ be an infinite-dimensional linear space. Show that the set of all finite-dimensional linear transformations in $\mathcal{L}[\mathcal{X}]$ is a proper left ideal of $\mathcal{L}[\mathcal{X}]$ with no identity. (*Hint*: Problem 2.25(b).)

(c) Show that, if $\mathcal{A}$ is an algebra and $\mathcal{I}$ is a proper ideal of $\mathcal{A}$, then the quotient space $\mathcal{A}/\mathcal{I}$ of $\mathcal{A}$ modulo $\mathcal{I}$ is an algebra. This is called the *quotient algebra* of $\mathcal{A}$ with respect to $\mathcal{I}$. If $\mathcal{A}$ has an identity 1, then the coset $1 + \mathcal{I}$ is the identity of $\mathcal{A}/\mathcal{I}$.

*Hint*: Recall that vector addition and scalar multiplication in the linear space $\mathcal{A}/\mathcal{I}$ are defined by

$$(x + \mathcal{I}) + (y + \mathcal{I}) = (x + y) + \mathcal{I},$$

$$\alpha(x + \mathcal{I}) = \alpha x + \mathcal{I},$$

for every $x$, $y \in \mathcal{A}$ and every scalar $\alpha$ (see Example 2H). Now show that the product of cosets in $\mathcal{A}/\mathcal{I}$ can be likewise defined by

$$(x + \mathcal{I})(y + \mathcal{I}) = xy + \mathcal{I}$$

for every $x, y \in \mathcal{A}$ (i.e., if $x' = x + u$ and $y' = y + v$, with $x, y \in \mathcal{A}$ and $u, v \in \mathcal{I}$, then there exists $z \in \mathcal{I}$ such that $x'y' + w = xy + z$ for any $w \in \mathcal{I}$, whenever $\mathcal{I}$ is a two-sided ideal of $\mathcal{A}$).

# 3
# Topological Structures

The basic concept behind the subject of point-set topology is the notion of "closeness" between two points in a set $X$. In order to get a numerical gauge of how close together two points in $X$ may be, we shall provide an extra structure to $X$, viz., a topological structure, that again goes beyond its purely set-theoretic structure. For most of our purposes the notion of closeness associated with a metric will be sufficient, and this leads to the concept of "metric space": a set upon which a "metric" is defined. The metric-space structure that a set acquires when a metric is defined on it is a special kind of topological structure. Metric spaces comprise the kernel of this chapter but general topological spaces are also introduced.

## 3.1   Metric Spaces

A *metric* (or *metric function*, or *distance function*) is a real-valued function on the Cartesian product of an arbitrary set with itself that has the following four properties, called the *metric axioms*.

**Definition 3.1.** Let $X$ be an arbitrary set. A real-valued function $d$ on the Cartesian product $X \times X$,

$$d \colon X \times X \to \mathbb{R},$$

is a *metric* on $X$ if the following conditions are satisfied for all $x$, $y$ and $z$ in $X$.

(i)   $d(x, y) \geq 0$   and   $d(x, x) = 0$   (*nonnegativeness*),
(ii)  $d(x, y) = 0$   only if   $x = y$   (*positiveness*),
(iii) $d(x, y) = d(y, x)$   (*symmetry*),
(iv)  $d(x, y) \leq d(x, z) + d(z, y)$   (*triangle inequality*).

A set $X$ equipped with a metric on it is a *metric space*.

A word on notation and terminology. The value of the metric $d$ on a pair of points of $X$ is called the *distance* between those points. According to the above definition a metric space actually is an ordered pair $(X, d)$ where $X$ is an arbitrary set, called the *underlying set* of the metric space $(X, d)$, and $d$ is a metric function defined on it. We shall often refer to a metric space in several ways. Sometimes we shall speak of $X$ itself as a metric space when the metric $d$ is either clear in the context or is immaterial. In this case we shall simply say "$X$ is a metric space". On the other hand, in order to avoid confusion among different metric spaces, we may occasionally insert a subscript on the metrics. For instance, $(X, d_X)$ and $(Y, d_Y)$ will stand for metric spaces where $X$ and $Y$ are the respective underlying sets, $d_X$ denotes the metric on $X$, and $d_Y$ the metric on $Y$. Moreover, if a set $X$ can be equipped with more than one metric, say $d_1$ and $d_2$, then $(X, d_1)$ and $(X, d_2)$ will represent *different* metric spaces with the same underlying set $X$. In brief, a metric space is an arbitrary set with an additional structure defined by means of a metric $d$. Such an additional structure is *the topological structure induced by the metric $d$*.

If $(X, d)$ is a metric space, and if $A$ is a subset of $X$, then it is easy to show that the restriction $d|_{A \times A} \colon A \times A \to \mathbb{R}$ of the metric $d$ to $A \times A$ is a metric on $A$ — the so-called *relative metric*. Equipped with the relative metric, $A$ is a *subspace* of $X$. We shall drop the subscript $A \times A$ from $d|_{A \times A}$ and say that $(A, d)$ is a subspace of $(X, d)$. Thus a subspace of a metric space $(X, d)$ is a subset $A$ of the underlying set $X$ equipped with the *relative metric*, which is itself a metric space. Roughly speaking, $A$ inherits the metric of $(X, d)$. If $(A, d)$ is a subspace of $(X, d)$ and $A$ is a proper subset of $X$, then $(A, d)$ is said to be a *proper subspace* of the metric space $(X, d)$.

**Example 3A.** The function $d \colon \mathbb{R} \times \mathbb{R} \to \mathbb{R}$ defined by

$$d(\alpha, \beta) = |\alpha - \beta|$$

for every $\alpha, \beta \in \mathbb{R}$ is a metric on $\mathbb{R}$. That is, it satisfies all the metric axioms in Definition 3.1, where $|\alpha|$ stands for the *absolute value* of $\alpha \in \mathbb{R}$: $|\alpha| = (\alpha^2)^{\frac{1}{2}}$. This is the *usual metric on $\mathbb{R}$*. The real line $\mathbb{R}$ equipped with its usual metric is the most important concrete metric space. If we refer to $\mathbb{R}$ as a metric space without specifying a metric on it, then it is understood that $\mathbb{R}$ has been equipped with its usual metric. Similarly, the function $d \colon \mathbb{C} \times \mathbb{C} \to \mathbb{R}$ given by $d(\xi, \upsilon) = |\xi - \upsilon|$ for every $\xi, \upsilon \in \mathbb{C}$ is a metric on $\mathbb{C}$. (Again, $|\xi|$ stands for the absolute value (or *modulus*) of a complex number $\xi$: $|\xi| = (\bar{\xi}\xi)^{\frac{1}{2}}$ with the upper bar denoting *complex conjugate*). This is the *usual metric on* $\mathbb{C}$. More generally, let $\mathbb{F}$ denote either the real field $\mathbb{R}$ or the complex field $\mathbb{C}$, and let $\mathbb{F}^n$ be the set of all ordered $n$-tuples of scalars in $\mathbb{F}$. For each real number $p \geq 1$, consider the function $d_p \colon \mathbb{F}^n \times \mathbb{F}^n \to \mathbb{R}$ defined by

$$d_p(x, y) = \left( \sum_{i=1}^{n} |\xi_i - \upsilon_i|^p \right)^{\frac{1}{p}},$$

and also the function $d_\infty \colon \mathbb{F}^n \times \mathbb{F}^n \to \mathbb{R}$ given by

$$d_\infty(x, y) = \max_{1 \leq i \leq n} |\xi_i - \upsilon_i|,$$

for every $x = (\xi_1, \dots, \xi_n)$ and $y = (\upsilon_1, \dots, \upsilon_n)$ in $\mathbb{F}^n$. These are metrics on $\mathbb{F}^n$. Indeed, all the metric axioms up to the triangle inequality are trivially verified. The triangle inequality follows from the Minkowski inequality (see Problem 3.4(a)). Note that $(\mathbb{Q}^n, d_p)$ is a subspace of $(\mathbb{R}^n, d_p)$ and $(\mathbb{Q}^n, d_\infty)$ is a subspace of $(\mathbb{R}^n, d_\infty)$. The special (very special, really) metric space $(\mathbb{R}^n, d_2)$ is called $n$-dimensional *Euclidean space* and $d_2$ is the *Euclidean metric* on $\mathbb{R}^n$. The metric space $(\mathbb{C}^n, d_2)$ is called $n$-dimensional *unitary space*. The singular role played by the metric $d_2$ will become clear in due course.

Recall that the notion of a bounded subset was defined for partially ordered sets in Section 1.5. In particular, boundedness is well-defined for subsets of the simply ordered set $(\mathbb{R}, \leq)$: the set of all real numbers $\mathbb{R}$ equipped with its natural ordering $\leq$ (see Section 1.6). Let us introduce a suitable and common notation for a subset of $\mathbb{R}$ that is bounded above. Since the simply ordered set $\mathbb{R}$ is a boundedly complete lattice (Example 1C), it follows that a subset $R$ of $\mathbb{R}$ is bounded above if and only if it has a supremum, sup $R$, in $\mathbb{R}$. In such a case we shall write sup $R < \infty$. Thus the notation sup $R < \infty$ simply means that $R$ is a subset of $\mathbb{R}$ which is bounded above. Otherwise (i.e., if $R \subseteq \mathbb{R}$ is not bounded above) we write sup $R = \infty$. With this in mind we shall extend the notion of boundedness from $(\mathbb{R}, \leq)$ to a metric space $(X, d)$ as follows. A nonempty subset $A$ of $X$ is a *bounded set* in the metric space $(X, d)$ if

$$\sup_{x, y \in A} d(x, y) < \infty.$$

That is, $A$ is bounded in $(X, d)$ if $\{d(x, y) \in \mathbb{R} \colon x, y \in A\}$ is a bounded subset of $(\mathbb{R}, \leq)$. Equivalently, if $\{d(x, y) \in \mathbb{R} \colon x, y \in A\}$ is bounded above in $\mathbb{R}$, since $0 \leq d(x, y)$ for every $x, y \in X$. An *unbounded set* is, of course, a set $A$ that is not bounded in $(X, d)$. The *diameter* of a nonempty bounded subset $A$ of $X$ (notation: diam$(A)$) is defined by

$$\mathrm{diam}(A) = \sup_{x, y \in A} d(x, y)$$

so that diam$(A) < \infty$ whenever a nonempty set $A$ is bounded in $(X, d)$. By convention the empty set $\varnothing$ is bounded and diam$(\varnothing) = 0$. If $A$ is unbounded we write

$\operatorname{diam}(A) = \infty$. Let $F$ be a function of a set $S$ to a metric space $(Y, d)$. $F$ is a *bounded function* if its range, $\mathcal{R}(F) = F(S)$, is a bounded subset in $(Y, d)$; that is, if

$$\sup_{s,t \in S} d(F(s), F(t)) < \infty.$$

Note that $R$ is bounded as a subset of the metric space $\mathbb{R}$ equipped with its usual metric if and only if $R$ is bounded as a subset of the simply ordered set $\mathbb{R}$ equipped with its natural ordering. Thus the notion of a *bounded subset of* $\mathbb{R}$ and the notion of a *bounded real-valued function* on an arbitrary set $S$ are both unambiguously defined.

**Proposition 3.2.** *Let $S$ be set and let $\mathbb{F}$ denote either the real field $\mathbb{R}$ or the complex field $\mathbb{C}$. Equip $\mathbb{F}$ with its usual metric. A function $\varphi \in \mathbb{F}^S$ (i.e., $\varphi \colon S \to \mathbb{F}$) is bounded if and only if*

$$\sup_{s \in S} |\varphi(s)| < \infty.$$

*Proof.* Consider a function $\varphi$ from a set $S$ to the field $\mathbb{F}$. Take $s$ and $t$ arbitrary in $S$, and let $d$ be the usual metric on $\mathbb{F}$ (see Example 3A). Since $\varphi(s), \varphi(t) \in \mathbb{F}$, it follows by Problem 3.1(a) that

$$\begin{aligned}
|\varphi(s)| - |\varphi(t)| &\leq \big||\varphi(s)| - |\varphi(t)|\big| = \big|d(\varphi(s), 0) - d(0, \varphi(t))\big| \\
&\leq d(\varphi(s), \varphi(t)) = |\varphi(s) - \varphi(t)| \leq |\varphi(s)| + |\varphi(t)|.
\end{aligned}$$

If $\sup_{s \in S}|\varphi(s)| < \infty$ (i.e., if $\{|\varphi(s)| \in \mathbb{R} \colon s \in S\}$ is bounded above) then

$$d(\varphi(s), \varphi(t)) \leq 2 \sup_{s \in S} |\varphi(s)|,$$

and hence $\sup_{s,t \in S} d(\varphi(s), \varphi(t)) \leq 2\sup_{s \in S}|\varphi(s)|$ so that the function $\varphi$ is bounded. On the other hand, if $\sup_{s,t \in S} d(\varphi(s), \varphi(t)) < \infty$, then

$$|\varphi(s)| \leq \sup_{s,t \in S} d(\varphi(s), \varphi(t)) + |\varphi(t)|,$$

and the real number $\sup_{s,t \in S} d(\varphi(s), \varphi(t)) + |\varphi(t)|$ is an upper bound for $\{|\varphi(s)| \in \mathbb{R} \colon s \in S\}$ for each $t \in S$. Thus $\sup_{s \in S}|\varphi(s)| < \infty$. $\qquad\square$

**Example 3B.** For each real number $p \geq 1$, let $\ell_+^p$, denote the set of all scalar-valued (real or complex) infinite sequences $\{\xi_k\}_{k \in \mathbb{N}}$ in $\mathbb{C}^{\mathbb{N}}$ (or in $\mathbb{C}^{\mathbb{N}_0}$) such that $\sum_{k=1}^{\infty}|\xi_k|^p < \infty$. We shall refer to this condition by saying that the elements of $\ell_+^p$ are *p-summable sequences*. Notation: $\sum_{k=1}^{\infty}|\xi_k|^p = \sup_{n \in \mathbb{N}}\sum_{k=1}^{n}|\xi_k|^p$. Thus, according to Proposition 3.2, $\sum_{k=1}^{\infty}|\xi_k|^p < \infty$ means that the nonnegative sequence $\{\sum_{k=1}^{n}|\xi_k|^p\}_{n \in \mathbb{N}}$ is bounded as a real-valued function on $\mathbb{N}$. Note that, if $\{\xi_k\}_{k \in \mathbb{N}}$ and $\{\upsilon_k\}_{k \in \mathbb{N}}$ are arbitrary sequences in $\ell_+^p$, then the Minkowski inequality (Problem

3.4(b)) ensures that $\sum_{k=1}^{\infty} |\xi_k - \upsilon_k|^p < \infty$. Hence we may consider the function $d_p \colon \ell_+^p \times \ell_+^p \to \mathbb{R}$ given by

$$d_p(x, y) = \left( \sum_{k=1}^{\infty} |\xi_k - \upsilon_k|^p \right)^{\frac{1}{p}}$$

for every $x = \{\xi_k\}_{k\in\mathbb{N}}$ and $y = \{\upsilon_k\}_{k\in\mathbb{N}}$ in $\ell_+^p$. We claim that $d_p$ is a metric on $\ell_+^p$. Indeed, as it happened in Example 3A, all the metric axioms up to the triangle inequality are readily verified; and the triangle inequality follows from the Minkowski inequality. Therefore $(\ell_+^p, d_p)$ is a metric space for each $p \geq 1$, and the metric $d_p$ is referred to as the usual metric on $\ell_+^p$. Now let $\ell_+^\infty$ denote the set of all scalar-valued *bounded sequences*; that is, the set of all real or complex-valued sequences $\{\xi_k\}_{k\in\mathbb{N}}$ such that $\sup_{k\in\mathbb{N}}|\xi_k| < \infty$. Again, the Minkowski inequality (Problem 3.4(b)) ensures that $\sup_{k\in\mathbb{N}}|\xi_k - \upsilon_k| < \infty$ whenever $\{\xi_k\}_{k\in\mathbb{N}}$ and $\{\upsilon_k\}_{k\in\mathbb{N}}$ lie in $\ell_+^\infty$, and hence we may consider the function $d_\infty \colon \ell_+^\infty \times \ell_+^\infty \to \mathbb{R}$ defined by

$$d_\infty(x, y) = \sup_{k\in\mathbb{N}} |\xi_k - \upsilon_k|$$

for every $x = \{\xi_k\}_{k\in\mathbb{N}}$ and $y = \{\upsilon_k\}_{k\in\mathbb{N}}$ in $\ell_+^\infty$. Proceeding as before (using the Minkowski inequality to verify the triangle inequality) it follows that $(\ell_+^\infty, d_\infty)$ is a metric space, and the metric $d_\infty$ is referred to as the usual metric on $\ell_+^\infty$. These metric spaces are the natural generalizations (for infinite sequences) of the metric spaces considered in Example 3A, and again the metric space $(\ell_+^2, d_2)$ will play a central role in the forthcoming chapters. There are counterparts of $\ell_+^p$ and $\ell_+^\infty$ for nets in $\mathbb{C}^{\mathbb{Z}}$. In fact, for each $p \geq 1$ let $\ell^p$ denote the set of all scalar-valued (real or complex) nets $\{\xi_k\}_{k\in\mathbb{Z}}$ such that $\sum_{k=-\infty}^{\infty}|\xi_k|^p < \infty$ (i.e., such that the nonnegative sequence $\{\sum_{k=-n}^{n}|\xi_k|^p\}_{n\in\mathbb{N}_0}$ is bounded), and let $\ell^\infty$ denote the set of all bounded nets in $\mathbb{C}^{\mathbb{Z}}$ (i.e., the set of all scalar-valued nets $\{\xi_k\}_{k\in\mathbb{Z}}$ such that $\sup_{k\in\mathbb{Z}}|\xi_k| < \infty$). The functions $d_p \colon \ell^p \times \ell^p \to \mathbb{R}$ and $d_\infty \colon \ell^\infty \times \ell^\infty \to \mathbb{R}$, given by

$$d_p(x, y) = \left( \sum_{k=-\infty}^{\infty} |\xi_k - \upsilon_k|^p \right)^{\frac{1}{p}}$$

for every $x = \{\xi_k\}_{k\in\mathbb{Z}}$ and $y = \{\upsilon_k\}_{k\in\mathbb{Z}}$ in $\ell^p$, and

$$d_\infty(x, y) = \sup_{k\in\mathbb{Z}} |\xi_k - \upsilon_k|$$

for every $x = \{\xi_k\}_{k\in\mathbb{Z}}$ and $y = \{\upsilon_k\}_{k\in\mathbb{Z}}$ in $\ell^\infty$, are metrics on $\ell^p$ (for each $p \geq 1$) and on $\ell^\infty$, respectively.

Let $(X, d)$ be a metric space. If $x$ is an arbitrary point in $X$, and $A$ is an arbitrary nonempty subset of $X$, then the *distance from $x$ to $A$* is the real number

$$d(x, A) = \inf_{a\in A} d(x, a).$$

If $A$ and $B$ are nonempty subsets of $X$, then the *distance between A and B* is the real number

$$d(A, B) = \inf_{a \in A, b \in B} d(a, b).$$

**Example 3C.** Let $S$ be a set and let $(Y, d)$ be a metric space. Let $B[S, Y]$ denote the subset of $Y^S$ consisting of all bounded mappings of $S$ into $(Y, d)$. According to Problem 3.6,

$$\sup_{s \in S} d(f(s), g(s)) \leq \operatorname{diam}(\mathcal{R}(f)) + \operatorname{diam}(\mathcal{R}(g)) + d(\mathcal{R}(f), \mathcal{R}(g))$$

so that $\sup_{s \in S} d(f(s), g(s)) \in \mathbb{R}$ for every $f, g \in B[S, Y]$. Thus we may consider the function $d_\infty \colon B[S, Y] \times B[S, Y] \to \mathbb{R}$ defined by

$$d_\infty(f, g) = \sup_{s \in S} d(f(s), g(s))$$

for each pair of mappings $f, g \in B[S, Y]$. This is a metric on $B[S, Y]$. Indeed, $d_\infty$ clearly satisfies conditions (i), (ii), and (iii) in Definition 3.1. To verify the triangle inequality (condition (iv)) proceed as follows. Take an arbitrary $s \in S$ and note that, if $f$, $g$, and $h$ are mappings in $B[S, Y]$, then (by the triangle inequality in $(Y, d)$)

$$d(f(s), g(s)) \leq d(f(s), h(s)) + d(h(s), g(s)) \leq d_\infty(f, h) + d_\infty(h, g).$$

Hence $d_\infty(f, g) \leq d_\infty(f, h) + d_\infty(f, g)$ and therefore $(B[S, Y], d_\infty)$ is a metric space. The metric $d_\infty$ is referred to as the *sup-metric* on $B[S, Y]$. Note that the metric spaces $(\ell_+^\infty, d_\infty)$ and $(\ell^\infty, d_\infty)$ of the previous example are particular cases of $(B[S, Y], d_\infty)$. Indeed, $\ell_+^\infty = B[\mathbb{N}, \mathbb{C}]$ and $\ell^\infty = B[\mathbb{Z}, \mathbb{C}]$.

**Example 3D.** The general concept of a continuous mapping between metric spaces will be defined in the next section. However, assuming that the reader is familiar with the particular notion of a real-valued continuous function of a real variable, we shall consider now the following example. Let $C[0, 1]$ denote the set of all scalar-valued (real or complex) continuous functions defined on the interval $[0, 1]$. For every $x, y \in C[0, 1]$ set

$$d_p(x, y) = \left( \int_0^1 |x(t) - y(t)|^p \, dt \right)^{\frac{1}{p}},$$

where $p$ is a real number such that $p \geq 1$, and

$$d_\infty = \sup_{t \in [0,1]} |x(t) - y(t)|.$$

These are metrics on the set $C[0, 1]$. That is, $d_p \colon C[0, 1] \times C[0, 1] \to \mathbb{R}$ and $d_\infty \colon C[0, 1] \times C[0, 1] \to \mathbb{R}$ are well-defined functions that satisfy all the conditions in Definition 3.1. Indeed, nonnegativeness and symmetry are trivially verified, positiveness for $d_p$ is ensured by the continuity of the elements in $C[0, 1]$, and the triangle inequality comes by the Minkowski inequality (Problem 3.4(c)): for every $x, y, z \in C[0, 1]$,

$$
\begin{aligned}
d_p(x, y) &= \left( \int_0^1 |x(t) - z(t) + z(t) - y(t)|^p \, dt \right)^{\frac{1}{p}} \\
&\leq \left( \int_0^1 |x(t) - z(t)|^p \, dt \right)^{\frac{1}{p}} + \left( \int_0^1 |z(t) - y(t)|^p \, dt \right)^{\frac{1}{p}} \\
&= d_p(x, z) + d_p(z, y), \\
d_\infty(x, y) &= \sup_{t \in [0,1]} |x(t) - z(t) + z(t) - y(t)| \\
&\leq \sup_{t \in [0,1]} |x(t) - z(t)| + \sup_{t \in [0,1]} |z(t) - y(t)| \\
&= d_\infty(x, z) + d_\infty(z, y).
\end{aligned}
$$

Let $B[0, 1]$ denote the set $B[S, Y]$ of Example 3C when $S = [0, 1]$ and $Y = \mathbb{F}$ (with $\mathbb{F}$ standing either for the real field $\mathbb{R}$ or for the complex field $\mathbb{C}$). Since $C[0, 1]$ is a subset of $B[0, 1]$ (reason: every scalar-valued continuous function defined on the interval $[0, 1]$ is bounded), it follows that $(C[0, 1], d_\infty)$ is a subspace of the metric space $(B[0, 1], d_\infty)$. The metric $d_\infty$ is called the *sup-metric* on $C[0, 1]$ and, as we shall see later, the "sup" in its definition in fact is a "max".

Let $X$ be an arbitrary set. A real-valued function $d$ on $X \times X$, $d \colon X \times X \to \mathbb{R}$, is a *pseudometric* on $X$ if it satisfies the axioms (i), (iii) and (iv) of Definition 3.1. A *pseudometric space* $(X, d)$ is a set $X$ equipped with a pseudometric $d$. The difference between a metric space and a pseudometric space is that a pseudometric does not necessarily satisfy the axiom (ii) in Definition 3.1 (i.e., it is possible for a pseudometric to vanish at a pair $(x, y)$ even though $x \neq y$). However, given a pseudometric space $(X, d)$, there exists a natural way to obtain a metric space $(\widetilde{X}, \widetilde{d})$ associated with $(X, d)$, where $\widetilde{d}$ is a metric on $\widetilde{X}$ associated with the pseudometric $d$ on $X$. Indeed, as we shall see next, a pseudometric $d$ induces an equivalence relation $\sim$ on $X$, and $\widetilde{X}$ is precisely the quotient space $X/\sim$ (i.e., the collection of all equivalence classes $[x]$ with respect to $\sim$ for every $x$ in $X$).

**Proposition 3.3.** *Let $d$ be a pseudometric on a set $X$ and consider the relation $\sim$ on $X$ defined as follows. If $x$ and $x'$ are elements of $X$, then*

$$
x' \sim x \quad \text{if} \quad d(x', x) = 0.
$$

*The relation $\sim$ is an equivalence relation on $X$ with the following property. For every $x$, $x'$, $y$ and $y'$ in $X$,*

$$
x' \sim x \quad \text{and} \quad y' \sim y \quad \text{imply} \quad d(x', y') = d(x, y).
$$

*Let $X/\sim$ be the quotient space of $X$ modulo $\sim$. For each pair $([x], [y])$ in $X/\sim \times X/\sim$ set*

$$
\widetilde{d}([x], [y]) = d(x, y)
$$

*for an arbitrary pair $(x, y)$ in $[x] \times [y]$. This defines a function*

$$\tilde{d} \colon X/{\sim} \times X/{\sim} \to \mathbb{R}$$

*which is a metric on the quotient space $X/{\sim}$.*

*Proof.* It is clear that the relation $\sim$ on $X$ is reflexive and symmetric because a pseudometric is nonnegative and symmetric. Transitivity comes from the triangle inequality: $0 \leq d(x, x'') \leq d(x, x') + d(x', x'')$ for every $x, x', x'' \in X$. Thus $\sim$ is an equivalence relation on $X$. Moreover, if $x' \sim x$ and $y' \sim y$ (i.e., if $x' \in [x]$ and $y' \in [y]$), then the triangle inequality in the pseudometric space $(X, d)$ ensures that

$$d(x, y) \leq d(x, x') + d(x', y') + d(y', y) = d(x', y')$$

and, similarly, $d(x', y') \leq d(x, y)$. Therefore

$$d(x', y') = d(x, y) \quad \text{whenever} \quad x' \sim x \text{ and } y' \sim y.$$

That is, given a pair of equivalence classes $[x] \subseteq X$ and $[y] \subseteq X$, the restriction of $d$ to $[x] \times [y] \subseteq X \times X$, $d|_{[x] \times [y]} \colon [x] \times [y] \to \mathbb{R}$, is a constant function. Thus, for each pair $([x], [y])$ in $X/{\sim} \times X/{\sim}$, set

$$\tilde{d}([x], [y]) = d|_{[x] \times [y]}(x, y) = d(x, y)$$

for any $x \in [x]$ and $y \in [y]$. This defines a function $\tilde{d} \colon X/{\sim} \times X/{\sim} \to \mathbb{R}$ which is nonnegative, symmetric, and satisfies the triangle inequality (along with $d$). The reason for defining equivalence classes is to ensure positiveness for $\tilde{d}$ from the nonnegativeness of the pseudometric $d$: if $\tilde{d}([x], [y]) = 0$, then $d(x, y) = 0$ so that $x \sim y$, and hence $[x] = [y]$. $\qquad\qquad\square$

**Example 3E.** The previous example exhibited different metric spaces with the same underlying set of all scalar-valued continuous functions on the interval $[0, 1]$. Here we shall allow discontinuous functions as well. Let $S$ be a nondegenerate interval of the real line $\mathbb{R}$ (typical examples: $S = [0, 1]$ or $S = \mathbb{R}$). For each real number $p \geq 1$, let $r^p(S)$ denote the set of all scalar-valued (real or complex) *p-integrable functions* on $S$. In this context, "$p$-integrable" means that a scalar-valued function $x$ on $S$ is Riemann integrable and $\int_S |x(s)|^p \, ds < \infty$ (i.e., the Riemann integral $\int_S |x(s)|^p \, ds$ exists as a number in $\mathbb{R}$). Consider the function $\delta_p \colon r^p(S) \times r^p(S) \to \mathbb{R}$ given by

$$\delta_p(x, y) = \left( \int_S |x(s) - y(s)|^p \, ds \right)^{\frac{1}{p}}$$

for every $x, y \in r^p(S)$. The Minkowski inequality (see Problem 3.4(c)) ensures that the function $\delta_p$ is well-defined, and also that it satisfies the triangle inequality. Moreover, nonnegativeness and symmetry are readily verified but positiveness fails. For instance, if $0$ denotes the null function on $S = [0, 1]$ (i.e., $0(s) = 0$ for all $s \in S$), and if $x(s) = 1$ for $s = \frac{1}{2}$ and zero elsewhere, then $\delta_p(x, 0) = 0$ although $x \neq 0$ (for $x(\frac{1}{2}) \neq 0(\frac{1}{2})$). Thus $\delta_p$ actually is a pseudometric on $r^p(S)$ rather than a metric, so that $(r^p(S), \delta_p)$ is a pseudometric space. However, if we "redefine" $r^p(S)$ by endowing it with a new notion of equality, different from the usual pointwise equality for functions, then perhaps we might make $\delta_p$ a metric on such a "redefinition" of $r^p(S)$. This in fact is the idea behind Proposition 3.3. Consider the equivalence relation $\sim$ on $r^p(S)$ defined as in Proposition 3.3: if $x$ and $x'$ are functions in $r^p(S)$, then $x' \sim x$ if $\delta_p(x', x) = 0$. Now set $R^p(S) = r^p(S)/\sim$, the collection of all equivalence classes $[x] = \{x' \in r^p(S): \ \delta_p(x', x) = 0\}$ for every $x \in r^p(S)$. Thus, according to Proposition 3.3, $(R^p(S), d_p)$ is a metric space where the metric $d_p: R^p(S) \times R^p(S) \to \mathbb{R}$ is defined by $d_p([x], [y]) = \delta_p(x, y)$ for arbitrary $x \in [x]$ and $y \in [y]$ for every $[x], [y] \in R^p(S)$. Note that equality in $R^p(S)$ is interpreted in the following way: if $[x]$ and $[y]$ are equivalence classes in $R^p(S)$, and if $x$ and $y$ are arbitrary functions in $[x]$ and $[y]$, respectively, then $[x] = [y]$ if and only if $\delta_p(x, y) = 0$. If $x$ is any element of $[x]$ then, in this context, it is usual to write $x$ for $[x]$ and hence $d_p(x, y)$ for $d_p([x], [y])$. Thus, following the common usage, we shall write $x \in R^p(S)$ instead of $[x] \in R^p(S)$, and also

$$d_p(x, y) = \left( \int_S |x(s) - y(s)|^p \, ds \right)^{\frac{1}{p}}$$

for every $x, y \in R^p(S)$ to represent the metric $d_p$ on $R^p(S)$. This is referred to as the usual metric on $R^p(S)$. Note that, according to this convention, $x = y$ in $R^p(S)$ if and only if $d_p(x, y) = 0$.


## 3.2    Convergence and Continuity

The notion of convergence, together with the notion of continuity, plays a central role in the theory of metric spaces.

**Definition 3.4.** Let $(X, d)$ be a metric space. An $X$-valued sequence $\{x_n\}$ (or a sequence in $X$ indexed by $\mathbb{N}$ or by $\mathbb{N}_0$) *converges* to a point $x$ in $X$ if for each real number $\varepsilon > 0$ there exists a positive integer $n_\varepsilon$ such that

$$n \geq n_\varepsilon \quad \text{implies} \quad d(x_n, x) < \varepsilon.$$

If $\{x_n\}$ converges to $x \in X$, then $\{x_n\}$ is said to be a *convergent sequence* and $x$ is said to be the *limit* of $\{x_n\}$ (notations: $\lim x_n = x$, $\lim_n x_n = x$, $\lim_{n \to \infty} x_n = x$, $x_n \to x$, or $x_n \to x$ as $n \to \infty$).

As defined above, convergence depends on the metric $d$ that equip the metric space $(X, d)$. To emphasize the rule played by the metric $d$, it is usual to refer to an $X$-valued convergent sequence $\{x_n\}$ by saying that $\{x_n\}$ *converges in* $(X, d)$. If an $X$-valued sequence $\{x_n\}$ does not converge in $(X, d)$ to the point $x \in X$, then we shall write $x_n \nrightarrow x$. Clearly, if $x_n \nrightarrow x$, then the sequence $\{x_n\}$ either converges in $(X, d)$ to another point different from $x$ or does not converge in $(X, d)$ to any $x$ in $X$. The notion of convergence in a metric space $(X, d)$ is a natural extension of the ordinary notion of convergence in the real line $\mathbb{R}$ (equipped with its usual metric). Indeed, let $(X, d)$ be a metric space, and consider an $X$-valued sequence $\{x_n\}$. Let $x$ be an arbitrary point in $X$ and consider the real-valued sequence $\{d(x_n, x)\}$. According to Definition 3.4,

$$x_n \to x \quad \text{if and only if} \quad d(x_n, x) \to 0.$$

This shows at once that *a convergent sequence in a metric space has a unique limit* (as we had anticipated in Definition 3.4 by referring to *the* limit of a convergent sequence). In fact, if $a$ and $b$ are points in $X$, then $0 \le d(a, b) \le d(a, x_n) + d(x_n, b)$ for every $n$. Thus, if $x_n \to a$ and $x_n \to b$ (i.e., $d(a, x_n) \to 0$ and $d(x_n, b) \to 0$), then $d(a, b) = 0$ (see Problem 3.10(c) and show that the sum of two convergent real-valued sequences $\{\alpha_n\}$ and $\{\beta_n\}$ is a convergent real-valued sequence with $\lim(\alpha_n + \beta_n) = \lim \alpha_n + \lim \beta_n$). Hence $a = b$.

**Example 3F.** Let $C[0, 1]$ denote the set of all scalar-valued continuous functions on the interval $[0, 1]$, and let $\{x_n\}$ be a $C[0, 1]$-valued sequence such that, for each integer $n \ge 1$, $x_n : [0, 1] \to \mathbb{R}$ is defined by

$$x_n(t) = \begin{cases} 1 - nt, & t \in [0, \frac{1}{n}], \\ 0, & t \in (\frac{1}{n}, 1]. \end{cases}$$

Consider the metric spaces $(C[0, 1], d_p)$ for $p \ge 1$ and $(C[0, 1], d_\infty)$ which were introduced in Example 3D. It is readily verified that the sequence $\{x_n\}$ converges in $(C[0, 1], d_p)$ to the null function $0 \in C[0, 1]$ for every $p \ge 1$. Indeed, take an arbitrary $p \ge 1$ and note that

$$d_p(x_n, 0) = \left( \int_0^1 |x_n(t)|^p \, dt \right)^{\frac{1}{p}} < \left( \tfrac{1}{n} \right)^{\frac{1}{p}}$$

for each $n \ge 1$. Since the sequence of real numbers $\{ (\frac{1}{n})^{\frac{1}{p}} \}$ converges to zero (when the real line $\mathbb{R}$ is equipped with its usual metric — apply Definition 3.4), it follows that $d_p(x_n, 0) \to 0$ as $n \to \infty$ (Problem 3.10(c)). That is,

$$x_n \to 0 \quad \text{in} \quad (C[0, 1], d_p).$$

However, $\{x_n\}$ does not converge in the metric space $(C[0, 1], d_\infty)$. Indeed, if there exists $x \in C[0, 1]$ such that $d_\infty(x_n, x) \to 0$, then it is easy to show that $x(0) = 1$ and $x(\varepsilon) = 0$ for all $\varepsilon \in (0, 1]$. Hence $x \notin C[0, 1]$, which is a contradiction. Conclusion: There is no $x \in C[0, 1]$ such that $x_n \to x$ in $(C[0, 1], d_\infty)$. Equivalently,

$$\{x_n\} \text{ does not converges in } (C[0, 1], d_\infty).$$

**Example 3G.** Consider the metric space $(B[S, Y], d_\infty)$ introduced in Example 3C, where $B[S, Y]$ denotes the set of all bounded functions of a set $S$ into a metric space $(Y, d)$, and $d_\infty$ is the sup-metric. Let $\{f_n\}$ be a $B[S, Y]$-valued sequence (i.e., a sequence of functions in $B[S, Y]$), and let $f$ be an arbitrary function in $B[S, Y]$. Since

$$0 \leq d(f_n(s), f(s)) \leq \sup_{s \in S} d(f_n(s), f(s)) = d_\infty(f_n, f)$$

for each index $n$ and all $s \in S$, it follows by Problem 3.10(c) that

$$f_n \to f \text{ in } (B[S, Y], d_\infty) \quad \text{implies} \quad f_n(s) \to f(s) \text{ in } (Y, d)$$

for every $s \in S$. If $f_n \to f$ in $(B[S, Y], d_\infty)$, then we say that the sequence $\{f_n\}$ of functions in $B[S, Y]$ *converges uniformly* to the function $f$ in $B[S, Y]$. If $f_n(s) \to f(s)$ in $(Y, d)$ for every $s \in S$, then we say that $\{f_n\}$ *converges pointwise* to $f$. Thus *uniform convergence implies pointwise convergence* (to the same limit), but the converse fails. For instance, set $S = [0, 1]$, $Y = \mathbb{F}$ (either the real field $\mathbb{R}$ or the complex field $\mathbb{C}$ equipped with their usual metric $d$), and set $B[0, 1] = B[[0, 1], \mathbb{F}]$. Recall that the metric space $(C[0, 1], d_\infty)$ of Example 3D is a subspace of $(B[0, 1], d_\infty)$. (Indeed, every scalar-valued continuous function defined on a bounded closed interval is a bounded function — we shall consider a generalized version of this well-known result later in this chapter). If $\{g_n\}$ is a sequence of functions in $C[0, 1]$ given by

$$g_n(s) = \frac{s^2}{s^2 + (1 - ns)^2}$$

for each integer $n \geq 1$ and every $s \in [0, 1]$, then it is easy to show (Definition 3.4) that

$$g_n(s) \to 0 \quad \text{in} \quad (\mathbb{R}, d)$$

for every $s \in [0, 1]$, so that the sequence $\{g_n\}$ of functions in $C[0, 1]$ converges pointwise to the null function $0 \in C[0, 1]$. On the other hand, note that $0 \leq g_n(s) \leq 1$ for all $s \in [0, 1]$, and $g_n(\frac{1}{n}) = 1$ for each $n \geq 1$. Therefore

$$d_\infty(g_n, 0) = \sup_{s \in [0,1]} |g_n(s)| = 1$$

for every integer $n \geq 1$. Thus $\{g_n\}$ does not converge uniformly to the null function, and hence it does not converge uniformly to any limit (for, if it converges uniformly, then it converges pointwise to the same limit). Conclusion: The $C[0, 1]$-valued sequence $\{g_n\}$ does not converge in the metric space $(C[0, 1], d_\infty)$. Briefly,

$$\{g_n\} \text{ does not converge in } (C[0, 1], d_\infty).$$

However, it does converge to the null function $0 \in C[0, 1]$ in the metric spaces $(C[0, 1], d_p)$ of Example 3D. That is,

$$g_n \to 0 \quad \text{in} \quad (C[0, 1], d_p)$$

for every $p \geq 1$. Indeed, since $g_n(0) = 0$ and $0 \leq g_n(s) = \frac{1}{1+(n-s^{-1})^2}$ for each $s \in (0, 1]$, it follows that $0 \leq g_n(s) \leq \frac{1}{1+n^2}$ for each $n \geq 1$ and every $s \in [0, 1]$. Therefore $\int_0^1 |g_n(s)|^p\, ds \leq \left(\frac{1}{1+n^2}\right)^p$, and hence

$$0 \leq d_p(g_n, 0) \leq \frac{1}{1 + n^2}$$

for each $n \geq 1$ and all $p \geq 1$. Since the sequence of positive numbers $\{\frac{1}{1+n^2}\}$ converges to zero in $(\mathbb{R}, d)$, it follows by Problem 3.10(c) that

$$d_p(g_n, 0) \to 0 \quad \text{as} \quad n \to \infty \quad \text{for every} \quad p \geq 1.$$

**Proposition 3.5.** *An $X$-valued sequence $\{x_n\}$ converges in a metric space $(X, d)$ to a limit $x \in X$ if and only if every subsequence of it converges in $(X, d)$ to $x$.*

*Proof.* If every subsequence converges to a fixed limit, then, in particular, the sequence itself converges to the same limit. On the other hand, suppose $x_n \to x$ in $(X, d)$. That is, for every $\varepsilon > 0$ there exists a positive integer $n_\varepsilon$ such that $n \geq n_\varepsilon$ implies $d(x_n, x) < \varepsilon$. Take an arbitrary subsequence $\{x_{n_k}\}_{k \in \mathbb{N}}$ of $\{x_n\}_{n \in \mathbb{N}}$. Since $k \leq n_k$ (reason: $\{n_k\}_{k \in \mathbb{N}}$ is a strictly increasing subsequence of the sequence $\{n\}_{n \in \mathbb{N}}$ — see Section 1.7), it follows that $k \geq n_\varepsilon$ implies $n_k \geq n_\varepsilon$ which in turn implies $d(x_{n_k}, x) < \varepsilon$. Therefore $x_{n_k} \to x$ in $(X, d)$ as $k \to \infty$. $\qquad\square$

As we saw in Section 1.7, nets constitute a natural generalization of (infinite) sequences. Thus it comes as no surprise that the concept of convergence can be generalized from sequences to nets in a metric space $(X, d)$. Indeed, an $X$-valued net $\{x_\gamma\}_{\gamma \in \Gamma}$ (or a net in $X$) indexed by a directed set $\Gamma$ *converges* to a point $x$ in $X$ if for each real number $\varepsilon > 0$ there exists an index $\gamma_\varepsilon$ in $\Gamma$ such that

$$\gamma \geq \gamma_\varepsilon \quad \text{implies} \quad d(x_\gamma, x) < \varepsilon.$$

If $\{x_\gamma\}_{\gamma \in \Gamma}$ converges to $x$, then it is said to be a *convergent net* and $x$ is said to be the *limit* of $\{x_\gamma\}_{\gamma \in \Gamma}$ (notations: $\lim x_\gamma = x$, $\lim_\gamma x_\gamma = x$, or $x_\gamma \to x$). Just as in the particular case of sequences, a convergent net in a metric space has a *unique* limit.

The notion of a real-valued continuous function on $\mathbb{R}$ is essential in classical analysis. One of the main reasons for investigating metric spaces is the generalization of the idea of continuity for maps between abstract metric spaces: *a map between metric spaces is continuous if it preserves closeness.*

**Definition 3.6.** Let $F: X \to Y$ be a function from a set $X$ to a set $Y$. Equip $X$ and $Y$ with metrics $d_X$ and $d_Y$, respectively, so that $(X, d_X)$ and $(Y, d_Y)$ are metric spaces. $F: (X, d_X) \to (Y, d_Y)$ is *continuous at the point* $x_0$ in $X$ if for each real number $\varepsilon > 0$ there exists a real number $\delta > 0$ (which certainly depends on $\varepsilon$ and may depend on $x_0$ as well) such that

$$d_X(x, x_0) < \delta \quad \text{implies} \quad d_Y(F(x), F(x_0)) < \varepsilon.$$

$F$ is *continuous* (or *continuous on* $X$) if it is continuous at every point of $X$; and *uniformly continuous* (on $X$) if for each real number $\varepsilon > 0$ there exists a real number $\delta > 0$ such that

$$d_X(x, x') < \delta \quad \text{implies} \quad d_Y(F(x), F(x')) < \varepsilon$$

for *all* $x$ and $x'$ in $X$.

It is clear that *a uniformly continuous mapping is continuous*, but the converse fails. The difference between continuity and uniform continuity is that if $F$ is uniformly continuous, then for each $\varepsilon > 0$ it is possible to take $\delta > 0$ (which depends only on $\varepsilon$) so as to ensure that the implication $\{d_X(x, x_0) < \delta \implies d_Y(F(x), F(x_0)) < \varepsilon\}$ holds for *all* points $x_0$ of $X$. We say that a mapping $F \colon (X, d_X) \to (Y, d_Y)$ is *Lipschitzian* if there exists a real number $\gamma > 0$ (called *Lipschitz constant*) such that

$$d_Y(F(x), F(x')) \leq \gamma\, d_X(x, x')$$

for all $x, x' \in X$ (which is referred to as the *Lipschitz condition*). It is readily verified that *every Lipschitzian mapping is uniformly continuous* but, again, the converse fails (see Problem 3.16). A *contraction* is a Lipschitzian mapping $F \colon (X, d_X) \to (Y, d_Y)$ with a Lipschitz constant $\gamma \leq 1$. That is, $F$ is a contraction if $d_Y(F(x), F(x')) \leq d_X(x, x')$ for all $x, x' \in X$ or, equivalently, if

$$\sup_{x \neq x'} \frac{d_Y(F(x), F(x'))}{d_X(x, x')} \leq 1.$$

$F$ is said to be a *strict contraction* if it is Lipschitzian with a Lipschitz constant $\gamma < 1$, which means that

$$\sup_{x \neq x'} \frac{d_Y(F(x), F(x'))}{d_X(x, x')} < 1.$$

Note that, if $d_Y(F(x), F(x')) < d_X(x, x')$ for all $x, x' \in X$, then $F$ is a contraction but not necessarily a strict contraction.

Consider a function $F$ from a metric space $(X, d_X)$ to a metric space $(Y, d_Y)$. If $F$ is continuous at a point $x_0 \in X$, then $x_0$ is said to be a *point of continuity* of $F$. Otherwise, if $F$ is not continuous at a point $x_0 \in X$, then $x_0$ is said to be a *point of discontinuity* of $F$, and $F$ is said to be *discontinuous at* $x_0$. $F$ is not continuous if there exists at least one point $x_0 \in X$ such that $F$ is discontinuous at $x_0$. According to Definition 3.6 a function $F$ is discontinuous at $x_0 \in X$ if and only if the following assertion holds true: there exists $\varepsilon > 0$ such that for every $\delta > 0$ there exists $x_\delta \in X$ with the property that

$$d_X(x_\delta, x_0) < \delta \quad \text{and} \quad d_Y(F(x_\delta), F(x_0)) \geq \varepsilon.$$

**Example 3H.** (a) Consider the set $R^2(\mathbb{R})$ defined in Example 3E. Put $Y = R^2(\mathbb{R})$ and let $X$ be the subset of $Y$ made up of all functions $x$ in $R^2(\mathbb{R})$ for which the formula

$$y(t) = \int_{-\infty}^{t} x(s)\,ds \quad \text{for each} \quad t \in \mathbb{R}$$

defines a function in $R^2(\mathbb{R})$. Briefly,

$$X = \left\{ x \in Y: \ \int_{-\infty}^{\infty} \left| \int_{-\infty}^{t} x(s)\,ds \right|^2 dt < \infty \right\}.$$

Recall that a "function" in $Y$ is, in fact, an equivalence class of functions as discussed in Example 3E. Thus consider the mapping $F: X \to Y$ that assigns to each function $x$ in $X$ the function $y = F(x)$ in $Y$ defined by the above formula. Now equip $R^2(\mathbb{R})$ with its usual metric $d_2$ (cf. Example 3E) so that $(X, d_2)$ is a subspace of the metric space $(Y, d_2)$. We claim that $F: (X, d_2) \to (Y, d_2)$ is *nowhere continuous*; that is, the mapping $F$ is discontinuous at every $x_0 \in X$ (see Problem 3.17(a)).

(b)  Now let $S$ be a (nondegenerate) closed and bounded interval of the real line $\mathbb{R}$ (typical example: $S = [0, 1]$), and consider the set $R^2(S)$ defined in Example 3E. If $x$ is a function in $R^2(S)$ (so that it is Riemann integrable), then set

$$y(t) = \int_{\min S}^{t} x(s)\,ds \quad \text{for each} \quad t \in S.$$

According to the Hölder inequality in Problem 3.3(c) it follows that $\int_S |x(s)|\,ds \leq (\int_S ds)^{\frac{1}{2}} (\int_S |x(s)|^2 ds)^{\frac{1}{2}}$ for every $x \in R^2(S)$. Therefore $|y(t)|^2 = |\int_0^t x(s)ds|^2 \leq (\int_S |x(s)|ds)^2 \leq \mathrm{diam}(S) \int_S |x(s)|^2 ds$ for each $t \in S$, and hence

$$\int_S |y(t)|^2\,dt \ \leq \ \mathrm{diam}(S)^2 \int_S |x(s)|^2\,ds \ < \ \infty$$

for every $x \in R^2(S)$. Thus the above formula defines a function $y$ in $R^2(S)$. Let $F$ be a mapping of $R^2(S)$ into itself that assigns to each function $x$ in $R^2(S)$ this function $y$ in $R^2(S)$, so that $y = F(x)$. Equip $R^2(S)$ with its usual metric $d_2$ (Example 3E). It is easy to show that $F: (R^2(S), d_2) \to (R^2(S), d_2)$ is uniformly continuous. As a matter of fact, the mapping $F$ is Lipschitzian (see Problem 3.17(b)). Comparing the example in item (a) with the present one we observe how different the metric spaces $R^2(\mathbb{R})$ and $R^2(S)$, both equipped with the usual metric $d_2$, can be: the "same" integral transformation $F$ that is nowhere continuous when defined on an appropriate subspace of $(R^2(\mathbb{R}), d_2)$ becomes Lipschitzian when defined on $(R^2(S), d_2)$.

The concepts of convergence and continuity are tightly intertwined. A particularly important result on the connection of these central concepts says that *a function is continuous if and only if it preserves convergence*. This supplies a useful necessary and sufficient condition for continuity in terms of convergence.

**Theorem 3.7.** *Consider a mapping* $F: (X, d_X) \to (Y, d_Y)$ *of a metric space* $(X, d_X)$ *into a metric space* $(Y, d_Y)$ *and let* $x_0$ *be a point in* $X$. *The following assertions are equivalent.*

(a)  $F$ *is continuous at* $x_0$.

(b)  *The* $Y$-*valued sequence* $\{F(x_n)\}$ *converges in* $(Y, d_Y)$ *to* $F(x_0) \in Y$ *whenever* $\{x_n\}$ *is an* $X$-*valued sequence that converges in* $(X, d_X)$ *to* $x_0 \in X$.

*Proof.* If $\{x_n\}$ is an $X$-valued sequence such that $x_n \to x_0$ in $(X, d_X)$ for some $x_0$ in $X$, then (Definition 3.4) for every $\delta > 0$ there exists a positive integer $n_\delta$ such that

$$n \geq n_\delta \quad \text{implies} \quad d_X(x_n, x_0) < \delta.$$

If $F: (X, d_X) \to (Y, d_Y)$ is continuous at $x_0$, then (Definition 3.6) for each $\varepsilon > 0$ there exists $\delta > 0$ such that

$$d_X(x_n, x_0) < \delta \quad \text{implies} \quad d_Y(F(x), F(x_0)) < \varepsilon.$$

Therefore, if $x_n \to x_0$ and $F$ is continuous at $x_0$, then for each $\varepsilon > 0$ there exists a positive integer $n_\varepsilon$ (e.g., $n_\varepsilon = n_\delta$) such that

$$n \geq n_\varepsilon \quad \text{implies} \quad d_Y(F(x_n), F(x_0)) < \varepsilon,$$

which means that (a)$\Rightarrow$(b). On the other hand, if $F$ is not continuous at $x_0$, then there exists $\varepsilon > 0$ such that for every $\delta > 0$ there exists $x_\delta \in X$ with the property that

$$d_X(x_\delta, x_0) < \delta \quad \text{and} \quad d_Y(F(x_\delta), F(x_0)) \geq \varepsilon.$$

In particular, for each positive integer $n$ there exists $x_n \in X$ such that

$$d_X(x_n, x_0) < \tfrac{1}{n} \quad \text{and} \quad d_Y(F(x_n), F(x_0)) \geq \varepsilon.$$

Thus $x_n \to x_0$ in $(X, d_X)$ (for $d_X(x_n, x_0) \to 0$) and $F(x_n) \not\to F(x_0)$ in $(Y, d_Y)$ (for $d_Y(F(x_n), F(x_0)) \not\to 0$). That is, the $Y$-valued sequence $\{F(x_n)\}$ does not converge to $F(x_0)$ (it may not converge in $(Y, d_Y)$ or, if it converges in $(Y, d_Y)$, then it does not converge to $F(x_0)$). Therefore, the denial of (a) implies the denial of (b). Equivalently, (b)$\Rightarrow$(a). $\qquad\square$

Note that the proof of (a)$\Rightarrow$(b) can be rewritten in terms of nets so that, *if the mapping* $F: (X, d_X) \to (Y, d_Y)$ *is continuous at* $x_0 \in X$, *and if* $\{x_\gamma\}_{\gamma \in \Gamma}$ *is an* $X$-*valued net that converges to* $x_0$, *then* $\{F(x_\gamma)\}_{\gamma \in \Gamma}$ *is a* $Y$-*valued net that converges to* $F(x_0)$. That is, if (b$'$) is the statement obtained from (b) by changing "sequence" to "net" in (b), then (a)$\Rightarrow$(b$'$). Conversely, since (b$'$)$\Rightarrow$(b) trivially (since (b) is a particular case of (b$'$)), it also follows that (b$'$)$\Rightarrow$(a) (because (b)$\Rightarrow$(a)).

We shall say that a mapping $F: (X, d_X) \to (Y, d_Y)$ of a metric space $(X, d_X)$ into a metric space $(Y, d_Y)$ *preserves convergence* if the $Y$-valued sequence $\{F(x_n)\}$

converges in $(Y, d_Y)$ whenever the $X$-valued sequence $\{x_n\}$ converges in $(X, d_X)$, and

$$F(\lim x_n) = \lim F(x_n).$$

**Corollary 3.8.** *A map between metric spaces is continuous if and only if it preserves convergence.*

*Proof.* Combine the above definition and the definition of a continuous function with Theorem 3.7. $\qquad\qquad\square$

**Example 3I.** Let $C[0, 1]$ denote the set of all scalar-valued (real or complex) continuous functions defined on the interval $[0, 1]$. Consider the map $\varphi \colon C[0, 1] \to \mathbb{C}$ defined by

$$\varphi(x) = \int_0^1 x(t)\, dt$$

for every $x$ in $C[0, 1]$. Equip $C[0, 1]$ with the sup-metric $d_\infty$ and $\mathbb{C}$ with its usual metric $d$. Take an arbitrary convergent sequence $\{x_n\}$ in $(C[0, 1], d_\infty)$ (i.e., an arbitrary $C[0, 1]$-valued sequence that converges in the metric space $(C[0, 1], d_\infty)$) and set $x_0 = \lim x_n \in C[0, 1]$. Note that for each positive integer $n$

$$
\begin{aligned}
0 \;\le\; d(\varphi(x_n), \varphi(x_0)) &= \left| \int_0^1 x_n(t)\, dt - \int_0^1 x_0(t)\, dt \right| \\[2mm]
&= \left| \int_0^1 (x_n(t) - x_0(t))\, dt \right| \le \int_0^1 |x_n(t) - x_0(t)|\, dt \\[2mm]
&\le \sup_{t \in [0,1]} |x_n(t) - x_0(t)| \int_0^1 dt = d_\infty(x_n, x_0)
\end{aligned}
$$

(by the Hölder inequality: Problem 3.3(c)). Since $d_\infty(x_n, x_0) \to 0$, it follows that $d(\varphi(x_n), \varphi(x_0)) \to 0$. Therefore $\varphi(x_n) \to \varphi(x_0)$ in $(\mathbb{C}, d)$ whenever $x_n \to x_0$ in $(C[0, 1], d_\infty)$, so that $\varphi \colon (C[0, 1], d_\infty) \to (\mathbb{C}, d)$ is continuous by Corollary 3.8. (In fact, $\varphi$ is a contraction, thus Lipschitzian, and hence uniformly continuous.)

## 3.3   Open Sets and Topology

Let $(X, d)$ be a metric space. For each point $x_0$ in $X$ and each nonnegative real number $\rho$ the set

$$B_\rho(x_0) = \{x \in X \colon \; d(x, x_0) < \rho\}$$

is the *open ball* with *center* $x_0$ and *radius* $\rho$ (or the open ball centered at $x_0$ with radius $\rho$). If $\rho = 0$, then $B_0(x_0)$ is empty; otherwise $B_\rho(x_0)$ always contains at least its center. The set

$$B_\rho[x_0] = \{x \in X \colon \; d(x, x_0) \le \rho\}$$

is the *closed ball* with center $x_0$ and radius $\rho$. It is clear that $B_\rho(x_0) \subseteq B_\rho[x_0]$. If $\rho = 0$, then $B_0[x_0] = \{x_0\}$ for every $x_0 \in X$.

**Definition 3.9.** A subset $U$ of a metric space $X$ is an *open set* in $X$ if it includes a nonempty open ball centered at each one of its points.

That is, $U$ is an open set in $X$ if and only if for every $u$ in $U$ there exists a positive number $\rho$ such that $B_\rho(x_0) \subseteq U$. Equivalently, $U \subseteq X$ is open in the metric space $(X, d)$ if and only if for *every* $u \in U$ there exist $\rho > 0$ such that

$$d(x, u) < \rho \quad \text{implies} \quad x \in U.$$

Thus, according to Definition 3.9, a subset $A$ of a metric space $(X, d)$ is not open if and only if there exists at least one point $a$ in $A$ such that every open ball with positive radius $\rho$ centered at $a$ contains a point of $X$ not in $A$. In other words, $A \subset X$ is not open in the metric space $(X, d)$ if and only if there exists at least one point $a \in A$ with the following property: for *every* $\rho > 0$ there exists $x \in X$ such that

$$d(x, a) < \rho \quad \text{and} \quad x \in X \backslash A.$$

This shows at once that the empty set $\varnothing$ is open in $X$ (reason: if a set is not open then it has at least one point); and also that the underlying set $X$ is always open in the metric space $(X, d)$ (reason: there is no point in $X \backslash X$).

**Proposition 3.10.** *An open ball is an open set.*

*Proof.* Let $B_\rho(x_0)$ be an open ball in a metric space $(X, d)$ with center at an arbitrary $x_0 \in X$ and with an arbitrary radius $\rho \geq 0$. Suppose $\rho > 0$ so that $B_\rho(x_0) \neq \varnothing$ (otherwise $B_\rho(x_0)$ is empty and hence trivially open). Take an arbitrary $u \in B_\rho(x_0)$, which means that $u \in X$ and $d(u, x_0) < \rho$. Set $\beta = \rho - d(u, x_0)$ so that $0 < \beta \leq \rho$, and let $x$ be a point in $X$. If $d(x, u) < \beta$, then the triangle inequality ensures that

$$d(x, x_0) \leq d(x, u) + d(u, x_0) < \beta + d(u, x_0) = \rho,$$

and hence $x \in B_\rho(x_0)$. Conclusion: For every $u \in B_\rho(x_0)$ there exists $\beta > 0$ such that

$$d(x, u) < \beta \quad \text{implies} \quad x \in B_\rho(x_0).$$

That is, $B_\rho(x_0)$ is an open set. $\qquad\square$

An *open neighborhood* of a point $x$ in a metric space is an open set containing $x$. In particular (see Proposition 3.10), every open ball with positive radius centered at a point $x$ in a metric space is an open neighborhood of $x$. A *neighborhood* of a point $x$ in a metric space $X$ is any subset of $X$ that includes an open neighborhood of $x$. Clearly, every open neighborhood of $x$ is a neighborhood of $x$.

Open sets can be used to give an alternative definition of continuity and convergence.

**Lemma 3.11.** *Consider a mapping $F : X \to Y$ of a metric space $X$ into a metric space $Y$ and let $x_0$ be a point in $X$. The following assertions are equivalent.*

(a)  *F is continuous at $x_0$.*

(b)  *The inverse image of every neighborhood of $F(x_0)$ is a neighborhood of $x_0$.*

*Proof.*  Consider the image $F(x_0) \in Y$ of $x_0 \in X$. Take an arbitrary neighborhood $N \subseteq Y$ of $F(x_0)$. Since $N$ includes an open neighborhood $U$ of $F(x_0)$, it follows that there exists an open ball $B_\varepsilon(F(x_0)) \subseteq U \subseteq N$ with center at $F(x_0)$ and radius $\varepsilon > 0$. If the mapping $F: X \to Y$ is continuous at $x_0$ (cf. Definition 3.6), then there exists $\delta > 0$ such that

$$d_Y(F(x), F(x_0)) < \varepsilon \quad \text{whenever} \quad d_X(x, x_0) < \delta,$$

where $d_X$ and $d_Y$ are the metrics on $X$ and $Y$, respectively. In other words, there exists $\delta > 0$ such that

$$x \in B_\delta(x_0) \quad \text{implies} \quad F(x) \in B_\varepsilon(F(x_0)).$$

Thus $B_\delta(x_0) \subseteq F^{-1}(B_\varepsilon(F(x_0))) \subseteq F^{-1}(U) \subseteq F^{-1}(N)$. Since the open ball $B_\delta(x_0)$ is an open neighborhood of $x_0$, and since $B_\delta(x_0) \subseteq F^{-1}(N)$, it follows that $F^{-1}(N)$ is a neighborhood of $x_0$. Hence (a) implies (b). Now suppose (b) holds true. Then, in particular, the inverse image $F^{-1}(B_\varepsilon(F(x_0)))$ of *each* open ball $B_\varepsilon(F(x_0))$ with center $F(x_0)$ and radius $\varepsilon > 0$ includes a neighborhood $N \subseteq X$ of $x_0$. This neighborhood $N$ includes an open neighborhood $U$ of $x_0$ which in turn includes an open ball $B_\delta(x_0)$ with center at $x_0$ and with a positive radius $\delta$ (cf. Definition 3.9). Therefore, for each $\varepsilon > 0$ there exists $\delta > 0$ such that

$$B_\delta(x_0) \subseteq U \subseteq N \subseteq F^{-1}(B_\varepsilon(F(x_0))).$$

Hence (see Problems 1.2(c,j))

$$F(B_\delta(x_0)) \subseteq B_\varepsilon(F(x_0)).$$

Equivalently, if $x \in B_\delta(x_0)$, then $F(x) \in B_\varepsilon(F(x_0))$. Thus for each $\varepsilon > 0$ there exists $\delta > 0$ such that

$$d_X(x, x_0) < \delta \quad \text{implies} \quad d_Y(F(x), F(x_0)) < \varepsilon,$$

where $d_X$ and $d_Y$ denote the metrics on $X$ and $Y$, respectively. That is, (a) holds true (Definition 3.6). □

**Theorem 3.12.** *A map between metric spaces is continuous if and only if the inverse image of each open set is an open set.*

*Proof.*  Let $F: X \to Y$ be a mapping of a metric space $X$ into a metric space $Y$.

(a)  Take any neighborhood $N \subseteq Y$ of $F(x) \in Y$ (for an arbitrary $x \in X$). Since $N$ includes an open neighborhood of $F(x)$, say $U$, it follows that $F(x) \in U \subseteq N$ which implies

$$x \in F^{-1}(U) \subseteq F^{-1}(N).$$

If the inverse image (under $F$) of each open set in $Y$ is an open set in $X$, then $F^{-1}(U)$ is open in $X$, and hence $F^{-1}(U)$ is an open neighborhood of $x$. Therefore, the inverse image $F^{-1}(N)$ is a neighborhood of $x$. Conclusion: The inverse image of every neighborhood of $F(x)$ (for any $x \in X$) is a neighborhood of $x$. Thus $F$ is continuous by Lemma 3.11.

(b)   Take an arbitrary open subset $U$ of $Y$. Suppose $\mathcal{R}(F) \cap U \neq \varnothing$ and take $x \in F^{-1}(U) \subseteq X$ arbitrary. Thus $F(x) \in U$ so that $U$ is an open neighborhood of $F(x)$. If $F$ is continuous, then it is continuous at $x$. Therefore, according to Lemma 3.11, $F^{-1}(U)$ is a neighborhood of $x$, and hence it includes a nonempty open ball $B_\delta(x)$ centered at $x$. Thus $B_\delta(x) \subseteq F^{-1}(U)$ so that $F^{-1}(U)$ is open in $X$ (reason: it includes a nonempty open ball of an arbitrary point of it). If $\mathcal{R}(F) \cap U = \varnothing$, then $F^{-1}(U) = \varnothing$ which is open. Conclusion: $F^{-1}(U)$ is open in $X$ for every open subset $U$ of $Y$. $\qquad\qquad\square$

**Corollary 3.13.** *The composition of two continuous functions is again a continuous function.*

*Proof.* Let $X$, $Y$ and $Z$ be metric spaces, and let $F: X \to Y$ and $G: Y \to Z$ be continuous functions. Take an arbitrary open set $U$ in $Z$. According to Theorem 3.12 the set $G^{-1}(U)$ is open in $Y$ so that $(GF)^{-1}(U) = F^{-1}(G^{-1}(U))$ is open in $X$. Thus, using Theorem 3.12 again, we conclude that $GF: X \to Z$ is continuous. $\qquad\qquad\square$

An $X$-valued sequence $\{x_n\}$ is said to be *eventually in* a subset $A$ of $X$ if there exists a positive integer $n_0$ such that

$$n \geq n_0 \quad \text{implies} \quad x_n \in A.$$

**Theorem 3.14.** *Let $\{x_n\}$ be a sequence in a metric space $X$, and let $x$ be a point in $X$. The following assertions are equivalent.*

(a)   $x_n \to x$ *in* $X$.

(b)   $\{x_n\}$ *is eventually in every neighborhood of* $x$.

*Proof.* If $x_n \to x$, then (definition of convergence) $\{x_n\}$ is eventually in every nonempty open ball centered at $x$. Hence it is eventually in every neighborhood of $x$ (cf. definitions of neighborhood and of open set). Conversely, if $\{x_n\}$ is eventually in every neighborhood of $x$ then, in particular, it is eventually in every nonempty open ball centered at $x$, which means that $x_n \to x$. $\qquad\qquad\square$

Observe that the above theorem is naturally extended from sequences to nets. *A net $\{x_\gamma\}_{\gamma \in \Gamma}$ in a metric space $X$ converges to $x \in X$ if and only if for every neighborhood $N$ of $x$ there exists an index $\gamma_0 \in \Gamma$ such that $x_\gamma \in N$ for every $\gamma \geq \gamma_0$.*

Given a metric space $X$, the collection of all open sets in $X$ is of paramount importance. The fundamental properties of it are stated in the next theorem.

**Theorem 3.15.** *If $X$ is a metric space, then*

(a) *the whole set $X$ and the empty set $\varnothing$ are open,*

(b) *the intersection of a finite collection of open sets is open,*

(c) *the union of an arbitrary collection of open sets is open.*

*Proof.* We have already verified that assertion (a) holds true. Let $\{U_n\}$ be a *finite* collection of open subsets of $X$. Suppose $\bigcap_n U_n \neq \varnothing$ (otherwise $\bigcap_n U_n$ is an open set). Take an arbitrary $u \in \bigcap_n U_n$ so that $u \in U_n$ for every index $n$. As each $U_n$ is an open subset of $X$, there exist open balls $B_{\alpha_n}(u) \subseteq U_n$ (with center at $u$ and radius $\alpha_n > 0$) for each index $n$. Consider the set $\{\alpha_n\}$ consisting of the radius of each $B_{\alpha_n}(u)$. Since $\{\alpha_n\}$ is a *finite* set of positive numbers, it follows that it has a positive minimum. Set $\alpha = \min\{\alpha_n\} > 0$ so that $B_\alpha(u) \subseteq \bigcap_n U_n$. Thus $\bigcap_n U_n$ is open in $X$ (i.e., for each $u \in \bigcap_n U_n$ there exists an open ball $B_\alpha(u) \subseteq \bigcap_n U_n$), which concludes the proof of (b). The proof of (c) goes as follows. Let $\mathcal{U}$ be an arbitrary collection of open subsets of $X$. Suppose $\mathcal{U}$ is nonempty (otherwise it is open by (a)) and take an arbitrary $u \in \bigcup \mathcal{U}$ so that $u \in U$ for some $U \in \mathcal{U}$. As $U$ is an open subset of $X$, there exists a nonempty open ball $B_\rho(u) \subseteq U \subseteq \bigcup \mathcal{U}$, which means that $\bigcup \mathcal{U}$ is open in $X$. $\qquad\square$

**Corollary 3.16.** *A subset of a metric space is open if and only if it is a union of open balls.*

*Proof.* The union of open balls in a metric space $X$ is an open set in $X$ because open balls are open sets (cf. Proposition 3.10 and Theorem 3.15). On the other hand, let $U$ be an open set in a metric space $X$. If $U$ is empty, then it coincides with the empty open ball. If $U$ is a nonempty open subset of $X$, then each $u \in U$ is the center of an open ball, say $B_{\rho_u}(u)$, included in $U$. Thus $U = \bigcup_{u \in U} \{u\} \subseteq \bigcup_{u \in U} B_{\rho_u}(u) \subseteq U$, and hence $U = \bigcup_{u \in U} B_{\rho_u}(u)$. $\qquad\square$

The collection $\mathcal{T}$ of all open sets in a metric space $X$ (which is a subcollection of the power set $\wp(X)$) is called the *topology* (or the *metric topology*) on $X$. As the elements of $\mathcal{T}$ are the open sets in the metric space $(X, d)$, and since the definition of an open set in $X$ depends on the particular metric $d$ that equips the metric space $(X, d)$, the collection $\mathcal{T}$ is also referred to as the *topology induced* (or *generated*, or *determined*) by the metric $d$.

Our starting point in this chapter was the definition of a metric space. A metric has been defined on a set $X$ as a real-valued function on $X \times X$ that satisfies the metric axioms of Definition 3.1. A possible and different approach is to define axiomatically an abstract notion of open sets (instead of an abstract notion of distance as we did in

Definition 3.1), and then to build up a theory based on it. Such a "new" beginning goes as follows.

**Definition 3.17.** A subcollection $\mathcal{T}$ of the power set $\wp(X)$ of a set $X$ is a *topology* on $X$ if it satisfies three axioms, viz.,

    (i)   The whole set $X$ and the empty set $\varnothing$ belong to $\mathcal{T}$.

    (ii)  The intersection of a finite collection of sets in $\mathcal{T}$ belongs to $\mathcal{T}$.

    (iii) The union of an arbitrary collection of sets in $\mathcal{T}$ belongs to $\mathcal{T}$.

A set $X$ equipped with a topology $\mathcal{T}$ is referred to as a *topological space* (denoted by $(X, \mathcal{T})$ or simply by $X$), and the elements of $\mathcal{T}$ are called the *open* subsets of $X$ with respect to $\mathcal{T}$. Thus a topology $\mathcal{T}$ on an underlying set $X$ is always identified with the collection of all open subsets of $X$: $U$ is open in $X$ with respect to $\mathcal{T}$ if and only if $U \in \mathcal{T}$. It is clear (see Theorem 3.15) that every metric space $(X, d)$ is a topological space, where the topology $\mathcal{T}$ (the metric topology, that is) is that induced by the metric. This topology $\mathcal{T}$ induced by the metric $d$, and the topological space $(X, \mathcal{T})$ obtained by equipping $X$ with $\mathcal{T}$, are said to be *metrized* by $d$. If $(X, \mathcal{T})$ is a topological space, and if there exists a metric $d$ on $X$ that metrizes $\mathcal{T}$, the topological space $(X, \mathcal{T})$ and the topology $\mathcal{T}$ are called *metrizable*. The notion of topological space is broader than the notion of metric space. Although every metric space is a topological space, the converse fails. There are topological spaces that are not metrizable.

**Example 3J.** Let $X$ be an arbitrary set and define a function $d : X \times X \to \mathbb{R}$ by

$$d(x, y) = \begin{cases} 0, & x = y, \\ 1, & x \neq y, \end{cases}$$

for every $x$ and $y$ in $X$. It is readily verified that $d$ is a metric on $X$, the so-called *discrete metric* on $X$. A set $X$ equipped with the discrete metric is called a *discrete space*. In a discrete space every open ball with radius $\rho \in (0, 1)$ is a singleton in $X$: $B_\rho(x_0) = \{x_0\}$ for every $x_0 \in X$ and every $\rho \in (0, 1)$. Thus, according to Definition 3.9, every subset of $X$ is an open set in the metric space $(X, d)$ equipped with the discrete metric $d$. That is, the metric topology coincides with the power set of $X$. Conversely, if $X$ is an arbitrary set, then the collection $\mathcal{T} = \wp(X)$ is a topology on $X$ (since $\mathcal{T}$ trivially satisfies the above three axioms), called the *discrete topology*, which is the *largest* topology on $X$ (any other topology on $X$ is a subcollection of the discrete topology). Summing up: The discrete topology

$$\mathcal{T} = \wp(X) \text{ is metrizable}$$

by the discrete metric. On the other extreme lies the topology $\mathcal{T} = \{\varnothing, X\}$, called the *indiscrete topology*, which is the *smallest* topology on $X$ (it is a subcollection of any other topology on $X$). If $X$ has more than one point, then the indiscrete topology

$$\mathcal{T} = \{\varnothing, X\} \text{ is not metrizable.}$$

Indeed, suppose there exists a metric $d$ on $X$ that induces the indiscrete topology. Take $u$ in $X$ arbitrary and consider the set $X \backslash \{u\}$. Since $\varnothing \neq X \backslash \{u\} \neq X$, it follows that this set is not open (with respect to the indiscrete topology). Thus there exists $v \in X \backslash \{u\}$ with the following property: for every $\rho > 0$ there exists $x \in X$ such that

$$d(x, v) < \rho \quad \text{and} \quad x \in X \backslash (X \backslash \{u\}) = \{u\}.$$

Hence $x = u$ so that $d(u, v) < \rho$ for every $\rho > 0$. Therefore $u = v$ (i.e., $d(u, v) = 0$), which is a contradiction (for $v \in X \backslash \{u\}$). Conclusion: There is no metric on $X$ that induces the indiscrete topology.

Continuity and convergence in a topological space can be defined as follows. A mapping $F \colon X \to Y$ of a topological space $(X, \mathcal{T}_X)$ into a topological space $(Y, \mathcal{T}_Y)$ is *continuous* if $F^{-1}(U) \in \mathcal{T}_X$ for every $U \in \mathcal{T}_Y$. An $X$-valued sequence $\{x_n\}$ *converges* in a topological space $(X, \mathcal{T})$ to a limit $x \in X$ if it is eventually in every $U \in \mathcal{T}$ that contains $x$. Carefully note that, for the particular case of metric spaces (or of metrizable topological spaces), the above definitions of continuity and convergence agree with Definitions 3.6 and 3.4, respectively, when the topological spaces are equipped with their metric topology. Indeed, these definitions are the topological-space versions of Theorems 3.12 and 3.14.

Many (but not all) of the theorems in the next sections hold for general topological spaces (metrizable or not), and we shall prove them by using a topological-space style (based on open sets rather than on open balls) whenever this is possible and convenient. However, as we had anticipated at the introduction of this chapter, our attention will focus mainly on metric spaces.

## 3.4   Equivalent Metrics and Homeomorphisms

Let $(X, d_1)$ and $(X, d_2)$ be two metric spaces with the same underlying set $X$. The metrics $d_1$ and $d_2$ are said to be *equivalent* (or $d_1$ and $d_2$ are *equivalent metrics* on $X$ — notation: $d_1 \sim d_2$) if they induce the same topology (i.e., a subset of $X$ is open in $(X, d_1)$ if and only if it is open in $(X, d_2)$). This notion of equivalence in fact is an equivalence relation on the collection of all metrics defined on a given set $X$. If $\mathcal{T}_1$ and $\mathcal{T}_2$ are the metric topologies on $X$ induced by the metrics $d_1$ and $d_2$, respectively, then

$$d_1 \sim d_2 \quad \text{if and only if} \quad \mathcal{T}_1 = \mathcal{T}_2.$$

If $\mathcal{T}_1 \subseteq \mathcal{T}_2$ (i.e., if every open set in $(X, d_1)$ is open in $(X, d_2)$), then $\mathcal{T}_2$ is said to be *stronger* than $\mathcal{T}_1$. In this case we also say that $\mathcal{T}_1$ is *weaker* than $\mathcal{T}_2$. The terms *finer* and *coarser* are also used as synonyms for "stronger" and "weaker", respectively. If either $\mathcal{T}_1 \subseteq \mathcal{T}_2$ or $\mathcal{T}_2 \subseteq \mathcal{T}_1$, then $\mathcal{T}_1$ and $\mathcal{T}_2$ are said to be *commensurable*. Otherwise (i.e., if neither $\mathcal{T}_1 \subseteq \mathcal{T}_2$ nor $\mathcal{T}_2 \subseteq \mathcal{T}_1$), the topologies are said to be *incommensurable*. As we shall see below, if $\mathcal{T}_2$ is stronger than $\mathcal{T}_1$, then continuity with respect to $\mathcal{T}_1$ implies continuity with respect to $\mathcal{T}_2$. On the other hand, if $\mathcal{T}_2$ is stronger than $\mathcal{T}_1$, then convergence with respect to $T_2$ implies convergence with respect to $T_1$. Briefly and roughly: "Strong convergence" implies "weak convergence" but "weak continuity" implies "strong continuity".

**Theorem 3.18.** *Let $d_1$ and $d_2$ be metrics on a set $X$, and consider the topologies $T_1$ and $T_2$ induced by $d_1$ and $d_2$, respectively. The following assertions are pairwise equivalent.*

(a) *$T_2$ is stronger than $T_1$    (i.e., $T_1 \subseteq T_2$).*

(b) *Every mapping $F : X \to Y$ that is continuous at $x_0 \in X$ as a mapping of $(X, d_1)$ into the metric space $(Y, d)$ is continuous at $x_0$ as a mapping of $(X, d_2)$ into $(Y, d)$.*

(c) *Every $X$-valued sequence that converges in $(X, d_2)$ to a limit $x \in X$ converges in $(X, d_1)$ to the same limit $x$.*

(d) *The identity map of $(X, d_2)$ onto $(X, d_1)$ is continuous.*

*Proof.* Consider the topologies $T_1$ and $T_2$ on $X$ induced by the metrics $d_1$ and $d_2$ on $X$, respectively. Let $T$ denote the topology on a set $Y$ induced by a metric $d$ on $Y$.

*Proof of* (a)$\Rightarrow$(b). If $F : (X, d_1) \to (Y, d)$ is continuous at $x_0 \in X$, then (Lemma 3.11) for every $U \in T$ that contains $F(x_0)$ there exists $U' \in T_1$ containing $x_0$ such that $U' \subseteq F^{-1}(U)$. If $T_1 \subseteq T_2$, then $U' \in T_2$: the inverse image (under $F$) of every open neighborhood of $F(x_0)$ in $T$ includes an open neighborhood of $x_0$ in $T_2$, which clearly implies that the inverse image (under $F$) of every neighborhood of $F(x_0)$ in $T$ is a neighborhood of $x_0$ in $T_2$. Thus, applying Lemma 3.11 again, $F : (X, d_2) \to (Y, d)$ is continuous at $x_0$.

*Proof of* (a)$\Rightarrow$(c). Let $\{x_n\}$ be an $X$-valued sequence. If $x_n \to x \in X$ in $(X, d_2)$, then (Theorem 3.14) $\{x_n\}$ is eventually in *every* open neighborhood of $x$ in $T_2$. If $T_1 \subseteq T_2$ then, in particular, $\{x_n\}$ is eventually in every neighborhood of $x$ in $T_1$. Therefore, applying Theorem 3.14 again, $x_n \to x$ in $(X, d_1)$.

*Proof of* (b)$\Rightarrow$(d). The identity map $I : (X, d_1) \to (X, d_1)$ of a metric space onto itself is trivially continuous. Thus, by setting $(Y, d) = (X, d_1)$ in (b), it follows that (b) implies (d).

*Proof of* (c)$\Rightarrow$(d). Corollary 3.8 ensures that (c) implies (d).

*Proof of* (d)$\Rightarrow$(a). According to Theorem 3.12 (d) implies (a) (i.e., if the identity $I : (X, d_2) \to (X, d_1)$ is continuous, then $U = I^{-1}(U)$ is open in $T_2$ whenever $U$ is open in $T_1$, and hence $T_1 \subseteq T_2$). $\qquad\square$

As the discrete topology is the strongest topology on $X$, the above theorem ensures that any function $F : X \to Y$ that is continuous in some topology on $X$ is continuous in the discrete topology. Actually, since every subset of $X$ is open in the discrete topology, it follows that the inverse image of every subset of $Y$ — no matter which topology equips the set $Y$ — is an open subset of $X$ when $X$ is equipped with the discrete topology. Therefore, *every function defined on a discrete topological space is continuous*. On the other hand, if an $X$-valued (infinite) sequence converges in the discrete topology, then it is *eventually constant* (i.e., it has only a finite number of entries not equal to its limit), and hence it converges in any topology on $X$.

**Corollary 3.19.** *Let $(X, d_1)$ and $(X, d_2)$ be metric spaces with the same underlying set $X$. The following assertions are pairwise equivalent.*

(a)  *$d_2$ and $d_1$ are equivalent metrics on $X$.*

(b)  *A mapping of $X$ into a set $Y$ is continuous at $x_0 \in X$ as a mapping of $(X, d_1)$ into the metric space $(Y, d)$ if and only if it is continuous at $x_0$ as a mapping of $(X, d_2)$ into $(Y, d)$.*

(c)  *An $X$-valued sequence converges in $(X, d_1)$ to $x \in X$ if and only if it converges in $(X, d_2)$ to $x$.*

(d)  *The identity map of $(X, d_1)$ onto $(X, d_2)$ and its inverse (i.e., the identity map of $(X, d_2)$ onto $(X, d_1)$) are both continuous.*

*Proof.* Recall that, by definition, two metrics $d_1$ and $d_2$ on a set $X$ are equivalent if the topologies $\mathcal{T}_1$ and $\mathcal{T}_2$ on $X$, induced by $d_1$ and $d_2$ respectively, coincide (i.e., if $\mathcal{T}_1 = \mathcal{T}_2$). Now apply Theorem 3.18.    $\square$

A one-to-one mapping $G$ of a metric space $X$ onto a metric space $Y$ is a *homeomorphism* if both $G \colon X \to Y$ and $G^{-1} \colon Y \to X$ are continuous. Equivalently, a homeomorphism between metric spaces is an invertible (i.e., injective and surjective) mapping that is continuous and has a continuous inverse. Obviously, $G$ is a homeomorphism from $X$ to $Y$ if and only if $G^{-1}$ is a homeomorphism from $Y$ to $X$. Two metric spaces are *homeomorphic* if there exists a homeomorphism between them. A function $F \colon X \to Y$ of a metric space $X$ into a metric space $Y$ is an *open map* (or an *open mapping*) if the image of each open set in $X$ is open in $Y$ (i.e., $F(U)$ is open in $Y$ whenever $U$ is open in $X$).

**Theorem 3.20.** *Let $X$ and $Y$ be metric spaces. If $G \colon X \to Y$ is invertible, then*

(a)  *$G$ is open if and only if $G^{-1}$ is continuous,*

(b)  *$G$ is continuous if and only if $G^{-1}$ is open,*

(c)  *$G$ is a homeomorphism if and only if $G$ and $G^{-1}$ are both open.*

*Proof.* If $G$ is invertible, then it is trivially verified that the inverse image of $B$ ($B \subseteq Y$) under $G$ coincides with the image of $B$ under the inverse of $G$ (tautologically:

$G^{-1}(B) = G^{-1}(B)$). Applying the same argument to the inverse $G^{-1}$ of $G$ (which is clearly invertible), $(G^{-1})^{-1}(A) = G(A)$ for each $A \subseteq X$. Thus the theorem is a straightforward combination of the definitions of open map and homeomorphism by using the alternative definition of continuity in Theorem 3.12. $\qquad\square$

Thus a homeomorphism provides *simultaneously* a one-to-one correspondence between the underlying sets $X$ and $Y$ (so that $X \leftrightarrow Y$, for a homeomorphism is injective and surjective) and between their topologies (so that $\mathcal{T}_X \leftrightarrow \mathcal{T}_Y$, for a homeomorphism puts the open sets of $\mathcal{T}_X$ into a one-to-one correspondence with the open sets of $\mathcal{T}_Y$). Indeed, if $\mathcal{T}_X$ and $\mathcal{T}_Y$ are the topologies on $X$ and $Y$, respectively, then a homeomorphism $G: X \to Y$ induces a map $\mathcal{G}: \mathcal{T}_X \to \mathcal{T}_Y$, defined by $\mathcal{G}(U) = G(U)$ for every $U \in \mathcal{T}_X$, which is injective and surjective according to Theorem 3.20. Therefore, any property of a metric space $X$ expressed entirely in terms of set operations and open sets is also possessed by each metric space homeomorphic to $X$. We call a property of a metric space a *topological property* or a *topological invariant* if whenever it is true for one metric space, say $X$, it is true for every metric space homeomorphic to $X$ (trivial examples: the cardinality of the underlying set and the cardinality of the topology). A map $F: X \to Y$ of a metric space $X$ into a metric space $Y$ is a *topological embedding* of $X$ into $Y$ if it establishes a homeomorphism of $X$ onto its range $\mathcal{R}(F)$ (i.e., $F: X \to Y$ is a topological embedding of $X$ into $Y$ if it is such that $F: X \to F(X)$ is a homeomorphism of $X$ onto the subspace $F(X)$ of $Y$).

**Example 3K.** Suppose $G: X \to Y$ is a homeomorphism of a metric space $X$ onto a metric space $Y$. Let $A$ be a subspace of $X$ and consider the subspace $G(A)$ of $Y$. According to Problem 3.30 the restriction $G|_A: A \to G(A)$ of $G$ to $A$ onto $G(A)$ is continuous. Similarly, the restriction $G^{-1}|_{G(A)}: G(A) \to A$ of the inverse of $G$ to $G(A)$ onto $G^{-1}(G(A)) = A$ is continuous as well. Since $G^{-1}|_{G(A)} = (G|_A)^{-1}$ (Problem 1.8), it follows that

$$G|_A: A \to G(A) \text{ is a homeomorphism,}$$

and hence $A$ and $G(A)$ are homeomorphic metric spaces (as subspaces of $X$ and $Y$, respectively). Thus, as we might expect, the restriction $G|_A: A \to Y$ of a homeomorphism $G: X \to Y$ to any subset $A$ of $X$ is a topological embedding of $A$ into $Y$.

The notions of homeomorphism, open map, topological invariant and topological embedding are germane to topological spaces in general (and to metric spaces in particular). For instance, both Theorem 3.20 and Example 3K can be likewise stated (and proved) in a topological-space setting. In other words, the metric has played no role in the above paragraph, and "metric space" can be replaced with "topological space" there. Next we shall consider a couple of concepts that only make sense in a metric space. A homeomorphism $G$ of a metric space $(X, d_X)$ onto a metric space $(Y, d_Y)$ is a *uniform homeomorphism* if both $G$ and $G^{-1}$ are uniformly continuous.

Two metric spaces are *uniformly homeomorphic* if there exists a uniform homeomorphism mapping one of them onto the other. An *isometry* between metric spaces is a map that preserves distance. Precisely, a mapping $J : (X, d_X) \to (Y, d_Y)$ of a metric space $(X, d_X)$ *into* a metric space $(Y, d_Y)$ is an *isometry* if

$$d_Y(J(x), J(x')) = d_X(x, x')$$

for every pair of points $x, x'$ in $X$. It is clear that *every isometry is an injective contraction*, and hence an injective and uniformly continuous mapping. Thus *every surjective isometry is a uniform homeomorphism* (the inverse of a surjective isometry is again a surjective isometry — trivial example: the identity mapping of a metric space into itself is a surjective isometry on that space). Two metric spaces are *isometric* (or *isometrically equivalent*) if there exists a surjective isometry between them, so that two *isometrically equivalent metric spaces are uniformly homeomorphic*. It is trivially verified that a composition of surjective isometries is a surjective isometry (transitivity), and this shows that the notion of isometrically equivalent metric spaces deserves its name: it is indeed an equivalence relation on any collection of metric spaces. If two metric spaces are isometrically equivalent, then they can be thought of as being essentially the same metric space — they may differ on the set-theoretic nature of their points but, as far as the metric space (topological) structure is concerned, they are indistinguishable. A surjective isometry not only preserves open sets (for it is a homeomorphism) but also distance.

Now consider two metric spaces $(X, d_1)$ and $(X, d_2)$ with the same underlying set $X$. According to Corollary 3.19 the metrics $d_1$ and $d_2$ are equivalent if and only if the identity map of $(X, d_1)$ onto $(X, d_2)$ is a homeomorphism (i.e., if and only if $I : (X, d_1) \to (X, d_2)$ and its inverse $I^{-1} : (X, d_2) \to (X, d_1)$ are both continuous). We say that the metrics $d_1$ and $d_2$ are *uniformly equivalent* if the identity map of $(X, d_1)$ onto $(X, d_2)$ is a uniform homeomorphism (i.e., $I : (X, d_1) \to (X, d_2)$ and its inverse $I^{-1} : (X, d_2) \to (X, d_1)$ are both uniformly continuous). For instance, if $I$ and $I^{-1}$ are both Lipschitzian, which means that there exist real numbers $\alpha > 0$ and $\beta > 0$ such that

$$\alpha\, d_1(x, x') \le d_2(x, x') \le \beta\, d_1(x, x')$$

for every $x, x'$ in $X$, then the metrics $d_1$ and $d_2$ are uniformly equivalent, and hence equivalent. Thus, if $d_1$ and $d_2$ are equivalent metrics on $X$, then $(X, d_1)$ and $(X, d_2)$ are homeomorphic metric spaces. However, the converse fails: there exist uniformly homeomorphic metric spaces with the same underlying set for which the identity is not a homeomorphism.

**Example 3L.** Take two metric spaces $(X, d_1)$ and $(X, d_2)$ with the same underlying set $X$ and consider the product spaces $(X \times X, d)$ and $(X \times X, d')$, where

$$\begin{aligned}
d((x, y), (u, v)) &= d_1(x, u) + d_2(y, v), \\
d'((x, y), (u, v)) &= d_2(x, u) + d_1(y, v),
\end{aligned}$$

for all ordered pairs $(x, y)$, $(u, v)$ in $X \times X$. That is, $(X \times X, d) = (X, d_1) \times (X, d_2)$ and $(X \times X, d') = (X, d_2) \times (X, d_1)$ — see Problem 3.9. Suppose the metrics $d_1$ and $d_2$ on $X$ are not equivalent so that either the identity map of $(X, d_1)$ onto $(X, d_2)$ or the identity map of $(X, d_2)$ onto $(X, d_1)$ (or both) is not continuous. Let $I \colon (X, d_1) \to (X, d_2)$ be the one that is not continuous. The identity map

$$I \colon (X \times X, d) \to (X \times X, d') \text{ is not continuous.}$$

Indeed, if it is continuous, then the restriction of it to any subspace of $X \times X, d$ is continuous (Problem 3.30). In particular, the restriction of it to $(X, d_1)$ — viewed as a subspace of $(X \times X, d) = (X, d_1) \times (X, d_2)$ — is continuous. But such a restriction is clearly identified with the identity map of $(X, d_1)$ onto $(X, d_2)$, which is not continuous. Thus $I \colon (X \times X, d) \to (X \times X, d')$ is not continuous, and hence the metrics $d$ and $d'$ on $X \times X$ are not equivalent. Now let $J \colon X \times X \to X \times X$ be the involution on $X \times X$ defined by $J((x, y)) = (y, x)$ for every $(x, y) \in X \times X$ (Problem 1.11). It is readily verified that $J \colon (X \times X, d) \to (X \times X, d')$ is a surjective isometry so that

$$J \colon (X \times X, d) \to (X \times X, d') \text{ is a uniform homeomorphism.}$$

Summing up: The metric spaces $(X \times X, d)$ and $(X \times X, d')$, with the same underlying set $X \times X$, are uniformly homeomorphic (more than that, they are isometrically equivalent) but the metrics $d$ and $d'$ on $X \times X$ are not equivalent.

Since two metric spaces with the same underlying set may be homeomorphic even if the identity between them is not a homeomorphism, it follows that a weaker version of Corollary 3.19 is obtained if we replace the homeomorphic identity with an arbitrary homeomorphism. This in fact can be formulated for arbitrary metric spaces (not necessarily with the same underlying set).

**Theorem 3.21.** *Let $X$ and $Y$ be metric spaces and let $G$ be an invertible mapping of $X$ onto $Y$. The following assertions are pairwise equivalent.*

(a)  *$G$ is a homeomorphism.*

(b)  *A mapping $F$ of $X$ into a metric space $Z$ is continuous if and only if the composition $F G^{-1} \colon Y \to Z$ is continuous.*

(c)  *An $X$-valued sequence $\{x_n\}$ converges in $X$ to a limit $x \in X$ if and only if the $Y$-valued sequence $\{G(x_n)\}$ converges in $Y$ to $G(x)$.*

*Proof.* Let $G \colon X \to Y$ be an invertible mapping of a metric space $X$ onto a metric space $Y$.

*Proof of* (a)⇒(b). Let $F: X \to Z$ be a mapping of $X$ into a metric space $Z$, and consider the commutative diagram

$$Y \xrightarrow{G^{-1}} X$$
$$H \searrow \quad \downarrow F$$
$$Z$$

so that $H = FG^{-1}: Y \to Z$. Suppose (a) holds true, and consider the following assertions.

(b$_1$)  $F: X \to Z$ is continuous.

(b$_2$)  $F^{-1}(U)$ is an open set in $X$ whenever $U$ is an open set in $Z$.

(b$_3$)  $G(F^{-1}(U))$ is an open set in $Y$ whenever $U$ is an open set in $Z$.

(b$_4$)  $(FG^{-1})^{-1}(U)$ is an open set in $Y$ whenever $U$ is an open set in $Z$.

(b$_5$)  $H = FG^{-1}: Y \to Z$ is continuous.

Theorem 3.12 says that (b$_1$) and (b$_2$) are equivalent. But (b$_2$) holds true if and only if (b$_3$) holds true by Theorem 3.20 (the homeomorphism $G: X \to Y$ puts the open sets of $X$ into a one-to-one correspondence with the open sets of $Y$). Now note that, as $G$ is invertible,

$$G(F^{-1}(A)) = \{G(x) \in Y: \ F(x) \in A\}$$
$$= \{y \in Y: \ F(G^{-1}(y)) \in A\} = (FG^{-1})^{-1}(A)$$

for every subset $A$ of $Z$. Thus (b$_3$) is equivalent to (b$_4$), which in turn is equivalent to (b$_5$) (cf. Theorem 3.12 again). Conclusion: (b$_1$)⇔(b$_5$) whenever (a) holds true.

*Proof of* (b)⇒(a). If (b) holds, then it holds in particular for $Z = X$ and for $Z = Y$. Thus (b) ensures that the following assertions hold true.

(b$'$)   If a mapping $F$ of $X$ into itself is continuous, then the mapping $H = FG^{-1}: Y \to X$ is continuous.

(b$''$)   A mapping $F$ of $X$ into $Y$ is continuous whenever the mapping $H = FG^{-1}: Y \to Y$ is continuous.

Since the identity of $X$ onto itself is a continuous mapping, (b$'$) implies that $G^{-1}: Y \to X$ is continuous. By setting $F = G$ in (b$''$) it follows that $G: X \to Y$ is continuous (because the identity $I = GG^{-1}: Y \to Y$ is continuous). Summing up: (b) implies that both $G$ and $G^{-1}$ are continuous, which means that (a) holds true.

*Proof of* (a)⇔(c). According to Corollary 3.8 an invertible mapping $G$ between metric spaces is continuous and has a continuous inverse if and only if both $G$ and $G^{-1}$ preserve convergence.                                                                □

# 3.5   Closed Sets and Closure

A subset $V$ of a metric space $X$ is *closed* in $X$ if its complement $X \backslash V$ is an open set in $X$.

**Theorem 3.22.** *If $X$ is a metric space, then*

   (a) *the whole set $X$ and the empty set $\varnothing$ are closed,*

   (b) *the union of a finite collection of closed sets is closed,*

   (c) *the intersection of an arbitrary collection of closed sets is closed.*

*Proof.* Apply the De Morgan laws to each item of Theorem 3.15. $\qquad\qquad$ □

Thus the concepts "closed" and "open" are dual of each other ($U$ is open in $X$ if and only if its complement $X \backslash U$ is closed in $X$, and $V$ is closed in $X$ if and only if its complement $X \backslash V$ is open in $X$); but they are neither exclusive (a set in a metric space may be both open and closed) nor exhaustive (a set in a metric space may be neither open nor closed).

**Theorem 3.23.** *A map between metric spaces is continuous if and only if the inverse image of each closed set is a closed set.*

*Proof.* Let $F: X \to Y$ be a mapping of a metric space $X$ into a metric space $Y$. Recall that $F^{-1}(Y \backslash B) = X \backslash F^{-1}(B)$ for every subset $B$ of $Y$ (Problem 1.2(b)). Suppose $F$ is continuous and take an arbitrary closed set $V$ in $Y$. Since $Y \backslash V$ is open in $Y$, it follows by Theorem 3.12 that $F^{-1}(Y \backslash V)$ is open in $X$. Thus $F^{-1}(V) = X \backslash F^{-1}(Y \backslash V)$ is closed in $X$. Therefore, the inverse image under $F$ of an arbitrary closed set $V$ in $Y$ is closed in $X$. Conversely, suppose the inverse image under $F$ of each closed set in $Y$ is a closed set in $X$ and take an arbitrary open set $U$ in $Y$. Thus $F^{-1}(Y \backslash U)$ is closed in $X$ (since $Y \backslash U$ is closed in $Y$) so that $F^{-1}(U) = X \backslash F^{-1}(Y \backslash U)$ is open in $X$. Conclusion: The inverse image under $F$ of an arbitrary open set $U$ in $Y$ is open in $X$. Therefore $F$ is continuous by Theorem 3.12. $\qquad$ □

A function $F: X \to Y$ of a metric space $X$ into a metric space $Y$ is a *closed map* (or a *closed mapping*) if the image of each closed set in $X$ is closed in $Y$ (i.e., $F(V)$ is closed in $Y$ whenever $V$ is closed in $X$). In general, a map $F: X \to Y$ may possess any combination of the attributes "continuous", "open" and "closed" (i.e., these are independent concepts). However, if $F: X \to Y$ is invertible (i.e., injective and surjective), then it is a closed map if and only if it is an open map.

**Theorem 3.24.** *Let $X$ and $Y$ be metric spaces. If a map $G: X \to Y$ is invertible, then*

   (a) *$G$ is closed if and only if $G^{-1}$ is continuous,*

(b)   *G is continuous if and only if $G^{-1}$ is closed,*

(c)   *G is a homeomorphism if and only if $G$ and $G^{-1}$ are both closed.*

*Proof.* Replace "open map" with "closed map" in the proof of Theorem 3.20 and use Theorem 3.23 instead of Theorem 3.12.    □

Let $A$ be a set in a metric space $X$ and let $\mathcal{V}_A$ be the collection of all closed subsets of $X$ that include $A$:

$$\mathcal{V}_A = \{V \in \mathscr{P}(X): \ V \text{ is closed in } X \text{ and } A \subseteq V\}.$$

The whole set $X$ always belongs to $\mathcal{V}_A$ so that $\mathcal{V}_A$ is never empty. The intersection of all sets in $\mathcal{V}_A$ is called the *closure* of $A$ in $X$, denoted by $A^-$ (i.e., $A^- = \bigcap \mathcal{V}_A$). According to Theorem 3.22(c) it follows that

$$A^- \text{ is closed in } X \quad \text{and} \quad A \subseteq A^-.$$

If $V \in \mathcal{V}_A$, then $A^- = \bigcap \mathcal{V}_A \subseteq V$. Thus, with respect to the inclusion ordering of $\mathscr{P}(X)$,

$$A^- \text{ is the smallest closed subset of } X \text{ that includes } A,$$

and hence (since $A^-$ is closed in $X$)

$$A \text{ is closed in } X \quad \text{if and only if} \quad A = A^-.$$

From the above displayed results it is readily verified that

$$\varnothing^- = \varnothing, \quad X^- = X, \quad (A^-)^- = A^-$$

and, if $B$ also is a set in $X$,

$$A \subseteq B \quad \text{implies} \quad A^- \subseteq B^-.$$

Moreover, since both $A$ and $B$ are subsets of $A \cup B$, it follows that $A^- \subseteq (A \cup B)^-$ and $B^- \subseteq (A \cup B)^-$ so that $A^- \cup B^- \subseteq (A \cup B)^-$. On the other hand, since $(A \cup B)^-$ is the smallest closed subset of $X$ that includes $A \cup B$, and since $A^- \cup B^-$ is closed (Theorem 3.22(b)) and includes $A \cup B$ (for $A \subseteq A^-$ and $B \subseteq B^-$ so that $A \cup B \subseteq A^- \cup B^-$), it follows that $(A \cup B)^- \subseteq A^- \cup B^-$. Therefore, if $A$ and $B$ are subsets of $X$, then

$$(A \cup B)^- = A^- \cup B^-.$$

It is easy to show by induction that the above identity holds for any finite collection of subsets of $X$. That is, *the closure of the union of a finite collection of subsets of $X$ coincides with the union of their closures.* In general (i.e., by allowing infinite collections as well) one has inclusion rather than equality. Indeed, if $\{A_\gamma\}_{\gamma \in \Gamma}$ is an arbitrary indexed family of subsets of $X$, then

$$\bigcup_\gamma A_\gamma^- \subseteq \left(\bigcup_\gamma A_\gamma\right)^-$$

because $A_\alpha \subseteq \bigcup_\gamma A_\gamma$ and hence $A_\alpha^- \subseteq (\bigcup_\gamma A_\gamma)^-$ for each index $\alpha \in \Gamma$. Similarly,

$$\left( \bigcap_\gamma A_\gamma \right)^- \subseteq \bigcap_\gamma A_\gamma^-$$

because $\bigcap_\gamma A_\gamma \subseteq \bigcap_\gamma A_\gamma^-$ and $\bigcap_\gamma A_\gamma^-$ is closed in $X$ by Theorem 3.22(c). However, the above two inclusions are not reversible in general so that equality does not hold.

**Example 3M.** Set $X = \mathbb{R}$ with its usual metric and consider the following subsets of $\mathbb{R}$. $A_n = [0, 1 - \frac{1}{n}]$, which is closed in $\mathbb{R}$ for each positive integer $n$, and $A = [0, 1)$, which is not closed in $\mathbb{R}$. Since

$$\bigcup_{n=1}^{\infty} A_n = A,$$

it follows that *the union of an infinite collection of closed sets is not necessarily closed* (see Theorem 3.22(b)). In particular, as $A_n^- = A_n$ for each $n$ and $A^- = [0, 1]$,

$$[0, 1) = \bigcup_{n=1}^{\infty} A_n^- \subset \left( \bigcup_{n=1}^{\infty} A_n \right)^- = [0, 1],$$

which is a *proper* inclusion. If $B = [1, 2]$ (so that $B^- = B$), then

$$\varnothing = (A \cap B)^- \subset A^- \cap B^- = \{1\},$$

so that the closure of any (even finite) intersection of sets may be a *proper* subset of the intersection of their closures.

A point $x$ in $X$ is *adherent* to $A$ (or an *adherent point* of $A$, or a *point of adherence* of $A$) if it belongs to the closure $A^-$ of $A$. It is clear that every point of $A$ is an adherent point of $A$ (i.e., $A \subseteq A^-$).

**Proposition 3.25.** *Let $A$ be a subset of a metric space $X$ and let $x$ be a point in $X$. The following assertions are pairwise equivalent.*

(a) *$x$ is a point of adherence of $A$.*

(b) *Every open set $U$ in $X$ that contains $x$ meets $A$ (i.e., if $U$ is open in $X$ and $x \in U$, then $A \cap U \neq \varnothing$).*

(c) *Every neighborhood $N$ of $x$ contains at least one point of $A$ (which may be $x$ itself).*

*Proof.* Suppose there exists an open set $U$ in $X$ containing $x$ for which $A \cap U = \varnothing$. Then $A \subseteq X \backslash U$, the set $X \backslash U$ is closed in $X$, and $x \notin X \backslash U$. Since $A^-$ is the smallest closed subset of $X$ that includes $A$, it follows that $A^- \subseteq X\backslash U$ so that $x \notin A^-$. Thus the denial of (b) implies the denial of (a), which means that (a) implies (b). Conversely, if $x \notin A^-$, then $x$ lies in the open set $X\backslash A^-$ which does not meet $A^-$ ($A^- \cap X\backslash A^- = \varnothing$). Therefore, the denial of (a) implies the denial of (b); that is, (b) implies (a). Finally note that (b) is equivalent to (c) as an obvious consequence of the definition of neighborhood. $\qquad\square$

A point $x$ in $X$ is a *point of accumulation* (or an *accumulation point*, or a *cluster point*) of $A$ if it is a point of adherence of $A\backslash\{x\}$. The set of all accumulation points of $A$ is the *derived set* of $A$, denoted by $A^\star$. Thus $x \in A^\star$ if and only if $x \in (A\backslash\{x\})^-$. It is clear that *every point of accumulation of $A$ is also a point of adherence of $A$*; that is, $A^\star \subseteq A^-$ (for $A\backslash\{x\} \subseteq A$ implies $(A\backslash\{x\})^- \subseteq A^-$). Actually,

$$A^- = A \cup A^\star.$$

Indeed, since $A^\star \subseteq A^-$ and $A \subseteq A^-$, it follows that $A \cup A^\star \subseteq A^-$. On the other hand, if $x \notin A \cup A^\star$, then $(A\backslash\{x\})^- = A^-$ (because $A\backslash\{x\} = A$ whenever $x \notin A$), and hence $x \notin A^-$ (for $x \notin A^\star$ so that $x \notin (A\backslash\{x\})^-$). Therefore, $x \in A^-$ implies $x \in A \cup A^\star$, which means $A^- \subseteq A \cup A^\star$. Hence $A^- = A \cup A^\star$. Thus

$$A = A^- \quad \text{if and only if} \quad A^\star \subseteq A.$$

That is, *$A$ is closed in $X$ if and only if it contains all its accumulation points*. It is trivially verified that

$$A \subseteq B \quad \text{implies} \quad A^\star \subseteq B^\star$$

whenever $A$ and $B$ are subsets of $X$. Also note that $A^- = \varnothing$ if and only if $A = \varnothing$ (for $\varnothing^- = \varnothing$ and $\varnothing \subseteq A \subseteq A^-$), and $A^\star = \varnothing$ whenever $A = \varnothing$ (because $A^\star \subseteq A^-$), but the converse fails (e.g., the derived set of a singleton is empty).

**Proposition 3.26.** *Let $A$ be a subset of a metric space $X$ and let $x$ be a point in $X$. The following assertions are pairwise equivalent.*

(a)  *$x$ is a point of accumulation of $A$.*

(b)  *Every open set $U$ in $X$ that contains $x$ also contains at least one point of $A$ other than $x$.*

(c)  *Every neighborhood $N$ of $x$ contains at least one point of $A$ distinct from $x$.*

*Proof.* Since $x \in X$ is a point of accumulation of $A$ if and only if it is a point of adherence of $A\backslash\{x\}$, it follows by Proposition 3.25 that the assertions (a), (b) and (c) are pairwise equivalent (replace $A$ with $A\backslash\{x\}$ in Proposition 3.25). $\qquad\square$

Everything that has been written so far in this section pertains to the realm of topological spaces (metrizable or not). However, the next results are typical of metric spaces.

**Proposition 3.27.** *Let $A$ be a subset of a metric space $(X, d)$ and let $x$ be a point in $X$. The following assertions are pairwise equivalent.*

(a)  *x is a point of adherence of* $A$.

(b)  *Every nonempty open ball centered at* $x$ *meets* $A$.

(c)  $A \neq \varnothing$ *and* $d(x, A) = 0$.

(d)  *There exists an* $A$-*valued sequence that converges to* $x$ *in* $(X, d)$.

*Proof.* The equivalence (a)$\Leftrightarrow$(b) follows by Proposition 3.25 (recall: a nonempty open ball centered at $x$ is a neighborhood of $x$ and, conversely, every neighborhood of $x$ includes a nonempty open ball centered at $x$, so that every nonempty open ball centered at $x$ meets $A$ if and only if every neighborhood of $x$ meets $A$). Clearly (b)$\Leftrightarrow$(c) (i.e., for each $\varepsilon > 0$ there exists $a \in A$ such that $d(x, a) < \varepsilon$ if and only if $A \neq \varnothing$ and $\inf_{a \in A} d(x, a) = 0$). Theorem 3.14 ensures that (d)$\Rightarrow$(b). On the other hand, if (b) holds true, then for each positive integer $n$ the open ball $B_{\frac{1}{n}}(x)$ meets $A$ (i.e., $B_{\frac{1}{n}}(x) \cap A \neq \varnothing$). Take $x_n \in B_{\frac{1}{n}}(x) \cap A$ so that $x_n \in A$ and $0 \leq d(x_n, x) < \frac{1}{n}$ for each $n$. Thus $\{x_n\}$ is an $A$-valued sequence such that $d(x_n, x) \to 0$. Therefore (b)$\Rightarrow$(d). $\qquad\qquad\square$

**Proposition 3.28.**  *Let* $A$ *be a subset of a metric space* $(X, d)$ *and let* $x$ *be a point in* $X$. *The following assertions are pairwise equivalent.*

(a)  *x is a point of accumulation of* $A$.

(b)  *Every nonempty open ball centered at* $x$ *contains a point of* $A$ *distinct from* $x$.

(c)  *Every nonempty open ball centered at* $x$ *contains infinitely many points of* $A$.

(d)  *There exists an* $A \backslash \{x\}$-*valued sequence of pairwise distinct points that converges to* $x$ *in* $(X, d)$.

*Proof.*  Note that (d)$\Rightarrow$(c) by Theorem 3.14, (c)$\Rightarrow$(b) trivially, and (d) $\Rightarrow$ (a) $\Rightarrow$ (b) by the previous proposition. To complete the proof it remains to show that (b)$\Rightarrow$(d). Let $B_\varepsilon(x)$ be open ball centered at $x \in X$ with radius $\varepsilon > 0$. We shall say that an $A$-valued sequence $\{x_k\}_{k \in \mathbb{N}}$ has Property $P_n$, for some integer $n \in \mathbb{N}$, if $x_k \in (B_{\frac{1}{k}}(x) \backslash \{x\})$ for each $k = 1, \dots, n{+}1$ and $d(x_{k+1}, x) < d(x_k, x)$ for every $k = 1, \dots, n$.

*Claim.*  If assertion (b) holds true, then there exists an $A$-valued sequence that has Property $P_n$ for every $n \in \mathbb{N}$.

*Proof.*  Suppose (b) holds true so that $(B_\varepsilon(x) \backslash \{x\}) \cap A \neq \varnothing$ for every $\varepsilon > 0$. Take an arbitrary $x_1 \in (B_1(x) \backslash \{x\}) \cap A$ and an arbitrary $x_2 \in (B_{\varepsilon_2}(x) \backslash \{x\}) \cap A$ where $\varepsilon_2 = \min\{\frac{1}{2}, d(x_1, x)\}$. Every $A$-valued sequence whose first two entries coincide with $x_1$ and $x_2$ has Property $P_1$. Suppose there exists an $A$-valued sequence that has Property $P_n$ for some integer $n \in \mathbb{N}$. Take any point from $(B_{\varepsilon_{n+2}}(x) \backslash \{x\}) \cap A$ where

$\varepsilon_{n+2} = \min\{\frac{1}{n+2}, d(x_{n+1}, x)\}$, and replace the $(n+2)$th entry of that sequence with this point. The resulting sequence has Property $P_{n+1}$. Thus there exists an $A$-valued sequence that has Property $P_{n+1}$ whenever there exists one that has Property $P_n$, and this concludes the proof by induction. $\square$

However, an $A$-valued sequence $\{x_k\}_{k\in\mathbb{N}}$ that has Property $P_n$ for every $n \in \mathbb{N}$ in fact is an $A\setminus\{x\}$-valued sequence of pairwise distinct points such that $0 < d(x_k, x) < \frac{1}{k}$ for every $k \in \mathbb{N}$. Therefore (b)$\Rightarrow$(d). $\qquad\square$

Recall that "point of adherence" and "point of accumulation" are concepts defined for sets, while "limit of a convergent sequence" is, of course, a concept defined for sequences. But the range of a sequence is a set, and it can have (many) accumulation points. Let $(X, d)$ be a metric space and let $\{x_n\}$ be an $X$-valued sequence. A point $x$ in $X$ is a *cluster point of the sequence* $\{x_n\}$ if some subsequence of $\{x_n\}$ converges to $x$. The cluster points of a sequence are precisely the accumulation points of its range (Proposition 3.28). If a sequence is convergent, then (Proposition 3.5) its range has only one point of accumulation which coincides with the unique limit of the sequence.

**Corollary 3.29.** *The derived set* $A^\star$ *of every subset* $A$ *of a metric space* $(X, d)$ *is closed in* $(X, d)$.

*Proof.* Let $A$ be an arbitrary subset of a metric space $(X, d)$. We want to show that $(A^\star)^- = A^\star$ (i.e., $A^\star$ is closed) or, equivalently, $(A^\star)^- \subseteq A^\star$ (recall: every set is included in its closure). If $A$ is empty, then the result is trivially verified ($\varnothing^\star = \varnothing = \varnothing^-$). Thus suppose $A$ is nonempty. Take an arbitrary $x^- \in (A^\star)^-$ and an arbitrary $\varepsilon > 0$. Proposition 3.27 ensures that $B_\varepsilon(x^-) \cap A^\star \neq \varnothing$. Take $x^\star \in B_\varepsilon(x^-) \cap A^\star$ and set $\delta = \varepsilon - d(x^\star, x^-)$. Note that $0 < \delta \leq \varepsilon$ (for $0 \leq d(x^\star, x^-) < \varepsilon$). Since $x^\star \in A^\star$, it follows by Proposition 3.28 that $B_\delta(x^\star) \cap A$ contains infinitely many points. Take $x \in B_\delta(x^\star) \cap A$ distinct from $x^-$ and from $x^\star$. Thus $0 < d(x, x^-) \leq d(x, x^\star) + d(x^\star, x^-) < \delta + d(x^\star, x^-) = \varepsilon$ by the triangle inequality. Therefore $x \in B_\varepsilon(x^-)$ and $x \neq x^-$. Conclusion: Every nonempty ball $B_\varepsilon(x^-)$ centered at $x^-$ contains a point $x$ of $A$ other than $x^-$. Thus $x^- \in A^\star$ by Proposition 3.28, and therefore $(A^\star)^- \subseteq A^\star$. $\qquad\square$

The above corollary does not hold in a general topological space. Indeed, if a set $X$ containing more than one point is equipped with the indiscrete topology (where the only open sets are $\varnothing$ and $X$), then the derived set $\{x\}^\star$ of a singleton $\{x\}$ is $X\setminus\{x\}$ which is not closed in that topology.

**Theorem 3.30.** (The Closed Set Theorem). *A subset* $A$ *of a set* $X$ *is closed in the metric space* $(X, d)$ *if and only if every* $A$-*valued sequence that converges in* $(X, d)$ *has its limit in* $A$.

*Proof.* (a) Take an arbitrary $A$-valued sequence $\{x_n\}$ that converges to $x \in X$ in $(X, d)$. By Theorem 3.14 $\{x_n\}$ is eventually in every neighborhood of $x$, and hence every neighborhood of $x$ contains a point of $A$. Thus $x$ is a point of adherence of

$A$ (Proposition 3.25); that is, $x \in A^-$. If $A = A^-$ (equivalently, if $A$ is closed in $(X, d)$), then $x \in A$.

(b) Take an arbitrary point $x \in A^-$ (i.e., an arbitrary point of adherence of $A$). According to Proposition 3.27, there exists an $A$-valued sequence that converges to $x$ in $(X, d)$. If every $A$-valued sequence that converges in $(X, d)$ has its limit in $A$, then $x \in A$. Thus $A^- \subseteq A$, and hence $A = A^-$ (for $A \subseteq A^-$ for every set $A$). That is, $A$ is closed in $(X, d)$. $\qquad\qquad\square$

This is a particularly useful result that will be often applied throughout this book. Note that part (a) of the proof holds for general topological spaces but not part (b). The counterpart of the above theorem for general (not necessarily metrizable) topological spaces is stated in terms of nets (instead of sequences).

**Example 3N.** Consider the set $B[X, Y]$ of all bounded mappings of a metric space $(X, d_X)$ into a metric space $(Y, d_Y)$, and let $BC[X, Y]$ denote the subset of $B[X, Y]$ consisting of all bounded continuous mappings of $(X, d_X)$ into $(Y, d_Y)$. Equip $B[X, Y]$ with the sup-metric $d_\infty$ as in Example 3C. We shall use the Closed Set Theorem to show that

$$BC[X, Y] \text{ is closed in } (B[X, Y], d_\infty).$$

Take an arbitrary $BC[X, Y]$-valued sequence $\{f_n\}$ that converges in $(B[X, Y], d_\infty)$ to a mapping $f \in B[X, Y]$. The triangle inequality in $(Y, d_Y)$ ensures that

$$d_Y(f(u), f(v)) \leq d_Y(f(u), f_n(u)) + d_Y(f_n(u), f_n(v)) + d_Y(f_n(v), f(v))$$

for each integer $n$ and every $u, v \in X$. Take an arbitrary real number $\varepsilon > 0$. Since $f_n \to f$ in $(B[X, Y], d_\infty)$, it follows that there exists a positive integer $n_\varepsilon$ such that $d_\infty(f_n, f) = \sup_{x \in X} d_Y(f_n(x), f(x)) < \frac{\varepsilon}{3}$, and hence $d_Y(f_n(x), f(x)) < \frac{\varepsilon}{3}$ for all $x \in X$, whenever $n \geq n_\varepsilon$ (uniform convergence — see Example 3G). Since each $f_n$ is continuous, it follows that there exists a real number $\delta_\varepsilon > 0$ (which may depend on $u$ and $v$) such that $d_Y(f_{n_\varepsilon}(u), f_{n_\varepsilon}(v)) < \frac{\varepsilon}{3}$ whenever $d_X(u, v) < \delta_\varepsilon$. Therefore $d_Y(f(u), f(v)) < \varepsilon$ whenever $d_X(u, v) < \delta_\varepsilon$, so that $f$ is continuous. That is, $f \in BC[X, Y]$. Thus, according to Theorem 3.30, $BC[X, Y]$ is a closed subset of the metric space $(B[X, Y], d_\infty)$. Particular case (see Examples 3D and 3G): $C[0, 1]$ is closed in $(B[0, 1], d_\infty)$.

## 3.6   Dense Sets and Separable Spaces

Let $A$ be a set in a metric space $X$ and let $\mathcal{U}_A$ be the collection of all open subsets of $X$ included in $A$:

$$\mathcal{U}_A = \big\{U \in \wp(X) : \; U \text{ is open in } X \text{ and } U \subseteq A\big\}.$$

The empty set $\varnothing$ of $X$ always belongs to $\mathcal{U}_A$ so that $\mathcal{U}_A$ is never empty. The union of all sets in $\mathcal{U}_A$ is called the *interior* of $A$ in $X$, denoted by $A^\circ$ (i.e., $A^\circ = \bigcup \mathcal{U}_A$). According to Theorem 3.15(c) it follows that

$$A^\circ \text{ is open in } X \quad \text{and} \quad A^\circ \subseteq A.$$

If $U \in \mathcal{U}_A$, then $U \subseteq \bigcup \mathcal{U}_A = A^\circ$. Thus, with respect to the inclusion ordering of $\wp(X)$,

$$A^\circ \text{ is the largest open subset of } X \text{ that is included in } A,$$

and hence (since $A^\circ$ is open in $X$)

$$A \text{ is open in } X \quad \text{if and only if} \quad A^\circ = A.$$

From the above displayed results it is readily verified that

$$\varnothing^\circ = \varnothing, \quad X^\circ = X, \quad (A^\circ)^\circ = A^\circ$$

and, if $B$ also is a set in $X$,

$$A \subseteq B \quad \text{implies} \quad A^\circ \subseteq B^\circ.$$

Moreover, since $A \cap B$ is a subset of both $A$ and $B$, it follows that $(A \cap B)^\circ \subseteq A^\circ \cap B^\circ$. On the other hand, since $(A \cap B)^\circ$ is the largest open subset of $X$ that is included in $A \cap B$, and since $A^\circ \cap B^\circ$ is open (Theorem 3.15(b)) and is included in $A \cap B$ (for $A^\circ \subseteq A$ and $B^\circ \subseteq B$ so that $A^\circ \cap B^\circ \subseteq A \cap B$), it follows that $A^\circ \cap B^\circ \subseteq (A \cap B)^\circ$. Therefore, if $A$ and $B$ are subsets of $X$, then

$$(A \cap B)^\circ = A^\circ \cap B^\circ.$$

It is easy to show by induction that the above identity holds for any finite collection of subsets of $X$. That is, *the interior of the intersection of a finite collection of subsets of $X$ coincides with the intersection of their interiors.* In general (i.e., by allowing infinite collections as well) one has inclusion rather than equality. Indeed, if $\{A_\gamma\}_{\gamma \in \Gamma}$ is an arbitrary indexed family of subsets of $X$, then

$$\left( \bigcap_\gamma A_\gamma \right)^\circ \subseteq \bigcap_\gamma A_\gamma^\circ$$

because $\bigcap_\gamma A_\gamma \subseteq A_\alpha$ and hence $\left( \bigcap_\gamma A_\gamma \right)^\circ \subseteq A_\alpha$ for each index $\alpha \in \Gamma$. Similarly,

$$\bigcup_\gamma A_\gamma^\circ \subseteq \left( \bigcup_\gamma A_\gamma \right)^\circ$$

because $\bigcup_\gamma A_\gamma^\circ \subseteq \bigcup_\gamma A_\gamma$ and $\bigcup_\gamma A_\gamma^\circ$ is open in $X$ by Theorem 3.15(c). However, the above two inclusions are not reversible in general so that equality does not hold.

**Example 3O.** This is the dual of Example 3M. Consider the setup of Example 3M and set $C_n = X \backslash A_n$, which is open in $\mathbb{R}$ for each positive integer $n$, and $C = X \backslash A$, which is not open in $\mathbb{R}$. Since

$$\bigcap_{n=1}^{\infty} C_n = \bigcap_{n=1}^{\infty} (X \backslash A_n) = X \backslash \bigcup_{n=1}^{\infty} A_n = X \backslash A = C,$$

it follows that *the intersection of an infinite collection of open sets is not necessarily open* (see Theorem 3.15(b)). In particular, as $C_n^{\circ} = C_n$,

$$(X \backslash A)^{\circ} = C^{\circ} = \left( \bigcap_{n=1}^{\infty} C_n \right)^{\circ} \subset \bigcap_{n=1}^{\infty} C_n^{\circ} = C = X \backslash A$$

which is a *proper* inclusion. Now set $D = X \backslash B = (-\infty, 1) \cup (2, \infty)$ (so that $D^{\circ} = D$). Thus $C^{\circ} \cup D^{\circ}$ is a *proper* subset of $(C \cup D)^{\circ}$:

$$\mathbb{R} \backslash \{1\} = C^{\circ} \cup D^{\circ} \subset (C \cup D)^{\circ} = \mathbb{R}.$$

Remark: The duality between "interior" and "closure" is clear:

$$(X \backslash A)^{-} = X \backslash A^{\circ} \quad \text{and} \quad (X \backslash A)^{\circ} = X \backslash A^{-}$$

for every subset $A$ of $X$. Indeed, $U \in \mathcal{U}_A$ if and only if $X \backslash U \in \mathcal{V}_{X \backslash A}$ (i.e., $U$ is open in $X$ and $U \subseteq A$ if and only if $X \backslash U$ is closed in $X$ and $X \backslash A \subseteq X \backslash U$) and, dually, $V \in \mathcal{V}_{X \backslash A}$ if and only if $X \backslash V \in \mathcal{U}_A$. Thus $A^{\circ} = \bigcup_{U \in \mathcal{U}_A} U = X \backslash \bigcap_{U \in \mathcal{U}_A} (X \backslash U) = X \backslash \bigcap_{V \in \mathcal{V}_{X \backslash A}} V = X \backslash (X \backslash A)^{-}$ and so $X \backslash A^{\circ} = (X \backslash A)^{-}$; which implies (swap $A$ and $X \backslash A$) that $X \backslash (X \backslash A)^{\circ} = A^{-}$ and hence $(X \backslash A)^{\circ} = X \backslash A^{-}$. This confirms the above identities and also their equivalent forms:

$$A^{\circ} = X \backslash (X \backslash A)^{-} \quad \text{and} \quad A^{-} = X \backslash (X \backslash A)^{\circ}.$$

A point $x \in X$ is an *interior point* of $A$ if it belongs to the interior $A^{\circ}$ of $A$. It is clear that every interior point of $A$ is a point of $A$ (i.e., $A^{\circ} \subseteq A$), and it is readily verified that $x \in A$ *is an interior point of $A$ if and only if there exists a neighborhood of $x$ included in $A$* (reason: $A^{\circ}$ is the largest open neighborhood of every interior point of $A$ that is included in $A$).

A subset $A$ of a metric space $X$ is called *dense* in $X$ (or *dense everywhere*) if its closure $A^{-}$ coincides with $X$ (i.e., if $A^{-} = X$). More generally, suppose $A$ and $B$ are subsets of a metric space $X$ such that $A \subseteq B$. $A$ is *dense in $B$* if $B \subseteq A^{-}$ or, equivalently, if $A^{-} = B^{-}$. (Why?) Clearly, if $A \subseteq B$ and $A^{-} = X$, then $B^{-} = X$. Note that the only closed set dense in $X$ is $X$ itself.

**Proposition 3.31.** *Let $A$ be a subset of a metric space $X$. The following assertions are pairwise equivalent.*

(a)  $A^- = X$    (i.e., *A is dense in X*).

(b)  *Every nonempty open subset of X meets A.*

(c)  $\mathcal{V}_A = \{X\}$.

(d)  $(X\backslash A)^\circ = \varnothing$    (i.e., *the complement of A has empty interior*).

*Proof.* Take any nonempty open subset $U$ of $X$, and take an arbitrary $u \in U \subseteq X$. If (a) holds true, then every point of $X$ is adherent to $A$. In particular, $u$ is adherent to $A$. Thus Proposition 3.25 ensures that $U$ meets $A$. Conclusion: (a)$\Rightarrow$(b). Now take an arbitrary proper closed subset $V$ of $X$, so that $\varnothing \neq X\backslash V$ is open in $X$. If (b) holds true, then $(X\backslash V) \cap A \neq \varnothing$. Thus $V$ does not include $A$, and hence $V \notin \mathcal{V}_A$. Therefore (b)$\Rightarrow$(c). Since $A^-$ surely belongs to $\mathcal{V}_A$, it follows that (c)$\Rightarrow$(a). The equivalence (a)$\Leftrightarrow$(d) is obvious from the identity $A^- = X\backslash(X\backslash A)^\circ$.    $\square$

The reader has probably observed that the concepts and results so far in this section apply to topological spaces in general. From now on the metric will play its role. Note that *a point in a subset A of a metric space X is an interior point of A if and only if it is the center of a nonempty open ball included in A* (reason: every nonempty open ball is a neighborhood and every neighborhood includes a nonempty open ball). We shall say that $(A, d)$ is a *dense subspace* of a metric space $(X, d)$ if the subset $A$ of $X$ is dense in $(X, d)$.

**Proposition 3.32.** *Let* $(X, d)$ *be a metric space and let A and B be subsets of X such that* $\varnothing \neq A \subseteq B \subseteq X$. *The following assertions are pairwise equivalent.*

(a)  $A^- = B^-$    (i.e., *A is dense in B*).

(b)  *Every nonempty open ball centered at any point b of B meets A.*

(c)  $\inf_{a \in A} d(b, a) = 0$ *for every* $b \in B$.

(d)  *For every point b in B there exists an A-valued sequence* $\{a_n\}$ *that converges in* $(X, d)$ *to b.*

*Proof.* Recall that $A^- = B^-$ if and only if $B \subseteq A^-$. Let $b$ be an arbitrary point in $B$. Thus assertion (a) can be rewritten as

(a$'$)  every point $b$ in $B$ is a point of adherence of $A$.

Now notice that assertions (a$'$), (a), (b) and (c) are pairwise equivalent by Proposition 3.27.    $\square$

**Corollary 3.33.** *Let F and G be continuous mappings of a metric space X into a metric space Y. If F and G coincide on a dense subset of X, then they coincide on the whole space X.*

*Proof.* Suppose $X$ is nonempty, to avoid trivialities, and let $A$ be a nonempty dense subset of $X$. Take an arbitrary $x \in X$ and let $\{a_n\}$ be an $A$-valued sequence that converges in $X$ to $x$ (whose existence is ensured by Proposition 3.32). If $F \colon X \to Y$ and $G \colon X \to Y$ are continuous mappings such that $F|_A = G|_A$, then

$$F(x) = F(\lim a_n) = \lim F(a_n) = \lim G(a_n) = G(\lim a_n) = G(x)$$

(Corollary 3.8). Thus $F(x) = G(x)$ for every $x \in X$; that is, $F = G$. $\qquad\square$

A metric space $(X, d)$ is *separable* if there exists a countable dense set in $X$. The density criteria in Proposition 3.32 (with $B = X$) are particularly useful to check separability.

**Example 3P.** Take an arbitrary integer $n \geq 1$, an arbitrary real $p \geq 1$, and consider the metric space $(\mathbb{R}^n, d_p)$ of Example 3A. Since the set of all rational numbers $\mathbb{Q}$ is dense in the real line $\mathbb{R}$ equipped with its usual metric, it follows that $\mathbb{Q}^n$ (the set of all rational $n$-tuples) is dense in $(\mathbb{R}^n, d_p)$. Indeed, $\mathbb{Q}^- = \mathbb{R}$ implies that $\inf_{v \in \mathbb{Q}} |\xi - v| = 0$ for every $\xi \in \mathbb{R}$, which in turn implies that

$$\inf_{y \in \mathbb{Q}^n} d_p(x, y) = \inf_{y=(v_1,\ldots,v_n) \in \mathbb{Q}^n} \left( \sum_{i=1}^{n} |\xi_i - v_i|^p \right)^{\frac{1}{p}} = 0$$

for every $x = (\xi_1, \ldots, \xi_n) \in \mathbb{R}^n$. Hence $(\mathbb{Q}^n)^- = \mathbb{R}^n$, according to Proposition 3.32. Moreover, since $\#\mathbb{Q}^n = \#\mathbb{Q} = \aleph_0$ (Problems 1.25(c) and 2.8), it follows that $\mathbb{Q}^n$ is countably infinite. Therefore $\mathbb{Q}^n$ is a countable dense subset of $(\mathbb{R}^n, d_p)$, and hence

$$(\mathbb{R}^n, d_p) \text{ is a separable metric space.}$$

Now consider the metric space $(\ell_+^p, d_p)$ for any $p \geq 1$ as in Example 3B, where $\ell_+^p$ is the set of all *real-valued* $p$-summable infinite sequences. Let $\ell_+^0$ be the subset of $\ell_+^p$ made up of all real-valued infinite sequences with a finite number of nonzero entries, and let $X$ be the subset of $\ell_+^0$ consisting of all rational-valued infinite sequences with a finite number of nonzero entries. The set $\ell_+^0$ is dense in $(\ell_+^p, d_p)$ — Problem 3.44(c). Since $\mathbb{Q}^- = \mathbb{R}$, it follows that $X$ is dense in $(\ell_+^0, d_p)$ — the proof is essentially the same as the proof that $\mathbb{Q}^n$ is dense in $(\mathbb{R}^n, d_p)$. Thus $X^- = (\ell_+^0)^- = \ell_+^p$, and hence $X$ is dense in $(\ell_+^p, d_p)$. Next we show that $X$ is countably infinite. In fact, $X$ is a linear space over the rational field $\mathbb{Q}$ for which $\dim X = \aleph_0$ (see Example 2J). Thus, according to Problem 2.8, $\#X = \max\{\#\mathbb{Q}, \dim X\} = \aleph_0$. Conclusion: $X$ is a countable dense subset of $(\ell_+^p, d_p)$, and so

$$(\ell_+^p, d_p) \text{ is a separable metric space.}$$

The same argument is readily extended to complex spaces so that $(\mathbb{C}^n, d_p)$ also is separable, as well as $(\ell_+^p, d_p)$ when $\ell_+^p$ is made up of all *complex-valued* $p$-summable infinite sequences. Finally we show that

$$(C[0, 1], d_\infty) \text{ is a separable metric space}$$

(see Example 3D). Actually, the set $P[0, 1]$ of all polynomials on $[0, 1]$ is dense in $(C[0, 1], d_\infty)$. This is the well-known *Weierstrass Theorem*, which says that *every continuous function in $C[0, 1]$ is the uniform limit of a sequence of polynomials in $P[0, 1]$* (i.e., for every $x \in C[0, 1]$ there exists a $P[0, 1]$-valued sequence $\{p_n\}$ such that $d_\infty(p_n, x) \to 0$). Moreover, it is easy to show that the set $X$ of all polynomials on $[0, 1]$ with rational coefficients is dense in $(P[0, 1], d_\infty)$, and hence $X$ is dense in $(C[0, 1], d_\infty)$. Since $X$ is a linear space over the rational field $\mathbb{Q}$, and since dim $X = \aleph_0$ (essentially the same proof as that in Example 2M), it follows by Problem 2.8 that $X$ is countable. Thus $X$ is a countable dense subset of $(C[0, 1], d_\infty)$.

A collection $\mathcal{B}$ of open subsets of a metric space $X$ is a *base* (or a *topological base*) for $X$ if every open set in $X$ is the union of some subcollection of $\mathcal{B}$. For instance, the collection of all open balls in a metric space (including the empty ball) is a base for $X$ (cf. Corollary 3.16). Note that the above definition of base forces the empty set $\varnothing$ of $X$ to be a member of any base for $X$.

**Proposition 3.34.** *Let $X$ be a metric space and let $\mathcal{B}$ be a collection of open subsets of $X$ that contains the empty set. The following assertions are pairwise equivalent.*

(a) *$\mathcal{B}$ is a base for $X$.*

(b) *For every nonempty open subset $U$ of $X$ and every point $x$ in $U$, there exists a set $B$ in $\mathcal{B}$ such that $x \in B \subseteq U$.*

(c) *For every $x$ in $X$ and every neighborhood $N$ of $x$, there exists a set $B$ in $\mathcal{B}$ such that $x \in B \subseteq N$.*

*Proof.* Take an arbitrary open subset $U$ of the metric space $X$ and set $\mathcal{B}_U = \{B \in \mathcal{B}: B \subseteq U\}$. If $\mathcal{B}$ is a base for $X$, then $U = \bigcup \mathcal{B}_U$ by definition of base. Thus, if $x \in U$, then $x \in B$ for some $B \in \mathcal{B}_U$ so that $x \in B \subseteq U$. That is, (a) implies (b). On the other hand, if (b) holds, then any open subset $U$ of $X$ clearly coincides with $\bigcup \mathcal{B}_U$, which shows that (a) holds true. Finally note that (b) and (c) are trivially equivalent: every neighborhood of $x$ includes an open set containing $X$, and every open set containing $x$ is a neighborhood of $x$. $\qquad \square$

**Theorem 3.35.** *A metric space $(X, d)$ is separable if and only if it has a countable base.*

*Proof.* Suppose $\mathcal{B}$ is a countable base for $X$. For each nonempty set $B_n$ in $\mathcal{B}$ take an arbitrary point $b_n$ in $B_n$. Now consider the set $\{b_n\}$ consisting of one point of each nonempty set $B_n$ in $\mathcal{B}$. The set $\{b_n\}$ is countable and every nonempty open subset $U$ of $X$ meets $\{b_n\}$ (since $U$ is the union of some subcollection of $\{B_n\}$ so that $U \cap \{b_n\} \neq \varnothing$). Therefore (cf. Proposition 3.31) $\{b_n\}$ is a countable dense subset of $X$, and hence $X$ is separable. On the other hand, suppose $X$ is separable. Then there exists a countable subset $A$ of $X$ that is dense in $X$. Consider the collection of nonempty open balls $\mathcal{B} = \{B_{\frac{1}{n}}(a): n \in \mathbb{N}$ and $a \in A\}$. Observe that $\mathcal{B} = \bigcup_{a \in A}\{B_{\frac{1}{n}}(a)\}_{n \in \mathbb{N}}$, a countable union of countable collections, so that $\mathcal{B}$ is itself a countable collection (Corollary 1.11).

*Claim.* For every $x \in X$ and every neighborhood $N$ of $x$ there exists a ball in $\mathcal{B}$ containing $x$ and included in $N$.

*Proof.* Take an arbitrary $x \in X$ and an arbitrary neighborhood $N$ of $x$. Let $B_\varepsilon(x)$ be an open ball of radius $\varepsilon > 0$, centered at $x$, and included in $N$. Now take a positive integer $n$ such that $\frac{1}{n} < \frac{\varepsilon}{2}$, and a point $a \in A$ such that $a \in B_{\frac{1}{n}}(x)$ (recall: since $A^- = X$, it follows by Proposition 3.32 that there exists $a \in A$ such that $a \in B_\rho(x)$ for every $x \in X$ and every $\rho > 0$). Obviously, $x \in B_{\frac{1}{n}}(a)$. Moreover, $B_{\frac{1}{n}}(a) \subseteq B_\varepsilon(x)$. Indeed, if $y \in B_{\frac{1}{n}}(a)$, then $d(x, y) \le d(x, a) + d(a, y) < \frac{2}{n} < \varepsilon$ so that $y \in B_\varepsilon(x)$. Thus $x \in B_{\frac{1}{n}}(a) \subseteq B_\varepsilon(x) \subseteq N$. $\square$

Therefore the countable collection $\mathcal{B} \cup \{\varnothing\}$ of open balls is a base for $X$ by Proposition 3.34. $\qquad\square$

**Corollary 3.36.** *Every subspace of a separable metric space is itself separable.*

*Proof.* Let $S$ be a subspace of a separable metric space $X$ and, according to Theorem 3.35, let $\mathcal{B}$ be a countable base for $X$. Set $\mathcal{B}_S = \{S \cap B: B \in \mathcal{B}\}$, which is a countable collection of subsets of $S$. Since the sets in $\mathcal{B}$ are open subsets of $X$, it follows that the sets in $\mathcal{B}_S$ are open relative to $S$ (see Problem 3.38(c)). Take an arbitrary nonempty relatively open subset $A$ of $S$ so that $A = S \cap U$ for some open subset $U$ of $X$ (Problem 3.38(c)). Since $U = \bigcup \mathcal{B}'$ for some subcollection $\mathcal{B}'$ of $\mathcal{B}$, it follows that $A = S \cap \bigcup_{B \in \mathcal{B}'} B = \bigcup_{B \in \mathcal{B}'} S \cap B = \bigcup \mathcal{B}'_S$, where $\mathcal{B}'_S - \{S \cap B: B \in \mathcal{B}'\}$ is a subcollection of $\mathcal{B}_S$. Thus $\mathcal{B}_S$ is a base for $S$. Therefore the subspace $S$ has a countable base, which means by the previous theorem that $S$ is separable. $\qquad\square$

Let $A$ be a subset of a metric space. An *isolated point* of $A$ is a point in $A$ that is not an accumulation point of $A$. That is, a point $x$ is an isolated point of $A$ if $x \in A \backslash A^\star$.

**Proposition 3.37.** *Let $A$ be a subset of a metric space $X$ and let $x$ be a point in $A$. The following assertions are pairwise equivalent.*

(a)  *$x$ is an isolated point of $A$.*

(b)  *There exists an open set $U$ in $X$ such that $A \cup U = \{x\}$.*

(c)  *There exists a neighborhood $N$ of $x$ such that $A \cup N = \{x\}$.*

(d)  *There exists a nonempty open ball $B_\rho(x)$ centered at $x$ such that $A \cup B_\rho(x) = \{x\}$.*

*Proof.* Assertion (a) is equivalent to assertion (b) by Proposition 3.26. Assertions (b), (c) and (d) are trivially pairwise equivalent.    □

A subset $A$ of $X$ consisting entirely of isolated points is a *discrete subset* of $X$. This means that in the subspace $A$ every set is open, and hence the subspace $A$ is homeomorphic to a discrete space (i.e., to a metric space equipped with the discrete metric). According to Theorem 3.35 and Corollary 3.36, *a discrete subset of a separable metric space is countable.* Thus, *if a metric space has an uncountable discrete subset, then it is not separable.*

**Example 3Q.** Let $S$ be a set, let $(Y, d)$ be a metric space, and consider the metric space $(B[S, Y], d_\infty)$ of all bounded mappings of $S$ into $(Y, d)$ equipped with the sup-metric $d_\infty$ (see Example 3C). Suppose $Y$ has more than one point, and let $y_0$ and $y_1$ be two distinct points in $Y$. As usual, let $2^S$ denote the set of all mappings on $S$ with values either $y_0$ or $y_1$ (i.e., the set of all mappings of $S$ into $\{y_0, y_1\}$ so that $2^S = \{y_0, y_1\}^S \subseteq B[S, Y]$). If $f, g \in 2^S$ and $f \neq g$ (i.e., if $f$ and $g$ are two distinct mappings on $S$ with values either $y_0$ or $y_1$), then

$$d_\infty(f, g) = \sup_{s \in S} d(f(s), g(s)) = d(y_0, y_1) \neq 0.$$

Therefore, any open ball $B_\rho(g) = \{f \in 2^S : d_\infty(f, g) < \rho\}$ centered at an arbitrary point $g$ of $2^S$ with radius $\rho = d(y_0, y_1)/2$ is such that $2^S \cap B_\rho(g) = \{g\}$. This means that every point of $2^S$ is an isolated point of it, and hence

$$2^S \text{ is a discrete set in } (B[S, Y], d_\infty).$$

If $S$ is an infinite set, then $2^S$ is an uncountable subset of $B[S, Y]$ (recall: if $S$ is infinite, then $\aleph_0 \leq \#S < \#2^S$ according to Theorems 1.4 and 1.5). Thus $(B[S, Y], d_\infty)$ *is not separable whenever* $2 \leq \#Y$ *and* $\aleph_0 \leq \#S$. Concrete example:

$$(\ell_+^\infty, d_\infty) \text{ is not a separable metric space.}$$

Indeed, set $S = \mathbb{N}$ and $Y = \mathbb{C}$ (or $Y = \mathbb{R}$) with its usual metric $d$, so that $(B[S, Y], d_\infty) = (\ell_+^\infty, d_\infty)$: the set of all scalar-valued bounded sequences equipped with the sup-metric, as introduced in Example 3B. The set $2^\mathbb{N}$, consisting of all sequences with values either 0 or 1, is an uncountable discrete subset of $(\ell_+^\infty, d_\infty)$.

In a discrete subset every point is isolated. The opposite notion is that of a set where every point is not isolated. A subset $A$ of a metric space $X$ is *dense in itself* if $A$ has no isolated point or, equivalently, if every point in $A$ is an accumulation point of $A$. That is, if $A \subseteq A^*$. Since $A^- = A \cup A^*$ for every subset $A$ of $X$, it follows that a set $A$ is dense in itself if and only if $A^* = A^-$. A subset $A$ of $X$ that is both closed in $X$ and dense in itself (i.e., such that $A^* = A$) is a *perfect set*: a closed set without isolated points. For instance, $\mathbb{Q} \cap [0, 1]$ is a countable perfect subset of the metric space $\mathbb{Q}$, but it is *not* perfect in the metric space $\mathbb{R}$ (since it is not closed in $\mathbb{R}$). As a matter of fact, every nonempty perfect subset of $\mathbb{R}$ is uncountable because $\mathbb{R}$ is a "complete" metric space, a concept that we shall define next.

# 3.7   Complete Spaces

Consider the metric space $(\mathbb{R}, d)$, where $d$ denotes the usual metric on the real line $\mathbb{R}$, and let $(A, d)$ be the subspace of $(\mathbb{R}, d)$ with $A = (0, 1]$. Let $\{\alpha_n\}_{n \in \mathbb{N}}$ be the $A$-valued sequence such that $\alpha_n = \frac{1}{n}$ for each $n \in \mathbb{N}$. Does $\{\alpha_n\}$ converge in the metric space $(A, d)$? It is clear that $\{\alpha_n\}$ converges to 0 in $(\mathbb{R}, d)$, and hence we might at a first glance think that it also converges in $(A, d)$. But the point 0 simply does not exist in $A$ so that it is nonsense to say that "$\alpha_n \to 0$ in $(A, d)$". In fact, $\{\alpha_n\}$ does not converge in the metric space $(A, d)$. However, the sequence $\{\alpha_n\}$ seems to possess a "special property" that makes it apparently convergent in spite of the particular underlying set $A$, and the metric space $(A, d)$ in turn seems to bear a "peculiar characteristic" that makes such a sequence fail to converge in it. The "special property" of the sequence $\{\alpha_n\}$ is that it is a *Cauchy sequence* in $(A, d)$ and the "peculiar characteristic" of the metric space $(A, d)$ is that it is not *complete*.

**Definition 3.38.** Let $(X, d)$ be a metric space. An $X$-valued sequence $\{x_n\}$ (indexed either by $\mathbb{N}$ or $\mathbb{N}_0$) is a *Cauchy sequence* in $(X, d)$ (or satisfies the *Cauchy criterion*) if for each real number $\varepsilon > 0$ there exists a positive integer $n_\varepsilon$ such that

$$n, m \geq n_\varepsilon \quad \text{implies} \quad d(x_m, x_n) < \varepsilon.$$

A usual notation for the Cauchy criterion is $\lim_{m,n} d(x_m, x_n) = 0$. Equivalently, an $X$-valued sequence $\{x_n\}$ is a Cauchy sequence if $\operatorname{diam}(\{x_k\}_{n \leq k}) \to 0$ as $n \to \infty$ (i.e., $\lim_n \operatorname{diam}(\{x_k\}_{n \leq k}) = 0$).

The basic facts about Cauchy sequences are stated in the proposition below. In particular, it shows that *every convergent sequence is bounded*, and that *a Cauchy sequence has a convergent subsequence if and only if every subsequence of it converges* (see Proposition 3.5).

**Proposition 3.39.** *Let $(X, d)$ be a metric space.*

(a) *Every convergent sequence in $(X, d)$ is a Cauchy sequence.*

(b) *Every Cauchy sequence in $(X, d)$ is bounded.*

(c) *If a Cauchy sequence in $(X, d)$ has a subsequence that converges in $(X, d)$, then it converges itself in $(X, d)$ and its limit coincides with the limit of that convergent subsequence.*

*Proof.* (a) Take an arbitrary $\varepsilon > 0$. If an $X$-valued sequence $\{x_n\}$ converges to a point $x \in X$, then there exists an integer $n_\varepsilon \geq 1$ such that $d(x_n, x) < \frac{\varepsilon}{2}$ whenever $n \geq n_\varepsilon$. Since $d(x_m, x_n) \leq d(x_m, x) + d(x, x_n)$ for every pair of indices $m, n$ (triangle inequality), it follows that $d(x_m, x_n) < \varepsilon$ whenever $m, n \geq n_\varepsilon$.

(b) If $\{x_n\}$ is a Cauchy sequence, then there exists an integer $n_1 > 1$ such that $d(x_m, x_n) < 1$ whenever $m, n \geq n_1$. Note that the set $\{d(x_m, x_n) \in \mathbb{R} : m, n < n_1\}$

has a maximum in $\mathbb{R}$, say $\beta$, because it is finite. Thus $d(x_m, x_n) \leq d(x_m, x_{n_1}) + d(x_{n_1}, x_n) \leq 2 \max\{1, \beta\}$ for every pair of indices $m, n$.

(c) Suppose $\{x_{n_k}\}$ is a subsequence of an $X$-valued Cauchy sequence $\{x_n\}$ that converges to a point $x \in X$ (i.e., $x_{n_k} \to x$ as $k \to \infty$). Take an arbitrary $\varepsilon > 0$. Since $\{x_n\}$ is a Cauchy sequence, it follows that there exists a positive integer $n_\varepsilon$ such that $d(x_m, x_n) < \frac{\varepsilon}{2}$ whenever $m, n \geq n_\varepsilon$. Since $\{x_{n_k}\}$ converges to $x$, it follows that there exists a positive integer $k_\varepsilon$ such that $d(x_{n_k}, x) < \frac{\varepsilon}{2}$ whenever $k \geq k_\varepsilon$. Thus, if $j$ is any integer with the property that $j \geq k_\varepsilon$ and $n_j \geq n_\varepsilon$ (for instance, if $j = \max\{n_\varepsilon, k_\varepsilon\}$), then $d(x_n, x) \leq d(x_n, x_{n_j}) + d(x_{n_j}, x) < \varepsilon$ for every $n \geq n_\varepsilon$, and therefore $\{x_n\}$ converges to $x$. $\qquad\square$

Although a convergent sequence always is a Cauchy sequence, the converse may fail. For instance, the $(0, 1]$-valued sequence $\{\frac{1}{n}\}_{n \in \mathbb{N}}$ is a Cauchy sequence in the metric space $((0, 1], d)$, where $d$ is the usual metric on $\mathbb{R}$, which does not converge in $((0, 1], d)$. There are however many metric spaces with the notable property that Cauchy sequences in it are convergent. Metric spaces possessing this property are so important that we give them a name. A metric space $X$ is *complete* if every Cauchy sequence in $X$ is a convergent sequence in $X$.

**Theorem 3.40.** *Let $A$ be a subset of a metric space $X$.*

(a) *If the subspace $A$ is complete, then $A$ is closed in $X$.*

(b) *If $X$ is complete and if $A$ is closed in $X$, then the subspace $A$ is complete.*

*Proof.* (a) Take an arbitrary $A$-valued sequence $\{a_n\}$ that converges in $X$. Since every convergent sequence is a Cauchy sequence, it follows that $\{a_n\}$ is a Cauchy sequence in $X$, and therefore a Cauchy sequence in the subspace $A$. If the subspace $A$ is complete, then $\{a_n\}$ converges in $A$. Conclusion: If $A$ is complete as a subspace of $X$, then every $A$-valued sequence that converges in $X$ has its limit in $A$. Thus, according to the Closed Set Theorem (Theorem 3.30), $A$ is closed in $X$.

(b) Take an arbitrary $A$-valued Cauchy sequence $\{a_n\}$. If $X$ is complete, then $\{a_n\}$ converges in $X$ to a point $a \in X$. If $A$ is closed in $X$, then Theorem 3.30 (the Closed Set Theorem again) ensures that $a \in A$, and hence $\{a_n\}$ converges in the subspace $A$. Conclusion: If $X$ is complete and $A$ is closed in $X$, then every Cauchy sequence in the subspace $A$ converges in $A$. That is, $A$ is complete as a subspace of $X$. $\qquad\square$

An important, although immediate, corollary of the above theorem says that "inside" a *complete metric space* the properties of being closed and complete coincide.

**Corollary 3.41.** *Let $X$ be a complete metric space. A subset $A$ of $X$ is closed in $X$ if and only if the subspace $A$ is complete.*

**Example 3R.** (a) A basic property of the real number system is that every bounded sequence of real numbers has a convergent subsequence. This and Proposition 3.39

ensure that the metric space $\mathbb{R}$ (equipped with its usual metric) is complete; and so is the metric space $\mathbb{C}$ of all complex numbers equipped with its usual metric (reason: if $\{\alpha_k\}$ is a Cauchy sequence in $\mathbb{C}$, then $\{\operatorname{Re}\alpha_k\}$ and $\{\operatorname{Im}\alpha_k\}$ are both Cauchy sequences in $\mathbb{R}$ so that they converge in $\mathbb{R}$, and hence $\{\alpha_k\}$ converges in $\mathbb{C}$). Since the set $\mathbb{Q}$ of all rational numbers is not closed in $\mathbb{R}$ (recall: $\mathbb{Q}^- = \mathbb{R}$), it follows by Corollary 3.41 that the metric $\mathbb{Q}$ is not complete More generally (but similarly), for every positive integer $n$,

$$\mathbb{R}^n \text{ and } \mathbb{C}^n \text{ are complete metric spaces}$$

when equipped with any of their metrics $d_p$ for $p \geq 1$ or $d_\infty$ (as in Example 3A); while

$$\mathbb{Q}^n \text{ is not a complete metric space.}$$

(b)   Now let $\mathbb{F}$ denote either the real field $\mathbb{R}$ or the complex field $\mathbb{C}$ equipped with their usual metrics. As we have just seen, $\mathbb{F}$ is a complete metric space. For each real number $p \geq 1$ let $(\ell_+^p, d_p)$ be the metric space of all $\mathbb{F}$-valued $p$-summable sequences equipped with its usual metric $d_p$ as in Example 3B. Take an arbitrary Cauchy sequence in $(\ell_+^p, d_p)$, say $\{x_n\}_{n\in\mathbb{N}}$. Recall that this is a sequence of sequences; that is, $x_n = \{\xi_n(k)\}_{k\in\mathbb{N}}$ is a sequence in $\ell_+^p$ for each integer $n \in \mathbb{N}$. The Cauchy criterion says: for every $\varepsilon > 0$ there exists an integer $n_\varepsilon \geq 1$ such that $d_p(x_m, x_n) < \varepsilon$ whenever $m, n \geq n_\varepsilon$. Thus

$$|\xi_m(k) - \xi_n(k)| \leq \left(\sum_{i=1}^\infty |\xi_m(i) - \xi_n(i)|^p\right)^{\frac{1}{p}} = d_p(x_m, x_n) < \varepsilon$$

for every $k \in \mathbb{N}$ whenever $m, n \geq n_\varepsilon$. Therefore, for each $k \in \mathbb{N}$ the scalar-valued sequence $\{\xi_n(k)\}_{n\in\mathbb{N}}$ is a Cauchy sequence in $\mathbb{F}$, and hence it converges in $\mathbb{F}$ (since $\mathbb{F}$ is complete) to, say, $\xi(k) \in \mathbb{F}$. Consider the scalar-valued sequence $x = \{\xi(k)\}_{k\in\mathbb{N}}$ consisting of those limits $\xi(k) \in \mathbb{F}$ for every $k \in \mathbb{N}$. First we show that $x \in \ell_+^p$. Since $\{x_n\}_{n\in\mathbb{N}}$ is a Cauchy sequence in $(\ell_+^p, d_p)$, it follows by Proposition 3.39 that it is bounded (i.e., $\sup_{m,n} d_p(x_m, x_n) < \infty$), and hence $\sup_m d_p(x_m, 0) < \infty$ where $0$ denotes the null sequence in $\ell_+^p$. (Indeed, for every $m \in \mathbb{N}$ the triangle inequality ensures that $d_p(x_m, 0) \leq \sup_{m,n} d_p(x_m, x_n) + d_p(x_n, 0)$ for an arbitrary $n \in \mathbb{N}$). Therefore,

$$\left(\sum_{k=1}^j |\xi_n(k)|^p\right)^{\frac{1}{p}} \leq \left(\sum_{k=1}^\infty |\xi_n(k)|^p\right)^{\frac{1}{p}} = d_p(x_n, 0) \leq \sup_m d_p(x_m, 0)$$

for every $n \in \mathbb{N}$ and each integer $j \geq 1$. Since $\xi_n(k) \to \xi(k)$ in $\mathbb{F}$ as $n \to \infty$ for each $k \in \mathbb{N}$, it follows that

$$\left(\sum_{k=1}^j |\xi(k)|^p\right)^{\frac{1}{p}} = \lim_n \left(\sum_{k=1}^j |\xi_n(k)|^p\right)^{\frac{1}{p}} \leq \sup_m d_p(x_m, 0)$$

for every $j \in \mathbb{N}$. Thus

$$\left(\sum_{k=1}^{\infty} |\xi(k)|^p\right)^{\frac{1}{p}} = \sup_{j}\left(\sum_{k=1}^{j} |\xi(k)|^p\right)^{\frac{1}{p}} \leq \sup_{m} d_p(x_m, 0),$$

which means that $x = \{\xi(k)\}_{k\in\mathbb{N}} \in \ell_+^p$. Next we show that $x_n \to x$ in $(\ell_+^p, d_p)$. Again, as $\{x_n\}_{n\in\mathbb{N}}$ is a Cauchy sequence in $(\ell_+^p, d_p)$, for any $\varepsilon > 0$ there exists an integer $n_\varepsilon \geq 1$ such that $d_p(x_m, x_n) < \varepsilon$ whenever $m, n \geq n_\varepsilon$. Thus

$$\sum_{k=1}^{j} |\xi_n(k) - \xi_m(k)|^p \leq \sum_{k=1}^{\infty} |\xi_n(k) - \xi_m(k)|^p < \varepsilon^p$$

for every integer $j \geq 1$ whenever $m, n \geq n_\varepsilon$. Since $\lim_m \xi_m(k) = \xi(k)$ for each $k \in \mathbb{N}$, it follows that $\sum_{k=1}^{j} |\xi_n(k) - \xi(k)|^p \leq \varepsilon^p$, and hence

$$d_p(x_n, x) = \left(\sum_{k=1}^{\infty} |\xi_n(k) - \xi(k)|^p\right)^{\frac{1}{p}} = \sup_{j}\left(\sum_{k=1}^{j} |\xi_n(k) - \xi(k)|^p\right)^{\frac{1}{p}} \leq \varepsilon,$$

whenever $n \geq n_\varepsilon$; which means that $x(n) \to x$ in $(\ell_+^p, d_p)$. Therefore

$$(\ell_+^p, d_p) \text{ is a complete metric space}$$

for every $p \geq 1$. Similarly (see Example 3B), for each $p \geq 1$,

$$(\ell^p, d_p) \text{ is a complete metric space.}$$

**Example 3S.** Let $S$ be a nonempty set, let $(Y, d)$ be a metric space, and consider the metric space $(B[S, Y], d_\infty)$ of all bounded mappings of $S$ into $(Y, d)$ equipped with the sup-metric $d_\infty$ (Example 3C). We claim that

$$(B[S, Y], d_\infty) \text{ is complete if and only if } (Y, d) \text{ is complete.}$$

(a) Indeed, if $\{f_n\}$ is a Cauchy sequence in $(B[S, Y], d_\infty)$, then $\{f_n(s)\}$ is a Cauchy sequence in $(Y, d)$ for every $s \in S$ (for $d(f_m(s), f_n(s)) \leq \sup_{s\in S} d(f_m(s), f_n(s)) = d_\infty(f_m, f_n)$ for each pair of integers $m, n$ and every $s \in S$), and hence $\{f_n(s)\}$ converges in $(Y, d)$ for every $s \in S$ whenever $(Y, d)$ is a complete metric space. For each $s \in S$ set $f(s) = \lim_n f_n(s)$ (i.e., $f_n(s) \to f(s)$ in $(Y, d)$), which defines a function $f$ of $S$ into $Y$. We shall show that $f \in B[S, Y]$ and that $f_n \to f$ in $(B[S, Y], d_\infty)$, thus proving that $(B[S, Y], d_\infty)$ is complete whenever $(Y, d)$ is complete. First note that, by the triangle inequality,

$$d(f(s), f(t)) \leq d(f(s), f_n(s)) + d(f_n(s), f_n(t)) + d(f_n(t), f(t))$$

for each positive integer $n$ and every pair of points $s, t$ in $S$. Now take an arbitrary $\varepsilon > 0$. Since $\{f_n\}$ is a Cauchy sequence in $(B[S, Y], d_\infty)$, it follows that there exists a positive integer $n_\varepsilon$ such that $d_\infty(f_m, f_n) = \sup_{s \in S} d(f_m(s), f_n(s)) < \varepsilon$, and hence

$$d(f_m(s), f_n(s)) \leq \varepsilon$$

for all $s \in S$, whenever $m, n \geq n_\varepsilon$. Moreover, since $f_m(s) \to f(s)$ in $(Y, d)$ for every $s \in S$, and since the metric is continuous (that is, $d(\cdot, y) \colon Y \to \mathbb{R}$ is a continuous function from the metric space $Y$ to the metric space $\mathbb{R}$ for every $y \in Y$), it also follows that $d(f(s), f_n(s)) = d(\lim_m f_m(s), f_n(s)) = \lim_m d(f_m(s), f_n(s))$ for each positive integer $n$ and every $s \in S$ (see Problem 3.14 or 3.34 and Corollary 3.8). Thus

$$d(f(s), f_n(s)) \leq \varepsilon$$

for all $s \in S$ whenever $n \geq n_\varepsilon$. Furthermore, as each $f_n$ lies in $B[S, Y]$, there exists a real number $\gamma_{n_\varepsilon}$ such that

$$\sup_{s,t \in S} d(f_{n_\varepsilon}(s), f_{n_\varepsilon}(t)) < \gamma_{n_\varepsilon}.$$

Summing up: For an arbitrary real number $\varepsilon > 0$ there exists a positive integer $n_\varepsilon$ such that

$$d(f(s), f(t)) \leq 2\varepsilon + \gamma_{n_\varepsilon}$$

for all $s, t \in S$, so that $f \in B[S, Y]$; and

$$d_\infty(f, f_n) = \sup_{s \in S} d(f(s), f_n(s)) \leq \varepsilon$$

whenever $n \geq n_\varepsilon$, so that $f_n \to f$ in $(B[S, Y], d_\infty)$.

(b) Conversely, take an arbitrary $Y$-valued sequence $\{y_n\}$. Suppose $S$ is nonempty and set $f_n(s) = y_n$ for each integer $n$ and all $s \in S$. This defines a sequence $\{f_n\}$ of constant mappings of $S$ into $Y$ which clearly lie in $B[S, Y]$ (a constant mapping is obviously bounded). Note that $d_\infty(f_m, f_n) = \sup_{s \in S} d(f_m(s), f_n(s)) = d(y_m, y_n)$ for every pair of integers $m, n$. Thus $\{f_n\}$ is a Cauchy sequence in $(B[S, Y], d_\infty)$ if and only if $\{y_n\}$ is a Cauchy sequence in $(Y, d)$. Moreover, $\{f_n\}$ converges in $(B[S, Y], d_\infty)$ if and only if $\{y_n\}$ converges in $(Y, d)$ (reason: if $d(y_n, y) \to 0$ for some $y \in Y$, then $d_\infty(f_n, f) \to 0$ where $f \in B[S, Y]$ is the constant mapping $f(s) = y$ for all $s \in S$; and if $d_\infty(f_n, f) \to 0$ for some $f \in B[S, Y]$, then $d(y_n, f(s)) = d(f_n(s), f(s))$ for each $n$ and every $s$, so that $d(y_n, f(s)) \to 0$ for all $s \in S$ — and hence $f$ must be a constant mapping). Now suppose $(Y, d)$ is not complete, which implies that there exists a Cauchy sequence in $(Y, d)$, say $\{y_n\}$, that fails to converge in $(Y, d)$. Thus the sequence $\{f_n\}$ of constant mappings $f_n(s) = y_n$ for each integer $n$ and all $s \in S$ is a Cauchy sequence in $(B[S, Y], d_\infty)$ that fails to converge in $(B[S, Y], d_\infty)$, and hence $(B[S, Y], d_\infty)$ is not complete. Conclusion: If $(B[S, Y], d_\infty)$ is complete, then $(Y, d)$ is complete.

(c) Concrete example: Set $S = \mathbb{N}$ or $S = \mathbb{Z}$ and $Y = \mathbb{F}$ (either the real field $\mathbb{R}$ or the complex field $\mathbb{C}$ equipped with their usual metric). Then

$$(\ell_+^\infty, d_\infty) \text{ and } (\ell^\infty, d_\infty) \text{ are complete metric spaces.}$$

**Example 3T.** Consider the set $B[X, Y]$ of all bounded mappings of a nonempty metric space $(X, d_X)$ into a metric space $(Y, d_Y)$ and equip it with the sup-metric $d_\infty$ as in the previous example. Let $BC[X, Y]$ be the set of all continuous mappings from $B[X, Y]$ (Example 3N), so that $(BC[X, Y], d_\infty)$ is the subspace of $(B[X, Y], d_\infty)$ made up of all bounded continuous mappings of $(X, d_X)$ into $(Y, d_Y)$. If $(Y, d_Y)$ is complete, then $(B[X, Y], d_\infty)$ is complete according to Example 3S. Since $BC[X, Y]$ is closed in $(B[X, Y], d_\infty)$ — Example 3N — it follows by Theorem 3.40 that $(BC[X, Y], d_\infty)$ is complete. On the other hand, the very same construction used in item (b) of the previous example shows that $(BC[X, Y], d_\infty)$ is not complete unless $(Y, d_Y)$ is. Conclusion:

$$(BC[X, Y], d_\infty) \text{ is complete if and only if } (Y, d_X) \text{ is complete.}$$

In particular (see Examples 3D, 3G and 3N),

$$(C[0, 1], d_\infty) \text{ is a complete metric space}$$

because the real line $\mathbb{R}$ or the complex plane $\mathbb{C}$ (equipped with their usual metrics, as always) are complete metric spaces (Example 3R). However, for any $p \geq 1$ (cf. Problem 3.58),

$$(C[0, 1], d_p) \text{ is not a complete metric space.}$$

The concept of completeness allows us to state and prove a useful result on contractions.

**Theorem 3.42.** (Contraction Mapping Theorem or Method of Successive Approximations). *A strict contraction $F$ of a nonempty complete metric space $(X, d)$ into itself has a unique fixed point $x \in X$, which is the limit in $(X, d)$ of every $X$-valued sequence of the form $\{F^n(x_0)\}_{n \in \mathbb{N}_0}$ for any $x_0 \in X$.*

*Proof.* Take an arbitrary point $x_0$ in $X$ and consider the $X$-valued sequence $\{x_n\}_{n \in \mathbb{N}_0}$ such that

$$x_n = F^n(x_0)$$

for each $n \in \mathbb{N}_0$. Recall that $F^n$ denotes the composition of $F : X \to X$ with itself $n$ times (and that $F^0$ is by convention the identity map on $X$). It is clear that the sequence $\{x_n\}_{n \in \mathbb{N}_0}$ satisfies the difference equation

$$x_{n+1} = F(x_n)$$

for every $n \in \mathbb{N}_0$. Conversely, if an $X$-valued sequence $\{x_n\}_{n \in \mathbb{N}_0}$ is recursively defined from any point $x_0 \in X$ onwards as $x_{n+1} = F(x_n)$ for every $n \in \mathbb{N}_0$, then it is of the form $x_n = F^n(x_0)$ for each $n \in \mathbb{N}_0$ (proof: induction). Now suppose $F \colon (X, d) \to (X, d)$ is a strict contraction and let $\gamma \in (0, 1)$ be a Lipschitz constant for $F$ so that

$$d(F(x), F(y)) \le \gamma \, d(x, y)$$

for every $x, y$ in $X$. A trivial induction shows that

$$d(F^n(x), F^n(y)) \le \gamma^n d(x, y)$$

for every nonnegative integer $n$ and every $x, y \in X$. Next take an arbitrary pair of nonnegative distinct integers, say $m < n$. Note that

$$x_n = F^n(x_0) = F^m(F^{n-m}(x_0)) = F^m(x_{n-m}),$$

and hence

$$d(x_m, x_n) = d(F^m(x_0), F^m(x_{n-m})) \le \gamma^m d(x_0, x_{n-m}).$$

By using the triangle inequality we get

$$d(x_0, x_{n-m}) \le \sum_{i=0}^{n-m-1} d(x_i, x_{i+1}),$$

and therefore

$$d(x_m, x_n) \le \gamma^m \sum_{i=0}^{n-m-1} d(x_i, x_{i+1}) \le \gamma^m \left( \sum_{i=0}^{n-m-1} \gamma^i \right) d(x_0, x_1).$$

Another trivial induction shows that $\sum_{i=0}^{k-1} \alpha^i = \frac{1-\alpha^k}{1-\alpha}$ for every real number $\alpha \neq 1$ and every integer $k \ge 1$. Thus, for any $\gamma \in (0, 1)$, $\sum_{i=0}^{n-m-1} \gamma^i = \frac{1-\gamma^{n-m}}{1-\gamma} < \frac{1}{1-\gamma}$ so that

$$d(x_m, x_n) < \frac{\gamma^m}{1 - \gamma} d(x_0, x_1) \quad \text{and} \quad \gamma^m \to 0,$$

and hence $\{x_n\}$ is a Cauchy sequence in $(X, d)$ (reason: for any $\varepsilon > 0$ there exists an integer $n_\varepsilon$ such that $\frac{\gamma^m}{1-\gamma} d(x_0, x_1) < \varepsilon$, which implies $d(x_m, x_n) < \varepsilon$, whenever $n > m \ge n_\varepsilon$). Thus $\{x_n\}$ converges in the complete metric space $(X, d)$. Set $x = \lim x_n \in X$. Since a contraction is continuous, it follows by Corollary 3.8 that $\{F(x_n)\}$ converges in $(X, d)$ and $F(\lim x_n) = \lim F(x_n)$. Therefore

$$x = \lim x_n = \lim x_{n+1} = \lim F(x_n) = F(\lim x_n) = F(x)$$

so that the limit of $\{x_n\}$ is a fixed point of $F$. Moreover, if $y$ is any fixed point of $F$, then $d(x, y) = d(F(x), F(y)) \le \gamma \, d(x, y)$, which implies that $d(x, y) = 0$ (because $\gamma \in (0, 1)$), and hence $x = y$. Conclusion: For any $x_0 \in X$ the sequence $\{F^n(x_0)\}$ converges in $(X, d)$ and its limit is the unique fixed point of $F$. $\qquad \square$

# 3.8    Continuous Extension and Completion

Recall that continuity preserves convergence (Corollary 3.8). Uniform continuity, as one might expect, goes beyond that. In fact, uniform continuity preserves Cauchy sequences too.

**Lemma 3.43.** *Let $F\colon X \to Y$ be a uniformly continuous mapping of a metric space $X$ into a metric space $Y$. If $\{x_n\}$ is a Cauchy sequence in $X$, then $\{F(x_n)\}$ is a Cauchy sequence in $Y$.*

*Proof.* The proof is straightforward by the definitions of Cauchy sequence and uniform convergence. Indeed, let $d_X$ and $d_Y$ denote the metrics on $X$ and $Y$, respectively, and take an arbitrary $X$-valued sequence $\{x_n\}$. If $F\colon X \to Y$ is uniformly continuous, then for every $\varepsilon > 0$ there exists $\delta_\varepsilon > 0$ such that

$$d_X(x_m, x_n) < \delta_\varepsilon \quad \text{implies} \quad d_Y(F(x_m), F(x_n)) < \varepsilon.$$

However, associated with $\delta_\varepsilon$ there exists a positive integer $n_\varepsilon$ such that

$$m, n \geq n_\varepsilon \quad \text{implies} \quad d_X(x_m, x_n) < \delta_\varepsilon$$

whenever $\{x_n\}$ is a Cauchy sequence in $X$. Hence, for every real number $\varepsilon > 0$ there exists a positive integer $n_\varepsilon$ such that

$$m, n \geq n_\varepsilon \quad \text{implies} \quad d_Y(F(x_m), F(x_n)) < \varepsilon,$$

which means that $\{F(x_n)\}$ is a Cauchy sequence in $Y$. $\qquad\qquad\square$

Thus, if $G\colon X \to Y$ is a uniform homeomorphism between two metric spaces $X$ and $Y$, then $\{x_n\}$ is a Cauchy sequence in $X$ if and only if $\{G(x_n)\}$ is a Cauchy sequence in $Y$, and therefore a uniform homeomorphism takes a complete metric space onto a complete metric space.

**Theorem 3.44.** *Take two uniformly homeomorphic metric spaces. One of them is complete if and only if the other is.*

*Proof.* Let $X$ and $Y$ be metric spaces and let $G\colon X \to Y$ be a uniform homeomorphism. Take an arbitrary Cauchy sequence $\{y_n\}$ in $Y$ and consider the sequence $\{x_n\}$ in $X$ such that $x_n = G^{-1}(y_n)$ for each $n$. Lemma 3.43 ensures that $\{x_n\}$ is a Cauchy sequence in $X$. If $X$ is complete, then $\{x_n\}$ converges in $X$ to, say, $x \in X$. Since $G$ is continuous, it follows by Corollary 3.8 that the sequence $\{y_n\}$, which is such that $y_n = G(x_n)$ for each $n$, converges in $Y$ to $y = G(x)$. Thus $Y$ is complete. $\qquad\square$

Carefully note that the above theorem does not hold if uniform homeomorphism is replaced by plain homeomorphism: if $X$ and $Y$ are homeomorphic metric spaces, then it is not necessarily true that $X$ is complete if and only if $Y$ is complete. In other words, *completeness is not a topological invariant* (continuity preserves convergence but not Cauchy sequences).

**Example 3U.** Let $\mathbb{R}$ be the real line with its usual metric. Set $A = (0, 1]$ and $B = [1, \infty)$, both subsets of $\mathbb{R}$. Consider the function $G \colon A \to B$ such that $G(\alpha) = \frac{1}{\alpha}$ for every $\alpha \in A$. As it is readily verified, $G$ is a homeomorphism of $A$ onto $B$, so that $A$ and $B$ are homeomorphic subspaces of $\mathbb{R}$. Now consider the $A$-valued sequence $\{\alpha_n\}$ with $\alpha_n = \frac{1}{n}$ for each $n \in \mathbb{N}$, which is a Cauchy sequence in $A$. However, $G(\alpha_n) = n$ for every $n \in \mathbb{N}$, and hence $\{G(\alpha_n)\}$ is certainly not a Cauchy sequence in $B$ (since it is not even bounded in $B$). Thus, according to Lemma 3.43, $G \colon A \to B$ (which is continuous) is not uniformly continuous. Actually, $B$ is a complete subspace of $\mathbb{R}$ (since $B$ is a closed subset of the complete metric space $\mathbb{R}$ — Corollary 3.41) and, as we have just seen, $A$ is not a complete subspace of $\mathbb{R}$ (the Cauchy sequence $\{\alpha_n\}$ does not converge in $A$ because its continuous image $\{G(\alpha_n)\}$ does not converge in $B$ — Corollary 3.8).

Lemma 3.43 also leads to an extremely useful result on extensions of uniformly continuous mappings of a *dense* subspace of a metric space into a *complete* metric space.

**Theorem 3.45.** *Every uniformly continuous mapping $F \colon A \to Y$ of a dense subspace $A$ of a metric space $X$ into a complete metric space $Y$ has a unique continuous extension over $X$, which in fact is uniformly continuous.*

*Proof.* Suppose $X$ is a nonempty metric space (otherwise the theorem is a triviality) and let $A$ be a dense subset of $X$. Take an arbitrary point $x$ in $X$. Since $A^- = X$, it follows by Proposition 3.32 that there exists an $A$-valued sequence $\{a_n\}$ that converges in $X$ to $x$, and hence $\{a_n\}$ is a Cauchy sequence in the metric space $X$ (Proposition 3.39) so that $\{a_n\}$ is a Cauchy sequence in the subspace $A$ of $X$. Now suppose $F \colon A \to Y$ is a uniformly continuous mapping of $A$ into a metric space $Y$. Thus, according to Lemma 3.43, $\{F(a_n)\}$ is a Cauchy sequence in $Y$. If $Y$ is a complete metric space, then the $Y$-valued sequence $\{F(a_n)\}$ converges in it. Let $y \in Y$ be the (unique) limit of $\{F(a_n)\}$ in $Y$:

$$y = \lim F(a_n).$$

We shall show now that $y$, which obviously depends on $x \in X$, does not depend on the $A$-valued sequence $\{a_n\}$ that converges in $X$ to $x$. Indeed, let $\{a_n'\}$ be an $A$-valued sequence converging in $X$ to $x$, and set

$$y' = \lim F(a_n').$$

Since both sequences $\{a_n\}$ and $\{a_n'\}$ converge in $X$ to the same limit $x$, it follows that $d_X(a_n, a_n') \to 0$ (see Problem 3.14(a)), where $d_X$ denotes the metric on $X$. Therefore, for every real number $\delta > 0$ there exists an index $n_\delta$ such that

$$n \geq n_\delta \quad \text{implies} \quad d_X(a_n, a_n') < \delta.$$

Moreover, since the mapping $F \colon A \to Y$ is uniformly continuous, for every real number $\varepsilon > 0$ there exists a real number $\delta_\varepsilon > 0$ such that

$$d_X(a, a') < \delta_\varepsilon \quad \text{implies} \quad d_Y(F(a), F(a')) < \varepsilon.$$

for all $a$ and $a'$ in $A$, where $d_Y$ denotes the metric on $Y$. Conclusion: Given an arbitrary $\varepsilon > 0$ there exists $\delta_\varepsilon > 0$, associated with which there exists an index $n_{\delta_\varepsilon}$, such that

$$n \geq n_{\delta_\varepsilon} \quad \text{implies} \quad d_Y(F(a_n), F(a'_n)) < \varepsilon.$$

Thus (cf. Problem 3.14(c)), $0 \leq d_Y(y, y') \leq \varepsilon$ for all $\varepsilon > 0$, and hence $d_Y(y, y') = 0$. That is, $y = y'$. Therefore, for each $x \in X$ set

$$\widehat{F}(x) = \lim F(a_n)$$

in $Y$, where $\{a_n\}$ is any $A$-valued sequence that converges in $X$ to $x$. This defines a mapping $\widehat{F} \colon X \to Y$ of $X$ into $Y$.

*Claim* 1. $\widehat{F}$ is an extension of $F$ over $X$.

*Proof.* Take an arbitrary $a$ in $A$ and consider the $A$-valued constant sequence $\{a_n\}$ such that $a_n = a$ for every index $n$. As the $Y$-valued sequence $\{F(a_n)\}$ is constant, it trivially converges in $Y$ to $F(a)$. Thus $\widehat{F}(a) = F(a)$ for every $a \in A$. That is, $\widehat{F}|_A = F$, which means that $F \colon A \to Y$ is a restriction of $\widehat{F} \colon X \to Y$ to $A \subseteq X$ or, equivalently, $\widehat{F}$ is an extension of $F$ over $X$. □

*Claim* 2. $\widehat{F}$ is uniformly continuous.

*Proof.* Take a pair of arbitrary points $x$ and $x'$ in $X$. Let $\{a_n\}$ and $\{a'_n\}$ be any pair of $A$-valued sequences converging in $X$ to $x$ and $x'$, respectively (recall: the existence of these sequences is ensured by Proposition 3.32 because $A$ is dense in $X$). Note that $d_X(a_n, a'_n) \leq d_X(a_n, x) + d_X(x, x') + d_X(x', a'_n)$ for every index $n$, by the triangle inequality in $X$. Thus, as $a_n \to x$ and $a'_n \to x'$ in $X$, for any $\delta > 0$ there exists an index $n_\delta$ such that (Definition 3.4)

$$d_X(x, x') < \delta \quad \text{implies} \quad d_X(a_n, a'_n) < 3\delta \ \text{ for every } n \geq n_\delta.$$

Since $F \colon A \to Y$ is uniformly continuous, it follows by Definition 3.6 that for every $\varepsilon > 0$ there exists $\delta_\varepsilon > 0$, which depends only on $\varepsilon$, such that

$$d_X(a_n, a'_n) < 3\delta_\varepsilon \quad \text{implies} \quad d_Y(F(a_n), F(a'_n)) < \varepsilon.$$

Thus, associated with each $\varepsilon > 0$ there exists $\delta_\varepsilon > 0$ (that depends only on $\varepsilon$), which in turn ensures the existence of an index $n_{\delta_\varepsilon}$, such that

$$d_X(x, x') < \delta_\varepsilon \quad \text{implies} \quad d_Y(F(a_n), F(a'_n)) < \varepsilon \ \text{ for every } n \geq n_{\delta_\varepsilon}.$$

Moreover, since $F(a_n) \to \widehat{F}(x)$ and $F(a'_n) \to \widehat{F}(x')$ in $Y$ by the very definition of $\widehat{F} \colon X \to Y$, it follows by Problem 3.14(c) that

$$d_Y(F(a_n), F(a'_n)) < \varepsilon \ \text{ for every } n \geq n_{\delta_\varepsilon} \quad \text{implies} \quad d_Y(\widehat{F}(x), \widehat{F}(x')) \leq \varepsilon.$$

Therefore, given an arbitrary $\varepsilon > 0$ there exists $\delta_\varepsilon > 0$ such that

$$d_X(x, x') < \delta_\varepsilon \quad \text{implies} \quad d_Y(\widehat{F}(x), \widehat{F}(x')) \leq \varepsilon$$

for all $x, x' \in X$. That is, $\widehat{F} \colon X \to Y$ is uniformly continuous (according to Definition 3.6). $\square$

Finally, since $\widehat{F} \colon X \to Y$ is continuous, it follows by Corollary 3.33 that if $\widehat{G} \colon X \to Y$ is a continuous extension of $F \colon A \to Y$ over $X$, then $\widehat{G} = \widehat{F}$ (because $A$ is dense in $X$ and $\widehat{G}|_A = \widehat{F}|_A = F$). Thus $\widehat{F}$ is the unique continuous extension of $F$ over $X$. $\hphantom{x}$ $\square$

**Corollary 3.46.** *Let $X$ and $Y$ be complete metric spaces, and let $A$ and $B$ be dense subspaces of $X$ and $Y$, respectively. If $G \colon A \to B$ is a uniform homeomorphism of $A$ onto $B$, then there exists a unique uniform homeomorphism $\widehat{G} \colon X \to Y$ of $X$ onto $Y$ that extends $G$ over $X$ (i.e., $\widehat{G}|_A = G$).*

*Proof.* Since $A$ is dense in $X$, $Y$ is complete, and $G \colon A \to B \subseteq Y$ is uniformly continuous, it follows by the previous theorem that $G$ has a unique uniformly continuous extension $\widehat{G} \colon X \to Y$. Similarly, the inverse $G^{-1} \colon B \to A$ of $G \colon A \to B$ has a unique uniformly continuous extension $\widehat{G^{-1}} \colon Y \to X$. Note that $(\widehat{G^{-1}}\widehat{G})|_A = G^{-1}G = I_A$ where $I_A \colon A \to A$ is the identity on $A$ (reason: $\widehat{G}|_A = G \colon A \to B$ and $\widehat{G^{-1}}|_B = G^{-1} \colon B \to A$). The identity $I_A$ is uniformly continuous (because its domain and range are subspaces of the same metric space $X$), and hence it has a unique continuous extension on $X$ (by the previous theorem) which clearly is $I_X \colon X \to X$, the identity on $X$ (recall: $I_X$ in fact is uniformly continuous because its domain and range are equipped with the same metric). Thus $\widehat{G^{-1}}\widehat{G} = I_X$, for $\widehat{G^{-1}}\widehat{G}$ is continuous (composition of continuous mappings) and is an extension of the uniformly continuous mapping $G^{-1}G = I_A$ over $X$. Similarly, $\widehat{G}\widehat{G^{-1}} = I_Y$ where $I_Y \colon Y \to Y$ is the identity on $Y$. Therefore $\widehat{G}^{-1} = \widehat{G^{-1}}$. Summing up: $\widehat{G} \colon X \to Y$ is an invertible uniformly continuous mapping with a uniformly continuous inverse (i.e., a uniform homeomorphism) which is the unique uniformly continuous extension of $G \colon A \to B$ over $X$. $\square$

Recall that every surjective isometry is a uniform homeomorphism. Suppose the uniform homeomorphism $G$ of the above corollary is a surjective isometry. Take an arbitrary pair of points $x$ and $x'$ in $X$ so that $\widehat{G}(x) = \lim G(a_n)$ and $\widehat{G}(x') = \lim G(a_n')$ in $Y$, where $\{a_n\}$ and $\{a_n'\}$ are $A$-valued sequences converging in $X$ to $x$ and $x'$, respectively (cf. proof of Theorem 3.45). Since $G$ is an isometry, it follows by Problem 3.14(b) that

$$d_Y(\widehat{G}(x), \widehat{G}(x')) = \lim d_Y(G(a_n), G(a_n')) = \lim d_X(a_n, a_n') = d_X(x, x').$$

Thus $\widehat{G}$ is an isometry as well, and hence a surjective isometry (since $\widehat{G}$ is a homeomorphism). This proves the following further corollary of Theorem 3.45.

**Corollary 3.47.** *Let A and B be dense subspaces of complete metric spaces X and Y, respectively. If $J: A \to B$ is a surjective isometry of A onto B, then there exists a unique surjective isometry $\widehat{J}: X \to Y$ of X onto Y that extends J over X. (i.e., $\widehat{J}|_A = J$).*

If a metric space $X$ is a subspace of a complete metric space $Z$, then its closure $X^-$ in $Z$ is a complete metric space by Theorem 3.40. In this case $X$ can be thought of as being "completed" by joining to it all its accumulation points from $Z$ (recall: $X^- = X \cup X^\star$), and $X^-$ can be viewed as a "completion" of $X$. However, if a metric space $X$ is not specified as being a subspace of a complete metric space $Z$, then the above approach of simply taking the closure of $X$ in $Z$ obviously collapses; but the idea of "completion" behind such an approach survives. To begin with, recall that two metric spaces, say $X$ and $\widetilde{X}$, are isometrically equivalent if there exists a surjective isometry of one of them onto the other (notation: $X \cong \widetilde{X}$). Isometrically equivalent metric spaces are regarded (as far as purely metric-space structure is concerned) as being essentially the same metric space. If $\widetilde{X}$ is a subspace of a complete metric space, then its closure $\widetilde{X}^-$ in that complete metric space is itself a complete metric space. With this in mind consider the following definition.

**Definition 3.48.** If the image of an isometry on a metric space $X$ is a dense subspace of a metric space $\widehat{X}$, then $X$ is said to be *densely embedded* in $\widehat{X}$. If a metric space $X$ is densely embedded in a complete metric space $\widehat{X}$, then $\widehat{X}$ is a *completion* of $X$.

Even if a metric space fails to be complete it can always be densely embedded in a complete metric space. Lemma 3.43 plays a central role in the proof of this statement.

**Theorem 3.49.** *Every metric space has a completion.*

*Proof.* Let $(X, d_X)$ be an arbitrary metric space and let $CS(X)$ denote the collection of all Cauchy sequences in $(X, d_X)$. Recall that, if $x = \{x_n\}$ and $y = \{y_n\}$ are sequences in $CS(X)$, then the real-valued sequence $\{d_X(x_n, y_n)\}$ converges in $\mathbb{R}$ (see Problem 3.53(a)). Thus, for each pair $(x, y)$ in $CS(X) \times CS(X)$ set

$$d(x, y) = \lim d_X(x_n, y_n).$$

This defines a function $d: CS(X) \times CS(X) \to \mathbb{R}$ which is a pseudometric on $CS(X)$. Indeed, nonnegativeness and symmetry are trivially verified and the triangle inequality in $(CS(X), d)$ follows at once by the triangle inequality in $(X, d_X)$. Consider a relation $\sim$ on $CS(X)$ defined as follows. If $x = \{x_n\}$ and $x' = \{x'_n\}$ are Cauchy sequences in $(X, d_X)$, then

$$x' \sim x \quad \text{if} \quad d(x', x) = 0.$$

Proposition 3.3 asserts that $\sim$ is an equivalence relation on $CS(X)$. Let $\widehat{X}$ be the collection of all equivalence classes $[x] \subseteq CS(X)$ with respect to $\sim$ for every sequence $x = \{x_n\}$ in $CS(X)$. In other words, set $\widehat{X} = CS(X)/\!\sim$, the quotient space of $CS(X)$ modulo $\sim$. For each pair $([x], [y])$ in $\widehat{X} \times \widehat{X}$ set

$$d_{\widehat{X}}([x], [y]) = d(x, y)$$

for an arbitrary pair $(x, y)$ in $[x] \times [y]$ (i.e., $d_{\widehat{X}}([x], [y]) = \lim d_X(x_n, y_n)$ where $\{x_n\}$ and $\{y_n\}$ are any Cauchy sequences from the equivalence classes $[x]$ and $[y]$, respectively). Proposition 3.3 also asserts that this actually defines a function $d_{\widehat{X}} : \widehat{X} \times \widehat{X} \to \mathbb{R}$ and, moreover, that such a function $d_{\widehat{X}}$ in fact is a metric on $\widehat{X}$. Thus

$$(\widehat{X}, d_{\widehat{X}}) \text{ is a metric space.}$$

Now consider the mapping $K : X \to \widehat{X}$ defined as follows. For each $x \in X$ take the constant sequence $x = \{x_n\} \in CS(X)$ such that $x_n = x$ for all indices $n$, and set $K(x) = [x] \in \widehat{X}$. That is, for each $x$ in $X$, $K(x)$ is the equivalence class in $\widehat{X}$ containing the constant sequence with entries equal to $x$. It is readily verified that

$$K : X \to \widehat{X} \text{ is an isometry.}$$

Indeed, take $x, y \in X$ arbitrary, let $x = \{x_n\}$ and $y = \{y_n\}$ be constant sequences such that $x_n = x$ and $y_n = y$ for all indices $n$, and note that $d_{\widehat{X}}(K(x), K(y)) = d_X(x_n, y_n) = d_X(x, y)$.

*Claim 1.*     $K(X)^- = \widehat{X}$.

*Proof.* Take an arbitrary $[x]$ in $\widehat{X}$ and an arbitrary $\{x_n\} \in [x]$ so that $\{x_n\}$ is a Cauchy sequence in $(X, d_X)$. Thus for each $\varepsilon > 0$ there exists an index $n_\varepsilon$ such that $d_X(x_n, x_{n_\varepsilon}) < \varepsilon$ for every $n \geq n_\varepsilon$. Put $[x_\varepsilon] = K(x_{n_\varepsilon}) \in K(X)$: the equivalence class in $\widehat{X}$ containing the constant sequence with entries equal to $x_{n_\varepsilon}$. Therefore, for each $[x] \in \widehat{X}$ and each $\varepsilon > 0$ there exists $[x_\varepsilon] \in K(X)$ such that $d_{\widehat{X}}([x_\varepsilon], [x]) = \lim d_X(x_n, x_{n_\varepsilon}) < \varepsilon$, and so $K(X)$ is dense in $\widehat{X}$ (Proposition 3.32). $\square$

*Claim 2.* The metric space $(\widehat{X}, d_{\widehat{X}})$ is complete.

*Proof.* Take an arbitrary Cauchy sequence $\{[x]_k\}_{k \geq 1}$ in $(\widehat{X}, d_{\widehat{X}})$. Recall that $K(X)$ is dense in $(\widehat{X}, d_{\widehat{X}})$, and hence for each positive integer $k$ there exists $[y]_k \in K(X)$ such that

$$d_{\widehat{X}}([x]_k, [y]_k) < \tfrac{1}{k}$$

(cf. Proposition 3.32). Then, since $\{[x]_k\}_{k \geq 1}$ is a Cauchy sequence in $(\widehat{X}, d_{\widehat{X}})$, and since

$$d_{\widehat{X}}([y]_j, [y]_k) \leq d_{\widehat{X}}([y]_j, [x]_j) + d_{\widehat{X}}([x]_j, [x]_k) + d_{\widehat{X}}([x]_k, [y]_k)$$

for every $j, k \geq 1$, it follows that the $K(X)$-valued sequence

$$\{[y]_k\}_{k \geq 1} \text{ is a Cauchy sequence in } (\widehat{X}, d_{\widehat{X}}).$$

Now take an arbitrary $k \geq 1$ and notice that, as $[y]_k$ lies in $K(X)$, there exists $y_k \in X$ such that the *constant* sequence $\boldsymbol{y}_k = \{y_k\}_{n \geq 1}$ belongs to the equivalence class $[\boldsymbol{y}]_k = K(y_k)$, and hence $y_k = K^{-1}([\boldsymbol{y}]_k)$. Indeed, $K : X \to K(X) \subseteq \widehat{X}$ is a surjective isometry of $X$ onto the subspace $K(X)$ of $\widehat{X}$ so that $K^{-1} : K(X) \to X$ is again a surjective isometry, thus uniformly continuous. Therefore, since $\{[\boldsymbol{y}]_k\}_{k \geq 1}$ is a Cauchy sequence in $(K(X), d_{\widehat{X}})$, it follows by Lemma 3.43 that

$$\{y_k\}_{k \geq 1} \text{ is a Cauchy sequence in } (X, d_X);$$

that is, $\boldsymbol{y} = \{y_k\}_{k \geq 1} \in CS(X)$ so that the equivalence class $[\boldsymbol{y}]$ lies in $\widehat{X}$. Next we show that

$$\{[\boldsymbol{x}]_k\}_{k \geq 1} \text{ converges in } (\widehat{X}, d_{\widehat{X}}) \text{ to } [\boldsymbol{y}] \in \widehat{X}.$$

In fact, for every $k \geq 1$,

$$0 \leq d_{\widehat{X}}([\boldsymbol{x}]_k, [\boldsymbol{y}]) \leq d_{\widehat{X}}([\boldsymbol{x}]_k, [\boldsymbol{y}]_k) + d_{\widehat{X}}([\boldsymbol{y}]_k, [\boldsymbol{y}]) \leq \tfrac{1}{k} + d_{\widehat{X}}([\boldsymbol{y}]_k, [\boldsymbol{y}]).$$

Take $\boldsymbol{y} = \{y_n\}_{n \geq 1} \in [\boldsymbol{y}]$ and, for each $k \geq 1$, take the *constant* sequence $\boldsymbol{y}_k = \{y_k\}_{n \geq 1} \in [\boldsymbol{y}]_k$ such that $y_k = K^{-1}([\boldsymbol{y}])$. By the definition of the metric $d_{\widehat{X}}$ on $\widehat{X}$, $d_{\widehat{X}}([\boldsymbol{y}]_k, [\boldsymbol{y}]) = \lim_n d_X(y_n, y_k)$ for every $k \geq 1$, and hence $\lim_k d_{\widehat{X}}([\boldsymbol{y}]_k, [\boldsymbol{y}]) = 0$ because $\{y_k\}_{k \geq 1}$ is a Cauchy sequence in $(X, d_X)$. Therefore

$$d_{\widehat{X}}([\boldsymbol{x}]_k, [\boldsymbol{y}]) \to 0 \quad \text{as} \quad k \to \infty.$$

Conclusion: Every Cauchy sequence in $(\widehat{X}, d_{\widehat{X}})$ converges in $(\widehat{X}, d_{\widehat{X}})$, so that $(\widehat{X}, d_{\widehat{X}})$ is a complete metric space. $\square$

Summing up: $X \cong K(X)$, $K(X)^- = \widehat{X}$, and $\widehat{X}$ is complete. That is, $X$ is densely embedded in a complete metric space $\widehat{X}$, which means that $\widehat{X}$ is a completion of $X$. $\square$

Corollary 3.47 leads to the proof that a completion of a metric space is essentially unique; that is, the completion of a metric space is unique up to a surjective isometry.

**Theorem 3.50.** *Any two completions of a metric space are isometrically equivalent.*

*Proof.* Let $X$ be a metric space and, according to Theorem 3.49, let $\widehat{X}$ and $\widehat{X}'$ be two completions of $X$. This means that there exist surjective isometries

$$J : \widetilde{X} \to X \quad \text{and} \quad J' : \widetilde{X}' \to X$$

where $\widetilde{X}$ is a dense subspace of $\widehat{X}$ and $\widetilde{X}'$ is a dense subspace of $\widehat{X}'$. Recall that a surjective isometry is invertible, and that its inverse is again a surjective isometry. Set

$$\widetilde{J} = J^{-1}J' : \widetilde{X}' \to \widetilde{X}$$

which, as a composition of surjective isometries, is a surjective isometry itself. Since $\widetilde{X}$ and $\widetilde{X}'$ are dense subspaces of the complete metric spaces $\widehat{X}$ and $\widehat{X}'$, respectively, it follows by Corollary 3.47 that there exists a unique surjective isometry

$$\widehat{J} : \widehat{X}' \to \widehat{X}$$

that extends $\widetilde{J}$ over $\widehat{X}'$. Thus $\widehat{X}'$ and $\widehat{X}$ are isometrically equivalent.     $\square$

According to Definition 3.48 a complete metric space $\widehat{X}$ is a completion of a metric space $X$ if there exists a dense subspace of $\widehat{X}$, say $\widetilde{X}$, which is isometrically equivalent to $X$. That is, if there exists a surjective isometry

$$J : \widetilde{X} \to X$$

of $\widetilde{X}$ onto $X$ for some dense subspace $\widetilde{X}$ of $\widehat{X}$. Now let $Y$ be another metric space, consider a mapping

$$F : X \to Y$$

of $X$ into $Y$, and let

$$\widehat{F} : \widehat{X} \to Y$$

be a mapping of a completion $\widehat{X}$ of $X$ into $Y$ such that $\widehat{F}(x) = F(J(x))$ for every $x \in \widetilde{X}$. That is, $\widehat{F}$ is an extension of the composition $FJ$ over $\widehat{X}$ or, equivalently, $FJ : \widetilde{X} \to Y$ is the restriction of $\widehat{F}$ to $\widetilde{X}$. It is usual to refer to $\widehat{F}$ as an *extension of F of over the completion* $\widehat{X}$ *of X* (which in fact is a slight abuse of terminology). The situation so far is illustrated by the following commutative diagram (recall: $\widehat{F}|_{\widetilde{X}} = FJ$).

$$\widetilde{X} \cong X \quad \xleftarrow{\;\;J\;\;} \quad \widetilde{X} \quad \subseteq \quad \widehat{X} = \widetilde{X}^{-}$$
$$F \searrow \quad \widehat{F}|_{\widetilde{X}} \downarrow \quad \swarrow \widehat{F}$$
$$Y$$

The next theorem says that, if $F$ is uniformly continuous and $Y$ is complete, then there exists an essentially unique continuous extension $\widehat{F}$ of $F$ over a completion $\widehat{X}$ of $X$.

**Theorem 3.51.** *Let $\widehat{X}$ be a completion of a metric space $X$ and let $Y$ be a complete metric space. Every uniformly continuous mapping $F : X \to Y$ has a uniformly continuous extension $\widehat{F} : \widehat{X} \to Y$ over the completion $\widehat{X}$ of $X$. Moreover, $\widehat{F}$ is unique up to a surjective isometry.*

*Proof. Existence.* Let $\widehat{X}$ be a completion of a metric space $X$. Thus there exists a dense subspace $\widetilde{X}$ of the metric space $\widehat{X}$, and a surjective isometry $J : \widetilde{X} \to X$ of $\widetilde{X}$ onto $X$. Suppose $F : X \to Y$ is a uniformly continuous mapping of $X$ into a metric space $Y$. Consider the composition $FJ : \widetilde{X} \to Y$, which is uniformly continuous as well (reason: $J$ is uniformly continuous). Since $\widetilde{X}$ is dense in $\widehat{X}$, $Y$ is complete, and $FJ : \widetilde{X} \to Y$ is uniformly continuous, it follows by Theorem 3.45 that there exists a unique continuous extension $\widehat{F} : \widehat{X} \to Y$ of $FJ : \widetilde{X} \to Y$ over $\widehat{X}$, which in fact is uniformly continuous. Thus $\widehat{F} : \widehat{X} \to Y$ is a uniformly continuous extension of $F : X \to Y$ over the completion $\widehat{X}$ of the metric space $X$.

*Uniqueness.* Suppose $\widehat{F}' \colon \widehat{X}' \to Y$ is another continuous extension of $F \colon X \to Y$ over some completion $\widehat{X}'$ of $X$ so that $\widehat{F}'|_{\widetilde{X}'} = FJ'$, where $\widetilde{X}'$ is a dense subspace of $\widehat{X}'$ and $J' \colon \widetilde{X}' \to X$ is a surjective isometry of $\widetilde{X}'$ onto $X$. Set $\widetilde{J} = J^{-1}J' \colon \widetilde{X}' \to \widetilde{X}$ as in the proof of Theorem 3.50, and let $\widehat{J} \colon \widehat{X}' \to \widehat{X}$ be the surjective isometry of $\widehat{X}'$ onto $\widehat{X}$ such that $\widehat{J}|_{\widetilde{X}'} = \widetilde{J}$. Thus, recalling that $\widehat{F}|_{\widetilde{X}} = FJ$, we get

$$\widehat{F}\widehat{J}|_{\widetilde{X}'} = F\widetilde{J} = FJJ^{-1}J' = \widehat{F}'|_{\widetilde{X}'}.$$

Therefore, the continuous mappings $\widehat{F}\widehat{J} \colon \widehat{X}' \to Y$ (composition of two continuous mappings) and $\widehat{F}' \colon \widehat{X}' \to Y$ coincide on a dense subset $\widetilde{X}'$ of $\widehat{X}'$, and hence $\widehat{F}' = \widehat{F}\widehat{J}$ by Corollary 3.33. Conclusion: If $\widehat{F}' \colon \widehat{X}' \to Y$ is another a continuous extension of $F$ over some completion $\widehat{X}'$ of $X$, then there exists a surjective isometry $\widehat{J} \colon \widehat{X}' \to \widehat{X}$ such that $\widehat{F}' = \widehat{F}\widehat{J}$. In other words, a continuous extension of $F$ over a completion of $X$ is unique up to a surjective isometry. $\qquad\qquad\square$

The commutative diagram, where $\bar{\subset}$ denotes dense inclusion,

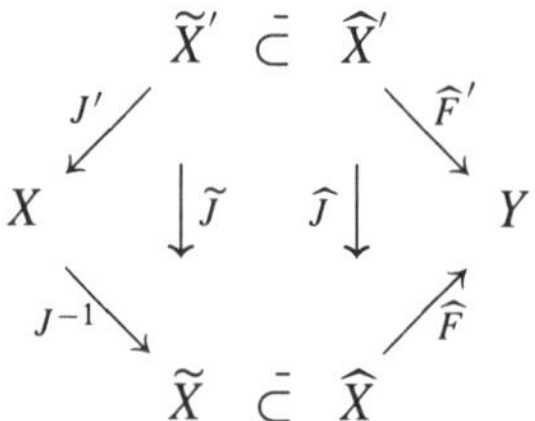

illustrates the uniqueness proofs of Theorems 3.50 and 3.51.


## 3.9   The Baire Category Theorem

We close our discussion on complete metric spaces with an important classification of subsets of a metric space into two categories. The basic notion behind such a classification is the following one. A subset $A$ of a metric space $X$ is *nowhere dense* (or *rare*) in $X$ if $(A^-)^\circ = \varnothing$ (i.e., if the interior of its closure is empty). Clearly, a closed subset of $X$ is nowhere dense in $X$ if and only if it has an empty interior. Note that $(A \backslash A^\circ)^\circ = \varnothing$ for every subset $A$ of a metric space $X$. Indeed $A^\circ$ is the largest open subset of $X$ that is included in $A$, so that the only open subset of $X$ that is included in $A \backslash A^\circ$ is the empty set $\varnothing$ of $X$. Therefore, *if $V$ is a closed subset of $X$, then $V \backslash V^\circ$ is nowhere dense in $X$.* (Reason: $V \backslash V^\circ = V^- \backslash V^\circ = \partial V$, and $\partial V$ is closed in $X$ — see Problem 3.41.) Dually, *if $U$ is an open subset of $X$, then $U^- \backslash U$ is nowhere dense in $X$* (recall: $U^- \backslash U = (X \backslash V)^- \backslash (X \backslash V) = (X \backslash V^\circ) \backslash (X \backslash V) = V \backslash V^\circ$). Carefully note that $\partial \mathbb{Q} = \mathbb{Q}^- \backslash \mathbb{Q}^\circ = (\mathbb{Q}^-)^\circ = \mathbb{R}$ in $\mathbb{R}$.

**Proposition 3.52.** *A singleton $\{x\}$ on a point $x$ in a metric space $X$ is nowhere dense in $X$ if and only if $x$ is not an isolated point of $X$.*

*Proof.* Recall that every singleton in a metric space $X$ is a closed subset of $X$ (Problem 3.37), and hence $\{x\} = \{x\}^-$ for every $x$ in $X$. According to Proposition 3.37 a point $x$ in $X$ is an isolated point of $X$ if and only if the singleton $\{x\}$ is an open set in $X$; that is, if and only if $\{x\}^\circ = \{x\}$. Thus a point $x$ in $X$ is not an isolated point of $X$ if and only if $\{x\}^\circ \subset \{x\}$ (i.e., $\{x\}^\circ \neq \{x\}$) or, equivalently, $\{x\}^\circ = \varnothing$ (since the empty set is the only proper subset of any singleton). But $\{x\}^\circ = \varnothing$ if and only if $(\{x\}^-)^\circ = \varnothing$ (because $\{x\} = \{x\}^-$ for every singleton $\{x\}$), which means that $\{x\}$ is nowhere dense in $X$. $\qquad\square$

The proposition below gives alternative characterizations of nowhere dense sets that will be required in the sequel.

**Proposition 3.53.** *Let $X$ be a metric space and let $A$ be a subset of $X$. The following assertions are pairwise equivalent.*

(a)  $(A^-)^\circ = \varnothing$    *(i.e., $A$ is nowhere dense in $X$).*

(b)  *For every nonempty open subset $U$ of $X$ there exists a nonempty open subset $U'$ of $X$ such that $U' \subset U$ and $U' \cap A = \varnothing$.*

(c)  *For every nonempty open subset $U$ of $X$ and every real number $\rho > 0$ there exists an open ball $B_\varepsilon$ with radius $\varepsilon \in (0, \rho)$ such that $B_\varepsilon^- \subset U$ and $B_\varepsilon^- \cap A = \varnothing$.*

*Proof.* Suppose $A$ is nonempty (otherwise the proposition is trivial).

*Proof of* (a)$\Leftrightarrow$(b). Take an arbitrary nonempty open subset $U$ of $X$. If $(A^-)^\circ = \varnothing$, then $U \backslash A^- \neq \varnothing$ (i.e., $A^-$ includes no nonempty open set), $U \backslash A^-$ is open in $X$ (since $U \backslash A^- = (X \backslash A^-) \cap U$), $U \backslash A^- \subset U$, and $(U \backslash A^-) \cap A = \varnothing$ Therefore (a) implies (b). Conversely, suppose $(A^-)^\circ \neq \varnothing$ so that there exists an open subset $U_0$ of $X$ such that $\varnothing \neq U_0 \subseteq A^-$. Then every point of $U_0$ is a point of adherence of $A$, and hence every nonempty open subset $U_0'$ of $U_0$ meets $A$ (cf. Proposition 3.25). Conclusion: The denial of (a) implies the denial of (b). That is, (b) implies (a).

*Proof of* (b)$\Leftrightarrow$(c). If (b) holds true, then (c) holds true for every open ball included in $U'$. Precisely, if (b) holds true and if $u$ is any point of the open set $U'$, then there exists a radius $\rho > 0$ such that $B_\rho(u) \subset U'$. Take an open ball $B_\varepsilon(u)$ with center at $u$ and radius $\varepsilon \in (0, \rho)$. Since $B_\varepsilon(u)^- \subseteq B_\rho(u)$, it follows that $B_\varepsilon(u)^- \subset U$ and $B_\varepsilon(u)^- \cap A = \varnothing$. Therefore (b) implies (c). On the other hand, suppose (c) holds true and set $U' = B_\varepsilon$, so that (c) implies (b). $\qquad\square$

By using Proposition 3.53 it is easy to show that $A \cup B$ is nowhere dense in $X$ whenever the sets $A$ and $B$ are both nowhere dense in $X$. Thus a trivial induction ensures that any *finite* union of nowhere dense subsets of a metric space $X$ is again nowhere dense in $X$. However, a countable union of nowhere dense subsets of $X$ does not need to be nowhere dense in $X$. A subset $A$ of a metric space $X$ is of *first category* (or *meagre*) in $X$ if it is a countable union of nowhere dense subsets of $X$.

That is, $A$ is of first category in $X$ if $A = \bigcup_{n \in \mathbb{N}} A_n$, where each $A_n$ is nowhere dense in $X$. The complement of a set of first category in $X$ is a *residual* (or *comeagre*) in $X$. A subset $B$ of $X$ is of *second category* (or *nonmeagre*) in $X$ if it is not of first category in $X$.

**Example 3V.** Let $X$ be a metric space. Recall that a subset of $X$ is dense in itself if and only if it has no isolated point. Thus, according to Proposition 3.52, if $X$ is nonempty and dense in itself, then every singleton in $X$ is nowhere dense in $X$. Moreover, a nonempty countable subset of $X$ is a countable union of singletons in $X$. Therefore, if $A$ is a nonempty countable subset of $X$, and if $X$ is dense in itself, then $A$ is a countable union of nowhere dense subsets of $X$. Summing up: *If a metric space $X$ is dense in itself, then every countable subset of it is of first category in $X$.* For instance, $\mathbb{Q}$ is a (dense) subset of first category in $\mathbb{R}$. Equivalently, *if a metric space $X$ has no isolated point, then every subset of second category in $X$ is uncountable.*

The following basic properties of sets of first category are (almost) immediate consequences of the definition. Note that assertions (a) and (b), but not assertion (c), in the proposition below still hold if we replace "sets of first category" by "nowhere dense sets".

**Proposition 3.54.** *Let $X$ be a metric space.*

(a) *A subset of a set of first category in $X$ is of first category in $X$.*

(b) *The intersection of an arbitrary collection of subsets of $X$ is of first category in $X$ if at least one of the subsets is of first category in $X$.*

(c) *The union of a countable collection of sets of first category in $X$ is of first category in $X$.*

*Proof.* If $B = \bigcup_n B_n$ and $A \subseteq B \subseteq X$, then $A = \bigcup_n A_n$ with $A_n = B_n \cap A \subseteq B_n$, so that $(A_n^-)^\circ \subseteq (B_n^-)^\circ$, for each $n$; and hence (a) holds true by the definitions of nowhere dense set and set of first category. Let $\{A_\gamma\}_{\gamma \in \Gamma}$ be an arbitrary collection of subsets of $X$. Since $\bigcap_{\gamma \in \Gamma} A_\gamma \subseteq A_\alpha$ for every $\alpha \in \Gamma$, it follows by item (a) that $\bigcap_{\gamma \in \Gamma} A$ is of first category in $X$ whenever at least one of the sets $A_\alpha$ in $\{A_\gamma\}_{\gamma \in \Gamma}$ is of first category in $X$. This proves assertion (b). If $\{A_n\}$ is a countable collection of subsets of $X$, and if each $A_n$ is a countable union of nowhere dense subsets of $X$, then $\bigcup_n A_n$ is itself a countable union of nowhere dense subsets of $X$ (recall: a countable union of a countable collection is again a countable collection — Corollary 1.11). Therefore $\bigcup_n A_n$ is a set of first category in $X$, which concludes the proof of assertion (c).                                    $\square$

Example 3V may suggest that sets of second category are particularly important. The next theorem, which plays a fundamental role in the theory of metric spaces, shows that they really are very important.

**Theorem 3.55.** (Baire Category Theorem). *Every nonempty open subset of a complete metric space $X$ is of second category in $X$.*

*Proof.* Let $\{A_n\}_{n\in\mathbb{N}}$ be an arbitrary countable collection of nowhere dense subsets of a metric space $X$ and set $A = \bigcup_{n\in\mathbb{N}} A_n \subseteq X$. Let $U$ be an arbitrary nonempty open subset of $X$.

*Claim.* For each integer $k \geq 1$ there exists a collection $\{B_n\}_{n=1}^{k+1}$ of open balls $B_n$ with radius $\varepsilon_n \in (0, \frac{1}{n})$ such that: $B_n^- \subset U$ and $B_n^- \cap A_n = \varnothing$ for each $n = 1, \ldots, k+1$, and $B_{n+1} \subset B_n$ for every $n = 1, \ldots, k$.

*Proof.* Since each $A_n$ is nowhere dense in $X$, it follows by Proposition 3.53 that there exist open balls $B_1$ and $B_2$ with center at $x_1$ and $x_2$ (both included in $U$) and positive radius $\varepsilon_1 < 1$ and $\varepsilon_2 < \frac{1}{2}$, respectively, such that $B_i^- \subset U$ and $B_i \cap A_i = \varnothing$, for $i = 1, 2$, and $B_2 \subset B_1$. Thus the claimed result holds for $k = 1$. Suppose it holds for some $k \geq 1$ and take a positive radius $\varepsilon_{k+1} < \min\{\varepsilon_k, \frac{1}{k+1}\}$. Proposition 3.53 ensures again the existence of an open ball $B_{k+1}$ with center at $x_{k+1} \in U$ and radius $\varepsilon_{k+1}$ such that $B_{k+1}^- \subset U$ and $B_{k+1} \cap A_{k+1} = \varnothing$. Clearly, $B_{k+1} \subset B_k$ so that the claimed result holds for $k + 1$ whenever it holds for some $k \geq 1$, which concludes the proof by induction. $\square$

Consider the collection $\{B_n\}_{n\in\mathbb{N}} = \bigcup_{k\in\mathbb{N}}\{B_n\}_{n=1}^{k+1}$. Since each open ball $B_n = B_{\varepsilon_n}(x_n) \subset U$ is such that $0 < \varepsilon_n < \frac{1}{n}$, it follows that the sequence $\{x_n\}_{n\in\mathbb{N}}$ of centers $x_n \in U$ is a Cauchy sequence in $X$ (reason: for each $\varepsilon > 0$ take a positive integer $n_\varepsilon > \frac{1}{\varepsilon}$ so that, if $n > m \geq n_\varepsilon$, then $B_{\varepsilon_n}(x_n) \subset B_{\varepsilon_m}(x_m)$ with $\varepsilon_m \leq \frac{1}{m} \leq \frac{1}{n_\varepsilon} < \varepsilon$, and hence $d(x_m, x_n) < \varepsilon$). Now suppose the metric space $X$ is complete. This ensures that the Cauchy sequence $\{x_n\}_{n\in\mathbb{N}}$ converges in $X$ to, say, $x \in X$. Take an arbitrary integer $i \geq 1$. Since $x$ is the limit of the sequence $\{x_n\}_{n\geq i}$ (it is a subsequence of $\{x_n\}_{n\in\mathbb{N}}$), and since $\{x_n\}_{n\geq i} \subset B_i$ ($B_{n+1} \subset B_n$ for every $n \in \mathbb{N}$), it follows that $x \in B_i^-$ (i.e., $x$ is an adherent point of $B_i$). Thus $x \notin A$, because $B_i^- \cap A_i = \varnothing$ for every $i \in \mathbb{N}$ and $A = \bigcup_{i\in\mathbb{N}} A_i$; and $x \in U$, because $B_i^- \subset U$ for all $i \in \mathbb{N}$. Hence $x \in U \backslash A$, and therefore $U \neq A$. Summing up: If $U$ is a nonempty open subset of a complete metric space $X$, and if $A$ is a set of first category in $X$ (i.e., if $A$ is a countable union of nowhere dense sets in $X$), then $U \neq A$. Conclusion: Every nonempty open subset of a complete metric space $X$ is not a set of first category in $X$. $\square$

In particular, as a metric space is always open in itself, we get at once the following corollary of Theorem 3.55.

**Corollary 3.56.** *A nonempty complete metric space is of second category in itself.*

**Corollary 3.57.** *The complement of a set of first category in a nonempty metric space is a dense subset of second category in that space.*

*Proof.* Let $X$ be a nonempty complete metric space. The union of two sets of first category in $X$ is again a set of first category in $X$ (Proposition 3.54). Since the union of a subset $A$ of $X$ and its complement $X\backslash A$ is the whole space $X$, it follows by Corollary 3.56 that $A$ and $A\backslash X$ cannot be both of first category in $X$. Thus $X\backslash A$ is of second category in $X$ whenever $A$ is of first category in $X$. Moreover, if $(X\backslash A)^- \neq X$, then $A^\circ$ is a nonempty open subset of $X$ (Proposition 3.31), and hence $A^\circ$ is a set of second category in $X$ (Theorem 3.55), which implies that $A$ is a set of second category in $X$ (reason: if $A$ is of first category in $X$, then $A^\circ$ is of first category in $X$ because $A^\circ \subseteq A$ — Proposition 3.54). Therefore, if $A$ is of first category in $X$, then $X\backslash A$ is dense in $X$. $\qquad\square$

In other words, *if $X \neq \varnothing$ is a complete metric space, then every residual in $X$ is both dense in $X$ and of second category in $x$.*

**Theorem 3.58.** *If a nonempty complete metric space is a countable union of closed sets, then at least one of them has nonempty interior.*

*Proof.* According to Corollary 3.56 a nonempty complete metric space is not a countable union of nowhere dense subsets of it. Thus, if a nonempty complete metric space $X$ is a countable union of subsets of $X$, then at least one of them is not nowhere dense in $X$ (i.e., the closure of at least one of them has nonempty interior). $\qquad\square$

This is a particularly useful version of the Baire Category Theorem. A further version of it, which is the dual statement of Theorem 3.58, reads as follows. (This in fact is the classical Theorem of Baire.)

**Theorem 3.59.** *Every countable intersection of open and dense subsets of a complete metric space $X$ is dense in $X$.*

*Proof.* Let $\{U_n\}$ be a countable collection of open and dense subsets of a nonempty complete metric space $X$ (if $X$ is empty the result is trivially verified). Set $V_n = X\backslash U_n$ for each $n$ so that $\{V_n\}$ is a countable collection of closed subsets of $X$ with empty interior (recall: $U_n^- = X$ means $(X\backslash U_n)^\circ = \varnothing$ by Proposition 3.31). If $(\bigcap_n U_n)^- \neq X$, then $X\backslash(\bigcap_n U_n)^- \neq \varnothing$, and hence $X\backslash\bigcap_n U_n \neq \varnothing$. However, $X\backslash\bigcap_n U_n = \bigcup_n (X\backslash U_n)$ — De Morgan laws — so that $\bigcup_n V_n \neq \varnothing$, which implies $(\bigcup_n V_n)^- \neq \varnothing$. Thus, according to Theorem 3.58, the nonempty subspace $(\bigcup_n V_n)^-$ of $X$ is not complete (reason: each $V_n$ is a closed subset of $(\bigcup_n V_n)^-$ — see Problem 3.38 — and $V_n^\circ = \varnothing$ for every $n$). On the other hand, Corollary 3.41 ensures that $(\bigcup_n V_n)^-$ is a complete subspace of the complete metric space $X$ (since $(\bigcup_n V_n)^-$ is a closed subset of $X$); which leads to a contradiction. Conclusion: $(\bigcap_n U_n)^- = X$. $\qquad\square$

The Baire Category Theorem is a nonconstructive existence theorem. For instance, for an arbitrary countable collection $\{A_n\}$ of nowhere dense sets in a nonempty complete metric space $X$, Corollary 3.57 asserts the existence of a dense set of points in $X$ with the property that none of them lies in $\{A_n\}$ for every $n$, but it does not tell us how to find those points. However, the unusual (and remarkable) fact about the Baire Category Theorem is that, while its hypothesis (completeness) has been defined in a metric space and is not a topological invariant (completeness is preserved by uniform homeomorphism but not by plain homeomorphism — see Example 3U), its conclusion is of purely topological nature and is a topological invariant. For instance, the conclusion in Theorem 3.55 (being of second category) is a topological invariant in a general topological space. Indeed, if $G: X \to Y$ is a homeomorphism between topological spaces $X$ and $Y$ and $A$ is an arbitrary subset of $X$, then it is easy to show that $G(A)^- = G(A^-)$ and $G(A)^\circ = G(A^\circ)$. Thus the property of being nowhere dense is a topological invariant (i.e., $(A^-)^\circ = \varnothing$ if and only if $(G(A)^-)^\circ = \varnothing$), and so is the property of being of first or second category (for $G(A) = G(\bigcup_n A_n) = \bigcup_n G(A_n)$ whenever $A = \bigcup_n A_n$). Such a purely topological conclusion suggests the following definition. A topological space is a *Baire space* if the conclusion of the classical Theorem of Baire holds on it. Precisely, a Baire space is a topological space $X$ on which every countable intersection of open and dense subsets of $X$ is dense in $X$. Thus Theorem 3.59 simply says that *every complete metric space is a Baire space*.

**Example 3W.** We shall now unfold three further consequences of the Baire Category Theorem, each resulting from one of the above three versions of it.

(a) Suppose $A$ is a set of first category in a complete metric space $X$. According to Corollary 3.56 $(X \backslash A)^- = X$ or, equivalently, $A^\circ = \varnothing$. Conclusion: *A set of first category in a complete metric space has empty interior*. Corollary: *A closed set of first category in a complete metric space is nowhere dense in that space* (i.e., if $A$ is a set of first category in a complete metric space, and if $A = A^-$, then $(A^-)^\circ = \varnothing$).

(b) Recall that a set without isolated points (i.e., dense in itself) in a complete metric space may be countable (example: $\mathbb{Q}$ in $\mathbb{R}$). Suppose $A$ is a nonempty perfect subset of a complete metric space $X$, which means that $A$ is closed in $X$ and dense in itself. If $A$ is a countable set, then it is the countable union of all singletons in it. Since every point in $A$ is not an isolated point of $A$ (for $A$ is dense in itself), it follows by Proposition 3.52 that every singleton in $A$ is nowhere dense in the subspace $A$, so that every singleton in $A$ has empty interior in $A$. (Recall: a singleton in a metric space $A$ is closed in $A$ — Problem 3.37.) Then $A$ is the countable union of closed sets in $A$, all of them with empty interior in $A$, and therefore the subspace $A$ is not complete according to Theorem 3.58. However, since $A$ is a closed subset of a complete metric space $X$, it follows by Corollary 3.41 that the subspace $A$ is complete. Thus the assumption that $A$ is countable leads to a contradiction. Conclusion: *A nonempty perfect set in a complete metric space is uncountable*.

(c) A subset of a metric space $X$ is a $G_\delta$ if it is a countable intersection of open subsets of $X$, and an $F_\sigma$ if it is a countable union of closed subsets of $X$. First observe that, if the complement $X \backslash A$ of a subset $A$ of $X$ is a countable union of subsets of $X$, then $A$ includes a $G_\delta$. In fact, if $X \backslash A = \bigcup_n C_n$, then $\bigcap_n (X \backslash C_n^-) \subseteq \bigcap_n (X \backslash C_n) = X \backslash \bigcup_n C_n = X \backslash (X \backslash A) = A$, and hence $A$ includes a $G_\delta$; viz., $\bigcap_n (X \backslash C_n^-)$. Moreover, if each $C_n$ is nowhere dense in $X$ (i.e., $(C_n^-)^\circ = \varnothing$), then $(X \backslash C_n^-)^- = X$ (see Proposition 3.31) so that $X \backslash C_n^-$ is open and dense in $X$ for every index $n$. Therefore, according to Theorem 3.59, $\bigcap_n (X \backslash C_n^-)$ is a dense $G_\delta$ in $X$ whenever $X$ is a complete metric space. Summing up: If $X \backslash A$ is of first category (i.e., a countable union of nowhere dense subsets of $X$) in a complete metric space $X$, then $A$ includes a dense $G_\delta$. Conversely, if a subset $A$ of a metric space $X$ includes a $G_\delta$, say $\bigcap_n U_n \subseteq A$ where each $U_n$ is open in $X$, then $X \backslash A \subseteq X \backslash \bigcap_n U_n = \bigcup_n X \backslash U_n$. If the $G_\delta$ is dense in $X$ (i.e., $(\bigcap_n U_n)^- = X$), then $(X \backslash \bigcap_n U_n)^\circ = \varnothing$ (see Proposition 3.31 again) so that $[(X \backslash U_n)^-]^\circ = (X \backslash U_n)^\circ \subseteq (\bigcup_m X \backslash U_m)^\circ = \varnothing$, and hence each $X \backslash U_n$ is nowhere dense in $X$. Thus $\bigcup_n X \backslash U_n$ is a set of first category in $X$, which implies that $X \backslash A$ is of first category in $X$ as well (since a subset of a set of first category is itself of first category — Proposition 3.54). Summing up: If $A$ includes a dense $G_\delta$ in a metric space $X$, then $X \backslash A$ is of first category in $X$. Conclusion: *A subset of a complete metric space is a residual* (i.e., its complement is of first category) *if and only if it includes a dense $G_\delta$.* (This generalizes Corollary 3.57.) Dually, *A subset of a complete metric space is of first category if and only if it is included in an $F_\sigma$ with empty interior.* (This generalizes the results of item (a).)

## 3.10   Compact Sets

Recall that a collection $\mathcal{A}$ of nonempty subsets of a set $X$ is a *covering* of $A \subseteq X$ (or $\mathcal{A}$ *covers* A) if $A \subseteq \bigcup \mathcal{A}$. If $\mathcal{A}$ is a covering of $A$, then any subcollection of $\mathcal{A}$ that also covers $A$ is a *subcovering* of $\mathcal{A}$. A covering of $A$ comprised only of open subsets of $X$ is called an *open covering*.

**Definition 3.60.** A metric space $X$ is *compact* if every open covering of $X$ includes a finite subcovering. A subset $A$ of a metric space $X$ is *compact* if it is compact as a subspace of $X$.

The notion of compactness plays an extremely important role in general topology. Note that any topology $\mathcal{T}$ on a metric space $X$ clearly is an open covering of $X$ which trivially has a finite subcovering; namely, the collection $\{X\}$ consisting of $X$ alone. However, the definition of a compact space demands that *every* open covering of it has a finite subcovering. The idea behind the definition of compact spaces is that even open coverings made up of "very small" open sets have a finite subcovering. Note that the definition of a compact subspace $A$ of a metric space $X$ is given in terms of the relative topology on $A$: an open covering of the subspace $A$ consists of relatively open subsets of $A$. The next elementary result says that this can be equally defined in terms of the topology on $X$.

**Proposition 3.61.** *A subset $A$ of a metric space $X$ is compact if and only if every covering of $A$ made up of open subsets of $X$ has a finite subcovering.*

*Proof.* If $\mathcal{U}$ is a covering of $A$ (i.e., $A \subseteq \bigcup \mathcal{U}$) consisting of open subsets of $X$, then $\{U \cap A : U \in \mathcal{U}\}$ is an open covering of the subspace $A$ (see Problem 3.38). Conversely, every open covering $\mathcal{U}_A$ of the subspace $A$ consisting of relatively open subsets of $A$ is of the form $\{U \cap A \colon U \in \mathcal{U}\}$ for some covering $\mathcal{U}$ of $A$ consisting of open subsets of $X$. (Reason: $U_A \in \mathcal{U}_A$ if and only if $U_A = A \cap U$ for some open subset $U$ of $X$ — see Problem 3.38 again.)     $\square$

The properties of being a closed subset and of being a compact subset of a metric space are certainly different from each other (trivial example: every metric space is closed in itself). However, "inside" a *compact metric space* these properties coincide.

**Theorem 3.62.** *Let $A$ be a subset of a metric space $X$.*

(a) *If $A$ is compact, then $A$ is closed in $X$.*

(b) *If $X$ is compact and if $A$ is closed in $X$, then $A$ is compact.*

*Proof.* (a) Let $A$ be a compact subset of a metric space $X$. If either $A = \varnothing$ or $A = X$, then $A$ is trivially closed in $X$. Thus suppose $\varnothing \neq X\backslash A \neq X$ and take an arbitrary point $x$ in $X\backslash A$. Since $x$ is distinct from every point in $A$, it follows that for each $a \in A$ there exists an open neighborhood $A_a$ of $a$ and an open neighborhood $X_a$ of $x$ such that $A_a \cap X_a = \varnothing$ (reason: every metric space is a Hausdorff space — see Problem 3.37). But $A \subseteq \bigcup_{a \in A} A_a$ so that $\{A_a\}_{a \in A}$ is a covering of $A$ consisting of nonempty open subsets of $X$. If $A$ is compact, then there exists a finite subset of $A$, say $\{a_i\}_{i=1}^n$, such that $A \subseteq \bigcup_{i=1}^n A_{a_i}$ (Proposition 3.61). Set $U_x = \bigcap_{i=1}^n X_{a_i}$, which is an open neighborhood of $x$ (recall: each $X_{a_i}$ is an open neighborhood of $x$). Since $A_{a_i} \cap U_x = \varnothing$ for each $i$, it follows that $(\bigcup_{i=1}^n A_{a_i}) \cap U_x = \varnothing$, and hence $A \cap U_x = \varnothing$. Therefore $U_x \subseteq X\backslash A$. Conclusion: $X\backslash A$ is open in $X$ (it includes an open neighborhood of each one of its points).

(b) Let $A$ be a closed subset of a compact metric space $X$. Take an arbitrary covering of $A$, say $\mathcal{U}_A$, consisting of open subsets of $X$. Thus $\mathcal{U}_A \cup \{X\backslash A\}$ is an open covering of $X$. As $X$ is compact, this covering includes a finite subcovering, say $\mathcal{U}$, so that $\mathcal{U}\backslash\{X\backslash A\} \subseteq \mathcal{U}_A$ is a finite subcovering of $\mathcal{U}_A$. Therefore, every covering of $A$ consisting of open subsets of $X$ has a finite subcovering, and hence (Proposition 3.61) $A$ is compact.     $\square$

**Corollary 3.63.** *Let $X$ be a compact metric space. A subset $A$ of $X$ is closed in $X$ if and only if it is compact.*

We say that a subset of a metric space $X$ is *relatively compact* (or *conditionally compact*) if it has a compact closure. It is clear by Corollary 3.63 that *every subset of a compact metric space is relatively compact*. Another important property of a compact set is that *the continuous image of a compact set is compact*.

**Theorem 3.64.** *Let $F \colon X \to Y$ be a continuous mapping of a metric space $X$ into a metric space $Y$.*

(a) *If $A$ is a compact subset of $X$, then $F(A)$ is compact in $Y$.*

(b) *If $X$ is compact, then $F$ is a closed mapping.*

(c) *If $X$ is compact and $F$ is injective, then $F$ is a homeomorphism of $X$ onto $F(X)$.*

*Proof.* (a) Let $\mathcal{U}$ be a covering of $F(A)$ (i.e., $F(A) \subseteq \bigcup_{U \in \mathcal{U}} U$) consisting of open subsets $U$ of $Y$. If $F$ is continuous, then $F^{-1}(U)$ is an open subset of $X$ for every $U \in \mathcal{U}$ according to Theorem 3.12. Set $\mathcal{F}^{-1}(\mathcal{U}) = \{F^{-1}(U)\colon U \in \mathcal{U}\}$; a collection of open subsets of $X$. Clearly (see Problem 1.2), $A \subseteq F^{-1}(F(A)) \subseteq F^{-1}(\bigcup_{U \in \mathcal{U}} U) = \bigcup_{U \in \mathcal{U}} F^{-1}(U)$ so that $\mathcal{F}^{-1}(\mathcal{U})$ is a covering of $A$ made up of open subsets of $X$. If $A$ is compact, then (cf. Proposition 3.61) there exists a finite subcollection of $\mathcal{F}^{-1}(\mathcal{U})$ covering $A$; that is, there exists $\{U_i\}_{i=1}^{n} \subseteq \mathcal{U}$ such that $A \subseteq \bigcup_{i=1}^{n} F^{-1}(U_i) \subseteq X$. Thus $F(A) \subseteq F(\bigcup_{i=1}^{n} F^{-1}(U_i)) \subseteq \bigcup_{i=1}^{n} U_i \subseteq Y$ (have another look at Problem 1.2), and hence $F(A)$ is compact by Proposition 3.61.

(b) If $X$ is compact and if $A$ is a closed subset of $X$, then $A$ is compact by Theorem 3.62(b). Hence $F(A)$ is a compact subset of $Y$ by item (a), so that $F(A)$ is closed in $Y$ according to Theorem 3.62(a).

(c) If $X$ is compact and $F$ is injective, then $F$ is a continuous invertible closed mapping of $X$ onto $F(X)$ by item (b), and therefore a homeomorphism of $X$ onto $F(X)$ (Theorem 3.24). $\qquad\square$

As compactness is preserved under continuous mappings, it is obviously preserved by homeomorphisms, and so *compactness is a topological invariant*. Moreover, a one-to-one continuous correspondence between compact metric spaces is a homeomorphism. These are straightforward corollaries of Theorem 3.64

**Corollary 3.65.** *If $X$ and $Y$ are homeomorphic metric spaces, then one is compact if and only if the other is.*

**Corollary 3.66.** *If $X$ and $Y$ are compact metric spaces, then every injective continuous mapping of $X$ onto $Y$ is a homeomorphism.*

Probably the reader has already noticed two important features: the metric has not played its role yet in this section, and the concepts of completeness and compactness share some common properties (e.g., compare Theorems 3.40 and 3.62). Indeed, the compactness proofs so far apply to topological spaces that are not necessarily metrizable. Actually they all apply to Hausdorff spaces (metrizable or not), and Theorems 3.62(b) and 3.64(a) do hold for general topological spaces (not necessarily Hausdorff). As for the connection between completeness and compactness, first note that the notion of completeness introduced in Section 3.7 needs a metric. Moreover, as we have just seen, compactness is a topological invariant while completeness is not preserved by plain homeomorphism (just by uniform homeomorphism — Theorem 3.44 and Example 3U). However, as we shall see in the next section, in a metric space compactness implies completeness.

Some of the most important results of mathematical analysis deal with continuous mappings on compact metric spaces. Theorem 3.64 is a special instance of such results that leads to many relevant corollaries (e.g., Corollary 3.85 in the next section is an extremely useful corollary of Theorem 3.64). Another important result in this line reads as follows.

**Theorem 3.67.** *Every continuous mapping of a compact metric space into an arbitrary metric space is uniformly continuous.*

*Proof.* Let $(X, d_X)$ and $(Y, d_Y)$ be metric spaces and take an arbitrary real number $\varepsilon > 0$. If $F : X \to Y$ is a continuous mapping, then for each $x \in X$ there exists a real number $\delta_\varepsilon(x) > 0$ such that

$$d_X(x', x) < 2\delta_\varepsilon(x) \quad \text{implies} \quad d_Y(F(x'), F(x)) < \varepsilon.$$

Let $B_{\delta_\varepsilon(x)}(x)$ be the open ball with center at the point $x$ and radius $\delta_\varepsilon(x)$. Consider the collection $\{B_{\delta_\varepsilon(x)}(x)\}_{x \in X}$, which surely covers $X$ (i.e., $X = \bigcup_{x \in X} B_{\delta_\varepsilon(x)}(x)$). If $X$ is compact (Definition 3.60), then this covering of $X$ includes a finite subcovering, say $\bigcup_{i=1}^n B_{\delta_\varepsilon(x_i)}(x_i)$ with $x_i \in X$ for each $i = 1, \ldots, n$. Take any $x' \in X$ so that

$$d_X(x_j, x') < \delta_\varepsilon(x_j)$$

for some $j = 1, \ldots, n$ (i.e., $X = \bigcup_{i=1}^n B_{\delta_\varepsilon(x_i)}(x_i)$ implies that every point of $X$ belongs to some ball $B_{\delta_\varepsilon(x_j)}(x_j)$). Therefore

$$d_Y(F(x_j), F(x')) < \varepsilon.$$

Set $\delta_\varepsilon = \min\{\delta_\varepsilon(x_i)\}_{i=1}^n$, which is a positive number. If

$$d_X(x', x) < \delta_\varepsilon,$$

then $d_X(x, x_j) \le d_X(x, x') + d_X(x', x_j) < \delta_\varepsilon + \delta_\varepsilon(x_j) \le 2\delta_\varepsilon(x_j)$ by the triangle inequality, and hence $d_Y(F(x), F(x_j)) < \varepsilon$. Thus, since $d_Y(F(x), F(x')) \le d_Y(F(x), F(x_j)) + d_Y(F(x_j), F(x'))$, it follows that

$$d_Y(F(x'), F(x)) < 2\varepsilon.$$

Conclusion: Given an arbitrary $\varepsilon > 0$ there exists $\delta_\varepsilon > 0$ such that

$$d_X(x', x) < \delta_\varepsilon \quad \text{implies} \quad d_Y(F(x'), F(x)) < 2\varepsilon$$

for all $x, x' \in X$. That is, $F : X \to Y$ is uniformly continuous. $\quad\square$

We shall now investigate alternative characterizations of compact sets that, unlike the fundamental concept posed in Definition 3.60, will be restricted to metrizable spaces.

**Definition 3.68.** Let $A$ be a subset of a metric space $(X, d)$. A subset $A_\varepsilon$ of $A$ is an **$\varepsilon$-net** for $A$ if for every point $x$ of $A$ there exists a point $y$ in $A_\varepsilon$ such that

$d(x, y) < \varepsilon$. A subset $A$ of $X$ is *totally bounded* in $(X, d)$ if for every real number $\varepsilon > 0$ there exists a finite $\varepsilon$-net for $A$.

**Proposition 3.69.** *Let $A$ be a subset of a metric space $X$. The following assertions are equivalent.*

> (a)  *$A$ is totally bounded.*

> (b)  *For every real number $\varepsilon > 0$ there exists a finite partition of $A$ into sets of diameter less than $\varepsilon$.*

*Proof.* Take an arbitrary $\varepsilon > 0$ and set $\rho = \frac{\varepsilon}{2}$. If there exists a finite $\rho$-net $A_\rho$ for $A$, then the finite collection of open balls $\{B_\rho(y)\}_{y \in A_\rho}$ covers $A$. That is, $A \subseteq \bigcup_{y \in A_\rho} B_\rho(y)$ because every $x \in A$ belongs to $B_\rho(y)$ for some $y \in A_\rho$ whenever $A_\rho$ is a $\rho$-net for $A$. Set $A_y = B_\rho(y) \cap A$ for each $y \in A_\rho$ so that $A = \bigcup_{y \in A_\rho} A_y$. A disjointification (Problem 1.18) of the finite collection $\{A_y\}_{y \in A_\rho}$ is a finite partition of $A$ into sets of diameter not greater than $\max_{y \in A_\rho} \mathrm{diam}(B_\rho(y) \cap A) \leq 2\rho = \varepsilon$. Thus (a) implies (b) according to Definition 3.68. On the other hand, if $\{A_i\}_{i=1}^n$ is a finite partition of $A$ into (nonempty) sets of diameter less than $\varepsilon$, then by taking one point $a_i$ of each set $A_i$ we get a finite set $\{a_i\}_{i=1}^n$ which is an $\varepsilon$-net for $A$. Therefore (b) implies (a).  $\square$

Note that *every finite subset of a metric space $X$ is totally bounded*: it is a finite $\varepsilon$-net for itself for every positive $\varepsilon$. In particular, the empty set of $X$ is totally bounded: for every positive $\varepsilon$ there is no point in $\varnothing$ within a distance greater than $\varepsilon$ for every point of $X$. It is also readily verified that *every subset of a totally bounded set is totally bounded* (indeed, $A_\varepsilon \cap B$ is an $\varepsilon$-net for $B \subseteq A$ whenever $A_\varepsilon$ is an $\varepsilon$-net for $A$). Moreover, *the closure of a totally bounded set is again totally bounded* (reason: $A_\varepsilon$ is a $2\varepsilon$-net for $A^-$ whenever $A_\varepsilon$ is an $\varepsilon$-net for $A$).

**Proposition 3.70.** *Let $A$ be a subset of a metric space $X$. If $A$ has a finite $\varepsilon$-net for some $\varepsilon > 0$, then $A$ is bounded in $X$.*

*Proof.* Suppose a nonempty subset $A$ of a metric space $(X, d)$ has a finite $\varepsilon$-net for some real number $\varepsilon > 0$. Take $x, y \in A$ arbitrary so that there exists $a, b \in A_\varepsilon$ for which $d(x, a) < \varepsilon$ and $d(x, b) < \varepsilon$. Thus $d(x, y) \leq d(x, a) + d(a, b) + d(b, y) < d(a, b) + 2\varepsilon$ by the triangle inequality. Therefore $\mathrm{diam}(A) \leq \mathrm{diam}(A_\varepsilon) + 2\varepsilon$. Since $\mathrm{diam}(A_\varepsilon) < \infty$ ($A_\varepsilon$ is a finite set), it follows that $\mathrm{diam}(A) < \infty$.  $\square$

**Corollary 3.71.** *Every totally bounded set is bounded.*

**Example 3X.** The converse fails. That is, a bounded set is not necessarily totally bounded. For instance, the closed unit ball $B_1[0]$ centered at the null sequence $0$ in the metric space $(\ell_+^2, d_2)$ of Example 3B is obviously bounded in $(\ell_+^2, d_2)$; actually, $\mathrm{diam}(B_1[0]) = 2$. We shall show that there is no finite $\varepsilon$-net for $B_1[0]$

with $\varepsilon \leq \frac{1}{2}\sqrt{2}$, and hence $B_1[0]$ is not totally bounded in $(\ell_+^2, d_2)$. Indeed, consider the countable subset $E = \{e_i\}_{i \in \mathbb{N}}$ of $B_1[0]$ made up of all scalar-valued sequences $e_i = \{\delta_{ik}\}_{k=1}^\infty$; that is, each sequence $e_i$ has just one nonzero entry (equal to one) at the $i$th position. If $A_\varepsilon$ is an $\varepsilon$-net for $B_1[0]$, then $A_\varepsilon$ contains a point within a distance less than $\varepsilon$ for each $e_i$ in $E$, and hence $A_\varepsilon$ must have a point in each open ball $B_\varepsilon(e_i)$. Since $d_2(e_i, e_j) = \sqrt{2}$ whenever $i \neq j$, it follows that $B_\varepsilon(e_i) \cap B_\varepsilon(e_j) = \varnothing$ (i.e., $B_\varepsilon(e_i) \backslash B_\varepsilon(e_j) = B_\varepsilon(e_i)$) whenever $i \neq j$ and $\varepsilon \leq \frac{1}{2}\sqrt{2}$. Thus, if $\varepsilon \leq \frac{1}{2}\sqrt{2}$, then for each $e_i \in E$ there exists $b_i \in A_\varepsilon \cap B_\varepsilon(e_i) \backslash B_\varepsilon(e_j)$ for every $j \neq i$. This establishes an injective function from $E$ to $A_\varepsilon$, so that $\#E \leq \#A_\varepsilon$. Therefore, $A_\varepsilon$ is at least countably infinite (i.e., $\aleph_0 \leq A_\varepsilon$). Conclusion: $B_1[0]$ is a closed and bounded set in the complete metric space $(\ell_+^2, d_2)$ that is not totally bounded in $(\ell_+^2, d_2)$.

**Proposition 3.72.** *A totally bounded metric space is separable.*

*Proof.* Suppose $X$ is a totally bounded metric space. For each positive integer $n$ let $X_n$ be a finite $\frac{1}{n}$-net for $X$. Set $A = \bigcup_{n \in \mathbb{N}} X_n$, which is a countable subset of $X$ (Corollary 1.11) and dense in $X$. Thus $X$ is separable. To verify that $A$ is dense in $X$, proceed as follows. Take $x \in X$ arbitrary. For each positive integer $n$ there exists $x_n \in X_n$ such that $d(x, x_n) < \frac{1}{n}$ (since $X_n$ is a $\frac{1}{n}$-net for $X$), and the $A$-valued sequence $\{x_n\}_{n \in \mathbb{N}}$ converges in $X$ to $x$. Thus $A^- = X$ by Proposition 3.32.     $\square$

**Lemma 3.73.** *A set $A$ in a metric space $X$ is totally bounded if and only if every $A$-valued sequence has a Cauchy subsequence.*

*Proof.* If $A$ is a finite set, then the result holds trivially (in this case every $A$-valued sequence has a constant subsequence). Thus suppose $A$ is an infinite set and let $d$ denote the metric on $X$.

(a) We shall say that an $A$-valued sequence $\{x_k\}_{k \in \mathbb{N}_0}$ has Property $P_n(\varepsilon)$, for some integer $n \in \mathbb{N}$, if there exists an $\varepsilon > 0$ such that $d(x_j, x_k) \geq \varepsilon$ for every pair $\{j, k\}$ of distinct integers $j, k = 0, 1, \dots, n$.

*Claim.* If $A$ is not totally bounded, then there exists an $\varepsilon > 0$ and an $A$-valued sequence that has Property $P_n(\varepsilon)$ for every $n \in \mathbb{N}$.

*Proof.* Suppose the infinite set $A$ is not totally bounded and let $\varepsilon$ be any positive real number for which there is no finite $\varepsilon$-net for $A$. In particular (and trivially), no singleton in $A$ is an $\varepsilon$-net for $A$, and hence there exists a pair of points in $A$, say $x_0$ and $x_1$, for which $d(x_0, x_1) \geq \varepsilon$. Thus every $A$-valued sequence whose first two entries coincide with $x_0$ and $x_1$ has Property $P_1(\varepsilon)$. Suppose there exists an $A$-valued sequence $\{x_k\}_{k \in \mathbb{N}_0}$ that has Property $P_n(\varepsilon)$ for some integer $n \in \mathbb{N}_0$, so that $d(x_j, x_k) \geq \varepsilon$ for every $j, k = 0, 1, \dots, n$ such that $j \neq k$. Since the set $\{x_k\}_{k=0}^n$ is not an $\varepsilon$-net for $A$ (recall: there is no finite $\varepsilon$-net for $A$), it follows that there exists a point in $A$, say $x'_{n+1}$, for which $d(x_k, x'_{n+1}) \geq \varepsilon$ for every $k = 0, 1, \dots, n$. Replace $x_{n+1}$ with $x'_{n+1}$ so that the resulting sequence has Property $P_{n+1}(\varepsilon)$. This concludes the proof by induction. $\square$

If an $A$-valued sequence $\{x_k\}_{k\in\mathbb{N}_0}$ has Property $P_n(\varepsilon)$ for every $n \in \mathbb{N}$, then $d(x_j, x_k) \geq \varepsilon$ for every nonnegative distinct integers $j$ and $k$, and hence it has no Cauchy subsequence. Conclusion: If $A$ is not totally bounded, then there exists an $A$-valued sequence that has no Cauchy subsequence. Equivalently, if every $A$-valued sequence has a Cauchy subsequence, then $A$ is totally bounded.

(b) Conversely, suppose $A$ is totally bounded and let $\{x_k\}_{k\in\mathbb{N}}$ be an arbitrary $A$-valued sequence. According to Proposition 3.69 there exists a finite partition $\mathcal{A}$ of $A$ into sets of diameter less than 1. Since $\mathcal{A}$ is a finite partition of $A$, it follows that at least one of its sets, say $A_1 \subseteq A$, has the property that the (infinite) $A$-valued sequence $x = \{x_k\}_{k\in\mathbb{N}}$ has an (infinite) subsequence, say $x_1 = \{x(1)_k\}_{k\in\mathbb{N}}$, whose entries lie in $A_1$. Note that $A_1$ is totally bounded (because $A$ is). Thus there exists a finite partition $\mathcal{A}_1$ of $A_1$, consisting of subsets of $A_1$ with diameter less than $\frac{1}{2}$, such that at least one of its sets, say $A_2 \subseteq A_1 \subseteq A$, has the property that the $A_1$-valued sequence $x_1 = \{x(1)_k\}_{k\in\mathbb{N}}$ has a subsequence, say $x_2 = \{x(2)_k\}_{k\in\mathbb{N}}$, whose entries lie in $A_2$. This leads to the inductive construction of a decreasing sequence $\{A_n\}_{n\in\mathbb{N}}$ of subsets of $A$ with $\operatorname{diam}(A_n) < \frac{1}{n}$, each including a subsequence $x_n = \{x(n)_k\}_{k\in\mathbb{N}}$ of the $A$-valued sequence $\{x_k\}_{k\in\mathbb{N}}$, for every $n \in \mathbb{N}$. Moreover, the sequence of sequences $\{x_n\}_{n\in\mathbb{N}}$ (i.e., the $A^{\mathbb{N}}$-valued sequence whose entries are the $A_n$-valued sequences $x_n$ for each $n \in \mathbb{N}$) has the property that $x_{n+1}$ is a subsequence of $x_n$ for each $n \in \mathbb{N}$. The kernel of the proof relies on the *diagonal procedure* (see Section 1.9). Consider the sequence $\{x(n)_n\}_{n\in\mathbb{N}}$ where, for each $n \in \mathbb{N}$, $x(n)_n$ is the $n$th entry of $x_n$. This sequence has the following properties: (1) it is an $A$-valued sequence (each $x(n)_n$ lies in $A_n \subseteq A$), (2) which in fact is a subsequence of $\{x_k\}_{k\in\mathbb{N}}$ (since $x_{n+1}$ is a subsequence of $x_n$ for each $n \in \mathbb{N}$ and $x_1$ is a subsequence of $x = \{x_k\}_{k\in\mathbb{N}}$), and (3) $d(x(m)_m, x(n)_n) < \frac{1}{m}$ whenever $m < n$ (since $x(m)_m \in A_m$, $x(n)_n \in A_n$, $A_n \subseteq A_m$, and $\operatorname{diam}(A_m) < \frac{1}{m}$ for every $m \in \mathbb{N}$). Therefore, the "diagonal" sequence $\{x(n)_n\}_{n\in\mathbb{N}}$ is a Cauchy subsequence of $\{x_k\}_{k\in\mathbb{N}}$.    $\square$

Total boundedness is not a topological invariant but it is preserved by uniform homeomorphism. In fact, the same example that shows that completeness is not preserved by plain homeomorphism (Example 3U) also shows that total boundedness is not preserved by plain homeomorphism (the sets $(0, 1]$ and $[1, \infty)$ are homeomorphic, but $(0, 1]$ is totally bounded — see Example 3Y below — while $[1, \infty)$ is not even bounded).

**Theorem 3.74.** *Let $F\colon X \to Y$ be a uniformly continuous mapping of a metric space $X$ into a metric space $Y$. If $A$ is a totally bounded subset of $X$, then $F(A)$ is totally bounded in $Y$.*

*Proof.* Let $A$ be a nonempty subset of $X$ and consider its image $F(A)$ in $Y$ under a mapping $F\colon X \to Y$ (if $A$ is empty the result is trivially verified). Take an arbitrary $F(A)$-valued sequence $\{y_n\}$ and consider any $A$-valued sequence $\{x_n\}$ such that $y_n = F(x_n)$ for every index $n$. If $A$ is totally bounded, then Lemma 3.73 ensures that $\{x_n\}$ has a Cauchy subsequence, say $\{x_{n_k}\}$. If $F\colon X \to Y$ is uniformly continuous, then $\{F(x_{n_k})\}$ is a Cauchy sequence in $Y$ (Lemma 3.43) which is a subsequence of $\{y_n\}$. Therefore, every $F(A)$-valued sequence has a Cauchy subsequence; that is, $F(A)$ is totally bounded (Lemma 3.73). $\qquad\square$

In particular, if $F: X \to Y$ is a surjective uniformly continuous mapping of a totally bounded metric space $X$ onto a metric space $Y$, then $Y$ is totally bounded.

**Corollary 3.75.** *If $X$ and $Y$ are uniformly homeomorphic metric spaces, then one is totally bounded if and only if the other is.*

Total boundedness is sometimes referred to as *precompactness*. Lemma 3.73 links completeness to compactness in a metric space. It actually leads to the proof that *a metric space is compact if and only if it is complete and totally bounded*. We shall prove this important assertion in the next section.

# 3.11   Sequential Compactness

The notion of compactness and total boundedness can be thought of as topological counterparts of the set-theoretic notion of "finiteness" in the sense that they may suggest "approximate finiteness" (see Propositions 3.61 and 3.69).

**Definition 3.76.** A metric space $X$ is *sequentially compact* if every $X$-valued sequence has a subsequence that converges in $X$. A subset $A$ of a metric space $X$ is *sequentially compact* if it is sequentially compact as a subspace of $X$.

**Proposition 3.77.** *Let $A$ be a subset of a metric space $X$. The following assertions are equivalent.*

  (a)   *$A$ is sequentially compact.*

  (b)   *Every infinite subset of $A$ has at least one accumulation point in $A$.*

*Proof.* If $A$ is empty, then the result holds trivially (it is sequentially compact because there is no $A$-valued sequence, and it satisfies condition (b) because it includes no infinite subset). Thus let $A$ be a nonempty set. Recall that the limits of the convergent subsequences of a given sequence are precisely the accumulation points of its range.

*Proof of* (a)$\Rightarrow$(b). If $B$ is an infinite subset of $A$, then $B$ includes a countably infinite set, and hence there exists a $B$-valued sequence $\{b_n\}_{n\in\mathbb{N}}$ of distinct points. If $A$ is sequentially compact, then this $A$-valued sequence has a subsequence that converges in $X$ to a point $a \in A$ (Definition 3.76). If $b_n \neq a$ for all $n \in \mathbb{N}$, then $\{b_n\}_{n\in\mathbb{N}}$ is a $B\backslash\{a\}$-valued sequence of distinct points that converges in $X$ to $a$, and therefore $a \in A$ is an accumulation point of $B$ (Proposition 3.28). If $b_m = a$ for some $m \in \mathbb{N}$, then remove this (unique) point from $\{b_n\}_{n\in\mathbb{N}}$ and get a $B\backslash\{a\}$-valued sequence of distinct points that converges in $X$ to $a$, so that $a \in A$ is again an accumulation point of $B$. Conclusion: (a) implies (b).

*Proof of* (b)$\Rightarrow$(a). Let $A$ be a subset of $X$ for which assertion (b) holds true. In particular, every countably infinite subset of $A$ has an accumulation point. Then every (infinite) $A$-valued sequence has a convergent subsequence, and hence $A$ is sequentially compact. That is, (b) implies (a).    $\square$

A subset $A$ of a metric space $X$ (which may be $X$ itself) is said to have the *Bolzano–Weierstrass property* if it satisfies condition (b) of Proposition 3.77. Thus *a metric space $X$ is sequentially compact if and only if it has the Bolzano–Weierstrass property*. Note that *every finite subset of a metric space $X$ is sequentially compact* (because it includes no infinite subset, and hence it has the Bolzano–Weierstrass property). In particular, the empty set is sequentially compact.

**Theorem 3.78.** *A metric space is sequentially compact if and only if it is totally bounded and complete.*

*Proof.* Let $\{x_n\}_{n\in\mathbb{N}}$ be an arbitrary $X$-valued sequence. If the metric space $X$ is both totally bounded and complete, then $\{x_n\}_{n\in\mathbb{N}}$ has a Cauchy subsequence (because $X$ is totally bounded — Lemma 3.73), which converges in $X$ (because $X$ is complete). Thus $X$ is sequentially compact (Definition 3.76). On the other hand, sequential compactness clearly implies total boundedness (see Definition 3.76, Proposition 3.39(a) and Lemma 3.73). Moreover, if $X$ is sequentially compact and if $\{x_n\}_{n\in\mathbb{N}}$ is an $X$-valued Cauchy sequence, then this sequence has a convergent subsequence (Definition 3.76), and hence $\{x_n\}_{n\in\mathbb{N}}$ converges in $X$ (Proposition 3.39(c)). Therefore, in a sequentially compact metric space $X$ every Cauchy sequence converges; that is, $X$ is complete.    $\square$

The next lemma establishes another necessary and sufficient condition for sequential compactness. We shall apply this lemma to prove the equivalence between the concepts of compactness and sequential compactness in a metric space.

**Lemma 3.79.** (Cantor). *A metric space $X$ is sequentially compact if and only if every decreasing sequence $\{V_n\}_{n\in\mathbb{N}}$ of nonempty closed subsets of $X$ has a nonempty intersection* (i.e., *is such that $\bigcap_{n\in\mathbb{N}} V_n \neq \varnothing$*).

*Proof.* (a) Let $\{V_n\}_{n\in\mathbb{N}}$ be a decreasing sequence (i.e., $V_{n+1} \subseteq V_n$ for every $n \in \mathbb{N}$) of nonempty closed subsets of a metric space $X$. For each $n \in \mathbb{N}$ let $v_n$ be a point of $V_n$ and consider the $X$-valued sequence $\{v_n\}_{n\in\mathbb{N}}$. If $X$ is sequentially compact, then $\{v_n\}_{n\in\mathbb{N}}$ has a subsequence that converges in $X$ to, say, $v \in X$ (Definition 3.76). Now take an arbitrary $m \in \mathbb{N}$ and note that this convergent subsequence is eventually in $V_m$ (because $\{V_n\}_{n\in\mathbb{N}}$ is decreasing), and hence it has a $V_m$-valued subsequence that converges in $X$ to $v$ (Proposition 3.5). Since $V_m$ is closed in $X$, it follows by the Closed Set Theorem that $v \in V_m$ (Theorem 3.20). Therefore, as $m$ is arbitrary, $v \in \bigcap_{m\in\mathbb{N}} V_m$.

(b)  Conversely, let $\{x_k\}_{k\in\mathbb{N}}$ be an arbitrary $X$-valued sequence and set $X_n = \{x_k \in X: k \geq n\}$ for each $n \in \mathbb{N}$. Observe that $\{X_n\}_{n\in\mathbb{N}}$ is a decreasing sequence of nonempty subsets of $X$, and so is the sequence of closed subsets of $X$, $\{X_n^-\}_{n\in\mathbb{N}}$, consisting of the closures of each $X_n$. If $\bigcap_{n\in\mathbb{N}}X_n^- \neq \varnothing$, then there exists $x \in X_n^-$ for all $n \in \mathbb{N}$. Take an arbitrary real number $\varepsilon > 0$ and consider the open ball $B_\varepsilon(x)$. Since $B_\varepsilon(x) \cap X_n \neq \varnothing$ for every $n \in \mathbb{N}$ (reason: put $A = X_n$ and $B = X_n^-$ in Proposition 3.32(b)), it follows that for every $n \in \mathbb{N}$ there exist integers $k \geq n$ for which $x_k \in B_\varepsilon(x)$. Thus every nonempty open ball centered at $x$ meets the range of the sequence $\{x_k\}_{k\in\mathbb{N}}$ infinitely often, and hence $x$ is the limit of some convergent subsequence of $\{x_k\}_{k\in\mathbb{N}}$ (Proposition 3.28). Conclusion: Every $X$-valued sequence has a convergent subsequence, which means that $X$ is sequentially compact.     $\square$

**Theorem 3.80.** *A metric space is compact if and only if it is sequentially compact.*

*Proof.* (a)  Suppose $X$ is a compact metric space (Definition 3.60). Let $\{V_n\}_{n=1}^\infty$ be an arbitrary decreasing sequence of nonempty closed subsets of $X$. Set $U_n = X \backslash V_n$ for each $n \in \mathbb{N}$ so that $\{U_n\}_{n=1}^\infty$ is an increasing sequence of proper open subsets of $X$. If $\{U_n\}_{n=1}^\infty$ covers $X$ (i.e., if $\bigcup_{n=1}^\infty U_n = X$), then $U_m = \bigcup_{n=1}^m U_n = X$ for some $m \in \mathbb{N}$ (because $\{U_n\}_{n=1}^\infty$ is increasing and $X$ is compact), which contradicts the fact that $U_n \neq X$ for every $n \in \mathbb{N}$. Outcome: $\bigcup_{n=1}^\infty U_n \neq X$ and hence (De Morgan laws) $\bigcap_{n=1}^\infty V_n \neq \varnothing$. Therefore $X$ is sequentially compact by Lemma 3.79.

(b)  On the other hand, suppose $X$ is a sequentially compact metric space. Since $X$ is separable (Proposition 3.72 and Theorem 3.78), it follows by Theorem 3.35 that $X$ has a countable base $\mathcal{B}$ of open subsets of $X$. Let $\mathcal{U}$ be an arbitrary open covering of $X$.

*Claim.*  There exists a countable subcollection $\mathcal{U}'$ of $\mathcal{U}$ that covers $X$.

*Proof.*  For each $U \in \mathcal{U}$ set $\mathcal{B}_\mathrm{U} = \{B \in \mathcal{B}: B \subseteq U\}$. Since $\mathcal{B}$ is a base for $X$, and since $U$ is an open subset of $X$, it follows by the very definition of base that $U = \bigcup \mathcal{B}_\mathrm{U}$. The collection $\mathcal{B}' = \bigcup_{U \in \mathcal{U}} \mathcal{B}_\mathrm{U}$ of open subsets of $X$ has the properties

$$\#\mathcal{B}' \leq \#\mathcal{B} \quad \text{and} \quad \bigcup \mathcal{U} \subseteq \bigcup \mathcal{B}'.$$

Indeed, since $\mathcal{B}_\mathrm{U} \subseteq \mathcal{B}$ for every $U \in \mathcal{U}$, it follows that $\bigcup_{U \in \mathcal{U}} \mathcal{B}_\mathrm{U} \subseteq \mathcal{B}$. Thus $\mathcal{B}' \subseteq \mathcal{B}$ so that $\#\mathcal{B}' \leq \#\mathcal{B}$. Moreover, if $U$ is an arbitrary set in $\mathcal{U}$, then $U = \bigcup \mathcal{B}_\mathrm{U} = \bigcup_{B \in \mathcal{B}_\mathrm{U}} B \subseteq \bigcup_{B \in \mathcal{B}'} B = \bigcup \mathcal{B}'$, and hence $\bigcup \mathcal{U} \subseteq \bigcup \mathcal{B}'$. Another property of the collection $\mathcal{B}'$ is that every set in $\mathcal{B}'$ is included in some set in $\mathcal{U}$ (reason: If $B' \in \mathcal{B}' = \bigcup_{U \in \mathcal{U}} \mathcal{B}_\mathrm{U}$, then $B' \in \mathcal{B}_{\mathrm{U}'} \subseteq \bigcup \mathcal{B}_{\mathrm{U}'} = U'$ for some $U' \in \mathcal{U}$). For each set $B'$ in $\mathcal{B}'$ take one set $U'$ in $\mathcal{U}$ that includes $B'$, and consider the subcollection $\mathcal{U}'$ of $\mathcal{U}$ consisting of all those sets $U'$. The very construction of $\mathcal{U}'$ establishes a surjective map of $\mathcal{B}'$ onto $\mathcal{U}'$ that embeds $\bigcup \mathcal{B}'$ in $\bigcup \mathcal{U}'$. Thus

$$\#\mathcal{U}' \leq \#\mathcal{B}' \quad \text{and} \quad \bigcup \mathcal{B}' \subseteq \bigcup \mathcal{U}'.$$

Therefore, by transitivity,

$$\#\mathcal{U}' \le \#\mathcal{B} \quad \text{and} \quad \bigcup \mathcal{U} \subseteq \bigcup \mathcal{U}'.$$

Conclusion: $\mathcal{U}'$ is a countable subcollection of $\mathcal{U}$ (because $\mathcal{B}$ is a countable base for $X$) which covers $X$ (because $\mathcal{U}$ covers $X$). $\square$

If $\mathcal{U}'$ is finite, then it is itself a finite subcovering of $\mathcal{U}$ so that $X$ is compact. If $\mathcal{U}'$ is countably infinite, then it can be indexed by $\mathbb{N}$ so that $\mathcal{U}' = \{U_n\}_{n=1}^{\infty}$, where each $U_n$ belongs to $\mathcal{U}$. For each $n \in \mathbb{N}$ set $V_n = X \backslash \bigcup_{i=1}^{n} U_i$ so that $\{V_n\}_{n=1}^{\infty}$ is a decreasing sequence of closed subsets of $X$. Since $\bigcup \mathcal{U}' = \bigcup_{n=1}^{\infty} U_n = X$ (recall: $\mathcal{U}'$ covers $X$), it follows that $\bigcap_{n=1}^{\infty} V_n = \varnothing$. Therefore, according to Lemma 3.79, at least one of the sets in $\{V_n\}_{n=1}^{\infty}$, say $V_m$, must be empty (because $X$ is sequentially compact). Thus $\bigcap_{n=1}^{m} V_n = \varnothing$, and hence $\bigcup_{n=1}^{m} U_n = X$. Conclusion: $\mathcal{U}$ includes a finite subcovering, viz. $\{U_n\}_{n=1}^{m}$, so that $X$ is compact. $\square$

The theorems have been proved. Let us now harvest the corollaries.

**Corollary 3.81.** *If $X$ is a metric space, then the following assertions are pairwise equivalent.*

(a)  *$X$ is compact.*

(b)  *$X$ is sequentially compact.*

(c)  *$X$ is complete and totally bounded.*

*Proof.*  Theorems 3.78 and 3.80. $\square$

As we have already observed, completeness and total boundedness are preserved by uniform homeomorphisms but not by plain homeomorphisms, whereas compactness is preserved by plain homeomorphisms. Thus completeness and total boundedness are not topological invariants. However, when taken together they mean compactness, which is a topological invariant.

**Corollary 3.82.** *Every compact subset of any metric space is closed and bounded.*

*Proof.*  Theorem 3.62(a) and Corollaries 3.71 and 3.81. $\square$

Recall that the converse fails. Indeed we exhibited in Example 3X a closed and bounded subset of a (complete) metric space that is not totally bounded, and hence not compact.

**Theorem 3.83.** (Heine–Borel). *A subset of $\mathbb{R}^n$ is compact if and only if it is closed and bounded.*

*Proof.*  The condition is clearly necessary by Corollary 3.82. We shall prove that it is also sufficient. Consider the real line $\mathbb{R}$ equipped with its usual metric. Let

$V_\rho$ be any nondegenerate closed and bounded interval, say $V_\rho = [\alpha, \ \alpha + \rho]$ for some real number $\alpha$ and some $\rho > 0$. Take an arbitrary real number $\varepsilon > 0$ and let $n_\varepsilon$ be a positive integer large enough so that $\rho < (n_\varepsilon + 1)\frac{\varepsilon}{2}$. For each integer $k = 0, 1, \ldots, n_\varepsilon$ consider the interval $A_k = [\alpha + k\frac{\varepsilon}{2}, \ \alpha + (k+1)\frac{\varepsilon}{2})$ of diameter $\frac{\varepsilon}{2}$. Since $A_j \cap A_k = \varnothing$ whenever $j \neq k$, and $V_\rho \subset [\alpha, \ \alpha + (n_\varepsilon + 1)\frac{\varepsilon}{2}) = \bigcup_{k=0}^{n_\varepsilon} A_k$, it follows that $\{A_k \cap V_\rho\}_{k=0}^{n_\varepsilon}$ is a finite partition of $V_\rho$ into sets of diameter less than $\varepsilon$. Thus every closed and bounded interval of the real line is totally bounded (Proposition 3.69). Now equip $\mathbb{R}^n$ with any of the metrics $d_\infty$ or $d_p$ for some $p \geq 1$ as in Example 3A (recall: these are uniformly equivalent metrics on $\mathbb{R}^n$ — Problem 3.33). Take an arbitrary bounded subset $B$ of $\mathbb{R}^n$ and consider the closed interval $V_\rho$ of diameter $\rho = \mathrm{diam}(B)$ such that $B \subseteq V_\rho^n$. Since $V_\rho$ is totally bounded in $\mathbb{R}$, it follows by Problem 3.64(a) that the Cartesian product $V_\rho^n$ is totally bounded in $\mathbb{R}^n$. Hence, as a subset of a totally bounded set, $B$ is totally bounded. Conclusion: *Every bounded subset of $\mathbb{R}^n$ is totally bounded*. Moreover, since $\mathbb{R}^n$ is a complete metric space (when equipped with any of these metrics — Example 3R(a)), it follows by Theorem 3.40(b) that every closed subset of $\mathbb{R}^n$ is a complete subspace of $\mathbb{R}^n$. Therefore every closed and bounded subset of $\mathbb{R}^n$ is compact (Corollary 3.81). $\quad\square$

The Heine–Borel Theorem is readily extended to $\mathbb{C}^n$ (again equipped with any of the uniformly equivalent metrics $d_\infty$ or $d_p$ for some $p \geq 1$ as in Example 3A): *A subset of $\mathbb{C}^n$ is compact if and only if it is closed and bounded in $\mathbb{C}^n$.*

**Corollary 3.84.** *Let $X$ be a complete metric space, and let $A$ be a subset of $X$.*

(a)   *$A$ is compact if and only if it is closed and totally bounded in $X$.*

(b)   *$A$ is relatively compact if and only if it is totally bounded in $X$.*

*Proof.* By Theorem 3.40(b) and Corollaries 3.81 and 3.82 we get the result in (a), which in turn leads to the result in (b) by recalling that $A^-$ is totally bounded if and only if $A$ is totally bounded. $\quad\square$

**Corollary 3.85.** *A continuous image of any compact set is closed and bounded.*

*Proof.* Theorem 3.64(a) and Corollary 3.82. $\quad\square$

**Theorem 3.86.** (Weierstrass). *If $\varphi\colon X \to \mathbb{R}$ is a continuous real-valued function on a metric space $X$, then $\varphi$ assumes both a maximum and a minimum value on each nonempty compact subset of $X$.*

*Proof.* If $\varphi$ is a continuous real-valued function defined on a (nonempty) compact metric space, then its (nonempty) range $\mathcal{R}(\varphi)$ is both closed and bounded in the real line $\mathbb{R}$ equipped with its usual metric (Corollary 3.85). Thus the bounded subset $\mathcal{R}(\varphi)$ of $\mathbb{R}$ has an infimum and a supremum in $\mathbb{R}$, which actually lie in $\mathcal{R}(\varphi)$ because $\mathcal{R}(\varphi)$ is closed in $\mathbb{R}$ (recall: a closed subset contains all its adherent points). $\quad\square$

**Example 3Y.** Consider the set $C[X, Y]$ of all continuous mappings of a metric space $X$ into a metric space $Y$ and let $B[X, Y]$ be the set of all bounded mappings of $X$ into $Y$. According to Corollary 3.85,

$$C[X, Y] \subseteq B[X, Y] \quad \text{whenever } X \text{ is compact.}$$

Thus the sup-metric $d_\infty$ on $B[X, Y]$ (see Example 3C) is inherited by $C[X, Y]$ if $X$ is compact, and hence $(C[X, Y], d_\infty)$ is a subspace of $(B[X, Y], d_\infty)$. In other words, if $X$ is compact, then the sup-metric $d_\infty$ is well-defined on $C[X, Y]$ so that, in this case, $(C[X, Y], d_\infty)$ is a metric space. In particular, $C[0, 1] \subset B[0, 1]$ because $[0, 1]$ is compact in $\mathbb{R}$ by the Heine–Borel Theorem (Theorem 3.83), and so $(C[0, 1], d_\infty)$ is a subspace of $(B[0, 1], d_\infty)$ as we had anticipated in Examples 3D and 3G, and used in Examples 3N and 3T. Moreover, since (1) the absolute value function $| \cdot |$: $[0, 1] \to \mathbb{R}$ is continuous, (2) $(x - y) \in C[0, 1]$ for every $x, y \in C[0, 1]$, (3) the interval $[0, 1]$ is compact, and since (4) the composition of continuous functions is continuous, it follows by the Weierstrass Theorem (Theorem 3.86) that

$$d_\infty(x, y) = \max_{t \in [0,1]} |x(t) - y(t)| \quad \text{for every} \quad x, y \in C[0, 1].$$

Now let $BC[X, Y]$ be the set of all bounded continuous mappings of $X$ into $Y$ and equip it with the sup-metric $d_\infty$ as in Example 3T. If $X$ is compact, then $C[X, Y] = BC[X, Y]$ and $(C[X, Y], d_\infty)$ is a metric space that coincides with the metric space $(BC[X, Y], d_\infty)$. Since $(BC[X, Y], d_\infty)$ is complete if and only if $Y$ is complete (Example 3T), it follows that

$$(C[X, Y], d_\infty) \text{ is complete if } X \text{ is compact and } Y \text{ is complete.}$$

**Example 3Z.** Suppose $(X, d_X)$ is a compact metric space, let $(Y, d_Y)$ be any metric space, and consider the metric space $(C[X, Y], d_\infty)$ of Example 3Y. Let $\Phi$ be a subset of $C[X, Y]$. We shall investigate a necessary and sufficient condition that $\Phi$ be totally bounded. To begin with let us pose the following definitions.

(i) A subset $\Phi$ of $C[X, Y]$ is *pointwise totally bounded* if for each $x$ in $X$ the set $\Phi(x) = \{f(x) \in Y : f \in \Phi\}$ is totally bounded in $Y$. Similarly, $\Phi$ is *pointwise bounded* if $\Phi(x)$ is a bounded in $Y$ for each $x \in X$ (i.e., if $\sup_{f,g \in \Phi} d_Y(f(x), g(x)) < \infty$ for each $x \in X$).

(ii) A subset $\Phi$ of $C[X, Y]$ is *equicontinuous at a point* $x_0 \in X$ if for each real number $\varepsilon > 0$ there exists a real number $\delta > 0$ such that $d_Y(f(x), f(x_0)) < \varepsilon$ whenever $d_X(x, x_0) < \delta$ for every $f \in \Phi$ (note: $\delta$ depends on $\varepsilon$ and may depend on $x_0$ but it does not depend on $f$ — hence the term "equicontinuity"). $\Phi$ is *equicontinuous* on $X$ if it is equicontinuous at every point of $X$.

Remark: If the subset $\Phi$ is equicontinuous on $X$, and if for each $\varepsilon > 0$ there exists a $\delta > 0$ (which depends only on $\varepsilon$) such that $d_X(x, x') < \delta$ implies

$d_Y(f(x), f(x')) < \varepsilon$ for all $x, x' \in X$ and every $f \in \Phi$, then $\Phi$ is *uniformly equicontinuous* on $X$. Uniform equicontinuity coincides with equicontinuity on a compact space (Theorem 3.67).

(a) Take $\varepsilon > 0$, $x_0 \in X$, and $f \in \Phi$ arbitrary. Let $\Phi_\varepsilon$ be an $\varepsilon$-net for $\Phi$. Thus there exists $g \in \Phi_\varepsilon$ such that $d_\infty(f, g) < \varepsilon$, and hence

$$d_Y(f(x_0), g(x_0)) \;\leq\; \sup_{x \in X} d_Y(f(x), g(x)) \;=\; d_\infty(f, g) \;<\; \varepsilon.$$

If $\Phi_\varepsilon$ is a finite $\varepsilon$-net for $\Phi$, then the set $\Phi_\varepsilon(x_0) = \{g(x_0) \in Y : g \in \Phi_\varepsilon\}$ is a finite $\varepsilon$-net for $\Phi(x_0)$. Therefore, if $\Phi$ is totally bounded, then $\Phi(x_0)$ is totally bounded for an arbitrary $x_0 \in X$. Moreover,

$$
\begin{aligned}
d_Y(f(x), f(x_0)) \;&\leq\; d_Y(f(x), g(x)) + d_Y(g(x), g(x_0)) + d_Y(g(x_0), f(x_0)) \\
&\leq\; 2\varepsilon + d_Y(g(x), g(x_0))
\end{aligned}
$$

for every $x \in X$ and every $g \in \Phi_\varepsilon$. Since each $g \in \Phi_\varepsilon$ is continuous, it follows that for each $g \in \Phi_\varepsilon$ there exists a $\delta_g = \delta_g(\varepsilon, x_0) > 0$ such that $d_X(x, x_0) < \delta_g$ implies $d_Y(g(x), g(x_0)) < \varepsilon$. If $\Phi_\varepsilon$ is a finite $\varepsilon$-net for $\Phi$, then set $\delta = \delta(\varepsilon, x_0) = \min\{\delta_g\}_{g \in \Phi_\varepsilon}$ so that $d_Y(g(x), g(x_0)) < \varepsilon$ whenever $d_X(x, x_0) < \delta$. Thus there exists a $\delta > 0$ (that does not depend on $f$) such that

$$d_X(x, x_0) < \delta \quad \text{implies} \quad d_Y(f(x), f(x_0)) < 3\varepsilon.$$

Therefore, if $\Phi$ is totally bounded, then $\Phi$ is equicontinuous at an arbitrary $x_0 \in X$. Summing up: if $\Phi$ is totally bounded, then $\Phi$ is pointwise totally bounded and equicontinuous on $X$

(b) Conversely, recall that $X$ is separable (because it is compact — Proposition 3.72 and Corollary 3.81) and take a countable dense subset $A$ of $X$. Consider the (infinite) $A$-valued sequence $\{a_i\}_{i \geq 1}$ consisting of an enumeration of all points of $A$ (followed by an arbitrary repetition of points of $A$ if $A$ is finite). Let $f = \{f_n\}_{n \geq 1}$ be an arbitrary $\Phi$-valued sequence, and suppose $\Phi$ is pointwise totally bounded so that $\Phi(x)$ is totally bounded in $Y$ for every $x \in X$. Thus, according o Lemma 3.73, for each $x \in X$ the $\Phi(x)$-valued sequence $\{f_n(x)\}_{n \geq 1}$ has a Cauchy subsequence. In particular, $\{f_n(a_1)\}_{n \geq 1}$ has a Cauchy subsequence, say $\{f_n^{(1)}(a_1)\}_{n \geq 1}$. Set $f_1 = \{f_n^{(1)}\}_{n \geq 1}$, which is a $\Phi$-valued subsequence of $f$ such that $\{f_n^{(1)}(a_1)\}_{n \geq 1}$ is a Cauchy sequence in $Y$. Now consider for each $x \in X$ the $\Phi(x)$-valued sequence $\{f_n^{(1)}(x)\}_{n \geq 1}$. Since $\Phi(x)$ is totally bounded for every $x \in X$, it follows by Lemma 3.73 that $\{f_n^{(1)}(x)\}_{n \geq 1}$ has a Cauchy subsequence for every $x \in X$. In particular, $\{f_n^{(1)}(a_2)\}_{n \geq 1}$ has a Cauchy sequence, say $\{f_n^{(2)}(a_2)\}_{n \geq 1}$. Set $f_2 = \{f_n^{(2)}\}_{n \geq 1}$, which is a $\Phi$-valued subsequence of $f_1$ such that $\{f_n^{(2)}(a_2)\}_{n \geq 1}$ is a Cauchy sequence in $Y$ and both $\{f_n^{(1)}(a_1)\}_{n \geq 1}$ and $\{f_n^{(1)}(a_2)\}_{n \geq 1}$ are Cauchy sequences in $Y$ as well (reason: $f_2$

is a subsequence of $f_1$, and hence $\{f_n^{(2)}(a_1)\}_{n\geq 1}$ is a subsequence of the Cauchy sequence $\{f_n^{(1)}(a_1)\}_{n\geq 1}$). This leads to the inductive construction of a sequence $\{f_k\}_{k\geq 1}$ of $\Phi$-valued subsequences of $f$ with the following properties.

**Property (1).** $f_{k+1} = \{f_n^{(k+1)}\}_{n\geq 1}$ is a subsequence of $f_k = \{f_n^{(k)}\}_{n\geq 1}$ for every $k \geq 1$.

**Property (2).** For each pair of integers $i \geq 1$ and $k \geq 1$, $\{f_n^{(k)}(a_i)\}_{n\geq 1}$ is a Cauchy sequence in $Y$ whenever $i \leq k$.

As it happened in part (b) of the proof of Lemma 3.73, the *diagonal procedure* plays a central role in this proof too. Take an arbitrary integer $i \geq 1$. By Property (1), the $Y$-valued sequence $\{f_n^{(n)}(a_i)\}_{n\geq i}$ is a subsequence of $\{f_n^{(i)}(a_i)\}_{n\geq 1}$, which in turn is a Cauchy sequence in $Y$ by Property (2). Thus $\{f_n^{(n)}(a_i)\}_{n\geq i}$ is a Cauchy sequence in $Y$, and so is $\{f_n^{(n)}(a_i)\}_{n\geq 1}$. Therefore, the "diagonal" sequence $\{f_n^{(n)}\}_{n\geq 1}$ is a subsequence of the $\Phi$-valued sequence $f = \{f_n\}_{n\geq 1}$ (cf. Property (1)) such that

$$\{f_n^{(n)}(a)\}_{n\geq 1} \text{ is a Cauchy sequence in } Y \text{ for every } a \in A.$$

Now take $\varepsilon > 0$ and $x \in X$ arbitrary. Suppose $\Phi$ is equicontinuous on $X$. Thus there exists a $\delta_\varepsilon = \delta_\varepsilon(x) > 0$ such that

$$d_X(x', x) < \delta_\varepsilon \quad \text{implies} \quad d_Y\big(f_n^{(n)}(x'), f_n^{(n)}(x)\big) < \varepsilon$$

for all $n$. Since $A$ is dense in $X$, it follows that there exists $a \in A$ such that $d_X(a, x) < \delta_\varepsilon$, and hence

$$d_Y\big((f_n^{(n)}(a), f_n^{(n)}(x)\big) < \varepsilon$$

for all $n$. But $\{f_n^{(n)}(a)\}_{n\geq 1}$ is a Cauchy sequence, which means that there exists a positive integer $n_\varepsilon = n_\varepsilon(a)$ such that

$$d_Y\big(f_m^{(m)}(a), f_n^{(n)}(a)\big) < \varepsilon$$

whenever $m, n \geq n_\varepsilon$. Hence, by the triangle inequality,

$$\begin{aligned} d_Y\big(f_m^{(m)}(x), f_n^{(n)}(x)\big) \;\leq\; & d_Y\big(f_m^{(m)}(x), f_m^{(m)}(a)\big) + d_Y\big(f_m^{(m)}(a), f_n^{(n)}(a)\big) \\ & + d_Y\big(f_n^{(n)}(a), f_n^{(n)}(x)\big) < 3\varepsilon \end{aligned}$$

whenever $m, n \geq n_\varepsilon$. Therefore, as $n_\varepsilon$ does not depend on $x$,

$$d_\infty\big(f_m^{(m)}, f_n^{(n)}\big) = \sup_{x\in X} d_Y\big(f_m^{(m)}(x), f_n^{(n)}(x)\big) < 3\varepsilon$$

whenever $m, n \geq n_\varepsilon$, which means that the subsequence $\{f_n^{(n)}\}_{n\geq 1}$ of $f = \{f_n\}_{n\geq 1}$ is a Cauchy sequence in $\Phi$. Thus $\Phi$ is totally bounded by Lemma 3.73. Summing up: If $\Phi$ is pointwise totally bounded and equicontinuous, then $\Phi$ is totally bounded.

(c) This is the *Arzelà–Ascoli Theorem*: *If $X$ is compact, then a subset of the metric space $(C[X, Y], d_\infty)$ is totally bounded if and only if it is pointwise totally bounded and equicontinuous.* The corollary below follows at once (cf. Example 3Y and Corollary 3.84): *If $X$ is compact and $Y$ is complete, then a subset of the metric space $(C[X, Y], d_\infty)$ is compact if and only if it is pointwise totally bounded, equicontinuous, and closed in $(C[X, Y], d_\infty)$.* Recall that total boundedness coincides with plain boundedness in the real line (see proof of Theorem 3.83). Thus we get the following particular case: *A subset $\Phi$ of the metric space $(C[0, 1], d_\infty)$ is compact if and only if it is pointwise bounded, equicontinuous and closed in $(C[0, 1], d_\infty)$.* Note: in this case pointwise boundedness means $\sup_{x \in \Phi} |x(t)| < \infty$ for each $t \in [0, 1]$.

## Suggested Reading

Bachman and Narici [1]
Brown and Pearcy [2]
Dieudonné [1]
Dugundji [1]
Goffman and Pedrick [1]
Kantorovich and Akilov [1]
Kelley [1]

Kolmogorov and Fomin [1]
Naylor and Sell [1]
Royden [1]
Schwartz [1]
Simmons [1]
Smart [1]
Sutherland [1]

# Problems

**Problem 3.1.** If $(X, d)$ is a metric space, then

$$\text{(a)} \qquad |d(x, y) - d(y, z)| \le d(x, z)$$

for every $x, y, z$ in $X$. (*Hint*: Use symmetry and the triangle inequality only.) Incidentally, the above inequality shows that the metric axioms (i) to (iv) in Definition 3.1 are not independent. For instance, the property $d(x, y) \ge 0$ in axiom (i) follows from symmetry and the triangle inequality. That is, "$d(x, y) \ge 0$ for every $x, y \in X$" in fact is a theorem derived by axioms (iii) and (iv). Moreover, show that

$$\text{(b)} \qquad |d(x, u) - d(v, y)| \le d(x, v) + d(u, y)$$

for every $u, v, x, y$ in $X$. (*Hint*: $d(x, u) \le d(x, v) + d(v, y) + d(y, u)$ and, similarly, $d(v, y) \le d(v, x) + d(x, u) + d(u, y)$ — use symmetry.)

**Problem 3.2.** Suppose $d_1 : X \times X \to \mathbb{R}$ and $d_2 : X \times X \to \mathbb{R}$ are metrics on a set $X$. Define the functions $d : X \times X \to \mathbb{R}$ and $d' : X \times X \to \mathbb{R}$ by

$$d(x, y) = d_1(x, y) + d_2(x, y) \quad \text{and} \quad d'(x, y) = \max \big\{ d_1(x, y), d_2(x, y) \big\}$$

for every $x, y \in X$. Show that both $d$ and $d'$ are metrics on $X$.

**Problem 3.3.** Let $p$ and $q$ be real numbers. If $p > 1$ and if $q = \frac{p}{p-1} > 1$ is the unique solution to the equation $\frac{1}{p} + \frac{1}{q} = 1$ (or, equivalently, the unique solution to the equation $p + q = pq$), then $p$ and $q$ are said to be *Hölder conjugates* of each other. Prove the following inequalities.

(a) If $p > 1$ and $q > 1$ are Hölder conjugates, and if $x = (\xi_1, \dots, \xi_n)$ and $y = (\upsilon_1, \dots, \upsilon_n)$ are arbitrary $n$-tuples in $\mathbb{C}^n$, then

$$\sum_{i=1}^{n} |\xi_i \upsilon_i| \le \left( \sum_{i=1}^{n} |\xi_i|^p \right)^{\frac{1}{p}} \left( \sum_{i=1}^{n} |\upsilon_i|^q \right)^{\frac{1}{q}}.$$

(*Hint*: Show that $\alpha \beta \le \frac{\alpha^p}{p} + \frac{\beta^q}{q}$ for every pair of positive real numbers $\alpha$ and $\beta$ whenever $p$ and $q$ are Hölder conjugates. Now set $\|x\|_p = \left( \sum_{i=1}^{n} |\xi_i|^p \right)^{\frac{1}{p}}$.) Moreover,

$$\sum_{i=1}^{n} |\xi_i \upsilon_i| \le \max_{1 \le i \le n} |\xi_i| \sum_{i=1}^{n} |\upsilon_i|.$$

These are the *Hölder inequalities for finite sums*.

(b) Let $x = \{\xi_k\}$ and $y = \{\upsilon_k\}$ be arbitrary complex-valued sequences (i.e., sequences in $\mathbb{C}^{\mathbb{N}}$). If $p > 1$ and $q > 1$ are Hölder conjugates, then

$$\sum_{k=1}^{\infty} |\xi_k \upsilon_k| \le \left( \sum_{k=1}^{\infty} |\xi_k|^p \right)^{\frac{1}{p}} \left( \sum_{k=1}^{\infty} |\upsilon_k|^q \right)^{\frac{1}{q}}$$

whenever $\sum_{k=1}^{\infty} |\xi_k|^p < \infty$ and $\sum_{k=1}^{\infty} |\upsilon_k|^q < \infty$; and

$$\sum_{k=1}^{\infty} |\xi_k \upsilon_k| \le \sup_{k \in \mathbb{N}} |\xi_k| \sum_{k=1}^{\infty} |\upsilon_k|$$

whenever $\sup_{k \in \mathbb{N}} |\xi_k| < \infty$ and $\sum_{k=1}^{\infty} |\upsilon_k| < \infty$. These are called *Hölder inequalities for infinite sums*.

(c) Finally, let $\Omega$ be a nonempty subset of $\mathbb{C}$, and let $x$ and $y$ be arbitrary complex-valued functions on $\Omega$ (i.e., functions in $\mathbb{C}^{\Omega}$). If $p > 1$ and $q > 1$ are Hölder conjugates, then

$$\int_{\Omega} |x \, y| \, d\omega \le \left( \int_{\Omega} |x|^p \, d\omega \right)^{\frac{1}{p}} \left( \int_{\Omega} |y|^q \, d\omega \right)^{\frac{1}{q}}$$

for all integrable functions $x$, $y$ in $\mathbb{C}^\Omega$ such that $\int_\Omega |x|^p \, d\omega < \infty$ and $\int_\Omega |y|^q \, d\omega < \infty$. Moreover, if $x$, $y \in \mathbb{C}^\Omega$ are integrable functions such that $\sup_{\omega \in \Omega} |x(\omega)| < \infty$ and $\int_\Omega |y| \, d\omega < \infty$, then

$$\int_\Omega |x \, y| \, d\omega \le \sup_{\omega \in \Omega} |x(\omega)| \int_\Omega |y| \, d\omega.$$

These are the *Hölder inequalities for integrals.*

**Problem 3.4.** Take any real number $p$ such that $p \ge 1$. Use the preceding problem to verify the following results.

(a) *Minkowski inequalities for finite sums.* If $x = (\xi_1, \dots, \xi_n)$ and $y = (\upsilon_1, \dots, \upsilon_n)$ are arbitrary $n$-tuples in $\mathbb{C}^n$, then

$$\left( \sum_{i=1}^n |\xi_i + \upsilon_i|^p \right)^{\frac{1}{p}} \le \left( \sum_{i=1}^n |\xi_i|^p \right)^{\frac{1}{p}} + \left( \sum_{i=1}^n |\upsilon_i|^p \right)^{\frac{1}{p}}$$

and

$$\max_{1 \le i \le n} |\xi_i + \upsilon_i| \le \max_{1 \le i \le n} |\xi_i| + \max_{1 \le i \le n} |\upsilon_i|.$$

*Hint*: Show that $|\xi + \upsilon| \le |\xi| + |\upsilon|$ for every pair $\{\xi, \upsilon\}$ of complex numbers, and also that $(\alpha + \beta)^p = (\alpha + \beta)^{p+1}\alpha + (\alpha + \beta)^{p+1}\beta$ for every pair $\{\alpha, \beta\}$ of nonnegative real numbers.

(b) *Minkowski inequalities for infinite sums.* If $x = \{\xi_k\}$ and $y = \{\upsilon_k\}$ are sequences in $\mathbb{C}^{\mathbb{N}}$ (i.e., complex-valued sequences) such that $\sum_{k=1}^\infty |\xi_k|^p < \infty$ and $\sum_{k=1}^\infty |\upsilon_k|^p < \infty$, then

$$\left( \sum_{k=1}^\infty |\xi_k + \upsilon_k|^p \right)^{\frac{1}{p}} \le \left( \sum_{k=1}^\infty |\xi_k|^p \right)^{\frac{1}{p}} + \left( \sum_{k=1}^\infty |\upsilon_k|^p \right)^{\frac{1}{p}}.$$

Moreover,

$$\sup_{k \in \mathbb{N}} |\xi_k + \upsilon_k| \le \sup_{k \in \mathbb{N}} |\xi_k| + \sup_{k \in \mathbb{N}} |\upsilon_k|$$

whenever $\sup_{k \in \mathbb{N}} |\xi_k| < \infty$ and $\sup_{k \in \mathbb{N}} |\upsilon_k| < \infty$.

(c) *Minkowski inequalities for integrals.* If $x$ and $y$ are integrable functions in $\mathbb{C}^\Omega$ (i.e., integrable complex-valued functions on $\Omega$) such that $\int_\Omega |x|^p \, d\omega < \infty$ and $\int_\Omega |y|^p \, d\omega < \infty$, where $\Omega$ is a nonempty subset of $\mathbb{C}$, then

$$\left( \int_\Omega |x + y|^p \, d\omega \right)^{\frac{1}{p}} \le \left( \int_\Omega |x|^p \, d\omega \right)^{\frac{1}{p}} + \left( \int_\Omega |y|^p \, d\omega \right)^{\frac{1}{p}}.$$

If $\sup_{\omega \in \Omega} |x(\omega)| < \infty$ and $\sup_{\omega \in \Omega} |y(\omega)| < \infty$, then

$$\sup_{\omega \in \Omega} |x(\omega) + y(\omega)| \le \sup_{\omega \in \Omega} |x(\omega)| + \sup_{\omega \in \Omega} |y(\omega)|.$$

**Problem 3.5.** Prove the *Jensen inequality*: If $p$ and $q$ are real numbers such that $0 < p < q$, then

$$\left( \sum_{k=1}^{\infty} |\xi_k|^q \right)^{\frac{1}{q}} \leq \left( \sum_{k=1}^{\infty} |\xi_k|^p \right)^{\frac{1}{p}}$$

for every $x = \{\xi_k\}$ in $\mathbb{C}^{\mathbb{N}}$ such that $\sum_{k=1}^{\infty} |\xi_k|^p < \infty$ (i.e., for every $p$-summable complex-valued sequence).

*Hint*: Show that $\sum_{k=1}^{\infty} \alpha_k^r \leq \left( \sum_{k=1}^{\infty} \alpha_k \right)^r$ for every $r \geq 1$ whenever the sequence of *nonnegative* real numbers $\{\alpha_k\}$ is such that $\sum_{k=1}^{\infty} \alpha_k < \infty$.

Now let $\ell_+^p$ and $\ell_+^\infty$ be the sets of all $p$-summable and bounded sequences from $\mathbb{F}^{\mathbb{N}}$, respectively, where either $\mathbb{F} = \mathbb{R}$ or $\mathbb{F} = \mathbb{C}$ (i.e., the sets of all $p$-summable and bounded scalar-valued sequences, respectively — see Example 3B). Verify that

$$1 \leq p < q \quad \text{implies} \quad \ell_+^p \subset \ell_+^q \subset \ell_+^\infty$$

(where $q$ is any real number greater than $p$), and show that these are in fact proper inclusions.

**Problem 3.6.** If $A$ and $B$ are nonempty and bounded subsets of a metric space $(X, d)$, then

$$\sup_{x \in A,\, y \in B} d(x, y) \leq \mathrm{diam}(A) + \mathrm{diam}(B) + d(A, B).$$

(*Hint*: $d(x, y) \leq d(x, a) + d(a, b) + d(b, y)$ for every $x, a \in A$ and every $y, b \in B$.) Now conclude that $A \cup B$ is bounded. Show that the union of a *finite* collection of bounded subsets of $X$ is a bounded subset of $X$.

**Problem 3.7.** Let $(X, d)$ be a metric space and define $d_i : X \times X \to \mathbb{R}$ for $i = 1, 2$ as follows.

$$d_1(x, y) = \frac{d(x, y)}{1 + d(x, y)} \quad \text{and} \quad d_2(x, y) = \min \{1, d(x, y)\}$$

for every $x, y \in X$. Show that $d_1$ and $d_2$ are metrics on $X$. Moreover, verify that every set in the metric spaces $(X, d_1)$ and $(X, d_2)$ is bounded.

**Problem 3.8.** Let $p, q$ and $r$ be positive numbers such that

$$\frac{1}{p} + \frac{1}{q} = \frac{1}{r}.$$

(a) Prove the following extension of the Hölder inequality for integrals.

$$\left( \int_\Omega |x\, y|^r \, d\omega \right)^{\frac{1}{r}} \leq \left( \int_\Omega |x|^p \, d\omega \right)^{\frac{1}{p}} \left( \int_\Omega |y|^q \, d\omega \right)^{\frac{1}{q}}$$

for all integrable functions $x, y$ in $\mathbb{C}^{\Omega}$ such that $\int_{\Omega} |x|^p \, d\omega < \infty$ and $\int_{\Omega} |y|^q \, d\omega < \infty$. (Note: Similar inequalities hold for finite and infinite sums.)

*Hint*: Verify that $\frac{p}{r}$ and $\frac{q}{r}$ are Hölder conjugates.

(b) Let $S$ be a bounded interval of the real line and let $d_p$ be the usual metric on $R^p(S)$ (see Example 3E). Show that

$$1 \le r < p \quad \text{implies} \quad R^p(S) \subset R^q(S).$$

*Hint*: Verify that $d_r(x, y) \le \mathrm{diam}(S)^{\frac{p-r}{r}} d_p(x, y)$ for every $x, y$ in $R^p(S)$ whenever $1 \le r < p$.

**Problem 3.9.** Let $\{(X_i, d_i)\}_{i=1}^{n}$ be a finite collection of metric spaces and set $Z = \prod_{i=1}^{n} X_i$, the Cartesian product of their underlying sets. Consider the functions $d_p$: $Z \times Z \to \mathbb{R}$ (for each real number $p \ge 1$) and $d_{\infty}$: $Z \times Z \to \mathbb{R}$ defined as follows.

$$d_p(x, y) = \left( \sum_{i=1}^{n} d_i(x_i, y_i)^p \right)^{\frac{1}{p}},$$

$$d_{\infty} = \max \left\{ d_k(x_i, y_i) \right\}_{i=1}^{n},$$

for every $x = (x_1, \dots, x_n)$ and $y = (y_1, \dots, y_n)$ in $Z = \prod_{i=1}^{n} X_i$. Show that these are metrics on $Z$.

Remark: Let $d: Z \times Z \to \mathbb{R}$ be any of the above metrics. The Cartesian product $\prod_{i=1}^{n} X_i$ equipped with $d$ is referred to as a *product space* of the metric spaces $\{(X_i, d_i)\}_{i=1}^{n}$, and the metric $d$ as a *product metric*. Sometimes, when the metric $d$ has been previously specified, it is convenient to denote the product space $(\prod_{i=1}^{n} X_i, d)$ by $\prod_{i=1}^{n}(X_i, d_i)$. This notation is particularly suitable for a product space of metric spaces with the same underlying set but different metrics. For instance, $(X, d_1) \times (X, d_2)$ and $(X, d_2) \times (X, d_1)$ are different metric spaces with the same underlying set $X \times X$.

**Problem 3.10.** Consider the real line $\mathbb{R}$ with its usual metric. Let $\{\alpha_n\}$ be a real-valued sequence (indexed by $\mathbb{N}$ or $\mathbb{N}_0$) and recall the following definitions.

(i)  $\{\alpha_n\}$ is bounded if $\sup_n |\alpha_n| < \infty$.

(ii)  $\{\alpha_n\}$ is increasing if $\alpha_n \le \alpha_{n+1}$ for every index $n$.

(iii)  $\{\alpha_n\}$ is decreasing if $\alpha_{n+1} \le \alpha_n$ for every index $n$.

(iv)  $\{\alpha_n\}$ is monotone if it is either increasing or decreasing.

Prove the next three statements.

(a) If $\{\alpha_n\}$ converges, then it is bounded.

(b) If $\{\alpha_n\}$ is monotone and bounded, then it converges.

Therefore, for a real-valued monotone sequence, boundedness becomes equivalent to convergence. Now suppose $\{\alpha_n\}$ is a nonnegative sequence (i.e., $0 \leq \alpha_n$ for every index $n$) and let $\{\beta_n\}$ be another real-valued sequence.

(c) If $0 \leq \alpha_n \leq \beta_n$ for every $n$ and $\beta_n \to 0$, then $\alpha_n \to 0$.

**Problem 3.11.** Consider again the real line $\mathbb{R}$ with its usual metric and let $\{\alpha_n\}$ be a sequence of *nonnegative* real numbers. For each integer $n \geq 1$ set

$$\sigma_n = \sum_{i=1}^{n} \alpha_i.$$

$\{\sigma_n\}$ is called the *sequence of partial sums* of $\{\alpha_n\}$, which clearly is nonnegative and increasing. Let us introduce the following usual notation and terminology.

(i)   If $\{\sigma_n\}$ converges, then we say that $\sum_{i=1}^{\infty} \alpha_i$ converges.

(ii)   If $\{\sigma_n\}$ is bounded, then we write $\sum_{i=1}^{\infty} \alpha_i < \infty$.

(iii)   For each index $n$ write $\sum_{i=n+1}^{\infty} \alpha_i = \sup_m \sigma_m - \sigma_n$.

The purpose of this problem is to show that the assertions

$$\sum_{i=1}^{\infty} \alpha_i \quad \text{converges},$$

$$\sum_{i=1}^{\infty} \alpha_i < \infty, \quad \text{and}$$

$$\sum_{i=n+1}^{\infty} \alpha_i \to 0 \quad \text{as} \quad n \to \infty$$

are pairwise equivalent. Prove the following propositions.

(a) If $\sup_m \sigma_m < \infty$, then $\sigma_n \to \sup_m \sigma_m$ as $n \to \infty$.

(b) If $\{\sigma_n\}$ converges, then $\lim_n \sigma_n = \sup_m \sigma_m$.

Obviously, the above convergences are all in $\mathbb{R}$ with its usual metric.

**Problem 3.12.** Let $\{\alpha_n\}$ be a sequence of *nonnegative* real numbers and equip the real line with its usual metric. Prove:

(a) If $\sum_{n=1}^{\infty} \alpha_n < \infty$, then $\lim_n \alpha_n = 0$.

(b) If $\{\alpha_n\}$ is decreasing and $\sum_{n=1}^{\infty} \alpha_n < \infty$, then $\lim_n n\alpha_n = 0$.

   *Hint*: Show that if $\lim_n \alpha_{2n} = \lim_n \alpha_{2n-1} = \alpha$, then $\lim_n \alpha_n = \alpha$.

Now exhibit a pair of nonnegative sequences $\{\beta_n\}$ and $\{\gamma_n\}$ with the following properties.

(c) $\sum_{n=1}^{\infty} \beta_n < \infty$ and $\sup_n n\beta_n = \infty$.

(d) $\{\gamma_n\}$ is decreasing, $\sum_{n=1}^{\infty} \gamma_n^2 < \infty$, and $\sup_n n\gamma_n = \infty$.

**Problem 3.13.** Let $\{\beta_n\}$ be a real-valued bounded sequence. Since the real line is a boundedly complete lattice in the natural ordering $\leq$ of $\mathbb{R}$ (Example 1C), it follows that both $\sup_{n \leq k} \beta_k = \sup\{\beta_k : n \leq k\}$ and $\inf_{n \leq k} \beta_k = \inf\{\beta_k : n \leq k\}$ exist in $\mathbb{R}$ for every index $n$. Set

$$\alpha_n = \inf_{n \leq k} \beta_k \quad \text{and} \quad \gamma_n = \sup_{n \leq k} \beta_k$$

for each $n$, consider the real-valued sequences $\{\alpha_n\}$ and $\{\gamma_n\}$, and equip the real line $\mathbb{R}$ with its usual metric.

(a) Show that both sequences $\{\alpha_n\}$ and $\{\gamma_n\}$ converge in $\mathbb{R}$.

*Hint*: These are bounded monotone sequences.

As they converge in $\mathbb{R}$, set

$$\alpha = \lim_n \alpha_n = \lim_n \inf_{n \leq k} \beta_k \quad \text{and} \quad \gamma = \lim_n \gamma_n = \lim_n \sup_{n \leq k} \beta_k.$$

The limits $\alpha \in \mathbb{R}$ and $\gamma \in \mathbb{R}$ are usually denoted by

$$\alpha = \liminf_n \beta_n \quad \text{and} \quad \gamma = \limsup_n \beta_n,$$

and called *limit inferior* (or *lower limit*) and *limit superior* (or *upper limit*) of the sequence $\{\beta_n\}$, respectively (see Problem 1.19).

(b) Show that $\liminf_n \beta_n \leq \limsup_n \beta_n$,

and prove the following propositions.

(c) If $\{\beta_n\}$ converges in $\mathbb{R}$, then $\liminf_n \beta_n = \limsup_n \beta_n = \lim_n \beta_n$.

*Hint*: Show that $|\alpha_n - \beta| \leq |\beta_n - \beta|$ for each $n$ and every $\beta \in \mathbb{R}$. Similarly, show that $|\beta_n - \gamma| \leq |\gamma_n - \gamma|$ for each $n$ and every $\gamma \in \mathbb{R}$.

(d) If $\liminf_n \beta_n = \limsup_n \beta_n = \beta$, then $\{\beta_n\}$ converges in $\mathbb{R}$ to $\beta$.

*Hint*: Show that $\alpha_n \leq \beta_n \leq \gamma_n$ for each $n$ and then apply the result of Problem 3.10(c).

Remark: If the real-valued sequence $\{\beta_n\}$ is not bounded above, then $\{\beta_k\}_{n \leq k}$ is not bounded above for every index $n$. In this case we write $\sup_{n \leq k} \beta_k = \infty$ for every $n$ and also $\limsup_n \beta_n = \infty$. Similarly, if $\{\beta_n\}$ is not bounded below, then $\{\beta_k\}_{n \leq k}$ is not bounded below for every index $n$. In this case we write $\inf_{n \leq k} \beta_k = -\infty$ for every $n$ and $\liminf_n \beta_n = -\infty$. Observe that the notations $\limsup_n \beta_n = \infty$ and $\liminf_n \beta_n = -\infty$ are mere formalisms; they simply say that $\{\beta_n\}$ is not bounded above or below, respectively. It should also be noticed that the concepts and results in this problem naturally generalize from sequences to nets of real numbers.

Warning: Let $\{\beta_n\}$ be an arbitrary real-valued sequence (bounded or not) and consider the subsets $B_n = \{\beta_k\}_{n \leq k}$ of $B = \{\beta_n\}$ for each index $n$. Since $\{B_n\}$ is a decreasing sequence of subsets of $B$ in the inclusion ordering, it follows by Problem 1.19 that

$$\lim_n B_n = \liminf_n B_n = \limsup_n B_n = \bigcap_n B_n,$$

which always exists as a set in the power set $\wp(\mathbb{R})$ (e.g., if $\beta_n = n$ for each $n \in \mathbb{N}$, then $\bigcap_{n \in \mathbb{N}} B_n = \varnothing$). Carefully note that the present problem is concerned with the natural ordering $\leq$ of $\mathbb{R}$, and not with the inclusion ordering $\subseteq$ of $\wp(\mathbb{R})$.

**Problem 3.14.** Let $\{x_n\}$ and $\{y_n\}$ be two convergent sequences (both indexed by $\mathbb{N}$ or $\mathbb{N}_0$) in a metric space $(X, d)$. Set $x = \lim x_n$ and $y = \lim y_n$ in $X$. Prove the following propositions.

(a) $d(x_n, u) \to d(x, u)$    and    $d(v, y_n) \to d(v, y)$    for each $u, v \in X$.

(b) $d(x_n, y_n) \to d(x, y)$    (*hint*: Problems 3.1(b) and 3.10(c)).

(c) If there exists $\alpha > 0$ and an integer $n_0$ such that $d(x_n, y_n) < \alpha$ for every $n \geq n_0$, then $d(x, y) \leq \alpha$.

   *Hint*: Use the triangle inequality to show that $d(x, y) < 2\varepsilon + \alpha$ for every $\varepsilon > 0$, so that $d(x, y) \leq \inf_{\varepsilon > 0}(2\varepsilon + \alpha)$.

**Problem 3.15.** Two (similarly indexed) sequences $\{x_n\}$ and $\{y_n\}$ in a metric space $(X, d)$ are *equiconvergent* if $\lim d(x_n, y_n) = 0$. Prove the following propositions.

(a) $\{x_n\}$ converges to $x$ in $(X, d)$ if and only if it is equiconvergent with the constant sequence $\{x'_n\}$ where $x'_n = x$ for all $n$.

(b) Two sequences that converge to the same limit are equiconvergent.

(c) If one of two equiconvergent sequences is convergent, then so is the other and both have the same limit.

**Problem 3.16.** Obviously, uniform continuity implies continuity.

(a)  Show that the function $f: \mathbb{R} \to \mathbb{R}$ defined by $f(x) = x^2$ for every $x \in \mathbb{R}$ is continuous but not uniformly continuous.

(b)  Prove that every Lipschitzian mapping is uniformly continuous.

(c)  Show that the function $g: [0, \infty) \to [0, \infty)$ defined by $g(x) = x^{\frac{1}{2}}$ for every $x \in [0, \infty)$ is uniformly continuous but not Lipschitzian.

**Problem 3.17.**  Consider the setup of Example 3H.

(a)  Show that the mapping $F: (X, d_2) \to (Y, d_2)$ of Example 3H(a) is nowhere continuous.

*Hint*: Take $x_0 \in X$ and $\delta > 0$ arbitrary. Let $\alpha$ be any positive real number such that $\frac{\delta^3}{18} \le \alpha^2 < \frac{\delta^3}{6}$ and consider the function $e_\delta: \mathbb{R} \to \mathbb{R}$ defined by

$$e_\delta(t) = \begin{cases} \alpha, & 0 \le t \le \frac{3}{\delta}, \\ -\alpha, & \frac{3}{\delta} < t \le \frac{6}{\delta}, \\ 0, & \text{otherwise.} \end{cases}$$

Verify that $e_\delta$ lies in $X$ so that $x_0 + e_\delta \in X$. Set $x_\delta = x_0 + e_\delta$ in $X$, $y_\delta = F(x_\delta)$ and $y_0 = F(x_0)$ in $Y$. Compute $y_\delta(t) - y_0(t)$ for $t \in \mathbb{R}$ and conclude that $d_2(x_\delta, x_0) < \delta$ and $d_2(F(x_\delta), F(x_0)) \ge 1$.

(b)  Show that the mapping $F: (R^2(S), d_2) \to (R^2(S), d_2)$ of Example 3H(b) is Lipschitzian (and hence uniformly continuous — *hint*: Hölder inequality).

**Problem 3.18.**  Let $C'[0, 1]$ be the subset of $C[0, 1]$ consisting of all differentiable functions from $C[0, 1]$ whose derivatives lie in $C[0, 1]$. Consider the subspace $(C'[0, 1], d_\infty)$ of the metric space $(C[0, 1], d_\infty)$ — see Example 3D. Let $D: C'[0, 1] \to C[0, 1]$ be the mapping that assigns to each $x \in C'[0, 1]$ its derivative in $C[0, 1]$.

(a)  Show that $D: (C'[0, 1], d_\infty) \to (C[0, 1], d_\infty)$ is nowhere continuous.

Now consider the function $d: C'[0, 1] \times C'[0, 1] \to \mathbb{R}$ defined by

$$d(x, y) = d_\infty(x, y) + d_\infty(D(x), D(y))$$

for every $x, y \in C'[0, 1]$, which is a metric on $C'[0, 1]$ (cf. Problem 3.2).

(b)  Show that $D: (C'[0, 1], d) \to (C[0, 1], d)$ is a contraction (thus Lipschitzian, and hence uniformly continuous).

**Problem 3.19.** Consider the real line $\mathbb{R}$ equipped with its usual metric. Set $B[0, \infty) = B[[0, \infty), \mathbb{R}]$ and $R^1[0, \infty) = R^1([0, \infty))$ as in Examples 3C and 3E, respectively. Let $d_\infty$ be the sup-metric on $B[0, \infty)$ and let $d_1$ be the usual metric on $R^1[0, \infty)$. Consider the set of all real-valued functions $x$ on $[0, \infty)$ that are 1-integrable (i.e., $x$ is Riemann integrable on $[0, \infty)$ and $\int_0^\infty |x(s)|\, ds < \infty$) and bounded (i.e., $\sup_{s \geq 0} |x(s)| < \infty$). Allowing a slight abuse of notation we write

$$X[0, \infty) = B[0, \infty) \cap R^1[0, \infty)$$

and set, for any real number $\alpha > 0$,

$$X[0, \alpha] = \{x \in X[0, \infty): \ x(s) = 0 \text{ for all } s > \alpha\}.$$

For each $x \in X[0, \infty)$ consider the function $y: [0, \infty) \to \mathbb{R}$ defined by the formula

$$y(t) = \int_0^t x(s)\, ds$$

for every $t \geq 0$. Let $BC[0, \infty) \subseteq B[0, \infty)$ denote the set of all real-valued bounded and continuous functions on $[0, \infty)$.

(a)  Show that $y \in BC[0, \infty)$.

Now consider the mapping $F: X[0, \infty) \to BC[0, \infty)$ that assigns to each function $x$ in $X[0, \infty)$ the function $y = F(x)$ in $BC[0, \infty)$ defined by the above formula. For simplicity we shall use the same notation $F$ for the restriction $F|_{X[0,\alpha]}$ of $F$ to $X[0, \alpha]$. By using the appropriate definitions, show that

(b)  $F: (X[0, \infty), d_1) \to (BC[0, \infty), d_\infty)$ is a contraction,

(c)  $F: (X[0, \infty), d_\infty) \to (BC[0, \infty), d_\infty)$ is nowhere continuous,

(d)  $F: (X[0, \alpha], d_\infty) \to (BC[0, \infty), d_\infty)$ is Lipschitzian.

**Problem 3.20.** Let $I$ be the identity map of $C[0, 1]$ onto itself and consider the metrics $d_\infty$ and $d_p$ (for any $p \geq 1$) on $C[0, 1]$ as in Example 3D. Verify that

(a)  $I: (C[0, 1]d_\infty) \to (C[0, 1], d_p)$ is a contraction,

(b)  $I: (C[0, 1]d_p) \to (C[0, 1], d_\infty)$ is nowhere continuous.

*Hint*: Consider the $C[0, 1]$-valued sequence $\{x_n\}$ of Example 3F and apply Theorem 3.7 to show that $I: (C[0, 1]d_p) \to (C[0, 1], d_\infty)$ is not continuous at $0 \in C[0, 1]$.

**Problem 3.21.** Recall that $\ell_+^p \subset \ell_+^\infty$ for every $p \geq 1$ (Problem 3.5) and consider the subspace $(\ell_+^p, d_\infty)$ of the metric space $(\ell_+^\infty, d_\infty)$. Let $I$ be the identity map of $\ell_+^p$ onto itself. Show that, for each $p \geq 1$,

(a)   $I: (\ell_+^p, d_p) \to (\ell_+^p, d_\infty)$ is a contraction,

(b)   $I: (\ell_+^p, d_\infty) \to (\ell_+^p, d_p)$ is nowhere continuous.

*Hint*: Use Theorem 3.7 to show that $I: (\ell_+^p, d_\infty) \to (\ell_+^p, d_p)$ is not continuous at $0 \in \ell_+^p$.

**Problem 3.22.** Let $\mathbb{F}$ denote either the real field $\mathbb{R}$ or the complex field $\mathbb{C}$ and let $\mathbb{F}^{\mathbb{N}}$ be the collection of all scalar-valued sequences indexed by $\mathbb{N}$. Consider the metric space $(\ell_+^p, d_p)$ of Example 3B for an arbitrary $p \geq 1$ and, for every $a = \{\alpha_k\} \in \mathbb{F}^{\mathbb{N}}$, consider the following subset of $\ell_+^p$.

$$X_a^p = \big\{ x = \{\xi_k\} \in \ell_+^p : \ \textstyle\sum_{k=1}^\infty |\alpha_k \xi_k|^p < \infty \big\}.$$

Let $D_a : (X_a^p, d_p) \to (\ell_+^p, d_p)$ be the *diagonal mapping* that assigns to each $x = \{\xi_k\}$ in $X_a^p$ the sequence $\{\alpha_k \xi_k\}$ in $\ell_+^p$; that is,

$$D_a(x) = \{\alpha_k \xi_k\} \quad \text{for every} \quad x = \{\xi_k\} \in X_a^p.$$

(a)   Show that $X_a^p = \ell_+^p$ and $D_a$ is Lipschitzian whenever $a \in \ell_+^\infty$ (i.e., whenever $\sup_k |\alpha_k| < \infty$).

(b)   If $\alpha_k = k$ for each $k \in \mathbb{N}$, then $D_a$ is not continuous. Prove this statement by using Theorem 3.7.

**Problem 3.23.** Take an arbitrary real number $\alpha \geq 0$ and set

$$\beta = \beta(\alpha) = \begin{cases} 1 - \alpha, & \alpha \leq 1, \\[2mm] \dfrac{\alpha - 1}{\alpha}, & \alpha \geq 1. \end{cases}$$

Consider the real-valued sequence $\{\xi_n\}$ recursively defined as follows.

$$\xi_0 = 0 \quad \text{and} \quad \xi_{n+1} = \tfrac{1}{2}(\beta + \xi_n^2)$$

for every $n \geq 0$. Verify that

(a)   $\{\xi_n\}$ is an increasing sequence,

(b)   $0 \leq \xi_n \leq \beta$ for all $n \geq 0$,

(c)   $\{\xi_n\}$ converges in $\mathbb{R}$, and

(d)   $\lim \xi_n = 1 - (1 - \beta)^{\frac{1}{2}}$.

*Hint*: According to Problem 3.16(a) the function $f : \mathbb{R} \to \mathbb{R}$ such that $f(x) = x^2$ for every $x \in \mathbb{R}$ is continuous. Use Theorem 3.7.

Thus conclude the *square root algorithm*: For every nonnegative real number $\alpha$,

$$\alpha^{\frac{1}{2}} = \begin{cases} 1 - \lim \xi_n, & \alpha \leq 1, \\ (1 - \lim \xi_n)^{-1}, & \alpha \geq 1, \end{cases}$$

where $\{\xi_n\}$ is recursively defined as above.

**Problem 3.24.** Take an arbitrary $C[0, 1]$-valued sequence $\{x_n\}$ that converges in $(C[0, 1], d_\infty)$ to $x \in C[0, 1]$. Take an arbitrary $[0,1]$-valued sequence $\{t_n\}$ that converges in $\mathbb{R}$ to $t \in [0, 1]$. Show that

$$x_n(t_n) \to x(t) \quad \text{in} \quad \mathbb{R}.$$

*Hint*: Recall that $\alpha_i + \beta_i \to \alpha + \beta$ whenever $\alpha_i \to \alpha$ and $\beta_i \to \beta$ in $\mathbb{R}$. Use Problem 3.10(c) and Theorem 3.7.

**Problem 3.25.** Consider the standard notion of curve length in the plane and let $D[0, 1]$ denote the subset of $C[0, 1]$ consisting of all real-valued functions on $[0, 1]$ whose graph has a finite length. Let $\varphi \colon D[0, 1] \to \mathbb{R}$ be the mapping that assigns to each function $x \in D[0, 1]$ the length of its graph (e.g., if $x \in C[0, 1]$ is given by $x(t) = (t - t^2)^{\frac{1}{2}}$ for every $t \in [0, 1]$, then $x \in D[0, 1]$ and $\varphi(x) = \frac{\pi}{2}$). Now consider the $D[0, 1]$-valued sequence $\{x_n\}$ defined as follows. For each $n \geq 1$ the graph of $x_n$ forms $n$ *equilateral* triangles of the same height when intercepted with the horizontal axis.

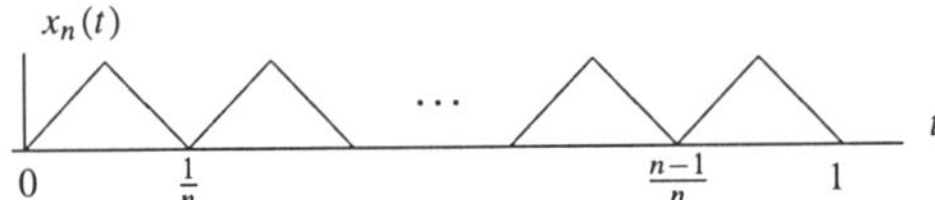


Use this sequence to show that the mapping $\varphi$ is not continuous when $D[0, 1]$ is equipped with the sup-metric $d_\infty$ and $\mathbb{R}$ is equipped with its usual metric. (*Hint*: Theorem 3.7.)

**Problem 3.26.** Let $C[0, \infty)$ denote the set of all real-valued continuous functions on $[0, \infty)$, set

$$XC[0, \infty) = X[0, \infty) \cap C[0, \infty) = B[0, \infty) \cap R^1[0, \infty) \cap C[0, \infty)$$

(with $X[0, \infty)$ defined as in Problem 3.19), and consider the mapping $\varphi \colon XC[0, \infty) \to \mathbb{R}$ given by

$$\varphi(x) = \int_0^\infty |x(t)|\, dt$$

for every $x \in XC[0, \infty)$. Let $\{x_n\}$ be the $XC[0, \infty)$-valued sequence such that, for each $n \geq 1$,

$$x_n(t) = \frac{1}{n^2} \begin{cases} t, & t \in [0, n], \\ 2n - t, & t \in [n, 2n], \\ 0, & \text{otherwise.} \end{cases}$$

Use this sequence to show that $\varphi \colon XC[0, \infty) \to \mathbb{R}$ is not continuous when $XC[0, \infty)$ is equipped with the sup-metric $d_\infty$ from $B[0, \infty)$ and $\mathbb{R}$ is equipped with its usual metric. (*Hint*: Theorem 3.7 — compare with Example 3I.)

**Problem 3.27.** Let $(X, d)$ be a metric space. A real-valued function $\varphi \colon X \to \mathbb{R}$ is *upper semicontinuous at a point* $x_0$ of $X$ if for each real number $\beta$ such that $\varphi(x_0) < \beta$ there exists a positive number $\delta$ such that

$$d(x, x_0) < \delta \quad \text{implies} \quad \varphi(x) < \beta.$$

It is *upper semicontinuous on X* if it is upper semicontinuous at every point of $X$. Similarly, a real-valued function $\psi \colon X \to \mathbb{R}$ is *lower semicontinuous at a point* $x_0$ of $X$ if for each real number $\alpha$ such that $\alpha < \psi(x_0)$ there exists a positive number $\delta$ such that

$$d(x, x_0) < \delta \quad \text{implies} \quad \alpha < \psi(x).$$

It is *lower semicontinuous on X* if it is lower semicontinuous at every point of $X$. Now equip the real line with its usual metric and prove the following proposition (which can be thought of as a real-valued semicontinuous version of Theorem 3.7).

(a)  $\varphi \colon X \to \mathbb{R}$ is upper semicontinuous at $x_0$ if and only if

$$\limsup_{n} \varphi(x_n) \leq \varphi(x_0)$$

for every $X$-valued sequence $\{x_n\}$ that converges in $(X, d)$ to $x_0$. Similarly, $\psi \colon X \to \mathbb{R}$ is lower semicontinuous at $x_0$ if and only if

$$\liminf_{n} \psi(x_n) \geq \psi(x_0)$$

for every $X$-valued sequence $\{x_n\}$ that converges in $(X, d)$ to $x_0$.

*Hint*: Take $\varepsilon > 0$ and $x_0$ arbitrary and set $\beta = \varphi(x_0) + \varepsilon$. Suppose $\varphi$ is upper semicontinuous at $x_0$ and show that there exists $\delta > 0$ such that

$$d(x, x_0) < \delta \quad \text{implies} \quad \varphi(x) < \varepsilon + \varphi(x_0).$$

If $x_n \to x_0$ in $(X, d)$, then show that there exists a positive integer $n_\delta$ such that

$$n \geq n_\delta \quad \text{implies} \quad \varphi(x_n) < \varepsilon + \varphi(x_0).$$

Now conclude: $\limsup_n \varphi(x_n) \leq \varphi(x_0)$. Conversely, if $\varphi\colon X \to \mathbb{R}$ is not upper semicontinuous at $x_0$, then verify that there exists $\beta > \varphi(x_0)$ such that for every $\delta > 0$ there exists $x_\delta \in X$ such that

$$d(x_\delta, x_0) < \delta \quad \text{and} \quad \varphi(x_\delta) \geq \beta.$$

Set $\delta_n = \frac{1}{n}$ and $x_n = x_{\delta_n}$ for each integer $n \geq 1$. Thus $x_n \to x_0$ in $(X, d)$ while $\varphi(x_n) \geq \beta$ for all $n$, and hence $\limsup_n \varphi(x_n) > \varphi(x_0)$.

(b)  Show that $\varphi\colon X \to \mathbb{R}$ is continuous if and only if it is both upper and lower semicontinuous on $X$.

*Hint*: Problem 3.13 and Theorem 3.7.

**Problem 3.28.**  Show that the composition of uniformly continuous mappings is again uniformly continuous.

**Problem 3.29.**  Let $X$, $Y$ and $Z$ be metric spaces. If $F\colon X \to Y$ is continuous at $x_0$, and if $G\colon Y \to Z$ is continuous at $F(x_0)$, then the composition $G \circ F\colon X \to Z$ is continuous at $x_0$. (*Hint*: Lemma 3.11.)

**Problem 3.30.**  The restriction of a continuous mapping to a subspace is continuous. That is, if $F\colon X \to Y$ is a mapping of a metric space $X$ into a metric space $Y$, and if $A$ is a subspace of $X$, then the restriction $F|_A\colon A \to Y$ of $F$ to $A$ is continuous. (*Hint*: $F|_A = F \circ J$ where $J\colon A \to X$ is the inclusion map — use Corollary 3.13.)

**Problem 3.31.**  Let $X$ be an arbitrary set. The largest (or strongest) topology on $X$ is the discrete topology $\wp(X)$ (where every subset of $X$ is open), and the smallest (or weakest) topology on $X$ is the indiscrete topology (where the only open subsets of $X$ are the empty set and the whole space). The collection of all topologies on $X$ is partially ordered by the inclusion ordering $\subseteq$. Recall that $\mathcal{T}_1 \subseteq \mathcal{T}_2$ (which means $\mathcal{T}_1$ is weaker than $\mathcal{T}_2$ or, equivalently, $\mathcal{T}_2$ is stronger than $\mathcal{T}_1$) if every element of $\mathcal{T}_1$ also is an element of $\mathcal{T}_2$. Prove the following propositions.

(a)  If $\mathcal{U} \subseteq \wp(X)$ is an arbitrary collection of subsets of $X$, then there exists a smallest (weakest) topology $\mathcal{T}$ on $X$ such that $\mathcal{U} \subseteq \mathcal{T}$.

*Hint*: The power set $\wp(X)$ is a topology on $X$. The intersection of a nonempty collection of topologies on $X$ is again a topology on $X$.

(b)  Show that the collection of all topologies on $X$ is a complete lattice in the inclusion ordering.

**Problem 3.32.**  Let $d_1$ and $d_2$ be two metrics on a set $X$. Show that $d_1$ and $d_2$ are equivalent if and only if for each $x_0 \in X$ and each $\varepsilon > 0$ the following two conditions hold.

(i)   There exists $\delta_1 > 0$ such that

$$d_1(x, x_0) < \delta_1 \quad \text{implies} \quad d_2(x, x_0) < \varepsilon.$$

(ii)  There exists $\delta_2 > 0$ such that

$$d_2(x, x_0) < \delta_2 \quad \text{implies} \quad d_1(x, x_0) < \varepsilon.$$

*Hint*: $d_1 \sim d_2$ if and only if the identity $I: (X, d_1) \to (X, d_2)$ is a homeomorphism.

**Problem 3.33.** Let $\{(X_i, d_i)\}_{i=1}^{n}$ be a finite collection of metric spaces and let $Z = \prod_{i=1}^{n} X_i$ be the Cartesian product of their underlying sets. Consider the metrics $d_p$ for each $p \geq 1$ and $d_\infty$ on $Z$ that were defined in Problem 3.9. Show that, for an arbitrary $p \geq 1$,

$$d_\infty(x, y) \leq d_p(x, y) \leq d_1(x, y) \leq n\, d_\infty(x, y)$$

for every $x = (x_1, \ldots, x_n)$ and $y = (y_1, \ldots, y_n)$ in $Z = \prod_{i=1}^{n} X_i$.

*Hint*: $d_p(x, y) \leq d_1(x, y)$ by the Jensen inequality (Problem 3.5).

Thus conclude that the metrics $d_\infty$ and $d_p$ for every $p \geq 1$ are all uniformly equivalent on $Z$, so that the product spaces $(\prod_{i=1}^{n} X_i, d_\infty)$ and $(\prod_{i=1}^{n} X_i, d_p)$ for every $p \geq 1$ are all uniformly homeomorphic.

**Problem 3.34.** Let $(X, d)$ be a metric space and equip the real line $\mathbb{R}$ with its usual metric. Take $u, v \in X$ arbitrary and note that both functions

$$d(\cdot, u): (X, d) \to \mathbb{R} \quad \text{and} \quad d(v, \cdot): (X, d) \to \mathbb{R}$$

preserve convergence (Problem 3.14(a)), and hence they are continuous by Corollary 3.8. Now consider the Cartesian product $X \times X$ equipped with the metric $d_1$ (see Problem 3.9: $d_1(z, w) = d(x, u) + d(y, v)$ for every $z = (x, y)$ and $w = (u, v)$ in $X \times X$).

(a)   Show that $d(\cdot, \cdot): (X \times X, d_1) \to \mathbb{R}$ is continuous.

   *Hint*: If $(x_n, y_n) \to (x, y)$ in $(X \times X, d_1)$, then $x_n \to x$ and $y_n \to y$ in $(X, d)$. Verify. Now use Problem 3.14(b) and Corollary 3.8.

Next let $d'$ denote any of the metrics $d_p$ (for an arbitrary $p \geq 1$) or $d_\infty$ on $X \times X$ as in Problem 3.9.

(b)   Show that $d(\cdot, \cdot): (X \times X, d') \to \mathbb{R}$ is continuous.

   *Hint*: See Problem 3.33 and Corollary 3.19. They ensure that the identity map $I: (X \times X, d') \to (X \times X, d_1)$ is a (uniform) homeomorphism. Now use item (a) and Corollary 3.13.

**Problem 3.35.** If $d$ and $d'$ are equivalent metrics on a set $X$, then an $X$-valued sequence $\{x_n\}$ converges in $(X, d)$ to $x \in X$ if and only if it converges in $(X, d')$ to the same limit $x$ (Corollary 3.19). If $d$ and $d'$ are not equivalent, then it may happen that an $X$-valued sequence $\{x_n\}$ converges to $x \in X$ in $(X, d)$ but does not converge (to any point) in $(X, d')$ (e.g., see Examples 3F and 3G). Can a sequence converge in $(X, d)$ to a point $x \in X$ and also converge in $(X, d')$ to a different point $x' \in X$? Yes, it can. We shall equip a set $X$ with two metrics $d$ and $d'$, and exhibit an $X$-valued sequence $\{x_n\}$ and a pair of distinct points $x$ and $x'$ in $X$ such that

$$x_n \to x \quad \text{in} \quad (X, d) \qquad \text{and} \qquad x_n \to x' \quad \text{in} \quad (X, d').$$

Consider the set $\mathbb{R}^2$ and let $d$ denote the Euclidean metric on it (or any of the metrics on $\mathbb{R}^2$ introduced in Example 3A, which are uniformly equivalent according to Problem 3.33). Set $v = (0, 1) \in \mathbb{R}^2$ and let $V$ be the vertical axis joining $v$ to the point $0 = (0, 0) \in \mathbb{R}^2$ (in the jargon of Chapter 2, set $V = \mathrm{span}\,\{v\}$). Now consider a function $d' \colon \mathbb{R}^2 \times \mathbb{R}^2 \to \mathbb{R}$ defined as follows. If $x$ and $y$ are both in $\mathbb{R}^2 \backslash V$ or both in $V$, then set $d'(x, y) = d(x, y)$. If one of them is in $V$ but not the other, then $d'(x, y) = d(x + v, y)$ if $x \in V$ and $y \in \mathbb{R}^2 \backslash V$, or $d'(x, y) = d(x, y + v)$ if $x \in \mathbb{R}^2 \backslash V$ and $y \in V$.

   (a)  Show that $d'$ is a metric on $\mathbb{R}^2$.

      *Hint*: If $x, y \in V$, then $d'(x, y) = d(x + v, y + v)$.

Next consider the $\mathbb{R}^2$-valued sequence $\{x_n\}$ where $x_n = (\frac{1}{n}, 1)$ for each $n \geq 1$. Show that

  (b)      $x_n \to v \quad \text{in} \quad (\mathbb{R}^2, d) \qquad \text{and} \qquad x_n \to 0 \quad \text{in} \quad (\mathbb{R}^2, d').$

(This construction was communicated by Ivo Fernandez Lopez).

**Problem 3.36.** Upper and lower semicontinuity were defined in Problem 3.27. Equip the real line with its usual metric, let $X$ be a metric space, and consider the following statement.

   (i)  Suppose $\varphi \colon X \to \mathbb{R}$ is an upper semicontinuous function on $X$ and suppose $\psi \colon X \to \mathbb{R}$ is a lower semicontinuous function on $X$. If $\varphi(x) \leq \psi(x)$ for every $x \in X$, then there exists a continuous function $\eta \colon X \to \mathbb{R}$ such that $\varphi(x) \leq \eta(x) \leq \psi(x)$ for every $x \in X$.

This is the *Hahn Interpolation Theorem*. Use it to prove the *Tietze Extension Theorem* which is stated below.

   (ii)  Let $A$ be a nonempty and closed subset of a metric space $X$. If $f \colon A \to \mathbb{R}$ is a bounded and continuous function on $A$, then it has a continuous extension $g \colon X \to \mathbb{R}$ over the whole space $X$. Moreover, $\inf_{x \in X} g(x) = \inf_{a \in A} f(a)$ and $\sup_{x \in X} g(x) = \sup_{a \in A} f(a)$.

*Hint*: Verify that the functions $\varphi : X \to \mathbb{R}$ and $\psi : X \to \mathbb{R}$ defined by

$$\varphi(x) = \begin{cases} f(x) & x \in A, \\ \inf_{a \in A} f(a), & x \in X \backslash A, \end{cases} \quad \text{and} \quad \psi(x) = \begin{cases} f(x), & x \in A, \\ \sup_{a \in A} f(a), & x \in X \backslash A, \end{cases}$$

extend $f$ over $X$ and satisfy the hypothesis of the Hahn Interpolation Theorem. To show that $\varphi$ is upper semicontinuous at an arbitrary point $a_0 \in A$, use the fact that $f$ is continuous on $A$ (take any $\beta > f(a_0)$ and set $\varepsilon = \beta - f(a_0)$). To show that $\varphi$ is upper semicontinuous at an arbitrary point $x_0 \in X \backslash A$, use the fact that $X \backslash A$ is open in $X$ and hence it includes an open ball $B_\delta(x_0)$ centered at $x_0$ for some radius $\delta > 0$ ($d(x, x_0) < \delta$ implies $\varphi(x) < \beta$ for every $\beta > \varphi(x_0)$).

**Problem 3.37.** We can define neighborhoods in a general topological space as we did in a metric space. Precisely, a *neighborhood* of a point $x$ in a topological space $X$ is any subset of $X$ that includes an open subset which contains $x$.

(a)  Show that a subset of a topological space $X$ is open in $X$ if and only if it is a neighborhood of each one of its points.

A topological space $X$ is a *Hausdorff space* if for every pair of distinct points $x$ and $y$ in $X$ there exist neighborhoods $N_x$ and $N_y$ of $x$ and $y$, respectively, such that $N_x \cap N_y = \varnothing$. Prove the following propositions.

(b)  Each singleton in a Hausdorff space is closed (i.e., $X \backslash \{x\}$ is open in $X$ for every $x \in X$).

(c)  Every metric space is a Hausdorff space (with respect to the metric topology).

(d)  For every pair of distinct points $x$ and $y$ in a metric space $X$ there exist nonempty open balls $B_\varepsilon(x)$ and $B_\rho(y)$ centered at $x$ and $y$, respectively, such that $B_\varepsilon(x) \cap B_\rho(y) = \varnothing$.

**Problem 3.38.** Let $(X, \mathcal{T}_X)$ be a topological space and let $A$ be a subset of $X$. A set $U' \subseteq A$ is said to be *open relative* to $A$ if $U' = A \cap U$ for some $U \in \mathcal{T}_X$.

(a)  Show that the collection $\mathcal{T}_A$ of all relatively open subsets of $A$ is a topology on $A$.

$\mathcal{T}_A$ is called the *relative topology* on $A$. When a subset $A$ of $X$ is equipped with this relative topology it is called a *subspace* of $X$; that is, $(A, \mathcal{T}_A)$ is a subspace of $(X, \mathcal{T}_X)$. If a subspace $A$ of $X$ is an open subset of $X$, then it is called an *open subspace* of $X$. Similarly, if it is a closed subset of $X$, then it is called a *closed subspace* of $X$. If $(Y, \mathcal{T}_Y)$ is a topological space, and if $F : X \to Y$ is a mapping of a set $X$ into $Y$, then the collection $\mathcal{F}^{-1}(\mathcal{T}_Y) = \{F^{-1}(U) : U \in \mathcal{T}_Y\}$ of all inverse images $F^{-1}(U)$ of open sets $U$ in $Y$ forms a topology on $X$. This is the topology *inversely induced* on $X$ by $F$, which is the weakest topology on $X$ that makes $F$ continuous.

(b)  Verify that the relative topology on $A$ is the topology inversely induced on $A$ by the inclusion map of $A$ into $X$. (Recall: the inclusion map of $A$ into $X$ is the function $J : A \to X$ defined by $J(x) = x$ for every $x \in A$.)

Now let $(X, d)$ be a metric space and let $\mathcal{T}_X$ be the metric topology on $X$. Suppose $A$ is a subset of $X$ and consider the (metric) subspace $(A, d)$ of the metric space $(X, d)$. Let $\mathcal{T}_A'$ be the metric topology induced on $A$ by the relative metric (i.e., let $\mathcal{T}_A'$ be the collection of all open sets in the metric space $(A, d)$).

(c)  Show that $U' \subseteq A$ is open in $(A, d)$ if and only if $U' = A \cap U$ for some $U \subseteq X$ open in $(X, d)$.

Thus the metric topology $\mathcal{T}_A'$ induced on $A$ by the relative metric coincides with the relative topology $\mathcal{T}_A$ induced on $A$ by the metric topology $\mathcal{T}_X$ on $X$; that is, $\mathcal{T}_A' = \mathcal{T}_A$, and hence the notion of *subspace* is unambiguously defined in a metric space. Let $A$ be a subspace of a metric space $X$.

(d)  Show that $V' \subseteq A$ is closed in $A$ (or *closed relative* to $A$) if and only if $V' = A \cap V$ for some closed subset $V$ of $X$.

   *Hint*: $A \backslash (A \cap B) = A \cap (X \backslash B)$ for arbitrary subsets $A$ and $B$ of $X$.

(e)  Open subsets of an open subspace of $X$ are open sets in $X$. Dually, closed subsets of a closed subspace of $X$ are closed sets in $X$. Prove.

Let $A$ be a subset of $B$ ($A \subseteq B \subseteq X$) and let $A^-$ and $B^-$ be the closures of $A$ and $B$ in $X$, respectively. Prove the following propositions.

(f)  $B \cap A^-$ coincides with the closure of $A$ in the subspace $B$.

(g)  $A$ is dense in the subspace $B$ if and only if $A^- = B^-$.

**Problem 3.39.**  A point $x$ in a metric space $X$ is a *condensation point* of a subset $A$ of $X$ if the intersection of $A$ with every nonempty open ball centered at $x$ is uncountable. Let $A'$ denote the set of all condensation points of an arbitrary set $A$ in $X$. Show that $A'$ is closed in $X$ and $A' \subseteq A^\star$, where $A^\star$ is the derived set of $A$.

**Problem 3.40.**  Take an arbitrary real number $p \geq 1$ and consider the metric space $(R^p[0, 1], d_p)$ of Example 3E. Let $C[0, 1]$ be the set of all scalar-valued continuous functions on the interval $[0, 1]$ as in Example 3D. Recall that $C[0, 1] \subset R^p[0, 1]$. This inclusion is interpreted in the following sense: if $x \in C[0, 1]$, then $[x] \in R^p[0, 1]$. Therefore, we are identifying $C[0, 1]$ with the collection of all equivalence classes $[x] = \{x' \in r^p[0, 1] : \delta_p(x', x) = 0\}$ that contain a continuous function $x \in C[0, 1]$ (see Example 3E). Use the Closed Set Theorem to show that $C[0, 1]$ is neither closed nor open in $(R^p[0, 1], d_p)$.

*Hint*: As usual, write $x$ for $[x]$. Consider the $C[0, 1]$-valued sequence $\{x_n\}_{n=1}^\infty$ and the $R^p[0, 1]\backslash C[0, 1]$-valued sequence $\{y_n\}_{n=1}^\infty$ defined by

$$x_n(t) = \begin{cases} 1, & t \in [0, \frac{1}{2}], \\ n+1-2nt, & t \in [\frac{1}{2}, \frac{n+1}{2n}], \\ 0, & t \in [\frac{n+1}{2n}, 1], \end{cases} \quad \text{and} \quad y_n(t) = \begin{cases} 1, & t \in [0, \frac{1}{n}), \\ 0, & t \in [\frac{1}{n}, 1]. \end{cases}$$

**Problem 3.41.** Let $A$ be a subset of a metric space $X$. The *boundary* of $A$ is the set $\partial A = A^- \backslash A^\circ$. A point $x \in X$ is a *boundary point* of $A$ if it belongs to $\partial A$. Prove the following propositions.

(a)  $\partial A = A^- \cap (X \backslash A)^- = X \backslash (A^\circ \cup (X \backslash A)^\circ) = \partial (X \backslash A)$.

(b)  $A^- = A^\circ \cup \partial A$, so that $A = A^-$ if and only if $\partial A \subseteq A$.

(c)  $\partial A$ is a closed subset of $X$ (i.e., $(\partial A)^- = \partial A$).

(d)  $A^\circ \cap \partial A = \varnothing$, so that $A = A^\circ$ if and only if $A \cap \partial A = \varnothing$.

(e)  The collection $\{A^\circ, \partial A, (X \backslash A)^\circ\}$ is a partition of $X$.

(f)  $\partial A \cap \partial B = \varnothing$ implies $(A \cup B)^\circ = A^\circ \cup B^\circ$ (for $B \subseteq X$).

**Problem 3.42.** Let $(X, d)$ be a metric space.

(a)  Show that a closed ball is a closed set.

Let $B_\rho(x)$ and $B_\rho[x]$ be arbitrary nonempty open and closed balls, respectively, both centered at the same point $x \in X$ and with the same radius $\rho > 0$. Prove the following propositions.

(b)  $B_\rho[x]^\circ = B_\rho(x)$    and    $\partial B_\rho[x] = \{y \in X : d(y, x) = \rho\}$.

(c)  $B_\rho(x)^- \subseteq B_\rho[x]$    and    $\partial B_\rho(x) \subseteq \partial B_\rho[x]$.

(d)  Show that the above inclusions may be proper.

 *Hint*: $X = [0, 1] \cup [2, 3]$, $x = 1$ and $\rho = 1$.

**Problem 3.43.** Let $A$ be an arbitrary subset of a metric space $(X, d)$. Show that

(a)  $\mathrm{diam}(A^\circ) \leq \mathrm{diam}(A) = \mathrm{diam}(A^-) = \mathrm{diam}(\partial A)$,

(b)  $d(x, A) = d(x, A^-)$    and    $d(x, A) = 0$ if and only if $x \in A^-$.

**Problem 3.44.** For an arbitrary $p \geq 1$ let $\ell_+^p$ be the set of all scalar-valued $p$-summable sequences as in Example 3B. Let $\ell_+^e$ be the set of all scalar-valued sequences $\{\xi_k\}_{k \in \mathbb{N}}$ for which there exist $\rho > 1$ and $\alpha \in (0, 1)$ such that $|\xi_k| \leq \rho \alpha^k$ for every $k \in \mathbb{N}$; and let $\ell_+^0$ be the set of all scalar-valued sequences with a finite number of nonzero entries.

(a) Prove: If $p$ and $q$ are real numbers such that $1 \leq p < q$, then

$$\ell_+^0 \subset \ell_+^e \subset \ell_+^p \subset \ell_+^q ,$$

where the above inclusions are all proper. (*Hint*: Problem 3.5.)

(b) Moreover, show that the sets $\ell_+^0$, $\ell_+^e$ and $\ell_+^p$ are all dense in the metric space $(\ell_+^q, d_q)$ of Example 3B. (*Hint*: Example 3P.)


**Problem 3.45.** Let $A$ and $B$ be subsets of a metric space $X$. Recall: $(A \cap B)^- \subseteq A^- \cap B^-$ and this inclusion may be proper. For instance, the sets in Example 3M are disjoint while their closures are not. We shall now exhibit a pair of sets $A$ and $B$ with the property

$$A \cap B \neq \varnothing \quad \text{and} \quad (A \cap B)^- \neq A^- \cap B^-.$$

Consider the metric space $(\ell_+^2, d_2)$ as in Example 3B. Set

$$A = \ell_+^1 \quad \text{and} \quad B = \{x_\beta \in \ell_+^2 : \ x_\beta = \beta\{\tfrac{1}{k}\}_{k \geq 1} \text{ for some } \beta \in \mathbb{C}\}.$$

Recall that $\ell_+^1 \subset \ell_+^2$ (Problem 3.5) and $(\ell_+^1)^- = \ell_+^2$ in $(\ell_+^2, d_2)$ (i.e., the set $\ell_+^1$ is dense in the metric space $(\ell_+^2, d_2)$ — Problem 3.44(b)). Show that $B = B^-$; that is, $B$ is closed in $(\ell_+^2, d_2)$. (*Hint*: Theorem 3.30.) Thus conclude that $A \cap B = \{0\}$, $(A \cap B)^- = \{0\}$ and $A^- \cap B^- = B$.


**Problem 3.46.** Suppose $F: X \to Y$ is a continuous mapping of a metric space $X$ into a metric space $Y$. Show that

(a) $$F(A^-) \subseteq F(A)^-$$

for every $A \subseteq X$. (*Hint*: Problem 1.2 and Theorem 3.23.) Now use the above result to conclude that, if $A \subseteq X$ and $C \subseteq Y$, then

(b) $$F(A) \subseteq C \quad \text{implies} \quad F(A^-) \subseteq C^-.$$

Finally, prove that

(c) $$A^- = B^- \quad \text{implies} \quad F(A)^- = F(B)^-$$

whenever $A \subseteq B \subseteq X$. (*Hint*: Proposition 3.32 and Corollary 3.8.) Thus, if $A$ is dense in $X$ and if $F$ is continuous and has a dense range (in particular, if $F$ is surjective), then $F(A)$ is dense in $Y$.

**Problem 3.47.** Consider the metric space $(\ell_+^p, d_p)$ for any $p \geq 1$, take an arbitrary scalar-valued sequence $a = \{\alpha_n\}_{n \geq 1}$ from $\ell_+^\infty$, and let $D_a : (\ell_+^p, d_p) \to (\ell_+^p, d_p)$ be the diagonal mapping defined in Problem 3.22. Suppose the bounded sequence $\{\alpha_n\}_{n \geq 1}$ is such that $\alpha_n \neq 0$ for every $n \geq 1$. Let $\ell_+^0$ denote the set of all scalar-valued sequences with a finite number of nonzero entries.

(a)  Show that $\mathcal{R}(D_a)^- = \ell_+^p$ (i.e., the range of $D_a$ is dense in $(\ell_+^p, d_p)$).

   *Hint*: Verify that $\ell_+^0 \subseteq \mathcal{R}(D_a) \subseteq \ell_+^p$ (see Problem 3.44).

(b)  Show that $D_a(\ell_+^0)^- = \ell_+^p$.

   *Hint*: Problems 3.22(a), 3.44(b) and 3.46(c).

**Problem 3.48.** Prove the following results.

(a)  If $X$ is a separable metric space and $F : X \to Y$ is a continuous and surjective mapping of $X$ onto a metric space $Y$, then $Y$ is separable (i.e., a continuous mapping preserves separability).

   *Hint*: Recall that, if there exists a surjective function of a set $A$ onto a set $B$, then $\#A \leq \#B$. Use Problems 3.46.

(b)  Separability is a topological invariant.

**Problem 3.49.** Verify the following propositions.

(a)  A metric space $X$ is separable if and only if there exists a countable subset $A$ of $X$ such that every nonempty open ball centered at each $x \in X$ meets $A$ (i.e., if and only if there exists a countable subset $A$ of $X$ such that every $x \in X$ is a point of adherence of $A$). (*Hint*: Propositions 3.27 and 3.32.)

(b)  The product space $(\prod_{i=1}^n X_i, d)$ of a finite collection $\{(X_i, d_i)\}_{i=1}^n$ of separable metric spaces is a separable metric space. (Note: $d$ is any of the metrics on $\prod_{i=1}^n X_i$ that were defined in Problem 3.9.)

(c)  If $\ell_+^{c_o}$ is the set of all scalar-valued sequences that converge to zero, then

$$(\ell_+^{c_o}, d_\infty) \text{ is a separable metric space.}$$

*Hint*: According to Proposition 3.39(b) $(\ell_+^{Co}, d_\infty)$ is a subspace of the non-separable metric space $(\ell_+^\infty, d_\infty)$ of Example 3Q. Show that the set of all rational-valued sequences with a finite number of nonzero entries is countable and dense in $(\ell_+^{Co}, d_\infty)$ (but not dense in $(\ell_+^\infty, d_\infty)$). See Example 3P. Note: We say that a complex number is "rational" if its real and imaginary parts are rational numbers.

(d)  The metric space $(\ell_+^{Co}, d_\infty)$ is not homeomorphic to $(\ell_+^\infty, d_\infty)$.

**Problem 3.50.** A subset $A$ of a topological space $X$ that is both closed and open is called *clopen* (or *closed-open*). A partition $\{A, B\}$ of $X$ into the union of two nonempty disjoint clopen sets $A$ and $B$ is a *disconnection* of $X$ (i.e., $\{A, B\}$ is a disconnection of $X = A \cup B$ if $A \cap B = \varnothing$ and $A$ and $B$ are both clopen subsets of $X$). If there exists a disconnection of $X$, then $X$ is called *disconnected*. Otherwise $X$ is called *connected*. In other words, *a topological space is connected if and only if the only clopen subsets of $X$ are the whole space $X$ and the empty set $\varnothing$*. A subset $A$ of $X$ is a *connected set* if, as a subspace of $X$, $A$ is a connected topological space. A topological space is *totally disconnected* if there is no connected subset of it containing two distinct points. Prove the following propositions.

(a)  $X$ is disconnected if and only if it is the union of two disjoint nonempty open sets.

(b)  $X$ is disconnected if and only if it is the union of two disjoint nonempty closed sets.

(c)  If $A$ is a connected set in a topological space $X$, and if $A \subseteq B \subseteq A^-$, then $B$ is connected. In particular, $A^-$ is connected whenever $A$ is connected.

Note: The closure $A^-$ of a set $A$ is defined in a topological space $X$ in the same way we have defined it in a metric space: the smallest closed subset of $X$ including $A$.

(d)  The continuous image of a connected set is a connected set.

(e)  Connectedness is a topological invariant.

Recall that a subset $A$ of a topological $X$ is discrete if it consists entirely of isolated points (i.e., if every point in $A$ does not belong to $(A\backslash\{x\})^-$, which means that in the subspace $A$ every set is open in $A$). Suppose $A$ is a discrete subset of $X$. If $B$ is an arbitrary subset of $A$ that contains more than one point, and if $x \in B$, then $B\backslash\{x\}$ and $\{x\}$ are both nonempty and open in $A$. Thus $B$ is disconnected, and hence $A$ is totally disconnected. Conclusion: *A discrete subset of a topological space is totally disconnected*.

(f)  Show that the converse of the above italicized proposition fails.

*Hint*: Verify that $\mathbb{Q}$ is a totally disconnected subset of $\mathbb{R}$ that is dense in itself (i.e., it has no isolated point).

**Problem 3.51.** Prove the following proposition.

(a) $\{x_n\}$ is a Cauchy sequence in a metric space $(X, d)$ if and only if

$$\limsup_{n \;\; k \geq 1} d(x_{n+k}, x_n) = 0$$

(i.e., if and only if $\{d(x_{n+k}, x_n)\}$ converges to zero uniformly in $k$).

(b) Show that the real-valued sequence $\{x_n\}$ such that $x_n = \log n$ for each $n \geq 1$ has the property

$$\lim_n d(x_{n+k}, x_n) = 0$$

for every $k \geq 1$, where $d$ is the usual metric on $\mathbb{R}$. (*Hint*: $\log \colon (0, \infty) \to \mathbb{R}$ is continuous — use Corollary 3.8.) However $\{x_n\}$ is not a Cauchy sequence (it is not even bounded).

**Problem 3.52.** Let $(X, d)$ be a metric space and let $\{x_n\}$ be an $X$-valued sequence. We say that $\{x_n\}$ is of *bounded variation* if

$$\sum_{n=1}^{\infty} d(x_{n+1}, x_n) < \infty.$$

If there exist real constants $\rho \geq 1$ and $\alpha \in (0, 1)$ such that

$$d(x_{n+1}, x_n) \leq \rho \alpha^n$$

for every $n$, then we say that $\{x_n\}$ has *exponentially decreasing increments*. Prove the following propositions.

(a) If a sequence in a metric space has exponentially decreasing increments, then it is of bounded variation.

(b) If a sequence in a metric space is of bounded variation, then it is a Cauchy sequence.

Thus, *if $(X, d)$ is a complete metric space, then every sequence of bounded variation converges in $(X, d)$, which implies that every sequence with exponentially decreasing increments converges in $(X, d)$.*

(c) Every Cauchy sequence in a metric space has a subsequence with exponentially decreasing increments (and therefore every Cauchy sequence in a metric space has a subsequence of bounded variation).

Now prove the converse of the above italicized statement:

(d)  If every sequence with exponentially decreasing increments in a metric space $(X, d)$ converges in $(X, d)$, then $(X, d)$ is complete.

*Hints*: (a) Verify that $\sum_{n=0}^{m} \alpha^n \to \frac{1}{1-\alpha}$ as $m \to \infty$ for any $\alpha \in (0, 1)$. (b) Use the triangle inequality and Problems 3.10 and 3.11. (c) Show that for each integer $k \geq 1$ there exists an integer $n_k \geq 1$ such that $d(x_n, x_{n_k}) \leq (\frac{1}{2})^k$ for every $n \geq n_k$ whenever $\{x_n\}$ is a Cauchy sequence. (d) Proposition 3.39(c).

**Problem 3.53.** If $\{x_n\}$ and $\{y_n\}$ are (similarly indexed) Cauchy sequences in a metric space $(X, d)$, then

(a)  the real sequence $\{d(x_n, y_n)\}$ converges in $\mathbb{R}$.

*Hint*: Use Problems 3.1(b) and 3.10(c) to show that $\{d(x_n, y_n)\}$ is a Cauchy sequence in $\mathbb{R}$.

Moreover, if $\{x_n'\}$ and $\{y_n'\}$ are Cauchy sequences in a metric space $(X, d)$ equiconvergent with $\{x_n\}$ and $\{y_n\}$, respectively, (i.e., if $\{x_n'\}$ and $\{y_n'\}$ are Cauchy sequences in $(X, d)$ such that $\lim d(x_n, x_n') = 0$ and $\lim d(y_n, y_n') = 0$ — see Problem 3.15), then

(b)  $$\lim d(x_n, y_n) \; = \; \lim d(x_n', y_n').$$

Hint: Set $\alpha_n = d(x_n, y_n)$, $\alpha_n' = d(x_n', y_n')$, $\alpha = \lim \alpha_n$, and $\alpha' = \lim \alpha_n'$. Use Problems 3.1(b) and 3.10(c) to show that $|\alpha_n - \alpha_n'| \to 0$. Now note that $0 \leq |\alpha - \alpha'| \leq |\alpha - \alpha_n| + |\alpha_n - \alpha_n'| + |\alpha_n' - \alpha'|$ for each $n$.

**Problem 3.54.** Suppose $\{x_n\}$ and $\{x_n'\}$ are two (similarly indexed) equiconvergent sequences in a metric space $X$ — see Problem 3.15.

(a)  Show that if one of them is a Cauchy sequence, then so is the other.

(b)  A metric space $X$ is complete whenever there exists a dense subset $A$ of $X$ such that every Cauchy sequence in $A$ converges in $X$. Prove.

**Problem 3.55.** Let $X$ be an arbitrary set. A function $d : X \times X \to \mathbb{R}$ is an *ultrametric* on $X$ if it satisfies conditions (i), (ii) and (iii) in Definition 3.1 and also the *ultrametric inequality*,

$$d(x, y) \; \leq \; \max \left\{ d(x, z), d(z, y) \right\}$$

for every $x$, $y$ and $z$ in $X$. Clearly, the ultrametric inequality implies the triangle inequality so that an ultrametric is a metric. Example: The discrete metric is an ultrametric. Let $d$ be an ultrametric on $X$ and let $x$, $y$ and $z$ be arbitrary points in $X$. Prove the following propositions.

(a)  If $d(x, z) \neq d(z, y)$, then $d(x, y) = \max\{d(x, z), d(z, y)\}$.

(b)  Every point in a nonempty open ball is a center of that ball. That is, if $\varepsilon > 0$ and $z \in B_\varepsilon(y)$, then $B_\varepsilon(y) = B_\varepsilon(z)$.

*Hint*: Suppose $z \in B_\varepsilon(y)$ and take any $x \in B_\varepsilon(y)$. First note that if $d(x, z) = d(z, y)$, then $x \in B_\varepsilon(z)$. Next use item (a) to show that if $d(x, z) \neq d(z, y)$, then $x \in B_\varepsilon(z)$. Thus conclude that $B_\varepsilon(y) \subseteq B_\varepsilon(z)$ whenever $z \in B_\varepsilon(y)$.

(c)  If two nonempty open balls meet, then one of then is included in the other. In particular, if two nonempty open balls of the same radius meet, then they coincide with each other.

(d)  Any nonempty open ball is a closed set. (*Hint*: Theorem 3.30.) Thus conclude that the metric space $(X, d)$ is totally disconnected. (*Hint*: Proposition 3.10 and Problem 3.50.)

(e)  A sequence $\{x_n\}$ in $(X, d)$ is a Cauchy sequence if and only if

$$\lim_n d(x_{n+1}, x_n) = 0.$$

Thus conclude that $\{d(x_{n+k}, x_n)\}$ converges to zero uniformly in $k$ if and only if it converges to zero for some integer $k \geq 1$; that is, $\lim_n \sup_{k \geq 1} d(x_{n+k}, x_n) = 0$ if and only if $\lim_n d(x_{n+k}, x_n) = 0$ for some $k \geq 1$. (Compare with Problem 3.51.)

**Problem 3.56.**  Let $S$ be a nonempty set and consider the collection $S^{\mathbb{N}}$ of all $S$-valued sequences. For any two distinct sequences $x = \{x_k\}$ and $y = \{y_k\}$ in $S^{\mathbb{N}}$ set $k(x, y) = \min\{k \in \mathbb{N}: x_k \neq y_k\}$ and define the function $d : S^{\mathbb{N}} \times S^{\mathbb{N}} \to \mathbb{R}$ by

$$d(x, y) = \begin{cases} 0, & x = y, \\ \frac{1}{k(x,y)}, & x \neq y. \end{cases}$$

(a)  Show that $d$ is an ultrametric on $S^{\mathbb{N}}$ and that $(S^{\mathbb{N}}, d)$ is a complete metric space.

This metric $d$ is called the *Baire metric* on $S^{\mathbb{N}}$. Now set $S = \mathbb{F}$ (where $\mathbb{F}$ denotes either the real field $\mathbb{R}$ or the complex field $\mathbb{C}$) and let $\ell_+^\infty$ be the set of all bounded scalar-valued sequences (i.e., $x = \{\xi_k\} \in \ell_+^\infty$ if and only if $\sup_k |\xi_k| < \infty$). Let $d$ be the Baire metric on $\mathbb{F}^{\mathbb{N}}$ and consider the subspace $(\ell_+^\infty, d)$ of the complete metric space $(\mathbb{F}^{\mathbb{N}}, d)$. Take the following $\ell_+^\infty$-valued sequence $\{x_n\}_{n \in \mathbb{N}}$: for each $n \in \mathbb{N}$, $x_n = \{\xi_n(k)\}_{k \in \mathbb{N}}$ where

$$\xi_n(k) = \begin{cases} n, & k = n, \\ 0, & k \neq n. \end{cases}$$

Let $d_\infty$ be the sup-metric on $\ell_+^\infty$ and recall that $(\ell_+^\infty, d_\infty)$ is a complete metric space (Example 3S(c)).

(b)  Show that $\{x_n\}_{n\in\mathbb{N}}$ converges to 0 (the null sequence) in $(\ell_+^\infty, d)$ but is unbounded in $(\ell_+^\infty, d_\infty)$.

(c)  Show that the metric space $(\ell_+^\infty, d)$ is not complete.

*Hint*: Consider the following $\ell_+^\infty$-valued sequence $\{y_n\}_{n\in\mathbb{N}}$: for each $n \in \mathbb{N}$, $y_n = \{\upsilon_n(k)\}_{i\in\mathbb{N}}$ where

$$\upsilon_n(k) = \begin{cases} k, & k \le n, \\ 0, & k > n. \end{cases}$$

Verify that $\{y_n\}_{n\in\mathbb{N}}$ converges in $(\mathbb{F}^\mathbb{N}, d)$ to $y = \{k\}_{k\in\mathbb{N}} \in \mathbb{F}^\mathbb{N}\backslash\ell_+^\infty$. Use item (a) and Theorems 3.30 and 3.40(a).

**Problem 3.57.** Recall: $(\ell_+^p, d_p)$ is a complete metric space for every $p \ge 1$ (Example 3R(b)) and $\ell_+^0 \subset \ell_+^p \subset \ell_+^q$ whenever $1 \le p < q$ (Problem 3.44(b)). Consider the subspace $(\ell_+^p, d_q)$ of $(\ell_+^q, d_q)$ and show that

$$(\ell_+^p, d_q) \text{ is not a complete metric space.}$$

Now consider the subspace $(\ell_+^0, d_p)$ of $(\ell_+^p, d_p)$ and show that

$$(\ell_+^0, d_p) \text{ is not a complete metric space.}$$

*Hint*: Problem 3.44(b) and Theorem 3.40(a).

**Problem 3.58.** Take an arbitrary real number $p \ge 1$ and consider the metric space $(C[0, 1], d_p)$ of Example 3D. Prove that

$$(C[0, 1], d_p) \text{ is not a complete metric space.}$$

Hint: Consider the $C[0, 1]$-valued sequence $\{x_n\}$ that was defined in Problem 3.40. First take an arbitrary pair of integers $m$ and $n$ such that $1 \le m < n$ and show that $d_p(x_m, x_n)^p < \frac{1}{2m}$. Then conclude that $\{x_n\}$ is a Cauchy sequence in $(C[0, 1], d_p)$. Next suppose there exists a function $x$ in $C[0, 1]$ such that $d_p(x_n, x) \to 0$. Show that

(i)  $\int_0^{\frac{1}{2}}|1 - x(t)|^p\, dt = 0$    and    (ii)  $\int_{\frac{n+1}{2n}}^{1}|x(t)|^p\, dt \to 0$  as  $n \to \infty$.

From (i) conclude that $x(t) = 1$ for all $t \in [0, \frac{1}{2}]$; in particular, $x(\frac{1}{2}) = 1$. From (ii) conclude that $x(t) = 0$ for all $t \in [\frac{n+1}{2n}, 1]$ and every $n \ge 1$; in particular, $x(\frac{n+1}{2n}) = 0$ for every $n \ge 1$ so that $x(\frac{1}{2}) = x(\lim \frac{n+1}{2n}) = \lim x(\frac{n+1}{2n}) = 0$ by Corollary 3.8. This leads to a contradiction (viz., $0 = 1$), and hence there is no function $x$ in $C[0, 1]$ such that $d_p(x_n, x) \to 0$. Thus the $C[0, 1]$-valued Cauchy sequence $\{x_n\}$ does not converge in $(C[0, 1], d_p)$.

**Problem 3.59.** Recall that $(\ell_+^\infty, d_\infty)$ is a complete metric space (Example 3S). Let $\ell_+^c$ denote the set of all scalar-valued convergent sequences (i.e., $x = \{\xi_i\} \in \ell_+^c$ if and only if $|\xi_i - \xi| \to 0$ for some scalar $\xi$) and let $\ell_+^{c_o}$ denote the subset of $\ell_+^c$ consisting of all sequences that converge to zero. Since every convergent sequence is bounded (Proposition 3.39), it follows that

$$\ell_+^0 \subset \ell_+^p \subset \ell_+^{c_o} \subset \ell_+^c \subset \ell_+^\infty,$$

with the sets $\ell_+^p$ $(p \geq 1)$ and $\ell_+^0$ defined as before (Problems 3.44 and 3.57). Use the Closed Set Theorem to verify the following propositions.

   (a) $(\ell_+^{c_o}, d_\infty)$ and $(\ell_+^c, d_\infty)$ are complete metric spaces.

   (b) $(\ell_+^0, d_\infty)$ and $(\ell_+^p, d_\infty)$ are not complete metric spaces.

*Hint*: (a) To show that $\ell_+^c$ is closed in $(\ell_+^\infty, d_\infty)$ proceed as follows. Take an arbitrary $\varepsilon > 0$. Let $\{x_n\}_{n\geq 1}$ be an $\ell_+^c$-valued sequence so that, for each $n \geq 1$, $x_n = \{\xi_n(k)\}_{k\geq 1}$ converges in $\mathbb{R}$. Thus for each $n \geq 1$ there exists an integer $k_{\varepsilon,n} \geq 1$ such that

$$|\xi_n(k) - \xi_n(j)| < \varepsilon$$

whenever $j, k \geq k_{\varepsilon,n}$. Suppose $\{x_n\}_{n\geq 1}$ converges in $(\ell_+^\infty, d_\infty)$ to $x = \{\xi(k)\}_{k\geq 1} \in \ell_+^\infty$ so that $\sup_k |\xi_n(k) - \xi(k)| \to 0$ as $n \to \infty$. Thus there exists an integer $n_\varepsilon \geq 1$ such that

$$|\xi_n(k) - \xi(k)| < \varepsilon$$

for every $k \geq 1$ whenever $n \geq n_\varepsilon$. Therefore,

$$|\xi(k) - \xi(j)| \leq |\xi(k) - \xi_{n_\varepsilon}(k)| + |\xi_{n_\varepsilon}(k) - \xi_{n_\varepsilon}(j)| + |\xi_{n_\varepsilon}(j) - \xi(j)| < 3\varepsilon$$

whenever $j, k \geq k_{\varepsilon,n_\varepsilon}$. Now conclude that $x$ lies in $\ell_+^c$. (b) To show that both sets $\ell_+^0$ and $\ell_+^p$ are not closed in the metric space $(\ell_+^\infty, d_\infty)$ set $x_n = \{1, (\frac{1}{2})^{\frac{1}{p}}, \ldots, (\frac{1}{n})^{\frac{1}{p}}, 0, 0, 0, \ldots\} \in \ell_+^0$ for each $n \geq 1$ so that the sequence $\{x_n\}_{n\geq 1}$ converges in $(\ell_+^\infty, d_\infty)$ to $x = \{(\frac{1}{k})^{\frac{1}{p}}\}_{k\geq 1} \in \ell_|^{c_o} \backslash \ell_|^p$.

Remark: Note that $\ell_+^p$ is not dense in $(\ell_+^\infty, d_\infty)$: if $y = \{v_k\}_{k\geq 1}$ is the constant sequence in $\ell_+^\infty$ with $v_k = 1$ for all $k \geq 1$, then $d_\infty(x, y) \geq 1$ for every $x \in \ell_+^p \subset \ell_+^{c_o}$. Hence $\ell_+^0$ is not dense in $(\ell_+^\infty, d_\infty)$ either.

**Problem 3.60.** Let $\{(X_i, d_i)\}_{i=1}^n$ be a finite collection of metric spaces and let $Z = \prod_{i=1}^n X_i$ be the Cartesian product of their underlying sets. Let $d$ denote any of the metrics $d_p$ (for an arbitrary $p \geq 1$) or $d_\infty$ on $Z = \prod_{i=1}^n X_i$ as in Problem 3.9. Show that the product space $(\prod_{i=1}^n X_i, d)$ is complete if and only if $(X_i, d_i)$ is a complete metric space for every $i = 1, \ldots, n$.

*Hint*: Consider the metric $d_1$ on $Z$ (see Problem 3.9) and show that $(\prod_{i=1}^n X_i, d_1)$ is complete if and only if each $(X_i, d_i)$ is complete. Now use Problem 3.33 and Lemma 3.43.

**Problem 3.61.** Take an arbitrary nondegenerate closed interval of the real line, say $I = [\alpha, \beta] \subset \mathbb{R}$ of positive length $\lambda = \beta - \alpha$. Consider the pair of closed intervals $\{[\alpha, \alpha + \frac{\lambda}{3}], [\beta - \frac{\lambda}{3}, \beta]\}$ consisting of the first and third closed subintervals of $I = [\alpha, \beta]$ of length $\frac{\lambda}{3}$, which will be referred to as the closed intervals *derived from $I$ by removal of the central open third of $I$*. If $\mathcal{I} = \{I_i\}_{i=1}^n$ is a finite disjoint collection of nondegenerate closed intervals in $\mathbb{R}$, then let $\mathcal{I}' = \{I_i'\}_{i=1}^{2n}$ be the corresponding finite collection obtained by replacing each closed interval $I_i$ in $\mathcal{I}$ with the pair of closed subintervals derived from $I_i$ by removal of the central open third of $I_i$. Now consider the unit interval $[0, 1]$ and set

$$\mathcal{I}_0 = \{[0, 1]\}.$$

The derived intervals from $[0, 1]$ by removal of the central open third are $[0, \frac{1}{3}]$ and $[\frac{2}{3}, 1]$. Set

$$\mathcal{I}_1 = \mathcal{I}_0' = \{[0, \tfrac{1}{3}], [\tfrac{2}{3}, 1]\}.$$

Similarly, replacing each closed interval in $\mathcal{I}_1$ by the pair of closed subintervals derived from it by removal of its central open third, set

$$\mathcal{I}_2 = \mathcal{I}_1' = \{[0, \tfrac{1}{9}], [\tfrac{2}{9}, \tfrac{1}{3}], [\tfrac{2}{3}, \tfrac{7}{9}], [\tfrac{8}{9}, 1]\}.$$

Take an arbitrary positive integer $n$. Suppose the collection of intervals $\mathcal{I}_k$ has already been defined for each $k = 0, 1, \ldots, n$, and set $\mathcal{I}_{n+1} = \mathcal{I}_n'$. This leads to an inductive construction of a disjoint collection $\mathcal{I}_n$ of $2^n$ nondegenerate closed subintervals of the unit interval $[0, 1]$, each having length $\frac{1}{3^n}$, for every $n \in \mathbb{N}_0$. Next set $C_n = \bigcup \mathcal{I}_n$ for each $n \in \mathbb{N}_0$ and note that $\{C_n\}_{n \in \mathbb{N}_0}$ is a decreasing sequence of subsets of the unit interval (i.e., $C_{n+1} \subset C_n \subset C_0 = [0, 1]$) such that each set $C_n$ is the union of $2^n$ disjoint nondegenerate subintervals of length $\frac{1}{3^n}$.

The *Cantor set* is the set

$$C = \bigcap_{n \in \mathbb{N}_0} C_n.$$

(a)  Show that the Cantor set is a *nonempty*, *closed* and *bounded* subset of the real line $\mathbb{R}$.

(b)  Show that the Cantor set has an *empty interior* and hence it is *nowhere dense*.

*Hint*: Recall that each set $C_n$ consists of $2^n$ intervals of length $\frac{1}{3^n}$. Take an arbitrary point $\gamma \in C$ and an arbitrary $\varepsilon > 0$. Verify that there exists a positive integer $n_\varepsilon$ such that the open ball $B_\varepsilon(\gamma)$ is not included in $C_{n_\varepsilon}$. Now conclude that the nonempty open ball $B_\varepsilon(\gamma)$ is not included in $C$, which means that $\gamma$ is not an interior point of $C$.

(c) Show that the Cantor set *has no isolated point* and hence it is a *perfect* subset of $\mathbb{R}$. Moreover, show that the Cantor set is *uncountable*.

   *Hint*: Consider the hint of item (b). Verify that there exists a positive integer $n_\varepsilon$ such that the open ball $B_\varepsilon(\gamma)$ includes an end point of some of the closed intervals of $C_{n_\varepsilon}$. Also see Example 3W(b), and recall that $\mathbb{R}$ is a complete metric space.

(d) Show that the Cantor set is *totally disconnected*.

   *Hint*: If $\alpha$ and $\gamma$ are two points of an arbitrary subset $A$ of $C$ such that $\alpha < \gamma$, and if $n$ is a positive integer such that $\frac{1}{3^n} < \gamma - \alpha$, then both $\alpha$ and $\gamma$ do not belong to any interval of length $\frac{1}{3^n}$. Thus $\alpha$ and $\gamma$ must belong to different intervals in $C_n$, so that there exists a real number $\beta$ such that $\alpha < \beta < \gamma$ and $\beta \notin C_n$. Verify that $\{A \cap (-\infty, \beta), A \cap (\beta, \infty)\}$ is a disconnection of the set $A$. See Problem 3.50.

Remark: Let $\mu(C_n)$ denote the length of each set $C_n$, which consists of $2^n$ *disjoint* intervals of length $\frac{1}{3^n}$. Thus $\mu(C_n) = (\frac{2}{3})^n$. If we agree that the length $\mu(A)$ of a subset $A$ of the real line can be defined somehow as to bear the property that $0 \le \mu(A) \le \mu(B)$ whenever $A \subseteq B \subseteq \mathbb{R}$ (provided the lengths $\mu(A)$ and $\mu(B)$ are "well-defined"), then the Cantor set $C$ is such that $0 \le \mu(C) = \mu(\bigcap_n C_n) \le \mu(C_m)$ for every $m$ (note: the length $\mu(\bigcap_n C_n)$ of $\bigcap_n C_n$ is "well-defined" whenever the length $\mu(C_n)$ is "well-defined" for every $C_n$). Hence $\mu(C) = 0$. That is, the Cantor set has *length zero*.

**Problem 3.62.** Consider the construction of the previous problem. Note that each set $C_n$ (for $n \ge 1$) is obtained from $C_{n-1}$ by removing $2^{n-1}$ central open subintervals, each of length $\frac{1}{3^n}$. Now, instead of removing at each iteration $2^{n-1}$ central open subintervals of length $\frac{1}{3^n}$, remove at each iteration $2^{n-1}$ central open subintervals of length $\frac{1}{4^n}$. Let $\{S_n\}_{n \in \mathbb{N}_0}$ be the resulting collection of closed subsets of the unit interval $S_0 = [0, 1]$, and note that the length of $S_n$ for each $n \in \mathbb{N}$ is

$$\mu(S_n) = 1 - \sum_{i=0}^{n-1} \frac{2^i}{4^{i+1}} = \frac{1}{2} + \frac{1}{2^{n+1}}.$$

Consider the sequence $\{x_n\}$ consisting of the characteristic functions of the subsets $S_n$ of $S_0$ for each $n \in \mathbb{N}$; that is

$$x_n(t) = \begin{cases} 1, & t \in S_n, \\ 0, & t \in S_0 \setminus S_n, \end{cases}$$

for every $n \geq 1$. Note that each $x_n$ belongs to $R^p[0, 1]$ for every $p \geq 1$ (see Example 3E) so that $\{x_n\}$ is an $R^p[0, 1]$-valued sequence.

(a) Equip $R^p[0, 1]$ with its usual metric $d_p$ and show that $\{x_n\}$ is a Cauchy sequence in $(R^p[0, 1], d_p)$.

(b) Show that $\{x_n\}$ does not converge in $(R^p[0, 1], d_p)$.

*Hint*: Suppose there exists $x$ in $R^p[0, 1]$ such that $d_p(x_n, x) \to 0$. Consider the null function $0 \in R^p[0, 1]$ and show that $d_p(x_n, 0) > \frac{1}{2}$ for all $n$. Use the triangle inequality to conclude that $d_p(x, 0) \geq \frac{1}{2}$. On the other hand, set $S = \bigcap_n S_n$ so that $S_0 \backslash S = \bigcup_n (S_0 \backslash S_n)$. Take an arbitrary positive integer $m$ and show that

$$\int_{\bigcup_{n=1}^{m}(S_0 \backslash S_n)} |x(t)|^p \, dt \leq \int_{S_0} |x(t) - x_k(t)|^p \, dt$$

for every $k \geq 1$. Finally, conclude that $\int_{S_0 \backslash S} |x(t)|^p \, dt = 0$, which implies that the lower Riemann integral of $|x|^p$ is zero. Since $|x|^p$ is Riemann integrable, it follows that $\int_{S_0} |x(t)|^p \, dt = 0$. This contradicts the fact that $\int_{S_0} |x(t)|^p \, dt \geq \frac{1}{2}$.

From (a) and (b) we conclude that, for any $p \geq 1$,

$$(R^p[0, 1], d_p) \text{ is not a complete metric space.}$$

Remark: The failure of $R^p[0, 1]$ to be complete when equipped with its usual metric $d_p$ is regarded as one of the defects in the definition of the Riemann integral. A more general concept of integral, viz., the *Lebesgue integral*, corrects this and other drawbacks of the Riemann integral. Let $L^p[0, 1]$ be the collection of all equivalence classes (as in Example 3E) of scalar-valued functions $x$ on $[0, 1]$ such that $\int_0^1 |x(t)|^p \, dt < \infty$, where the integral now is the Lebesgue integral. Since any Riemann integrable function is Lebesgue integrable, it follows that $R^p[0, 1] \subset L^p[0, 1]$. Moreover, $R^p[0, 1]$ is dense in the metric space $(L^p[0, 1], d_p)$ so that

$$(L^p[0, 1], d_p) \text{ is a completion of } (R^p[0, 1], d_p).$$

**Problem 3.63.** A metric space is complete if and only if every decreasing sequence $\{V_n\}_{n=1}^{\infty}$ of nonempty closed subsets of $X$ for which $\text{diam}(V_n) \to 0$ is such that $\bigcap_{n=1}^{\infty} V_n \neq \varnothing$.

*Hint*: This result, likewise Lemma 3.79, is also attributed to Cantor. Its proof follows closely the proof of Lemma 3.79. Consider the same $X$-valued sequence $\{v_n\}_{n \in \mathbb{N}}$ that was defined in part (a) of the proof of Lemma 3.79. Show that $\{v_n\}_{n \in \mathbb{N}}$ is a Cauchy sequence if $\text{diam}(V_n) \to 0$. Suppose $X$ is complete, set $v = \lim v_n$, and verify that $v \in V_n$ for an arbitrary $m \in \mathbb{N}$ so that $\bigcap_{m \in \mathbb{N}} V_m \neq \varnothing$. On the other hand, let $\{x_n\}_{n\in\mathbb{N}}$ be an arbitrary $X$-valued Cauchy sequence and consider the decreasing sequence $\{V_m^-\}_{m\in\mathbb{N}}$ of nonempty closed subsets of $X$ that was defined in part (b) of the proof of Lemma 3.79. Show that $V_n^- \to 0$. If $\bigcap_{m\in\mathbb{N}} V_m^- \neq \varnothing$, then there exists $v \in V_m^-$ for all $m \in \mathbb{N}$. Verify that $x_n \to v$ and conclude that $X$ is complete.

**Problem 3.64.** Let $\{(X_i, d_i)\}_{i=1}^n$ be a finite collection of metric spaces and let $Z = \prod_{i=1}^n X_i$ be the Cartesian product of their underlying sets. Let $d$ denote any of the metrics $d_p$ (for an arbitrary $p \geq 1$) or $d_\infty$ on $Z = \prod_{i=1}^n X_i$ as in Problem 3.9.

(a) Show that the product space $(\prod_{i=1}^n X_i, d)$ is totally bounded if and only if $(X_i, d_i)$ is totally bounded for every $i = 1, \ldots, n$.

*Hint*: First use Lemma 3.73 to show that $(\prod_{i=1}^n X_i, d_1)$ is totally bounded if and only if each $(X_i, d_i)$ is totally bounded. Then apply Problems 3.33 and Corollary 3.81.

(b) Show that $(\prod_{i=1}^n X_i, d)$ is compact if and only if $(X_i, d_i)$ is compact for every $i = 1, \ldots, n$.

*Hint*: Item (a), Problem 3.60, and Corollary 3.81.

Remark: Let $\{X_\gamma\}_{\gamma\in\Gamma}$ be an indexed family of nonempty topological spaces and let $Z = \prod_{\gamma\in\Gamma} X_\gamma$ be the Cartesian product of the underlying sets $\{X_\gamma\}_{\gamma\in\Gamma}$. The *product topology* on $Z$ is the topology inversely induced on $Z$ by the family $\{\pi_\gamma\}_{\gamma\in\Gamma}$ of projections of $X$ onto each $X_\gamma$ (i.e., the weakest topology on $Z$ that makes each projection $\pi_\gamma : Z \to X_\gamma$ continuous). Compactness in a topological space is defined as in Definition 3.60. An important result, the *Tikhonov Theorem*, says that $\prod_{\gamma\in\Gamma} X_\gamma$ *is compact if and only if* $X_\gamma$ *is compact for every* $\gamma \in \Gamma$.

**Problem 3.65.** Every closed ball of positive radius in $(\ell_+^p, d_p)$ is not totally bounded (and hence not compact).

*Hint*: Take an arbitrary $p \geq 1$ and consider the metric space $(\ell_+^p, d_p)$ of Example 3B. Let $B_\rho[x_0]$ be a closed ball of radius $\rho > 0$ centered at an arbitrary $x_0 \in \ell_+^p$. Consider the $\ell_+^p$ valued sequence $\{e_i\}_{i\in\mathbb{N}}$ of Example 3X and set $x_i = \rho e_i + x_0$ for each $i \in \mathbb{N}$. Instead of following the approach of Example 3X, show that $\{x_i\}_{i\in\mathbb{N}}$ is a $B_\rho[x_0]$-valued sequence that has no Cauchy subsequence. Then apply Lemma 3.73.

**Problem 3.66.** Prove the following propositions.

(a) Every closed ball of positive radius in $(C[0, 1], d_\infty)$ is not compact.

*Hint*: Consider the metric space $(C[0, 1], d_\infty)$ of Example 3D. Let $B_\rho[x_0]$ be the closed ball of radius $\rho > 0$ centered at an arbitrary $x_0 \in C[0, 1]$. Consider the mapping $\varphi : B_\rho[x_0] \to \mathbb{R}$ defined by

$$\varphi(x) = \int_0^1 |x(t) - x_0(t)|\, dt - |x(0) - x_0(0)|$$

for every $x \in B_\rho[x_0]$. Equip $B_\rho[x_0]$ with the sup-metric $d_\infty$ and the real line with its usual metric. Show that $\varphi$ is continuous, $\varphi(x) < \rho$ for all $x \in B_\rho[x_0]$, and $\sup_{x \in B_\rho[x_0]} \varphi(x) = \rho$. Now use the Weierstrass Theorem (Theorem 3.86) to verify that $B_\rho[x_0]$ is not compact.

(b) Every closed ball of positive radius in $(C[0, 1], d_\infty)$ is not totally bounded.

*Hint*: Problems 3.42(a), Theorems 3.40 and 3.81, and item (a).

**Problem 3.67.** A topological space $X$ is *locally compact* if every point of $X$ has a compact neighborhood. Prove the following assertions.

(a) A metric space $X$ is locally compact if and only if there exists a compact closed ball of positive radius centered at each point of $X$.

(b) $\mathbb{R}^n$ and $\mathbb{C}^n$ (equipped with any of their uniformly equivalent metrics of Example 3A) are locally compact.

(c) $(\ell_+^p, d_p)$ and $(C[0, 1], d_\infty)$ are not locally compact.

(d) Every open subspace and every closed subspace of a locally compact metric space is locally compact.

**Problem 3.68.** Consider the metric space $(\ell_+^p, d_p)$ for some $p \geq 1$.

(a) Prove that a subset $A$ of $\ell_+^p$ is totally bounded if and only if

$$\sup_{\{\xi_k\} \in A} \sum_{k=1}^\infty |\xi_k|^p < \infty \quad \text{and} \quad \lim_n \sup_{\{\xi_k\} \in A} \sum_{k=n}^\infty |\xi_k|^p = 0$$

(i.e., $A$ is bounded and $\sum_{k=n}^\infty |\xi_k|^p \to 0$ as $n \to \infty$ uniformly on $A$).

(b) Show that a subset $A$ of $\ell_+^p$ is compact if and only if it is closed and satisfies the above conditions.

**Problem 3.69.** Let $\{\xi_k(0)\}$ be an arbitrary point in $\ell_+^p$ and set

$$S_0 = \left\{ \{\xi_k\} \in \ell_+^p : \ |\xi_k| \leq |\xi_k(0)| \ \text{for every} \ k \right\}.$$

Use the preceding problem to show that $S_0$ is a compact subset of the metric space $(\ell_+^p, d_p)$. In particular, the set

$$S = \left\{ \{\xi_k\}_{k \geq 1} \in \ell_+^2 : \ |\xi_k| \leq \tfrac{1}{k} \ \text{for every} \ k \geq 1 \right\},$$

which is known as the *Hilbert* cube, is compact in $(\ell_+^2, d_2)$. Show that the Hilbert cube has an empty interior (*hint*: verify that $(\ell_+^2 \backslash S)^- = \ell_+^2$) and then conclude that it is nowhere dense.

**Problem 3.70.** Suppose $X$ is a compact metric space, let $Y$ be any metric space, and consider the metric space $(C[X, Y], d_\infty)$ of Example 3Y. Take an arbitrary real number $\gamma > 0$ and let $C_\gamma[X, Y]$ denote the subset of $C[X, Y]$ consisting of all Lipschitzian mappings of $X$ into $Y$ that have a Lipschitz constant less than or equal to $\gamma$.

(a) Show that $C_\gamma[X, Y]$ is equicontinuous and closed in $(C[X, Y], d_\infty)$.

*Hint*: Set $\delta = \frac{\varepsilon}{\gamma}$ for equicontinuity. Use the Closed Set Theorem: if $f_n \in C_\gamma[X, Y]$ and if $f_n \to f \in C[X, Y]$, then

$$\begin{aligned} d_Y(f(x), f(y)) &\le d_Y(f(x), f_n(x)) + d_Y(f_n(x), f_n(y)) + d_Y(f_n(y), f(y)) \\ &\le 2d_\infty(f_n, f) + \gamma d_X(x, y). \end{aligned}$$

From now on suppose the space $Y$ is compact. Thus $Y$ is complete (Corollary 3.81), and hence $(C[X, Y], d_\infty)$ is complete (Example 3Y).

(b) Show that $C_\gamma[X, Y]$ is pointwise totally bounded and conclude that $C_\gamma[X, Y]$ is a compact subset of the metric space $(C[X, Y], d_\infty)$.

Particular case ($\gamma = 1$): The set $C_1[X, Y]$ of all contractions of a compact metric space $X$ into a compact metric space $Y$ is a compact subset of $(C[X, Y], d_\infty)$. Let $I[X, Y]$ denote the set of all isometries of a compact metric space $X$ into a compact metric space $Y$ so that $I[X, Y] \subset C_1[X, Y]$.

(c) Show that $I[X, Y]$ is closed in $(C[X, Y], d_\infty)$ and conclude that $I[X, Y]$ is compact in $(C[X, Y], d_\infty)$.

*Hint*. Apply the Closed Set Theorem: if $\{f_n\}$ is an $I[X, Y]$-valued sequence that converges to $f \in C[X, Y]$, then (Problem 3.1(b))

$$\begin{aligned} |d_X(x, y) - d_Y(f(x), f(y))| &= |d_Y(f_n(x), f_n(y)) - d_Y(f(x), f(y))| \\ &\le d_Y(f_n(x), f(x)) + d_Y(f_n(y), f(y)) \\ &\le d_\infty(f_n, f). \end{aligned}$$

# 4

# Banach Spaces

Our purpose now is to put algebra and topology to work together. For instance, from algebra we get the notion of finite sums (either ordinary or direct sums of vectors, linear manifolds, or linear transformations), and from topology the notion of convergent sequences. If algebraic and topological structures are suitably laid on the same underlying set, then we may consider the concept of infinite sums and convergent series. More importantly, as continuity plays a central role in the theory of topological spaces, and linear transformation plays a central role in the theory of linear spaces, when algebra and topology are properly combined they yield the concept of *continuous linear transformation*; the very central theme of this book.

## 4.1   Normed Spaces

To begin with let us point out, once and for all, that throughout this chapter $\mathbb{F}$ will denote either the real field $\mathbb{R}$ or the complex field $\mathbb{C}$, both equipped with their usual topologies induced by their usual metrics.

If we intend to combine algebra and topology so that a given set is endowed with both algebraic and topological structures, then we might simply equip a linear space with some metric, and hence it would become a linear space that is also a metric space. However, an arbitrary metric on a linear space may induce a topological structure that has nothing to do with the algebraic structure (i.e., these structures may live apart on the same underlying set). A richer and more useful structure is obtained when the metric recognizes the operations of vector addition and scalar multiplication that come with the linear space, and incorporate these operations in its own definition. With this in mind, let us first define a couple of concepts. A metric (or a pseudometric) $d$ on a linear space $\mathcal{X}$ over $\mathbb{F}$ is said to be *additively invariant* if

$$d(x, y) = d(x + z, y + z)$$

for every $x$, $y$ and $z$ in $\mathcal{X}$ (which means that the translation mapping $\mathcal{X} \to \mathcal{X}$ defined by $x \mapsto x + z$ for any $z \in \mathcal{X}$ is an isometry). If $d$ is such that

$$d(\alpha x, \alpha y) = |\alpha| d(x, y)$$

for every $x$ and $y$ in $\mathcal{X}$ and every $\alpha$ in $\mathbb{F}$, then the metric $d$ is called *absolutely homogeneous*. A program for equipping a linear space with a metric that has the above "linear-like" properties goes as follows. Let $p \colon \mathcal{X} \to \mathbb{R}$ be a real-valued functional on a linear space over $\mathbb{F}$ (recall: $\mathbb{F}$ is either $\mathbb{R}$ or $\mathbb{C}$ so that $\mathbb{R}$ is always embedded in $\mathbb{F}$). It is *nonnegative homogeneous* if

$$p(\alpha x) = \alpha p(x)$$

for every $x$ in $\mathcal{X}$ and every *nonnegative* (real) scalar $\alpha$ in $\mathbb{F}$, and *subadditive* if

$$p(x + y) \leq p(x) + p(y)$$

for every $x$ and $y$ in $\mathcal{X}$. If $p$ is both nonnegative homogeneous and subadditive, then it is called a *sublinear* functional. If

$$p(\alpha x) = |\alpha| p(x)$$

for every $x$ in $\mathcal{X}$ and $\alpha$ in $\mathbb{F}$, then $p$ is *absolutely homogeneous*. A subadditive absolutely homogeneous functional is a *convex* functional. (Note that this includes the classical definition of a convex functional: if $p \colon \mathcal{X} \to \mathbb{R}$ is convex, then $p(\alpha x + \beta x) \leq \alpha p(x) + \beta p(x)$ for every $x, y \in \mathcal{X}$ and every $\alpha \in [0, 1]$ with $\beta = 1 - \alpha$.) If

$$p(x) \geq 0$$

for all $x$ in $\mathcal{X}$, then $p$ is *nonnegative*. A nonnegative convex functional is a *seminorm* (or a *pseudonorm* — i.e., a nonnegative absolutely homogeneous subadditive functional). If

$$p(x) > 0 \quad \text{whenever} \quad x \neq 0,$$

then $p$ is called *positive*. A positive seminorm is a *norm* (i.e., a positive absolutely homogeneous subadditive functional). Summing up: A norm is a real-valued functional on a linear space with the following four properties, called the *norm axioms*.

**Definition 4.1.** Let $\mathcal{X}$ be a linear space over $\mathbb{F}$. A real-valued function

$$\| \ \| \colon \mathcal{X} \to \mathbb{R}$$

is a *norm* on $\mathcal{X}$ if the following conditions are satisfied for all vectors $x$ and $y$ in $X$ and all scalars $\alpha$ in $\mathbb{F}$.

(i)   $\|x\| \geq 0$    (*nonnegativeness*),

(ii)   $\|x\| > 0$   if   $x \neq 0$    (*positiveness*),

(iii)   $\|\alpha x\| = |\alpha|\|x\|$    (*absolute homogeneity*),

(iv)   $\|x + y\| \leq \|x\| + \|y\|$    (*subadditivity – triangle inequality*).

A linear space $\mathcal{X}$ equipped with a norm on it is a *normed space* (synonyms: *normed linear space* or *normed vector space*). If $\mathcal{X}$ is a real or complex linear space (so that $\mathbb{F} = \mathbb{R}$ or $\mathbb{F} = \mathbb{C}$) equipped with a norm on it, then it is referred to as a *real* or *complex normed space*, respectively.

Note that these are not independent axioms. For instance, axiom (i) can be derived from axioms (ii) and (iii): an absolutely homogeneous positive functional is necessarily nonnegative. Indeed, for $\alpha = 0$ in (iii) we get $\|0\| = 0$ and, conversely, $x = 0$ whenever $\|x\| = 0$ by positiveness in (ii). Therefore, if $\| \; \|: \mathcal{X} \to \mathbb{R}$ is a norm, then

$$\|x\| = 0 \quad \text{if and only if} \quad x = 0.$$

**Proposition 4.2.** *If $\| \; \|: \mathcal{X} \to \mathbb{R}$ is a norm on a linear space $\mathcal{X}$, then the function $d: \mathcal{X} \times \mathcal{X} \to \mathbb{R}$, defined by*

$$d(x, y) = \|x - y\|$$

*for every $x, y \in \mathcal{X}$, is a metric on $\mathcal{X}$.*

*Proof.* From (i) and (ii) in Definition 4.1 we get the metric axiom (i) of Definition 3.1. Positiveness (ii) and absolute homogeneity (iii) of Definition 4.1 imply positiveness (ii) and symmetry of Definition 3.1, respectively. Finally, the triangle inequality (iv) of Definition 3.1 follows from the triangle inequality (iv) and absolute homogeneity (iii) of Definition 4.1. $\qquad\qquad\square$

A word on notation and terminology. According to Definition 4.1 a normed space actually is an ordered pair $(\mathcal{X}, \| \; \|)$, where $\mathcal{X}$ is a linear space and $\| \; \|$ is a norm on $\mathcal{X}$. As in the case of a metric space, we shall refer to a normed space in several ways. We may speak of $\mathcal{X}$ itself as a normed space when the norm $\| \; \|$ is either clear in the context or immaterial and, in this case, we shall simply say "$\mathcal{X}$ is a normed space". However, in order to avoid confusion among different normed spaces, we may occasionally insert a subscript on the norms (e.g., $(\mathcal{X}, \| \; \|_X)$ and $(\mathcal{Y}, \| \; \|_Y)$). If a linear space $\mathcal{X}$ can be equipped with more than one norm, say $\| \; \|_1$ and $\| \; \|_2$, then $(\mathcal{X}, \| \; \|_1)$ and $(\mathcal{X}, \| \; \|_2)$ will represent *different* normed spaces with the same linear space $\mathcal{X}$. The metric $d$ of Proposition 4.2 is the *metric generated* by the norm $\| \; \|$, and so a normed space is a special kind of linear metric space. Whenever we refer to the topological (metric) structure of a normed space $(\mathcal{X}, \| \; \|)$ it will always be understood that such a topology on $\mathcal{X}$ is that induced by the metric $d$ generated by the norm $\| \ \|$. This is the so-called *norm topology*. Note that the norm $\|x\|$ of every vector $x$ in a normed space $(\mathcal{X}, \| \ \|)$ is precisely the distance (in the norm topology) between $x$ and the origin 0 of the linear space $\mathcal{X}$ (i.e., $\|x\| = d(x, 0)$, where $d$ is the metric generated by the norm $\| \ \|$). Proposition 4.2 says that every norm on a linear space generates a metric, but an arbitrary metric on a linear space may not be generated by any norm on it. The next proposition tells us when a metric on a linear space is generated by a norm.

**Proposition 4.3.** *Let $\mathcal{X}$ be a linear space. A metric on $\mathcal{X}$ is generated by a norm on $\mathcal{X}$ if and only if it is additively invariant and absolutely homogeneous. Moreover, for each additively invariant and absolutely homogeneous metric on $\mathcal{X}$ there exists a unique norm on $\mathcal{X}$ that generates it.*

*Proof.* If $d_{\|\|}$ is a metric on a normed space $\mathcal{X}$ generated by a norm $\| \ \|$ on $\mathcal{X}$, then it is additively invariant $(d_{\|\|}(x, y) = \|x - y\| = \|x+z-(y+z)\| = d_{\|\|}(x+z, y+z)$ for every $x$, $y$ and $z$ in $\mathcal{X}$) and absolutely homogeneous $(d_{\|\|}(\alpha x, \alpha y) = \|\alpha x - \alpha y\| = \|\alpha(x - y)\| = |\alpha|\|x - y\| = |\alpha| d_{\|\|}(x, y)$ for every $x$ and $y$ in $\mathcal{X}$ and every scalar $\alpha$). Conversely, if $d$ is an additively invariant and absolutely homogeneous metric on a linear space $\mathcal{X}$, then the function $\| \ \|_d \colon \mathcal{X} \to \mathbb{R}$ defined by $\|x\|_d = d(x, 0)$ for every $x$ in $\mathcal{X}$ is a norm on $\mathcal{X}$. Indeed, properties (i) and (ii) of Definition 4.1 are trivially verified by the first two metric axioms in Definition 3.1. Properties (iii) and (iv) of Definition 4.1 follow from absolute homogeneity $(\|\alpha x\|_d = d(\alpha x, 0) = |\alpha| d(x, 0) = |\alpha|\|x\|_d$ for every $x$ in $\mathcal{X}$ and every scalar $\alpha$) and additive invariance $(\|x + y\|_d = d(x + y, 0) = d(x, -y) \leq d(x, 0)+d(0, -y) = d(x, 0)+d(0, y) = \|x\| + \|y\|$ for every $x$ and $y$ in $\mathcal{X}$). This norm $\| \ \|_d$ on $\mathcal{X}$ clearly generates the metric $d$ (for $\|x - y\|_d = d(x - y, 0) = d(x, y)$ for every $x$ and $y$ in $\mathcal{X}$). Uniqueness is straightforward: if $\| \ \|_1$ and $\| \ \|_2$ generate $d$, then $\|x\|_1 = d(x, 0) = \|x\|_2$ for all $x$ in $\mathcal{X}$. $\qquad\square$

Let $(\mathcal{X}, \| \ \|)$ be a normed space. By Proposition 4.2 and Problem 3.1 it follows at once that

$$\big|\,\|x\| - \|y\|\,\big| \leq \|x - y\|$$

for every $x, y \in \mathcal{X}$. Thus *the norm* $\| \ \| \colon \mathcal{X} \to \mathbb{R}$ *is a continuous mapping* with respect to the norm topology of $\mathcal{X}$ (see Problem 3.34). In fact, the above inequality says that every norm is a contraction (thus Lipschitzian and hence uniformly continuous). Therefore (cf. Corollary 3.8 and Lemma 3.43), a norm preserves convergence: *if $x_n \to x$ in the norm topology of $\mathcal{X}$, then $\|x_n\| \to \|x\|$ in $\mathbb{R}$*; and also preserves Cauchy sequences: *if $\{x_n\}$ is a Cauchy sequence in $\mathcal{X}$ with respect to the metric generated by the norm on $\mathcal{X}$, then $\{\|x_n\|\}$ is a Cauchy sequence in $\mathbb{R}$*.

A *Banach space* is a complete normed space. Obviously, completeness refers to the norm topology: a Banach space is a normed space that is complete as a metric space with respect to the metric generated by the norm. A *real* or *complex Banach space* is a complete real or complex normed space, respectively.

Let $\mathcal{X}$ be a linear space and let $\{x_n\}$ be an $\mathcal{X}$-valued sequence indexed by $\mathbb{N}$ (or by $\mathbb{N}_0$). For each $n \geq 1$ set

$$y_n = \sum_{i=1}^{n} x_i$$

in $\mathcal{X}$ so that $\{y_n\}_{n=1}^{\infty}$ is again an $\mathcal{X}$-valued sequence. This is called the *sequence of partial sums* of $\{x_n\}_{n=1}^{\infty}$. Now equip $\mathcal{X}$ with a norm $\|\ \|$. If the sequence of partial sums $\{y_n\}_{n=1}^{\infty}$ converges in the normed space $\mathcal{X}$ to a point $y$ in $\mathcal{X}$ (i.e., if $\|y_n - y\| \to 0$), then we say that $\{x_n\}_{n=1}^{\infty}$ is a *summable sequence* (or that the *infinite series* $\sum_{i=1}^{\infty} x_i$ converges in $\mathcal{X}$ to $y$ — notation: $y = \sum_{i=1}^{\infty} x_i$). If the real-valued sequence $\{\|x_i\|\}_{i=1}^{\infty}$ is summable (i.e., if the infinite series $\sum_{i=1}^{\infty} \|x_i\|$ converges in $\mathbb{R}$ or, equivalently, if $\sum_{i=1}^{\infty} \|x_i\| < \infty$ — see Problem 3.11), then we say that $\{x_n\}_{n=1}^{\infty}$ is an *absolutely summable sequence* (or that the infinite series $\sum_{i=1}^{\infty} x_i$ is *absolutely convergent*).

**Proposition 4.4.** *A normed space is a Banach space if and only if every absolutely summable sequence is summable.*

*Proof.* Let $(\mathcal{X}, \|\ \|)$ be a normed space and let $\{x_n\}_{n=0}^{\infty}$ be an arbitrary $\mathcal{X}$-valued sequence.

(a) Consider the sequence $\{y_n\}_{n=0}^{\infty}$ of partial sums of $\{x_n\}_{n=0}^{\infty}$,

$$y_n = \sum_{i=0}^{n} x_i$$

in $\mathcal{X}$ for each $n \geq 0$. It is readily verified by induction (with a little help from the triangle inequality) that

$$\|y_{n+k} - y_n\| = \left\| \sum_{i=n+1}^{n+k} x_i \right\| < \sum_{i=n+1}^{n+k} \|x_i\|$$

for every pair of integers $n \geq 0$ and $k \geq 1$. Suppose $\{x_n\}_{n=0}^{\infty}$ is an absolutely summable sequence (i.e., $\sum_{i=0}^{\infty} \|x_i\| < \infty$) so that

$$0 \leq \sup_{k \geq 1} \|y_{n+k} - y_n\| \leq \sum_{i=n+1}^{\infty} \|x_i\| \to 0 \quad \text{as} \quad n \to \infty,$$

and hence $\lim_n \sup_{k \geq 1} \|y_{n+k} - y_n\| = 0$ (Problems 3.10(c) and 3.11). Equivalently, $\{y_n\}_{n=0}^{\infty}$ is a Cauchy sequence in $\mathcal{X}$ (Problem 3.51). Therefore, if $\mathcal{X}$ is a Banach space, then $\{y_n\}_{n=0}^{\infty}$ converges in $\mathcal{X}$, which means that $\{x_n\}_{n=0}^{\infty}$ is a summable sequence. Conclusion: An arbitrary absolutely summable sequence in a Banach space is summable.

(b) Conversely, suppose $\{x_n\}_{n=0}^{\infty}$ is a Cauchy sequence. According to Problem 3.52(c) $\{x_n\}_{n=0}^{\infty}$ has a subsequence $\{x_{n_k}\}_{k=0}^{\infty}$ of bounded variation. Set $z_0 = x_{n_0}$ and

$$z_{k+1} = x_{n_{k+1}} - x_{n_k}$$

in $\mathcal{X}$, so that $x_{n_{k+1}} = x_{n_k} + z_{k+1}$, for every $k \geq 0$. Thus

$$x_{n_k} = \sum_{i=0}^{k} z_i$$

for every $k \geq 0$ (see Problem 2.19). Since $\{x_{n_k}\}_{k=0}^{\infty}$ is of bounded variation (i.e., $\sum_{k=0}^{\infty} \|x_{n_{k+1}} - x_{n_k}\| < \infty$), it follows that $\{z_k\}_{k=0}^{\infty}$ is an absolutely summable sequence in $\mathcal{X}$. If every absolutely summable sequence in $\mathcal{X}$ is summable, then $\{z_k\}_{k=0}^{\infty}$ is a summable sequence, which implies that the subsequence $\{x_{n_k}\}_{k=0}^{\infty}$ converges in $\mathcal{X}$. Thus (see Proposition 3.39(c)) the Cauchy sequence $\{x_n\}_{n=0}^{\infty}$ converges in $\mathcal{X}$. Conclusion: Every Cauchy sequence in $\mathcal{X}$ converges in $\mathcal{X}$, which means that the normed space $\mathcal{X}$ is complete (i.e., $\mathcal{X}$ is a Banach space). $\qquad\square$

## 4.2   Examples

Many of the examples of metric spaces exhibited in Chapter 3 are in fact examples of normed spaces: linear spaces equipped with an additively invariant and absolutely homogeneous metric.

**Example 4A.** Let $\mathbb{F}^n$ be the linear space over $\mathbb{F}$ of Example 2D (with either $\mathbb{F} = \mathbb{R}$ or $\mathbb{F} = \mathbb{C}$). Consider the functions $\| \; \|_p \colon \mathbb{F}^n \to \mathbb{R}$ (for each real number $p \geq 1$) and $\| \; \|_\infty \colon \mathbb{F}^n \to \mathbb{R}$ defined by

$$\|x\|_p = \left( \sum_{i=1}^{n} |\xi_i|^p \right)^{\frac{1}{p}} \quad \text{and} \quad \|x\|_\infty = \max_{1 \leq i \leq n} |\xi_i|$$

for every $x = (\xi_1, \ldots, \xi_n)$ in $\mathbb{F}^n$. It is easy to verify that these are norms on $\mathbb{F}^n$ (the triangle inequality follows from the Minkowski inequality of Problem 3.4(a)), and also that the metrics generated by each of them are precisely the metrics $d_p$ (for $p \geq 1$) and $d_\infty$ of Example 3A. Since $\mathbb{F}^n$, when equipped with any of these metrics, is a complete metric space (Example 3.R(a)), it follows that

$$\mathbb{F}^n \text{ is a Banach space}$$

when equipped with any of the norms $\| \; \|_p$ or $\| \; \|_\infty$. In particular, for $n = 1$ all of these norms are reduced to the absolute value function $| \; | \colon \mathbb{F} \to \mathbb{R}$, which is the *usual norm on* $\mathbb{F}$. The norm $\| \; \|_2$ plays a special role. On $\mathbb{R}^n$ it is the *Euclidean norm*, and the real Banach space $(\mathbb{R}^n, \| \ \|_2)$ is the $n$-dimensional *Euclidean space*. The complex Banach space $(\mathbb{C}^n, \| \ \|_2)$ is the $n$-dimensional *unitary space*.

**Example 4B.** According to Example 2E the set $\mathbb{F}^{\mathbb{N}}$ (or $\mathbb{F}^{\mathbb{N}_0}$) of all scalar-valued sequences is a linear space over $\mathbb{F}$. Now consider the subsets $\ell_+^p$ and $\ell_+^\infty$ of $\mathbb{F}^{\mathbb{N}}$ defined as in Example 3B. These are linear manifolds of $\mathbb{F}^{\mathbb{N}}$ (vector addition and scalar multiplication — pointwise defined — of $p$-summable or bounded sequences are again $p$-summable or bounded sequences, respectively), and hence $\ell_+^p$ and $\ell_+^\infty$ are linear spaces over $\mathbb{F}$. For each $p \geq 1$ consider function $\| \ \|_p \colon \ell_+^p \to \mathbb{R}$ defined by

$$\|x\|_p = \left( \sum_{k=1}^\infty |\xi_k|^p \right)^{\frac{1}{p}}$$

for every $x = \{\xi_k\}_{k \in \mathbb{N}}$ in $\ell_+^p$, and the function $\| \ \|_\infty \colon \ell_+^\infty \to \mathbb{R}$ given by

$$\|x\|_\infty = \sup_{k \in \mathbb{N}} |\xi_k|$$

for every $x = \{\xi_k\}_{k \in \mathbb{N}}$ in $\ell_+^\infty$. It is readily verified that $\| \ \|_p$ is a norm on $\ell_+^p$ and $\| \ \|_\infty$ is a norm on $\ell_+^\infty$ (as before, the Minkowski inequality leads to the triangle inequality). Moreover, the norm $\| \ \|_p$ generates the metric $d_p$ and the norm $\| \ \|_\infty$ generates the metric $d_\infty$ of Example 3B. These are the usual norms on $\ell_+^p$ and $\ell_+^\infty$. Since $(\ell_+^p, d_p)$ is a complete metric space, and since $(\ell_+^\infty, d_\infty)$ also is a complete metric space (see Examples 3.R(b) and 3S), it follows that

$$(\ell_+^p, \| \ \|_p) \text{ and } (\ell_+^\infty, \| \ \|_\infty) \text{ are Banach spaces.}$$

Similarly (see Examples 2E, 3B, 3R(b) and 3S again),

$$(\ell^p, \| \ \|_p) \text{ and } (\ell^\infty, \| \ \|_\infty) \text{ are Banach spaces,}$$

where the functions $\| \ \|_p \colon \ell^p \to \mathbb{R}$ and $\| \ \|_\infty \colon \ell^\infty \to \mathbb{R}$, defined by

$$\|x\|_p = \left( \sum_{k=-\infty}^\infty |\xi_k|^p \right)^{\frac{1}{p}} \quad \text{and} \quad \|x\|_\infty = \sup_{k \in \mathbb{Z}} |\xi_k|$$

for $x = \{\xi_k\}_{k \in \mathbb{Z}}$ in $\ell^p$ or in $\ell^\infty$, respectively, are the usual norms on the linear manifolds $\ell^p$ and $\ell^\infty$ of the linear space $\mathbb{F}^{\mathbb{Z}}$.

Let $\mathcal{X}$ be a linear space. A real-valued function $\| \ \| \colon \mathcal{X} \to \mathbb{R}$ is a *seminorm* (or a *pseudonorm*) on $\mathcal{X}$ if it satisfies the three axioms (i), (iii) and (iv) of Definition 4.1. It is worth noticing that the inequality $\big| \|x\| - \|y\| \big| \leq \|x - y\|$ for every $x, y$ in $\mathcal{X}$ still holds for a seminorm. The difference between a norm and a seminorm is that a seminorm does not necessarily satisfy the axiom (ii) of Definition 4.1 (i.e., a seminorm surely vanishes at the origin but it may also vanish at a nonzero vector).

A seminorm generates a pseudometric as in Proposition 4.2: if $\| \ \|$ is a seminorm on $\mathcal{X}$, then $d(x, y) = \|x - y\|$ for every $x, y \in \mathcal{X}$ defines an additively invariant and absolutely homogeneous pseudometric on $\mathcal{X}$. Moreover, if a pseudometric is additively invariant and absolutely homogeneous, then it is generated by a seminorm as in Proposition 4.3: if $d$ is an additively invariant and absolutely homogeneous pseudometric on $\mathcal{X}$, then $\|x\| = d(x, 0)$ for every $x \in \mathcal{X}$ defines a seminorm on $\mathcal{X}$ such that $d(x, y) = \|x - y\|$ for every $x, y \in \mathcal{X}$.

**Proposition 4.5.** *Let* $\| \ \|$ *be a seminorm on a linear space* $\mathcal{X}$. *The set* $\mathcal{N}$ *of all vectors* $x$ *in* $\mathcal{X}$ *for which* $\|x\| = 0$,

$$\mathcal{N} = \{x \in \mathcal{X}: \ \|x\| = 0\},$$

*is a linear manifold of* $\mathcal{X}$. *Consider the quotient space* $\mathcal{X}/\mathcal{N}$ *and set*

$$\|[x]\|_{\sim} = \|x\|$$

*for every coset* $[x]$ *in* $\mathcal{X}/\mathcal{N}$, *where* $x$ *is an arbitrary vector in* $[x]$. *This defines a norm on the linear space* $\mathcal{X}/\mathcal{N}$ *so that* $(\mathcal{X}/\mathcal{N}, \| \ \|_{\sim})$ *is a normed space.*

*Proof.* Indeed, $\mathcal{N}$ is a linear manifold of $\mathcal{X}$ (if $u, v \in \mathcal{N}$, then $0 \leq \|u + v\| \leq \|u\| + \|v\| = 0$ and $0 \leq \|\alpha u\| \leq |\alpha| \|u\| = 0$ so that $u + v \in \mathcal{N}$ and $\alpha u \in \mathcal{N}$ for every scalar $\alpha$). Now consider the quotient space $\mathcal{X}/\mathcal{N}$ of $\mathcal{X}$ modulo $\mathcal{N}$ as in Example 2H, which is a linear space over the same scalar field of $\mathcal{X}$. Take an arbitrary coset

$$[x] = x + \mathcal{N} = \{x' \in \mathcal{X}: \ x' = x + z \text{ for some } z \in \mathcal{N}\}$$

in $\mathcal{X}/\mathcal{N}$ and note that $\|u\| = \|v\|$ for every $u$ and $v$ in $[x]$ (if $u, v \in [x]$, then $u - v \in \mathcal{N}$ and $0 \leq \big| \|u\| - \|v\| \big| \leq \|u - v\| = 0$). Thus set

$$\|[x]\|_{\sim} = \|x\|$$

for an arbitrary $x \in [x]$ (i.e., for an arbitrary representative of the equivalence class $[x]$), which defines a function from $\mathcal{X}/\mathcal{N}$ to $\mathbb{R}$,

$$\| \ \|_{\sim} : \mathcal{X}/\mathcal{N} \to \mathbb{R}.$$

It is clear that $\|[x]\|_{\sim} \geq 0$. If $\|[x]\|_{\sim} = 0$, then $[x] = \mathcal{N} = [0]$, the origin of the linear space $\mathcal{X}/\mathcal{N}$ (reason: $\|[x]\|_{\sim} = 0$ implies that every $x'$ in $[x]$ belongs to $\mathcal{N}$ and also that every $u$ in $\mathcal{N}$ belongs to $[x]$). Moreover,

$$\|\alpha[x]\|_{\sim} = \|[\alpha x]\|_{\sim} = \|\alpha x\| = |\alpha| \|x\| = |\alpha| \|[x]\|_{\sim},$$

$$\|[x] + [y]\|_{\sim} = \|[x + y]\|_{\sim} = \|x + y\| \leq \|x\| + \|y\| = \|[x]\|_{\sim} + \|[y]\|_{\sim},$$

for every $[x], [y] \in \mathcal{X}/\mathcal{N}$ and every scalar $\alpha$ (Example 2H). Therefore (Definition 4.1), $\| \ \|_{\sim}$ is a norm on the linear space $\mathcal{X}/\mathcal{N}$. $\qquad\square$

Remark: Note that the relation $\sim$ on $\mathcal{X}$ defined by

$$x' \sim x \quad \text{if} \quad \|x' - x\| = 0$$

or, equivalently,

$$x' \sim x \quad \text{if} \quad x' - x \in \mathcal{N}$$

is a linear equivalence relation on the linear space $\mathcal{X}$ in the sense of Example 2G. Consider the quotient space of $\mathcal{X}$ modulo $\sim$, $\mathcal{X}/\sim$. If vector addition and scalar multiplication are defined in $\mathcal{X}/\sim$ as in Example 2H, then $X/\sim$ is a linear space that coincides with $X/\mathcal{N}$. In this case (i.e., if $\mathcal{X}$ is a linear space and $\| \; \|$ is a seminorm on $\mathcal{X}$), then the metric $d_\sim$ on $X/\mathcal{N}$ generated by the norm $\| \; \|_\sim$ is precisely the metric $\tilde{d}$ on $\mathcal{X}/\sim$ of Proposition 3.3 obtained by the pseudometric $d$ on $\mathcal{X}$ generated by the seminorm $\| \; \|$.

**Example 4C.** Take an arbitrary real number $p \geq 1$ and consider the setup of Example 3E: $r^p(S)$ is the set of all $\mathbb{F}$-valued Riemann $p$-integrable functions on a nondegenerate interval $S$ of the real line $\mathbb{R}$. This is a linear manifold of the linear space $\mathbb{F}^S$ (see Example 2E), and hence $r^p(S)$ is a linear space over $\mathbb{F}$. Indeed, addition and scalar multiplication of Riemann $p$-integrable functions on $S$ are again Riemann $p$-integrable functions on $S$ (Minkowski inequality). It is clear that the pseudometric $\delta_p$ on $r^p(S)$ defined in Example 3E is additively invariant and absolutely homogeneous. Thus $\delta_p$ is generated by a seminorm $| \; |_p$ on $r^p(S)$,

$$|x|_p = \delta_p(x, 0) = \left( \int_S |x(s)|^p \, ds \right)^{\frac{1}{p}}$$

for every $x \in r^p(S)$, so that $\delta_p(x, y) = |x - y|_p$ for every $x, y \in r^p(S)$. Now consider the linear manifold $\mathcal{N} = \{x \in r^p(S) : |x|_p = 0\}$ and set $R^p(S) = r^p(S)/\mathcal{N}$, the quotient space of $r^p(S)$ modulo $\mathcal{N}$ (i.e., the collection of all equivalence classes $[x] = \{x' \in r^p(S) : |x' - x|_p = 0\}$ for every $x \in r^p(S)$). By Proposition 4.5 the function $\| \; \|_p \colon R^p(S) \to \mathbb{R}$, defined by $\|[x]\|_p = |x|_p$ for every $[x]$ in $R^p(S)$, is a norm on $R^p(S)$ (where $x$ is an arbitrary representative of the equivalence class $[x]$). This is the usual norm on $R^p(S)$. Note that $R^p(S)$ is precisely the quotient space $r^p(S)/\sim$ of Example 3E (see the remark that follows Proposition 4.5). Moreover, $\| \; \|_p$ is the norm on $R^p(S)$ that generates the usual metric $d_p$ of Example 3E:

$$d_p([x], [y]) = \|[x] - [y]\|_p = \|[x - y]\|_p = |x - y|_p = \delta_p(x, y)$$

for every $[x], [y] \in R^p(S)$, where $x$ and $y$ are arbitrary vectors in $[x]$ and $[y]$, respectively. According to common usage we shall write $x \in R^p(S)$ instead of $[x] \in R^p(S)$, and also

$$\|x\|_p = d_p(x, 0) = \left( \int_S |x(s)|^p \, ds \right)^{\frac{1}{p}}$$

for every $x \in R^p(S)$ to represent the norm $\| \ \|_p$ on $R^p(S)$. Therefore,

$$(R^p(S), \| \ \|_p) \text{ is a normed space}$$

but not a Banach space (Problem 3.62). Its completion, of course, is:

$$(L^p(S), \| \ \|_p) \text{ is a Banach space.}$$

(We shall dicuss the completion of a normed space in Section 4.7.)

**Example 4D.** Let $C[0, 1]$ be the set of all $\mathbb{F}$-valued continuous functions on the interval $[0, 1]$ as in Example 3D. Again, this is a linear manifold of the linear space $\mathbb{F}^{[0,1]}$ of Example 2E (addition and scalar multiplication of continuous functions are continuous functions), and hence $C[0, 1]$ is a linear space over $\mathbb{F}$. In fact, $C[0, 1]$ is a linear manifold of the linear space $r^p[0, 1]$ of the previous example (every continuous function on $[0, 1]$ is Riemann $p$-integrable). For each $p \geq 1$ consider the function $\| \ \|_p : C[0, 1] \to \mathbb{R}$ defined by

$$\|x\|_p = \left( \int_0^1 |x(t)|^p \, dt \right)^{\frac{1}{p}}$$

for every $x \in C[0, 1]$. This is the norm on $C[0, 1]$ that generates the metric $d_p$ of Example 3D so that

$$(C[0, 1], \| \ \|_p) \text{ is a normed space.}$$

According to Problem 3.58 $(C[0, 1], \| \ \|_p)$ is not a Banach space. Recall that $C[0, 1]$ can be viewed as a subset of $R^p[0, 1]$ (in the sense of Problem 3.40) and, as such, it can be shown that $C[0, 1]$ is dense in $(R^p[0, 1], \| \ \|_p)$. Therefore (see Problems 3.38(g) and 3.62), the Banach space $(L^p[0, 1], \| \ \|_p)$ is a completion of $(C[0, 1], \| \ \|_p)$.

Let $\{\mathcal{X}_\gamma\}_{\gamma \in \Gamma}$ be an indexed family of linear spaces over the same field $\mathbb{F}$. The set $\bigoplus_{\gamma \in \Gamma} \mathcal{X}_\gamma$ of all indexed families $\{x_\gamma\}_{\gamma \in \Gamma}$, where $x_\gamma \in \mathcal{X}_\gamma$ for each $\gamma \in \Gamma$, becomes a linear space over $\mathbb{F}$ if vector addition and scalar multiplication are defined on $\bigoplus_{\gamma \in \Gamma} \mathcal{X}_\gamma$ as

$$\{x_\gamma\}_{\gamma \in \Gamma} \oplus \{y_\gamma\}_{\gamma \in \Gamma} = \{x_\gamma + y_\gamma\}_{\gamma \in \Gamma} \quad \text{and} \quad \alpha\{x_\gamma\}_{\gamma \in \Gamma} = \{\alpha x_\gamma\}_{\gamma \in \Gamma}$$

for every $\{x_\gamma\}_{\gamma \in \Gamma}$ and $\{y_\gamma\}_{\gamma \in \Gamma}$ in $\bigoplus_{\gamma \in \Gamma} \mathcal{X}_\gamma$ and every $\alpha$ in $\mathbb{F}$. This is the *direct sum* (or the *full direct sum*) of the family $\{\mathcal{X}_\gamma\}_{\gamma \in \Gamma}$. The underlying set of the linear space $\bigoplus_{\gamma \in \Gamma} \mathcal{X}_\gamma$ is the Cartesian product $\prod_{\gamma \in \Gamma} \mathcal{X}_\gamma$ of the underlying sets of each linear space $\mathcal{X}_\gamma$ (cf. Section 2.8).

**Example 4E.** Let $\{(\mathcal{X}_i, \| \ \|_i)\}_{i=1}^n$ be a finite collection of normed spaces, where the linear spaces $\mathcal{X}_i$ are all over the same field $\mathbb{F}$, and let $\bigoplus_{i=1}^n \mathcal{X}_i$ be the direct sum of the family $\{\mathcal{X}_i\}_{i=1}^n$. Consider the functions $\|\ \|_p \colon \bigoplus_{i=1}^n \mathcal{X}_i \to \mathbb{R}$ (for each real number $p \geq 1$) and $\|\ \|_\infty \colon \bigoplus_{i=1}^n \mathcal{X}_i \to \mathbb{R}$ defined by

$$\|x\|_p = \Big( \sum_{i=1}^n \|x_i\|_i^p \Big)^{\frac{1}{p}} \quad \text{and} \quad \|x\|_\infty = \max_{1 \leq i \leq n} \|x_i\|_i$$

for every $x = (x_1, \dots, x_n)$ in $\bigoplus_{i=1}^n \mathcal{X}_i$. It is easy to verify that these are norms on the direct sum $\bigoplus_{i=1}^n \mathcal{X}_i$. For instance, the triangle inequality for the norm $\|\ \|_p$ comes from the Minkowski inequality (Problem 3.4): for every $x = (x_1, \dots, x_n)$ and $y = (y_1, \dots, y_n)$ in $\bigoplus_{i=1}^n \mathcal{X}_i$,

$$\begin{aligned}
\|x \oplus y\|_p &= \Big( \sum_{i=1}^n \|x_i + y_i\|_i^p \Big)^{\frac{1}{p}} \leq \Big( \sum_{i=1}^n (\|x_i\|_i + \|y_i\|_i)^p \Big)^{\frac{1}{p}} \\
&\leq \Big( \sum_{i=1}^n \|x_i\|_i^p \Big)^{\frac{1}{p}} + \Big( \sum_{i=1}^n \|y_i\|_i^p \Big)^{\frac{1}{p}} = \|x\|_p + \|y\|_p.
\end{aligned}$$

Moreover, these norms generate the metrics $d_p$ and $d_\infty$ of Problem 3.9 (recall: the underlying set of the linear space $\bigoplus_{i=1}^n \mathcal{X}_i$ is the Cartesian product $\prod_{i=1}^n \mathcal{X}_i$ of the underlying sets of each linear space $\mathcal{X}_i$), and therefore (Problem 3.60)

$$\Big( \bigoplus_{i=1}^n \mathcal{X}_i, \|\ \|_p \Big) \quad \text{and} \quad \Big( \bigoplus_{i=1}^n \mathcal{X}_i, \|\ \|_\infty \Big) \quad \text{are Banach spaces}$$

if and only if each $(\mathcal{X}_i, \|\ \|_i)$ is a Banach space. If the normed spaces $(\mathcal{X}_i, \|\ \|_i)$ coincide with a fixed normed space $(\mathcal{X}, \|\ \|)$, then $\bigoplus_{i=1}^n \mathcal{X}$ is the direct sum of $n$ copies of $\mathcal{X}$ (a linear space whose underlying set is the Cartesian product $\prod_{i=1}^n \mathcal{X} = \mathcal{X}^n$ of $n$ copies of the underlying set of the linear space $\mathcal{X}$ — Section 1.7). We can identify $\bigoplus_{i=1}^n \mathcal{X}$ with $\mathcal{X}^n$ (where the linear operations on $\mathcal{X}^n$ are defined coordinatewise) so that

$$(\mathcal{X}^n, \|\ \|_p) \quad \text{and} \quad (\mathcal{X}^n, \|\ \|_\infty) \quad \text{are Banach spaces}$$

whenever $(\mathcal{X}, \|\ \|)$ is a Banach space. Note that this generalizes the Banach spaces of Example 4A.

**Example 4F.** Let $\{(\mathcal{X}_k, \|\ \|_k)\}$ be a countably infinite collection of normed spaces, where the linear spaces $\mathcal{X}_k$ are all over the same field $\mathbb{F}$, and let $\bigoplus_k \mathcal{X}_k$ be the direct sum of $\{\mathcal{X}_k\}$. For each $p \geq 1$ consider the subset $\big[ \bigoplus_k \mathcal{X}_k \big]_p$ of $\bigoplus_k \mathcal{X}_k$ consisting of all *p-summable families* $\{x_k\}$ in $\bigoplus_k \mathcal{X}_k$. That is, $\{x_k\} \in \big[ \bigoplus_k \mathcal{X}_k \big]_p$ if and only if $\sum_k \|x_k\|_k^p < \infty$, where $\sum_k \|x_k\|_k^p$ is the supremum of the set of all finite sums of positive numbers from $\{\|x_k\|_k^p\}$. This is a linear manifold of the linear space $\bigoplus_k \mathcal{X}_k$ and so $\big[ \bigoplus_k \mathcal{X}_k \big]_p$ is a linear space over $\mathbb{F}$. As in Example 4E, it is easy to show that the function $\|\ \|_p \colon \big[ \bigoplus_k \mathcal{X}_k \big]_p \to \mathbb{R}$, defined by

$$\|x\|_p = \Big( \sum_k \|x_k\|_k^p \Big)^{\frac{1}{p}}$$

for every $x = \{x_k\} \in \left[\bigoplus_k \mathcal{X}_k\right]_p$, is a norm on $\left[\bigoplus_k \mathcal{X}_k\right]_p$. Now consider the subset $\left[\bigoplus_k \mathcal{X}_k\right]_\infty$ of $\bigoplus_k \mathcal{X}_k$ consisting of all *bounded families* $\{x_k\}$ in $\bigoplus_k \mathcal{X}_k$ (i.e., $\{x_k\} \in \left[\bigoplus_k \mathcal{X}_k\right]_\infty$ if and only if $\sup_k \|x_k\|_k < \infty$). This again is a linear manifold of the linear space $\bigoplus_k \mathcal{X}_k$ so that $\left[\bigoplus_k \mathcal{X}_k\right]_\infty$ is itself a linear space over $\mathbb{F}$. It is readily verified that the function $\| \ \|_\infty \colon \left[\bigoplus_k \mathcal{X}_k\right]_\infty \to \mathbb{R}$, defined by

$$\|x\|_\infty = \sup_k \|x_k\|_k$$

for every $x = \{x_k\} \in \left[\bigoplus_k \mathcal{X}_k\right]_\infty$, is a norm on $\left[\bigoplus_k \mathcal{X}_k\right]_\infty$. Moreover, it can also be shown that

$$\left(\left[\textstyle\bigoplus_k \mathcal{X}_k\right]_p, \| \ \|_p\right) \ \text{and} \ \left(\left[\textstyle\bigoplus_k \mathcal{X}_k\right]_\infty, \| \ \|_\infty\right) \ \text{are Banach spaces}$$

if and only if each $(\mathcal{X}_k, \| \ \|_k)$ is a Banach space (hint: Example 3R(b)). Again, if the normed spaces $(\mathcal{X}_k, \| \ \|_k)$ coincide with a fixed normed space $(\mathcal{X}, \| \ \|)$, then $\bigoplus_k \mathcal{X}$ is the direct sum of countably infinite copies of $\mathcal{X}$. (Recall that $\bigoplus_k \mathcal{X}$ is a linear space whose underlying set is the Cartesian product $\prod_k \mathcal{X}$ of countably infinite copies of the underlying set of the linear space $\mathcal{X}$, which coincide with $\mathcal{X}^\mathbb{N}$, $\mathcal{X}^{\mathbb{N}_0}$ or $\mathcal{X}^\mathbb{Z}$ if the indices $k$ run over $\mathbb{N}$, $\mathbb{N}_0$ or $\mathbb{Z}$, respectively — Section 1.7). As before, we shall adopt the identifications $\bigoplus_{k \in \mathbb{N}} \mathcal{X} = \mathcal{X}^\mathbb{N}$, $\bigoplus_{k \in \mathbb{N}_0} \mathcal{X} = \mathcal{X}^{\mathbb{N}_0}$ and $\bigoplus_{k \in \mathbb{Z}} \mathcal{X} = \mathcal{X}^\mathbb{Z}$ (the linear operations on $\mathcal{X}^\mathbb{N}$, $\mathcal{X}^{\mathbb{N}_0}$ and $\mathcal{X}^\mathbb{Z}$ are defined as in Example 2F). It is usual to denote $\left[\bigoplus_k \mathcal{X}\right]_p$ in $\mathcal{X}^\mathbb{N}$ (or in $\mathcal{X}^{\mathbb{N}_0}$) by $\ell_+^p(\mathcal{X})$: the linear manifold of $\mathcal{X}^\mathbb{N}$ consisting of all $p$-summable $\mathcal{X}$-valued sequences; and $\left[\bigoplus_k \mathcal{X}\right]_\infty$ in $\mathcal{X}^\mathbb{N}$ (or in $\mathcal{X}^{\mathbb{N}_0}$) by $\ell_+^\infty(\mathcal{X})$: the linear manifold of $\mathcal{X}^\mathbb{N}$ consisting of all bounded $\mathcal{X}$-valued sequences. That is,

$$\ell_+^p(\mathcal{X}) = \left\{\{x_k\} \in \mathcal{X}^\mathbb{N} \colon \ \textstyle\sum_{k=1}^\infty \|x_k\|^p < \infty\right\},$$

$$\ell_+^\infty(\mathcal{X}) = \left\{\{x_k\} \in \mathcal{X}^\mathbb{N} \colon \ \sup_{k \in \mathbb{N}} \|x_k\| < \infty\right\}.$$

The norms $\|x\|_p = \left(\sum_{k=1}^\infty \|x_k\|^p\right)^{\frac{1}{p}}$ and $\|x\|_\infty = \sup_{k \in \mathbb{N}} \|x_k\|$, for every $x = \{x_k\}$ either in $\ell_+^p(\mathcal{X})$ or in $\ell_+^\infty(\mathcal{X})$, are the usual norms on $\ell_+^p(\mathcal{X})$ and $\ell_+^\infty(\mathcal{X})$, respectively, and

$$(\ell_+^p(\mathcal{X}), \| \ \|_p) \ \text{and} \ (\ell_+^\infty(\mathcal{X}), \| \ \|_\infty) \ \text{are Banach spaces}$$

whenever $(\mathcal{X}, \| \ \|)$ is a Banach space. Similarly, $\left[\bigoplus_k \mathcal{X}\right]_p$ in $\mathcal{X}^\mathbb{Z}$ is denoted by $\ell^p(\mathcal{X})$ and $\left[\bigoplus_k \mathcal{X}\right]_\infty$ in $\mathcal{X}^\mathbb{Z}$ is denoted by $\ell^\infty(\mathcal{X})$ and, when equipped with their usual norms $\| \ \|_p$ and $\| \ \|_\infty$,

$$(\ell^p(\mathcal{X}), \| \ \|_p) \ \text{and} \ (\ell^\infty(\mathcal{X}), \| \ \|_\infty) \ \text{are Banach spaces}$$

whenever $(\mathcal{X}, \| \ \|)$ is a Banach space. This example generalizes the Banach spaces of Example 4B.

# 4.3   Subspaces and Quotient Spaces

If $(\mathcal{X}, \| \ \|)$ is a normed space, and if $\mathcal{M}$ is a linear manifold of the linear space $\mathcal{X}$, then it is easy to show that the restriction $\| \ \|_{\mathrm{M}} \colon \mathcal{M} \to \mathbb{R}$ of the norm $\| \ \| \colon \mathcal{X} \to \mathbb{R}$ to $\mathcal{M}$ is a norm on $\mathcal{M}$ so that $(\mathcal{M}, \| \ \|_{\mathrm{M}})$ is a normed space. Moreover, the metric $d_{\mathrm{M}} \colon \mathcal{M} \times \mathcal{M} \to \mathbb{R}$ generated by the norm $\| \ \|_{\mathrm{M}}$ on $\mathcal{M}$ coincides with the restriction to $\mathcal{M} \times \mathcal{M}$ of the metric $d \colon \mathcal{X} \times \mathcal{X} \to \mathbb{R}$ generated by the norm $\| \ \|$ on $\mathcal{X}$. Thus $(\mathcal{M}, d_{\mathrm{M}})$ is a subspace of the metric space $(\mathcal{X}, d)$. If a linear manifold of a normed space is regarded as a normed space, then it will be understood that the norm on it is the restricted norm $\| \ \|_{\mathrm{M}}$. We shall drop the subscripts and write $(\mathcal{M}, \| \ \|)$ and $(\mathcal{M}, d)$ instead of $(\mathcal{M}, \| \ \|_{\mathrm{M}})$ and $(\mathcal{M}, d_{\mathrm{M}})$, respectively, and often refer to the normed space $(\mathcal{M}, \| \ \|)$ by simply saying that "$\mathcal{M}$ is a linear manifold of $\mathcal{X}$".

**Proposition 4.6.** *Let $\mathcal{M}$ be a linear manifold of a normed space $\mathcal{X}$. If $\mathcal{M}$ is open in $\mathcal{X}$, then $\mathcal{M} = \mathcal{X}$.*

*Proof.* Since $\mathcal{M}$ is a linear manifold of the linear space $\mathcal{X}$, the origin $0$ of $\mathcal{X}$ lies in $\mathcal{M}$. If $\mathcal{M}$ is open in $\mathcal{X}$ (in the norm topology, of course), then $\mathcal{M}$ includes a nonempty open ball with center at the origin. That is, $B_\varepsilon(0) = \{y \in \mathcal{X} \colon \|y\| < \varepsilon\} \subset \mathcal{M}$ for some $\varepsilon > 0$. Take an arbitrary nonzero vector $x$ in $\mathcal{X}$ and set $z = \varepsilon(2\|x\|)^{-1}x \in \mathcal{X}$ so that $\|z\| = \frac{\varepsilon}{2}$, and hence $z \in B_\varepsilon(0) \subset \mathcal{M}$. Thus $x = (2\|x\|)\varepsilon^{-1}z$ lies in $\mathcal{M}$ (since $\mathcal{M}$ is a linear space). Therefore, every nonzero vector in $\mathcal{X}$ also lies in $\mathcal{M}$. Conclusion: $\mathcal{X} \subseteq \mathcal{M}$ so that $\mathcal{M} = \mathcal{X}$ (because $\mathcal{M} \subseteq \mathcal{X}$). $\qquad\square$

This shows that a normed space $\mathcal{X}$ is itself the only open linear manifold of $\mathcal{X}$. On the other hand, the closed linear manifolds of a normed space are far more interesting. They in fact are so important that we give then a name. A *closed* linear manifold of a normed space is called a *subspace* of $\mathcal{X}$. (Warning: A subspace of a metric space $(X, d)$ is simply a subset of $X$ equipped with the "same" metric $d$, while a subspace of a normed space $(\mathcal{X}, \| \ \|)$ is a linear manifold of $\mathcal{X}$ equipped with the "same" norm $\| \ \|$ that is *closed* in $(\mathcal{X}, \| \ \|)$).

**Proposition 4.7.** *A linear manifold of a Banach space $\mathcal{X}$ is itself a Banach space if and only if it is a subspace of $\mathcal{X}$.*

*Proof.* Corollary 3.41. $\qquad\square$

**Example 4G.** As usual, let $Y^S$ denote the collection of all functions of a nonempty set $S$ into a set $Y$. If $\mathcal{Y}$ is a linear space over $\mathbb{F}$, then $\mathcal{Y}^S$ is a linear space over $\mathbb{F}$ (Example 2F). Suppose $\mathcal{Y}$ is a normed space and consider the subset $B[S, \mathcal{Y}]$ of $\mathcal{Y}^S$ consisting of all bounded mappings of $S$ into $\mathcal{Y}$. This is a linear manifold of $\mathcal{Y}^S$ (vector addition and scalar multiplication of bounded mappings are bounded mappings), and hence $B[S, \mathcal{Y}]$ is a linear space over $\mathbb{F}$. Now consider the function $\| \ \|_\infty \colon B[S, \mathcal{Y}] \to \mathbb{R}$ defined by

$$\|f\|_\infty = \sup_{s \in S} \|f(s)\|$$

for every $f \in B[S, \mathcal{Y}]$, where $\| \ \| : \mathcal{Y} \to \mathbb{R}$ is the norm on $\mathcal{Y}$. It is easy to verify that $(B[S, \mathcal{Y}], \| \ \|_\infty)$ is a normed space. The norm $\| \ \|_\infty$ on $B[S, \mathcal{Y}]$, which is referred to as the *sup-norm*, generates the sup-metric $d_\infty$ of Example 3C. Thus, according to Example 3S,

$$(B[S, \mathcal{Y}], \| \ \|_\infty) \text{ is a Banach space if and only if } \mathcal{Y} \text{ is a Banach space.}$$

Moreover, if $X$ is a nonempty metric space and if $BC[X, \mathcal{Y}]$ is the subset of $B[X, \mathcal{Y}]$ made up of all continuous mappings from $B[X, \mathcal{Y}]$, then $BC[X, \mathcal{Y}]$ is a linear manifold of the linear space $B[X, \mathcal{Y}]$ (addition and scalar multiplication of continuous functions are continuous functions), and hence a linear space over $\mathbb{F}$. According to Example 3N the linear manifold $BC[X, \mathcal{Y}]$ is closed in $(B[X, \mathcal{Y}], \| \ \|_\infty)$ and so $(BC[X, \mathcal{Y}], \| \ \|_\infty)$ is a subspace of the normed space $(B[X, \mathcal{Y}], \| \ \|_\infty)$. Thus

$$(BC[X, \mathcal{Y}], \| \ \|_\infty) \text{ is a Banach space if and only if } \mathcal{Y} \text{ is a Banach space}$$

(Example 3T — also see Proposition 4.7). If the metric space $X$ is compact, then $C[X, \mathcal{Y}] = BC[X, \mathcal{Y}]$, where $C[X, \mathcal{Y}]$ stands for the set of all continuous mappings of the compact metric space $X$ into the normed space $\mathcal{Y}$ (Example 3Y). Therefore,

$$(C[X, \mathcal{Y}], \| \ \|_\infty) \text{ is a Banach space if } X \text{ is compact and } \mathcal{Y} \text{ is Banach.}$$

By setting $X = [0, 1]$ equipped with the usual metric on $\mathbb{R}$ (which is compact — Heine–Borel Theorem), it follows that

$$(C[0, 1], \| \ \|_\infty) \text{ is a Banach space,}$$

where $C[0, 1] = C([0, 1], \mathbb{F})$ is the linear space over $\mathbb{F}$ of all $\mathbb{F}$-valued continuous functions defined on the closed interval $[0, 1]$ (recall: $(\mathbb{F}, | \ |)$ is a Banach space — Example 4A), and

$$\|x\|_\infty = \sup_{0 \le t \le 1} |x(t)| = \max_{0 \le t \le 1} |x(t)| \quad \text{for every} \quad x \in C[0, 1]$$

(cf., Examples 3T and 3Y) is the *sup-norm* on $C[0, 1]$.

In any normed space $\mathcal{X}$ the zero linear manifold $\{0\}$ and the whole space $\mathcal{X}$ are subspaces of $\mathcal{X}$. If a subspace $\mathcal{M}$ is a proper subset of $\mathcal{X}$, then it is said to be a *proper subspace*. A *nontrivial subspace* $\mathcal{M}$ of a normed space $\mathcal{X}$ is a nonzero proper subspace of it ($\{0\} \neq \mathcal{M} \neq \mathcal{X}$). The next proposition shows that, if the dimension of the linear space $\mathcal{X}$ is greater than one, then there are many nontrivial subspaces of the normed space $\mathcal{X}$.

**Proposition 4.8.** *Let $\mathcal{X}$ be a normed space.*

(a) *The closure $\mathcal{M}^-$ of a linear manifold $\mathcal{M}$ of $\mathcal{X}$ is a subspace of $\mathcal{X}$.*

(b) *The intersection of an arbitrary nonempty collection of subspaces of $\mathcal{X}$ is again a subspace of $\mathcal{X}$.*

*Proof.* (a) The closure $\mathcal{M}^-$ of a linear manifold $\mathcal{M}$ of a normed space $\mathcal{X}$ is clearly a closed subset of $\mathcal{X}$. What is left to be shown is that this closed subset of $\mathcal{X}$ is also a linear manifold of $\mathcal{X}$. Take two arbitrary points $x$ and $y$ in $\mathcal{M}^-$. According to Proposition 3.27 there exist $\mathcal{M}$-valued sequences $\{x_n\}$ and $\{y_n\}$ that converge to $x$ and $y$, respectively, in the norm topology of $\mathcal{X}$. Therefore (cf. Problem 4.1), the $\mathcal{M}$-valued sequence $\{x_n + y_n\}$ converges in $\mathcal{X}$ to $x + y$, and hence $x + y \in \mathcal{M}^-$ (see Proposition 3.27 again).

(b) The intersection of an arbitrary nonempty collection of linear manifolds of a linear space is a linear manifold (cf. Section 2.2), and the intersection of an arbitrary collection of closed subsets of a metric space is a closed set (cf. Theorem 3.22). Thus the intersection of an arbitrary nonempty collection of closed linear manifolds of a normed space is a closed linear manifold. $\qquad\qquad\square$

Let $A$ be a subset of a normed space $\mathcal{X}$. The (linear) span of $A$, span $A$, was defined in Section 2.2 as the intersection of all linear manifolds of $\mathcal{X}$ that include $A$, which coincides with the smallest linear manifold of $\mathcal{X}$ that includes $A$. Recall that the smallest closed subset of $\mathcal{X}$ that includes span $A$ is precisely its closure $(\text{span } A)^-$ in $\mathcal{X}$ (by the very definition of closure). According to Proposition 4.8(a) $(\text{span } A)^-$ is a subspace of $\mathcal{X}$. This is the smallest closed linear manifold of $\mathcal{X}$ that includes $A$. Set

$$\bigvee A = (\text{span } A)^-,$$

which is called the *subspace spanned* by $A$. If a subspace $\mathcal{M}$ of $\mathcal{X}$ (which may be $\mathcal{X}$ itself) is such that $\mathcal{M} = \bigvee A$ for some subset $A$ of $\mathcal{X}$, then we say that $A$ *spans* $\mathcal{M}$ or that $A$ is a *spanning set* for $\mathcal{M}$ (warning: the same terminology of Section 2.2 but now with a different meaning); or still that $A$ is a *total set* for $\mathcal{M}$. Also note that the intersection of all subspaces of $\mathcal{X}$ that include $A$ is the smallest subspace of $\mathcal{X}$ (see Proposition 4.8(b)) that includes $A$. Summing up: $\bigvee A$ *is the smallest subspace of $\mathcal{X}$ that includes $A$*, which coincides with *the intersection of all subspaces of $\mathcal{X}$ that include $A$.* It is readily verified that $\bigvee \varnothing = \{0\}$, $\bigvee \mathcal{M} = \mathcal{M}^-$ for every linear manifold $\mathcal{M}$ of $\mathcal{X}$, and $A \subseteq \bigvee A = \bigvee(\bigvee A)$ for every subset $A$ of $\mathcal{X}$. Moreover, if $A$ and $B$ are subsets of $\mathcal{X}$, then

$$A \subseteq B \quad \text{implies} \quad \bigvee A \subseteq \bigvee B.$$

**Proposition 4.9.** *Let $\mathcal{X}$ be a normed space.*

(a) *A set $A$ spans $\mathcal{X}$ if and only if every linear manifold of $\mathcal{X}$ that includes $A$ is dense in $\mathcal{X}$.*

(b) *$\mathcal{X}$ is separable if and only if it is spanned by a countable set.*

*Proof.* (a) Let $A$ be a subset of a normed space $\mathcal{X}$. Take an arbitrary linear manifold $\mathcal{M}$ of $\mathcal{X}$ such that $A \subseteq \mathcal{M}$. Recall: $\bigvee A \subseteq \bigvee \mathcal{M} = \mathcal{M}^-$. Thus $\bigvee A = \mathcal{X}$ implies $\mathcal{M}^- = \mathcal{X}$. Conversely, span $A$ is a linear manifold of $\mathcal{X}$ that includes $A$. If every linear manifold of $\mathcal{X}$ that includes $A$ is dense in $\mathcal{X}$, then $\bigvee A = (\text{span } A)^- = \mathcal{X}$.

(b) Let $\mathcal{X}$ be a normed space. If $\mathcal{X}$ is separable as a metric space, then (by definition) there exists a countable set, say $A \subseteq \mathcal{X}$, such that $A^- = \mathcal{X}$. Since $A \subseteq \bigvee A = (\bigvee A)^-$, it follows that $\bigvee A = \mathcal{X}$. Conversely, suppose $(\text{span } A)^- = \mathcal{X}$ for some countable subset $A$ of $\mathcal{X}$. Recall that span $A$ is the set of all (finite) linear combinations of vectors in $A$ (Proposition 2.2). Let $\mathcal{M}$ denote the set of all (finite) linear combinations of vectors in $A$ with rational coefficients (note: we say that a complex number is "rational" if its real and imaginary parts are rational numbers). Since $\mathbb{Q}^- = \mathbb{R}$, it follows that $\mathcal{M}^- = (\text{span } A)^-$ (see Example 3P), and hence $\mathcal{M}^- = \mathcal{X}$. Moreover, it is readily verified that $\mathcal{M}$ is a linear space over the rational field $\mathbb{Q}$ and so $\mathcal{M} = \text{span } A$ (over $\mathbb{Q}$). Thus $\mathcal{M}$ has a Hamel basis included in $A$ (Theorem 2.6) so that $\dim \mathcal{M} \leq \#A \leq \aleph_0$. Therefore, $\#\mathcal{M} = \max\{\#\mathbb{Q}, \dim \mathcal{M}\} = \aleph_0$ (cf. Problem 2.8). Conclusion: $\mathcal{M}$ is a countable dense subset of $\mathcal{X}$. Outcome: $\mathcal{X}$ is separable. $\qquad\square$

In Section 2.2 we considered the subcollection $\mathcal{L}at(\mathcal{X})$ of the power set $\wp(\mathcal{X})$ of a linear space $\mathcal{X}$ consisting of all linear manifolds of $\mathcal{X}$. Recall that $\mathcal{L}at(\mathcal{X})$ was shown to be a complete lattice (in the inclusion ordering of $\wp(\mathcal{X})$): if $\{\mathcal{M}_\gamma\}_{\gamma \in \Gamma}$ is an arbitrary nonempty subcollection of $\mathcal{L}at(\mathcal{X})$ (i.e., an arbitrary nonempty indexed family of linear manifolds of $\mathcal{X}$), then $\inf\{\mathcal{M}_\gamma\}_{\gamma \in \Gamma} = \bigcap_{\gamma \in \Gamma} \mathcal{M}_\gamma \in \mathcal{L}at(\mathcal{X})$ and $\sup\{\mathcal{M}_\gamma\}_{\gamma \in \Gamma} = \text{span}\left(\bigcup_{\gamma \in \Gamma} \mathcal{M}_\gamma\right) = \sum_{\gamma \in \Gamma} \mathcal{M}_\gamma \in \mathcal{L}at(\mathcal{X})$. In particular, if $\{\mathcal{M}, \mathcal{N}\}$ is a pair of linear manifolds of $\mathcal{X}$, then $\mathcal{M} \wedge \mathcal{N} = \mathcal{M} \cap \mathcal{N}$ and $\mathcal{M} \vee \mathcal{N} = \text{span}(\mathcal{M} \cup \mathcal{N}) = \mathcal{M} + \mathcal{N}$ are both linear manifolds of $\mathcal{X}$.

Now shift from linear manifolds to subspaces. Let $\text{Lat}(\mathcal{X})$ denote the subcollection of the power set $\wp(\mathcal{X})$ of a normed space $\mathcal{X}$ made up of all *subspaces* of $\mathcal{X}$. Clearly, $\text{Lat}(\mathcal{X}) \subseteq \mathcal{L}at(\mathcal{X})$. If $\{\mathcal{M}_\gamma\}_{\gamma \in \Gamma}$ is an arbitrary nonempty subcollection of $\text{Lat}(\mathcal{X})$ (i.e., an arbitrary nonempty indexed family of subspaces of the normed space $\mathcal{X}$), then $\bigcap_{\gamma \in \Gamma} \mathcal{M}_\gamma \in \text{Lat}(\mathcal{X})$ (by Proposition 4.8(b)) and $\bigcap_{\gamma \in \Gamma} \mathcal{M}_\gamma \subseteq \mathcal{M}_\alpha$ for every $\mathcal{M}_\alpha \in \{\mathcal{M}_\gamma\}_{\gamma \in \Gamma}$, so that $\bigcap_{\gamma \in \Gamma} \mathcal{M}_\gamma$ is a lower bound for $\{\mathcal{M}_\gamma\}_{\gamma \in \Gamma}$. Moreover, if $\mathcal{V}$ in $\text{Lat}(\mathcal{X})$ is a lower bound for $\{\mathcal{M}_\gamma\}_{\gamma \in \Gamma}$ (i.e., if $\mathcal{V} \subseteq \mathcal{M}_\gamma$ for all $\gamma \in \Gamma$), then $\mathcal{V} \subseteq \bigcap_{\gamma \in \Gamma} \mathcal{M}_\gamma$. Thus $\bigcap_{\gamma \in \Gamma} \mathcal{M}_\gamma = \inf\{\mathcal{M}_\gamma\}_{\gamma \in \Gamma}$. If we adopt the usual notation $\bigwedge_{\gamma \in \Gamma} \mathcal{M}_\gamma = \inf\{\mathcal{M}_\gamma\}_{\gamma \in \Gamma}$, then

$$\bigwedge_{\gamma \in \Gamma} \mathcal{M}_\gamma = \bigcap_{\gamma \in \Gamma} \mathcal{M}_\gamma.$$

Similarly, $\mathcal{M}_\alpha \subseteq \bigvee\left(\bigcap_{\gamma \in \Gamma} \mathcal{M}_\gamma\right) \in \text{Lat}(\mathcal{X})$ for every $\mathcal{M}_\alpha \in \{\mathcal{M}_\gamma\}_{\gamma \in \Gamma}$ and, if $\mathcal{U}$ in $\text{Lat}(\mathcal{X})$ is an upper bound for $\{\mathcal{M}_\gamma\}_{\gamma \in \Gamma}$ (i.e., if $\mathcal{M}_\gamma \subseteq \mathcal{U}$ for all $\gamma \in \Gamma$ so that $\bigcup_{\gamma \in \Gamma} \mathcal{M}_\gamma \subseteq \mathcal{U}$), then $\bigvee\left(\bigcup_{\gamma \in \Gamma} \mathcal{M}_\gamma\right) \subseteq \mathcal{U}$. Thus $\bigvee\left(\bigcup_{\gamma \in \Gamma} \mathcal{M}_\gamma\right) = \sup\{\mathcal{M}_\gamma\}_{\gamma \in \Gamma}$.

Again, if we take up the usual notation $\bigvee_{\gamma \in \Gamma} \mathcal{M}_\gamma = \sup\{\mathcal{M}_\gamma\}_{\gamma \in \Gamma}$, then

$$\bigvee_{\gamma \in \Gamma} \mathcal{M}_\gamma = \bigvee\left(\bigcup_{\gamma \in \Gamma} \mathcal{M}_\gamma\right) = \left(\sum_{\gamma \in \Gamma} \mathcal{M}_\gamma\right)^-.$$

This is the *topological sum* of $\{\mathcal{M}_\gamma\}_{\gamma \in \Gamma}$. Conclusion: $\mathrm{Lat}(\mathcal{X})$ is a complete lattice. *The collection of all subspaces of a normed space is a complete lattice in the inclusion ordering.* If $\mathcal{M}$ and $\mathcal{N}$ are subspaces of $\mathcal{X}$, then $\mathcal{M} \wedge \mathcal{N} = \mathcal{M} \cap \mathcal{N}$ and $\mathcal{M} \vee \mathcal{N} = \bigvee(\mathcal{M} \cup \mathcal{N}) = (\mathcal{M} + \mathcal{N})^-$ lie in $\mathrm{Lat}(\mathcal{X})$. However (and this is rather important) it may happen that $\mathcal{M} + \mathcal{N} \neq (\mathcal{M} + \mathcal{N})^-$: the (algebraic) sum of subspaces is not necessarily a subspace (it is a linear manifold but not necessarily a *closed* linear manifold). We shall see later (next chapter) an example of a couple of subspaces (of a Banach space) whose sum is not closed.

Next we consider the quotient space of a normed space modulo a subspace. Suppose $\mathcal{M}$ is a subspace of a normed space $(\mathcal{X}, \| \ \|_X)$. Let

$$[x] = x + \mathcal{M} = \left\{x' \in \mathcal{X}: \ x' = x + u \text{ for some } u \in \mathcal{M}\right\}$$

be the coset (of $x$ modulo $\mathcal{M}$) in the quotient space $\mathcal{X}/\mathcal{M}$ (of $\mathcal{X}$ modulo $\mathcal{M}$), which is a linear space over the same field of the linear space $\mathcal{X}$ (see Example 2H). Consider the function $\| \ \|: \mathcal{X}/\mathcal{M} \to \mathbb{R}$ given by

$$\|[x]\| = \inf_{x' \in [x]} \|x'\|_X = \inf_{u \in \mathcal{M}} \|x + u\|_X$$

for every $[x]$ in $\mathcal{X}/\mathcal{M}$, where $x$ is any vector of $\mathcal{X}$ in $[x]$. This defines a function from $\mathcal{X}/\mathcal{M}$ to $\mathbb{R}$, for $\|[x]\|$ depends only on the coset $[x]$ and not on a particular representative vector $x$ in $[x]$. Note that $\inf_{u \in \mathcal{M}}\|x + u\|_X = \inf_{u \in \mathcal{M}}\|x - u\|_X$ because $\mathcal{M}$ is a linear manifold. Thus, with $d_X$ denoting the metric on $\mathcal{X}$ generated by the norm $\| \ \|_X$,

$$\|[x]\| = \inf_{u \in \mathcal{M}} d_X(x, u) = d_X(x, \mathcal{M})$$

for every $[x] \in \mathcal{X}/\mathcal{M}$, where $x$ is any vector of $\mathcal{X}$ in $[x]$. We claim that $\| \ \|$ is a norm on the quotient space $\mathcal{X}/\mathcal{M}$. Indeed, nonnegativeness is trivially verified:

$$\|[x]\| \geq 0$$

for every $[x] \in \mathcal{X}/\mathcal{M}$. To verify positiveness proceed as follows. Take an arbitrary $x \in [x]$. If $\|[x]\| = 0$, then $d_X(x, \mathcal{M}) = 0$ so that $x \in \mathcal{M}^- = \mathcal{M}$ (cf. Problem 3.43(b) — recall: $\mathcal{M}$ is closed in $\mathcal{X}$). This means that $x \in [0]$ (reason: the origin $[0]$ of $\mathcal{X}/\mathcal{M}$ is $\mathcal{M}$ — see Example 2H), and hence $[x] = [0]$. Equivalently,

$$\|[x]\| > 0 \quad \text{whenever} \quad [x] \neq 0.$$

Absolute homogeneity and subadditivity (i.e., the triangle inequality) are also easily verified. Recall that (Example 2H)

$$\alpha[x] = [\alpha x] \quad \text{and} \quad [x] + [y] = [x + y]$$

for every $[x]$, $[y]$ in $\mathcal{X}/\mathcal{M}$ and every scalar $\alpha$. Since $\mathcal{M}$ is a linear manifold, it follows by absolute homogeneity and subadditivity of the norm $\| \ \|_X$ that

$$
\begin{aligned}
\|\alpha[x]\| \ &= \ \|[\alpha x]\| \ = \ \inf_{u\in\mathcal{M}} \|\alpha x + u\|_X \ = \ \inf_{u\in\mathcal{M}} \|\alpha x + \alpha u\|_X \\
&= \ |\alpha| \inf_{u\in\mathcal{M}} \|x + u\|_X \ = \ |\alpha| \, \|[x]\|, \\
\|[x] + [y]\| \ &= \ \|[x + y]\| \ = \ \inf_{u\in\mathcal{M}} \|x + y + u\|_X \ = \ \inf_{u\in\mathcal{M}} \|x + u + y + u\|_X \\
&\leq \ \inf_{u\in\mathcal{M}} \|x + u\|_X + \inf_{u\in\mathcal{M}} \|y + u\|_X \ = \ \|[x]\| + \|[y]\|,
\end{aligned}
$$

for every $[x]$, $[y]$ in $\mathcal{X}/\mathcal{M}$ and every scalar $\alpha$. Such a norm is referred to as the *quotient norm* and, whenever a quotient space $\mathcal{X}/\mathcal{M}$ (of a *normed* space $\mathcal{X}$ modulo a *subspace* $\mathcal{M}$) is regarded as a normed space, it is this quotient norm that will be assumed to equip it. Note that $\inf_{u\in\mathcal{M}}\|x + u\|_X \leq \|x\|_X + \inf_{u\in\mathcal{M}}\|u\|_X = \|x\|_X$, so that

$$
\|[x]\| \leq \|x\|_X
$$

for every $[x] \in \mathcal{X}/\mathcal{M}$ and every $x \in [x]$. Thus

$$
\|[x] - [y]\| \ = \ \|[x - y]\| \ \leq \ \|x - y\|_X
$$

for every $x, y \in \mathcal{X}$. Therefore the natural mapping $\pi$ of $\mathcal{X}$ onto $\mathcal{X}/\mathcal{M}$, $\pi(x) = [x]$ for every $x \in \mathcal{X}$, is uniformly continuous (a contraction, actually), and hence it preserves convergence and Cauchy sequences (Corollary 3.8 and Lemma 3.43). *If $\{x_n\}$ converges in $\mathcal{X}$ to $x \in \mathcal{X}$, then $\{[x_n]\}$ converges in $\mathcal{X}/\mathcal{M}$ to $[x] \in \mathcal{X}/\mathcal{M}$; if $\{x_n\}$ is a Cauchy sequence in $\mathcal{X}$, then $\{[x_n]\}$ is a Cauchy sequence in $\mathcal{X}/\mathcal{M}$.*

**Proposition 4.10.** *If $\mathcal{M}$ is a subspace of a Banach space $\mathcal{X}$, then the quotient space $\mathcal{X}/\mathcal{M}$ is a Banach space.*

*Proof.* Let $\mathcal{M}$ be a subspace of a normed space $(\mathcal{X}, \| \ \|_X)$. Consider the quotient space $\mathcal{X}/\mathcal{M}$ equipped with the quotient norm $\| \ \|$ and let $\{[x_n]\}_{n=0}^{\infty}$ be an arbitrary Cauchy sequence in $\mathcal{X}/\mathcal{M}$. According to Problem 3.52(c) $\{[x_n]\}_{n=0}^{\infty}$ has a subsequence $\{[x_{n_k}]\}_{k=0}^{\infty}$ of bounded variation, so that

$$
\sum_{k=0}^{\infty} \|[x_{n_{k+1}} - x_{n_k}]\| \ = \ \sum_{k=0}^{\infty} \|[x_{n_{k+1}}] - [x_{n_k}]\| \ < \ \infty.
$$

Note: If $x \in \mathcal{X}$, then there exists an $\varepsilon > 0$ and a vector $u_{\varepsilon,x} \in \mathcal{M}$ such that

$$
\|x + u_{\varepsilon,x}\|_X \ \leq \ \inf_{u\in\mathcal{M}} \|x + u\|_X + \varepsilon \ = \ \|[x]\| + \varepsilon.
$$

In particular, for each integer $k \geq 0$ there exists a vector $u_k \in \mathcal{M}$ such that

$$
\|(x_{n_{k+1}} - x_{n_k}) + u_k\|_X \ \leq \ \|[x_{n_{k+1}} - x_{n_k}]\| + (\tfrac{1}{2})^k.
$$

Therefore, as $\sum_{k=0}^{\infty}(\frac{1}{2})^k < \infty$,

$$\sum_{k=0}^{\infty} \|(x_{n_{k+1}} - x_{n_k}) + u_k\|_X < \infty.$$

In words, the $\mathcal{X}$-valued sequence $\{(x_{n_{k+1}} - x_{n_k}) + u_k\}_{k=0}^{\infty}$ is absolutely summable. Now consider the sequence $\{y_k\}_{k=0}^{\infty}$ in $\mathcal{X}$ of partial sums of $\{(x_{n_{k+1}} - x_{n_k}) + u_k\}_{k=0}^{\infty}$. A trivial induction shows that

$$y_k = \sum_{i=0}^{k}(x_{n_{i+1}} - x_{n_i}) + u_i = (x_{n_{k+1}} - x_{n_0}) + \sum_{i=0}^{k} u_i$$

for each $k \geq 0$. If $\mathcal{X}$ is a Banach space, then the absolutely summable sequence $\{(x_{n_{k+1}} - x_{n_k}) + u_k\}_{k=0}^{\infty}$ is summable (Proposition 4.4), which means that $\{y_k\}_{k=0}^{\infty}$ converges in $\mathcal{X}$. Hence $\{[y_k]\}_{k=0}^{\infty}$ converges in $\mathcal{X}/\mathcal{M}$ (for the natural mapping of $\mathcal{X}$ onto $\mathcal{X}/\mathcal{M}$, $x \mapsto [x]$, is continuous). But $[x_{n_{k+1}}] - [x_{n_0}] = [x_{n_{k+1}} - x_{n_0}] = [y_k]$ for each $k \geq 0$ (because $\sum_{i=0}^{k} u_i$ lies in $\mathcal{M}$ for every $k \geq 0$). Thus the subsequence $\{[x_{n_k}]\}_{k=0}^{\infty}$ converges in $\mathcal{X}/\mathcal{M}$, and therefore the Cauchy sequence $\{[x_n]\}_{n=0}^{\infty}$ converges in $\mathcal{X}/\mathcal{M}$ (Proposition 3.39). Conclusion: Every Cauchy sequence in $\mathcal{X}/\mathcal{M}$ converges in $\mathcal{X}/\mathcal{M}$; that is, $\mathcal{X}/\mathcal{M}$ is a Banach space. $\qquad\square$

## 4.4  Bounded Linear Transformations

Let $\mathcal{X}$ and $\mathcal{Y}$ be normed spaces. A *continuous linear transformation* of $\mathcal{X}$ into $\mathcal{Y}$ is a linear transformation of the linear space $\mathcal{X}$ into the linear space $\mathcal{Y}$ that is continuous with respect to the norm topologies of $\mathcal{X}$ and $\mathcal{Y}$. (Note that $\mathcal{X}$ and $\mathcal{Y}$ are necessarily linear spaces over the same scalar field.) This is the most important concept that results from the combination of algebraic and topological structures.

**Definition 4.11.** A linear transformation $T$ of a normed space $\mathcal{X}$ into a normed space $\mathcal{Y}$ is *bounded* if there exists a constant $\beta \geq 0$ such that

$$\|Tx\| \leq \beta \|x\|$$

for every $x \in \mathcal{X}$. (The norm on the left-hand side is the norm on $\mathcal{Y}$ and that on the right-hand side is the norm on $\mathcal{X}$).

**Proposition 4.12.** *A linear transformation of a normed space $\mathcal{X}$ into a normed space $\mathcal{Y}$ is bounded if and only if it maps bounded subsets of $\mathcal{X}$ into bounded subsets of $\mathcal{Y}$.*

*Proof.* Let $A$ be a bounded subset of $\mathcal{X}$ so that $\sup_{a \in A} \|a\| < \infty$ (Problem 4.5). If $T$ is bounded, then there exists a real number $\beta > 0$ such that $\|Tx\| \leq \beta \|x\|$ for every $x \in \mathcal{X}$. Therefore

$$\sup_{y \in T(A)} \|y\| = \sup_{a \in A} \|Ta\| \le \beta \sup_{a \in A} \|a\| < \infty,$$

and hence $T(A)$ is bounded in $\mathcal{Y}$. Conversely, suppose $T(A)$ is bounded in $\mathcal{Y}$ for every bounded set $A$ in $\mathcal{X}$. In particular, $T(B_1[0])$ is bounded in $\mathcal{Y}$: the closed unit ball centered at the origin of $\mathcal{X}$, $B_1[0]$, is certainly bounded in $\mathcal{X}$. Thus

$$\sup_{\|b\| \le 1} \|Tb\| = \sup_{b \in B_1[0]} \|Tb\| = \sup_{y \in T(B_1[0])} \|y\| < \infty.$$

Take an arbitrary nonzero $x$ in $\mathcal{X}$. Since $\left\| \frac{x}{\|x\|} \right\| = 1$, it follows that

$$\|Tx\| = \|x\| \left\| T\!\left(\tfrac{x}{\|x\|}\right) \right\| \le \sup_{\|b\| \le 1} \|Tb\| \, \|x\|$$

for every $0 \ne x \in \mathcal{X}$, and hence $T$ is bounded (since the inequality $\|Tx\| \le \sup_{\|b\| \le 1} \|Tb\| \, \|x\|$ holds trivially for $x = 0$).     $\square$

The next elementary result is extremely important.

**Proposition 4.13.** *Let $T$ be a bounded linear transformation of a normed space $\mathcal{X}$ into a normed space $\mathcal{Y}$. The null space $\mathcal{N}(T)$ of $T$ is a subspace of $\mathcal{X}$.*

*Proof.* The null space (or kernel) of $T$,

$$\mathcal{N}(T) = \{x \in \mathcal{X}: \ Tx = 0\} = T^{-1}(\{0\}),$$

is a linear manifold of $\mathcal{X}$ (Problem 2.10). The Closed Set Theorem (Theorem 3.30) ensures that it is closed in $\mathcal{X}$. Indeed, if $\{x_n\}$ is an $\mathcal{N}(T)$-valued sequence (i.e., $Tx_n = 0$ for every $n$) that converges in $\mathcal{X}$ to $x \in \mathcal{X}$, then

$$0 \le \|Tx\| = \|Tx_n - Tx\| = \|T(x_n - x)\| \le \beta \|x_n - x\| \to 0$$

as $n \to \infty$ (because $T$ is linear and bounded), and hence $x \in \mathcal{N}(T)$.     $\square$

**Theorem 4.14.** *Let $T$ be a linear transformation of a normed space $\mathcal{X}$ into a normed space $\mathcal{Y}$. The following assertions are pairwise equivalent.*

  (a)  *$T$ is bounded.*

  (b)  *$T$ is Lipschitzian.*

  (c)  *$T$ is uniformly continuous.*

  (d)  *$T$ is continuous.*

  (e)  *$T$ is continuous at some point $x_0$ of $\mathcal{X}$.*

(f)   *T is continuous at the origin $0 \in \mathcal{X}$.*

*Proof.* Let $T: \mathcal{X} \to \mathcal{Y}$ be a linear transformation. If $T$ is bounded, then there exists $\beta \geq 0$ such that, for every $x_1, x_2 \in \mathcal{X}$,

$$\|Tx_1 - Tx_2\| = \|T(x_1 - x_2)\| \leq \beta\|x_1 - x_2\|,$$

and hence (a)$\Rightarrow$(b). Recall that (b)$\Rightarrow$(c)$\Rightarrow$(d)$\Rightarrow$(e) trivially. Now suppose $T$ is continuous at $x_0 \in \mathcal{X}$. Then for every $\varepsilon > 0$ there exists $\delta > 0$ such that $\|Tx - T0\| = \|Tx\| = \|T(x + x_0) - Tx_0\| < \varepsilon$ whenever $\|x - 0\| = \|x\| = \|(x + x_0) - x_0\| < \delta$, and so $T$ is continuous at $0 \in \mathcal{X}$. Therefore (e)$\Rightarrow$(f). Next suppose $T$ is continuous at $0 \in \mathcal{X}$. Thus (for $\varepsilon = 1$) there exists $\delta > 0$ such that $\|Tx\| = \|Tx - T0\| < 1$ whenever $\|x\| = \|x - 0\| < \delta$. Since $\left\|\frac{\delta}{2\|x\|}x\right\| < \delta$ for every nonzero $x$ in $\mathcal{X}$,

$$\|Tx\| = \frac{2\|x\|}{\delta}\left\|\frac{\delta}{2\|x\|}Tx\right\| = \frac{2\|x\|}{\delta}\left\|T\left(\frac{\delta}{2\|x\|}x\right)\right\| < \frac{2}{\delta}\|x\|$$

for every $0 \neq x \in \mathcal{X}$, and hence $T$ is bounded. Thus (f)$\Rightarrow$(a).     $\square$

Observe from Theorem 4.14 that the terms "*bounded linear transformation*" and "*continuous linear transformation*" are synonyms, and as such we shall use them interchangeably. If $\mathcal{X}$ and $\mathcal{Y}$ are normed spaces (over the same field $\mathbb{F}$), then the collection of all bounded linear transformations of $\mathcal{X}$ into $\mathcal{Y}$ will be denoted by $\mathcal{B}[\mathcal{X}, \mathcal{Y}]$. It is easy to verify (triangle inequality and homogeneity for the norm on $\mathcal{Y}$) that $\mathcal{B}[\mathcal{X}, \mathcal{Y}]$ is a linear manifold of the linear space $\mathcal{L}[\mathcal{X}, \mathcal{Y}]$ of all linear transformations of $\mathcal{X}$ into $\mathcal{Y}$ (see Section 2.5), and hence $\mathcal{B}[\mathcal{X}, \mathcal{Y}]$ *is a linear space over* $\mathbb{F}$. The origin of $\mathcal{B}[\mathcal{X}, \mathcal{Y}]$ is the *null transformation*, denoted by $O$ ($Ox = 0$ for every $x \subset \mathcal{X}$). For each $T \subset \mathcal{B}[\mathcal{X}, \mathcal{Y}]$ set

$$\|T\| = \inf\left\{\beta \geq 0: \ \|Tx\| \leq \beta\|x\| \text{ for every } x \in \mathcal{X}\right\}.$$

(Recall: if $T \in \mathcal{B}[\mathcal{X}, \mathcal{Y}]$, then there exists $\beta > 0$ such that $\|Tx\| < \beta\|x\|$ for every $x \in \mathcal{X}$, so that the nonnegative number $\|T\|$ exists for every $T \in \mathcal{B}[\mathcal{X}, \mathcal{Y}]$.) If $x$ is any vector in $\mathcal{X}$, then $\|Tx\| \leq (\|T\| + \varepsilon)\|x\|$ for every $\varepsilon > 0$ so that $\|Tx\| \leq \inf_{\varepsilon > 0}(\|T\| + \varepsilon)\|x\|$. Thus, for every $x \in \mathcal{X}$,

$$\|Tx\| \leq \|T\|\|x\|,$$

and hence $\|T\| = \min\left\{\beta \geq 0: \ \|Tx\| \leq \beta\|x\| \text{ for every } x \in \mathcal{X}\right\}$. It is also easy to show that

$$\|T\| = \sup_{\|x\| \leq 1}\|Tx\| = \sup_{\|x\| < 1}\|Tx\| = \sup_{\|x\| = 1}\|Tx\| = \sup_{x \neq 0}\frac{\|Tx\|}{\|x\|},$$

where the last two expressions make sense only if $\mathcal{X} \neq \{0\}$. Hence

$$\|T\| \geq 0 \quad \text{and} \quad \|T\| = 0 \text{ if and only if } T = O.$$

Moreover, for any $\alpha$ in $\mathbb{F}$ and any $S$ in $\mathcal{B}[\mathcal{X}, \mathcal{Y}]$, $\|(\alpha T)x\| = \|\alpha(Tx)\| = |\alpha|\|Tx\| \leq |\alpha|\|T\|\|x\|$ and $\|(T+S)x\| = \|Tx + Sx\| \leq \|Tx\| + \|Sx\| \leq (\|T\| + \|S\|)\|x\|$ for every $x \in \mathcal{X}$. Therefore,

$$\|\alpha T\| \leq |\alpha|\|T\| \quad \text{and} \quad \|T + S\| \leq \|T\| + \|S\|$$

for every $T, S \in \mathcal{B}[\mathcal{X}, \mathcal{Y}]$ and every scalar $\alpha \in \mathbb{F}$. Conclusion: The function $\mathcal{B}[\mathcal{X}, \mathcal{Y}] \to \mathbb{R}$ defined by $T \mapsto \|T\|$ is a norm on the linear space $\mathcal{B}[\mathcal{X}, \mathcal{Y}]$. Thus $\mathcal{B}[\mathcal{X}, \mathcal{Y}]$ *is a normed space*, and $\|T\|$ is referred to as the *norm* of $T \in \mathcal{B}[\mathcal{X}, \mathcal{Y}]$ (or the *usual norm* on $\mathcal{B}[\mathcal{X}, \mathcal{Y}]$, or still as the *induced uniform norm* on $\mathcal{B}[\mathcal{X}, \mathcal{Y}]$). This is the norm that will be assumed to equip $\mathcal{B}[\mathcal{X}, \mathcal{Y}]$ whenever $\mathcal{B}[\mathcal{X}, \mathcal{Y}]$ is regarded as a normed space. A *contraction* in $\mathcal{B}[\mathcal{X}, \mathcal{Y}]$ is a bounded linear transformation $T \in \mathcal{B}[\mathcal{X}, \mathcal{Y}]$ such that $\|T\| \leq 1$. Clearly,

$$\|T\| \leq 1 \quad \Longleftrightarrow \quad \|Tx\| \leq \|x\| \text{ for every } x \in \mathcal{X}.$$

If $\mathcal{X} \neq \{0\}$, then a transformation $T \in \mathcal{B}[\mathcal{X}, \mathcal{Y}]$ is a contraction if and only if $\sup_{x \neq 0}(\|Tx\|/\|x\|) \leq 1$. A *strict contraction* in $\mathcal{B}[\mathcal{X}, \mathcal{Y}]$ is a bounded linear transformation $T \in \mathcal{B}[\mathcal{X}, \mathcal{Y}]$ such that $\|T\| < 1$. It is obvious that every strict contraction is a contraction. $T \in \mathcal{B}[\mathcal{X}, \mathcal{Y}]$ is a strict contraction if and only if $\mathcal{X} \neq \{0\}$ and $\sup_{x \neq 0}(\|Tx\|/\|x\|) < 1$. Carefully note that, if $T \in \mathcal{B}[\mathcal{X}, \mathcal{Y}]$ and $\mathcal{X} \neq \{0\}$, then

$$\|T\| < 1 \quad \Longrightarrow \quad \|Tx\| < \|x\| \text{ for every } 0 \neq x \in \mathcal{X} \quad \Longrightarrow \quad \|T\| \leq 1.$$

**Example 4H.** The converses of the above implications fail. Indeed, Let $\ell_+^p$ be the Banach space of Example 4B for some $p \geq 1$. We shall drop the subscript "$p$" from the norm $\|\ \|_p$ on $\ell_+^p$ and write simply $\|\ \|$. Now consider the *diagonal mapping* $D_a : \ell_+^p \to \ell_+^p$ of Problem 3.22 for some bounded sequence $a = \{\alpha_k\}_{k=0}^\infty \in \ell_+^\infty$:

$$D_a x = \{\alpha_k \xi_k\}_{k=0}^\infty \quad \text{for every} \quad x = \{\xi_k\}_{k=0}^\infty \in \ell_+^p$$

(i.e., $D_a(\xi_0, \xi_1, \xi_2, \dots) = (\alpha_0 \xi_0, \alpha_1 \xi_1, \alpha_2 \xi_2, \dots)$). Common notations for a diagonal mapping of $\ell_+^p$ into itself include the usual representation as an *infinite diagonal matrix*:

$$D_a = \mathrm{diag}(\{\alpha_k\}_{k=0}^\infty) = \mathrm{diag}(\alpha_0, \alpha_1, \alpha_2, \dots) = \begin{pmatrix} \alpha_0 & & & \\ & \alpha_1 & & \\ & & \alpha_2 & \\ & & & \ddots \end{pmatrix}.$$

This bounded linear transformation is called a *diagonal operator*. Linearity is trivially verified and boundedness (i.e., continuity) follows from Problem 3.22(a). Indeed, if $a = \{\alpha_k\}_{k=0}^\infty \in \ell_+^\infty$, then

$$\|D_a x\| = \left(\sum_{k=0}^\infty |\alpha_k \xi_k|^p\right)^{\frac{1}{p}} \leq \sup_k |\alpha_k| \left(\sum_{k=0}^\infty |\xi_k|^p\right)^{\frac{1}{p}} = \sup_k |\alpha_k| \|x\|$$

for every $x = \{\xi_k\}_{k=0}^{\infty} \in \ell_+^p$, so that $\|D_a\| \leq \sup_k |\alpha_k| = \|a\|_{\infty}$. On the other hand, consider the $\ell_+^p$-valued sequence $\{e_i\}_{i=0}^{\infty}$ where, for each $i \geq 0$, $e_i$ is a scalar-valued sequence with just one nonzero entry (equal to one) at the $i$th position (i.e., $e_i = \{\delta_{ik}\}_{k=0}^{\infty}$ for every $i \geq 0$, with $\delta_{ik} = 1$ if $k = i$ and $\delta_{ik} = 0$ if $k \neq i$). Since $D_a e_i = \alpha_i e_i$ and $\|e_i\| = 1$, it follows that $\|D_a e_i\| = |\alpha_i|$ and hence $\|D_a\| = \sup_{\|x\|=1} \|D_a x\| \geq \|D_a e_i\| = |\alpha_i|$, for every $i \geq 0$. Thus $\|D_a\| \geq \sup_i |\alpha_i| = \|a\|_{\infty}$. Therefore,

$$\|D_a\| = \|a\|_{\infty}.$$

If $a = \{\alpha_k\}_{k=0}^{\infty}$ is a constant sequence, say $\alpha_k = \alpha$ for all $k \geq 0$, then $D_a = \mathrm{diag}(\alpha, \alpha, \alpha, \dots) = \alpha I$, a multiple of the identity $I$ on $\ell_+^p$. In this case $D_a$ is a *scalar operator* (see Problem 4.19). It is clear that $\|\alpha I x\| = |\alpha| \|x\|$ for every $x \in \ell_+^p$ and $\|\alpha I\| = |\alpha|$. If $a = \{\alpha_k\}_{k=0}^{\infty}$ is the increasing sequence with $\alpha_k = \frac{k+1}{k+2}$ for every $k \geq 0$, then $D_a = \mathrm{diag}(\frac{1}{2}, \frac{2}{3}, \frac{3}{4}, \dots)$ is such that $\|D_a x\| < \|x\|$ for every $0 \neq x \in \ell_+^p$ and $\|D_a\| = 1$.

**Proposition 4.15.** $\mathcal{B}[\mathcal{X}, \mathcal{Y}]$ *is a Banach space if* $\mathcal{Y}$ *is a Banach space.*

*Proof.* Let $\{T_n\}$ be an arbitrary Cauchy sequence in $\mathcal{B}[\mathcal{X}, \mathcal{Y}]$. For each $x \in \mathcal{X}$ the sequence $\{T_n x\}$ is a Cauchy sequence in $\mathcal{Y}$ (reason: $\|T_m x - T_n x\| \leq \|T_m - T_n\| \|x\|$ for every $x \in \mathcal{X}$ and every pair of indices $m$ and $n$). Since $\mathcal{Y}$ is complete, $\{T_n x\}$ converges in $\mathcal{Y}$. Thus for every $x \in \mathcal{X}$ there exists a vector $y_x \in \mathcal{Y}$ such that $T_n x \to y_x$ in $\mathcal{Y}$: the (unique) limit of $\{T_n x\}$. Let $T$ be the mapping that assigns to each $x$ in $\mathcal{X}$ this vector $y_x$ in $\mathcal{Y}$; that is, $T x = \lim_n T_n x$ for every $x \in \mathcal{X}$.

*Claim 1.*     $T \colon \mathcal{X} \to \mathcal{Y}$ is linear.

*Proof.* Since $T_n$ is linear for each $n$, and since the linear operations are continuous (Problem 4.1), it follows that

$$
\begin{aligned}
T(\alpha_1 x_1 + \alpha_2 x_2) &= \lim_n T_n(\alpha_1 x_1 + \alpha_2 x_2) = \lim_n (T_n \alpha_1 x_1 + T_n \alpha_2 x_2) \\
&= \alpha_1 \lim_n T_n x_1 + \alpha_2 \lim_n T_n x_2 = \alpha_1 T x_1 + \alpha_2 T x_2
\end{aligned}
$$

for every $x_1, x_2 \in \mathcal{X}$ and every pair of scalars $\alpha_1$ and $\alpha_2$. $\square$

*Claim 2.*     $T \colon \mathcal{X} \to \mathcal{Y}$ is bounded.

*Proof.* Take an arbitrary real number $\varepsilon > 0$. Since $\{T_n\}$ is a Cauchy sequence in $\mathcal{B}[\mathcal{X}, \mathcal{Y}]$, it follows that there exists an integer $n_\varepsilon \geq 1$ such that $\|T_m - T_n\| < \varepsilon$ whenever $m, n \geq n_\varepsilon$. Thus $\|T_m x - T_n x\| \leq \|T_m - T_n\| \|x\| < \varepsilon \|x\|$ for every $x \in \mathcal{X}$ and every $m, n \geq n_\varepsilon$. Therefore,

$$
\begin{aligned}
\|(T_m - T)x\| &= \|T_m x - T x\| = \left\| T_m x - \lim_n T_n x \right\| \\
&= \left\| \lim_n (T_m x - T_n x) \right\| = \lim_n \left\| T_m x - T_n x \right\| < \varepsilon \|x\|
\end{aligned}
$$

for every $x \in \mathcal{X}$ and every $m \geq n_\varepsilon$ (reason: every norm is continuous — see Corollary 3.8). Hence the linear transformation $T_m - T$ is bounded for each $m \geq n_\varepsilon$, and so is $T = T_m - (T_m - T)$. $\square$

*Claim 3.*    $T_m \to T$ in $\mathcal{B}[\mathcal{X}, \mathcal{Y}]$.

*Proof.* For every $\varepsilon > 0$ there exists an integer $n_\varepsilon \geq 1$ such that

$$m \geq n_\varepsilon \quad \text{implies} \quad \|T_m - T\| = \sup_{\|x\| \leq 1} \|(T_m - T)x\| < \varepsilon$$

according to the above displayed inequality. $\square$

Conclusion: Every Cauchy sequence $\{T_n\}$ in $\mathcal{B}[\mathcal{X}, \mathcal{Y}]$ converges in $\mathcal{B}[\mathcal{X}, \mathcal{Y}]$ to $T \in \mathcal{B}[\mathcal{X}, \mathcal{Y}]$, which means that the normed space $\mathcal{B}[\mathcal{X}, \mathcal{Y}]$ is complete; that is $\mathcal{B}[\mathcal{X}, \mathcal{Y}]$ is a Banach space. $\qquad\qquad\square$

Comparing the above proposition with Example 4G we might expect that, if $\mathcal{X} \neq \{0\}$, then $\mathcal{B}[\mathcal{X}, \mathcal{Y}]$ is a Banach space if and only if $\mathcal{Y}$ is a Banach space. This indeed is the case, and we shall prove the converse of Proposition 4.15 in Section 4.10. Recall that bounded linear transformations can be multiplied (where multiplication means composition). The resulting transformation is linear as well as bounded.

**Proposition 4.16.** *Let $\mathcal{X}$, $\mathcal{Y}$ and $\mathcal{Z}$ be normed spaces over the same scalar field. If $T \in \mathcal{B}[\mathcal{X}, \mathcal{Y}]$ and $S \in \mathcal{B}[\mathcal{Y}, \mathcal{Z}]$, then $ST \in \mathcal{B}[\mathcal{X}, \mathcal{Z}]$ and*

$$\|ST\| \leq \|S\|\|T\|.$$

*Proof.* The composition $ST : \mathcal{X} \to \mathcal{Z}$ is linear (Problem 2.15) and $\|STx\| \leq \|S\|\|Tx\| \leq \|S\|\|T\|\|x\|$ for every $x$ in $\mathcal{X}$. $\qquad\qquad\square$

The above inequality is a rather important additional property shared by the (induced uniform) norm of bounded linear transformations. We shall refer to it as the *operator norm property.*

Let $\mathcal{X}$ be a normed space and set $\mathcal{B}[\mathcal{X}] = \mathcal{B}[\mathcal{X}, \mathcal{X}]$ for short. The elements of $\mathcal{B}[\mathcal{X}]$ are called *operators*. In other words, by an *operator* (or a *bounded linear operator*) we simply mean a bounded linear transformation of a normed space $\mathcal{X}$ into itself, so that $\mathcal{B}[\mathcal{X}]$ is the normed space of all operators on $\mathcal{X}$.

**Example 4I.** Let $\{(\mathcal{X}_k, \|\ \|_k)\}_{k \in \mathbb{I}}$ be a countable (either finite or infinite) indexed family of normed spaces (over the same scalar field), and consider the direct sum $\bigoplus_{k \in \mathbb{I}} \mathcal{X}_k$ of the linear spaces $\{\mathcal{X}_k\}_{k \in \mathbb{I}}$. (To avoid trivialities we assume that each $\mathcal{X}_k$ is nonzero.) Equip the linear space $\bigoplus_{k \in \mathbb{I}} \mathcal{X}_k$ with any of the norms $\|\ \|_\infty$ or $\|\ \|_p$ (for any $p \geq 1$) as in Examples 4E and 4F. If the index set $\mathbb{I}$ is countably infinite, then write $\left(\bigoplus_{k \in \mathbb{I}} \mathcal{X}_k, \|\ \|_\infty\right)$ for the normed space $\left([\bigoplus_{k \in \mathbb{I}} \mathcal{X}_k]_\infty, \|\ \|_\infty\right)$

and $\left(\bigoplus_{k\in\mathbb{I}}\mathcal{X}_k, \|\ \|_p\right)$ for the normed space $\left([\bigoplus_{k\in\mathbb{I}}\mathcal{X}_k]_p, \|\ \|_p\right)$. For each $i \in \mathbb{I}$ define a mapping $P_i : \bigoplus_{k\in\mathbb{I}}\mathcal{X}_k \to \bigoplus_{k\in\mathbb{I}}\mathcal{X}_k$ by

$$P_i x = x_i \quad \text{for every} \quad x = \{x_k\}_{k\in\mathbb{I}} \in \bigoplus_{k\in\mathbb{I}}\mathcal{X}_k,$$

where we are identifying each vector $x_i \in \mathcal{X}_i$ with the indexed family $\{x_i(k)\}_{k\in\mathbb{I}} \in \bigoplus_{k\in\mathbb{I}}\mathcal{X}_k$ that has just one nonzero entry (equal to $x_i$) at the $i$th position (i.e., set $x_i(k) = \delta_{ik}x_k$ so that $x_i(k) = 0_k \in \mathcal{X}_k$, the origin of $\mathcal{X}_k$, if $k \neq i$ and $x_i(i) = x_i$). Briefly, we are writing $x_i$ for $(0_1, \dots, 0_{i-1}, x_i, 0_{i+1}, \dots)$. Each $P_i$ is a linear transformation. Indeed,

$$\begin{aligned}
P_i(\alpha u \oplus \beta v) &= P_i\big(\alpha\{u_k\}_{k\in\mathbb{I}} \oplus \beta\{v_k\}_{k\in\mathbb{I}}\big) \\
&= P_i\big(\{\alpha u_k + \beta v_k\}_{k\in\mathbb{I}}\big) = \alpha u_i + \beta v_i = \alpha P_i u \oplus \beta P_i v
\end{aligned}$$

for every $u = \{u_k\}_{k\in\mathbb{I}}$ and $v = \{v_k\}_{k\in\mathbb{I}}$ in $\bigoplus_{k\in\mathbb{I}}\mathcal{X}_k$ and every pair of scalars $\{\alpha, \beta\}$. Moreover,

$$\mathcal{R}(P_i) = \mathcal{X}_i$$

if we identify each normed space $\mathcal{X}_i$ with the linear manifold $\bigoplus_{k\in\mathbb{I}}\mathcal{X}_i(k)$ of $\bigoplus_{k\in\mathbb{I}}\mathcal{X}_k$, such that $\mathcal{X}_i(k) = \{0_k\}$ for $k \neq i$ and $\mathcal{X}_i(i) = \mathcal{X}_i$, equipped with any of the norms $\|\ \|_\infty$ or $\|\ \|_p$. (To identify $\mathcal{X}_i$ with $\bigoplus_{k\in\mathbb{I}}\mathcal{X}_i(k)$ simply means that these normed spaces are *isometrically isomorphic*, a concept that will be introduced in Section 4.7.) Note that $\{x_i(k)\}_{k\in\mathbb{I}}$ lies in $\bigoplus_{k\in\mathbb{I}}\mathcal{X}_i(k)$ for every $x_i$ in $\mathcal{X}_i$ and $\|x_i\|_i = \|\{x_i(k)\}_{k\in\mathbb{I}}\|_\infty = \|\{x_i(k)\}_{k\in\mathbb{I}}\|_p$ for any $p \geq 1$. Also note that each $P_i$ is a projection (i.e., an idempotent linear transformation):

$$P_i^2 x = P_i\{x_i(k)\}_{k\in\mathbb{I}} = \{x_i(k)\}_{k\in\mathbb{I}} = P_i x$$

for every $x = \{x_k\}_{k\in\mathbb{I}}$ in $\bigoplus_{k\in\mathbb{I}}\mathcal{X}_k$ so that $P_i = P_i^2$ for each $i \in \mathbb{I}$. In fact, each $P_i$ is a continuous projection. To verify that the mapping $P_i : \bigoplus_{k\in\mathbb{I}}\mathcal{X}_k \to \bigoplus_{k\in\mathbb{I}}\mathcal{X}_k$ is continuous, when $\bigoplus_{k\in\mathbb{I}}\mathcal{X}_k$ is equipped with any of the norms $\|\ \|_\infty$ or $\|\ \|_p$, observe that

$$\|P_i x\|_\infty = \|\{x_i(k)\}_{k\in\mathbb{I}}\|_\infty = \|x_i\|_i \leq \sup_{k\in\mathbb{I}} \|x_k\|_i = \|x\|_\infty$$

for every $x = \{x_k\}_{i\in\mathbb{I}}$ in $\left(\bigoplus_{k\in\mathbb{I}}\mathcal{X}_k, \|\ \|_\infty\right)$, and

$$\|P_i x\|_p^p = \|\{x_i(k)\}_{k\in\mathbb{I}}\|_p^p = \|x_i\|_i^p \leq \sum_{k\in\mathbb{I}} \|x_k\|_k^p = \|x\|_p^p$$

for every $x = \{x_k\}_{k\in\mathbb{I}}$ in $\left(\bigoplus_{k\in\mathbb{I}}\mathcal{X}_k, \|\ \|_p\right)$. Thus $P_i$ is a bounded linear transformation, and hence continuous (Theorem 4.14). In other words, each $P_i$ is a *projection operator* in $\mathcal{B}[\bigoplus_{k\in\mathbb{I}}\mathcal{X}_k]$. Actually, if we use the same symbols $\|\ \|_\infty$ and $\|\ \|_p$ to denote the induced uniform norms on $\mathcal{B}[\bigoplus_{k\in\mathbb{I}}\mathcal{X}_k]$ whenever $\bigoplus_{k\in\mathbb{I}}\mathcal{X}_k$ is equipped with either $\|\ \|_\infty$ or $\|\ \|_p$, respectively, then the above inequalities ensure that $\|P_i\|_\infty \leq 1$ and $\|P_i\|_p \leq 1$. Thus each $P_i$ is a *contraction* with respect to any of the norms $\| \ \|_\infty$ or $\| \ \|_p$. On the other hand, since $P_i = P_i^2$, it follows that $\|P_i\|_\infty = \|P_i^2\|_\infty \leq \|P_i\|_\infty^2$ and $\|P_i\|_p = \|P_i^2\|_p \leq \|P_i\|_p^2$ (operator norm property). Hence $1 \leq \|P_i\|_\infty$ and $1 \leq \|P_i\|_\infty$ (for $P_k \neq O$) so that

$$\|P_i\|_\infty = \|P_i\|_p = 1.$$

Summing up: If $\bigoplus_{k \in \mathbb{I}} \mathcal{X}_k$ is equipped with either $\| \ \|_\infty$ or $\| \ \|_p$, then $\{P_i\}_{i \in \mathbb{I}}$ is a family of projections with unit norm in $\mathcal{B}[\bigoplus_{k \in \mathbb{I}} \mathcal{X}_k]$. When we identify $\bigoplus_{k \in \mathbb{I}} \mathcal{X}_i(k) = \mathcal{R}(P_i)$ with $\mathcal{X}_i$, as we actually did, then each map $P_i \colon \bigoplus_{k \in \mathbb{I}} \mathcal{X}_k \to \bigoplus_{k \in \mathbb{I}} \mathcal{X}_i(k) \subseteq \bigoplus_{k \in \mathbb{I}} \mathcal{X}_k$ can be viewed as a function from $\bigoplus_{k \in \mathbb{I}} \mathcal{X}_k$ onto $\mathcal{X}_i$, and hence we write $P_i \colon \bigoplus_{k \in \mathbb{I}} \mathcal{X}_k \to \mathcal{X}_i$. This is called the *natural projection* of $\bigoplus_{k \in \mathbb{I}} \mathcal{X}_k$ onto $\mathcal{X}_i$.

**Definition 4.17.** A normed space $\mathcal{A}$ that is simultaneously an algebra (as in Problem 2.30) with respect to a product operation $\mathcal{A} \times \mathcal{A} \to \mathcal{A}$, say $(A, B) \mapsto AB$, such that $\|AB\| \leq \|A\| \|B\|$ is called a *normed algebra*. A *Banach algebra* is a normed algebra that is complete as a normed space. If a normed algebra possesses an identity $I$ such that $\|I\| = 1$, then it is a *unital normed algebra* (or a *normed algebra with identity*). If a unital normed algebra is complete as a normed space, then it is a *unital Banach algebra* (or a *Banach algebra with identity*).

Let $\mathcal{L}[\mathcal{X}]$ be the linear space of all linear transformations on $\mathcal{X}$ (i.e., $\mathcal{L}[\mathcal{X}] = \mathcal{L}[\mathcal{X}, \mathcal{X}]$: the linear space of all linear transformations of a linear space $\mathcal{X}$ into itself — Problem 2.13). We have already noticed that $\mathcal{B}[\mathcal{X}]$ is a linear manifold of $\mathcal{L}[\mathcal{X}]$. If $\mathcal{X} \neq \{0\}$, then $\mathcal{L}[\mathcal{X}]$ is much more than a mere linear space; it actually is an algebra with identity, where the product in $\mathcal{L}[\mathcal{X}]$ is interpreted as composition (Problem 2.30). The operator norm property ensures that $\mathcal{B}[\mathcal{X}]$ is a subalgebra (in the algebraic sense) of $\mathcal{L}[\mathcal{X}]$, and hence an algebra in its own right. Moreover, the operator norm property also ensures that $\mathcal{B}[\mathcal{X}]$ in fact is a normed algebra. Let $I$ denote the *identity operator* in $\mathcal{B}[\mathcal{X}]$ (i.e., $Ix = x$ for every $x \in \mathcal{X}$), which is the identity of the algebra $\mathcal{B}[\mathcal{X}]$. Of course, $\|I\| = 1$ by the very definition of norm (the induced uniform norm, that is) on $\mathcal{B}[\mathcal{X}]$. Thus $\mathcal{B}[\mathcal{X}]$ *is a normed algebra with identity* and, according to Proposition 4.15,

$\mathcal{B}[\mathcal{X}]$ is a Banach algebra with identity if $\mathcal{X} \neq \{0\}$ is a Banach space.

We know from Problem 4.1 that addition and scalar multiplication in $\mathcal{B}[\mathcal{X}]$ are continuous (as they are in any normed space). The operator norm property allows us to conclude that multiplication is continuous too. Indeed, if $\{S_n\}$ and $\{T_n\}$ are $\mathcal{B}[\mathcal{X}]$-valued sequences that converge in $\mathcal{B}[\mathcal{X}]$ to $S \in \mathcal{B}[\mathcal{X}]$ and $T \in \mathcal{B}[\mathcal{X}]$, respectively, then

$$\|S_n T_n - ST\| = \|S_n(T_n - T) + (S_n - S)T\| \leq \|S_n\| \|T_n - T\| + \|S_n - S\| \|T\|$$

for every $n$. Since $\sup_n \|S_n\| < \infty$ (reason: every convergent sequence is bounded), it follows that $\{S_n T_n\}$ converges in $\mathcal{B}[\mathcal{X}]$ to $ST$. Therefore the product operation $\mathcal{B}[\mathcal{X}] \times \mathcal{B}[\mathcal{X}] \to \mathcal{B}[\mathcal{X}]$ given by $(S, T) \mapsto ST$ is continuous (Corollary 3.8), where the topology on $\mathcal{B}[\mathcal{X}] \times \mathcal{B}[\mathcal{X}]$ is that induced by any of the equivalent metrics of Problems 3.9 and 3.33.

## 4.5 The Open Mapping Theorem and Continuous Inverses

Recall that a function $F : X \to Y$ from a set $X$ to a set $Y$ has an inverse on its range $\mathcal{R}(F)$ if there exists a (unique) function $F^{-1} : \mathcal{R}(F) \to X$ (called the *inverse of* $F$ *on* $\mathcal{R}(F)$) such that $F^{-1}F = I_X$, where $I_X$ stands for the identity on $X$. Moreover, $F$ has an inverse on its range if and only if it is injective. If $F$ is injective and surjective, then it is called *invertible* (in the set-theoretic sense) and $F^{-1} : Y \to X$ is the *inverse* of $F$. Furthermore, $F : X \to Y$ is invertible if and only if there exists a unique function $F^{-1} : Y \to X$ (the inverse of it) such that $F^{-1}F = I_X$ and $FF^{-1} = I_Y$, where $I_Y$ stands for the identity on $Y$. (See Section 1.3 and Problems 1.5 to 1.8). Now let $T$ be a linear transformation of a linear space $X$ into a linear space $Y$ (i.e., $T \in \mathcal{L}[X, Y]$). According to Theorems 2.8 and 2.10 $\mathcal{N}(T) = \{0\}$ if and only if $T$ has a linear inverse on its range $\mathcal{R}(T)$. That is,

$$\mathcal{N}(T) = \{0\} \iff T \text{ is injective} \iff \text{There exists } T^{-1} \in \mathcal{L}[\mathcal{R}(T), X].$$

Clearly, $\mathcal{N}(T) = \{0\}$ if and only if $0 < \|Tx\|$ for every nonzero $x \in X$ (whenever $Y$ is a normed space). This leads to the proposition below.

**Proposition 4.18.** *Take $T \in \mathcal{L}[X, Y]$, where $X$ is a linear space and $Y$ is a normed space. The following assertions are equivalent.*

  (a) *There exists $T^{-1} \in \mathcal{L}[\mathcal{R}(T), X]$  (i.e., $T$ has an inverse on its range).*

  (b) $0 < \|Tx\|$ *for every nonzero $x$ in $X$.*

**Definition 4.19.** A linear transformation $T$ of a normed space $X$ into a normed space $Y$ is *bounded below* if there exists a constant $\alpha > 0$ such that

$$\alpha \|x\| \le \|Tx\|$$

for every $x \in X$ (where the norm on the left-hand side is the norm on $X$ and that on the right-hand side is the norm on $Y$).

Note: $T \in \mathcal{B}[X, Y]$ actually means that $T \in \mathcal{L}[X, Y]$ is *bounded above*. The next result is a continuous version of the previous proposition.

**Proposition 4.20.** *Take $T \in \mathcal{L}[X, Y]$, where $X$ and $Y$ are normed spaces. The following assertions are equivalent.*

  (a) *There exists $T^{-1} \in \mathcal{B}[\mathcal{R}(T), X]$  (i.e., $T$ has a continuous inverse on its range).*

  (b) *$T$ is bounded below.*

*Moreover, if $\mathcal{X}$ is a Banach space, then each of the above equivalent assertions implies that $\mathcal{R}(T)^- = \mathcal{R}(T)$    (i.e., $\mathcal{R}(T)$ is closed in $\mathcal{Y}$).*

*Proof.* (i) If (a) holds true, then there exists a constant $\beta > 0$ such that $\|T^{-1}y\| \leq \beta\|y\|$ for every $y \in \mathcal{R}(T)$ (Definition 4.11). Take an arbitrary $x \in \mathcal{X}$ so that $Tx \in \mathcal{R}(T)$. Thus $\|x\| = \|T^{-1}Tx\| \leq \beta\|Tx\|$, and hence $\frac{1}{\beta}\|x\| \leq \|Tx\|$. Therefore (a)$\Rightarrow$(b).

(ii) Conversely, if (b) holds true, then $0 < \|Tx\|$ for every nonzero $x$ in $\mathcal{X}$, which implies that there exists $T^{-1} \in \mathcal{L}[\mathcal{R}(T), \mathcal{X}]$ by Proposition 4.18. Take an arbitrary $y \in \mathcal{R}(T)$ so that $y = Tx$ for some $x \in \mathcal{X}$. Thus $\|T^{-1}y\| = \|T^{-1}Tx\| = \|x\| \leq \frac{1}{\alpha}\|Tx\| = \frac{1}{\alpha}\|y\|$, which means that $T^{-1}$ is bounded (Definition 4.11). That is, (b)$\Rightarrow$(a).

(iii) Now take an arbitrary $\mathcal{R}(T)$-valued convergent sequence $\{y_n\}$. Since $y_n$ lies in $\mathcal{R}(T)$ for every $n$, it follows that there exists an $\mathcal{X}$-valued sequence $\{x_n\}$ such that $y_n = Tx_n$ for each $n$. Thus $\{Tx_n\}$ converges in $\mathcal{Y}$, and hence $\{Tx_n\}$ is a Cauchy sequence in $\mathcal{Y}$. If $T$ is bounded below, then there exists $\alpha > 0$ such that

$$0 \leq \alpha\|x_m - x_n\| \leq \|T(x_m - x_n)\| = \|Tx_m - Tx_n\|$$

for every pair of indices $m$ and $n$ (recall: $T$ is linear), so that $\{x_n\}$ also is a Cauchy sequence (in $\mathcal{X}$), and therefore it converges in $\mathcal{X}$ to, say, $x \in \mathcal{X}$ (whenever $\mathcal{X}$ is a Banach space). But $T$ is continuous and so $y_n = Tx_n \to Tx$ (Corollary 3.8), which implies that the (unique) limit of $\{y_n\}$ lies in $\mathcal{R}(T)$. Conclusion: $\mathcal{R}(T)$ is closed in $\mathcal{Y}$ by the Closed Set Theorem (Theorem 3.30).    $\square$

**Example 4J.** Take a sequence $a = \{\alpha_k\}_{k=0}^{\infty}$ in $\ell_+^{\infty}$ and consider the diagonal operator $D_a$ in $\mathcal{B}[\ell_+^p]$ (for some $p \geq 1$) of Example 4H. If $\alpha_k \neq 0$ for every $k \geq 0$ and $D_a x = 0$ for some $x = \{\xi_k\}_{k=0}^{\infty}$ in $\ell_+^p$, then $x = 0$ (i.e., if $\alpha_k \neq 0$ and $\alpha_k \xi_k = 0$ for every $k \geq 0$, then $\xi_k = 0$ for every $k \geq 0$). Thus $\alpha_k \neq 0$ for every $k \geq 0$ implies $\mathcal{N}(D_a) = \{0\}$. Conversely, if $\mathcal{N}(D_a) = \{0\}$, then $D_a x \neq 0$ for every nonzero $x$ in $\ell_+^p$. In particular, $D_a e_i = \alpha_i e_i \neq 0$ so that $\|D_a e_i\| = |\alpha_i| \neq 0$, and hence $\alpha_i \neq 0$, for every $i \geq 0$ (see Example 4H). Conclusion:

$$\mathcal{N}(D_a) = \{0\} \text{ if and and only } \alpha_k \neq 0 \text{ for every } k \geq 0.$$

Therefore, since a linear transformation has a (linear) inverse on its range if and only if its null space is zero (i.e., if and only if it is injective), we may conclude:

There exists $D_a^{-1} \in \mathcal{L}[\mathcal{R}(D_a), \ell_+^p]$ if and only if $\alpha_k \neq 0$ for every $k \geq 0$.

In this case, the linear (but not necessarily bounded) transformation $D_a^{-1}$ is again a diagonal mapping,

$$D_a^{-1}y = \{\alpha_k^{-1}\upsilon_k\}_{k=0}^{\infty} \quad \text{for every} \quad y = \{\upsilon_k\}_{k=0}^{\infty} \in \mathcal{R}(D_a) \subseteq \ell_+^p,$$

whose domain $\mathcal{D}(D_a^{-1})$ is precisely the range $\mathcal{R}(D_a)$ of $D_a$ (reason: such a mapping is a diagonal, as diagonals were defined in Problem 3.22, and $D_a^{-1}D_a$ is the identity on $\ell_+^p$). Recall that $\mathcal{D}(D_a^{-1}) = \ell_+^p$ whenever $\sup_k |\alpha_k^{-1}| < \infty$ (Problem 3.22(a)), and hence $\mathcal{R}(D_a) = \ell_+^p$ whenever $\inf_k |\alpha_k| > 0$ (since $\inf_k |\alpha_k| > 0$ implies $\sup_k |\alpha_k^{-1}| < \infty$). Moreover,

$$\|D_a x\| = \Big( \sum_{k=0}^{\infty} |\alpha_k \xi_k|^p \Big)^{\frac{1}{p}} \geq \inf_k |\alpha_k| \Big( \sum_{k=0}^{\infty} |\xi_k|^p \Big)^{\frac{1}{p}} = \inf_k |\alpha_k| \, \|x\|$$

for every $x = \{\xi_k\}_{k=0}^{\infty} \in \ell_+^p$, so that $D_a$ is bounded below whenever $\inf_k |\alpha_k| > 0$. Conversely, if $D_a$ is bounded below, then there exists $\alpha > 0$ such $\alpha \leq \|D_a e_i\| = |\alpha_i|$ for every $i \geq 0$ so that $\inf_i |\alpha_i| \geq \alpha > 0$. Outcome (cf. Proposition 4.20):

There exists $D_a^{-1} \in \mathcal{B}[\ell_+^p]$ if and only if $\inf_k |\alpha_k| > 0$.

That is, the inverse of a diagonal operator $D_a$ in $B[\ell_+^p]$ exists and is itself a diagonal operator in $B[\ell_+^p]$ if and only if the bounded sequence $a = \{\alpha_k\}_{k=0}^{\infty}$ is *bounded away from zero* (see Problem 4.5).

*If $T \in \mathcal{B}[\mathcal{X}, \mathcal{Y}]$, $\mathcal{X}$ is a Banach space, and there exists $T^{-1} \in \mathcal{B}[\mathcal{Y}, \mathcal{X}]$, then $\mathcal{Y}$ is a Banach space.* Proof: Theorem 4.14, Lemma 3.43 and Corollary 3.8 (note: this supplies another proof for part of Proposition 4.20). Question: Does the converse hold? That is, if $\mathcal{X}$ and $\mathcal{Y}$ are Banach spaces, and if $T \in \mathcal{B}[\mathcal{X}, \mathcal{Y}]$ is invertible in the set-theoretic sense (i.e., injective and surjective), does it follow that $T^{-1} \in \mathcal{B}[\mathcal{Y}, \mathcal{X}]$? Yes, it does. This is the Inverse Mapping Theorem, but to prove it we need a fundamental result in the theory of bounded linear transformations between Banach spaces, namely, the Open Mapping Theorem. Recall that a function $F \colon X \to Y$ of a metric space $X$ into a metric space $Y$ is an open mapping if $F(U)$ is open in $Y$ whenever $U$ is open in $X$ (Section 3.4). The Open Mapping Theorem says that *every continuous linear transformation of a Banach space onto a Banach space is an open mapping.* In other words, *a surjective bounded linear transformation between Banach spaces maps open sets into open sets.*

**Theorem 4.21.** (The Open Mapping Theorem). *If $\mathcal{X}$ and $\mathcal{Y}$ are Banach spaces and $T \in \mathcal{B}[\mathcal{X}, \mathcal{Y}]$ is surjective, then $T$ is an open mapping.*

*Proof.* Let $T$ be a surjective bounded linear transformation of a Banach space $\mathcal{X}$ onto a Banach space $\mathcal{Y}$. The nonempty open ball with center at the origin of $\mathcal{X}$ and radius $\varepsilon$ will de denoted by $X_\varepsilon$, and the nonempty open ball with center at the origin of $\mathcal{Y}$ and radius $\delta$ will be denoted by $Y_\delta$.

*Claim 1.* For every $\varepsilon > 0$ there exists $\delta > 0$ such that $Y_\delta \subseteq T(X_\varepsilon)^-$ (i.e., $T(X_\varepsilon)^-$ is a neighborhood of 0 in $\mathcal{Y}$).

*Proof.* Since each $x$ in $\mathcal{X}$ belongs to $X_n$ for every integer $n > \|x\|$, it follows that the sequence of open balls $\{X_n\}$ covers $\mathcal{X}$. That is, $\mathcal{X} = \bigcup_{n=1}^{\infty} X_n$, and hence

$T(\mathcal{X}) = \bigcup_{n=1}^{\infty} T(X_n)$ (see Problem 1.2(e)). Thus the countable family $\{T(X_n)\}$ of subsets of $\mathcal{Y}$ covers $T(\mathcal{X})$. If $T(\mathcal{X}) = \mathcal{Y}$ and $\mathcal{Y}$ is complete, then the Baire Category Theorem (Theorem 3.58) ensures the existence of a positive integer $m$ such that $T(X_m)^-$ has nonempty interior. Since $T$ is linear, $T(X_\varepsilon) = \frac{\varepsilon}{m} T(X_m)$ and so $T(X_\varepsilon)^- = \frac{\varepsilon}{m} T(X_m)^-$ has nonempty interior for every $\varepsilon > 0$ (because multiplication by a nonzero scalar is a homeomorphism). Take an arbitrary $\varepsilon > 0$ and an arbitrary $y_0 \in [T(X_{\frac{\varepsilon}{2}})^-]^\circ$ so that there exists a nonempty open ball with center at $y_0$, say $B_\delta(y_0)$, such that $B_\delta(y_0) \subseteq T(X_{\frac{\varepsilon}{2}})^-$. If $y$ is an arbitrary point of $Y_\delta$, then $\|y + y_0 - y_0\| = \|y\| < \delta$ and hence $y + y_0 \in B_\delta(y_0)$. Thus both $y + y_0$ and $y_0$ lie in $T(X_{\frac{\varepsilon}{2}})^-$, which means that

$$\inf_{u \in X_{\frac{\varepsilon}{2}}} \|Tu - y - y_0\| = 0 \quad \text{and} \quad \inf_{v \in X_{\frac{\varepsilon}{2}}} \|Tv - y_0\| = 0.$$

Therefore (recall: $u, v \in X_{\frac{\varepsilon}{2}}$ implies $u - v \in X_\varepsilon$),

$$\begin{aligned}
\inf_{x \in X_\varepsilon} \|Tx - y\| \;&\leq\; \inf_{u,v \in X_{\frac{\varepsilon}{2}}} \|T(u - v) - y\| \\
&=\; \inf_{u,v \in X_{\frac{\varepsilon}{2}}} \|Tu - y - y_0 + y_0 - Tv\| \\
&\leq\; \inf_{u \in X_{\frac{\varepsilon}{2}}} \|Tu - y - y_0\| + \inf_{v \in X_{\frac{\varepsilon}{2}}} \|Tv - y_0\| = 0,
\end{aligned}$$

and hence $y \in T(X_\varepsilon)^-$. Conclusion: $Y_\delta \subseteq T(X_\varepsilon)^-$. $\square$

*Claim* 2. For every $\varepsilon > 0$ there exists $\delta > 0$ such that $Y_\delta \subseteq T(X_\varepsilon)$ (i.e., we may even erase the closure symbol from Claim 1).

*Proof.* Take an arbitrary $\varepsilon > 0$, and set $\varepsilon_n = \frac{\varepsilon}{2^n}$ for every $n \in \mathbb{N}$. According to Claim 1, for each $n \in \mathbb{N}$ there exists $\delta_n \in (0, \varepsilon_n)$ such that

$$Y_{\delta_n} \subseteq T(X_{\varepsilon_n})^-.$$

If $y_n \in Y_{\delta_n} \subseteq T(X_{\varepsilon_n})^-$ for some $n \in \mathbb{N}$, then $\inf_{x \in X_{\varepsilon_n}} \|y_n - Tx\| = 0$ and so there exists $x_n \in X_n$ such that $\|y_n - Tx_n\| < \delta_{n+1}$. Set $\delta = \delta_1$ and take an arbitrary $y \in Y_\delta$. We claim that there exists an $\mathcal{X}$-valued sequence $\{x_n\}$ with the properties

$$x_n \in X_{\varepsilon_n} \quad \text{and} \quad y - \sum_{k=1}^{n} Tx_n \in Y_{\delta_{n+1}}$$

for every $n \in \mathbb{N}$. Indeed, $y \in Y_{\delta_1}$ implies that there exists $x_1 \in X_{\varepsilon_1}$ such that $\|y - Tx_1\| < \delta_2$, and hence the above properties hold for $x_1$ (i.e., they hold for $n = 1$). Now suppose they hold for some $n \geq 1$, so that there exists $x_k \in X_{\varepsilon_k}$ for each $k = 1, \ldots, n$ such that $y - \sum_{k=1}^{n} Tx_k \in Y_{\delta_{n+1}}$. This implies that there exists $x_{n+1} \in X_{\varepsilon_{n+1}}$ such that $\|y - \sum_{k=1}^{n} Tx_k - Tx_{n+1}\| < \delta_{n+2}$, and hence there exists $x_k \in X_{\varepsilon_k}$ for each $k = 1, \ldots, n+1$ such that $y - \sum_{k=1}^{n+1} Tx_k \in Y_{\delta_{n+2}}$. Therefore, assuming that the above properties hold for some $n \geq 1$ we conclude that they hold for $n + 1$, which completes the induction argument. Now we shall use this sequence $\{x_n\}$ to show that

$$y = Tx \quad \text{for some} \quad x \in X_\varepsilon.$$

Since $\|x_n\| < \frac{\varepsilon}{2^n}$ for each $n \in \mathbb{N}$, $\sum_{k=1}^\infty \|x_k\| < \varepsilon \sum_{k=1}^\infty \frac{1}{2^k} = \varepsilon$ (for $\sum_{k=0}^\infty \frac{1}{2^k} = 2$) and so $\{x_n\}$ is absolutely summable. Since $X$ is a Banach space, this implies that $\{x_n\}$ is summable (Proposition 4.4), which means that $\{\sum_{k=1}^n x_k\}$ converges in $X$. That is, there exists $x \in X$ such that

$$\sum_{k=1}^n x_k \to x \ \text{ in } \ X \ \text{ as } \ n \to \infty.$$

Moreover, $\|\sum_{k=1}^n x_k\| \leq \sum_{k=1}^n \|x_k\| < \varepsilon$ for all $n$ so that

$$\|x\| = \left\| \lim_n \sum_{k=1}^n x_k \right\| = \lim_n \left\| \sum_{k=1}^n x_k \right\| \leq \sup_n \left\| \sum_{k=1}^n x_k \right\| < \varepsilon$$

(recall: every norm is continuous), and hence $x \in X_\varepsilon$. Since $T$ is continuous and linear, and since

$$\sum_{k=1}^n Tx_k \to y \ \text{ in } \ \mathcal{Y} \ \text{ as } \ n \to \infty$$

(for $\|y - \sum_{k=1}^n Tx_k\| < \delta_{n+1} < \varepsilon_{n+1} = \frac{\varepsilon}{2^{n+1}}$ for every $n$), it follows that

$$Tx = T \lim_n \sum_{k=1}^n x_k = \lim_n T \sum_{k=1}^n x_k = \lim_n \sum_{k=1}^n Tx_k = y$$

and so $y \in T(X_\varepsilon)$. That is, an arbitrary point of $Y_\delta$ lies in $T(X_\varepsilon)$. Therefore $Y_\delta \subseteq T(X_\varepsilon)$. $\square$

Let $U$ be any open subset of $X$ and take an arbitrary $z \in T(U)$. Thus there exists $u \in U$ such that $z = Tu$. But $u$ is an interior point of $U$ (because $U$ is open in $X$), which means that there exists a nonempty open ball included in $U$ with center at $u$, say $B_\varepsilon(u) \subseteq U$. Translate this ball to 0 in $X$ and note that the translated ball coincides with the open ball of radius $\varepsilon$ about the origin of $X$:

$$X_\varepsilon = B_\varepsilon(u) - u = \left\{ x \in X : \ x = v - u \text{ for some } v \in B_\varepsilon(u) \right\}.$$

According to Claim 2, there exists a nonempty open ball $Y_\delta$ about the origin of $\mathcal{Y}$ such that $Y_\delta \subseteq T(X_\varepsilon)$. Translate this ball to the point $z$ and get the open ball with center at $z$ and radius $\delta$:

$$B_\delta(z) = Y_\delta + z = \left\{ w \in \mathcal{Y} : \ w = y + z \text{ for some } y \in Y_\delta \right\}.$$

Then, since $T$ is linear,

$$B_\delta(z) \subseteq T(X_\varepsilon) + Tu = T(X_\varepsilon + u) = T(B_\varepsilon(u)) \subseteq T(U),$$

and hence $z$ is an interior point of $T(U)$. Conclusion: Every point of $T(U)$ is interior, which means that $T(U)$ is open in $\mathcal{Y}$.    $\square$

A straightforward application of the Open Mapping Theorem says that *an injective and surjective bounded linear transformation between Banach spaces has a bounded inverse.*

**Theorem 4.22.** (The Inverse Mapping Theorem or The Banach Continuous Inverse Theorem). *If $\mathcal{X}$ and $\mathcal{Y}$ are Banach spaces and $T \in \mathcal{B}[\mathcal{X}, \mathcal{Y}]$ is injective and surjective, then $T^{-1} \in \mathcal{B}[\mathcal{Y}, \mathcal{X}]$.*

*Proof.* Theorems 2.10, 3.20 and 4.21.    $\square$

If $\mathcal{X}$ and $\mathcal{Y}$ are normed spaces, then an element $T$ of $\mathcal{B}[\mathcal{X}, \mathcal{Y}]$ is called *invertible* if it is an invertible mapping in the set-theoretic sense (i.e., injective and surjective) and its inverse $T^{-1}$ lies in $\mathcal{B}[\mathcal{Y}, \mathcal{X}]$. According to Theorem 2.10 this can be briefly stated as: $T \in \mathcal{B}[\mathcal{X}, \mathcal{Y}]$ is *invertible* if it has a bounded inverse. We shall denote the set of all invertible elements of $\mathcal{B}[\mathcal{X}, \mathcal{Y}]$ by $\mathcal{G}[\mathcal{X}, \mathcal{Y}]$. The Inverse Mapping Theorem says that, if $\mathcal{X}$ and $\mathcal{Y}$ are Banach spaces, then both meanings of the term "invertible" coincide, and $T^{-1} \in \mathcal{G}[\mathcal{Y}, \mathcal{X}]$ whenever $T \in \mathcal{G}[\mathcal{X}, \mathcal{Y}]$ (for $(T^{-1})^{-1} = T$ — Problem 1.8). Note that $\mathcal{G}[\mathcal{X}, \mathcal{Y}]$ is not a linear manifold of $\mathcal{B}[\mathcal{X}, \mathcal{Y}]$: addition is not a binary operation on $\mathcal{G}[\mathcal{X}, \mathcal{Y}]$ (e.g., $\mathrm{diag}(\{\frac{-k}{k+1}\}_{k=1}^\infty) + I = \mathrm{diag}(\{\frac{1}{k+1}\}_{k=1}^\infty)$ — see Example 4J). On the other hand, the composition (product) of invertible bounded linear transformations is again an invertible bounded linear transformation.

**Corollary 4.23.** *If $T \in \mathcal{G}[\mathcal{X}, \mathcal{Y}]$ and $S \in \mathcal{G}[\mathcal{Y}, \mathcal{Z}]$, then $ST \in \mathcal{G}[\mathcal{X}, \mathcal{Z}]$ and $(ST)^{-1} = T^{-1}S^{-1}$ whenever $\mathcal{X}$, $\mathcal{Y}$ and $\mathcal{Z}$ are Banach spaces.*

*Proof.* Problem 1.10, Proposition 4.16 and Theorem 4.22.    $\square$

**Corollary 4.24.** *Let $\mathcal{X}$ and $\mathcal{Y}$ be Banach spaces and take $T$ in $\mathcal{B}[\mathcal{X}, \mathcal{Y}]$. The following assertions are pairwise equivalent.*

(a) *There exists $T^{-1} \in \mathcal{B}[\mathcal{R}(T), \mathcal{X}]$    (i.e., $T$ has a continuous inverse on its range).*

(b) *$T$ is bounded below.*

(c) *$\mathcal{N}(T) = \{0\}$ and $\mathcal{R}(T)^- = \mathcal{R}(T)$    (i.e., $T$ is injective and has a closed range).*

*Proof.* According to Proposition 4.20, assertions (a) and (b) are equivalent and each of them implies $\mathcal{R}(T)^- = \mathcal{R}(T)$ whenever $\mathcal{X}$ is a Banach space. Moreover,

(b) trivially implies $\mathcal{N}(T) = \{0\}$. Thus any of the equivalent assertions (a) or (b) implies (c). On the other hand, if $\mathcal{N}(T) = \{0\}$, then $T$ is injective. If, in addition, the linear manifold $\mathcal{R}(T)$ is closed in the Banach space $\mathcal{Y}$, then it is itself a Banach space (Proposition 4.7) so that $T : \mathcal{X} \to \mathcal{R}(T)$ is an injective and surjective bounded linear transformation of the Banach space $\mathcal{X}$ onto the Banach space $\mathcal{R}(T)$. Therefore, its inverse $T^{-1}$ lies in $\mathcal{B}[\mathcal{R}(T), \mathcal{X}]$ by the Inverse Mapping Theorem. That is, (c) implies (a). $\qquad\square$

Let $\mathcal{X} \neq \{0\}$ be a normed space. Consider the algebra $\mathcal{L}[\mathcal{X}]$ (of all linear transformations on $\mathcal{X}$), and also the normed algebra $\mathcal{B}[\mathcal{X}]$ (of all operators on $\mathcal{X}$), which is a subalgebra (in the purely algebraic sense) of $\mathcal{L}[\mathcal{X}]$. Let $\mathcal{G}[\mathcal{X}]$ denote the set of all invertible operators in $\mathcal{B}[\mathcal{X}]$ (i.e., set $\mathcal{G}[\mathcal{X}] = \mathcal{G}[\mathcal{X}, \mathcal{X}]$). Suppose $\mathcal{X}$ is a Banach space, and let $T \in \mathcal{B}[\mathcal{X}]$ be an invertible element of the algebra $\mathcal{L}[\mathcal{X}]$ (i.e., there exists $T^{-1} \in \mathcal{L}[\mathcal{X}]$ such that $T^{-1}T = TT^{-1} = I$, where $I$ stands for the identity of $\mathcal{L}[\mathcal{X}]$). Thus, as we had already observed, the Inverse Mapping Theorem ensures that the concept of an invertible element of the Banach algebra $\mathcal{B}[\mathcal{X}]$ is unambiguously defined (for $T^{-1} \in \mathcal{B}[\mathcal{X}]$). In other words, the set-theoretic inverse of an operator on a Banach space is again an operator on the same Banach space. Moreover (since the inverse of an invertible operator is itself invertible), the set $\mathcal{G}[\mathcal{X}]$ of all invertible operators in $\mathcal{B}[\mathcal{X}]$ forms a *group* under multiplication (every operator in $\mathcal{G}[\mathcal{X}] \subset \mathcal{B}[\mathcal{X}]$ has an inverse in $\mathcal{G}[\mathcal{X}]$) whenever $\mathcal{X}$ is a Banach space.

We close this section with another important application of the Open Mapping Theorem. Let $\mathcal{X}$ and $\mathcal{Y}$ be normed spaces (over the same scalar field) and let $\mathcal{X} \oplus \mathcal{Y}$ be the direct sum of them, which is a linear space whose underlying set is the Cartesian product $\mathcal{X} \times \mathcal{Y}$ of the underlying sets of the linear spaces $\mathcal{X}$ and $\mathcal{Y}$. Recall that the graph of a transformation $T : \mathcal{X} \to \mathcal{Y}$ is the set

$$G_T = \big\{(x, y) \in \mathcal{X} \times \mathcal{Y} : \ y = Tx\big\} = \big\{(x, Tx) \in \mathcal{X} \oplus \mathcal{Y} : \ x \in \mathcal{X}\big\}$$

(see Section 1.2). If $T$ is linear, then $G_T$ is a linear manifold of the linear space $\mathcal{X} \oplus \mathcal{Y}$ (for $\alpha(u, Tu) \oplus \beta(v, Tv) = (\alpha u + \beta v, \ T(\alpha u + \beta v))$ for every $u, v \in \mathcal{X}$ and every pair if scalars $\{\alpha, \beta\}$). Equip the linear space $\mathcal{X} \oplus \mathcal{Y}$ with any of the norms of Example 4E and consider the normed space $\mathcal{X} \oplus \mathcal{Y}$.

**Theorem 4.25.** (The Closed Graph Theorem). *If $\mathcal{X}$ and $\mathcal{Y}$ are Banach spaces and $T \in \mathcal{L}[\mathcal{X}, \mathcal{Y}]$, then $T$ is continuous (i.e., $T \in \mathcal{B}[\mathcal{X}, \mathcal{Y}]$) if and only if $G_T$ is closed in $\mathcal{X} \oplus \mathcal{Y}$.*

*Proof.* Let $P_X : \mathcal{X} \oplus \mathcal{Y} \to \mathcal{X}$ and $P_Y : \mathcal{X} \oplus \mathcal{Y} \to \mathcal{Y}$ be defined by

$$P_X(x, y) = x \quad \text{and} \quad P_Y(x, y) = y$$

for every $(x, y) \in \mathcal{X} \oplus \mathcal{Y}$. Consider the restriction $P_X|_{G_T} : G_T \to \mathcal{X}$ of $P_X$ to the linear manifold $G_T$ of the normed space $\mathcal{X} \oplus \mathcal{Y}$. Observe that $P_X$ and $P_Y$ are the natural projections of $\mathcal{X} \oplus \mathcal{Y}$ onto $\mathcal{X}$ and $\mathcal{Y}$, respectively, which are both linear and bounded (see Example 4I). Thus $P_Y \in \mathcal{B}[\mathcal{X} \oplus \mathcal{Y}, \mathcal{Y}]$ and $P_X|_{G_T} \in \mathcal{B}[G_T, \mathcal{X}]$

(Problems 2.14 and 3.30). Moreover, $P_X|_{G_T}$ is clearly surjective and injective (if $P_X(x, Tx) = 0$, then $x = 0$ and hence $(x, Tx) = (0, 0) \in G_T$; that is, $\mathcal{N}(P_X|_{G_T}) = \{0\}$).

(a)  Recall that $\mathcal{X} \oplus \mathcal{Y}$ is a Banach space whenever $\mathcal{X}$ and $\mathcal{Y}$ are Banach spaces (Example 4E). If $G_T$ is closed in $\mathcal{X} \oplus \mathcal{Y}$, then $G_T$ is itself a Banach space (Proposition 4.7), and the Inverse Mapping Theorem (Theorem 4.22) ensures that the inverse $(P_X|_{G_T})^{-1}$ of $P_X|_{G_T}$ lies in $\mathcal{B}[\mathcal{X}, G_T]$. Since $T = P_Y(P_X|_{G_T})^{-1}$ (for $T P_X|_{G_T} = P_Y|_{G_T}$), it follows by Proposition 4.16 that $T$ is bounded.

(b)  Conversely, take an arbitrary sequence $\{(x_n, Tx_n)\}$ in $G_T$ that converges in $\mathcal{X} \oplus \mathcal{Y}$ to, say, $(x, y) \in \mathcal{X} \oplus \mathcal{Y}$. Since $P_X$ and $P_Y$ are continuous, it follows by Corollary 3.8 that

$$\lim x_n = \lim P_X(x_n, Tx_n) = P_X \lim(x_n, Tx_n) = P_X(x, y) = x,$$

$$\lim Tx_n = \lim P_Y(x_n, Tx_n) = P_Y \lim(x_n, Tx_n) = P_Y(x, y) = y.$$

If $T$ is continuous, then (Corollary 3.8 again)

$$y = \lim Tx_n = T \lim x_n = Tx,$$

and hence $(x, y) = (x, Tx) \in G_T$. Therefore $G_T$ is closed in $\mathcal{X} \oplus \mathcal{Y}$ by the Closed Set Theorem (Theorem 3.30).    $\square$


## 4.6   Equivalence and Finite-Dimensional Spaces

Recall that two sets are said to be *equivalent* if there exists a one-to-one correspondence (i.e., an injective and surjective mapping or, equivalently, an invertible mapping) between them (Chapter 1). Two linear spaces are *isomorphic* if there exists an isomorphism (i.e., an invertible linear transformation) of one of them onto the other (recall: the inverse of an invertible linear transformation is again a linear transformation). An isomorphism (or a linear-space isomorphism) is then a one-to-one correspondence that preserves the linear operations between the linear spaces, and hence it preserves the algebraic structure (Chapter 2). Two topological spaces are *homeomorphic* if there exists a homeomorphism (i.e., an invertible continuous mapping whose inverse also is continuous) of one of them onto the other. A homeomorphism provides a one-to-one correspondence between the topologies on the respective spaces, thus preserving the topological structure. In particular, a homeomorphism preserves convergence. A uniform homeomorphism between metric spaces is a homeomorphism where continuity (in both senses) is strengthened to uniform continuity, so that a uniform homeomorphism also preserves Cauchy sequences. Two metric spaces are *uniformly homeomorphic* if there exists a uniform homeomorphism of one of them onto the other (Chapter 3).

Now, as one would expect, we shall be interested in preserving both algebraic and topological structures between two normed spaces. A mapping of a normed space $\mathcal{X}$ onto a normed space $\mathcal{Y}$ that is (simultaneously) a homeomorphism and an isomorphism is called a *topological isomorphism* (or an *equivalence*), and $\mathcal{X}$ and $\mathcal{Y}$ are said to be *topologically isomorphic* (or *equivalent*) if there exists a topological isomorphism between them. Clearly, continuity refers to the norm topologies: a topological isomorphism is a mapping of a normed space $\mathcal{X}$ onto a normed space $\mathcal{Y}$, which is a homeomorphism when $\mathcal{X}$ and $\mathcal{Y}$ are viewed as metric spaces (equipped with the metrics generated by their respective norms), and also is an isomorphism between the linear spaces $\mathcal{X}$ and $\mathcal{Y}$. Since an isomorphism is just an injective and surjective linear transformation between linear spaces, it follows that *a topological isomorphism is simply a linear homeomorphism between normed spaces*. Thus the topological isomorphisms between $\mathcal{X}$ and $\mathcal{Y}$ are precisely the elements of $\mathcal{G}[\mathcal{X}, \mathcal{Y}]$: if $\mathcal{X}$ and $\mathcal{Y}$ are normed spaces, then $W : \mathcal{X} \to \mathcal{Y}$ is a topological isomorphism if and only if $W$ is an invertible element of $\mathcal{B}[\mathcal{X}, \mathcal{Y}]$. Therefore, $\mathcal{X}$ *and* $\mathcal{Y}$ *are topologically isomorphic if and only if there exists a linear homeomorphism between them* or, equivalently, *if and only if* $\mathcal{G}[\mathcal{X}, \mathcal{Y}] \neq \varnothing$. Conversely, the elements of $\mathcal{G}[\mathcal{X}, \mathcal{Y}]$ are also characterized as those linear-space isomorphisms that are bounded above and below (Theorem 4.14 and Proposition 4.20). Thus $\mathcal{X}$ *and* $\mathcal{Y}$ *are topologically isomorphic if and only if there exists an isomorphism* $W \in \mathcal{L}[\mathcal{X}, \mathcal{Y}]$ *and a pair of positive constants* $\alpha$ *and* $\beta$ *such that*

$$\alpha \|x\| \leq \|Wx\| \leq \beta \|x\| \quad \text{for every} \quad x \in \mathcal{X}.$$

The Inverse Mapping Theorem says that an injective and surjective bounded linear transformation of a Banach space onto a Banach space is a homeomorphism, and hence a topological isomorphism: *if* $\mathcal{X}$ *and* $\mathcal{Y}$ *are Banach spaces, then they are topologically isomorphic if and only if there exists an isomorphism in* $\mathcal{B}[\mathcal{X}, \mathcal{Y}]$.

Two norms on the same linear space are said to be *equivalent* if the metrics generated by them are equivalent. In other words (see Section 3.4), $\| \ \|_1$ and $\| \ \|_2$ are equivalent norms on a linear space $\mathcal{X}$ if and only if they induce the same norm topology on $\mathcal{X}$.

**Proposition 4.26.** *Let* $\| \ \|_1$ *and* $\| \ \|_2$ *be two norms on the same linear space* $\mathcal{X}$. *The following assertions are pairwise equivalent.*

(a)  $\| \ \|_1$ *and* $\| \ \|_2$ *are equivalent norms.*

(b)  *The identity map between the normed spaces* $(\mathcal{X}, \| \ \|_1)$ *and* $(\mathcal{X}, \| \ \|_2)$ *is a topological isomorphism.*

(c)  *There exist real constants* $\alpha > 0$ *and* $\beta > 0$ *such that*

$$\alpha \|x\|_1 \leq \|x\|_2 \leq \beta \|x\|_1 \quad \text{for every} \quad x \in \mathcal{X}.$$

*Proof.* Recall that the identity obviously is an isomorphism of a linear space onto itself, and hence it is a topological isomorphism between the normed spaces $(\mathcal{X}, \|\ \|_1)$ and $(\mathcal{X}, \|\ \|_2)$ if and only if it is a homeomorphism. Thus assertions (a) and (b) are equivalent by Corollary 3.19. Assertion (c) simply says that the identity $I : (\mathcal{X}, \|\ \|_1) \to (\mathcal{X}, \|\ \|_2)$ is bounded above and below or, equivalently, that it is continuous and its inverse $I^{-1} : (\mathcal{X}, \|\ \|_2) \to (\mathcal{X}, \|\ \|_1)$ also is continuous (Theorem 4.14 and Proposition 4.20), which means that it is a homeomorphism. That is, assertions (b) and (c) are equivalent.                                        $\square$

It is worth noticing that metrics generated by norms are equivalent if and only if they are uniformly equivalent. Indeed, continuity coincides with uniform continuity for linear transformations between normed spaces (Theorem 4.14), and hence the identity $I : (\mathcal{X}, \|\ \|_1) \to (\mathcal{X}, \|\ \|_2)$ is a homeomorphism if and only if it is a uniform homeomorphism. Observe that, according to Problem 3.33, all norms of Example 4E are equivalent. In particular, all norms of Example 4A are equivalent.

**Theorem 4.27.** *If $\mathcal{X}$ is a finite-dimensional linear space, then any two norms on $\mathcal{X}$ are equivalent.*

*Proof.* Let $B = \{e_i\}_{i=1}^n$ be a Hamel basis for an $n$-dimensional linear space $\mathcal{X}$ over $\mathbb{F}$ so that every vector in $\mathcal{X}$ has a unique expansion on $B$,

$$x = \sum_{i=1}^{n} \xi_i e_i,$$

where $\{\xi_i\}_{i=1}^n$ is a family of scalars: the coordinates of $x$ with respect to $B$. It is easy to show that the function $\|\ \|_0 : \mathcal{X} \to \mathbb{R}$, defined by

$$\|x\|_0 = \max_{1 \le i \le n} |\xi_i|$$

for every $x \in \mathcal{X}$, is a norm on $\mathcal{X}$. Now take an arbitrary norm $\|\ \|$ on $\mathcal{X}$. We shall verify that $\|\ \|_0$ and $\|\ \|$ are equivalent. First observe that

$$\|x\| = \left\| \sum_{i=1}^{n} \xi_i e_i \right\| \le \sum_{i=1}^{n} |\xi_i| \|e_i\| \le \left( \sum_{i=1}^{n} \|e_i\| \right) \|x\|_0$$

for every $x \in \mathcal{X}$. To show the other inequality consider the linear space $\mathbb{F}^n$ equipped with the norm $\|\ \|_\infty$ of Example 4A; that is,

$$\|a\|_\infty = \max_{1 \le i \le n} |\alpha_i|$$

for every vector $a = (\alpha_1, \dots, \alpha_n) \in \mathbb{F}^n$; and consider the transformation $L : (\mathbb{F}^n, \|\ \|_\infty) \to (\mathcal{X}, \|\ \|)$ defined by

$$La = \sum_{i=1}^{n} \alpha_i e_i$$

for every $a = (\alpha_1, \cdots, \alpha_n) \in \mathbb{F}^n$. It is readily verified that $L$ is linear and bounded. Indeed,

$$\|La\| = \Big\| \sum_{i=1}^{n} \alpha_i e_i \Big\| \leq \Big( \sum_{i=1}^{n} \|e_i\| \Big) \|a\|_\infty$$

so that the transformation $L$ is continuous (Theorem 4.14), and so is the real-valued function $\varphi \colon (\mathbb{F}^n, \|\ \|_\infty) \to \mathbb{R}$ such that

$$\varphi(a) = \|La\|$$

for every $a = (\alpha_1, \dots, \alpha_n) \in \mathbb{F}^n$ (recall: $\|\ \| \colon \mathcal{X} \to \mathbb{R}$ is continuous, and composition of continuous functions is a continuous function). The unit sphere $S = \partial B_1[0] = \{a \in \mathbb{F}^n \colon \|a\|_\infty\}$ is compact in $(\mathbb{F}^n, \|\ \|_\infty)$ by Theorem 3.83, and hence the continuous function $\varphi$ assumes a minimum value on $S$ (Theorem 3.86). Moreover, since $\{e_i\}_{i=1}^{n}$ is linearly independent, it follows that the linear transformation $L$ is injective ($L(a) = 0$ if and only if $a = 0$), so that this minimum value of $\varphi$ on $S$ is positive (for $\varphi(a) = \|L(a)\| > 0$ for every $a \in S$). Summing up: there exists $a_m \in S$ such that

$$0 < \varphi(a_m) \leq \varphi(a)$$

for all $a \in S$. Let $x = (\xi_1, \dots, \xi_n) \in \mathbb{F}^n$ be the $n$-tuple whose entries are the coordinates $\{\xi_i\}_{i=1}^{n}$ of an arbitrary $x = \sum_{i=1}^{n} \xi_i e_i$ in $\mathcal{X}$ with respect to the basis $B$. Note that

$$\|x\|_0 = \|x\|_\infty \neq 0$$

whenever $x \neq 0$, and hence

$$\varphi(a_m)\|x\|_0 \leq \varphi\Big(\frac{x}{\|x\|_\infty}\Big)\|x\|_\infty = \Big\| L\Big(\frac{x}{\|x\|_\infty}\Big) \Big\| \|x\|_\infty$$

$$= \|Lx\| = \Big\| \sum_{i=1}^{n} \xi_i e_i \Big\| = \|x\|$$

for every nonzero $x$ in $\mathcal{X}$. Therefore, by setting $\alpha = \varphi(a_m) > 0$ and $\beta = \sum_{i=1}^{n} \|e_i\| > 0$ (which do not depend on $x$), it follows that

$$\alpha\|x\|_0 \leq \|x\| \leq \beta\|x\|_0$$

for every $x \in \mathcal{X}$. Conclusion: Every norm on $\mathcal{X}$ is equivalent to the norm $\|\ \|_0$, and hence every two norms on $\mathcal{X}$ are equivalent. $\qquad\square$

**Corollary 4.28.** *A finite-dimensional normed space is a Banach space.*

*Proof.* Let $\mathcal{X}$ be an $n$-dimensional linear space over $\mathbb{F}$ and let $B = \{e_i\}_{i=1}^{n}$ be a Hamel basis for $\mathcal{X}$. Take an arbitrary $\mathcal{X}$-valued sequence $\{x_k\}_{k\geq 1}$ so that

$$x_k = \sum_{i=1}^{n} \xi_k(i) e_i,$$

where $\{\xi_k(i)\}_{i=1}^n$ is a family of scalars (the coordinates of $x_k$ with respect to the Hamel basis $B$) for each integer $k \geq 1$. Equip $\mathcal{X}$ with a norm $\|\ \|$ and suppose $\{x_k\}_{k\geq 1}$ is a Cauchy sequence in $(\mathcal{X}, \|\ \|)$. Consequently, for every $\varepsilon > 0$ there exists an integer $k_\varepsilon \geq 1$ such that $\|x_j - x_k\| < \varepsilon$ whenever $j, k \geq k_\varepsilon$. Now consider the equivalent norm $\|\ \|_0$ on $\mathcal{X}$ (which was defined in the proof of Theorem 4.27) so that, for every $j, k \geq k_\varepsilon$,

$$\max_{1 \leq i \leq n} |\xi_j(i) - \xi_k(i)| = \|x_j - x_k\|_0 \leq \tfrac{1}{\alpha}\|x_j - x_k\| < \tfrac{\varepsilon}{\alpha}$$

for some positive constant $\alpha$. Hence $\{\xi_k(i)\}_{k\geq 1}$ is a Cauchy sequence in $(\mathbb{F}, |\ |)$ for each $i = 1, \ldots, n$, where $|\ |$ stands for the usual norm on $\mathbb{F}$. As $(\mathbb{F}, |\ |)$ is a Banach space (Example 4A), each sequence $\{\xi_k(i)\}_{k\geq 1}$ converges in $(\mathbb{F}, |\ |)$ to, say, $\xi(i) \in \mathbb{F}$. That is, for every $\varepsilon > 0$ there exists an integer $k_{i,\varepsilon} \geq 1$ for each $i = 1, \ldots, n$ such that $|\xi_k(i) - \xi(i)| < \varepsilon$ whenever $k \geq k_{i,\varepsilon}$. By setting $x = \sum_{i=1}^n \xi(i) e_i$ in $\mathcal{X}$ we get

$$\|x_k - x\| \leq \sum_{i=1}^n |\xi_k(i) - \xi(i)|\,\|e_i\| \leq \left(\sum_{i=1}^n \|e_i\|\right)\varepsilon$$

whenever $k \geq k_\varepsilon = \max\{k_{i,\varepsilon}\}_{i=1}^n$, and hence $x_k \to x$ in $(\mathcal{X}, \|\ \|)$ as $k \to \infty$. Conclusion: If $\mathcal{X}$ is a finite-dimensional linear space and $\|\ \|$ is any norm on $\mathcal{X}$, then every Cauchy sequence in $(\mathcal{X}, \|\ \|)$ converges in $(\mathcal{X}, \|\ \|)$, and therefore $(\mathcal{X}, \|\ \|)$ is a Banach space. $\qquad\square$

**Corollary 4.29.** *Every finite-dimensional linear manifold of any normed space $\mathcal{X}$ is a subspace of $\mathcal{X}$.*

*Proof.* Corollary 4.28 and Theorem 3.40(a). $\qquad\square$

**Corollary 4.30.** *Let $\mathcal{X}$ and $\mathcal{Y} \neq \{0\}$ be normed spaces. Every linear transformation of $\mathcal{X}$ into $\mathcal{Y}$ is continuous if and only if $\mathcal{X}$ is finite-dimensional.*

*Proof.* The statement can be rewritten as follows. If $\mathcal{Y} \neq \{0\}$, then

$$\dim \mathcal{X} < \infty \quad \text{if and only if} \quad \mathcal{L}[\mathcal{X}, \mathcal{Y}] = \mathcal{B}[\mathcal{X}, \mathcal{Y}].$$

(a) If $\dim \mathcal{X} = 0$, then $\mathcal{X} = \{0\}$ and hence $\mathcal{L}[\mathcal{X}, \mathcal{Y}] = \{O\}$, so that $\mathcal{L}[\mathcal{X}, \mathcal{Y}]$ trivially coincides with $\mathcal{B}[\mathcal{X}, \mathcal{Y}]$. Suppose $\dim \mathcal{X} = n$ for some positive integer $n$ and let $B = \{e_i\}_{i=1}^n$ be a Hamel basis for $\mathcal{X}$. Take an arbitrary $x \in \mathcal{X}$ and consider its (unique) expansion on $B$,

$$x = \sum_{i=1}^n \xi_i e_i,$$

where $\{\xi_i\}_{i=1}^n$ is a family of scalars consisting of the coordinates of $x$ with respect to the basis $B$. Let $\|\ \|_X$ and $\|\ \|_Y$ denote the norms on $\mathcal{X}$ and $\mathcal{Y}$, respectively. If $T \in \mathcal{L}[\mathcal{X}, \mathcal{Y}]$, then

$$\|Tx\|_Y = \left\|\sum_{i=1}^n \xi_i T e_i\right\|_Y \leq \sum_{i=1}^n |\xi_i|\,\|T e_i\|_Y \leq \left(\sum_{i=1}^n \|T e_i\|_Y\right)\max_{1 \leq i \leq n} |\xi_i|.$$

The norm $\|\ \|_0$ on $\mathcal{X}$ that was defined in the proof of Theorem 4.27 is equivalent to $\|\ \|_X$. Hence there exists $\alpha > 0$ such that

$$\|Tx\|_Y \le \left( \sum_{i=1}^{n} \|Te_i\|_Y \right) \|x\|_0 \le \left( \tfrac{1}{\alpha} \sum_{i=1}^{n} \|Te_i\|_Y \right) \|x\|_X,$$

and so $T \in \mathcal{B}[\mathcal{X}, \mathcal{Y}]$. Thus $\mathcal{L}[\mathcal{X}, \mathcal{Y}] \subseteq \mathcal{B}[\mathcal{X}, \mathcal{Y}]$ (i.e., $\mathcal{L}[\mathcal{X}, \mathcal{Y}] = \mathcal{B}[\mathcal{X}, \mathcal{Y}]$).

(b) Conversely, suppose $\mathcal{X}$ is infinite-dimensional and let $\{e_\gamma\}_{\gamma \in \Gamma}$ be an indexed Hamel basis for $\mathcal{X}$. Since $\{e_\gamma\}_{\gamma \in \Gamma}$ is an infinite set, it has a countably infinite subset, say $\{e_k\}_{k \in \mathbb{N}}$. Set $\mathcal{M} = \mathrm{span}\,\{e_k\}_{k \in \mathbb{N}}$, a linear manifold of $\mathcal{X}$ for which $\{e_k\}_{k \in \mathbb{N}}$ is a Hamel basis. Thus every vector $x$ in $\mathcal{M} \subseteq \mathcal{X}$ has a unique representation as a (finite) linear combination of vectors in $\{e_k\}_{k \in \mathbb{N}}$. This means that there exists a unique (similarly indexed) family of scalars $\{\xi_k\}_{k \in \mathbb{N}}$ such that $\xi_k = 0$ for all but a finite set of indices $k$, and

$$x = \sum_{k \in \mathbb{N}} \xi_k e_k$$

(see Section 2.4). Take an arbitrary nonzero vector $y$ in $\mathcal{Y}$ and consider the mapping $L_{\mathrm{M}} \colon \mathcal{M} \to \mathcal{Y}$ defined by

$$L_{\mathrm{M}} x = \sum_{k \in \mathbb{N}} k\, \xi_k \|e_k\| y$$

for every $x \in \mathcal{M}$. It is readily verified that $L_{\mathrm{M}}$ is linear (recall: the above sums are finite and unique). Extend $L_{\mathrm{M}}$ to $\mathcal{X}$, according to Theorem 2.9, and get a linear transformation $L \colon \mathcal{X} \to \mathcal{Y}$ such that $L|_{\mathcal{M}} = L_{\mathrm{M}}$. Since $Le_k = L_{\mathrm{M}} e_k = k\|e_k\| y$ and $e_k \ne 0$ for each $k \in \mathbb{N}$, it follows that

$$k\|y\| = \frac{\|Le_k\|}{\|e_k\|} \le \sup_{x \ne 0} \frac{\|Lx\|}{\|x\|}$$

for every $k \in \mathbb{N}$. Outcome: There is no constant $\beta \ge 0$ for which $\|Lx\| \le \beta\|x\|$ for all $x \in \mathcal{X}$, and therefore the linear transformation $L \colon \mathcal{X} \to \mathcal{Y}$ is not bounded; that is, $L \in \mathcal{L}[\mathcal{X}, \mathcal{Y}] \backslash \mathcal{B}[\mathcal{X}, \mathcal{Y}]$. Conclusion: If $\mathcal{L}[\mathcal{X}, \mathcal{Y}] = \mathcal{B}[\mathcal{X}, \mathcal{Y}]$, then $\dim \mathcal{X} < \infty$. $\qquad\square$

Corollary 4.30 holds, in particular, if $\mathcal{X}$ is a Banach space. In this case, the above construction (proof of Corollary 4.30(b) exhibits (via Theorem 2.9) our first example of an unbounded linear transformation $L$ whose domain $\mathcal{X}$ is an infinite-dimensional Banach space and the codomain $\mathcal{Y}$ is any nonzero normed space.

**Corollary 4.31.** *Two finite-dimensional normed spaces are topologically isomorphic if and only if they have the same dimension.*

*Proof.* Let $\mathcal{X}$ and $\mathcal{Y}$ be finite-dimensional normed spaces. Corollaries 4.28 and 4.30 ensure that $\mathcal{X}$ and $\mathcal{Y}$ are Banach spaces and $\mathcal{B}[\mathcal{X}, \mathcal{Y}] = \mathcal{L}[\mathcal{X}, \mathcal{Y}]$. Thus $\mathcal{X}$ and $\mathcal{Y}$ are topologically isomorphic if and only if they are isomorphic linear spaces (Theorem 4.22), which means that dim $\mathcal{X}$ = dim $\mathcal{Y}$ (Theorem 2.12).    $\square$

Recall that every compact subset of any metric space is closed and bounded. The Heine–Borel Theorem asserts that every closed and bounded subset of $\mathbb{F}^n$ is compact. This equivalence (i.e., in $\mathbb{F}^n$ a set is compact if and only if it is closed and bounded) actually is a property that characterizes finite-dimensional normed spaces.

**Corollary 4.32.** *If $\mathcal{X}$ is a finite-dimensional normed space, then every closed and bounded subset of $\mathcal{X}$ is compact.*

*Proof.* Let $\mathcal{X}$ be a finite-dimensional normed space over $\mathbb{F}$ and let $A$ be a subset of $\mathcal{X}$ that is closed and bounded (in the norm topology of $\mathcal{X}$). According to Corollary 4.31 $\mathcal{X}$ is topologically isomorphic to $\mathbb{F}^n$, where $n = \dim \mathcal{X}$ and $\mathbb{F}^n$ is equipped with any norm (which are all equivalent by Theorem 4.27). Take a topological isomorphism $W \in \mathcal{G}[\mathcal{X}, \mathbb{F}^n]$ and note that $W(A)$ is closed (because $W$ is a closed map — Theorem 3.24) and bounded (for $\sup_{a \in A} \|Wa\| \le \|W\| \sup_{a \in A} \|a\|$ — see Problem 4.5) in $\mathbb{F}^n$. Thus $W(A)$ is compact in $\mathbb{F}^n$ (Theorem 3.83). Therefore, since $W^{-1} \in \mathcal{G}[\mathbb{F}^n, \mathcal{X}]$ is a homeomorphism, and since compactness is a topological invariant (Theorem 3.64), it follows that $A = W^{-1}(W(A))$ is compact in $\mathcal{X}$.    $\square$

To establish the converse (i.e., *if every closed and bounded subset of a normed space $\mathcal{X}$ is compact, then $\mathcal{X}$ is finite-dimensional*) we need the following result.

**Lemma 4.33.** (Riesz). *Let $\mathcal{M}$ be a proper subspace of a normed space $\mathcal{X}$. For each real number $\alpha \in (0, 1)$ there exists a vector $x_\alpha$ in $\mathcal{X}$ such that $\|x_\alpha\| = 1$ and*

$$\alpha \le d(x_\alpha, \mathcal{M}) \le 1.$$

*Proof.* The second inequality holds trivially because $\mathcal{M}$ is a linear manifold of $\mathcal{X}$, and hence it contains the origin of $\mathcal{X}$ (i.e., $d(x, \mathcal{M}) = \inf_{u \in \mathcal{M}} \|x - u\| \le \|x - 0\| = \|x\|$ for every $x \in \mathcal{X}$). Since $\mathcal{M}$ is a proper closed subset of $\mathcal{X}$, it follows that $\mathcal{X} \backslash \mathcal{M}$ is nonempty and open in $\mathcal{X}$. Take any vector $z \in \mathcal{X} \backslash \mathcal{M}$ such that $0 < d(z, \mathcal{M})$ (e.g., take the center of an arbitrary nonempty open ball included in the open set $\mathcal{X} \backslash \mathcal{M}$). Thus for each $\varepsilon > 0$ there exists a vector $v_\varepsilon \in \mathcal{M}$ such that

$$0 < d(z, \mathcal{M}) = \inf_{u \in \mathcal{M}} \|z - u\| \le \|z - v_\varepsilon\| \le (1 + \varepsilon)\, d(z, \mathcal{M}).$$

Set $x_\varepsilon = \|z - v_\varepsilon\|^{-1} (z - v_\varepsilon)$ in $\mathcal{X}$ so that $\|x_\varepsilon\| = 1$ and

$$\begin{aligned}
\|z - v_\varepsilon\|\, d(x_\varepsilon, \mathcal{M}) &= \inf_{u \in \mathcal{M}} \big\| \|z - v_\varepsilon\| x_\varepsilon - \|z - v_\varepsilon\| u \big\| \\
&= \inf_{v \in \mathcal{M}} \|z - v_\varepsilon - v\| = d(z, \mathcal{M}).
\end{aligned}$$

Hence

$$\tfrac{1}{1+\varepsilon} \le d(x_\varepsilon, \mathcal{M}),$$

which concludes the proof by setting $\alpha = \frac{1}{1+\varepsilon}$. □

This result is sometimes referred to as the "Almost Orthogonality Lemma". (Why? — Draw a picture.) It is worth remarking that the lemma does not ensure the existence of a vector $x_1$ in $\mathcal{X}$ with $\|x_1\| = 1$ and $d(x_1, \mathcal{M}) = 1$. Such a vector may not exist in a Banach space but it certainly exists in a Hilbert space (next chapter).

**Corollary 4.34.** *Let $\mathcal{X}$ be a normed space. If the closed unit ball $B_1[0] = \{x \in \mathcal{X} : \|x\| \le 1\}$ is compact, then $\mathcal{X}$ is finite-dimensional.*

*Proof.* If $B_1[0]$ is compact, then it is totally bounded (Corollary 3.81) so that there exists a finite $\frac{1}{2}$-net for $B_1[0]$ (Definition 3.68), say $\{u_i\}_{i=1}^{n} \subset B_1[0]$. Set $\mathcal{M} = \mathrm{span}\,\{u_i\}_{i=1}^{n}$, which is a finite-dimensional subspace of $\mathcal{X}$ (reason: span $\{u_i\}_{i=1}^{n}$ is a finite-dimensional linear manifold of $\mathcal{X}$ because there exists a Hamel basis for it included in $\{u_i\}_{i=1}^{n}$, by Theorem 2.6, and hence it is a subspace of $\mathcal{X}$ according to Corollary 4.29). If $\mathcal{X}$ is infinite-dimensional, then $\mathcal{M} \ne \mathcal{X}$ and, in this case, Lemma 4.33 ensures the existence of a vector $x \in B_1[0]$ such that $\frac{1}{2} < d(x, \mathcal{M})$. Thus $\frac{1}{2} < \|x - u\|$ for every $u \in \mathcal{M}$ and, in particular, $\frac{1}{2} < \|x - u_i\|$ for every $i = 1, \dots, n$, which contradicts the fact that $\{u_i\}_{i=1}^{n}$ is a $\frac{1}{2}$-net for $B_1[0]$. Conclusion: If $B_1[0]$ is compact in $\mathcal{X}$, then $\mathcal{X}$ is finite-dimensional. □

## 4.7  Continuous Linear Extension and Completion

We shall now transfer the extension results of Section 3.8 on uniformly continuous mappings between metric spaces to bounded linear transformations between normed spaces. The normed-space versions of Theorem 3.45 and Corollary 3.46 read as follows.

**Theorem 4.35.** *Let $\mathcal{M}$ be a dense linear manifold of a normed space $\mathcal{X}$ and let $\mathcal{Y}$ be a Banach space. Every $T \in \mathcal{B}[\mathcal{M}, \mathcal{Y}]$ has a unique extension $\widehat{T}$ in $\mathcal{B}[\mathcal{X}, \mathcal{Y}]$. Moreover $\|\widehat{T}\| = \|T\|$.*

*Proof.* Let $\mathcal{M}$ and $\mathcal{Y}$ be normed spaces. Recall that every $T \in \mathcal{B}[\mathcal{M}, \mathcal{Y}]$ is uniformly continuous (Theorem 4.14). If $\mathcal{M}$ is a dense linear manifold of a normed space $\mathcal{X}$, and $\mathcal{Y}$ is a Banach space (i.e., a complete normed space), then Theorem 3.45 ensures that there exists a unique continuous extension $\widehat{T} : \mathcal{X} \to \mathcal{Y}$ of $T$ over $\mathcal{X}$. It remains to verify that

$$\widehat{T} \text{ is linear and } \|\widehat{T}\| = \|T\|.$$

Let $u$ and $v$ be arbitrary vectors in $\mathcal{X}$. Since $\mathcal{M}^- = \mathcal{X}$, it follows by Proposition 3.32 that there exist $\mathcal{M}$-valued sequences $\{u_n\}$ and $\{v_n\}$ converging in $\mathcal{X}$ to $u$ and $v$ respectively. Then, by Corollary 3.8,

$$\widehat{T}(u + v) = \widehat{T}(\lim u_n + \lim v_n) = \widehat{T}(\lim(u_n + v_n)) = \lim \widehat{T}(u_n + v_n)$$

because addition is a continuous operation (Problem 4.1) and $\widehat{T}$ is a continuous mapping. But $T = \widehat{T}|_{\mathcal{M}}$, $\mathcal{M}$ is a linear space, and $T$ is a linear transformation. Hence

$$\widehat{T}(u_n + v_n) = T(u_n + v_n) = Tu_n + Tv_n = \widehat{T}u_n + \widehat{T}v_n$$

for every $n$. The same continuity argument (Problem 4.1 and Corollary 3.8) makes the backward path:

$$\lim \widehat{T}(u_n + v_n) = \lim(\widehat{T}u_n + \widehat{T}v_n) = \lim \widehat{T}u_n + \lim \widehat{T}v_n = \widehat{T}u + \widehat{T}v.$$

This shows that $\widehat{T}$ is additive. Similarly, for every scalar $\alpha$,

$$\begin{aligned}
\widehat{T}(\alpha u) &= \widehat{T}(\alpha \lim u_n) = \widehat{T}(\lim \alpha u_n) = \lim \widehat{T}(\alpha u_n) \\
&= \lim T\alpha u_n = \lim \alpha T u_n = \lim \alpha \widehat{T}u_n = \alpha \lim \widehat{T}u_n = \alpha \widehat{T}u
\end{aligned}$$

so that $\widehat{T}$ is also homogeneous. That is, $\widehat{T}$ is linear. Thus $T \in \mathcal{B}[\mathcal{X}, \mathcal{Y}]$. Moreover,

$$\|\widehat{T}u_n\| = \|Tu_n\| \leq \|T\|\,\|u_n\|$$

for every $n$. Applying once again the same continuity argument, and recalling that the norm is continuous, it follows that $\lim \|u_n\| = \|u\|$ and

$$\|\widehat{T}u\| = \|\widehat{T}\lim u_n\| = \lim \|\widehat{T}u_n\| \leq \|T\| \lim \|u_n\| = \|T\|\,\|u\|,$$

and hence $\|\widehat{T}\| = \sup_{\|u\|=1} \|\widehat{T}u\| \leq \|T\|$. On the other hand,

$$\|T\| = \sup_{0\neq w\in\mathcal{M}} \frac{\|Tw\|}{\|w\|} = \sup_{0\neq w\in\mathcal{M}} \frac{\|\widehat{T}w\|}{\|w\|} \leq \sup_{0\neq x\in\mathcal{X}} \frac{\|\widehat{T}x\|}{\|x\|} = \|\widehat{T}\|. \qquad \square$$

**Corollary 4.36.** *Let $\mathcal{X}$ and $\mathcal{Y}$ be Banach spaces. Let $\mathcal{M}$ and $\mathcal{N}$ be dense linear manifolds of $\mathcal{X}$ and $\mathcal{Y}$, respectively. If $W \in \mathcal{G}[\mathcal{M}, \mathcal{N}]$, then there exists a unique $\widehat{W} \in \mathcal{G}[\mathcal{X}, \mathcal{Y}]$ that extends $W$ over $\mathcal{X}$.*

*Proof.* Recall that every $W \in \mathcal{G}[\mathcal{M}, \mathcal{N}]$ is a linear homeomorphism, and hence a uniform homeomorphism (Theorem 4.14). Thus Corollary 3.46 ensures that there exists a unique homeomorphism $\widehat{W} \colon \mathcal{X} \to \mathcal{Y}$ that extends $W$ over $\mathcal{X}$. On the other hand, the previous theorem says that there exists a unique continuous linear extension of $W \colon \mathcal{M} \to \mathcal{Y}$ over $\mathcal{X}$, say $\widehat{W}' \in \mathcal{B}[\mathcal{X}, \mathcal{Y}]$. Thus $\widehat{W}$ and $\widehat{W}'$ are both continuous mappings of $\mathcal{X}$ into $\mathcal{Y}$ that coincide on a dense subset $\mathcal{M}$ of $\mathcal{X}$: $\widehat{W}|_{\mathcal{M}} = W = \widehat{W}'|_{\mathcal{M}}$. Therefore $\widehat{W} = \widehat{W}'$ by Corollary 3.33, and hence the homeomorphism $\widehat{W}$ of $\mathcal{X}$ onto $\mathcal{Y}$ lies in $\mathcal{B}[\mathcal{X}, \mathcal{Y}]$. Conclusion: $W$ is a linear homeomorphism, which means a topological isomorphism; that is, $W \in \mathcal{G}[\mathcal{X}, \mathcal{Y}]$. $\qquad \square$

An even stronger form of homeomorphism is that of a surjective isometry (a surjective mapping that preserves distance — recall: every isometry is an injective contraction, and the inverse of a surjective isometry is again a surjective isometry).

If there exists a surjective isometry between two metric spaces, then they are said to be *isometric* or *isometrically equivalent* (Chapter 3). Thus a linear isometry of a normed space $\mathcal{X}$ into a normed space $\mathcal{Y}$ (i.e., a linear transformation of $\mathcal{X}$ into $\mathcal{Y}$ that is also an isometry) is necessarily an element of $\mathcal{B}[\mathcal{X}, \mathcal{Y}]$, since an isometry is continuous. A surjective isometry in $\mathcal{B}[\mathcal{X}, \mathcal{Y}]$ (i.e., a linear surjective isometry or, equivalently, a linear-space isomorphism that is also an isometry) is called an *isometric isomorphism*. Two normed spaces are *isometrically isomorphic* if there exists an isometric isomorphism between them. The next proposition places linear isometries in a normed-space setting.

**Proposition 4.37.** *Take $V \in \mathcal{L}[\mathcal{X}, \mathcal{Y}]$, where $\mathcal{X}$ and $\mathcal{Y}$ are normed spaces. The following assertions are equivalent.*

   (a)   *$V$ is an isometry*     *(i.e., $\|Vu - Vv\| = \|u - v\|$ for every $u, v \in \mathcal{X}$).*

   (b)   *$\|Vx\| = \|x\|$ for every $x \in \mathcal{X}$.*

*If $\mathcal{Y} = \mathcal{X}$, then each of the above assertions also is equivalent to*

   (c)   *$\|V^n x\| = \|x\|$ for every $x \in \mathcal{X}$ and every integer $n \geq 1$.*

*Proof.* Assertions (a) and (b) are clearly equivalent because $V$ is linear. Indeed, set $v = 0$ in (a) to get (b) and, conversely, set $x = u - v$ in (b) to get (a). Moreover, take $V \in \mathcal{L}[\mathcal{X}]$ and suppose (b) holds true. Then (c) holds trivially for $n = 1$. If (c) holds for some $n \geq 1$, then $\|V^{n+1} x\| = \|V^n V x\| = \|Vx\| = \|x\|$ for every $x \in \mathcal{X}$. Conclusion: (b) implies (c) by induction. Finally note that (c) trivially implies (b).     $\square$

It is clear that every isometric isomorphism is a topological isomorphism, and that the identity $I$ of $\mathcal{B}[\mathcal{X}]$ is an isometric isomorphism. It is also clear that the inverse of a topological (isometric) isomorphism is itself a topological (isometric) isomorphism, and that the composition (product) of two topological (isometric) isomorphisms is again a topological (isometric) isomorphism. Therefore, these concepts (topological isomorphism and isometric isomorphism, that is) have the defining properties of an equivalence relation (viz., reflexivity, symmetry and transitivity).

**Corollary 4.38.** *Let $\mathcal{M}$ and $\mathcal{N}$ be dense linear manifolds of Banach spaces $\mathcal{X}$ and $\mathcal{Y}$, respectively. If $U : \mathcal{M} \to \mathcal{N}$ is an isometric isomorphism of $\mathcal{M}$ onto $\mathcal{N}$, then there exists a unique isometric isomorphism $\widehat{U} : \mathcal{X} \to \mathcal{Y}$ of $\mathcal{X}$ onto $\mathcal{Y}$ that extends $U$ over $\mathcal{X}$.*

*Proof.* According to Corollary 4.36 there exists a unique topological isomorphism $\widehat{U} : \mathcal{X} \to \mathcal{Y}$ that extends $U : \mathcal{M} \to \mathcal{N}$ over $\mathcal{X}$. Note that the mappings $\| \; \| : \mathcal{X} \to \mathbb{R}$ and $\|\widehat{U}( \; )\| : \mathcal{X} \to \mathbb{R}$ are continuous (composition of continuous functions is again a continuous function), and also that they coincide on a dense subset $\mathcal{M}$ of $\mathcal{X}$ ($\|\widehat{U}u\| = \|Uu\| = \|u\|$ for every $u \in \mathcal{M}$ because $\widehat{U}|_{\mathcal{M}} = U$ and $U$

is an isometry — Proposition 4.37). Then, according to Corollary 3.33, they coincide on the whole space $\mathcal{X}$; that is, $\|\widehat{U}x\| = \|x\|$ for every $x \in \mathcal{X}$. Therefore (Proposition 4.37 again), the topological isomorphism $\widehat{U}$ is an isometry, and hence an isometric isomorphism. $\qquad\qquad\square$

Since every normed space is a metric space, every normed space has a completion in the sense of Definition 3.48. The question that naturally arrises is whether a normed space has a completion that is a Banach space. In other words, is a completion (in the norm topology, of course) of a normed space a complete *normed space*? We shall first redefine the concept of completion in a normed-space setting.

**Definition 4.39.** If the image of a *linear* isometry on a normed space $\mathcal{X}$ is a dense linear manifold of a Banach space $\widehat{\mathcal{X}}$, then $\widehat{\mathcal{X}}$ is a *completion* of $\mathcal{X}$. Equivalently, if a normed space $\mathcal{X}$ is isometrically isomorphic to a dense linear manifold of a Banach space $\widehat{\mathcal{X}}$, then $\widehat{\mathcal{X}}$ is a completion of $\mathcal{X}$. (In particular, any Banach space is a completion of every dense linear manifold of it, for the inclusion map is a linear isometry.)

**Theorem 4.40.** *Every normed space has a completion.*

*Proof.* Let $(\mathcal{X}, \|\ \|_\mathrm{X})$ be a normed space. Consider the linear space $\mathcal{X}^{\mathbb{N}}$ of all $\mathcal{X}$-valued sequences. It is readily verified that the collection $CS(\mathcal{X})$ of all Cauchy sequences in $(\mathcal{X}, \|\ \|_\mathrm{X})$ is a linear manifold of $\mathcal{X}^{\mathbb{N}}$, and hence a linear space (over the same field of the linear space $\mathcal{X}$). The program is to rewrite the proof of Theorem 3.49 in terms of the metric $d_\mathrm{X}$ generated by the norm $\|\ \|_\mathrm{X}$. Thus

$$\|x\| = \lim \|x_n\|_\mathrm{X}$$

defines a seminorm on the linear space $CS(\mathcal{X})$, where $x = \{x_n\}$ is an arbitrary sequence in $CS(\mathcal{X})$. Set

$$\mathcal{N} = \{x \in CS(\mathcal{X}): \ \|x\| = 0\}.$$

Proposition 4.5 ensures that $\mathcal{M}$ is a linear manifold of $CS(\mathcal{X})$, and also that

$$\|[x]\|_{\widehat{\mathrm{X}}} = \|x\|$$

defines a norm on the quotient space $\widehat{\mathcal{X}} = CS(\mathcal{X})/\mathcal{N}$, where $x$ is an arbitrary element of an arbitrary coset $[x] = x + \mathcal{N}$ in $\widehat{\mathcal{X}}$, so that

$$(\widehat{\mathcal{X}}, \|\ \|_{\widehat{\mathrm{X}}}) \text{ is a normed space.}$$

Now consider the mapping $K : \mathcal{X} \to \widehat{\mathcal{X}}$ that assigns to each vector $x \in \mathcal{X}$ the coset $[x] \in \widehat{\mathcal{X}}$ containing the constant sequence $x = \{x_n\} \in CS(\mathcal{X})$ such that $x_n = x$ for all $n$. It is easy to show that

$$K : \mathcal{X} \to \widehat{\mathcal{X}} \text{ is a linear isometry.}$$

Moreover, according to Claims 1 and 2 in the proof of Theorem 3.49, the range of $K$ (which is clearly a linear manifold of $\widehat{\mathcal{X}}$) is dense in $(\widehat{\mathcal{X}}, \| \; \|_{\widehat{X}})$, and $(\widehat{\mathcal{X}}, \| \; \|_{\widehat{X}})$ is a complete normed space. That is,

$$K(\mathcal{X})^- = \widehat{\mathcal{X}} \quad \text{and} \quad \widehat{\mathcal{X}} \text{ is a Banach space.}$$

Summing up: $\mathcal{X}$ is isometrically isomorphic to $K(\mathcal{X})$, which is a dense linear manifold of $\widehat{\mathcal{X}}$, which in turn is a Banach space. Therefore $\widehat{\mathcal{X}}$ is a completion of $\mathcal{X}$. $\qquad\square$

The completion of a normed space is essentially unique; that is, the completion of a normed space is unique up to an isometric isomorphism (i.e., up to a linear surjective isometry).

**Theorem 4.41.** *Any two completions of a normed space are isometrically isomorphic.*

*Proof.* Replace "metric space", "surjective isometry" and "subspace", respectively, with "normed space", "isometric isomorphism" and "linear manifold" in the proof of Theorem 3.50. $\qquad\square$

According to Definition 4.39, a Banach space $\widehat{\mathcal{X}}$ is a completion of a normed space $\mathcal{X}$ if there exists a dense linear manifold of $\widehat{\mathcal{X}}$, say $\widetilde{\mathcal{X}}$, which is isometrically isomorphic to $\mathcal{X}$. That is, if there exists a surjective isometry $U_X \in \mathcal{G}[\widetilde{\mathcal{X}}, \mathcal{X}]$ for some dense linear manifold $\widetilde{\mathcal{X}}$ of $\widehat{\mathcal{X}}$. Let $\widehat{\mathcal{Y}}$ be a completion of a normed space $\mathcal{Y}$, so that $\mathcal{Y}$ is isometrically isomorphic to a dense linear manifold $\widetilde{\mathcal{Y}}$ of the Banach space $\widehat{\mathcal{Y}}$, and let $U_Y \in \mathcal{G}[\mathcal{Y}, \widetilde{\mathcal{Y}}]$ be the surjective isometry between $\widetilde{\mathcal{Y}}$ and $\mathcal{Y}$. Take $T \in \mathcal{B}[\mathcal{X}, \mathcal{Y}]$ and set $\widetilde{T} = U_Y T U_X \in \mathcal{B}[\widetilde{\mathcal{X}}, \widetilde{\mathcal{Y}}]$ as in the following commutative diagram.

$$
\begin{array}{ccc}
\mathcal{X} & \xrightarrow{\;T\;} & \mathcal{Y} \\[2pt]
{\scriptstyle U_X}\uparrow & & \downarrow{\scriptstyle U_Y} \\[2pt]
\widetilde{\mathcal{X}} & \xrightarrow{\;\widetilde{T}\;} & \widetilde{\mathcal{Y}}
\end{array}
$$

Define $\widehat{T} \in \mathcal{B}[\widehat{\mathcal{X}}, \widehat{\mathcal{Y}}]$ as the extension of $\widetilde{T} \in \mathcal{B}[\widetilde{\mathcal{X}}, \widehat{\mathcal{Y}}]$ over $\widehat{\mathcal{X}}$ (Theorem 4.35) so that $\|\widehat{T}\| = \|U_Y T U_X\| = \|T\|$ (see Problem 4.41). It is again usual to refer to $\widehat{T}$ as the *extension of $T$ over the completion $\widehat{\mathcal{X}}$ of $\mathcal{X}$ into the completion $\widehat{\mathcal{Y}}$ of $\mathcal{Y}$*. Moreover, since any pair of completions of $\mathcal{X}$ and $\mathcal{Y}$, say $\{\widehat{\mathcal{X}}, \widehat{\mathcal{X}}'\}$ and $\{\widehat{\mathcal{Y}}, \widehat{\mathcal{Y}}'\}$, are isometrically isomorphic (i.e., since there exist surjective isometries $\widehat{U}_X \in \mathcal{G}[\widehat{\mathcal{X}}', \widehat{\mathcal{X}}]$ and $\widehat{U}_Y \in \mathcal{G}[\widehat{\mathcal{Y}}, \widehat{\mathcal{Y}}']$), it follows (see proof of Theorem 3.51) that any two extensions $\widehat{T} \in \mathcal{B}[\widehat{\mathcal{X}}, \widehat{\mathcal{Y}}]$ and $\widehat{T}' \in \mathcal{B}[\widehat{\mathcal{X}}', \widehat{\mathcal{Y}}']$ of $T \in \mathcal{B}[\mathcal{X}, \mathcal{Y}]$ over the completions $\widehat{\mathcal{X}}$ and $\widehat{\mathcal{X}}'$ into the completions $\widehat{\mathcal{Y}}$ and $\widehat{\mathcal{Y}}'$, respectively, are unique up to isometric isomorphisms in the sense that $\widehat{T}' = \widehat{U}_Y \widehat{T} \widehat{U}_X$ (and hence $\|\widehat{T}'\| = \|\widehat{T}\| = \|T\|$). The commutative diagrams

$$\begin{array}{ccccccc}
\mathcal{Y} & \xrightarrow{U_Y} & \widetilde{\mathcal{Y}} & \subset & \widehat{\mathcal{Y}} & \xrightarrow{\widehat{U}_Y} & \widehat{\mathcal{Y}} \\
T \uparrow & & \uparrow \widetilde{T} & & \widehat{T} \uparrow & & \uparrow \widehat{T}' \\
\mathcal{X} & \xleftarrow{U_X} & \widetilde{\mathcal{X}} & \subset & \widehat{\mathcal{X}} & \xleftarrow{\widehat{U}_X} & \widehat{\mathcal{X}}
\end{array}$$

illustrate such an extension program, which is finally stated as follows.

**Theorem 4.42.** *Let the Banach spaces $\widehat{\mathcal{X}}$ and $\widehat{\mathcal{Y}}$ be completions of the normed spaces $\mathcal{X}$ and $\mathcal{Y}$, respectively. Every $T \in \mathcal{B}[\mathcal{X}, \mathcal{Y}]$ has an extension $\widehat{T} \in \mathcal{B}[\widehat{\mathcal{X}}, \widehat{\mathcal{Y}}]$ over the completion $\widehat{\mathcal{X}}$ of $\mathcal{X}$ into the completion $\widehat{\mathcal{Y}}$ of $\mathcal{Y}$. Moreover, $\widehat{T}$ is unique up to isometric isomorphisms and $\|\widehat{T}\| = \|T\|$.*

## 4.8   The Banach–Steinhaus Theorem and Operator Convergence

Let $\mathcal{X}$ and $\mathcal{Y}$ be normed spaces and let $\Theta$ be a subset of $\mathcal{B}[\mathcal{X}, \mathcal{Y}]$. We shall say that $\Theta$ is *strongly bounded* (or *pointwise bounded*) if for each $x \in \mathcal{X}$ the set $\Theta x = \{Tx \in \mathcal{Y}: T \in \Theta\}$ is bounded in $\mathcal{Y}$ (with respect to the norm topology of $\mathcal{Y}$), and *uniformly bounded* if it is itself bounded in $\mathcal{B}[\mathcal{X}, \mathcal{Y}]$ (with respect to the uniform norm topology of $\mathcal{B}[\mathcal{X}, \mathcal{Y}]$). It is clear that uniform boundedness implies strong boundedness; that is,

$$\sup_{T \in \Theta} \|T\| < \infty \quad \text{implies} \quad \sup_{T \in \Theta} \|Tx\| < \infty \ \text{ for every } x \in \mathcal{X},$$

since $\|Tx\| \le \|T\|\|x\|$ for every $x \in \mathcal{X}$. The Banach–Steinhaus Theorem ensures the converse whenever $\mathcal{X}$ is a Banach space:

$$\sup_{T \in \Theta} \|Tx\| < \infty \ \text{for every } x \text{ in a Banach space } \mathcal{X} \text{ implies} \ \sup_{T \in \Theta} \|T\| < \infty.$$

That is, *a collection $\Theta$ of bounded linear transformations of a Banach space $\mathcal{X}$ into a normed space $\mathcal{Y}$ is uniformly bounded if and only if it is strongly bounded.* This will be our second important application of the Baire Category Theorem (the first was the Open Mapping Theorem).

**Theorem 4.43.** (The Banach–Steinhaus Theorem or The Uniform Boundedness Principle). *Let $\{T_\gamma\}_{\gamma \in \Gamma}$ be an arbitrary indexed family of bounded linear transformations of a Banach space $\mathcal{X}$ into a normed space $\mathcal{Y}$. If $\sup_{\gamma \in \Gamma} \|T_\gamma x\| < \infty$ for every $x \in \mathcal{X}$, then $\sup_{\gamma \in \Gamma} \|T_\gamma\| < \infty$.*

*Proof.* For each $n \in \mathbb{N}$ set $A_n = \{x \in \mathcal{X}: \|T_\gamma x\| \le n \text{ for all } \gamma \in \Gamma\}$.

*Claim 1.*    $A_n$ is closed in $\mathcal{X}$.

*Proof.* Let $\{x_k\}$ be an $A_n$-valued sequence that converges in $\mathcal{X}$ to, say, $x \in \mathcal{X}$. Take an arbitrary $\gamma \in \Gamma$ and note that

$$
\begin{aligned}
\|T_\gamma x\| &= \|T_\gamma(x - x_k) + T_\gamma x_k\| \\
&\le \|T_\gamma(x - x_k)\| + \|T_\gamma x_k\| \;\le\; \|T_\gamma\|\|x - x_k\| + n
\end{aligned}
$$

for every integer $k$. Since $\|x - x_k\| \to 0$ as $k \to \infty$, it follows that $\|T_\gamma x\| \le n$. Thus, as $\gamma$ is an arbitrary index in $\Gamma$, $\|T_\gamma x\| \le n$ for all $\gamma \in \Gamma$, and hence $x \in A_n$. Conclusion: $A_n$ is closed in $\mathcal{X}$ by the Closed Set Theorem (Theorem 3.30). $\square$

*Claim 2.* If $\sup_{\gamma \in \Gamma}\|T_\gamma x\| < \infty$ for every $x \in \mathcal{X}$, then $\{A_n\}$ covers $\mathcal{X}$.

*Proof.* Take an arbitrary $x \in \mathcal{X}$. If $\sup_{\gamma \in \Gamma}\|T_\gamma x\| < \infty$, then there exists an integer $n_x \in \mathbb{N}$ such that $\sup_{\gamma \in \Gamma}\|T_\gamma x\| \le n_x$, and hence $x \in A_{n_x}$. Therefore, $\mathcal{X} \subseteq \bigcup_{n \in \mathbb{N}} A_n$ (i.e., $\mathcal{X} = \bigcup_{n \in \mathbb{N}} A_n$). $\square$

Then, by the Baire Category Theorem (Theorem 3.58), at least one of the sets $A_n$, say $A_{n_0}$, has a nonempty interior. This means that there exists a vector $x_0$ in $\mathcal{X}$ and a radius $\rho > 0$ such that the open ball $B_\rho(x_0)$ is included in $A_{n_0}$ (and hence $\|T_\gamma x_0\| < n_0$ for every $\gamma \in \Gamma$). Now take any nonzero vector $x$ in $\mathcal{X}$ and set $x' = \rho\|x\|^{-1}x \in \mathcal{X}$. Since $x' \in B_\rho(0)$, it follows that $x' + x_0 \in B_\rho(x_0) \subset A_{n_0}$. Thus, for every $\gamma \in \Gamma$, $\|T_\gamma(x' + x_0)\| < n_0$ and

$$
\begin{aligned}
\rho\frac{\|T_\gamma x\|}{\|x\|} = \|T_\gamma x'\| &= \|T_\gamma(x' + x_0) - T_\gamma x_0\| \\
&\le \|T_\gamma(x' + x_0)\| + \|T_\gamma x_0\| \;<\; 2n_0.
\end{aligned}
$$

Therefore $\|T_\gamma\| = \sup_{x \ne 0}\frac{\|T_\gamma x\|}{\|x\|} \le \frac{2n_0}{\rho}$ for all $\gamma \in \Gamma$. $\qquad\square$

Let $\{L_n\}$ be a sequence of linear transformations of a normed space $\mathcal{X}$ into a normed space $\mathcal{Y}$ (i.e., an $\mathcal{L}[\mathcal{X}, \mathcal{Y}]$-valued sequence). $\{L_n\}$ is *pointwise convergent* if for every $x \in \mathcal{X}$ the $\mathcal{Y}$-valued sequence $\{L_n x\}$ converges in $\mathcal{Y}$. That is, if for each $x \in \mathcal{X}$ there exists $y_x \in \mathcal{Y}$ such that $L_n x \to y_x$ in $\mathcal{Y}$ as $n \to \infty$. As the limit is unique, this actually defines a mapping $L : \mathcal{X} \to \mathcal{Y}$ given by $L(x) = y_x$ for every $x \in \mathcal{X}$. Moreover, according to Problem 4.1, it is readily verified that $L$ is linear. Therefore, *an $\mathcal{L}[\mathcal{X}, \mathcal{Y}]$-valued sequence $\{L_n\}$ is pointwise convergent if and only if there exists $L \in \mathcal{L}[\mathcal{X}, \mathcal{Y}]$ such that $\|(L_n - L)x\| \to 0$ as $n \to \infty$ for every $x \in \mathcal{X}$.* Now consider a sequence $\{T_n\}$ of bounded linear transformations of a normed space $\mathcal{X}$ into a normed space $\mathcal{Y}$ (i.e., a $\mathcal{B}[\mathcal{X}, \mathcal{Y}]$-valued sequence).

**Proposition 4.44.** *If $\mathcal{X}$ is a Banach space and $\mathcal{Y}$ is a normed space, then a $\mathcal{B}[\mathcal{X}, \mathcal{Y}]$-valued sequence $\{T_n\}$ is pointwise convergent if and only if there exists $T \in \mathcal{B}[\mathcal{X}, \mathcal{Y}]$ such that $\|(T_n - T)x\| \to 0$ for every $x \in \mathcal{X}$.*

*Proof.* According to the above italicized result (about pointwise convergence of $\mathcal{L}[\mathcal{X}, \mathcal{Y}]$-valued sequences) all that remains to be verified is that $T$ is bounded whenever $\|(T_n - T)x\| \to 0$ for every $x \in \mathcal{X}$. Since the sequence $\{T_n x\}$ converges in $\mathcal{Y}$, it follows that it is bounded in $\mathcal{Y}$ (Proposition 3.39), which means that $\sup_n \|T_n x\| < \infty$ for every $x \in \mathcal{X}$ (Problem 4.5). Hence $\sup_n \|T_n\| < \infty$ by the Banach–Steinhaus Theorem (Theorem 4.43) because $\mathcal{X}$ is a Banach space. Therefore,

$$\|Tx\| = \lim_n \|T_n x\| \le \left( \limsup_n \|T_n\| \right) \|x\| \le \left( \sup_n \|T_n\| \right) \|x\|$$

for every $x \in \mathcal{X}$ so that $T \in \mathcal{B}[\mathcal{X}, \mathcal{Y}]$ and $\|T\| \le \limsup_n \|T_n\|$. $\qquad\qquad\square$

**Definition 4.45.** Let $\mathcal{X}$ and $\mathcal{Y}$ be normed spaces and let $\{T_n\}$ be a $\mathcal{B}[\mathcal{X}, \mathcal{Y}]$-valued sequence. If $\{T_n\}$ converges in $\mathcal{B}[\mathcal{X}, \mathcal{Y}]$, that is, if there exists $T \in \mathcal{B}[\mathcal{X}, \mathcal{Y}]$ such that

$$\|T_n - T\| \to 0,$$

then we say that $\{T_n\}$ *converges uniformly* (or *converges in the uniform topology*, or *in the operator norm topology*) to $T$. In this case, $T \in \mathcal{B}[\mathcal{X}, \mathcal{Y}]$ is called the *uniform limit* of $\{T_n\}$. Notation: $T_n \overset{u}{\longrightarrow} T$ (or $T_n - T \overset{u}{\longrightarrow} O$). If $\{T_n\}$ does not converge uniformly to $T$, then we write $T_n \overset{u}{\nrightarrow} T$. If there exists $T \in \mathcal{B}[\mathcal{X}, \mathcal{Y}]$ such that

$$\|(T_n - T)x\| \to 0 \quad \text{for every } x \in \mathcal{X},$$

then we say that $\{T_n\}$ *converges strongly* (or *converges in the strong (operator) topology*) to $T$, which is called the *strong limit* of $\{T_n\}$. Notation: $T_n \overset{s}{\longrightarrow} T$ (or $T_n - T \overset{s}{\longrightarrow} O$). Again, if $\{T_n\}$ does not converge strongly to $T$, then we write $T_n \overset{s}{\nrightarrow} T$.

Uniform convergence implies strong convergence (to the same limit). Indeed, for each $n$, $0 \le \|(T_n - T)x\| \le \|T_n - T\| \|x\|$ for every $x \in \mathcal{X}$, and hence

$$T_n \overset{u}{\longrightarrow} T \quad \Longrightarrow \quad T_n \overset{s}{\longrightarrow} T.$$

The next proposition says that on a finite-dimensional normed space $\mathcal{X}$ the concepts of strong and uniform convergence coincide.

**Proposition 4.46.** *Let $\mathcal{X}$ and $\mathcal{Y}$ be normed spaces. Consider a $\mathcal{B}[\mathcal{X}, \mathcal{Y}]$-valued sequence $\{T_k\}$ and let $T$ be a transformation in $\mathcal{B}[\mathcal{X}, \mathcal{Y}]$. If $\mathcal{X}$ is finite-dimensional, then $T_k \overset{s}{\longrightarrow} T$ if and only if $T_k \overset{u}{\longrightarrow} T$.*

*Proof.* Suppose $(\mathcal{X}, \| \ \|_{\mathrm{X}})$ is a finite-dimensional normed space and let $B = \{e_i\}_{i=1}^{n}$ be a Hamel basis for the linear space $\mathcal{X}$. Take an arbitrary $x$ in $\mathcal{X}$ and consider its unique expansion on $B$,

$$x = \sum_{i=1}^{n} \xi_i e_i,$$

where the scalars $\{\xi_i\}_{i=1}^n$ are the coordinates of $x$ with respect to the basis $B$. Set

$$\|x\|_0 = \max_{1\le i\le n} |\xi_i|,$$

which defines a norm on $\mathcal{X}$ as we saw in the proof of Theorem 4.27. If $\{T_k\}$ converges strongly to $T \in \mathcal{B}[\mathcal{X}, \mathcal{Y}]$, then $\|(T_k - T)e_i\|_Y \to 0$ as $k \to \infty$ for each $e_i \in B$. Thus for each $i = 1, \dots, n$ and for every $\varepsilon > 0$ there exists a positive integer $k_{i,\varepsilon}$ such that $\|(T_k - T)e_i\|_Y < \varepsilon$ whenever $k \ge k_{i,\varepsilon}$. Therefore,

$$\|(T_k - T)x\|_Y = \left\| \sum_{i=1}^n \xi_i (T_k - T)e_i \right\|_Y \le \sum_{i=1}^n |\xi_i| \, \|(T_k - T)e_i\|_Y$$

$$\le n \max_{1\le i\le n} |\xi_i| \max_{1\le i\le n} \|(T_k - T)e_i\|_Y < n\|x\|_0 \varepsilon$$

whenever $k \ge k_\varepsilon = \max\{k_{i,\varepsilon}\}_{i=1}^n$. However, $\|\ \|_0$ and $\|\ \|_X$ are equivalent norms on $\mathcal{X}$ (Theorem 4.27) so that there exists a constant $\alpha > 0$ such that $\alpha\|x\|_0 \le \|x\|_X$. Since $n$ and $\alpha$ do not depend on $x$, it follows that

$$\|(T_k - T)x\|_Y < \tfrac{n}{\alpha} \varepsilon \|x\|_X$$

for every $x \in \mathcal{X}$ whenever $k \ge k_\varepsilon$. Hence, for every $\varepsilon > 0$ there exists a positive integer $k_\varepsilon$ such that

$$k \ge k_\varepsilon \quad \text{implies} \quad \|(T_k - T)\| = \sup_{\|x\|\le 1} \|(T_k - T)x\|_Y < \tfrac{n}{\alpha}\varepsilon,$$

and therefore $\{T_k\}$ converges uniformly to $T$. Summing up: $T_k \xrightarrow{\ s\ } T$ implies $T_k \xrightarrow{\ u\ } T$ whenever $\mathcal{X}$ is a finite-dimensional normed space. This concludes the proof once uniform convergence always implies strong convergence. $\qquad\square$

**Proposition 4.47.** *If a Cauchy sequence in $\mathcal{B}[\mathcal{X}, \mathcal{Y}]$ is strongly convergent, then it is uniformly convergent.*

*Proof.* Let $\mathcal{X}$ and $\mathcal{Y}$ be normed spaces, let $T$ be an operator in $\mathcal{B}[\mathcal{X}, \mathcal{Y}]$, and let $\{T_n\}$ be a $\mathcal{B}[\mathcal{X}, \mathcal{Y}]$-valued sequence. Thus, for each pair of indices $m$ and $n$,

$$\|(T_n - T)x\| = \|(T_n - T_m)x + (T_m - T)x\| \le \|T_n - T_m\| \, \|x\| + \|(T_m - T)x\|$$

for every $x \in \mathcal{X}$. Now take an arbitrary $\varepsilon > 0$. If $\{T_n\}$ is a Cauchy sequence in $\mathcal{B}[\mathcal{X}, \mathcal{Y}]$, then there exists a positive integer $n_\varepsilon$ such that $\|T_n - T_m\| < \tfrac{\varepsilon}{2}$, and hence

$$\|(T_n - T)x\| < \tfrac{\varepsilon}{2}\|x\| + \|(T_m - T)x\|$$

for every $x \in \mathcal{X}$, whenever $m, n \ge n_\varepsilon$. If $\{T_n\}$ converges strongly to $T$, then for each $x \in \mathcal{X}$ there exists an integer $m_{\varepsilon,x} \ge n_\varepsilon$ such that $\|(T_m - T)x\| < \tfrac{\varepsilon}{2}$ whenever $m \ge m_{\varepsilon,x}$. Hence

$$\|(T_n - T)x\| < \tfrac{\varepsilon}{2}\|x\| + \tfrac{\varepsilon}{2}$$

for every $x \in \mathcal{X}$, and so

$$\|T_n - T\| = \sup_{\|x\| \leq 1} \|(T_n - T)x\| < \varepsilon,$$

whenever $n \geq n_\varepsilon$. Conclusion: For every $\varepsilon > 0$ there exists a positive integer $n_\varepsilon$ such that

$$n \geq n_\varepsilon \quad \text{implies} \quad \|T_n - T\| < \varepsilon,$$

which means that $\{T_n\}$ converges in $\mathcal{B}[\mathcal{X}, \mathcal{Y}]$ to $T$. $\qquad\qquad$ $\square$

Outcome: If $\mathcal{X}$ and $\mathcal{Y}$ are normed spaces and $\{T_n\}$ is a $\mathcal{B}[\mathcal{X}, \mathcal{Y}]$-valued sequence, then $\{T_n\}$ *is uniformly convergent if and only if it is a strongly convergent Cauchy sequence in $\mathcal{B}[\mathcal{X}, \mathcal{Y}]$* (recall that every convergent sequence is a Cauchy sequence). If $\{T_n\}$ is strongly convergent, then it is strongly bounded (Proposition 3.39). Now suppose $\mathcal{X}$ is a Banach space. In this case Proposition 4.44 asserts that pointwise convergence coincides with strong convergence, and strong boundedness coincides with uniform boundedness (Theorem 4.43). Hence

$$T_n \xrightarrow{\;s\;} T \quad \Longrightarrow \quad \sup_n \|T_n\| < \infty \text{ whenever } \mathcal{X} \text{ is Banach.}$$

Take $T \in \mathcal{B}[\mathcal{X}]$ and consider the power sequence $\{T^n\}$ in the normed algebra $\mathcal{B}[\mathcal{X}]$. The operator $T$ is called *uniformly stable* if the power sequence $\{T^n\}$ converges uniformly to the null operator; that is, if $T^n \xrightarrow{\;u\;} O$ (equivalently, $\|T^n\| \to 0$). $T$ is *strongly stable* if the power sequence $\{T^n\}$ converges strongly to the null operator; that is, if $T^n \xrightarrow{\;s\;} O$ (equivalently, $\|T^n x\| \to 0$ for every $x \in \mathcal{X}$). It is called *power bounded* if $\{T^n\}$ is a bounded sequence in $\mathcal{B}[\mathcal{X}]$; that is, if $\sup_n \|T^n\| < \infty$. If $\mathcal{X}$ is a Banach space, then the Banach–Steinhaus Theorem ensures that $T$ is power bounded if and only if $\sup_n \|T^n x\| < \infty$ for every $x \in \mathcal{X}$. Clearly, uniform stability implies strong stability,

$$T^n \xrightarrow{\;u\;} O \quad \Longrightarrow \quad T^n \xrightarrow{\;s\;} O,$$

which in turn implies power boundedness whenever $\mathcal{X}$ is a Banach space:

$$T^n \xrightarrow{\;s\;} O \quad \Longrightarrow \quad \sup_n \|T^n\| < \infty \text{ if } \mathcal{X} \text{ is Banach.}$$

However, the converses fail.

**Example 4K.** Consider the diagonal operator $D_a \in \mathcal{B}[\ell_+^p]$ of Example 4H (for some $p \geq 1$, where $a = \{\alpha_k\}_{k=0}^\infty$ lies in $\ell_+^\infty$), and recall that $\|D_a\| = \sup_k |\alpha_k| = \|a\|_\infty$. It is readily verified by induction that the $n$th power of $D_a$, $D_a^n$, is again a diagonal operator in $\mathcal{B}[\ell_+^p]$. Indeed,

$$D_a^n x = \{\alpha_k^n \xi_k\}_{k=0}^\infty \quad \text{for every} \quad x = \{\xi_k\}_{k=0}^\infty \in \ell_+^p$$

so that $\|D_a^n\| = \sup_k |\alpha_k|^n = \|a\|_\infty^n$ for every $n \geq 0$. Hence $D_a \in \mathcal{B}[\ell_+^p]$ is uniformly stable if and only if $\|a\|_\infty < 1$; that is,

$$D_a^n \xrightarrow{u} O \quad \text{if and only if} \quad \sup_k |\alpha_k| < 1.$$

Next we shall investigate strong stability. If $\|D_a^n x\| \to 0$ for every $x \in \ell_+^p$ then, in particular (see Example 4H), $\|D_a^n e_i\| = |\alpha_i|^n \to 0$ as $n \to \infty$, and hence $|\alpha_i| < 1$, for every $i \geq 0$. On the other hand, let $x = \{\xi_k\}_{k=0}^\infty$ be an arbitrary point in $\ell_+^p$ and note that

$$
\begin{aligned}
0 \leq \|D_a^n x\|^p &= \sum_{k=0}^\infty |\alpha_k|^{np} |\xi_k|^p = \sum_{k=0}^m |\alpha_k|^{np} |\xi_k|^p + \sum_{k=m+1}^\infty |\alpha_k|^{np} |\xi_k|^p \\
&\leq \max_{0 \leq k \leq m} |\alpha_k|^{np} \sum_{k=0}^m |\xi_k|^p + \sup_{k>m} |\alpha_k|^{np} \sum_{k=m+1}^\infty |\xi_k|^p
\end{aligned}
$$

for every pair of integers $m, n \geq 0$. If $|\alpha_k| < 1$ for every $k \geq 0$, then $\sup_k |\alpha_k|^{np} \leq 1$ for all $n \geq 0$. Thus, as $\sum_{k=0}^m |\xi_k|^p \leq \|x\|^p$ for all $m \geq 0$,

$$0 \leq \|D_a^n x\|^p \leq \max_{0 \leq k \leq m} |\alpha_k|^{np} \|x\|^p + \sum_{k=m+1}^\infty |\xi_k|^p$$

for every $m, n \geq 0$. Now take an arbitrary $\varepsilon > 0$. Since $\sum_{k=m+1}^\infty |\xi_k|^p \to 0$ as $m \to \infty$ (for $\sum_{k=0}^\infty |\xi_k|^p = \|x\|^p < \infty$ — Problem 3.11), it follows that there exists a positive integer $m_\varepsilon$ such that

$$m \geq m_\varepsilon \quad \text{implies} \quad \sum_{k=m+1}^\infty |\xi_k|^p < \frac{\varepsilon^p}{2}.$$

If $|\alpha_k| < 1$ for every $k \geq 0$, then $\max\{|\alpha_k|^p\}_{k=0}^m < 1$ so that $\lim_n \max_{0 \leq k \leq m} |\alpha_k|^{np} = \lim_n (\max_{0 \leq k \leq m} |\alpha_k|^p)^n = 0$ for every nonnegative integer $m$. In particular, $\lim_n \max_{0 \leq k \leq m_\varepsilon} |\alpha_k|^{np} = 0$. Then there exists a positive integer $n_\varepsilon$ such that

$$n \geq n_\varepsilon \quad \text{implies} \quad \max_{0 \leq k \leq m_\varepsilon} |\alpha_k|^{np} < \frac{\varepsilon^p}{2}.$$

But $0 \leq \|D_a^n x\|^p \leq \max_{0 \leq k \leq m_\varepsilon} |\alpha_k|^{np} \|x\|^p + \sum_{k=m_\varepsilon+1}^\infty |\xi_k|^p$, and hence

$$n \geq n_\varepsilon \quad \text{implies} \quad 0 \leq \|D_a^n x\| < \varepsilon.$$

Therefore, if $|\alpha_k| < 1$ for every $k \geq 0$, then $\|D_a^n x\| \to 0$ for every $x \in \ell_+^p$. Conclusion: $D_a \in \mathcal{B}[\ell_+^p]$ is strongly stable if and only if $|\alpha_k| < 1$ for every $k \geq 0$; that is,

$$D_a^n \xrightarrow{s} O \quad \text{if and only if} \quad |\alpha_k| < 1 \text{ for every } k \geq 0.$$

For instance, the diagonal operator $D_a = \mathrm{diag}\big(\{\frac{k+1}{k+2}\}_{k=0}^{\infty}\big) \in \mathcal{B}[\ell_+^p]$ of Example 4H is strongly stable but not uniformly stable.

**Example 4L.** Consider the mapping $S_- : \ell_+^p \to \ell_+^p$ defined by

$$S_- = \{\xi_{k+1}\}_{k=0}^{\infty} \quad \text{for every} \quad x = \{\xi_k\}_{k=0}^{\infty} \in \ell_+^p$$

(i.e., $S_-(\xi_0, \xi_1, \xi_2, \dots) = (\xi_1, \xi_2, \xi_3, \dots)$), which is also represented as an infinite matrix

$$S_- = \begin{pmatrix} 0 & 1 & & & \\ & 0 & 1 & & \\ & & 0 & 1 & \\ & & & & \ddots \end{pmatrix},$$

where every entry above the main diagonal is equal to one and the remaining entries are all zero. This is the *backward unilateral shift* on $\ell_+^p$. It is readily verified that $S_-$ is linear and bounded so that $S_- \in \mathcal{B}[\ell_+^p]$. Actually, $\|S_- x\|^p = \sum_{k=1}^{\infty} |\xi_k|^p \leq \sum_{k=0}^{\infty} |\xi_k|^p = \|x\|^p$ for every $x = \{\xi_k\}_{k=0}^{\infty} \in \ell_+^p$ so that $S_-$ is, in fact, a contraction (i.e., $\|S_-\| \leq 1$). Consider the power sequence $\{S_-^n\}$ in $\mathcal{B}[\ell_+^p]$. A trivial induction shows that, for each $n \in \mathbb{N}_0$,

$$S_-^n x = \{\xi_{k+n}\}_{k=0}^{\infty} \quad \text{for every} \quad x = \{\xi_k\}_{k=0}^{\infty} \in \ell_+^p.$$

Hence $\|S_-^n x\|^p = \sum_{k=n}^{\infty} |\xi_k|^p \to 0$ as $n \to \infty$ for every $x = \{\xi_k\}_{k=0}^{\infty} \in \ell_+^p$ (cf. Problem 3.11) so that $S_-$ is strongly stable; that is,

$$S_-^n \xrightarrow{\ s\ } O.$$

However, $S_-$ is not uniformly stable as we shall see next. Indeed, $\|S_-^n\| \leq \|S_-\|^n \leq 1$ for every $n \geq 0$ (see Problem 4.47(a)). On the other hand, consider the $\ell_+^p$-valued sequence $\{e_i\}_{i=0}^{\infty}$ where, for each integer $i \geq 0$, $e_i$ is a scalar-valued sequence with just one nonzero entry (equal to one) at the $i$th position (i.e., $e_i = \{\delta_{ik}\}_{k=0}^{\infty} \in \ell_+^p$ for every $i \geq 0$). Note that $\|e_i\| = 1$ for every $i \geq 0$ and $S_-^n e_{n+1} = e_1$ for every $n \geq 0$. Thus $\|S_-^n\| = \sup_{\|x\|=1} \|S_-^n x\| \geq \|S_-^n e_{n+1}\| = \|e_1\| = 1$ for every nonnegative integer $n$. Therefore,

$$\|S_-^n\| = 1 \quad \text{for every} \quad n \geq 0$$

so that $S_-^n \xrightarrow{\ u\ }\!\!\!\!\!/\ \ O$. Conclusion: The power sequence

$$\{S_-^n\} \text{ does not converge uniformly.}$$

Reason: Since $S_-^n \xrightarrow{\ s\ } O$, and since uniform convergence implies strong convergence to the same limit, it follows that either $S_-^n \xrightarrow{\ u\ } O$ or $\{S_-^n\}$ does not converge uniformly.

Let $\mathcal{X}$ and $\mathcal{Y}$ be normed spaces and let $\Theta$ be a subset of $\mathcal{B}[\mathcal{X}, \mathcal{Y}]$. According to the Closed Set Theorem, $\Theta$ is closed in $\mathcal{B}[\mathcal{X}, \mathcal{Y}]$ if and only if every $\Theta$-valued sequence that converges in $\mathcal{B}[\mathcal{X}, \mathcal{Y}]$ has its limit in $\Theta$. Equivalently, every uniformly convergent sequence $\{T_n\}$ of bounded linear transformations in $\Theta \subseteq \mathcal{B}[\mathcal{X}, \mathcal{Y}]$ has its (uniform) limit $T$ in $\Theta$. In this case the set $\Theta \subseteq \mathcal{B}[\mathcal{X}, \mathcal{Y}]$ is also called *uniformly closed* (or *closed in the uniform topology of* $\mathcal{B}[\mathcal{X}, \mathcal{Y}]$). We say that a subset $\Theta$ of $\mathcal{B}[\mathcal{X}, \mathcal{Y}]$ is *strongly closed* in $\mathcal{B}[\mathcal{X}, \mathcal{Y}]$ if every $\Theta$-valued strongly convergent sequence $\{T_n\}$ has its (strong) limit $T$ in $\Theta$.

**Proposition 4.48.** *If* $\Theta \subseteq \mathcal{B}[\mathcal{X}, \mathcal{Y}]$ *is strongly closed in* $\mathcal{B}[\mathcal{X}, \mathcal{Y}]$, *then it is (uniformly) closed in* $\mathcal{B}[\mathcal{X}, \mathcal{Y}]$.

*Proof.* Take an arbitrary $\Theta$-valued uniformly convergent sequence, say $\{T_n\}$, and let $T \in \mathcal{B}[\mathcal{X}, \mathcal{Y}]$ be its (uniform) limit. Since uniform convergence implies strong convergence to the same limit, it follows that $\{T_n\}$ converges strongly to $T$. If every $\Theta$-valued strongly convergent sequence has its (strong) limit in $\Theta$, then $T \in \Theta$. Conclusion: Every $\Theta$-valued uniformly convergent sequence has its (uniform) limit in $\Theta$. $\qquad\qquad\square$

Remark: If $\mathcal{X}$ is finite-dimensional, then strong convergence coincides with uniform convergence (Proposition 4.46), and hence the concepts of strongly closed and uniformly closed in $\mathcal{B}[\mathcal{X}, \mathcal{Y}]$ coincide as well whenever $\mathcal{X}$ is a finite-dimensional normed space.

**Example 4M.** Take an arbitrary $p \geq 1$ and let $\mathcal{D}$ be the collection of all diagonal operators in $\mathcal{B}[\ell_+^p]$. That is,

$$\mathcal{D} = \big\{ D_a \in \mathcal{L}[\ell_+^p] \colon \ D_a = \operatorname{diag}(\{\alpha_k\}_{k=0}^{\infty}) \ \text{and} \ a = \{\alpha_k\}_{k=0}^{\infty} \in \ell_+^{\infty} \big\}$$

(see Example 4H). Set

$$\mathcal{D}_\infty = \big\{ D_a \in \mathcal{D} \colon \ a \in \ell_+^{c_o} \big\},$$

the collection of all diagonal operators $D_a = \operatorname{diag}(\{\alpha_k\}_{k=0}^{\infty})$ in $\mathcal{B}[\ell_+^p]$ such that the scalar-valued sequence $a = \{\alpha_k\}_{k=0}^{\infty}$ converges to zero. As a matter of fact, both $\mathcal{D}$ and $\mathcal{D}_\infty$ are subalgebras of the Banach algebra $\mathcal{B}[\ell_+^p]$. Let $\{D_{a_n}\}_{n\geq 1}$ be an arbitrary $\mathcal{D}_\infty$-valued uniformly convergent sequence. Hence $D_{a_n} \overset{u}{\longrightarrow} D \in \mathcal{B}[\ell_+^p]$ with each $D_{a_n} = \operatorname{diag}(\{\alpha_k(n)\}_{k=0}^{\infty})$ in $\mathcal{D}_\infty$ so that each $a_n = \{\alpha_k(n)\}_{k=0}^{\infty}$ lies in $\ell_+^{c_o}$. This implies that $D_{a_n} \overset{s}{\longrightarrow} D$ and, according to Problem 4.51, $D$ is a diagonal operator; that is, $D = D_a = \operatorname{diag}(\{\alpha_k\}_{k=0}^{\infty}) \in \mathcal{D}$ for some $a = \{\alpha_k\}_{k=0}^{\infty} \in \ell_+^{\infty}$. Thus (see Example 4H)

$$\|D_{a_n} - D_a\| = \|a_n - a\|_\infty = \sup_k |\alpha_k(n) - \alpha_k|.$$

Now take an arbitrary $\varepsilon > 0$. Since $\sup_k |\alpha_k(n) - \alpha_k| \to 0$ as $n \to \infty$ (because $\|D_{a_n} - D_a\| \to 0$), there exists a positive integer $n_\varepsilon$ such that

$$\sup_k |\alpha_k(n_\varepsilon) - \alpha_k| < \varepsilon.$$

Moreover, since $\alpha_{k(n_\varepsilon)} \to 0$ as $k \to \infty$ (for $a_{n_\varepsilon} \in \ell_+^{c_o}$), there exists a positive integer $k_\varepsilon$ such that

$$|\alpha_{k(n_\varepsilon)}| < \varepsilon \quad \text{whenever} \quad k \geq k_\varepsilon.$$

Finally recall that $\big||\alpha_k| - |\alpha_{k(n_\varepsilon)}|\big| \leq |\alpha_k - \alpha_{k(n_\varepsilon)}|$, and hence

$$|\alpha_k| \leq \sup_k |\alpha_k - \alpha_{k(n_\varepsilon)}| + |\alpha_{k(n_\varepsilon)}|$$

for every $k$. Therefore,

$$k \geq k_\varepsilon \quad \text{implies} \quad |\alpha_k| < 2\varepsilon$$

so that $\alpha_k \to 0$. Thus $a = \{\alpha_k\}_{k=0}^\infty \in \ell_+^{c_o}$ and so $D_a \in \mathcal{D}_\infty$. Conclusion:

$$\mathcal{D}_\infty \text{ is closed in } \mathcal{B}[\ell_+^p].$$

Next consider a $\mathcal{D}_\infty$-valued sequence $\{D_n\}_{n \geq 1}$, where each $D_n$ is a diagonal operator whose only nonzero entries are the first $n$ entries in the main diagonal, which are all equal to one:

$$D_n = \mathrm{diag}(1, \dots, 1, 0, 0, 0, \dots) \in \mathcal{D}_\infty.$$

Observe that $\|D_n x - x\|^p = \sum_{k=n+1}^\infty |\xi_k|^p \to 0$ as $n \to \infty$ for every $x = \{\xi_k\}_{k=0}^\infty$ (because $\sum_{k=0}^\infty |\xi_k|^k = \|x\| < \infty$ — Problem 3.11). Hence

$$D_n \xrightarrow{\ s\ } I$$

but the identity is a diagonal operator that does not lie in $\mathcal{D}_\infty$ (i.e., $I \in \mathcal{D} \backslash \mathcal{D}_\infty$). Outcome:

$$\mathcal{D}_\infty \text{ is not strongly closed in } \mathcal{B}[\ell_+^p].$$

## 4.9   Compact Operators

Let $\mathcal{X}$ and $\mathcal{Y}$ be normed spaces. A linear transformation $T \in \mathcal{L}[\mathcal{X}, \mathcal{Y}]$ is *compact* (or *completely continuous*) if it maps bounded subsets of $\mathcal{X}$ into relatively compact subsets of $\mathcal{Y}$. That is, if $T(A)^-$ is compact in $\mathcal{Y}$ whenever $A$ is bounded in $\mathcal{X}$. Equivalently, $T \in \mathcal{L}[\mathcal{X}, \mathcal{Y}]$ is compact if $T(A)$ lies in a compact subset of $\mathcal{Y}$ whenever $A$ is bounded in $\mathcal{X}$ (see Theorem 3.62).

**Theorem 4.49.** *If* $T \in \mathcal{L}[\mathcal{X}, \mathcal{Y}]$ *is compact, then* $T \in \mathcal{B}[\mathcal{X}, \mathcal{Y}]$.

*Proof.* Take an arbitrary bounded subset $A$ of $\mathcal{X}$. If $T \in \mathcal{L}[\mathcal{X}, \mathcal{Y}]$ is compact, then $T(A)^-$ is a compact subset of $\mathcal{Y}$. Thus $T(A)^-$ is totally bounded in $\mathcal{Y}$ (Corollary 3.81), and so $T(A)^-$ is bounded in $\mathcal{Y}$ (Corollary 3.71). Hence $T(A)$ is a bounded subset of $\mathcal{Y}$, and therefore the linear transformation $T$ is bounded by Proposition 4.12. $\qquad\qquad\square$

In other words, *every compact linear transformation is continuous*. The converse is clearly false, for the identity $I$ of an infinite-dimensional normed space $\mathcal{X}$ into itself is not compact by Corollary 4.34. Recall that $T \in \mathcal{L}[\mathcal{X}, \mathcal{Y}]$ is of *finite rank* (or a *finite-dimensional linear transformation*) if it has a finite-dimensional range (see Problem 2.18).

**Proposition 4.50.** *Let $\mathcal{X}$ and $\mathcal{Y}$ be normed spaces. If $T \in \mathcal{B}[\mathcal{X}, \mathcal{Y}]$ is of finite rank, then it is compact.*

*Proof.* Take any bounded subset $A$ of $\mathcal{X}$. If $T \in \mathcal{B}[\mathcal{X}, \mathcal{Y}]$, then $T(A)$ is a bounded subset of $\mathcal{Y}$ (Proposition 4.12). Thus $T(A)^-$ is closed and bounded in $\mathcal{Y}$. If $T$ is of finite rank, then the range $\mathcal{R}(T)$ of $T$ is a finite-dimensional subspace of $\mathcal{Y}$ (Corollary 4.29), and hence $\mathcal{R}(T)$ is a closed subset of $\mathcal{Y}$. Thus $T(A)^-$ is closed and bounded in $\mathcal{R}(T)$, according to Problem 3.38(d), and therefore a compact subset of $\mathcal{R}(T)$ by Corollary 4.30. Since the metric space $\mathcal{R}(T)$ is a subspace of the metric space $\mathcal{Y}$, it follows that $T(A)^-$ is compact in $\mathcal{Y}$. $\qquad\square$

The assumption "$T$ is bounded" cannot be removed from the statement of Proposition 4.50. Actually, we have already exhibited an unbounded linear transformation $L$ of an arbitrary infinite-dimensional normed space $\mathcal{X}$ into an arbitrary nonzero normed space $\mathcal{Y}$. If $\dim \mathcal{Y} = 1$, then $L \in \mathcal{L}[\mathcal{X}, \mathcal{Y}]$ is a rank-one linear transformation that is not even bounded (see proof of Corollary 4.30, part (b)).

**Corollary 4.51.** *If $\mathcal{X}$ is a finite-dimensional normed space and $\mathcal{Y}$ is any normed space, then every linear transformation $T \in \mathcal{L}[\mathcal{X}, \mathcal{Y}]$ is of finite rank and compact.*

*Proof.* If $\mathcal{X}$ is finite-dimensional, then $T \in \mathcal{B}[\mathcal{X}, \mathcal{Y}]$ (Corollary 4.30) and $\dim \mathcal{R}(T) < \infty$ (Problems 2.17 and 2.18). Thus $T$ is bounded and of finite rank, and hence compact by the preceding proposition. $\qquad\square$

**Theorem 4.52.** *Let $T$ be a linear transformation of a normed space $\mathcal{X}$ into a normed space $\mathcal{Y}$. The following four assertions are pairwise equivalent.*

(a)  *$T$ is compact (i.e., $T$ maps bounded sets into relatively compact sets).*

(b)  *$T$ maps the unit ball $B_1[0]$ into a relatively compact set.*

(c)  *$T$ maps every $B_1[0]$-valued sequence into a sequence that has a convergent subsequence.*

(d)  *$T$ maps bounded sequences into sequences that have a convergent subsequence.*

*Moreover, each of the above equivalent assertions implies that*

(e)  *$T$ maps bounded sets into totally bounded sets,*

*which in turn implies that*

(f)   *T maps the unit ball $B_1[0]$ into a totally bounded set.*

*Furthermore, if $\mathcal{Y}$ is a Banach space, then these six assertions are all pairwise equivalent.*

*Proof.* First note that (a)$\Rightarrow$(b) trivially. Hence, in order to verify that (a), (b), (c) and (d) are pairwise equivalent, it is enough to show that (b)$\Rightarrow$(c) $\Rightarrow$ (d)$\Rightarrow$(a).

*Proof of* (b)$\Rightarrow$(c). Take an arbitrary $B_1[0]$-valued sequence $\{x_n\}$. If (b) holds, then $\{Tx_n\}$ lies in a compact subset of $\mathcal{Y}$ or, equivalently, in a sequentially compact subset of $\mathcal{Y}$ (Corollary 3.81). Thus, according to Definition 3.76, the $\mathcal{Y}$-valued sequence $\{Tx_n\}$ has a convergent subsequence. Therefore (b) implies (c).

*Proof of* (c)$\Rightarrow$(d). Take an arbitrary $\mathcal{X}$-valued bounded sequence $\{x_n\}$ so that there exists a nonnegative real number $\beta \geq \sup_n \|x_x\|$ for which $\{\beta^{-1}x_n\}$ is a $B_1[0]$-valued sequence. If assertion (c) holds, then $\{T(\beta^{-1}x_n)\} = \beta^{-1}\{Tx_n\}$ has a convergent subsequence, and so $\{Tx_n\}$ has a convergent subsequence. Thus (c) implies (d).

*Proof of* (d)$\Rightarrow$(a). If (d) holds, then the image of every bounded subset of $\mathcal{X}$ contains a subsequence that converges in $\mathcal{Y}$. Then the closure $T(A)^-$ of the image $T(A)$ of any bounded subset $A$ of $\mathcal{X}$ contains a convergent subsequence by Theorem 3.30, which means that $T(A)^-$ is sequentially compact (Definition 3.76). Therefore (d) implies (a) by Corollary 3.81.

Moreover, observe that (a) implies (e). Indeed, if $T(A)^-$ is a compact subset of $\mathcal{Y}$ whenever $A$ is a bounded subset of $\mathcal{X}$, then $T(A)^-$ (and hence $T(A)$) is totally bounded in $\mathcal{Y}$ whenever $A$ is bounded in $\mathcal{X}$, according to Corollary 3.81. Also note that (e) trivially implies (f). Conversely, suppose (f) holds true so that $T(B_1[0])$ is a totally bounded subset of $\mathcal{Y}$. If $\mathcal{Y}$ is a Banach space, then Corollary 3.84(b) ensures that $T(B_1[0])$ is relatively compact in $\mathcal{Y}$. Therefore (f) implies (b) whenever $\mathcal{Y}$ is a Banach space. $\qquad\qquad\square$

We shall denote the collection of all compact linear transformation of a normed space $\mathcal{X}$ into a normed space $\mathcal{Y}$ by $\mathcal{B}_\infty[\mathcal{X}, \mathcal{Y}]$. Recall that $\mathcal{B}_\infty[\mathcal{X}, \mathcal{Y}] \subseteq \mathcal{B}[\mathcal{X}, \mathcal{Y}]$ by Theorem 4.49 (and $\mathcal{B}_\infty[\mathcal{X}, \mathcal{Y}] = \mathcal{B}[\mathcal{X}, \mathcal{Y}]$ whenever $\mathcal{X}$ is finite-dimensional by Corollary 4.51). Accordingly, we shall write $\mathcal{B}_\infty[\mathcal{X}]$ for $\mathcal{B}_\infty[\mathcal{X}, \mathcal{X}]$: the collection of all compact operators on a normed space $\mathcal{X}$.

**Theorem 4.53.** *Let $\mathcal{X}$ and $\mathcal{Y}$ be normed spaces.*

(a)   $\mathcal{B}_\infty[\mathcal{X}, \mathcal{Y}]$ *is a linear manifold of $\mathcal{B}[\mathcal{X}, \mathcal{Y}]$.*

(b)   *If $\mathcal{Y}$ is a Banach space, then $\mathcal{B}_\infty[\mathcal{X}, \mathcal{Y}]$ is a subspace of $\mathcal{B}[\mathcal{X}, \mathcal{Y}]$.*

*Proof.* It is trivially verified that $\alpha T \in \mathcal{B}_\infty[\mathcal{X}, \mathcal{Y}]$ for every scalar $\alpha$ whenever $T \in \mathcal{B}_\infty[\mathcal{X}, \mathcal{Y}]$ (Theorem 4.52(d)) . In order to verify that $S + T \in \mathcal{B}_\infty[\mathcal{X}, \mathcal{Y}]$ for every $S, T \in \mathcal{B}_\infty[\mathcal{X}, \mathcal{Y}]$ proceed as follows. Take $S$ and $T$ in $\mathcal{B}_\infty[\mathcal{X}, \mathcal{Y}] \subseteq$

$\mathcal{B}[\mathcal{X}, \mathcal{Y}]$, and let $\{x_n\}_{n\geq 1}$ be an arbitrary $\mathcal{X}$-valued bounded sequence. Theorem 4.52(d) ensures that there exists a subsequence of $\{Tx_n\}_{n\geq 1}$, say $\{Tx_{n_k}\}_{k\geq 1}$, that converges in $\mathcal{Y}$. Now consider the subsequence $\{x_{n_k}\}_{k\geq 1}$ of $\{x_n\}_{n\geq 1}$, which is clearly bounded. Then (Theorem 4.52(d) again) the sequence $\{Sx_{n_k}\}_{k\geq 1}$ has a subsequence, say $\{Sx_{n_{k_j}}\}_{j\geq 1}$, that converges in $\mathcal{Y}$. Since $\{Tx_{n_{k_j}}\}_{j\geq 1}$ is a subsequence of the convergent sequence $\{Tx_{n_k}\}_{k\geq 1}$, it follows that $\{Tx_{n_{k_j}}\}_{j\geq 1}$ also converges in $\mathcal{Y}$ (Proposition 3.5). Therefore $\{(S+T)x_{n_{k_j}}\}_{j\geq 1} = \{Sx_{n_{k_j}}\}_{j\geq 1} + \{Tx_{n_{k_j}}\}_{j\geq 1}$ is a convergent subsequence of $\{(S+T)x_n\}_{n\geq 1}$, and hence $S+T \in \mathcal{B}[\mathcal{X}, \mathcal{Y}]$ is compact by Theorem 4.52(d). Thus $S + T \in \mathcal{B}_\infty[\mathcal{X}, \mathcal{Y}]$. Conclusion: $\mathcal{B}_\infty[\mathcal{X}, \mathcal{Y}]$ is a linear manifold of $\mathcal{B}[\mathcal{X}, \mathcal{Y}]$.

*Claim.* If $\mathcal{Y}$ is a Banach space, then $\mathcal{B}_\infty[\mathcal{X}, \mathcal{Y}]$ is closed in $\mathcal{B}[\mathcal{X}, \mathcal{Y}]$.

*Proof.* Take an arbitrary $\mathcal{B}_\infty[\mathcal{X}, \mathcal{Y}]$-valued sequence $\{T_n\}$ that converges (uniformly) in $\mathcal{B}[\mathcal{X}, \mathcal{Y}]$ to, say, $T \in \mathcal{B}[\mathcal{X}, \mathcal{Y}]$. Thus for each $\varepsilon > 0$ there exists a positive integer $n_\varepsilon$ such that

$$\|(T - T_{n_\varepsilon})x\| \leq \|T - T_{n_\varepsilon}\|\|x\| < \tfrac{\varepsilon}{2}\|x\|$$

for every $x \in \mathcal{X}$. Since $T_{n_\varepsilon}$ is compact, it follows by Theorem 4.52(f) that the image $T_{n_\varepsilon}(B_1[0])$ of the unit ball $B_1[0]$ is totally bounded in $\mathcal{Y}$, and hence $T_{n_\varepsilon}(B_1[0])$ has a finite $\tfrac{\varepsilon}{2}$-net, say $Y_\varepsilon$ (Definition 3.68). Therefore, for each $x \in B_1[0]$ there exists $y \in Y_\varepsilon$ such that $\|T_{n_\varepsilon}x - y\| < \tfrac{\varepsilon}{2}$, and so

$$\begin{aligned}
\|Tx - y\| &= \|Tx - T_{n_\varepsilon}x + T_{n_\varepsilon}x - y\| \\
&\leq \|(T - T_{n_\varepsilon})x\| + \|T_{n_\varepsilon}x - y\| < \tfrac{\varepsilon}{2}\|x\| + \tfrac{\varepsilon}{2} < \varepsilon.
\end{aligned}$$

That is, $Y_\varepsilon$ is a finite $\tfrac{\varepsilon}{2}$-net for $T(B_1[0])$, which means that $T(B_1[0])$ is totally bounded in $\mathcal{Y}$. Thus, if $\mathcal{Y}$ is a Banach space, then $T$ is compact by Theorem 4.52. Conclusion: Every $\mathcal{B}_\infty[\mathcal{X}, \mathcal{Y}]$-valued sequence that converges in $\mathcal{B}[\mathcal{X}, \mathcal{Y}]$ has its limit in $\mathcal{B}_\infty[\mathcal{X}, \mathcal{Y}]$. Hence $\mathcal{B}_\infty[\mathcal{X}, \mathcal{Y}]$ is closed in $\mathcal{B}[\mathcal{X}, \mathcal{Y}]$ by the Closed Set Theorem (Theorem 3.30). $\square$

Outcome: If $\mathcal{Y}$ is a Banach space, then $\mathcal{B}_\infty[\mathcal{X}, \mathcal{Y}]$ is a subspace of $\mathcal{B}[\mathcal{X}, \mathcal{Y}]$ (which are Banach spaces by Propositions 4.15 and 4.7). $\qquad\square$

Recall that a *two-sided ideal* $\mathcal{I}$ of an algebra $\mathcal{A}$ is a subalgebra of $\mathcal{A}$ such that the product (both left and right products) of every element of $\mathcal{I}$ with any element of $\mathcal{A}$ is again an element of $\mathcal{I}$ (see Problem 2.30).

**Proposition 4.54.** *If $\mathcal{X}$ is a normed space, then $\mathcal{B}_\infty[\mathcal{X}]$ is a two-sided ideal of the normed algebra $\mathcal{B}[\mathcal{X}]$.*

*Proof.* $\mathcal{B}_\infty[\mathcal{X}]$ is a linear manifold of $\mathcal{B}[\mathcal{X}]$ by Theorem 4.53. Take $S \in \mathcal{B}_\infty[\mathcal{X}]$ and $T \in \mathcal{B}[\mathcal{X}]$ arbitrary.

*Claim.* Both $ST$ and $TS$ lie in $\mathcal{B}_\infty[\mathcal{X}]$.

*Proof.* Let $A$ be any bounded subset of $\mathcal{X}$ so that $T(A)$ is bounded by Proposition 4.12. Since $S$ is compact, it follows (by definition) that $S(T(A))^-$ is compact. Thus the composition $ST$ maps bounded sets into relatively compact sets, which means that $ST$ is compact. Moreover, $S(A)^-$ is compact as well. Since $T$ is continuous, it follows that $T(S(A)^-)$ is compact by Theorem 3.64 (and hence $T(S(A)^-)$ is closed — Theorem 3.62), and also that $T(S(A))^- = T(S(A)^-)^- = T(S(A)^-)$ according to Problem 3.46. Therefore $T(S(A))^-$ is compact. Thus the composition $TS$ maps bounded sets into relatively compact sets, which means that $TS$ is compact. $\square$

Conclusion: $\mathcal{B}_\infty[\mathcal{X}]$ is a two-sided ideal of $\mathcal{B}[\mathcal{X}]$.    $\square$

Let $\mathcal{B}_0[\mathcal{X}, \mathcal{Y}]$ denote the collection of all finite-rank bounded linear transformations of a normed space $\mathcal{X}$ into a normed space $\mathcal{Y}$. By Proposition 4.50 it follows that $\mathcal{B}_0[\mathcal{X}, \mathcal{Y}] \subseteq \mathcal{B}_\infty[\mathcal{X}, \mathcal{Y}]$ (and $\mathcal{B}_0[\mathcal{X}, \mathcal{Y}] = \mathcal{B}_\infty[\mathcal{X}, \mathcal{Y}] = \mathcal{B}[\mathcal{X}, \mathcal{Y}] = \mathcal{L}[\mathcal{X}, \mathcal{Y}]$ whenever $\mathcal{X}$ is finite dimensional — Corollary 4.51). We shall write $\mathcal{B}_0[\mathcal{X}]$ for $\mathcal{B}_0[\mathcal{X}, \mathcal{X}]$, the collection of all finite-rank operators on $\mathcal{X}$. It is readily verified that both $ST$ and $TS$ lie in $\mathcal{B}_0[\mathcal{X}]$ for every $S \in \mathcal{B}_0[\mathcal{X}]$ and every $T \in \mathcal{B}[\mathcal{X}]$. Indeed, it is clear that $ST$ is of finite rank (for the range of $ST$ is trivially included in the range of $S$). Moreover, $TS$ is of finite rank because $TS = T|_{\mathcal{R}(S)}S$ and the domain of $T|_{\mathcal{R}(S)}$ is the finite-dimensional range of $S$ (and so its own range is finite-dimensional as well — see Problem 2.17). Therefore $\mathcal{B}_0[\mathcal{X}]$ *is also a two-sided ideal of* $\mathcal{B}[\mathcal{X}]$.

**Corollary 4.55.** *Let $\mathcal{X}$ and $\mathcal{Y}$ be normed spaces. If $\mathcal{Y}$ is a Banach space, then every $\mathcal{B}_0[\mathcal{X}, \mathcal{Y}]$-valued sequence that converges (uniformly) in $\mathcal{B}[\mathcal{X}, \mathcal{Y}]$ has its limit in $\mathcal{B}_\infty[\mathcal{X}, \mathcal{Y}]$.*

*Proof.* This is a straightforward application of Theorem 4.53(b) for $\mathcal{B}_0[\mathcal{X}, \mathcal{Y}] \subseteq \mathcal{B}_\infty[\mathcal{X}, \mathcal{Y}]$. Indeed, since $\mathcal{B}_\infty[\mathcal{X}, \mathcal{Y}]$ is closed in $\mathcal{B}[\mathcal{X}, \mathcal{Y}]$, the Closed Set Theorem ensures that every $\mathcal{B}_\infty[\mathcal{X}, \mathcal{Y}]$-valued sequence (and hence every $\mathcal{B}_0[\mathcal{X}, \mathcal{Y}]$-valued sequence) that converges in $\mathcal{B}[\mathcal{X}, \mathcal{Y}]$ has its limit in $\mathcal{B}_\infty[\mathcal{X}, \mathcal{Y}]$.    $\square$

**Example 4N.** Consider the diagonal operator $D_a \in \mathcal{B}[\ell_+^p]$ of Example 4K (for some $p \geq 1$, where $a = \{\alpha_k\}_{k=0}^\infty$ lies in $\ell_+^\infty$). We shall show that

$$D_a \text{ is compact if and only if } \alpha_k \to 0 \text{ as } k \to \infty.$$

Consider the $\ell_+^p$-valued sequence $\{e_i\}_{i=0}^\infty$ with $e_i = \{\delta_{ik}\}_{k=0}^\infty \in \ell_+^p$ for each $i \geq 0$ (just one nonzero entry equal to one at the $i$th position) as in Example 4H.

(a) For each nonnegative integer $n$ set $D_{a_n} = \mathrm{diag}(\alpha_0, \dots, \alpha_n, 0, 0, 0, \dots)$ in $\mathcal{B}[\ell_+^p]$. It is readily verified that each $D_{a_n}$ is of finite rank (proof: $y \in \mathcal{R}(D_{a_n})$ if and only if $y = \sum_{i=0}^n \alpha_i \xi_i e_i$ for some $x = \{\xi_i\}_{i=0}^\infty \in \ell_+^p$, so that $\mathcal{R}(D_{a_n}) \subseteq$ span $\{e_i\}_{i=0}^n$, and hence $\dim \mathcal{R}(D_{a_n}) \leq n$ by Theorem 2.6). Moreover (Problem 4H again), $\|D_{a_n} - D_a\| = \sup_{k \geq n+1} |\alpha_k|$ for every $n \geq 0$. If $\lim_n |\alpha_n| = 0$, then $\limsup_n |\alpha_n| = \lim_n \sup_{k \geq n} |\alpha_k| = 0$ (Problem 3.13), and hence $\lim_n \|D_{a_n} - D_a\| = 0$. Thus $\alpha_n \to 0$ implies $D_{a_n} \xrightarrow{u} D_a$, which in turn implies that $D_a$ is compact by Corollary 4.55.

(b) Conversely, if $\alpha_i \not\to 0$, then $\{\alpha_i\}_{i=0}^{\infty}$ has a subsequence, say $\{\alpha_{i_n}\}_{n=0}^{\infty}$, such that $\inf_n |\alpha_{i_n}| > 0$. Set $\varepsilon = \inf_n |\alpha_{i_n}| > 0$, and note that $\|D_a e_{i_m} - D_a e_{i_n}\|^p = \|\alpha_{i_m} e_{i_m} - \alpha_{i_n} e_{i_n}\|^p = |\alpha_{i_m}|^p + |\alpha_{i_n}|^p \geq 2\varepsilon^p$ whenever $m \neq n$. Then every subsequence of $\{D_a e_{i_n}\}_{n=0}^{\infty}$ is not a Cauchy sequence, and hence does not converge in $\ell_+^p$. Therefore, the bounded sequence $\{e_{i_n}\}_{n=0}^{\infty}$ is such that its image under $D_a$, $\{D_a e_{i_n}\}_{n=0}^{\infty}$, has no convergent subsequence. Hence $D_a$ is not compact by Theorem 4.52(c). Conclusion: If $D_a$ is compact, then $\alpha_i \to 0$ as $i \to \infty$.

**Example 4O.** According to Theorem 4.53 and Corollary 4.55, it follows that

$$\mathcal{B}_0[\mathcal{X}, \mathcal{Y}] \text{ is closed in } \mathcal{B}_\infty[\mathcal{X}, \mathcal{Y}],$$

$$\mathcal{B}_\infty[\mathcal{X}, \mathcal{Y}] \text{ is closed in } \mathcal{B}[\mathcal{X}, \mathcal{Y}],$$

and hence

$$\mathcal{B}_0[\mathcal{X}, \mathcal{Y}] \text{ is closed in } \mathcal{B}[\mathcal{X}, \mathcal{Y}],$$

whenever $\mathcal{Y}$ is a Banach space. Now consider the setup of Examples 4M and 4N, and note that $\mathcal{D}_\infty = \{D_a \in \mathcal{D} : a \in \ell_+^{c_o}\}$ is precisely the collection of all compact diagonal operators on the Banach space $\ell_+^p$. It was shown in Example 4M that $\mathcal{D}_\infty$ is not strongly closed in $\mathcal{B}[\ell_+^p]$ by exhibiting a sequence $\{D_n\}$ of finite-rank diagonal operators (and hence a sequence of compact diagonal operators) that converges strongly to the identity $I$ on $\ell_+^p$, which is not even compact. Conclusion:

$$\mathcal{B}_0[\ell_+^p] \text{ is not strongly closed in } \mathcal{B}_\infty[\ell_+^p],$$

which implies that

$$\mathcal{B}_0[\ell_+^p] \text{ and } \mathcal{B}_\infty[\ell_+^p] \text{ are not strongly closed in } \mathcal{B}[\ell_+^p].$$

It may be tempting to think that the converse of Corollary 4.55 holds. It in fact does hold whenever the Banach space $\mathcal{Y}$ has a Schauder basis (Problem 4.11), and it also holds whenever $\mathcal{Y}$ is a Hilbert space (next Chapter). In these cases, *every $T$ in $\mathcal{B}_\infty[\mathcal{X}, \mathcal{Y}]$ is the uniform limit of a $\mathcal{B}_0[\mathcal{X}, \mathcal{Y}]$-valued sequence.* But such an italicized result fails in general (see Problem 4.58). However, every compact linear transformation comes close to having a finite-dimensional range in the following sense.

**Proposition 4.56.** *Let $\mathcal{X}$ and $\mathcal{Y}$ be normed spaces and take $T \in \mathcal{L}[\mathcal{X}, \mathcal{Y}]$. If $T$ is compact, then for each $\varepsilon > 0$ there exists a finite-dimensional subspace $\mathcal{R}_\varepsilon$ of the range $\mathcal{R}(T)$ of $T$ such that*

$$d(Tx, \mathcal{R}_\varepsilon) \leq \varepsilon \|x\| \quad \text{for every} \quad x \in \mathcal{X}.$$

*Proof.* Take an arbitrary $\varepsilon > 0$. If $T \in \mathcal{L}[\mathcal{X}, \mathcal{Y}]$ is compact, then the image $T(B_1[0])$ of the closed unit ball $B_1[0]$ is totally bounded in the normed space $\mathcal{R}(T)$ by Theorem

4.52(b). Thus there exists a finite $\varepsilon$-net for $T(B_1[0])$, say $\{v_i\}_{i=1}^{n_\varepsilon} \subset \mathcal{R}(T)$. That is, for every $y \in T(B_1[0])$ there exists $v_y \in \{v_i\}_{i=1}^{n_\varepsilon}$ such that $\|y - v_y\| < \varepsilon$. Set $\mathcal{R}_\varepsilon = \text{span}\,\{v_i\}_{i=1}^{n_\varepsilon} \subseteq \mathcal{R}(T)$. $\mathcal{R}_\varepsilon$ is a finite-dimensional subspace of $\mathcal{R}(T)$ (Theorem 2.6, Proposition 4.7 and Corollary 4.28), and

$$
\begin{aligned}
\frac{1}{\|x\|} d(Tx, \mathcal{R}_\varepsilon) \;&=\; \frac{1}{\|x\|} \inf_{u \in \mathcal{R}_\varepsilon} \|Tx - u\| \;=\; \inf_{u \in \mathcal{R}_\varepsilon} \left\| \frac{Tx}{\|x\|} - \frac{u}{\|x\|} \right\| \\[2mm]
&=\; \inf_{v \in \mathcal{R}_\varepsilon} \left\| T\!\left(\tfrac{x}{\|x\|}\right) - v \right\| \;\le\; \inf_{v \in \{v_i\}_{i=1}^{n_\varepsilon}} \left\| T\!\left(\tfrac{x}{\|x\|}\right) - v \right\| \;<\; \varepsilon
\end{aligned}
$$

for every nonzero $x$ in $\mathcal{X}$, which concludes the proof (for $0 \in \mathcal{R}_\varepsilon$). $\qquad\square$

**Proposition 4.57.** *The range of a compact linear transformation is separable.*

*Proof.* Let $\mathcal{X}$ be a normed space. Consider the collection $\{B_n(0)\}_{n \in \mathbb{N}}$ of all open balls with center at the origin of $\mathcal{X}$ and radius $n$, which clearly covers $\mathcal{X}$. That is

$$
\mathcal{X} = \bigcup_n B_n(0).
$$

Let $\mathcal{Y}$ be a normed space and let $T \colon \mathcal{X} \to \mathcal{Y}$ be any mapping of $\mathcal{X}$ into $\mathcal{Y}$ so that (see Problem 1.2(e))

$$
\mathcal{R}(T) = T(\mathcal{X}) = T\!\left(\bigcup_n B_n(0)\right) = \bigcup_n T(B_n(0)).
$$

If $T \in \mathcal{B}_\infty[\mathcal{X}, \mathcal{Y}]$, then each $T(B_n(0))$ is separable (Theorem 4.52 and Proposition 3.72), and hence each $T(B_n(0))$ has a countable dense subset. That is, for each $n \in \mathbb{N}$ there exists a countable set $A_n \subseteq T(B_n(0))$ such that $A_n^- = T(B_n(0))^-$ (cf. Problem 3.38(g)). Therefore,

$$
\bigcup_n A_n \subseteq \bigcup_n T(B_n(0)) \subseteq \bigcup_n A_n^- \subseteq \left(\bigcup_n A_n\right)^-
$$

(see Section 3.5). Thus $\left(\bigcup_n A_n\right)^- = \mathcal{R}(T)^-$ so that $\bigcup_n A_n$ is dense in $\mathcal{R}(T)$ (cf. Problem 3.38(g) again). Moreover, since each $A_n$ is countable, it follows by Corollary 1.11 that the countable union $\bigcup_n A_n$ is countable as well. Outcome: $\bigcup_n A_n$ is a countable dense subset of $\mathcal{R}(T)$, which means that $\mathcal{R}(T)$ is separable. $\qquad\square$

If $T \colon \mathcal{X} \to \mathcal{Y}$ is a compact linear transformation of a normed space $\mathcal{X}$ into a normed space $\mathcal{Y}$, then the restriction $T|_\mathcal{M} \colon \mathcal{M} \to \mathcal{Y}$ of $T$ to a linear manifold $\mathcal{M}$ of $\mathcal{X}$ is a linear transformation (Problem 2.14). Moreover, it is clear by the very definition of compact linear transformations that $T|_\mathcal{M}$ is compact as well: *the restriction of a compact linear transformation $T \colon \mathcal{X} \to \mathcal{Y}$ to a linear manifold of $\mathcal{X}$ is again a compact linear transformation* (i.e., $T|_\mathcal{M}$ lies in $\mathcal{B}_\infty[\mathcal{M}, \mathcal{Y}]$ whenever $T$ lies in $\mathcal{B}_\infty[\mathcal{X}, \mathcal{Y}]$). On the other hand, if $\mathcal{M}$ is a linear manifold of an infinite-dimensional Banach space $\mathcal{X}$, and $T \colon \mathcal{M} \to \mathcal{Y}$ is a compact linear transformation of $\mathcal{M}$ into a Banach space $\mathcal{Y}$, then it is easy to show that an arbitrary bounded linear extension of $T$ over $\mathcal{X}$ may not be compact (see Problem 4.60). However, if $\mathcal{M}$ is dense in $\mathcal{X}$, then the extension of $T$ over $\mathcal{X}$ must be compact. In fact, *the extension of a compact linear transformation* $T: \mathcal{X} \to \mathcal{Y}$ *over a completion of* $\mathcal{X}$ *into a completion of* $\mathcal{Y}$ *is again a compact linear transformation.* Recall that every bounded linear transformation $T$ of a normed space $\mathcal{X}$ into a normed space $\mathcal{Y}$ has a unique (up to an isometric isomorphism) bounded linear extension $\widehat{T}$ over the (essentially unique) completion $\widehat{\mathcal{X}}$ of $\mathcal{X}$ into the (essentially unique) completion $\widehat{\mathcal{Y}}$ of $\mathcal{Y}$ (Theorems 4.40 to 4.42). The next theorem says that $\widehat{T} \in \mathcal{B}[\widehat{\mathcal{X}}, \widehat{\mathcal{Y}}]$ is compact whenever $T \in \mathcal{B}[\mathcal{X}, \mathcal{Y}]$ is compact.

**Theorem 4.58.** *Let the Banach spaces $\widehat{\mathcal{X}}$ and $\widehat{\mathcal{Y}}$ be completions of the normed spaces $\mathcal{X}$ and $\mathcal{Y}$, respectively. If $T$ lies in $\mathcal{B}_\infty[\mathcal{X}, \mathcal{Y}]$, then its bounded linear extension $\widehat{T}: \widehat{\mathcal{X}} \to \widehat{\mathcal{Y}}$ lies in $\mathcal{B}_\infty[\widehat{\mathcal{X}}, \widehat{\mathcal{Y}}]$.*

*Proof.* Let $\widehat{\mathcal{X}}$ and $\widehat{\mathcal{Y}}$ be completions of $\mathcal{X}$ and $\mathcal{Y}$. Thus there exist dense linear manifolds $\widetilde{\mathcal{X}}$ and $\widetilde{\mathcal{Y}}$ of $\widehat{\mathcal{X}}$ and $\widehat{\mathcal{Y}}$ that are isometrically isomorphic to $\mathcal{X}$ and $\mathcal{Y}$, respectively (Definition 4.39 and Theorem 4.40). Let $U_X \in \mathcal{G}[\widetilde{\mathcal{X}}, \mathcal{X}]$ and $U_Y \in \mathcal{G}[\mathcal{Y}, \widetilde{\mathcal{Y}}]$ denote such isometric isomorphisms. Take $T \in \mathcal{B}[\mathcal{X}, \mathcal{Y}]$ and set $\widetilde{T} = U_Y T U_X \in \mathcal{B}[\widetilde{\mathcal{X}}, \widetilde{\mathcal{Y}}]$ so that the diagram

$$
\begin{array}{ccc}
\mathcal{Y} & \xrightarrow{\ U_Y\ } & \widetilde{\mathcal{Y}} \\[4pt]
{\scriptstyle T}\uparrow & & \uparrow{\scriptstyle \widetilde{T}} \\[4pt]
\mathcal{X} & \xleftarrow{\ U_X\ } & \widetilde{\mathcal{X}}
\end{array}
$$

commutes. Now take an arbitrary bounded $\widehat{\mathcal{X}}$-valued sequence $\{\widehat{x}_n\}$. Since $\widetilde{\mathcal{X}}^- = \widehat{\mathcal{X}}$ (i.e., since $\inf\{\|\widehat{x} - \widetilde{x}\|: \widetilde{x} \subset \widetilde{\mathcal{X}}\} = 0$ for every $\widehat{x} \in \widehat{\mathcal{X}}$ — see Proposition 3.32), it follows that there exists an $\widetilde{\mathcal{X}}$-valued sequence $\{\widetilde{x}_n\}$ equiconvergent with $\{\widehat{x}_n\}$ (i.e., such that $\|\widehat{x}_n - \widetilde{x}_n\| \to 0$; for instance, for each integer $n$ take $\widetilde{x}_n$ in $\widetilde{\mathcal{X}}$ such that $\|\widehat{x}_n - \widetilde{x}_n\| < \frac{1}{n+1}$), which is bounded (for $\|\widetilde{x}_n\| \leq \|\widehat{x}_n - \widetilde{x}_n\| + \|\widehat{x}_n\|$ for every $n$). Consider the $\mathcal{X}$-valued sequence $\{x_n\}$ such that $x_n = U_X \widetilde{x}_n$ for each $n$, which is bounded too: $\|x_n\| = \|U_X \widetilde{x}_n\| = \|\widetilde{x}_n\|$ for every $n$. If $T$ is compact, then the $\mathcal{Y}$-valued sequence $\{Tx_n\}$ has a convergent subsequence (Theorem 4.52), say, $\{Tx_{n_k}\}$. Thus $\{U_Y Tx_{n_k}\}$ converges in $\widetilde{\mathcal{Y}}$ (because $U_Y$ is a homeomorphism). Therefore $\{\widetilde{T}\widetilde{x}_{n_k}\}$ converges in $\widetilde{\mathcal{Y}}$ (for $\widetilde{T}\widetilde{x}_{n_k} = U_Y T U_X \widetilde{x}_{n_k} = U_Y Tx_{n_k}$ for each $k$) to, say, $\widetilde{y} \in \widetilde{\mathcal{Y}} \subseteq \widehat{\mathcal{Y}}$. Hence $\{\widehat{T}\widehat{x}_{n_k}\}$ converges in $\widehat{\mathcal{Y}}$. Indeed,

$$
\begin{aligned}
\|\widehat{T}\widehat{x}_{n_k} - \widetilde{y}\| &= \|\widehat{T}(\widehat{x}_{n_k} - \widetilde{x}_{n_k}) + \widehat{T}\widetilde{x}_{n_k} - \widetilde{y}\| \\
&\leq \|\widehat{T}\|\,\|\widehat{x}_{n_k} - \widetilde{x}_{n_k}\| + \|\widetilde{T}\widetilde{x}_{n_k} - \widetilde{y}\|
\end{aligned}
$$

for every $k$ because $\widetilde{T} = \widehat{T}|_{\widetilde{\mathcal{X}}}$, so that $\widehat{T}\widehat{x}_{n_k} \to \widetilde{y}$ (reason: $\|\widetilde{T}\widetilde{x}_{n_k} - \widetilde{y}\| \to 0$ and $\|\widehat{x}_{n_k} - \widetilde{x}_{n_k}\| \to 0$ — see Proposition 3.5) as $k \to \infty$. Conclusion: $\widehat{T}$ maps bounded sequences into sequences that have a convergent subsequence; that is, $\widehat{T}$ is compact (Theorem 4.52). $\qquad\square$

# 4.10   The Hahn–Banach Theorem and Dual Spaces

Three extremely important results on continuous (i.e., bounded) linear transformations, which yield a solid foundation for a large portion of modern analysis, are the *Open Mapping Theorem*, the *Banach–Steinhaus Theorem* and the *Hahn–Banach Theorem*. The Hahn–Banach Theorem is concerned with the existence of bounded linear extensions for bounded linear functionals (i.e., for scalar-valued bounded linear transformations), and it is the basis for several existence results that are often applied in functional analysis. In particular, the Hahn–Banach Theorem ensures the existence of a large supply of continuous linear functionals on a normed space $\mathcal{X}$, and hence it is of fundamental importance for introducing the *dual space* of $\mathcal{X}$ (the collection of all continuous linear functionals on $\mathcal{X}$).

Let $\mathcal{M}$ be any linear manifold of a linear space $\mathcal{X}$ and consider a linear transformation $L\colon \mathcal{M} \to \mathcal{Y}$ of $\mathcal{M}$ into a linear space $\mathcal{Y}$. From a purely algebraic point of view, a plain linear extension $\widehat{L}\colon \mathcal{X} \to \mathcal{Y}$ of $L$ over $\mathcal{X}$ has already been investigated in Theorem 2.9. On the other hand, if $\mathcal{M}$ is a dense linear manifold of the normed space $\mathcal{X}$ and $T\colon \mathcal{M} \to \mathcal{Y}$ is a bounded linear transformation of $\mathcal{M}$ into a Banach space $\mathcal{Y}$, then $T$ has a unique bounded linear extension $\widehat{T}\colon \mathcal{X} \to \mathcal{Y}$ over $\mathcal{X}$ (Theorem 4.35). In particular, every bounded linear functional on a dense linear manifold $\mathcal{M}$ of a normed space $\mathcal{X}$ has a bounded linear extension over $\mathcal{X}$. The results of Section 3.8 (and also of Section 4.7), that ensure the existence of a uniformly continuous extension over a metric space $X$ of a uniformly continuous mapping on a dense subset of $X$, are called *extension by continuity*. What the Hahn–Banach Theorem does is to ensure the existence of a bounded linear extension $\widehat{f}\colon \mathcal{X} \to \mathbb{F}$ over $\mathcal{X}$ for every bounded linear functional $f\colon \mathcal{M} \to \mathbb{F}$ on *any* linear manifold $\mathcal{M}$ of the normed space $\mathcal{X}$. (Here $\mathcal{M}$ is not necessarily dense in $\mathcal{X}$ so that extension by continuity collapses.)

We shall approach the Hahn–Banach Theorem step-by-step. The first steps are purely algebraic and, as such, could have been introduced in Chapter 2. To begin with, we consider the following lemma on linear functionals acting on a linear manifold of a real linear space, which are dominated by a sublinear functional (i.e., by a nonnegative homogeneous and subadditive functional.)

**Lemma 4.59.** *Let $\mathcal{M}_0$ be a proper linear manifold of a real linear space $\mathcal{X}$. Take $x_1 \in \mathcal{X}\backslash\mathcal{M}_0$ and consider the linear manifold $\mathcal{M}_1$ of $\mathcal{X}$ generated by $\mathcal{M}_0$ and $x_1$,*

$$\mathcal{M}_1 = \mathcal{M}_0 + \mathrm{span}\,\{x_1\}.$$

*Let $p\colon \mathcal{X} \to \mathbb{R}$ be a sublinear functional on $\mathcal{X}$. If $f_0\colon \mathcal{M}_0 \to \mathbb{R}$ is a linear functional on $\mathcal{M}_0$ such that*

$$f_0(x) \leq p(x) \quad \text{for every} \quad x \in \mathcal{M}_0,$$

*then there exists a linear extension $f_1\colon \mathcal{M}_1 \to \mathbb{R}$ of $f_0$ over $\mathcal{M}_1$ such that*

$$f_1(x) \leq p(x) \quad \text{for every} \quad x \in \mathcal{M}_1.$$

*Proof.* Take an arbitrary vector $x_1$ in $\mathcal{X} \backslash \mathcal{M}_0$.

*Claim.* There exists a real number $c_1$ such that

$$-p(-w - x_1) - f_0(w) \leq c_1 \leq p(w + x_1) - f_0(w)$$

for every $w \in \mathcal{M}_0$.

*Proof.* Since the linear functional $f_0 \colon \mathcal{M}_0 \to \mathbb{R}$ is dominated by a subadditive functional $p \colon \mathcal{X} \to \mathbb{R}$, it follows that

$$
\begin{aligned}
f_0(v) - f_0(u) = f_0(v - u) &\leq p(v - u) = p(v + x_1 - u - x_1) \\
&\leq p(v + x_1) + p(-u - x_1)
\end{aligned}
$$

for every $u, v \in \mathcal{M}_0$. Therefore,

$$-p(-u - x_1) - f_0(u) \leq p(v + x_1) - f_0(v)$$

for every $u \in \mathcal{M}_0$ and every $v \in \mathcal{M}_0$. Set

$$a_1 = \sup_{u \in \mathcal{M}_0} \left(-p(-u - x_1) - f_0(u)\right) \quad \text{and} \quad b_1 = \inf_{v \in \mathcal{M}_0} \left(p(v + x_1) - f_0(v)\right).$$

The above inequality ensures that $a_1$ and $b_1$ are real numbers, and also that $a_1 \leq b_1$. Thus the claimed result holds for any $c_1 \in [a_1, b_1]$. $\square$

Recall that every $x$ in $\mathcal{M}_1 = \mathcal{M}_0 + \operatorname{span}\{x_1\}$ can be *uniquely* written as $x = x_0 + \alpha x_1$ with $x_0$ in $\mathcal{M}_0$ and $\alpha$ in $\mathbb{R}$. Now consider the functional $f_1 \colon \mathcal{M}_1 \to \mathbb{R}$ defined by the formula

$$f_1(x) = f_0(x_0) + \alpha c_1$$

for every $x \in \mathcal{M}_1$, where the pair $(x_0, \alpha)$ in $\mathcal{M}_0 \times \mathbb{R}$ stands for the unique representation of $x$ in $\mathcal{M}_0 + \operatorname{span}\{x_1\}$. It is readily verified that $f_1$ is, in fact, a linear extension of $f_0$ over $\mathcal{M}_1$ (i.e., $f_1$ inherits the linearity of $f_0$ and $f_1|_{\mathcal{M}_0} = f_0$). Finally, we show that $p$ also dominates $f_1$. Take an arbitrary $x = x_0 + \alpha x_1$ in $\mathcal{M}_1$ and consider the three possibilities, viz., $\alpha = 0$, $\alpha > 0$, or $\alpha < 0$. If $\alpha = 0$, then $f_1(x) \leq p(x)$ trivially (in this case, $f(x) = f_0(x_0) \leq p(x_0) = p(x)$). Next recall that $p$ is nonnegative homogeneous (i.e., $p(\gamma z) = \gamma p(z)$ for every $z \in \mathcal{X}$ and every $\gamma \geq 0$), and consider above claimed inequalities. If $\alpha > 0$, then

$$
\begin{aligned}
f_1(x) = f_0(x_0) + \alpha c_1 &\leq f_0(x_0) + \alpha p\left(\tfrac{x_0}{\alpha} + x_1\right) - \alpha f_0\left(\tfrac{x_0}{\alpha}\right) \\
&= p(x_0 + \alpha x_1) = p(x).
\end{aligned}
$$

On the other hand, if $\alpha < 0$, then

$$
\begin{aligned}
f_1(x) = f_0(x_0) + \alpha c_1 &= f_0(x_0) - |\alpha| c_1 \\
&\leq f_0(x_0) + |\alpha| p\left(\tfrac{-x_0}{\alpha} - x_1\right) + |\alpha| f_0\left(\tfrac{-x_0}{\alpha}\right) \\
&= p(x_0 - |\alpha| x_1) = p(x_0 + \alpha x_1) = p(x),
\end{aligned}
$$

which concludes the proof.    □

**Theorem 4.60.** (Real Hahn–Banach Theorem). *Let $\mathcal{M}$ be a linear manifold of a real linear space $\mathcal{X}$ and let $p\colon \mathcal{X} \to \mathbb{R}$ be a sublinear functional on $\mathcal{X}$. If $f\colon \mathcal{M} \to \mathbb{R}$ is a linear functional on $\mathcal{M}$ such that*

$$f(x) \leq p(x) \quad \text{for every} \quad x \in \mathcal{M},$$

*then there exists a linear extension $\widehat{f}\colon \mathcal{X} \to \mathbb{R}$ of $f$ over $\mathcal{X}$ such that*

$$\widehat{f}(x) \leq p(x) \quad \text{for every} \quad x \in \mathcal{X}.$$

*Proof.* First note that, except for the dominance condition that travels from $f$ to $\widehat{f}$, this could be viewed as a particular case of Theorem 2.9, whose kernel's proof is Zorn's Lemma. Let

$$\mathcal{K} = \left\{ \varphi \in \mathcal{L}[\mathcal{N}, \mathbb{R}] \colon \ \mathcal{N} \in \mathcal{L}at(\mathcal{X}), \ \mathcal{M} \subseteq \mathcal{N} \text{ and } f = \varphi|_{\mathcal{M}} \right\}$$

be the collection of all linear functionals on linear manifolds of the real linear space $\mathcal{X}$ which extend the linear functional $f\colon \mathcal{M} \to \mathbb{R}$, and set

$$\mathcal{K}' = \left\{ \varphi \in \mathcal{K} \colon \ \varphi(x) \leq p(x) \text{ for every } x \in \mathcal{N} = \mathcal{D}(\varphi) \right\}.$$

Note that $\mathcal{K}'$ is not empty (for $f \in \mathcal{K}'$). Following the proof of Theorem 2.9, $\mathcal{K}$ is partially ordered (in the extension ordering, and so is its subcollection $\mathcal{K}'$) and every chain in $\mathcal{K}$ has a supremum in $\mathcal{K}$. Then every chain $\{\varphi_\gamma\}$ in $\mathcal{K}'$ has a supremum $\bigvee_\gamma \varphi_\gamma$ in $\mathcal{K}$, which actually lies in $\mathcal{K}'$. Indeed, since each $\varphi_\gamma$ is such that $\varphi_\gamma(x) \leq p(x)$ for every $x$ in the domain $\mathcal{D}(\varphi_\gamma)$ of $\varphi_\gamma$, and since $\{\varphi_\gamma\}$ is a chain, it follows that $\left(\bigvee_\gamma \varphi_\gamma\right)(x) \leq p(x)$ for every $x$ in the domain $\bigcup_\gamma \mathcal{D}(\varphi_\gamma)$ of $\bigvee_\gamma \varphi_\gamma$. (In fact, if $x \in \bigcup_\gamma \mathcal{D}(\varphi_\gamma)$, then $x \in \mathcal{D}(\varphi_\mu)$ for some $\varphi_\mu \in \{\varphi_\gamma\}$, and hence $\left(\bigvee_\gamma \varphi_\gamma\right)(x) = \varphi_\mu(x) \leq p(x)$ because $\left(\bigvee_\gamma \varphi_\gamma\right)|_{\mathcal{D}(\varphi_\mu)} = \varphi_\mu$.) Therefore, every chain in $\mathcal{K}'$ has a supremum (and so an upper bound) in $\mathcal{K}'$. Thus, according to Zorn's Lemma, $\mathcal{K}'$ has a maximal element, say, $\varphi_0\colon \mathcal{N}_0 \to \mathbb{R}$. Now we shall apply Lemma 4.59 to show that $\mathcal{N}_0 = \mathcal{X}$. Suppose $\mathcal{N}_0 \neq \mathcal{X}$. Take $x_1 \in \mathcal{X}\backslash\mathcal{N}_0$ and consider the linear manifold $\mathcal{N}_1$ of $\mathcal{X}$ generated by $\mathcal{N}_0$ and $x_1$,

$$\mathcal{N}_1 = \mathcal{N}_0 + \operatorname{span}\{x_1\},$$

which properly includes $\mathcal{N}_0$. Since $\varphi_0 \in \mathcal{K}'$, it follows that $\varphi_0(x) \leq p(x)$ for every $x \in \mathcal{N}_0$. Thus, according to Lemma 4.59, there is a linear extension $\varphi_1\colon \mathcal{N}_1 \to \mathbb{R}$ of $\varphi_0$ over $\mathcal{N}_1$ such that $\varphi_1(x) \leq p(x)$ for every $x \in \mathcal{N}_1$. Therefore $\varphi_0 \leq \varphi_1 \in \mathcal{K}'$, which contradicts the fact that $\varphi_0$ is maximal in $\mathcal{K}'$ (for $\varphi_0 \neq \varphi_1$). Conclusion: $\mathcal{N}_0 = \mathcal{X}$. Outcome: $\varphi_0$ is a linear extension of $f$ over $\mathcal{X}$ which is dominated by $p$.    □

**Theorem 4.61.** (Hahn–Banach Theorem). *Let $\mathcal{M}$ be a linear manifold of a linear space $\mathcal{X}$ over $\mathbb{F}$ and let $p\colon \mathcal{X} \to \mathbb{R}$ be a seminorm on $\mathcal{X}$. If $f\colon \mathcal{M} \to \mathbb{F}$ is a linear functional on $\mathcal{M}$ such that*

$$|f(x)| \leq p(x) \quad \text{for every} \quad x \in \mathcal{M},$$

*then there exists a linear extension $\widehat{f}\colon \mathcal{X} \to \mathbb{F}$ of $f$ over $\mathcal{X}$ such that*

$$|\widehat{f}(x)| \leq p(x) \quad \text{for every} \quad x \in \mathcal{X}.$$

*Proof.* As we have agreed in the introduction to this chapter, $\mathbb{F}$ denotes either the complex field $\mathbb{C}$ or the real field $\mathbb{R}$. Recall that a seminorm is a nonnegative convex functional (i.e., a nonnegative absolutely homogeneous subadditive functional).

(a)  If $\mathbb{F} = \mathbb{R}$, then this is an easy corollary of the previous theorem. Indeed, if $\mathbb{F} = \mathbb{R}$, then the condition $|f| \leq p$ trivially implies $f \leq p$ on $\mathcal{M}$. As a seminorm is a sublinear functional, Theorem 4.60 ensures the existence of a linear extension $\widehat{f}\colon \mathcal{X} \to \mathbb{R}$ of $f$ over $\mathcal{X}$ such that $\widehat{f} \leq p$ on $\mathcal{X}$. Since $\widehat{f}$ is linear and $p$ is absolutely homogeneous, it follows that $-\widehat{f}(x) = \widehat{f}(-x) \leq p(-x) = |{-1}|\, p(x) = p(x)$ for every $x \in \mathcal{X}$. Hence $-p \leq \widehat{f}$, and therefore $|\widehat{f}| \leq |p| = p$ on $\mathcal{X}$ (for $p$ is nonnegative).

(b) Suppose $\mathbb{F} = \mathbb{C}$ and note that the complex linear space $\mathcal{X}$ can also be viewed as a real linear space (where scalar multiplication now means multiplication by *real* scalars only). Moreover, if $\mathcal{M}$ is a linear manifold of the (complex) linear space $\mathcal{X}$, then it is also a (real) linear manifold of $\mathcal{X}$ when $\mathcal{X}$ is regarded as a real linear space. Furthermore, if $f\colon \mathcal{M} \to \mathbb{C}$ is a complex-valued functional on $\mathcal{M}$, and if $g\colon \mathcal{M} \to \mathbb{R}$ and $h\colon \mathcal{M} \to \mathbb{R}$ are defined by $g(x) = \operatorname{Re} f(x)$ and $h(x) = \operatorname{Im} f(x)$ for every $x \in \mathcal{X}$, then

$$f = g + ih.$$

Now recall that $f\colon \mathcal{M} \to \mathbb{C}$ is linear. Thus, for an arbitrary $\alpha \in \mathbb{R}$,

$$g(\alpha x) + ih(\alpha x) = f(\alpha x) = \alpha f(x) = \alpha g(x) + i\alpha h(x),$$

and hence $g(\alpha x) = \alpha g(x)$ and $h(\alpha x) = \alpha h(x)$ for every $x \in \mathcal{M}$ (because $g$ and $h$ are real-valued). Similarly, for every $x, y \in \mathcal{M}$,

$$g(x+y) + ih(x+y) = f(x+y) = f(x) + f(y) = g(x) + ih(x) + g(y) + ih(y),$$

and so $g(x + y) = g(x) + g(y)$ and $h(x + y) = h(x) + h(y)$.

*Conclusion:* $g\colon \mathcal{M} \to \mathbb{R}$ and $h\colon \mathcal{M} \to \mathbb{R}$ are *linear* functionals on $\mathcal{M}$ when $\mathcal{M}$ is regarded as a real linear space.

Observe that $f(ix) = if(x)$, and hence $g(ix) + ih(ix) = ig(x) - h(x)$, for every $x \in \mathcal{M}$. Since $g(x), g(ix), h(x)$ and $h(ix)$ are real numbers, it follows that $h(x) = -g(ix)$, and therefore

$$f(x) = g(x) - ig(ix),$$

for every $x \in \mathcal{M}$. If $|f| < p$ on $\mathcal{M}$, then $g = \mathrm{Re}\, f \leq p$ on $\mathcal{M}$. Since $g$ is a linear functional on the (real) linear manifold $\mathcal{M}$, and since $p$ is a sublinear functional on the (real) linear space $\mathcal{X}$ (because it is a seminorm on the complex linear space $\mathcal{X}$), it follows by Theorem 4.60 that there exists a real-valued linear extension $\widehat{g}$ of $g$ over the (real) linear space $\mathcal{X}$ such that $\widehat{g} < p$ on $\mathcal{X}$. Consider the functional $\widehat{f}\colon \mathcal{X} \to \mathbb{C}$ (on the complex linear space $\mathcal{X}$) defined by

$$\widehat{f}(x) = \widehat{g}(x) - i\,\widehat{g}(ix)$$

for every $x \in \mathcal{X}$. It is clear that $\widehat{f}$ extends $f$ over $\mathcal{X}$ (reason: If $x \in \mathcal{M}$ — so that $ix \in \mathcal{M}$ — then $\widehat{f}(x) = g(x) - ig(x) = f(x)$). It is also readily verified that $\widehat{f}$ is a linear functional on the complex space $\mathcal{X}$. Indeed, additivity and (real) homogeneity are trivially verified (because $\widehat{g}$ is additive and homogeneous on the real linear space $\mathcal{X}$). Thus it suffices to verify that $\widehat{f}(ix) = i\,\widehat{f}(x)$ for every $x \in \mathcal{X}$. In fact, $\widehat{f}(ix) = \widehat{g}(ix) - i\,\widehat{g}(-x) = \widehat{g}(ix) + i\,\widehat{g}(x) = i\,\widehat{f}(x)$ for every $x \in \mathcal{X}$. Therefore, $\widehat{f}$ *is a linear extension of* $f$ *over* $\mathcal{X}$. Finally we show that

$$|\widehat{f}(x)| \leq p(x)$$

for every $x \in \mathcal{X}$. Take an arbitrary $x$ in $\mathcal{X}$ and write the complex number $\widehat{f}(x)$ in polar form: $\widehat{f}(x) = \rho e^{i\theta}$ (if $\widehat{f}(x) = 0$, then $\rho = 0$ and $\theta$ is any number between $0$ and $2\pi$). Since $\widehat{f}$ is a linear functional on the complex space $\mathcal{X}$, it follows that $\widehat{f}(e^{-i\theta}x) = \rho = |\widehat{f}(x)|$, which is a real number. Then $\widehat{f}(e^{-i\theta}x) = \widehat{g}(e^{-i\theta}x)$, and hence

$$|\widehat{f}(x)| = \widehat{g}(e^{-i\theta}x) \leq p(e^{-i\theta}x) = |e^{-i\theta}|p(x) = p(x),$$

since $p\colon \mathcal{X} \to \mathbb{R}$ is absolutely homogeneous on the complex space $\mathcal{X}$.    □

Theorems 4.60 and 4.61 are called Dominated Extension Theorems (in which no topology is involved). The next one is the Continuous Extension Theorem.

**Theorem 4.62.** (Hahn–Banach Theorem in Normed Space). *Let* $\mathcal{M}$ *be a linear manifold of a normed space* $\mathcal{X}$. *Every bounded linear functional* $f\colon \mathcal{M} \to \mathbb{F}$ *defined on* $\mathcal{M}$ *has a bounded linear extension* $\widehat{f}\colon \mathcal{X} \to \mathbb{F}$ *over the whole space* $\mathcal{X}$ *such that* $\|\widehat{f}\| = \|f\|$.

*Proof.* Take an arbitrary $f \in \mathcal{B}[\mathcal{M}, \mathbb{F}]$ so that $|f(x)| \leq \|f\|\,\|x\|$ for each $x \in \mathcal{M}$. Set $p(x) = \|f\|\,\|x\|$ for every $x \in \mathcal{X}$, which defines a seminorm on $\mathcal{X}$ (since $p\colon \mathcal{X} \to \mathbb{R}$ is a multiple of a norm $\|\ \|\colon \mathcal{X} \to \mathbb{R}$ on $\mathcal{X}$ — in fact, $p$ is a norm on $\mathcal{X}$ whenever $f \neq 0$). Since $|f(x)| \leq p(x)$ for every $x \in \mathcal{M}$, it follows by the previous theorem that there exists a linear extension $\widehat{f}\colon \mathcal{X} \to \mathbb{F}$ of $f$ over $\mathcal{X}$ such that

$$|\widehat{f}(x)| \leq p(x) = \|f\|\,\|x\|$$

for every $x \in \mathcal{X}$. Thus $\widehat{f}$ is bounded (i.e., $\widehat{f} \in \mathcal{B}[\mathcal{X}, \mathbb{F}]$) and $\|\widehat{f}\| \leq \|f\|$. On the other hand, $f(x) = \widehat{f}(x)$ for every $x \in \mathcal{M}$ (because $f = \widehat{f}|_{\mathcal{M}}$), and hence

$$\|f\| = \sup_{x \in \mathcal{M}} |\widehat{f}(x)| \leq \sup_{x \in \mathcal{X}} |\widehat{f}(x)| = \|\widehat{f}\|.$$

Therefore $\|\widehat{f}\| = \|f\|$. $\qquad\qquad\qquad\qquad\qquad\qquad\qquad\qquad\qquad\qquad\qquad\square$

Here are some consequences of the Hahn–Banach Theorem that are particularly useful.

**Corollary 4.63.** *Let $\mathcal{M}$ be a proper subspace of a normed space $\mathcal{X}$. If $x_0 \in \mathcal{X}\backslash\mathcal{M}$, then there exists a bounded linear functional $f: \mathcal{X} \to \mathbb{F}$ such that $f(x_0) = 1$, $f(\mathcal{M}) = \{0\}$, and $\|f\| = d(x_0, \mathcal{M})^{-1}$.*

*Proof.* First note that $d(x_0, \mathcal{M}) \neq 0$: the distance from $x_0$ to $\mathcal{M}$ is strictly positive because $x_0 \in \mathcal{X}\backslash\mathcal{M}$ and $\mathcal{M} = \mathcal{M}^-$ (see Problem 3.43(b)). Hence $d(x_0, \mathcal{M})^{-1}$ is well-defined. Now consider the linear manifold $\mathcal{M}_0$ of $\mathcal{X}$ generated by $\mathcal{M}$ and $x_0$,

$$\mathcal{M}_0 = \mathcal{M} + \mathrm{span}\,\{x_0\},$$

so that every $x$ in $\mathcal{M}_0$ can be uniquely written as $x = u + \alpha x_0$ with $u$ in $\mathcal{M}$ and $\alpha$ in $\mathbb{F}$. Let $f_0: \mathcal{M}_0 \to \mathbb{F}$ be a functional on $\mathcal{M}_0$ defined by

$$f_0(u + \alpha x_0) = \alpha$$

for every $x = u + \alpha x_0 \in \mathcal{M}_0$. It is easy to verify that $f_0$ is linear, $f_0(\mathcal{M}) = \{0\}$, and $f_0(x_0) = 1$. Next we show that $f_0$ is bounded (so that $f_0 \in \mathcal{B}[\mathcal{M}_0, \mathbb{F}]$) and $\|f_0\| = d(x_0, \mathcal{M})^{-1}$. Consider the set

$$S = \big\{(u, \alpha) \in \mathcal{M}{\times}\mathbb{F}:\ (u, \alpha) \neq (0, 0)\big\}$$

and its partition $\{S_1, S_2\}$, where

$$S_1 = \big\{(u, \alpha) \in \mathcal{M}{\times}\mathbb{F}:\ \alpha \neq 0\big\},$$

$$S_2 = \big\{(u, \alpha) \in \mathcal{M}{\times}\mathbb{F}:\ u \neq 0 \text{ and } \alpha = 0\big\}.$$

Observe that $x = u + \alpha x_0 = 0$ (in $\mathcal{M}_0$) if and only if $u = 0$ (in $\mathcal{M}$) and $\alpha = 0$ (in $\mathbb{F}$) —reason: $\mathcal{M}$ is a linear manifold of $\mathcal{X}$ and $x_0 \subset \mathcal{X}\backslash\mathcal{M}$, so that $\mathrm{span}\,\{x_0\}{\cap}\mathcal{M} = \{0\}$. Hence $x = u + \alpha x_0 \neq 0$ in $\mathcal{M}_0$ if and only if $(u, \alpha) \in S$, which implies that

$$\|f_0\| = \sup_{0 \neq x \in \mathcal{M}_0} \frac{|f_0(x)|}{\|x\|} = \sup_{(u,\alpha)\in S} \frac{|\alpha|}{\|u + \alpha x_0\|} = \sup_{(u,\alpha)\in S_1} \frac{|\alpha|}{\|u + \alpha x_0\|},$$

since $\sup_{(u,\alpha)\in S_2} \frac{|\alpha|}{\|u + \alpha x_0\|} = 0$ and $S = S_1 \cup S_2$. However, $\inf_{v\in\mathcal{M}}\|v + x_0\| = \inf_{v\in\mathcal{M}}\|x_0 - v\| = d(x_0, \mathcal{M}) \neq 0$, and so (see Problem 4.5)

$$\sup_{(u,\alpha)\in S_1} \frac{|\alpha|}{\|u + \alpha x_0\|} = \sup_{(u,\alpha)\in S_1} \frac{1}{\|\alpha^{-1}u + x_0\|} = \sup_{v\in\mathcal{M}} \frac{1}{\|v + x_0\|} = \frac{1}{d(x_0,\mathcal{M})}.$$

Summing up: $f_0$ is a bounded linear functional on the linear manifold $\mathcal{M}_0$ of $\mathcal{X}$ such that $f_0(x_0) = 1$, $f_0(\mathcal{M}) = \{0\}$ and $\|f_0\| = d(x_0, \mathcal{M})^{-1}$. Therefore, according to Theorem 4.62, $f_0: \mathcal{M}_0 \to \mathbb{F}$ has a bounded linear extension $f: \mathcal{X} \to \mathbb{F}$ over

$\mathcal{X}$ such that $\|f\| = \|f_0\| = d(x_0, \mathcal{M})^{-1}$. Moreover, since $f|_{\mathcal{M}_0} = f_0$, it also follows that $f(x_0) = f_0(x_0) = 1$ and $f(\mathcal{M}) = f_0(\mathcal{M}) = \{0\}$, which concludes the proof. $\qquad\square$

Recall that $\{0\}$ is a proper subspace of any nonzero normed space $\mathcal{X}$. Thus, according to Corollary 4.63, for each $x_0 \neq 0$ in $\mathcal{X} \neq \{0\}$ there exists a bounded linear functional $f \colon \mathcal{X} \to \mathbb{F}$ such that $f(x_0) = 1$ and $\|f\| = d(x_0, \{0\})^{-1} = \|x_0\|^{-1}$. Now set $f_0 = \|x_0\| f$, which is again a bounded linear functional on $\mathcal{X}$, so that $f_0(x_0) = \|x_0\|$ and $\|f_0\| = 1$. Moreover, if $x_0 = 0$, then take $x_1 \neq 0$ in $\mathcal{X}$ and a bounded linear functional $f_1$ on $\mathcal{X}$ such that $f_1(x_1) = \|x_1\|$ and $\|f_1\| = 1$. Since $f_1$ is linear, $f_1(x_0) = f_1(0) = 0 = \|x_0\|$. This proves the following corollary.

**Corollary 4.64.** *For each vector $x_0$ in a normed space $\mathcal{X} \neq \{0\}$ there exists a bounded linear functional $f \colon \mathcal{X} \to \mathbb{F}$ such that $\|f\| = 1$ and $f(x_0) = \|x_0\|$. Consequently, there exist nonzero bounded linear functionals defined on every nonzero normed space.*

Let $\mathcal{X}$ and $\mathcal{Y}$ be normed spaces over the same field, and consider the normed space $\mathcal{B}[\mathcal{X}, \mathcal{Y}]$ of all bounded linear transformations of $\mathcal{X}$ into $\mathcal{Y}$. If $\mathcal{X} \neq \{0\}$, then Corollary 4.64 ensures the existence of $f \neq 0$ in $\mathcal{B}[\mathcal{X}, \mathbb{F}]$. Suppose $\mathcal{Y} \neq \{0\}$, take any $y \neq 0$ in $\mathcal{Y}$, and set

$$T(x) = f(x) y$$

for every $x \in \mathcal{X}$. This defines a nonzero mapping $T \colon \mathcal{X} \to \mathcal{Y}$ which certainly is linear and bounded. Conclusion: *There exists $T \neq O$ in $\mathcal{B}[\mathcal{X}, \mathcal{Y}]$ whenever $\mathcal{X}$ and $\mathcal{Y}$ are nonzero normed spaces.*

**Example 4P.** Proposition 4.15 says that $\mathcal{B}[\mathcal{X}, \mathcal{Y}]$ *is a Banach space whenever $\mathcal{Y}$ is a Banach space.* If $\mathcal{X} = \{0\}$, then $\mathcal{B}[\mathcal{X}, \mathcal{Y}] = \{O\}$, which is a trivial Banach space regardless whether $\mathcal{Y}$ is Banach or not. Thus the converse of Proposition 4.15 should read as follows.

If $\mathcal{X} \neq \{0\}$ and $\mathcal{B}[\mathcal{X}, \mathcal{Y}]$ is a Banach space, then $\mathcal{Y}$ is a Banach space.

Corollary 4.64 asserts that there exists $f \neq 0$ in $\mathcal{B}[\mathcal{X}, \mathbb{F}]$ whenever $\mathcal{X} \neq \{0\}$. Take an arbitrary Cauchy sequence $\{y_n\}$ in $\mathcal{Y}$ and consider the $\mathcal{B}[\mathcal{X}, \mathcal{Y}]$-valued sequence $\{T_n\}$ such that, for each $n$,

$$T_n x = f(x) y_n$$

for every $x \in \mathcal{X}$. Each $T_n$ in fact lies in $\mathcal{B}[\mathcal{X}, \mathcal{Y}]$ because $f$ lies in $\mathcal{B}[\mathcal{X}, \mathbb{F}]$. Indeed, for each integer $n$,

$$\|T_n\| = \sup_{\|x\| \leq 1} \|T_n x\| = \sup_{\|x\| \leq 1} |f(x)| \|y_n\| = \|f\| \|y_n\|$$

and, for any pair of integers $m$ and $n$,

$$\|T_m - T_n\| = \sup_{\|x\| \leq 1} \|(T_m - T_n)x\| = \|f\| \|y_m - y_n\|.$$

Hence $\{T_n\}$ is a Cauchy sequence in $\mathcal{B}[\mathcal{X}, \mathcal{Y}]$. If $\mathcal{B}[\mathcal{X}, \mathcal{Y}]$ is complete, then $\{T_n\}$ converges in $\mathcal{B}[\mathcal{X}, \mathcal{Y}]$ to, say, $T \in \mathcal{B}[\mathcal{X}, \mathcal{Y}]$. Since $f \neq 0$, there exists $x_0 \in \mathcal{X}$ such that $f(x_0) \neq 0$. Therefore,

$$y_n = f(x_0)^{-1} T_n(x_0) \ \to\ f(x_0)^{-1} T(x_0) \quad \text{in} \quad \mathcal{Y}$$

(uniform convergence implies strong convergence to the same limit), and so $\{y_n\}$ converges in $\mathcal{Y}$. Conclusion: *If $\mathcal{X} \neq \{0\}$ and $\mathcal{B}[\mathcal{X}, \mathcal{Y}]$ is complete, then $\mathcal{Y}$ is complete.*

The *dual space* (or *conjugate space*) of a normed space $\mathcal{X}$, denoted by $\mathcal{X}^*$, is the normed space of all continuous linear functionals on $\mathcal{X}$ (i.e., $\mathcal{X}^* = \mathcal{B}[\mathcal{X}, \mathbb{F}]$, where $\mathbb{F}$ stands either for the real field $\mathbb{R}$ or the complex field $\mathbb{C}$, whether $\mathcal{X}$ is a real or complex normed space, respectively). Obviously, $\mathcal{X}^* = \{0\}$ whenever $\mathcal{X} = \{0\}$. Corollary 4.64 ensures the converse: $\mathcal{X}^* \neq \{0\}$ whenever $\mathcal{X} \neq \{0\}$. Indeed, if $f(x) = 0$ for all $f \in \mathcal{X}^*$, then $x = 0$. As a matter of fact, if $x \neq y$ in $\mathcal{X}$, then there exists $f \in \mathcal{X}^*$ such that $f(x) - f(y) = f(x - y) = \|x - y\| \neq 0$ (Corollary 4.64 again), and hence $f(x) \neq f(y)$. This is usually expressed by saying that $\mathcal{X}^*$ *separates the points of* $\mathcal{X}$. Still from Corollary 4.64, for each nonzero $x \in \mathcal{X}$ there exists $f_0 \in \mathcal{X}^*$ such that $\|f_0\| = 1$ and $\|x\| = |f_0(x)|$. Therefore,

$$\|x\| = \frac{|f_0(x)|}{\|f_0\|} \leq \left\{ \begin{array}{c} \sup_{f \in \mathcal{X}^*,\, \|f\| \leq 1} |f(x)| \\[4pt] \sup_{0 \neq f \in \mathcal{X}^*} \frac{|f(x)|}{\|f\|} \end{array} \right\} \leq \|x\|$$

(recall: $|f(x)| \leq \|f\|\|x\|$ for every $x$ in $\mathcal{X}$ and every $f$ in $\mathcal{X}^*$), which shows a symmetry in the definitions of the norms in $\mathcal{X}$ and $\mathcal{X}^*$:

$$\|x\| = \sup_{f \in \mathcal{X}^*,\, \|f\| \leq 1} |f(x)| = \sup_{0 \neq f \in \mathcal{X}^*} \frac{|f(x)|}{\|f\|}$$

for every $x \in \mathcal{X}$. Observe that, according to Proposition 4.15, $\mathcal{X}^*$ *is a Banach space for every normed space $\mathcal{X}$* (reason: $\mathcal{X}^* = \mathcal{B}[\mathcal{X}, \mathbb{F}]$ and $(\mathbb{F}, |\ |)$ is a Banach space).

**Proposition 4.65.** *If the dual space $\mathcal{X}^*$ of a normed space $\mathcal{X}$ is separable, then $\mathcal{X}$ itself is separable.*

*Proof.* If $\mathcal{X}^* = \{0\}$, then $\mathcal{X} = \{0\}$ and the result holds trivially. Thus suppose $\mathcal{X}^* \neq \{0\}$ is separable and consider the unit sphere about the origin of $\mathcal{X}^*$, viz., $S_1 = \{f \in \mathcal{X}^*\colon \|f\| = 1\}$. Since every subset of a separable metric space is separable (Corollary 3.36), it follows that $S_1$ includes a countable dense subset, say, $\{f_n\}$. For each $f_n$ there exists $x_n \in \mathcal{X}$ such that

$$\|x_n\| = 1 \quad \text{and} \quad \tfrac{1}{2} < |f_n(x_n)|.$$

Indeed, $\sup_{\|x\|=1} |f_n(x)| = \|f_n\| = 1$ because $f_n \in S_1$. Consider the countable subset $\{x_n\}$ of $\mathcal{X}$ and put

$$\mathcal{M} = \operatorname{span} \{x_n\}.$$

If $\mathcal{M}^- \neq \mathcal{X}$, then Corollary 4.63 ensures that there exists $0 \neq f \in \mathcal{X}^*$ such that $f(\mathcal{M}) = \{0\}$. Set $f_0 = \|f\|^{-1} f$ in $\mathcal{X}^*$ so that

$$f_0 \in S_1 \quad \text{and} \quad f_0(x_n) = 0 \ \text{ for every } n.$$

Hence $|f_n(x_n)| \leq |(f_n - f_0)(x_n)| \leq \|f_n - f_0\|$, and therefore

$$\tfrac{1}{2} \leq \|f_n - f_0\|$$

for every $n$. But this implies that the set $\{f_n\}$ is not dense in $S_1$ (see Proposition 3.32), which contradicts the very definition of $\{f_n\}$. Outcome: $\mathcal{M}^- = \mathcal{X}$, and so $\mathcal{X}$ is separable by Proposition 4.9(b). $\qquad\square$

If $\mathcal{X} \neq \{0\}$, then $\mathcal{X}^* \neq \{0\}$ and hence $(\mathcal{X}^*)^*$, the dual of $\mathcal{X}^*$, is again a nonzero Banach space. We shall write $\mathcal{X}^{**}$ instead of $(\mathcal{X}^*)^*$, which is called the *second dual* (or *bidual*) of $\mathcal{X}$. It is clear that $\mathcal{X}$, $\mathcal{X}^*$ and $\mathcal{X}^{**}$ are normed spaces over the same scalar field. The next result shows that $\mathcal{X}$ can be identified with a linear manifold of $\mathcal{X}^{**}$, so that $\mathcal{X}$ is naturally embedded in its second dual $\mathcal{X}^{**}$.

**Theorem 4.66.** *Every normed space $\mathcal{X}$ is isometrically isomorphic to a linear manifold of $\mathcal{X}^{**}$.*

*Proof.* Suppose $\mathcal{X} \neq \{0\}$ (otherwise the result is trivially verified), take an arbitrary $x$ in $\mathcal{X}$, and consider the functional $\varphi_x : \mathcal{X}^* \to \mathbb{F}$ defined on the dual $\mathcal{X}^*$ of $\mathcal{X}$ by

$$\varphi_x(f) = f(x)$$

for every $f$ in $\mathcal{X}^*$. Since $\mathcal{X}^*$ is a linear space,

$$\varphi_x(\alpha f + \beta g) = (\alpha f + \beta g)(x) = \alpha f(x) + \beta g(x) = \alpha \varphi_x(f) + \beta \varphi_x(g)$$

for every $f, g \in \mathcal{X}^*$ and every $\alpha, \beta \in \mathbb{F}$, so that $\varphi_x$ is linear. Moreover, since the elements of $\mathcal{X}^*$ are bounded and linear, it also follows that

$$|\varphi_x(f)| = |f(x)| \leq \|f\| \|x\|$$

for every $f \in \mathcal{X}^*$, and hence $\varphi_x$ is bounded. Thus $\varphi_x \in \mathcal{X}^{**}$. Indeed,

$$\|\varphi_x\| = \|x\|,$$

since Corollary 4.64 ensures the existence of $f_0 \in \mathcal{X}^*$ such that $\|f_0\| = 1$ and $|f_0(x)| = \|x\|$, and therefore

$$\|\varphi_x\| = \sup_{\|f\| \leq 1} |f(x)| \leq \|x\| = |f_0(x)| = |\varphi_x(f_0)| \leq \|\varphi_x\| \|f_0\| = \|\varphi_x\|.$$

Let $\Phi : \mathcal{X} \to \mathcal{X}^{**}$ be the mapping that assigns to each vector $x$ in $\mathcal{X}$ the functional $\varphi_x$ in $\mathcal{X}^{**}$; that is,

$$\Phi(x) = \varphi_x$$

for every $x \in \mathcal{X}$. It is easy to verify that $\Phi$ is linear. Since $\|\Phi(x)\| = \|x\|$ for every $x \in \mathcal{X}$, it follows that $\Phi$ is a linear isometry of $\mathcal{X}$ into $\mathcal{X}^{**}$ (see Proposition 4.37). Hence $\Phi\colon \mathcal{X} \to \mathcal{R}(\Phi) \subseteq \mathcal{X}^{**}$ is an isometric isomorphism of $\mathcal{X}$ onto $\mathcal{R}(\Phi) = \Phi(\mathcal{X})$, the range of $\Phi$. Thus the range of $\Phi$ is a linear manifold of $\mathcal{X}^{**}$ isometrically isomorphic to $\mathcal{X}$.    $\square$

This linear isometry $\Phi\colon \mathcal{X} \to \mathcal{X}^{**}$ is known as the *natural embedding* of the normed space $\mathcal{X}$ into its second dual $\mathcal{X}^{**}$. If $\Phi$ is surjective (i.e., if $\Phi(\mathcal{X}) = \mathcal{X}^{**}$), then we say that $\mathcal{X}$ is *reflexive*. Equivalently, $\mathcal{X}$ is reflexive if and only if the *natural embedding* $\Phi\colon \mathcal{X} \to \mathcal{X}^{**}$ is an isometric isomorphism of $\mathcal{X}$ onto $\mathcal{X}^{**}$. Thus, if $\mathcal{X}$ is reflexive, then $\mathcal{X}$ and $\mathcal{X}^{**}$ are isometrically isomorphic (notation: $\mathcal{X} \cong \mathcal{X}^{**}$). The converse, however, fails: $\mathcal{X} \cong \mathcal{X}^{**}$ clearly implies $\Phi(\mathcal{X}) \cong \mathcal{X}^{**}$ (for composition of isometric isomorphisms is again an isometric isomorphism) but does not imply $\Phi(\mathcal{X}) = \mathcal{X}^{**}$. Since $\mathcal{X}^{**}$ (the dual of $\mathcal{X}^{*}$) always is a Banach space, it follows by Problem 4.37 that *every reflexive normed space is a Banach space*. Again, the converse fails (i.e., there exist nonreflexive Banach spaces, as we shall see in Example 4S below). Recall that separability is a topological invariant (see Problem 3.48) so that the converse of Proposition 4.65 holds for reflexive Banach spaces. Indeed, if $\mathcal{X} \cong \mathcal{X}^{**}$, then $\mathcal{X}$ is separable if and only if $\mathcal{X}^{**}$ is separable, which implies that $\mathcal{X}^{*}$ is separable by Proposition 4.65. Therefore, if $\mathcal{X}$ is separable and $\mathcal{X}^{*}$ is not separable, then $\mathcal{X} \not\cong \mathcal{X}^{**}$ and hence $\mathcal{X}$ is not reflexive: *a separable Banach space with a nonseparable dual is not reflexive*. This provides a necessary condition for reflexivity. Here is an equivalent condition.

**Proposition 4.67.** *A Banach space $\mathcal{X}$ is reflexive if and only for each $\varphi \in \mathcal{X}^{**}$ there exists $x \in \mathcal{X}$ such that*

$$\varphi(f) - f(x) \quad \text{for every} \quad f \in \mathcal{X}^{*}.$$

*Proof.* Let $\Phi\colon \mathcal{X} \to \mathcal{X}^{**}$ be the natural embedding of $\mathcal{X}$ into $\mathcal{X}^{**}$ and take an arbitrary $\varphi \in \mathcal{X}^{**}$. There exists $x \in \mathcal{X}$ such that $\varphi(f) = f(x)$ for every $f \in \mathcal{X}^{*}$ if and only if $\varphi = \varphi_x$, which means that $\varphi \in \mathcal{R}(\Phi)$. Equivalently, if and only if $\Phi$ is surjective.    $\square$

**Example 3Q.** If $\mathcal{X}$ is a finite-dimensional normed space, then $\dim \mathcal{X} = \dim \mathcal{X}^{*}$ by Problem 4.64. Thus $\dim \mathcal{X}^{*} = \dim \mathcal{X}^{**}$ because $\mathcal{X}^{*}$ is finite-dimensional (Problem 4.64 again), and hence $\dim \mathcal{X} = \dim \mathcal{X}^{**}$. Let $\Phi\colon \mathcal{X} \to \mathcal{X}^{**}$ be the natural embedding of $\mathcal{X}$ into $\mathcal{X}^{**}$. Since $\Phi(\mathcal{X})$ is a linear manifold of the finite-dimensional linear space $\mathcal{X}^{**}$, it follows by Problem 2.7 that $\Phi(\mathcal{X})$ also is a finite-dimensional linear space. Therefore, as $\mathcal{X}$ and $\Phi(\mathcal{X})$ are topologically isomorphic finite-dimensional normed spaces, Corollary 4.31 ensures that $\dim \Phi(\mathcal{X}) = \dim \mathcal{X}$. Then $\Phi(\mathcal{X})$ is a linear manifold of the finite-dimensional space $\mathcal{X}^{**}$ and $\dim \Phi(\mathcal{X}) = \dim \mathcal{X}^{**}$. This implies that $\Phi(\mathcal{X}) = \mathcal{X}^{**}$ (Problem 2.7 again). Conclusion:

Every finite-dimensional normed space is reflexive.

**Example 4R.** Take an arbitrary pair $\{p, q\}$ of Hölder conjugates (i.e., take real numbers $p > 1$ and $q > 1$ such that $\frac{1}{p} + \frac{1}{q} = 1$), and consider the Banach spaces $\ell_+^p$ and $\ell_+^q$ of Example 4B. It can be shown that there exists a natural isometric isomorphism $J_p \colon \ell_+^p \to (\ell_+^q)^*$ of $\ell_+^p$ onto the dual of $\ell_+^q$. Thus, symmetrically, there exists a natural isometric isomorphism $J_q \colon \ell_+^q \to (\ell_+^p)^*$ of $\ell_+^q$ onto the dual of $\ell_+^p$. It then follows by Problem 4.65 that there exists an isometric isomorphism $J_q^* \colon (\ell_+^q)^* \to (\ell_+^p)^{**}$ of the dual of $\ell_+^q$ onto the second dual of $\ell_+^p$. Therefore, the composition $J_q^* J_p \colon \ell_+^p \to (\ell_+^p)^{**}$ is an isometric isomorphism of $\ell_+^p$ onto its second dual so that $\ell_+^p \cong (\ell_+^p)^{**}$. Moreover, it can also be shown that this isometric isomorphism actually coincides with the natural embedding $\Phi \colon \ell_+^p \to (\ell_+^p)^{**}$. Conclusion:

$$\ell_+^p \text{ is a reflexive Banach space for every } p > 1.$$

In particular, the very special space $\ell_+^2$, besides being reflexive, is also isometrically equivalent to its own dual (actually, as we shall see in Section 5.11, the *real* space $\ell_+^2$ is isometrically isomorphic to $(\ell_+^2)^*$).

**Example 4S.** There are, however, nonreflexive Banach spaces. For instance, consider the linear spaces $\ell_+^\infty$ and $\ell_+^1$ equipped with their usual norms ($\| \ \|_\infty$ and $\| \ \|_1$, respectively). Since $\ell_+^{c_o}$ is a linear manifold of the linear space $\ell_+^\infty$, equip it with the sup-norm as well. Recall that $\ell_+^{c_o}$ and $\ell_+^1$ are separable Banach space but the Banach space $\ell_+^\infty$ is not separable (see Examples 3P to 3S and Problems 3.49 and 3.59). It is not very difficult to check that $(\ell_+^{c_o})^* \cong \ell_+^1$ and $(\ell_+^1)^* \cong \ell_+^\infty$, and so $(\ell_+^{c_o})^{**} \cong \ell_+^\infty$ (see Problem 4.65 again). Thus $\ell_+^1$ is a separable Banach space with a nonseparable dual (reason: $(\ell_+^1)^*$ is not separable because $(\ell_+^1)^* \cong \ell_+^\infty$ and separability is a topological invariant). Hence,

$$\ell_+^1 \text{ is a nonreflexive Banach space}.$$

Since $\ell_+^{c_o}$ is not even homeomorphic to $\ell_+^\infty$ (Problem 3.49) and $(\ell_+^{c_o})^{**} \cong \ell_+^\infty$, it follows that $\ell_+^{c_o} \not\cong (\ell_+^{c_o})^{**}$. Therefore,

$$\ell_+^{c_o} \text{ is a nonreflexive Banach space}.$$

## Suggested Reading


Bachman and Narici [1]  
Banach [1]  
Beauzamy [1]  
Berberian [2]  
Brown and Pearcy [1]  
Conway [1]  
Kantorovich and Akilov [1]  
Kolmogorov and Fomin [1]  
Kreyszig [1]  
Maddox [1]  
Naylor and Sell [1]  
Reed and Simon [1]  

Douglas [1]

Dunford and Schwartz [1]

Goffman and Pedrick [1]

Goldberg [1]

Hille and Phillips [1]

Istrăţescu [1]

Robertson and Robertson [1]

Rudin [1]

Schwartz [1]

Simmons [1]

Taylor and Lay [1]

Yosida [1]

# Problems

**Problem 4.1.** We shall say that a topology $\mathcal{T}$ on a linear space $\mathcal{X}$ over a field $\mathbb{F}$ is *compatible with the linear structure* of $\mathcal{X}$ if vector addition and scalar multiplication are continuous mappings of $\mathcal{X} \times \mathcal{X}$ into $\mathcal{X}$ and of $\mathbb{F} \times \mathcal{X}$ into $\mathcal{X}$, respectively. In this case $\mathcal{T}$ is said to be a *compatible topology* (or a *linear topology*) on $\mathcal{X}$. When we refer to continuity of the mappings $\mathcal{X} \times \mathcal{X} \to \mathcal{X}$ and $\mathbb{F} \times \mathcal{X} \to \mathcal{X}$ defined by $(x, y) \mapsto x+y$ and $(\alpha, x) \mapsto \alpha x$, respectively, it is understood that $\mathcal{X} \times \mathcal{X}$ and $\mathbb{F} \times \mathcal{X}$ are equipped with their product topology. If $\mathcal{X}$ is a metric space, then these are the topologies induced by any of the uniformly equivalent metrics of Problems 3.9 and 3.33. If $\mathcal{X}$ is a general topological space, then these are the product topologies (cf. remark in Problem 3.64). A *topological vector space* (or *topological linear space*) is a linear space $\mathcal{X}$ equipped with a compatible topology.

(a) Show that, for each $y$ in a topological vector space $\mathcal{X}$ and each $\alpha$ in $\mathbb{F}$, the translation mapping $x \mapsto x + y$ and the scaling mapping $x \mapsto \alpha x$ are homeomorphisms of $\mathcal{X}$ onto itself.

(b) Show that every normed space is a topological vector space (a metrizable topological vector space, that is).

In other words, show that vector addition and scalar multiplication are continuous mappings of $\mathcal{X} \times \mathcal{X}$ into $\mathcal{X}$ and of $\mathbb{F} \times \mathcal{X}$ into $\mathcal{X}$, respectively, with respect to the norm topology on $\mathcal{X}$.

**Problem 4.2.** Consider the definitions of convex set and convex hull in a linear space (Problem 2.2). Recall that the closure of a subset of a topological space is the intersection of all closed subsets that include it.

(a) Show that in a topological vector space the closure of a convex set is convex.

The intersection of all closed and convex subsets that include a subset $A$ of a topological vector space is called the *closed convex hull* of $A$.

(b) Show that in a topological vector space the closed convex hull of a set $A$ coincides with $\mathrm{co}(A)^-$ (i.e., it coincides with the closure of the convex hull of $A$).

A subset $A$ of a linear space $\mathcal{X}$ is *balanced* if $\alpha x \in A$ for every vector $x \in A$ and every scalar $\alpha$ such that $|\alpha| \le 1$ (i.e., if $\alpha A \subseteq A$ whenever $|\alpha| \le 1$). A subset of a linear space is *absolutely convex* if it is both convex and balanced.

(c) Show that a subset $A$ of a linear space is absolutely convex if and only if $\alpha x + \beta y \in A$ for every $x, y \in A$ and all scalars $\alpha, \beta$ such that $|\alpha| + |\beta| \le 1$ (hence the term "absolutely convex").

(d) Show that the interior $A^\circ$ of an absolutely convex set $A$ contains the origin whenever $A^\circ$ is nonempty.

(e) Show that in a topological vector space the closure of a balanced set is balanced, and therefore the closure of an absolutely convex set is absolutely convex.

A subset $A$ of a linear space is *absorbing* (or *absorbent*) if for each vector $x \in \mathcal{X}$ there exists $\varepsilon > 0$ such that $\alpha x \in A$ for every scalar $\alpha$ with $|\alpha| \le \varepsilon$. Equivalently, if for each $x \in \mathcal{X}$ there exists $\lambda > 0$ such that $x \in \mu A$ for every scalar $\mu$ with $|\mu| \ge \lambda$.

(f) Show that in a topological vector space every neighborhood of the origin is absorbing.

A subset $A$ of a linear space $\mathcal{X}$ *absorbs* a subset $B$ of $\mathcal{X}$ (or $B$ is *absorbed* by $A$) if there exists $\beta > 0$ such that $x \in B$ implies $x \in \mu A$ for every scalar $\mu$ with $|\mu| \ge \beta$ (i.e., if $B \subseteq \mu A$ whenever $|\mu| \ge \beta$). In particular, $A$ is absorbing if and only if it absorbs every singleton $\{x\}$ in $\mathcal{X}$. A subset $B$ of a topological vector space is said to be *bounded* if it is absorbed by every neighborhood of the origin.

**Problem 4.3.** Let $\mathcal{X}$ be a linear space over a field $\mathbb{F}$ (either $\mathbb{F} = \mathbb{C}$ or $\mathbb{F} = \mathbb{R}$). A *quasinorm* on $\mathcal{X}$ is a real-valued positive subadditive functional $\| \ \| \colon \mathcal{X} \to \mathbb{R}$ on $\mathcal{X}$ that satisfies the axioms (i), (ii) and (iv) of Definition 4.1 but, instead of axiom (iii), it satisfies the following ones.

$$(\mathrm{iii}') \quad \|\alpha x\| \le \|x\| \quad \text{whenever } \alpha \le 1,$$
$$(\mathrm{iii}'') \quad \|\alpha_n x\| \to 0 \quad \text{whenever } \alpha_n \to 0,$$

for every $x \in \mathcal{X}$ (as usual, $\alpha$ stands for a scalar in $\mathbb{F}$ and $\{\alpha_n\}$ for a scalar-valued sequence). A linear space $\mathcal{X}$ equipped with a quasinorm is called a *quasinormed space*. Consider the mapping $d \colon \mathcal{X} \times \mathcal{X} \to \mathbb{R}$ defined by $d(x, y) = \|x - y\|$ for every $x, y \in \mathcal{X}$, where $\| \ \| \colon \mathcal{X} \to \mathbb{R}$ is a quasinorm on $\mathcal{X}$.

(a) Show that $d$ is an additively invariant metric on $\mathcal{X}$ that also satisfies the condition
$$d(\alpha x, \alpha y) \le d(x, y)$$
for every $x, y \in \mathcal{X}$ and every $\alpha \in \mathbb{F}$ such that $|\alpha| \le 1$.

This is called the *metric generated* by the quasinorm $\|\ \|$.

    (b)  Show that a norm on $\mathcal{X}$ is a quasinorm on $\mathcal{X}$, so that every normed space is a quasinormed space.

A quasinormed space that is complete as a metric space (with respect to the metric generated by the quasinorm) is called an *F-space*.

    (c)  Verify that every Banach space is an $F$-space.

**Problem 4.4.** A *neighborhood base* at a point $x$ in a topological vector space is a collection $\mathcal{N}$ of neighborhoods of $x$ with the property that every neighborhood of $x$ includes some neighborhood in $\mathcal{N}$. A *locally convex space* (or simply a *convex space*) is a topological vector space that has a neighborhood base at the origin consisting of convex sets.

    (a)  Show that every normed space is locally convex.

A *barrel* is a subset of a locally convex space that is convex, balanced, absorbing and closed. It can be shown that every locally convex space has a neighborhood base at the origin consisting of barrels. A locally convex space is called *barreled* if every barrel is a neighborhood of the origin. Barreled spaces can be thought of as a generalization of Banach spaces. Indeed, a sequence $\{x_n\}$ in a locally convex space is a *Cauchy sequence* if for each neighborhood $N$ at the origin there exists an integer $n_N$ such that $x_m - x_n \in N$ for all $m, n \geq n_N$. It can be verified that every convergent sequence in a metrizable locally convex space $\mathcal{X}$ (i.e., every sequence that is eventually in every open neighborhood of a point $x$ in $\mathcal{X}$) is a Cauchy sequence. A set $A$ in a metrizable locally convex space is *complete* if every Cauchy sequence in $A$ converges to a point of $A$. A complete metrizable locally convex space is called a *Fréchet space*. Recall the definition of $F$-space (Problem 4.3) and also that every Banach space is an $F$-space.

    (b)  Show that every $F$-space is a Fréchet space.

    (c)  Show that every Fréchet space is a barreled space.

        *Hint*: Note that a Fréchet space $\mathcal{X}$ is a complete metric space. Take an arbitrary barrel $B$ in $\mathcal{X}$ and show that the countable collection $\{nB\}_{n\geq 1}$ of closed sets covers $\mathcal{X}$. Now apply the Baire Category Theorem (Theorem 3.58) — see Problem 4.2(d).

We shall return to barreled spaces in Problem 4.44.

**Problem 4.5.** Consider the definition of a bounded subset of a metric space: $A$ is bounded if and only if $\mathrm{diam}(A) < \infty$ (Section 3.1).

(a) Show that a subset $A$ of a normed space $\mathcal{X}$ is bounded if and only if $\sup_{x \in A} \|x\| < \infty$. (By convention, $\sup_{x \in \varnothing} \|x\| = 0$.)

Now consider the definition of a bounded subset of a topological vector space as given in Problem 4.2.

(b) Show that a set $A$ is bounded as a subset of a normed space $\mathcal{X}$ if and only if it is bounded as a subset of the topological vector space $\mathcal{X}$. That is, $\sup_{x \in A} \|x\| < \infty$ if and only if $A$ is absorbed by every neighborhood of the origin of $\mathcal{X}$. In other words, the notion of a bounded subset of a normed space is unambiguously defined.

Let $A$ be a subset of a normed space. Suppose $A \backslash \{0\} \neq \varnothing$ and prove the following propositions.

(c)  $\sup_{x \in A} \|x\| < \infty$  implies  $\inf_{x \in A \backslash \{0\}} \|x\|^{-1} = \left( \sup_{x \in A} \|x\| \right)^{-1}$.

(d)  $\inf_{x \in A} \|x\| \neq 0$  implies  $\sup_{x \in A \backslash \{0\}} \|x\|^{-1} = \left( \inf_{x \in A} \|x\| \right)^{-1}$.

Clearly, $\inf_{x \in A} \|x\| \neq 0$ if and only if $\inf_{x \in A} \|x\| > 0$ (since $\|x\| \geq 0$ for all $x$ in any normed space). It is also clear that $\inf_{x \in A} \|x\| \leq \sup_{x \in A} \|x\|$, and $\inf_{x \in A} \|x\| < \infty$ even if $A$ is unbounded. Show that

(e)  $\inf_{x \in A} \|x\| \neq 0$  if and only if  $\sup_{x \in A \backslash \{0\}} \|x\|^{-1} < \infty$.

A nonempty subset $A$ of a normed space is *bounded away from zero* if $\inf_{x \in A} \|x\| \neq 0$. Accordingly, a mapping $F$ of a nonempty set $S$ into a normed space $\mathcal{X}$ is bounded if and only if $\sup_{s \in S} \|F(s)\| < \infty$; and bounded away from zero if and only if $\inf_{s \in S} \|F(s)\| \neq 0$. In particular, an $\mathcal{X}$-valued sequence $\{x_n\}$ is bounded if and only if $\sup_n \|x_n\| < \infty$; and bounded away from zero if and only if $\inf_n \|x_n\| \neq 0$.

**Problem 4.6.** This problem is entirely based on the triangle inequality. Consider the spaces $(\ell_+^1, \| \; \|_1)$ and $(\ell_+^\infty, \| \; \|_\infty)$ of Example 4B. Take $x = \{\xi_k\}_{k=1}^\infty \in \ell_+^\infty$ and let $\{x_n\}_{n=1}^\infty$ be an $\ell_+^1$-valued sequence (i.e., each $x_n = \{\xi_n(k)\}_{k=1}^\infty$ lies in $\ell_+^1$). Recall: $\ell_+^1 \subset \ell_+^\infty$. Show that

(a) if $\|x_n - x\|_\infty \to 0$ and $\sup_n \|x_n\|_1 < \infty$, then $x \in \ell_+^1$.

Hint: $\sum_{k=1}^m |\xi_k| \leq m \|x_n - x\|_\infty + \sup_n \|x_n\|_1$ for each $m \geq 1$.

Now suppose $x \in \ell_+^1$ and show that

(b) if $\|x_n - x\|_\infty \to 0$ and $\|x_n\|_1 \to \|x\|_1$, then $\|x_n - x\|_1 \to 0$.

*Hint:* If $x = \{\xi_k\}_{k=1}^{\infty}$ and $z = \{\zeta_k\}_{k=1}^{\infty}$ are in $\ell_+^1$ then, for each $m \geq 1$,

$$\|z\|_1 + \|x\|_1 \leq \|z - x\|_1 + 2m\|z\|_\infty + 2\sum_{k=m+1}^{\infty}|\xi_k|.$$

(Note that $\|z\|_1 + \|x\|_1 = \sum_{k=1}^{m}(|\xi_k| - |\zeta_k|) + \sum_{k=m+1}^{\infty}(|\zeta_k| - |\xi_k|) + 2(\sum_{k=1}^{m}|\zeta_k| + \sum_{k=m+1}^{\infty}|\xi_k|)$ and $\big||\xi_k| - |\zeta_k|\big| \leq |\xi_k - \zeta_k|$.) Prove the above auxiliary inequality and conclude: if $x,\, y \in \ell_+^1$, then

$$\|y - x\|_1 \leq \big|\,\|y\|_1 - \|x\|_1\big| + 2m\|y - x\|_\infty + 2\sum_{k=m+1}^{\infty}|\xi_k|$$

for each $m \geq 1$. Show that, under the assumptions of (b),

$$\limsup_n \|x_n - x\|_1 \leq 2\sum_{k=m+1}^{\infty}|\xi_k| \quad \text{for all} \quad m \geq 1.$$

Next suppose the sequence $\{x_n\}_{n=1}^{\infty}$ is dominated by a vector $y$ in $\ell_+^1$. That is, $|\xi_n(k)| \leq |\upsilon_k|$ for each $k \geq 1$ and all $n \geq 1$, for some $y = \{\upsilon_k\}_{k=1}^{\infty}$ in $\ell_+^1$. Equivalently, $\sum_{k=1}^{\infty}\sup_n|\xi_n(k)| < \infty$. Show that

(c)  if $\|x_n - x\|_\infty \to 0$ and $\sum_{k=1}^{\infty}\sup_n|\xi_n(k)| < \infty$, then $\|x_n - x\|_1 \to 0$.

*Hint:* Item (a) and the dominance condition ensure that $x \in \ell_+^1$ ($\sup_n\|x_n\|_1 \leq \|y\|_1$ for some $y \in \ell_+^1$). Moreover, for each $m \geq 1$,

$$\|x_n - x\|_1 \leq m\|x_n - x\|_\infty + \sum_{k=m+1}^{\infty}|\xi_n(k)| + \sum_{k=m+1}^{\infty}|\xi_k|.$$

Extend these results to $\ell_+^1(\mathcal{X})$ and $\ell_+^\infty(\mathcal{X})$ as in Example 4F.

**Problem 4.7.** Let $\{x_i\}_{i=1}^{\infty}$ be a sequence in a normed space $\mathcal{X}$ and consider the sequence $\{y_n\}_{n=1}^{\infty}$ of partial sums of $\{x_i\}_{i=1}^{\infty}$; that is, $y_n = \sum_{i=1}^{n}x_i$ for each $n \geq 1$. Prove that the following assertions are pairwise equivalent.

(a)  $\{y_n\}_{n=1}^{\infty}$ is a Cauchy sequence in $\mathcal{X}$.

(b)  The real sequence $\{\|\sum_{i=n}^{n+k}x_i\|\}_{n=1}^{\infty}$ converges to zero uniformly in $k$; that is,

$$\limsup_{n \ \ k\geq 0} \left\|\sum_{i=n}^{n+k}x_i\right\| = 0.$$

(c)  For each real number $\varepsilon > 0$ there exists an integer $n_\varepsilon \geq 1$ such that

$$\left\|\sum_{i=m}^{n}x_i\right\| < \varepsilon \quad \text{whenever} \quad n_\varepsilon \leq m \leq n.$$

(*Hint*: Problem 3.51). Observe that, according to item (b), $x_n \to 0$ in $\mathcal{X}$ whenever $\{\sum_{i=1}^{n} x_i\}_{n=1}^{\infty}$ is a Cauchy sequence in $\mathcal{X}$ and, in particular, whenever the infinite series $\sum_{i=1}^{\infty} x_i$ converges in $\mathcal{X}$. That is, $x_n \to 0$ in $\mathcal{X}$ *for every summable sequence* $\{x_i\}_{i=1}^{\infty}$ *in* $\mathcal{X}$. Now suppose $\mathcal{X}$ is a Banach space. As $\mathcal{X}$ is complete, then (by the very definition of completeness) each of the above equivalent assertions also is equivalent to the following one.

(d)   The infinite series $\sum_{i=1}^{\infty} x_i$ converges in $\mathcal{X}$ (i.e., the sequence $\{x_i\}_{i=1}^{\infty}$ is summable).

If $\mathcal{X}$ is a Banach space, then condition (c) (or condition (b)) is referred to as the *Cauchy Criterion* for convergent infinite series.

**Problem 4.8.**   Proposition 4.4 says that every absolutely summable sequence in a Banach space is summable.

(a)   Consider the $\ell_+^2$-valued sequence $\{x_i\}_{i=1}^{\infty}$ where, for each positive integer $i$, $x_i = \frac{1}{i} e_i$ with $e_i = \{\delta_{ik}\}_{k=1}^{\infty} \in \ell_+^2$ (just one nonzero entry equal to one at the $i$th position). Show that $\{x_i\}_{i=1}^{\infty}$ is a summable sequence in the Banach space $\ell_+^2$ but is not absolutely summable.

Problem 4.7 says that $\{x_i\}_{i=1}^{\infty}$ is a summable sequence in a Banach space if and only if $\lim_n \sup_{k \geq 0} \|\sum_{i=n}^{n+k} x_i\| = 0$, which clearly implies that $\lim_n \|\sum_{i=n}^{n+k} x_i\| = 0$ for every integer $k \geq 0$.

(b)   Give an example of a nonsummable sequence (or, equivalently, an example of a nonconvergent series) in a Banach space such that $\lim_n \|\sum_{i=n}^{n+k} x_i\| = 0$ for every integer $k \geq 0$. (*Hint*: $\xi_i = \frac{1}{i}$ in $\mathbb{R}$.)

**Problem 4.9.**   Prove the following propositions.

(a)   If $\{x_i\}_{i=0}^{\infty}$ and $\{y_i\}_{i=0}^{\infty}$ are summable sequences in a normed space $\mathcal{X}$, then $\{\alpha x_i + \beta y_i\}_{i=0}^{\infty} = \alpha\{x_i\}_{i=0}^{\infty} + \beta\{y_i\}_{i=0}^{\infty}$ is again a summable sequence in $\mathcal{X}$, and

$$\sum_{i=0}^{\infty}(\alpha x_i + \beta y_i) = \alpha \sum_{i=0}^{\infty} x_i + \beta \sum_{i=0}^{\infty} y_i,$$

for every pair of scalars $\alpha, \beta \in \mathbb{F}$. This shows that the collection of all $\mathcal{X}$-valued summable sequences is a linear manifold of the linear space $\mathcal{X}^{\mathbb{N}_0}$ (see Example 2F), and hence is a linear space itself.

(b)   If $\{x_i\}_{i=0}^{\infty}$ is a summable sequences in a normed space $\mathcal{X}$, then

$$\sum_{i=n}^{\infty} x_i \to 0 \quad \text{in} \quad \mathcal{X} \quad \text{as} \quad n \to \infty.$$

*Hint:* $\sum_{i=0}^{m} x_i = \sum_{i=0}^{n-1} x_i + \sum_{i=n}^{m} x_i$ for every $1 \leq n \leq m$. Now use item (a) to verify that the series $\sum_{i=n}^{\infty} x_i = \sum_{i=0}^{\infty} x_{n-i}$ converges in $\mathcal{X}$, and $\sum_{i=0}^{\infty} x_i = \sum_{i=0}^{n-1} x_i + \sum_{i=n}^{\infty} x_i$ for every $n \geq 1$. Thus the result follows by uniqueness of the limit.

**Problem 4.10.** Let $x = \{x_i\}_{i=0}^{\infty}$ be a sequence of linearly independent vectors in a normed space $\mathcal{X}$, and let $\mathcal{A}_x$ be the collection of all scalar-valued sequences $a = \{\alpha_i\}_{i=0}^{\infty}$ for which the series $\sum_{i=0}^{\infty} \alpha_i x_i$ converges in $\mathcal{X}$ (i.e., such that $\{\alpha_i x_i\}_{i=0}^{\infty}$ is a summable sequence in $\mathcal{X}$). Show that

(a) $\qquad\qquad \mathcal{A}_x$ is a linear manifold of the linear space $\mathbb{F}^{\mathbb{N}_0}$.

Recall: $\mathbb{F} = \mathbb{C}$ or $\mathbb{F} = \mathbb{R}$, whether $\mathcal{X}$ is a complex or real linear space, respectively. Verify that the function $\| \, \| \colon \mathcal{A}_x \to \mathbb{R}$, defined by

(b)
$$\|a\| = \sup_n \left\| \sum_{i=0}^{n} \alpha_i x_i \right\|_X$$

for every $a = \{\alpha_i\}_{i=0}^{\infty}$ in $\mathcal{A}_x$, is a norm on $\mathcal{A}_x$. Take an arbitrary integer $i \geq 0$. Show that

$$|\alpha_i| \, \|x_i\|_X \leq 2\|a\|$$

for every $a = \{\alpha_i\}_{i=0}^{\infty}$ in $\mathcal{A}_x$ (*hint:* $\alpha_i x_i = \sum_{j=0}^{i} \alpha_j x_j - \sum_{j=0}^{i-1} \alpha_j x_j$ for each $i \geq 1$), and hence

(c)
$$|\alpha_i - \beta_i| \, \|x_i\|_X \leq 2\|a - b\|$$

for every $a = \{\alpha_i\}_{i=0}^{\infty}$ and $b = \{\beta_i\}_{i=0}^{\infty}$ in $\mathcal{A}_x$. (Reason: $\mathcal{A}_x$ is a linear space by item (a).) Now let $\{a_k\}_{k=0}^{\infty}$ be a Cauchy sequence in the normed space $\mathcal{A}_x$. That is, each $a_k = \{\alpha_k(i)\}_{i=0}^{\infty}$ lies in $\mathcal{A}_x$ and the $\mathcal{A}_x$-valued sequence $\{a_k\}_{k=0}^{\infty}$ is Cauchy in $(\mathcal{A}_x, \| \, \|)$, where $\| \, \|$ is the norm in (b). According to (c), $|\alpha_k(i) - \alpha_\ell(i)| \leq 2\|x_i\|_X^{-1} \|a_k - a_\ell\|$ for each $i \geq 0$ and every $k, \ell \geq 0$ (recall: $\{x_i\}_{i=0}^{\infty}$ is linearly independent so that $\|x_i\|_X \neq 0$ for every $i \geq 0$). Thus the scalar-valued sequence $\{\alpha_k(i)\}_{k=0}^{\infty}$ is Cauchy in $\mathbb{F}$, and so convergent in $\mathbb{F}$, for each $i \geq 0$. Set

$$\widetilde{\alpha}_i = \lim_k \alpha_k(i)$$

in $\mathbb{F}$ for each integer $i \geq 0$ and consider the sequence $\widetilde{a} = \{\widetilde{\alpha}_i\}_{i=0}^{\infty}$ in $\mathbb{F}^{\mathbb{N}_0}$. Take an arbitrary $\varepsilon > 0$. Show that there exists an integer $k_\varepsilon \geq 0$ such that

$$\left\| \sum_{i=m}^{n} (\alpha_k(i) - \alpha_\ell(i)) x_i \right\|_X < 2\varepsilon$$

for every $0 \le m \le n$, whenever $k, \ell \ge k_\varepsilon$ (*hint:* $\sum_{i=m}^{n}(\alpha_{k(i)} - \alpha_{\ell(i)})x_i = \sum_{i=0}^{n}(\alpha_{k(i)} - \alpha_{\ell(i)})x_i - \sum_{i=0}^{m-1}(\alpha_{k(i)} - \alpha_{\ell(i)})x_i$ and $\{a_k\}_{k=0}^{\infty}$ is a Cauchy sequence in $\mathcal{A}_x$), and hence

$$(d) \qquad \Big\| \sum_{i=m}^{n} (\alpha_{k(i)} - \widetilde{\alpha}_i)x_i \Big\|_X < 2\varepsilon$$

for every pair of nonnegative integers $m \le n$, whenever $k \ge k_\varepsilon$. (*Hint:* Note that $\|\sum_{i=m}^{n}(\alpha_{k(i)} - \lim_\ell \alpha_{\ell(i)})x_i\|_X = \lim_\ell \|\sum_{i=m}^{n}(\alpha_{k(i)} - \alpha_{\ell(i)})x_i\|_X$. Why?) Next prove the following claims.

(e) \qquad\qquad\qquad If $\mathcal{X}$ is a Banach space, then $\widetilde{a} \in \mathcal{A}_x$.

(*Hint:* (d) implies that the infinite series $\sum_{i=0}^{\infty}(\alpha_{k(i)} - \widetilde{\alpha}_i)x_i$ converges in the Banach space $\mathcal{X}$ for every $k \ge k_\varepsilon$ — see Problem 4.7: Cauchy Criterion — and hence $a_k - \widetilde{a} \in \mathcal{A}_x$ for each $k \ge k_\varepsilon$.)

(f) \qquad\qquad\qquad If $\widetilde{a} \in \mathcal{A}_x$, then $a_k \to \widetilde{a}$ in $\mathcal{A}_x$.

(*Hint:* $\widetilde{a} \in \mathcal{A}_x$ implies $\|a_k - \widetilde{a}\| < 2\varepsilon$ whenever $k \ge k_\varepsilon$ by setting $m = 0$ in (d).) Finally conclude from (e) and (f):

If $(\mathcal{X}, \| \ \|_X)$ is a Banach space, then $(\mathcal{A}_x, \| \ \|)$ is a Banach space.

**Problem 4.11.** A sequence $\{x_i\}_{i=0}^{\infty}$ of vectors in a normed space $\mathcal{X}$ is a *Schauder basis* for $\mathcal{X}$ if for each $x$ in $\mathcal{X}$ there exists a *unique* (similarly indexed) sequence of scalars $\{\alpha_i\}_{i=0}^{\infty}$ such that

$$x = \sum_{i=0}^{\infty} \alpha_i x_i$$

(i.e., $x = \lim_n \sum_{i=0}^{n} \alpha_i x_i$). The entries of the scalar-valued sequence $\{\alpha_i\}_{i=0}^{\infty}$ are called the *coefficients* of $x$ with respect to the Schauder basis $\{x_i\}_{i=0}^{\infty}$, and the (convergent) series $\sum_{i=0}^{\infty} \alpha_i x_i$ is the *expansion* of $x$ with respect to $\{x_i\}_{i=0}^{\infty}$. Prove the following assertions.

(a) Every Schauder basis for a normed space $\mathcal{X}$ is a sequence of linearly independent vectors in $\mathcal{X}$ that spans $\mathcal{X}$. That is, if $\{x_i\}_{i=0}^{\infty}$ is a Schauder basis for $\mathcal{X}$, then

    (i) $\ \ \{x_i\}_{i=0}^{\infty}$ is linearly independent, and

    (ii) $\ \ \bigvee \{x_i\}_{i=0}^{\infty} = \mathcal{X}$.

(b) If a normed space has a Schauder basis, then it is separable.

*Hint*: Proposition 4.9(b).

Remark: An infinite sequence of linearly independent vectors exists only in an infinite-dimensional linear space. It is readily verified that finite-dimensional normed spaces are separable Banach spaces (see Problem 4.37 below) but, in this case, the purely algebraic notion of a (finite) Hamel basis is enough. Thus, when the concept of a Schauder basis is under discussion, only infinite-dimensional spaces are considered. Does every separable Banach space have a Schauder basis? This is a famous question, raised by Banach himself in the early thirties, that remained open for a long period. Each separable Banach space that ever came up in analysis during that period (and this includes all classical examples) had a Schauder basis. The surprising negative answer to that question was given by Enflo, who constructed in the early seventies a separable Banach space that has no Schauder basis. See also the remark in Problem 4.58 below.

**Problem 4.12.** As usual, for each integer $i \geq 0$ let $e_i$ be a scalar-valued sequence with just one nonzero entry (equal to one) at the $i$th position (i.e., $e_i = \{\delta_{ik}\}_{k=0}^{\infty}$ for every $i \geq 0$). Consider the Banach spaces $\ell_+^{\infty}$ and $\ell_+^p$ for every $p \geq 1$ as in Example 4B. Show that the sequence $\{e_i\}_{i=0}^{\infty}$ is a Schauder basis for each $\ell_+^p$, and verify that $\ell_+^{\infty}$ has no Schauder basis. (*Hint*: Example 3Q.)

**Problem 4.13.** Let $\mathcal{M}$ be a subspace of a normed space $\mathcal{X}$. If $\mathcal{X}$ is a Banach space, then $\mathcal{M}$ and $\mathcal{X}/\mathcal{M}$ are both Banach spaces (Propositions 4.7 and 4.10). Conversely, *if $\mathcal{M}$ and $\mathcal{X}/\mathcal{M}$ are Banach spaces, then $\mathcal{X}$ is a Banach space.* Indeed, suppose $\mathcal{M}$ and $\mathcal{X}/\mathcal{M}$ are Banach spaces, and let $\{x_n\}_{n \in \mathbb{N}}$ be an arbitrary Cauchy sequence in $\mathcal{X}$.

(a)  Show that $\{[x_n]\}$ is a Cauchy sequence in $\mathcal{X}/\mathcal{M}$ and conclude that $\{[x_n]\}$ converges in $\mathcal{X}/\mathcal{M}$ to, say, $[x] \in \mathcal{X}/\mathcal{M}$.

(b)  Take any $x$ in $[x]$. For each $n$ there exists $\widetilde{x}_n$ in $[x_n - x]$ such that

$$0 \leq \|\widetilde{x}_n\|_X \leq \|[x_n - x]\| + \tfrac{1}{n} = \|[x_n] - [x]\| + \tfrac{1}{n}.$$

Prove the above assertion and conclude: $\widetilde{x}_n \to 0$ in $\mathcal{X}$.

Observe that $\widetilde{x}_n - x_n + x$ lies in $\mathcal{M}$ for each $n$. In fact, $[\widetilde{x}_n] = [x_n - x]$ (for $\widetilde{x}_n \in [x_n - x]$), and so $[\widetilde{x}_n - x_n + x] = [\widetilde{x}_n] - [x_n - x] = [0] = \mathcal{M}$.

(c)  Set $u_n = \widetilde{x}_n - x_n + x$ in $\mathcal{M}$ and show that $\{u_n\}$ is a Cauchy sequence in $\mathcal{M}$. Thus conclude that $\{u_n\}$ converges in $\mathcal{M}$ (and hence in $\mathcal{X}$) to, say, $u \in \mathcal{M}$.

Since $x_n = \widetilde{x}_n + x - u_n$ for each integer $n$, it follows that $\{x_n\}$ converges in $\mathcal{X}$ to $x - u \in \mathcal{X}$ (for vector addition and scalar multiplication are continuous mappings — see Problem 4.1). Outcome: Every Cauchy sequence in $\mathcal{X}$ converges in $\mathcal{X}$, which means that $\mathcal{X}$ is a Banach space.

**Problem 4.14.** Consider the linear space $\ell_+^{c_o}$ and let $\{a_k\}_{k=1}^{\infty}$ be an $\ell_+^{c_o}$-valued sequence. That is, each $a_k = \{\alpha_k(n)\}_{n=1}^{\infty}$ is a scalar-valued sequence that converges to zero:

$$\lim_n |\alpha_k(n)| = 0 \quad \text{for every} \quad k \geq 1.$$

Suppose further that there exists a real number $\alpha \geq 0$ such that $|\alpha_k(n)| \leq \alpha$ for all $k, n \geq 1$ or, equivalently, suppose

$$\sup_k \sup_n |\alpha_k(n)| < \infty.$$

Take an arbitrary $x = \{\xi_k\}_{k=1}^{\infty}$ in $\ell_+^1$ so that

$$\sum_{k=1}^{\infty} |\xi_k| < \infty.$$

Consider the above assumptions and prove the following proposition.

(a) $\qquad \lim_n \sup_k |\alpha_k(n)||\xi_k| = 0 \quad \text{and} \quad \sum_{k=1}^{\infty} \sup_n |\alpha_k(n)||\xi_k| < \infty.$

> *Hint*: $\sup_k |\alpha_k(n)||\xi_k| \leq \max\{\max_{1 \leq k \leq m} |\alpha_k(n)||\xi_k|, \alpha \sum_{k=m+1}^{\infty} |\xi_k|\}$ for every $m, n \geq 1$, which implies
>
> $$\lim_n \sup_k |\alpha_k(n)||\xi_k| \leq \alpha \sum_{k=m+1}^{\infty} |\xi_k| \quad \text{for every} \quad m \geq 1.$$

Next use the dominated convergence of Problem 4.6(c) to show that

(b) $\qquad\qquad\qquad \lim_n \sum_{k=1}^{\infty} |\alpha_k(n)||\xi_k| = 0.$

Now conclude: For each integer $n \geq 1$ the infinite series $\sum_{k=1}^{\infty} \alpha_k(n)\xi_k$ converges (in the Banach space $\mathbb{F}$) and the scalar-valued sequence $\{\sum_{k=1}^{\infty} \alpha_k(n)\xi_k\}_{n=1}^{\infty}$ is an element of $\ell_+^{c_o}$. Therefore, every infinite matrix $A = [\alpha_k(n)]_{k,n \geq 1}$ whose rows $a_k$ satisfy the above assumptions represents a mapping of $\ell_+^1$ into $\ell_+^{c_o}$. Equip $\ell_+^1$ and $\ell_+^{c_o}$ with their usual norms ($\| \; \|_1$ on $\ell_+^1$ and $\| \; \|_\infty$ on $\ell_+^{c_o}$). Observe that the assumptions on $A$ simply say that $\{a_k\}_{k=1}^{\infty}$ is a bounded (i.e., $\sup_k \|a_k\|_\infty < \infty$) $\ell_+^{c_o}$-valued sequence. Show that, under these assumptions, such a mapping in fact is a bounded linear transformation of $\ell_+^1$ into $\ell_+^{c_o}$:

$$A \in \mathcal{B}[\ell_+^1, \ell_+^{c_o}] \quad \text{and} \quad \|A\| = \sup_k \|a_k\|_\infty.$$

**Problem 4.15.** If $\{x_k\}_{k=0}^{\infty}$ is a summable sequence in a normed space $\mathcal{X}$ and $T$ is a continuous linear transformation of $\mathcal{X}$ into a normed space $\mathcal{Y}$ (i.e., if $T \in \mathcal{B}[\mathcal{X}, \mathcal{Y}]$), then show that $\{T x_k\}_{k=0}^{\infty}$ is a summable sequence in $\mathcal{Y}$ and

$$\sum_{k=0}^{\infty} T x_k = T\Big(\sum_{k=0}^{\infty} x_k\Big).$$

**Problem 4.16.** Let $(\mathcal{X}_1, \|\ \|_1)$ and $(\mathcal{X}_2, \|\ \|_2)$ be normed spaces and consider the normed space $\mathcal{X}_1 \oplus \mathcal{X}_2$ equipped with any of the norms of Example 4E, which we shall denote simply by $\|\ \|$. That is, for every $(x_1, x_2)$ in $\mathcal{X}_1 \oplus \mathcal{X}_2$, either $\|(x_1, x_2)\|^p = \|x_1\|_1^p + \|x_2\|_2^p$ for some $p \geq 1$ or $\|(x_1, x_2)\| = \max\{\|x_1\|_1, \|x_2\|_2\}$. Now let $T_1$ and $T_2$ be operators on $\mathcal{X}_1$ and $\mathcal{X}_2$, respectively (i.e., $T_1 \in \mathcal{B}[\mathcal{X}_1]$ and $T_2 \in \mathcal{B}[\mathcal{X}_2]$). Consider the direct sum $T \in \mathcal{L}[\mathcal{X}_1 \oplus \mathcal{X}_2]$ of $T_1$ and $T_2$ (defined in Section 2.9):

$$T = T_1 \oplus T_2 = \begin{pmatrix} T_1 & O \\ O & T_2 \end{pmatrix},$$

where $T_1 = T|_{\mathcal{X}_1}$ and $T_2 = T|_{\mathcal{X}_2}$ are the direct summands of $T$.

(a) Show that $T \in \mathcal{B}[\mathcal{X}_1 \oplus \mathcal{X}_2]$ and $\|T\| = \max\big\{\|T_1\|, \|T_2\|\big\}$.

*Hint*: Whatever is the norm $\|\ \|$ that equips $\mathcal{X}_1 \oplus \mathcal{X}_2$ (among those of Example 4E), $\|T(x_1, x_2)\| \leq \max\big\{\|T_1\|, \|T_2\|\big\} \|(x_1, x_2)\|$ for $(x_1, x_2) \in \mathcal{X}_1 \oplus \mathcal{X}_2$.

Generalize to a countable direct sum. That is, let $\{\mathcal{X}_k\}$ be an indexed countable family of normed spaces and consider the normed space $\bigoplus_k \mathcal{X}_k$ of Examples 4E or 4F (equipped with any of those norms, so that either $\bigoplus_k \mathcal{X}_k = [\bigoplus_k \mathcal{X}_k]_p$ or $\bigoplus_k \mathcal{X}_k = [\bigoplus_k \mathcal{X}_k]_\infty$ in case of a countably infinite family as in Example 4F). Let $\{T_k\}$ be a similarly indexed countable family of operators on $\mathcal{X}_k$ (each $T_k$ lying in $\mathcal{B}[\mathcal{X}_k]$) such that $\sup_k \|T_k\| < \infty$. Set

$$T\{x_k\} = \{T_k x_k\} \quad \text{for each} \quad \{x_k\} \in \bigoplus_k \mathcal{X}_k.$$

(b) Show that this actually defines a bounded linear transformation $T$ of $\bigoplus_k \mathcal{X}_k$ into itself. Such an operator is usually denoted by

$$T = \bigoplus_k T_k \quad \text{in} \quad \mathcal{B}\big[\bigoplus_k \mathcal{X}_k\big],$$

and referred to as the *direct sum* of $\{T_k\}$. Moreover, verify that $T_k = T|_{\mathcal{X}_k}$ (in the sense of Section 2.9) for each $k$. These are the *direct summands* of $T$. Finally, show that

$$\|T\| = \sup_k \|T_k\|.$$

If $\mathcal{X}_k = \mathcal{X}$ for a single normed space $\mathcal{X}$, then $T = \bigoplus_k T_k$ is sometimes referred to as a *block diagonal operator* acting on $\ell_+^p(\mathcal{X})$ or on $\ell_+^\infty(\mathcal{X})$.

**Problem 4.17.** Let $\mathcal{X}_i$ and $\mathcal{Y}_i$ be normed spaces for $i = 1, 2$ and consider the normed spaces $\mathcal{X}_1 \oplus \mathcal{X}_2$ and $\mathcal{Y}_1 \oplus \mathcal{Y}_2$ equipped with any of the norms of Example 4E.

(a) Take $T_{ij}$ in $\mathcal{B}[\mathcal{X}_j, \mathcal{X}_i]$ for $i, j = 1, 2$ so that $T_{11}x_1 + T_{12}x_2$ lies in $\mathcal{Y}_1$ and $T_{21}x_1 + T_{22}x_2$ lies in $\mathcal{Y}_2$ for every $(x_1, x_2)$ in $\mathcal{X}_1 \oplus \mathcal{X}_2$. Set

$$T(x_1, x_2) = (T_{11}x_1 + T_{12}x_2, \; T_{21}x_2 + T_{22}x_2) \quad \text{in} \quad \mathcal{Y}_1 \oplus \mathcal{Y}_2.$$

Show that this defines a mapping $T : \mathcal{X}_1 \oplus \mathcal{X}_2 \to \mathcal{Y}_1 \oplus \mathcal{Y}_2$, which in fact lies in $\mathcal{B}[\mathcal{X}_1 \oplus \mathcal{X}_2, \mathcal{Y}_1 \oplus \mathcal{Y}_2]$, and

$$\max_{\substack{i=1,2 \\ j=1,2}} \left\{ \|T_{ij}\| \right\} \leq \|T\| \leq 4 \max_{\substack{i=1,2 \\ j=1,2}} \left\{ \|T_{ij}\| \right\}.$$

(b) Conversely, suppose $T \in \mathcal{B}[\mathcal{X}_1 \oplus \mathcal{X}_2, \mathcal{Y}_1 \oplus \mathcal{Y}_2]$. If $x_1$ is an arbitrary vector in $\mathcal{X}_1$, then

$$T(x_1, 0) = (T_{11}x_1, \; T_{21}x_1)$$

in $\mathcal{Y}_1 \oplus \mathcal{Y}_2$, where $T_{11}$ is a mapping of $\mathcal{X}_1$ into $\mathcal{Y}_1$ and $T_{21}$ is a mapping of $\mathcal{X}_1$ into $\mathcal{Y}_2$. Similarly, if $x_2$ is any vector in $\mathcal{X}_2$, then

$$T(0, x_2) = (T_{12}x_2, \; T_{22}x_2)$$

in $\mathcal{Y}_1 \oplus \mathcal{Y}_2$, where $T_{12}$ is a mapping of $\mathcal{X}_2$ into $\mathcal{Y}_1$ and $T_{22}$ is a mapping of $\mathcal{X}_2$ into $\mathcal{Y}_2$. Show that $T_{ij} \in \mathcal{B}[\mathcal{X}_j, \mathcal{Y}_i]$ and $\|T_{ij}\| \leq \|T\|$ for every $i = 1, 2$ and $j = 1, 2$.

Consider the bounded linear transformation $T \in \mathcal{B}[\mathcal{X}_1 \oplus \mathcal{X}_2, \mathcal{Y}_1 \oplus \mathcal{Y}_2]$ of item (b). Since $T(x_1, x_2) = T(x_1, 0) \oplus T(0, x_2)$ in $\mathcal{Y}_1 \oplus \mathcal{Y}_2$, it follows that $T(x_1, x_2) = (T_{11}x_1 + T_{12}x_2, \; T_{21}x_1 + T_{22}x_2)$ for every $(x_1, x_2)$ in $\mathcal{X}_1 \oplus \mathcal{X}_2$ as in item (a). This establishes a one-to-one correspondence between each $T$ in $\mathcal{B}[\mathcal{X}_1 \oplus \mathcal{X}_2, \mathcal{Y}_1 \oplus \mathcal{Y}_2]$ and the $2 \times 2$ matrix of bounded linear transformations $[T_{ij}]$, called the *operator matrix* for $T$, which we shall represent by the same symbol $T$ (instead of, for instance, $[T]$) and write

$$T = \begin{pmatrix} T_{11} & T_{12} \\ T_{21} & T_{22} \end{pmatrix}.$$

Note that, if $\mathcal{Y}_i = \mathcal{X}_i$ for $i = 1, 2$, then $T$ is the direct sum $T_{11} \oplus T_{22}$ in $\mathcal{B}[\mathcal{X}_1 \oplus \mathcal{X}_2]$ of the previous problem if and only if $T_{12} = O$ in $\mathcal{B}[\mathcal{X}_2, \mathcal{X}_1]$ and $T_{21} = O$ in $\mathcal{B}[\mathcal{X}_1, \mathcal{X}_2]$.

**Problem 4.18.** Let $T$ be an operator on a normed space $\mathcal{X}$ (i.e., $T \in \mathcal{B}[\mathcal{X}]$). Recall that a subset $A$ of $\mathcal{X}$ is $T$-invariant (or $A$ is an invariant subset for $T$) if $T(A) \subseteq A$ (i.e., $Tx \in A$ whenever $x \in A$). If $\mathcal{M}$ is a linear manifold (or a subspace) of $\mathcal{X}$ and, as a subset of $\mathcal{X}$, is $T$-invariant, then we say that $\mathcal{M}$ is an *invariant linear manifold* (or an *invariant subspace*) for $T$. Prove the following assertions.

(a)  If $A$ is $T$-invariant, then $A^-$ is $T$-invariant.

(b)  If $\mathcal{M}$ is an invariant linear manifold for $T$, then $\mathcal{M}^-$ is an invariant subspace for $T$.

(c)  $\{0\}$ and $\mathcal{X}$ are invariant subspaces for every $T$ in $\mathcal{B}[\mathcal{X}]$.

**Problem 4.19.**  Let $\mathcal{X}$ be a normed space and let $\mathrm{Lat}(\mathcal{X})$ be the lattice of all subspaces of $\mathcal{X}$. Recall that $\{0\}$ and $\mathcal{X}$ are subspaces of $\mathcal{X}$ (so that they are elements of $\mathrm{Lat}(\mathcal{X})$). These are the trivial elements of $\mathrm{Lat}(\mathcal{X})$: a subspace in $\mathrm{Lat}(\mathcal{X})$ is nontrivial if it is a proper nonzero subspace of $\mathcal{X}$ (i.e., $\mathcal{M} \in \mathrm{Lat}(\mathcal{X})$ is nontrivial if $\{0\} \neq \mathcal{M} \neq \mathcal{X}$).

(a)  Check that there exist nontrivial subspaces in $\mathrm{Lat}(\mathcal{X})$ if and only if the dimension of $\mathcal{X}$ is greater than 1 (i.e., $\mathrm{Lat}(\mathcal{X}) \neq \big\{\{0\},\,\mathcal{X}\big\}$ if and only if $\dim \mathcal{X} > 1$).

Let $\mathcal{B}[\mathcal{X}]$ be the unital algebra of all operators on a normed space $\mathcal{X}$ and let $T$ be an operator in $\mathcal{B}[\mathcal{X}]$. A *nontrivial invariant subspace* for $T$ is a nontrivial element of $\mathrm{Lat}(\mathcal{X})$ which is invariant for $T$ (i.e., a subspace $\mathcal{M} \in \mathrm{Lat}(\mathcal{X})$ such that $\{0\} \neq \mathcal{M} \neq \mathcal{X}$ and $T(\mathcal{M}) \subseteq \mathcal{M}$). An element of $\mathcal{B}[\mathcal{X}]$ is a *scalar operator* if it is a multiple if the identity, say, $\alpha I$ for some scalar $\alpha$.

(b)  Verify that every subspace in $\mathrm{Lat}(\mathcal{X})$ is invariant for any scalar operator in $\mathcal{B}[\mathcal{X}]$, and hence every scalar operator has a nontrivial invariant subspace whenever $\dim \mathcal{X} > 1$.

**Problem 4.20.**  Let $\mathcal{X}$ be a normed space and take $T \in \mathcal{B}[\mathcal{X}]$. Prove the following propositions.

(a)  $\mathcal{N}(T)$ and $\mathcal{R}(T)^-$ are invariant subspaces for $T$.

(b)  If $T$ has no nontrivial invariant subspace, then $\mathcal{N}(T) = \{0\}$ and $\mathcal{R}(T)^- = \mathcal{X}$.

Take $S$ and $T$ in $\mathcal{B}[\mathcal{X}]$. We say that $S$ and $T$ *commute* if $ST = TS$.

(c)  Show that if $S$ and $T$ commute, then $\mathcal{N}(S)$, $\mathcal{N}(T)$, $\mathcal{R}(S)^-$ and $\mathcal{R}(T)^-$ are invariant subspaces for both $S$ and $T$.

**Problem 4.21.**  Let $S \in \mathcal{B}[\mathcal{X}]$ and $T \in \mathcal{B}[\mathcal{X}]$ be nonzero operators on a normed space $\mathcal{X}$. Suppose $ST = O$ and show that

(a)  $T(\mathcal{N}(S)) \subseteq T(\mathcal{X}) = \mathcal{R}(T) \subseteq \mathcal{N}(S)$,

(b)  $\{0\} \neq \mathcal{N}(S) \neq \mathcal{X}$    and    $\{0\} \neq \mathcal{R}(T)^- \neq \mathcal{X}$,

(c)  $S(\mathcal{R}(T)^-) \subseteq S(\mathcal{R}(T))^- \subseteq \mathcal{R}(T)^-$.

Conclusion: *If $S \neq O$, $T \neq O$ and $ST = O$, then $\mathcal{N}(S)$ and $\mathcal{R}(T)^-$ are nontrivial invariant subspaces for both $S$ and $T$.*

**Problem 4.22.** Let $\mathcal{X}$ be a normed space and take $T \in \mathcal{B}[\mathcal{X}]$. Every nonzero polynomial $p(T)$ of $T$ (defined as in Problem 2.20) lies in $\mathcal{B}[\mathcal{X}]$. (Reason: $\mathcal{B}[\mathcal{X}]$ is an algebra.) Show that

(a) $\mathcal{N}(p(T))$ and $\mathcal{R}(p(T))^-$ are invariant subspaces for $T$.

Recall that an operator in $\mathcal{B}[\mathcal{X}]$ is nilpotent if $T^n = O$ for some positive integer $n$, and algebraic if $p(T) = O$ for some nonzero polynomial $p$ (see Problem 2.20 again).

(b) Show that every nilpotent operator in $\mathcal{B}[\mathcal{X}]$ (with dim $\mathcal{X} > 1$) has a nontrivial invariant subspace.

(c) Suppose $\mathcal{X}$ is a *complex* normed space and dim $\mathcal{X} > 1$. Show that every algebraic operator in $\mathcal{B}[\mathcal{X}]$ has a nontrivial invariant subspace.

*Hint*: A polynomial (in one complex variable and with complex coefficients) of degree $n \geq 1$ is the product of a polynomial of degree $n - 1$ and a polynomial of degree 1.

**Problem 4.23.** Let $\mathrm{Lat}(T)$ denote the subcollection of $\mathrm{Lat}(\mathcal{X})$ made up of *all invariant subspaces for $T \in \mathcal{B}[\mathcal{X}]$*, where $\mathcal{X}$ is a normed space. Clearly (see Problems 4.18 and 4.19), $T$ has no nontrivial invariant subspace if and only if $\mathrm{Lat}(T) = \{\{0\}, \mathcal{X}\}$.

(a) Show that $\mathrm{Lat}(T)$ is a complete lattice in the inclusion ordering.

*Hint*: Intersection and closure of sum of invariant subspaces are again invariant subspaces. See Section 4.3.

Take any operator $T$ in $\mathcal{B}[\mathcal{X}]$ and an arbitrary vector $x$ in $\mathcal{X}$. Consider the $\mathcal{X}$-valued power sequence $\{T^n x\}_{n \geq 0}$. The range of $\{T^n x\}_{n \geq 0}$ is called the *orbit of $x$ under $T$*.

(b) Show that the (linear) span of the orbit of $x$ under $T$ is the set of the images of all nonzero polynomials of $T$ at $x$; that is,

$$\mathrm{span}\,\{T^n x\}_{n \geq 0} = \{p(T)x \in \mathcal{X}: \ p \text{ is a nonzero polynomial}\}.$$

Since $\mathrm{span}\,\{T^n x\}_{n \geq 0}$ is a linear manifold of $\mathcal{X}$, it follows that its closure, $(\mathrm{span}\,\{T^n x\}_{n \geq 0})^- = \bigvee\{T^n x\}_{n \geq 0}$, is a subspace of $\mathcal{X}$ (Proposition 4.8(b)). That is, $\bigvee\{T^n x\}_{n \geq 0} \in \mathrm{Lat}(\mathcal{X})$.

(c) Show that $\bigvee\{T^n x\}_{n \geq 0} \in \mathrm{Lat}(T)$.

These are the *cyclic subspaces* in $\mathrm{Lat}(T)$: $\mathcal{M} \in \mathrm{Lat}(T)$ is cyclic for $T$ if $\mathcal{M} = \bigvee \{T^n x\}_{n \geq 0}$ for some $x \in \mathcal{X}$. If $\bigvee \{T^n x\}_{n \geq 0} = \mathcal{X}$, then $x$ is said to be a *cyclic vector* for $T$. We say that a linear manifold $\mathcal{M}$ of $\mathcal{X}$ is *totally cyclic* for $T$ if every nonzero vector in $\mathcal{M}$ is cyclic for $T$.

(d)  Verify that $T$ has no nontrivial invariant subspace if and only if $\mathcal{X}$ is totally cyclic for $T$.

**Problem 4.24.**  Let $\mathcal{X}$ and $\mathcal{Y}$ be normed spaces. A bounded linear transformation $X \in \mathcal{B}[\mathcal{X}, \mathcal{Y}]$ *intertwines* an operator $T \in \mathcal{B}[\mathcal{X}]$ to an operator $S \in \mathcal{B}[\mathcal{Y}]$ if

$$XT = SX.$$

If there exists an $X$ intertwining $T$ to $S$, then we say $T$ is *intertwined* to $S$. Suppose $XT = SX$. Show by induction that

(a) $$XT^n = S^n X$$

for every positive integer $n$. Hence verify that

(b) $$Xp(T) = p(S)X$$

for every polynomial $p$. Now use Problem 4.23(b) to prove that

(c) $$X\big(\mathrm{span}\,\{T^n x\}_{n \geq 0}\big) = \mathrm{span}\,\{S^n Xx\}_{n \geq 0}$$

for each $x \in \mathcal{X}$, and therefore (see Problem 3.46(a))

(d) $$X\big(\bigvee \{T^n x\}_{n \geq 0}\big) \subseteq \bigvee \{S^n Xx\}_{n \geq 0}$$

for every $x \in \mathcal{X}$. An operator $T$ in $\mathcal{B}[\mathcal{X}]$ is *densely intertwined* to an operator $S$ in $\mathcal{B}[\mathcal{Y}]$ if there exists a bounded linear transformation $X$ in $\mathcal{B}[\mathcal{X}, \mathcal{Y}]$ with dense range intertwining $T$ to $S$. If $XT = SX$ and $\mathcal{R}(X)^- = \mathcal{Y}$, then show that

(e) $$\bigvee \{T^n x\}_{n \geq 0} = \mathcal{X} \quad \text{implies} \quad \mathcal{Y} = \bigvee \{S^n Xx\}_{n \geq 0}.$$

Conclusion: *Suppose $T$ in $\mathcal{B}[\mathcal{X}]$ is densely intertwined to $S$ in $\mathcal{B}[\mathcal{Y}]$ and let $X$ in $\mathcal{B}[\mathcal{X}, \mathcal{Y}]$ be a transformation with dense range intertwining $T$ to $S$. If $x \in \mathcal{X}$ is a cyclic vector for $T$, then $Xx \in \mathcal{Y}$ is a cyclic vector for $S$. Therefore, if a linear manifold $\mathcal{M}$ of $\mathcal{X}$ is totally cyclic for $T$, then the linear manifold $X(\mathcal{M})$ of $\mathcal{Y}$ is totally cyclic for $S$.*

**Problem 4.25.**  Here is a sufficient condition for transferring nontrivial invariant subspaces from $S$ to $T$ whenever $T$ is densely intertwined to $S$. Let $\mathcal{X}$ and $\mathcal{Y}$ be normed spaces and take $T \in \mathcal{B}[\mathcal{X}]$, $S \in \mathcal{B}[\mathcal{Y}]$ and $X \in \mathcal{B}[\mathcal{X}, \mathcal{Y}]$ such that

$$XT = SX.$$

Prove the following assertions.

(a) If $\mathcal{M} \subseteq \mathcal{Y}$ is an invariant subspace for $S$, then the inverse image of $\mathcal{M}$ under $X$, $X^{-1}(\mathcal{M}) \subseteq \mathcal{X}$, is an invariant subspace for $T$.

(b) If, in addition, $\mathcal{M} \neq \mathcal{Y}$, $\mathcal{R}(X) \cap \mathcal{M} \neq \{0\}$ and $\mathcal{R}(X)^- = \mathcal{Y}$, then $\{0\} \neq X^{-1}(\mathcal{M}) \neq \mathcal{X}$.

*Hint*: Problems 1.2 and 2.11, and Theorem 3.23.

Conclusion: *If $T$ is densely intertwined to $S$, then the inverse image under the intertwining transformation $X$ of a nontrivial invariant subspace $\mathcal{M}$ for $S$ is a nontrivial invariant subspace for $T$, provided that the range of $X$ is not (algebraically) disjoint with $\mathcal{M}$.*

Show that the condition $\mathcal{R}(X) \cap \mathcal{M} \neq \{0\}$ in (b) is not redundant. That is, if $\mathcal{M}$ is a subspace of $\mathcal{Y}$, then show that

(c) $\{0\} \neq \mathcal{M} \neq \mathcal{Y}$ and $\mathcal{R}(X)^- = \mathcal{Y}$ does not imply $\mathcal{R}(X) \cap \mathcal{M} \neq \{0\}$.

However, if $X$ is surjective, then the condition $\mathcal{R}(X) \cap \mathcal{M} \neq \{0\}$ in (b) is trivially satisfied whenever $\mathcal{M} \neq \{0\}$. Actually, with the assumption $XT = SX$ still in force, check the proposition below.

(d) If $S$ has a nontrivial invariant subspace, and if $\mathcal{R}(X) = \mathcal{Y}$, then $T$ has a nontrivial invariant subspace.

**Problem 4.26.** Let $\mathcal{X}$ be a normed space. The *commutant* of an operator $T$ in $\mathcal{B}[\mathcal{X}]$ is the set $\{T\}'$ consisting of all operators in $\mathcal{B}[\mathcal{X}]$ that commute with $T$. That is,

$$\{T\}' = \{C \in \mathcal{B}[\mathcal{X}]: \ CT = TC\}.$$

In other words, the commutant of an operator is the set of all operators intertwining it to itself.

(a) Show that $\{T\}'$ is an operator algebra that contains the identity (i.e., $\{T\}'$ is a unital subalgebra of the normed algebra $\mathcal{B}[\mathcal{X}]$).

A linear manifold (or a subspace) of $\mathcal{X}$ is *hyperinvariant* for $T \in \mathcal{B}[\mathcal{X}]$ if it is invariant for every $C \in \{T\}'$; that is, if it is an invariant linear manifold (or an invariant subspace) for every operator in $\mathcal{B}[\mathcal{X}]$ that commutes with $T$. As $T \in \{T\}'$, every hyperinvariant linear manifold (subspace) for $T$ obviously is an invariant linear manifold (subspace) for $T$. Take an arbitrary $T \in \mathcal{B}[\mathcal{X}]$ and, for each $x \in \mathcal{X}$, set

$$\mathcal{T}_x = \bigcup_{C \in \{T\}'} Cx = \{y \in \mathcal{X}: \ y = Cx \text{ for some } C \in \{T\}'\}.$$

It is clear that $\mathcal{T}_x$ is never empty (for instance, $x \in \mathcal{T}_x$ because $I \in \{T\}'$). In fact, $0 \in \mathcal{T}_x$ for every $x \in \mathcal{X}$, and $\mathcal{T}_x = \{0\}$ if and only if $x = 0$. Prove the following proposition.

(b)  For each $x \in \mathcal{X}$, $\mathcal{T}_x^-$ is a hyperinvariant subspace for $T$.

> *Hint*: As an algebra, $\{T\}'$ is a linear space. This implies that $\mathcal{T}_x$ is a linear manifold of $\mathcal{X}$. If $y = C_0 x$ for some $C_0 \in \{T\}'$, then $Cy = CC_0 x \in \mathcal{T}_x$ for every $C \in \{T\}'$ (i.e., $\mathcal{T}_x$ is hyperinvariant for $T$ because $\{T\}'$ is an algebra). See Problem 4.18(b).

**Problem 4.27.**  Let $\mathcal{X}$ and $\mathcal{Y}$ be any normed spaces. Take $T \in \mathcal{B}[\mathcal{X}]$, $S \in \mathcal{B}[\mathcal{Y}]$, $X \in \mathcal{B}[\mathcal{X}, \mathcal{Y}]$, and $Y \in \mathcal{B}[\mathcal{Y}, \mathcal{X}]$ such that

$$XT = SX \quad \text{and} \quad YS = TY.$$

Show that if $C \in \mathcal{B}[\mathcal{X}]$ commutes with $T$, then $XCY$ commutes with $S$. That is (see Problem 4.26), show that

(a)  $XCY \in \{S\}'$ for every $C \in \{T\}'$.

Now consider the subspace $\mathcal{T}_x^-$ of $\mathcal{X}$ that, according to Problem 4.26, is nonzero and hyperinvariant for $T$ for every nonzero $x$ in $\mathcal{X}$. Under the above assumptions on $T$ and $S$, prove the following propositions.

(b)  Suppose $\mathcal{M}$ is a nontrivial hyperinvariant subspace for $S$. If $\mathcal{R}(X)^- = \mathcal{Y}$ and $\mathcal{N}(Y) \cap \mathcal{M} = \{0\}$, then $Y(\mathcal{M}) \neq \{0\}$ and $\mathcal{T}_x^- \neq \mathcal{X}$ for every nonzero $x$ in $Y(\mathcal{M})$. Consequently, $\mathcal{T}_x^-$ is a nontrivial hyperinvariant subspace for $T$ whenever $x$ is a nonzero vector in $Y(\mathcal{M})$.

> *Hint*: Since $\mathcal{M}$ is hyperinvariant for $S$, it follows from (a) that $\mathcal{M}$ is invariant for $XCY$ whenever $C \subset \{T\}'$. Use this fact to show that $X(\mathcal{T}_x) \subseteq \mathcal{M}$ for every $x \in Y(\mathcal{M})$, and hence $X(\mathcal{T}_x^-) \subseteq \mathcal{M}^- = \mathcal{M}$ (see Problem 3.46(a)). Now verify that $\mathcal{T}_x^- = \mathcal{X}$ implies $\mathcal{R}(X)^- = X(\mathcal{X})^- = X(\mathcal{T}_x^-)^- \subseteq \mathcal{M}$. Therefore, if $\mathcal{M} \neq \mathcal{Y}$ and $\mathcal{R}(X)^- = \mathcal{Y}$, then $\mathcal{T}_x^- \neq \mathcal{X}$ for every vector $x$ in $Y(\mathcal{M})$. Next observe that if $Y(\mathcal{M}) = \{0\}$ (i.e., if $\mathcal{M} \subseteq \mathcal{N}(Y)$), then $\mathcal{N}(Y) \cap \mathcal{M} = \mathcal{M}$. Thus conclude: if $\mathcal{M} \neq \{0\}$ and $\mathcal{N}(Y) \cap \mathcal{M} = \{0\}$, then $Y(\mathcal{M}) \neq \{0\}$. Finally recall that $\mathcal{T}_x \neq \{0\}$ for every $x \neq 0$ in $\mathcal{X}$, and hence $\{0\} \neq \mathcal{T}_x^- \neq \mathcal{X}$ for every nonzero vector $x$ in $Y(\mathcal{M})$.

(c)  If $S$ has a nontrivial hyperinvariant subspace, and if $\mathcal{R}(X)^- = \mathcal{Y}$ and $\mathcal{N}(Y) = \{0\}$, then $T$ has a nontrivial hyperinvariant subspace.

**Problem 4.28.**  A bounded linear transformation $X$ of a normed space $\mathcal{X}$ into a normed space $\mathcal{Y}$ is *quasiinvertible* (or a *quasiaffinity*) if it is injective and has a dense range (i.e., $\mathcal{N}(X) = \{0\}$ and $\mathcal{R}(X)^- = \mathcal{Y}$). An operator $T \in \mathcal{B}[\mathcal{X}]$ is a *quasiaffine transform* of an operator $S \in \mathcal{B}[\mathcal{Y}]$ if there exists a quasiinvertible transformation $X \in \mathcal{B}[\mathcal{X}, \mathcal{Y}]$ intertwining $T$ to $S$. Two operators are *quasisimilar* if they are quasiaffine transforms of each other. In other words, $T \in \mathcal{B}[\mathcal{X}]$ and $S \in \mathcal{B}[\mathcal{Y}]$ are quasisimilar (notation: $T \sim S$) if there exists $X \in \mathcal{B}[\mathcal{X}, \mathcal{Y}]$ and $Y \in \mathcal{B}[\mathcal{Y}, \mathcal{X}]$ such that

$$\mathcal{N}(X) = \{0\}, \quad \mathcal{R}(X)^- = \mathcal{Y}, \quad \mathcal{N}(Y) = \{0\}, \quad \mathcal{R}(Y)^- = \mathcal{X},$$

$$XT = SX \quad \text{and} \quad YS = TY.$$

(a) Show that quasisimilarity has the defining properties of an equivalence relation.

(b) If two operators are quasisimilar and if one of them has a nontrivial hyperinvariant subspace, them so has the other. Prove.


**Problem 4.29.** Let $\mathcal{X}$ and $\mathcal{Y}$ be normed spaces. Two operators $T \in \mathcal{B}[\mathcal{X}]$ and $S \in \mathcal{B}[\mathcal{Y}]$ are *similar* (notation: $T \approx S$) if there exists an injective and surjective bounded linear transformation $X$ of $\mathcal{X}$ onto $\mathcal{Y}$, with a bounded inverse $X^{-1}$ of $\mathcal{Y}$ onto $\mathcal{X}$, that intertwines $T$ to $S$. That is, $T \in \mathcal{B}[\mathcal{X}]$ and $S \in \mathcal{B}[\mathcal{Y}]$ are similar if there exists $X \in \mathcal{B}[\mathcal{X}, \mathcal{Y}]$ such that $\mathcal{N}(X) = \{0\}$, $\mathcal{R}(X) = \mathcal{Y}$, $X^{-1} \in \mathcal{B}[\mathcal{Y}, \mathcal{X}]$ and $XT = SX$.

(a) Let $T$ be an operator on $\mathcal{X}$ and let $S$ be an operator on $\mathcal{Y}$. If $X$ is a bounded linear transformation of $\mathcal{X}$ onto $\mathcal{Y}$ with a bounded inverse $X^{-1}$ of $\mathcal{Y}$ onto $\mathcal{X}$ (which is always linear), then check that

$$XT = SX \iff T = X^{-1}SX \iff S = XTX^{-1} \iff X^{-1}S = TX^{-1}.$$

Now prove the following assertions.

(b) If $T$ and $S$ are similar, then they are quasisimilar.

(c) Similarity has the defining properties of an equivalence relation.

(d) If two operators are similar, and if one of them has a nontrivial invariant subspace, then so has the other. (Hint: Problem 4.25).

Note that we are using the same terminology of Section 2.7, namely "similar", but now with a different meaning. The linear transformation $X \colon \mathcal{X} \to \mathcal{Y}$ in fact is a (linear) isomorphism so that $\mathcal{X}$ and $\mathcal{Y}$ are isomorphic linear spaces, and hence the concept of similarity defined above implies the purely algebraic homonymous concept defined in Section 2.7. However, we are now imposing that all linear transformations involved are continuous (or, equivalently, bounded) viz., $T$, $S$, $X$ and also the inverse $X^{-1}$ of $X$.

**Problem 4.30.** Let $\{x_i\}_{i=0}^{\infty}$ be a Schauder basis for a (separable) Banach space $\mathcal{X}$ (see Problem 4.11) so that every $x \in \mathcal{X}$ has a unique expansion

$$x = \sum_{i=0}^{\infty} \alpha_i(x) x_i$$

with respect to the basis $\{x_i\}_{i=0}^{\infty}$. For each integer $i \geq 0$ consider the functional $\varphi_i : \mathcal{X} \to \mathbb{F}$ that assigns to each $x \in \mathcal{X}$ its *unique* coefficient $\alpha_i(x)$ in the above expansion:

$$\varphi_i(x) = \alpha_i(x)$$

for every $x \in \mathcal{X}$. Show that $\varphi_i$ is a bounded linear functional (i.e., $\varphi_i \in \mathcal{B}[\mathcal{X}, \mathbb{F}]$ for each $i \geq 0$). In other words, each coefficient in a Schauder basis expansion for a vector $x$ in a Banach space $\mathcal{X}$ is a bounded linear functional on $\mathcal{X}$.

*Hint*: Let $\mathcal{A}_{\mathbf{x}}$ be the Banach space defined in Problem 4.10. Consider the mapping $\Phi : \mathcal{A}_{\mathbf{x}} \to \mathcal{X}$ given by

$$\Phi(a) = \sum_{i=0}^{\infty} \alpha_i x_i$$

for every $a = \{\alpha_i\}_{i=0}^{\infty}$ in $\mathcal{A}_{\mathbf{x}}$. Verify that $\Phi$ is linear, injective, surjective and bounded (actually, $\Phi$ is a contraction: $\Phi(a) \leq \|a\|$ for every $a \in \mathcal{A}_{\mathbf{x}}$). Now apply Theorem 4.22 to conclude that $\Phi \in \mathcal{G}[\mathcal{A}_{\mathbf{x}}, \mathcal{X}]$. For each integer $i \geq 0$ consider the functional $\psi_i : \mathcal{A}_{\mathbf{x}} \to \mathbb{F}$ given by

$$\psi_i(a) = \alpha_i$$

for every $a = \{\alpha_i\}_{i \geq 0} \in \mathcal{A}_{\mathbf{x}}$. Show that each $\psi_i$ is linear and bounded. Finally, observe that the following diagram commutes.

$$\mathcal{X} \xrightarrow{\ \Phi^{-1}\ } \mathcal{A}_{\mathbf{x}}$$
$$\varphi_i \searrow \quad \downarrow \psi_i$$
$$\mathbb{F}$$

**Problem 4.31.** Let $\mathcal{X}$ and $\mathcal{Y}$ be normed spaces (over the same scalar field) and let $\mathcal{M}$ be a linear manifold of $\mathcal{X}$. Equip the direct sum of $\mathcal{M}$ and $\mathcal{Y}$ with any of the norms of Example 4E and consider the normed space $\mathcal{M} \oplus \mathcal{Y}$. A linear transformation $L : \mathcal{M} \to \mathcal{Y}$ is called *closed* if its graph is closed in $\mathcal{M} \oplus \mathcal{Y}$. Since a subspace simply means a closed linear manifold, and recalling that the graph of any linear transformation of $\mathcal{M}$ into $\mathcal{Y}$ is a linear manifold of the linear space $\mathcal{M} \oplus \mathcal{Y}$, such a definition can be rewritten as follows. A linear transformation $L : \mathcal{M} \to \mathcal{Y}$ is closed if its graph is a subspace of the normed space $\mathcal{M} \oplus \mathcal{Y}$. Take an arbitrary $L \in \mathcal{L}[\mathcal{X}, \mathcal{Y}]$ and prove that the assertions below are equivalent.

(i)   $L$ is closed.

(ii)   If $\{u_n\}$ is an $\mathcal{M}$-valued sequence that converges in $\mathcal{X}$, and if its image under $L$ converges in $\mathcal{Y}$, then

$$\lim u_n \in \mathcal{M} \quad \text{and} \quad \lim L u_n = L \lim u_n.$$

Symbolically, *L is closed if and only if*

$$\left\{ \begin{array}{c} u_n \in \mathcal{M} \to u \in \mathcal{X} \\ L u_n \to y \in \mathcal{Y} \end{array} \right\} \quad \Longrightarrow \quad \left\{ \begin{array}{c} u \in \mathcal{M} \\ y = L u \end{array} \right\}.$$

*Hint*: Apply the Closed Set Theorem. Use the one-norm on $\mathcal{M} \oplus \mathcal{Y}$ (i.e., $\|(u, y)\|_1 = \|u\|_X + \|y\|_Y$ for every $(u, y) \in \mathcal{M} \oplus \mathcal{Y}$.

**Problem 4.32.** Consider the setup of the previous problem and prove the following propositions.

(a) If $L \in \mathcal{B}[\mathcal{M}, \mathcal{Y}]$ and $\mathcal{M}$ is closed in $\mathcal{X}$, then $L$ is closed.

Every bounded linear transformation defined on a subspace of a normed space is closed. In particular (set $\mathcal{M} = \mathcal{X}$), if $L \in \mathcal{B}[\mathcal{X}, \mathcal{Y}]$ then $L$ is closed.

(b) If $\mathcal{M}$ and $\mathcal{Y}$ are Banach spaces and $L \in \mathcal{L}[\mathcal{M}, \mathcal{Y}]$ is closed, then $L \in \mathcal{B}[\mathcal{M}, \mathcal{Y}]$.

Every closed linear transformation between Banach spaces is bounded.

(c) If $\mathcal{Y}$ is a Banach space and $L \in \mathcal{B}[\mathcal{M}, \mathcal{Y}]$ is closed, then $\mathcal{M}$ is closed in $\mathcal{X}$.

Every closed and bounded linear transformation into a Banach space has a closed domain.

Hint: Closed Graph Theorem and Closed Set Theorem.

Recall that continuity means convergence preservation in the sense of Theorem 3.7, and also that the notions of "bounded" and "continuous" coincide for a linear transformation between normed spaces (Theorem 4.14). Compare Corollary 3.8 with Problem 4.31 and prove the next proposition.

(d) If $\mathcal{M}$ and $\mathcal{Y}$ are Banach spaces, then $L \in \mathcal{L}[\mathcal{M}, \mathcal{Y}]$ is continuous if and only if it is closed.

**Problem 4.33.** Let $\mathcal{X}$ and $\mathcal{Y}$ be Banach spaces and let $\mathcal{M}$ be a linear manifold of $\mathcal{X}$. Take $L \in \mathcal{L}[\mathcal{M}, \mathcal{Y}]$ and consider the following assertions.

    (i)   $\mathcal{M}$ is closed in $\mathcal{X}$ (so that $\mathcal{M}$ is a Banach space).

    (ii)   $L$ is a closed linear transformation.

    (iii)   $L$ is bounded (i.e., $L$ is continuous).

According to Problem 4.32 these three assertions are related as follows: *each pair of them implies the other.*

  (a)  Exhibit a bounded linear transformation that is not closed.

Hint: $\ell_+^1$ is a dense linear manifold of $(\ell_+^2, \| \ \|_2)$. Take the inclusion map of $(\ell_+^1, \| \ \|_2)$ into $(\ell_+^2, \| \ \|_2)$.

The classical example of a closed linear transformation that is not bounded is the differential mapping $D\colon C'[0, 1] \to C[0, 1])$ defined in Problem 3.18. It is easy to show that $C'[0, 1]$, the set of all differentiable functions in $C[0, 1]$ whose derivatives lie in $C[0, 1]$, is a linear manifold of the Banach space $C[0, 1]$ equipped with the sup-norm. It is also readily verified that $D$ is linear. Moreover, according to Problem 3.18(a), $D$ is not continuous (and hence unbounded). However, if $\{u_n\}$ is a uniformly convergent sequence of continuously differentiable functions whose derivative sequence $\{Du_n\}$ also converges uniformly, then $\lim Du_n = D(\lim u_n)$. This is a standard result from advanced calculus. Thus $D$ is closed by Problem 4.31.

  (b)  Give another example of a closed linear transformation that is not bounded.

*Hint:* $\mathcal{X} = \mathcal{Y} = \ell_+^1$, $\mathcal{M} = \left\{ x = \{\xi_k\}_{k=1}^\infty \in \ell_+^1 \colon \sum_{k=1}^\infty k|\xi_k| < \infty \right\}$, and $D = \mathrm{diag}(\{k\}_{k=1}^\infty) = \mathrm{diag}(1, 2, 3, \ldots)\colon \mathcal{M} \to \ell_+^1$. Verify that $\mathcal{M}$ is a linear manifold of $\ell_+^1$. Use $x_n = \frac{1}{n^2}(1, \ldots, 1, 0, 0, 0, \ldots) \in \mathcal{M}$ (the first $n$ entries are all equal to $\frac{1}{n^2}$; the rest is zero) to show that $D$ is not continuous (Corollary 3.8). Suppose $u_n \to u \in \ell_+^1$, with $u_n = \{\xi_n(k)\}_{k=1}^\infty$ in $\mathcal{M}$, and $Du_n \to y = \{v(k)\}_{k=1}^\infty \in \ell_+^1$. Set $\xi(k) = \frac{1}{k}v(k)$ so that $x = \{\xi(k)\}_{k=1}^\infty$ lies in $\mathcal{M}$. Now show that $\|u_n - x\|_1 \le \|Du_n - y\|_1$, and hence $u_n \to x$. Therefore, $u = x \in \mathcal{M}$ (uniqueness of the limit) and $y = Du$ (for $y = Dx$). Apply Problem 4.31 to conclude that $D$ is closed. Generalize to injective diagonals with unbounded entries.

**Problem 4.34.** Let $\mathcal{M}$ and $\mathcal{N}$ be subspaces of a normed space $\mathcal{X}$. If $\mathcal{M}$ and $\mathcal{N}$ are algebraic complements of each other (i.e., if $\mathcal{M} + \mathcal{N} = \mathcal{X}$ and $\mathcal{M} \cap \mathcal{N} = \{0\}$), then we say that $\mathcal{M}$ and $\mathcal{N}$ are *complementary subspaces* in $\mathcal{X}$. According to Theorem 2.14 the natural mapping $\Phi\colon \mathcal{M} \oplus \mathcal{N} \to \mathcal{M} + \mathcal{N}$, defined by $\Phi((u, v)) = u + v$ for every $(u, v) \in \mathcal{M} \oplus \mathcal{N}$, is an isomorphism between the linear spaces $\mathcal{M} \oplus \mathcal{N}$

and $\mathcal{M} + \mathcal{N}$ whenever $\mathcal{M} \cap \mathcal{N} = \{0\}$. Consider the direct sum $\mathcal{M} \oplus \mathcal{N}$ equipped with any of the norms of Example 4E and prove the following claim.

*If $\mathcal{M}$ and $\mathcal{N}$ are complementary subspaces in a Banach space $\mathcal{X}$, then the natural mapping $\Phi: \mathcal{M} \oplus \mathcal{N} \to \mathcal{M} + \mathcal{N}$ is a topological isomorphism.*

*Hint*: Show that the isomorphism $\Phi$ is a contraction when $\mathcal{M} \oplus \mathcal{N}$ is equipped with the norm $\| \ \|_1$. Recall that $\mathcal{M}$ and $\mathcal{N}$ are Banach spaces (Proposition 4.7) and conclude that $\mathcal{M} \oplus \mathcal{N}$ is again a Banach space (Example 4E). Apply the Inverse Mapping Theorem to prove that $\Phi$ is a topological isomorphism when $\mathcal{M} \oplus \mathcal{N}$ is equipped with the norm $\| \ \|_1$. Also recall that the norms of Example 4E are equivalent (see the remarks that follow Proposition 4.26).

**Problem 4.35.** Prove the following propositions.

(a) If $P: \mathcal{X} \to \mathcal{X}$ is a continuous projection on a normed space $\mathcal{X}$, then $\mathcal{R}(P)$ and $\mathcal{N}(P)$ are complementary subspaces in $\mathcal{X}$

   *Hint*: Recall that $\mathcal{R}(P) = \mathcal{N}(I - P)$. Apply Theorem 2.12 and Proposition 4.13.

(b) Conversely, if $\mathcal{M}$ and $\mathcal{N}$ are complementary subspaces in a *Banach* space $\mathcal{X}$, then the unique projection $P: \mathcal{X} \to \mathcal{X}$ with $\mathcal{R}(P) = \mathcal{M}$ and $\mathcal{N}(P) = \mathcal{N}$ of Theorem 2.20 is continuous and $\|P\| \geq 1$.

   *Hint*: Consider the natural mapping $\Phi: \mathcal{M} \oplus \mathcal{N} \to \mathcal{M} + \mathcal{N}$ of the direct sum $\mathcal{M} \oplus \mathcal{N}$ (equipped with any of the norms of Example 4E) onto $\mathcal{X} = \mathcal{M} + \mathcal{N}$. Let $P_M: \mathcal{M} \oplus \mathcal{N} \to \mathcal{M} \subseteq \mathcal{X}$ be the map defined by $P_M(u, v) = u$ for every $(u, v) \in \mathcal{M} \oplus \mathcal{N}$. Recall from Example 4I that $P_M$ is a contraction (indeed, $\|P_M\| = 1$). Apply the previous problem to verify that the diagram

$$\mathcal{M} \oplus \mathcal{N} \xleftarrow{\ \Phi^{-1}\ } \mathcal{M} + \mathcal{N}$$

$$P_M \searrow \qquad \downarrow P$$

$$\mathcal{X}$$

   commutes. Thus show that $P$ is continuous (note: $Pu = u$ for every $u \in \mathcal{M} = \mathcal{R}(P)$).

Remarks: $P_M$ is, in fact, a continuous projection of $\mathcal{M} \oplus \mathcal{N}$ into itself whose range is $\mathcal{R}(P_M) = \mathcal{M} \oplus \{0\}$. If we identify $\mathcal{M} \oplus \{0\}$ with $\mathcal{M}$ (as we did in Example 4I), then $P_M: \mathcal{M} \oplus \mathcal{N} \to \mathcal{M} \oplus \{0\} \subseteq \mathcal{M} \oplus \mathcal{N}$ can be viewed as a map from $\mathcal{M} \oplus \mathcal{N}$ onto $\mathcal{M}$, and hence we wrote $P_M: \mathcal{M} \oplus \mathcal{N} \to \mathcal{M} \subset \mathcal{X}$; the continuous natural projection of $\mathcal{M} \oplus \mathcal{N}$ onto $\mathcal{M}$. It is also worth noticing that the above propositions hold for the *complementary projection* $E = (I - P): \mathcal{X} \to \mathcal{X}$ as well, for $\mathcal{N}(E) = \mathcal{R}(P)$ and $\mathcal{R}(E) = \mathcal{N}(P)$.

**Problem 4.36.** Consider a bounded linear transformation $T \in \mathcal{B}[\mathcal{X}, \mathcal{Y}]$ of a Banach space $\mathcal{X}$ into a Banach space $\mathcal{Y}$. Let $\mathcal{M}$ be a complementary subspace of $\mathcal{N}(T)$ in $\mathcal{X}$. That is, $\mathcal{M}$ is a subspace of $\mathcal{X}$ that is also an algebraic complement of the null space $\mathcal{N}(T)$ of $T$:

$$\mathcal{M} = \mathcal{M}^-, \quad \mathcal{X} = \mathcal{M} + \mathcal{N}(T) \quad \text{and} \quad \mathcal{M} \cap \mathcal{N}(T) = \{0\}.$$

Set $T_M = T|_{\mathcal{M}} \colon \mathcal{M} \to \mathcal{Y}$, the restriction of $T$ to $\mathcal{M}$, and verify the following propositions.

(a) $T_M \in \mathcal{B}[\mathcal{M}, \mathcal{Y}]$, $\mathcal{R}(T_M) = \mathcal{R}(T)$ and $\mathcal{N}(T_M) = \{0\}$.

    *Hint*: Problems 2.14 and 3.30.

(b) $\mathcal{R}(T_M) = \mathcal{R}(T_M)^-$ if and only if there exists $T_M^{-1} \in \mathcal{B}[\mathcal{R}(T_M), \mathcal{M}]$.

    *Hint*: Proposition 4.7 and Corollary 4.24.

(c) If $A \subseteq \mathcal{R}(T)$ and $T_M^{-1}(A)^- = \mathcal{M}$, then $T^{-1}(A)^- = \mathcal{X}$.

    *Hint*: Take an arbitrary $x = u + v \in \mathcal{X} = \mathcal{M} + \mathcal{N}(T)$, with $u$ in $\mathcal{M}$ and $v$ in $\mathcal{N}(T)$. Verify that there exists a $T_M^{-1}(A)$-valued sequence $\{u_n\}$ that converges to $u$. Set $x_n = u_n + v$ in $\mathcal{X}$ and show that $\{x_n\}$ is a $T_M^{-1}(A)$-valued sequence that converges to $x$. Apply Proposition 3.32.

Now use the above results to prove the following assertion.

(d) If $A \subseteq \mathcal{R}(T)$ and $A^- = \mathcal{R}(T) = \mathcal{R}(T)^-$, then $T^{-1}(A)^- = \mathcal{X}$.

That is, the inverse image under $T$ of a dense subset of the range of $T$ is dense in $\mathcal{X}$ whenever $\mathcal{X}$ and $\mathcal{Y}$ are Banach spaces and $T \in \mathcal{B}[\mathcal{X}, \mathcal{Y}]$ has a closed range and a null space with a complementary subspace in $\mathcal{X}$. This can be viewed as a converse to Problem 3.46(c).

**Problem 4.37.** Prove the following propositions.

(a) Every finite-dimensional normed space is a separable Banach space.

    *Hint*: Example 3P, Problem 3.48, and Corollaries 4.28 and 4.31.

(b) If $\mathcal{X}$ and $\mathcal{Y}$ are topologically isomorphic normed spaces, and if one of them is a (separable) Banach space, then so is the other.

    *Hint*: Theorems 3.44 and 4.14.

**Problem 4.38.** Let $\mathcal{X}$ and $\mathcal{Y}$ be normed spaces and take $T \in \mathcal{L}[\mathcal{X}, \mathcal{Y}]$. If either $\mathcal{X}$ or $\mathcal{Y}$ is finite-dimensional, then $T$ is of finite rank (Problems 2.6 and 2.17). $\mathcal{R}(T)$ is a subspace of $\mathcal{Y}$ whenever $T$ is of finite rank (Corollary 4.29). If $T$ is injective and of finite rank, then $\mathcal{X}$ is finite-dimensional (Theorem 2.8 and Problems 2.6 and 2.17). Use Problem 2.7 and Corollaries 4.24 and 4.28 to prove the following assertions.

(a)  If $\mathcal{Y}$ is a Banach space and $T \in \mathcal{B}[\mathcal{X}, \mathcal{Y}]$ is of finite rank and injective, then $T$ has a bounded inverse on its range.

(b)  If $\mathcal{X}$ is finite-dimensional, then every injective operator in $\mathcal{B}[\mathcal{X}]$ is invertible.

(c)  If $\mathcal{X}$ is finite-dimensional and $T \in \mathcal{L}[\mathcal{X}]$, then $\mathcal{N}(T) = \{0\}$ if and only if $T \in \mathcal{G}[\mathcal{X}]$.

In other words, a linear transformation of a finite-dimensional normed space into itself is a topological isomorphism if and only if it is injective.

(d)  If $\mathcal{X}$ is finite-dimensional, then every linear isometry of $\mathcal{X}$ into itself is an isometric isomorphism.

That is, every linear isometry of a finite-dimensional normed space into itself is surjective

**Problem 4.39.**  The previous problem says that nonsurjective isometries in $\mathcal{B}[\mathcal{X}]$ may exist only if the normed space $\mathcal{X}$ is infinite-dimensional. Here is an example. Let $(\ell_+, \|\ \|)$ denote either the normed space $(\ell_+^p, \|\ \|_p)$ for some $p \geq 1$ or $(\ell_+^\infty, \|\ \|_\infty)$. Consider the mapping $S_+ : \ell_+ \to \ell_+$ defined as follows. For every $x = \{\xi_k\}_{k=0}^\infty \in \ell_+$,

$$S_+ x = \{\upsilon_k\}_{k=0}^\infty \qquad \text{with} \qquad \upsilon_k = \begin{cases} 0, & k = 0, \\ \xi_{k-1}, & k \geq 1, \end{cases}$$

(i.e., $S_+(\xi_0, \xi_1, \xi_2, \dots) = (0, \xi_0, \xi_1, \xi_2, \dots ))$ which is also represented by the infinite matrix

$$S_+ = \begin{pmatrix} 0 & & & \\ 1 & 0 & & \\ & 1 & 0 & \\ & & 1 & \ddots \\ & & & \end{pmatrix},$$

where every entry below the main diagonal is equal to one and the remaining entries are all zero. This is the *unilateral shift* on $\ell_+$.

(a)  Show that $S_+$ is a linear nonsurjective isometry.

Since $S_+$ is a linear isometry, it follows by Proposition 4.37 that $\| S_+^n x \| = \| x \|$ for every $x \in \ell_+$ and all $n \geq 1$, and hence

$$S_+ \in \mathcal{B}[\ell_+] \qquad \text{with} \qquad \| S_+^n \| = 1 \ \text{ for all } \ n \geq 0.$$

Consider the backward unilateral shift $S_-$ of Example 4L, now acting either on $\ell_+^p$ or on $\ell_+^\infty$. Recall that $S_- \in \mathcal{B}[\ell_+]$ and $\| S_-^n \| = 1$ for all $n \geq 0$ (this has been verified in Example 4L for $(\ell_+, \|\ \|) = (\ell_+^p, \|\ \|_p)$ but the same argument ensures that it holds for $(\ell_+, \|\ \|) = (\ell_+^\infty, \|\ \|_\infty)$ as well).

(b)  Show that $S_- S_+ = I : \ell_+ \to \ell_+$, the identity on $\ell_+$.

Therefore, $S_- \in \mathcal{B}[\ell_+]$ is a left inverse of $S_+ \in \mathcal{B}[\ell_+]$.

(c)  Conclude that $S_-$ is surjective but not injective.


**Problem 4.40.**  Let $(\ell, \| \; \|)$ denote either the normed space $(\ell^p, \| \; \|_p)$ for some $p \geq 1$ or $(\ell^\infty, \| \; \|_\infty)$. Consider the mapping $S \colon \ell \to \ell$ defined by

$$Sx = \{\xi_{k-1}\}_{k=-\infty}^{\infty} \quad \text{for every} \quad x = \{\xi_k\}_{k=-\infty}^{\infty} \in \ell$$

(i.e., $S(\dots, \xi_{-2}, \xi_{-1}, (\xi_0), \xi_1, \xi_2, \dots) = (\dots, \xi_{-3}, \xi_{-2}, (\xi_{-1}), \xi_0, \xi_1, \dots))$ which is also represented by the following (doubly) infinite matrix (with the inner parenthesis indicating the zero-zero position).

$$S = \begin{pmatrix} \ddots & & & & \\ & 1 & 0 & & \\ & & 1 & (0) & \\ & & & 1 & 0 \\ & & & & \ddots \end{pmatrix},$$

where every entry below the main diagonal is equal to one and the remaining entries are all zero. This is the *bilateral shift* on $\ell$.

(a)  Show that $S$ is a linear surjective isometry.

That is, $S$ is an isometric isomorphism, and hence

$$S \in \mathcal{G}[\ell] \quad \text{with} \quad \|S^n\| = 1 \text{ for all } n \geq 0.$$

Its inverse $S^{-1}$ is then again an isometric isomorphism, so that

$$S^{-1} \in \mathcal{G}[\ell] \quad \text{with} \quad \|(S^{-1})^n\| = 1 \text{ for all } n \geq 0.$$

(b)  Verify that the inverse $S^{-1}$ of $S$ is given by the formula

$$S^{-1}x = \{\xi_{k+1}\}_{k=-\infty}^{\infty} \quad \text{for every} \quad x = \{\xi_k\}_{k=-\infty}^{\infty} \in \ell$$

(i.e., $S^{-1}(\dots, \xi_{-2}, \xi_{-1}, (\xi_0), \xi_1, \xi_2, \dots) = (\dots, \xi_{-1}, \xi_0, (\xi_1), \xi_2, \xi_3, \dots))$ which is also represented by a (doubly) infinite matrix

$$S^{-1} = \begin{pmatrix} \ddots & 1 & & & \\ & 0 & 1 & & \\ & & (0) & 1 & \\ & & & 0 & \\ & & & & \ddots \end{pmatrix},$$

where every entry above the main diagonal is equal to one and the remaining entries are all zero. This called the *backward bilateral shift* on $\ell$.

**Problem 4.41.** Prove the following propositions.

(a) Let $\mathcal{W}$, $\mathcal{X}$, $\mathcal{Y}$ and $\mathcal{Z}$ be normed spaces (over the same scalar field). If $V \in \mathcal{B}[\mathcal{X}, \mathcal{Y}]$ is an isometry, then

$$\|T V\| = \|T\| \quad \text{and} \quad \|V S\| = \|S\|$$

for every $T \in \mathcal{B}[\mathcal{Y}, \mathcal{Z}]$ and every $S \in \mathcal{B}[\mathcal{W}, \mathcal{X}]$.

Remark: This extends the following result. $\|V^n\| = 1$ for every $n \geq 1$ whenever $V$ is an isometry in $\mathcal{B}[\mathcal{X}]$ (Proposition 4.37(c)).

(b) Every linear isometry of a Banach space into a normed space has a closed range.

*Hint*: Proposition 4.20 and 4.37.

**Problem 4.42.** Let $\mathcal{X}$ and $\mathcal{Y}$ be normed spaces. Verify that $T$ in $\mathcal{B}[\mathcal{X}]$ and $S$ in $\mathcal{B}[\mathcal{Y}]$ are similar (in the sense of Problem 4.29) if and only if there exists a topological isomorphism intertwining $T$ to $S$. That is, if and only if there exists $W$ in $\mathcal{G}[\mathcal{X}, \mathcal{Y}]$ such that
$$W T = S W.$$

Clearly, $\mathcal{X}$ and $\mathcal{Y}$ must be topologically isomorphic normed spaces if there are similar elements in $\mathcal{B}[\mathcal{X}]$ and $\mathcal{B}[\mathcal{Y}]$. A stronger form of similarity is obtained when there exists an isometric isomorphism, say $U$ in $\mathcal{G}[\mathcal{X}, \mathcal{Y}]$, intertwining $T$ to $S$; that is,

$$U T = S U.$$

If this happens, then we say that $T$ and $S$ are *isometrically equivalent* (notation: $T \cong S$). Again, $\mathcal{X}$ and $\mathcal{Y}$ must be isometrically isomorphic normed spaces if there are isometrically equivalent elements in $\mathcal{B}[\mathcal{X}]$ and $\mathcal{B}[\mathcal{Y}]$. As in the case of similarity, show that isometric equivalence has the defining properties of an equivalence relation. An important difference between similarity and isometric equivalence is that isometric equivalence is norm-preserving: if $T$ and $S$ are isometrically equivalent, then $\|T\| = \|S\|$. Prove this identity and show that it may fail if $T$ and $S$ are simply similar. Now let $\mathcal{X}$ and $\mathcal{Y}$ be Banach spaces. Show that, in this case, $T$ in $\mathcal{B}[\mathcal{X}]$ and $S$ in $\mathcal{B}[\mathcal{Y}]$ are similar if and only if there exists an injective and surjective bounded linear transformation in $\mathcal{B}[\mathcal{X}, \mathcal{Y}]$ intertwining $T$ to $S$.

**Problem 4.43.** Let $X$ be a normed space. Verify that the following three conditions are pairwise equivalent.

(a) $\mathcal{X}$ is separable (as a metric space).

(b) There exists a countable subset of $\mathcal{X}$ that spans $\mathcal{X}$.

(c) There exists a dense linear manifold $\mathcal{M}$ of $\mathcal{X}$ such that $\dim \mathcal{M} \leq \aleph_0$.

*Hint*: Proposition 4.9.

(d) Moreover, show also that a completion $\widehat{\mathcal{X}}$ of a separable normed space $\mathcal{X}$ is itself separable.

**Problem 4.44.** In many senses barreled spaces in a locally-convex-space setting plays a role similar to Banach spaces in a normed-space setting. In fact, as we saw in Problem 4.4, a Banach space is barreled. Barreled spaces actually are the spaces where the Banach–Steinhaus Theorem holds in a locally-convex-space setting: *Every pointwise bounded collection of continuous linear transformations of a barreled space into a locally convex space is equicontinuous*. To see that this is exactly the locally-convex-space version of Theorem 4.43, we need the notion of *equicontinuity* in a locally convex space. Let $\mathcal{X}$ and $\mathcal{Y}$ be topological vector spaces. A subset $\Theta$ of $\mathcal{L}[\mathcal{X}, \mathcal{Y}]$ is *equicontinuous* if for each neighborhood $N_Y$ of the origin of $\mathcal{Y}$ there exists a neighborhood $N_X$ of the origin of $\mathcal{X}$ such that $T(N_X) \subseteq N_Y$ for all $T \in \Theta$.

(a) Show that if $\mathcal{X}$ and $\mathcal{Y}$ are normed spaces, then $\Theta \subseteq \mathcal{L}[\mathcal{X}, \mathcal{Y}]$ is equicontinuous if and only if $\Theta \subseteq \mathcal{B}[\mathcal{X}, \mathcal{Y}]$ and $\sup_{T \in \Theta} \|T\| < \infty$.

The notion of a bounded set in a topological vector space (and, in particular, in a locally convex space) was defined in Problem 4.2. Moreover, it was shown in Problem 4.5(b) that this in fact is the natural extension to topological vector spaces of the usual notion of a bounded set in a normed space.

(b) Show that the Banach–Steinhaus Theorem can be stated as follows: *Every pointwise bounded collection of continuous linear transformations of a Banach space into a normed space is equicontinuous*.

**Problem 4.45.** Let $\{T_n\}$ be a sequence in $\mathcal{B}[\mathcal{X}, \mathcal{Y}]$, where $\mathcal{X}$ and $\mathcal{Y}$ are normed spaces. Prove the following results.

(a) If $T_n \xrightarrow{s} T$ for some $T \in \mathcal{B}[\mathcal{X}, \mathcal{Y}]$, then $\|T_n x\| \to \|T x\|$ for every $x \in \mathcal{X}$ and $\|T\| \leq \liminf_n \|T_n\|$.

(b) If $\sup_n \|T_n\| < \infty$ and $\{T_n a\}$ is a Cauchy sequence in $\mathcal{Y}$ for every $a$ in a dense subset $A$ of $\mathcal{X}$, then $\{T_n x\}$ is a Cauchy sequence in $\mathcal{Y}$ for every $x$ in $\mathcal{X}$.

$\quad$ *Hint*: $T_n x - T_m x = T_n x - T_n a_i + T_n a_i - T_m a_i + T_m a_i - T_m x$.

(c) If there exists $T \in \mathcal{B}[\mathcal{X}, \mathcal{Y}]$ such that $T_n a \to T a$ for every $a$ in a dense subset $A$ of $\mathcal{X}$, and if $\sup_n \|T_n\| < \infty$, then $T_n \xrightarrow{s} T$.

$\quad$ *Hint*: $(T_n - T)x = (T_n - T)(x - a_\varepsilon) + (T_n - T)a_\varepsilon$.

(d) If $\mathcal{X}$ is a Banach space and $\{T_n x\}$ is a Cauchy sequence for every $x \in \mathcal{X}$, then $\sup_n \|T_n\| < \infty$.

(e) If $\mathcal{X}$ and $\mathcal{Y}$ are Banach spaces and $\{T_n x\}$ is a Cauchy sequence for every $x \in \mathcal{X}$, then $T_n \xrightarrow{s} T$ for some $T \in \mathcal{B}[\mathcal{X}, \mathcal{Y}]$.

**Problem 4.46.** Let $\{T_n\}$ be a sequence in $\mathcal{B}[\mathcal{X}, \mathcal{Y}]$ and let $\{S_n\}$ be a sequence in $\mathcal{B}[\mathcal{Y}, \mathcal{Z}]$, where $\mathcal{X}$, $\mathcal{Y}$ and $\mathcal{Z}$ are normed spaces. Suppose

$$T_n \xrightarrow{s} T \quad \text{and} \quad S_n \xrightarrow{s} S$$

for $T \in \mathcal{B}[\mathcal{X}, \mathcal{Y}]$ and $S \in \mathcal{B}[\mathcal{Y}, \mathcal{Z}]$. Prove the following propositions.

(a) If $\sup_n \|S_n\| < \infty$, then $S_n T_n \xrightarrow{s} S T$.

(b) If $\mathcal{Y}$ is a Banach space, then $S_n T_n \xrightarrow{s} S T$.

(c) If $S_n \xrightarrow{u} S$, then $S_n T_n \xrightarrow{s} S T$.

(d) If $S_n \xrightarrow{u} S$ and $T_n \xrightarrow{u} T$, then $S_n T_n \xrightarrow{u} S T$.

Finally, show that addition of strongly (uniformly) convergent sequences of bounded linear transformations is again a strongly (uniformly) convergent sequence of bounded linear transformations whose strong (uniform) limit is the sum of the strong (uniform) limits of each summand.

**Problem 4.47.** Let $\mathcal{X}$ be a Banach space and let $T$ be an operator in $\mathcal{B}[\mathcal{X}]$. If $\lambda$ is any scalar such that $\|T\| < |\lambda|$, then $(\lambda I - T)$ is an invertible element of $\mathcal{B}[\mathcal{X}]$ (i.e., $(\lambda I - T) \in \mathcal{G}[\mathcal{X}]$) and the series $\sum_{i=0}^{\infty} \frac{T^i}{\lambda^{i+1}}$ converges in $\mathcal{B}[\mathcal{X}]$ to $(\lambda I - T)^{-1}$. That is,

$$\|T\| < |\lambda| \quad \text{implies} \quad (\lambda I - T)^{-1} = \frac{1}{\lambda} \sum_{i=0}^{\infty} \left(\tfrac{T}{\lambda}\right)^i \in \mathcal{B}[\mathcal{X}].$$

This is a rather important result, known as the *von Neumann expansion*. The purpose of this problem is to prove it. Take $T \in \mathcal{B}[\mathcal{X}]$ and $0 \neq \lambda \in \mathbb{F}$ arbitrary. Show by induction that, for each integer $n \geq 0$,

(a)
$$\|T^n\| \leq \|T\|^n, \quad \text{and}$$

(b)
$$(\lambda I - T)\frac{1}{\lambda} \sum_{i=0}^{n} \left(\tfrac{T}{\lambda}\right)^i = \frac{1}{\lambda} \sum_{i=0}^{n} \left(\tfrac{T}{\lambda}\right)^i (\lambda I - T) = I - \left(\tfrac{T}{\lambda}\right)^{n+1}.$$

From now on suppose $\|T\| < |\lambda|$ and consider the power sequence $\left\{\left(\frac{T}{\lambda}\right)^n\right\}_{n=0}^{\infty}$ in $\mathcal{B}[\mathcal{X}]$. Use the result in (a) to show that

(c)  $\qquad\qquad \left\{\left(\frac{T}{\lambda}\right)^n\right\}_{n=0}^{\infty}$ is absolutely summable.

Thus conclude that (cf. Problem 3.12)

(d)  $\qquad\qquad\qquad\qquad \left(\frac{T}{\lambda}\right)^n \xrightarrow{u} O$

(i.e., $\frac{T}{\lambda}$ is uniformly stable), and also that (see Proposition 4.4)

(e)  $\qquad\qquad\qquad \left\{\left(\frac{T}{\lambda}\right)^n\right\}_{n=0}^{\infty}$ is summable.

This means that the series $\sum_{i=0}^{\infty}\left(\frac{T}{\lambda}\right)^i$ converges in $\mathcal{B}[\mathcal{X}]$ or, equivalently, that there exists an operator in $\mathcal{B}[\mathcal{X}]$, say $\sum_{i=0}^{\infty}\left(\frac{T}{\lambda}\right)^i$, such that

$$\sum_{i=0}^{n}\left(\frac{T}{\lambda}\right)^i \xrightarrow{u} \sum_{i=0}^{\infty}\left(\frac{T}{\lambda}\right)^i.$$

Apply the results in (b) and (d) to check the following convergences.

(f)  $\qquad (\lambda I - T)\frac{1}{\lambda}\sum_{i=0}^{n}\left(\frac{T}{\lambda}\right)^i \xrightarrow{u} I \quad$ and $\quad \frac{1}{\lambda}\sum_{i=0}^{n}\left(\frac{T}{\lambda}\right)^i(\lambda I - T) \xrightarrow{u} I.$

Now use (e) and (f) to show that

(g)  $\qquad (\lambda I - T)\frac{1}{\lambda}\sum_{i=0}^{\infty}\left(\frac{T}{\lambda}\right)^i = \frac{1}{\lambda}\sum_{i=0}^{\infty}\left(\frac{T}{\lambda}\right)^i(\lambda I - T) = I.$

Hence $\frac{1}{\lambda}\sum_{i=0}^{\infty}\left(\frac{T}{\lambda}\right)^i \in \mathcal{B}[\mathcal{X}]$ is the inverse of $(\lambda I - T) \in \mathcal{B}[\mathcal{X}]$ (Problem 1.7). Therefore, $(\lambda I - T) \in \mathcal{B}[\mathcal{X}]$ is an invertible element of $\mathcal{B}[\mathcal{X}]$ (i.e., $(\lambda I - T) \in \mathcal{G}[\mathcal{X}]$) whose inverse $(\lambda I - T)^{-1}$ is the (uniform) limit of the sequence $\left\{\sum_{i=0}^{n}\frac{T^i}{\lambda^{i+1}}\right\}_{n=0}^{\infty}$; that is,

$$(\lambda I - T)^{-1} = \frac{1}{\lambda}\sum_{i=0}^{\infty}\left(\frac{T}{\lambda}\right)^i \in \mathcal{B}[\mathcal{X}].$$

Finally, verify that

(h)  $\qquad \|(\lambda I - T)^{-1}\| \le \frac{1}{|\lambda|}\sum_{i=0}^{\infty}\left(\frac{\|T\|}{|\lambda|}\right)^i = (|\lambda| - \|T\|)^{-1}.$

Remark: Exactly the same proof applies if, instead of $\mathcal{B}[\mathcal{X}]$, we were working on an abstract unital Banach algebra $\mathcal{A}$.

**Problem 4.48.** If $T$ is a strict contraction on a Banach space $\mathcal{X}$, then

$$(I - T)^{-1} = \sum_{i=0}^{\infty} T^i \in \mathcal{B}[\mathcal{X}].$$

This is the special case of Problem 4.47 for $\lambda = 1$. Use it to prove the following assertions.

(a) Every operator in the open unit ball centered at the identity $I$ of $\mathcal{B}[\mathcal{X}]$ is invertible (i.e., if $\|I - S\| < 1$, then $S \in \mathcal{G}[\mathcal{X}]$).

(b) Let $\mathcal{X}$ and $\mathcal{Y}$ be Banach spaces. Centered at each $T \in \mathcal{G}[\mathcal{X}, \mathcal{Y}]$ there exists a nonempty open ball $B_\varepsilon(T) \subset \mathcal{G}[\mathcal{X}, \mathcal{Y}]$ such that $\sup_{S \in B_\varepsilon(T)} \|S^{-1}\| < \infty$. In particular, $\mathcal{G}[\mathcal{X}, \mathcal{Y}]$ is open in $\mathcal{B}[\mathcal{X}, \mathcal{Y}]$.

  *Hint*: Suppose $\|T - S\| < \varepsilon < \|T^{-1}\|^{-1}$ so that

$$\|I_X - T^{-1}S\| = \|T^{-1}(T - S)\| < \|T^{-1}\|\varepsilon < 1.$$

  Thus $T^{-1}S = I_X - (I_X - T^{-1}S)$ lies in $\mathcal{G}[\mathcal{X}]$ by (a). Hence $S = T T^{-1}S$ lies in $\mathcal{G}[\mathcal{X}, \mathcal{Y}]$ (Corollary 4.23). Moreover, $\|S^{-1}\| = \|S^{-1}TT^{-1}\| \leq \|T^{-1}\|\|S^{-1}T\|$. But

$$\|S^{-1}T\| = \|(T^{-1}S)^{-1}\| = \|[I_X - (I_X - T^{-1}S)]^{-1}\|$$
$$\leq (1 - \|I_X - T^{-1}S\|)^{-1}$$

  (cf. Problem 4.47(h)). Conclude: $\|S^{-1}\| \leq \|T^{-1}\|(1 - \|T^{-1}\|\varepsilon)^{-1}$.

(c) Inversion is a continuous mapping. That is, if $\mathcal{X}$ and $\mathcal{Y}$ are Banach spaces, then the map $T \mapsto T^{-1}$ of $\mathcal{G}[\mathcal{X}, \mathcal{Y}]$ into $\mathcal{G}[\mathcal{Y}, \mathcal{X}]$ is continuous.

  *Hint*: $T^{-1} - S^{-1} = T^{-1}(S - T)S^{-1}$. If $\{T_n\}$ in $\mathcal{G}[\mathcal{X}, \mathcal{Y}]$ converges in $\mathcal{B}[\mathcal{X}, \mathcal{Y}]$ to $S \in \mathcal{G}[\mathcal{X}, \mathcal{Y}]$, then $\sup_n \|T_n^{-1}\| < \infty$, and so $T_n^{-1} \to S^{-1}$.

**Problem 4.49.** Let $\mathcal{X}$ be a normed space. Show that the collection of all strict contractions in $\mathcal{B}[\mathcal{X}]$ is not closed in $\mathcal{B}[\mathcal{X}]$ (and hence not strongly closed in $\mathcal{B}[\mathcal{X}]$).
*Hint*: Each $D_n = \mathrm{diag}(\frac{1}{2}, \dots, \frac{n}{n+1}, 0, 0, 0, \dots)$, with just the first $n$ entries different from zero, is a strict contraction in any $\mathcal{B}[\ell_+^p]$ (and in $\mathcal{B}[\ell_+^\infty]$).

On the other hand, show that the collection of all contractions in $\mathcal{B}[\mathcal{X}]$ is strongly closed in $\mathcal{B}[\mathcal{X}]$, and hence (uniformly) closed in $\mathcal{B}[\mathcal{X}]$.
*Hint*: $\big|\|T_n x\| - \|Tx\|\big| \leq \|(T_n - T)x\|$.

**Problem 4.50.** Show that the strong limit of a sequence of linear isometries is again a linear isometry. In other words, the collection of all isometries from $\mathcal{B}[\mathcal{X}, \mathcal{Y}]$

is strongly closed, and hence uniformly closed ($\mathcal{X}$ and $\mathcal{Y}$ are, of course, normed spaces).

*Hint*: Proposition 4.37 and Problem 4.45(a).

**Problem 4.51.** Take an arbitrary $p \geq 1$ and consider the normed space $\ell_+^p$ of Example 4B. Let $\{D_n\}$ be a sequence of diagonal operators in $\mathcal{B}[\ell_+^p]$. If $\{D_n\}$ converges strongly to $D \in \mathcal{B}[\ell_+^p]$, then $D$ is a diagonal operator. Prove.

**Problem 4.52.** Let $\{P_k\}_{k=1}^{\infty}$ be a sequence of diagonal operators in $\mathcal{B}[\ell_+^p]$ such that, for each $k \geq 1$,

$$P_k = \mathrm{diag}(e_k) = \mathrm{diag}(0, \dots, 0, 1, 0, 0, 0, \dots)$$

(the only nonzero entry is equal to one and lies at the $k$th position). Set

$$E_n = \sum_{k=1}^{n} P_k = \mathrm{diag}(1, \dots, 1, 0, 0, 0, \dots) \in \mathcal{B}[\ell_+^p]$$

for every integer $n \geq 1$. Show that $E_n \xrightarrow{s} I$, the identity in $B[\ell_+^p]$, but $\{E_n\}_{n=1}^{\infty}$ does not converge uniformly because $\|E_n - I\| = 1$ for all $n \geq 1$.

**Problem 4.53.** Let $a = \{\alpha_k\}_{k=0}^{\infty}$ be a scalar-valued sequence and consider a sequence $\{D_n\}_{n=0}^{\infty}$ of diagonal mappings of the Banach space $\ell_+^p$ into itself such that

$$D_n = \mathrm{diag}(\alpha_0, \dots, \alpha_n, 0, 0, 0, \dots)$$

for each integer $n \geq 0$, where the entries in the main diagonal are all null except perhaps for the first $n+1$. It is clear that $D_n \in \mathcal{B}[\ell_+^p]$ for every $n \geq 0$. (Reason: $\mathcal{B}_0[\ell_+^p] \subset \mathcal{B}[\ell_+^p]$.) If $a \in \ell_+^{\infty}$, then consider the diagonal operator $D_a = \mathrm{diag}(\{\alpha_k\}_{k=0}^{\infty}) \in \mathcal{B}[\ell_+^p]$ of Examples 4H and 4K. Now prove the following assertions.

(a) If $\sup_k |\alpha_k| < \infty$, then $D_n \xrightarrow{s} D_a$.

   *Hint*: $\|D_n x - D_a x\|^p \leq \sup_k |\alpha_k|^p \sum_{k=n}^{\infty} |\xi_k|^p$ for $x = \{\xi_k\}_{k=0}^{\infty} \in \ell_+^p$.

(b) Conversely, if $\{D_n x\}_{n=0}^{\infty}$ converges in $\ell_+^p$ for every $x \in \ell_+^p$, then $\sup_k |\alpha_k| < \infty$, and hence $D_n \xrightarrow{s} D_a$.

   *Hint*: Proposition 3.39, Theorem 4.43 and Example 4H.

(c) If $\lim_k |\alpha_k| = 0$, then $D_n \xrightarrow{u} D_a$.

   *Hint*: Verify that $\|(D_n - D_a)x\|^p \leq \sup_{n \leq k} |\alpha_k|^p \|x\|^p$ and then conclude: $\lim_n \|D_n - D_a\| \leq \lim_n \sup_{n \leq k} |\alpha_k| = \lim \sup_n |\alpha_k|$.

(d) Conversely, if $\{D_n\}_{n=0}^{\infty}$ converges uniformly, then $\lim_k |\alpha_k| = 0$, and hence $D_n \xrightarrow{u} D_a$.

*Hint*: Uniform convergence implies strong convergence. Apply (c). Compute $(D_n - D_a)e_k$ for every $k, n$.

**Problem 4.54.** Take any $\alpha \in \mathbb{C}$ such that $\alpha \neq 1$. Consider the operators $A$ and $P$ in $\mathcal{B}[\mathbb{C}^2]$ identified with the matrices

$$A = \begin{pmatrix} 0 & 1 \\ -\alpha & 1+\alpha \end{pmatrix} \quad \text{and} \quad P = \tfrac{1}{\alpha-1}\begin{pmatrix} \alpha & -1 \\ \alpha & -1 \end{pmatrix}$$

(i.e., these matrices are the representations of $A$ and $P$ with respect to the canonical basis for $\mathbb{C}^2$).

(a) Show that $PA = AP = P = P^2$.

(b) Prove by induction that

$$A^{n+1} = \alpha A^n + (1-\alpha)P,$$

and hence (see Problem 2.19 or supply another induction)

$$A^n = \alpha^n(I - P) + P,$$

for every integer $n \geq 0$.

(c) Finally, show that

$$|\alpha| < 1 \quad \text{implies} \quad A^n \xrightarrow{u} P \quad \text{and} \quad \|P\| > 1,$$

where $\| \ \|$ denotes the norm on $\mathcal{B}[\mathbb{C}^2]$ induced by any of the norms $\| \ \|_p$ (for $p \geq 1$) or $\| \ \|_\infty$ on the linear space $\mathbb{C}^2$ as in Example 4A.

*Hint*: $1 < \|Pe_1\|_\infty \leq \|Pe_1\|_p$ (cf. Problem 3.33).

**Problem 4.55.** Take $T \in \mathcal{L}[\mathcal{X}]$, a linear transformation on a normed space $\mathcal{X}$. Suppose the power sequence $\{T^n\}$ is pointwise convergent, which means that there exists $P \in \mathcal{L}[\mathcal{X}]$ such that $T^n x \to Px$ in $\mathcal{X}$ for every $x \in \mathcal{X}$. Show that

(a) $PT^k = T^k P = P = P^k$ for every integer $k \geq 1$,

(b) $(T - P)^n = T^n - P$ for every integer $n \geq 1$.

Now suppose $T$ lies in $\mathcal{B}[\mathcal{X}]$ and prove the following propositions.

(c) If $T^n \xrightarrow{s} P \in \mathcal{B}[\mathcal{X}]$, then $P$ is a projection and $(T - P)^n \xrightarrow{s} O$.

(d) If $T^n x \to Px$ for every $x \in \mathcal{X}$, then $P \in \mathcal{L}[\mathcal{X}]$ is a projection. If, in addition, $\mathcal{X}$ is a Banach space, then $P$ is a continuous projection and $(T - P) \in \mathcal{B}[\mathcal{X}]$ is strongly stable.

**Problem 4.56.** Let $F\colon X \to X$ be a mapping of a set $X$ into itself. Recall that $F$ is injective if and only if it has a left inverse $F^{-1}\colon \mathcal{R}(X) \to X$ on its range. Thus, if $F$ is injective and idempotent (i.e., $F = F^2$), then $F = F^{-1}FF = F^{-1}F = I$, and hence

(a) the unique idempotent injective mapping is the identity.

This is a purely set-theoretic result (no algebra or topology is involved). Now let $X$ be a metric space and recall that every isometry is injective. Therefore,

(b) the unique idempotent isometry is the identity.

In particular, if $\mathcal{X}$ is a normed space and $F\colon \mathcal{X} \to \mathcal{X}$ is a projection (i.e., an idempotent linear transformation) and an isometry, then $F = I$: the identity is the unique isometry that also is a projection. Show that

(c) the unique linear isometry that has a strongly convergent power sequence is the identity.

*Hint*: Problems 4.50 and 4.55.

**Problem 4.57.** Let $\{T_n\}$ be a sequence of bounded linear transformations in $\mathcal{B}[\mathcal{Y}, \mathcal{Z}]$, where $\mathcal{Y}$ is a Banach space and $\mathcal{Z}$ is a normed space, and take $T \in \mathcal{B}[\mathcal{Y}, \mathcal{Z}]$. Use Proposition 4.46 to show that, if $\mathcal{M}$ is a finite-dimensional subspace of $\mathcal{Y}$, then

(a) $$T_n \xrightarrow{s} T \quad \text{implies} \quad (T_n - T)|_{\mathcal{M}} \xrightarrow{u} O.$$

Now let $K$ be a compact linear transformation of a normed space $\mathcal{X}$ into $\mathcal{Y}$ (i.e., $K \in \mathcal{B}_\infty[\mathcal{X}, \mathcal{Y}]$). Prove that

(b) $$T_n \xrightarrow{s} T \quad \text{implies} \quad T_n K \xrightarrow{u} T K.$$

*Hint*: Take an arbitrary $x \in \mathcal{X}$. Use Proposition 4.56 to show that for each $\varepsilon > 0$ there exists a finite-dimensional subspace $\mathcal{R}_\varepsilon$ of $\mathcal{Y}$ and a vector $r_{\varepsilon,x}$ in $\mathcal{R}_\varepsilon$ such that

$$\|Kx - r_{\varepsilon,x}\| < 2\varepsilon \|x\| \quad \text{and} \quad \|r_{\varepsilon,x}\| < (2\varepsilon + \|K\|)\|x\|.$$

Then verify that

$$
\begin{aligned}
\|(T_n K - TK)x\| &\le \|T_n - T\|\|Kx - r_{\varepsilon,x}\| + \left\|(T_n - T)|_{\mathcal{R}_\varepsilon}\right\|\|r_{\varepsilon,x}\| \\
&< \left(2\varepsilon\|T_n - T\| + (2\varepsilon + \|K\|)\left\|(T_n - T)|_{\mathcal{R}_\varepsilon}\right\|\right)\|x\|.
\end{aligned}
$$

Now apply the Banach–Steinhaus Theorem (Theorem 4.43) to ensure that $\sup_n \|T_n - T\| < \infty$ and conclude from item (a): for every $\varepsilon > 0$

$$\limsup_n \|T_n K - TK\| < \left(2 \sup_n \|T_n - T\|\right)\varepsilon.$$

**Problem 4.58.** Prove the converse of Corollary 4.55 under the assumption that the Banach space $\mathcal{Y}$ has a Schauder basis. In other words, prove the following proposition.

*If $\mathcal{Y}$ is a Banach space with a Schauder basis and $\mathcal{X}$ is a normed space, then every compact linear transformation in $\mathcal{B}_\infty[\mathcal{X}, \mathcal{Y}]$ is the uniform limit of a sequence of finite-rank linear transformations in $\mathcal{B}_0[\mathcal{X}, \mathcal{Y}]$.*

*Hint*: Suppose $\mathcal{Y}$ is infinite-dimensional (otherwise the result is trivially verified) and has a Schauder basis. Take an arbitrary $K$ in $\mathcal{B}_\infty[\mathcal{X}, \mathcal{Y}]$. $\mathcal{R}(K)^-$ is a Banach space possessing a Schauder basis, say $\{y_i\}_{i=0}^\infty$, so that every $y \in \mathcal{R}(K)^-$ has a unique expansion $y = \sum_{i=0}^\infty \alpha_i(y)\, y_i$ (Problem 4.11). For each nonnegative integer $n$ consider the mapping $E_n \colon \mathcal{R}(K)^- \to \mathcal{R}(K)^-$ defined by $E_n y = \sum_{i=0}^n \alpha_i(y)\, y_i$. Show that each $E_n$ is bounded and linear (Problem 4.30), and also of finite rank (for $\mathcal{R}(E_n) \subseteq \bigvee\{y_i\}_{i=0}^n$). Moreover, show that $\{E_n\}_{n=0}^\infty$ converges strongly to the identity operator $I$ on $\mathcal{R}(K)^-$ (Problem 4.9(b)). Use the previous problem to conclude that $E_n K \xrightarrow{u} K$. Finally, check that $E_n K$ lies in $\mathcal{B}_0[\mathcal{X}, \mathcal{Y}]$ for each $n$.

Remark: Consider the remark in Problem 4.11. There we commented on the construction of a separable Banach space that has no Schauder basis. Actually, such a breakthrough was achieved by Enflo in 1973 when he exhibited a separable (and reflexive) Banach space $\mathcal{X}$ for which $\mathcal{B}_0[\mathcal{X}]$ is not dense in $\mathcal{B}_\infty[\mathcal{X}]$, so that there exist compact operators on $\mathcal{X}$ that are not the (uniform) limit of finite-rank operators (and hence the converse of Corollary 4.55 fails in general). Thus, according to the above proposition, such an $\mathcal{X}$ is a separable Banach space without a Schauder basis.

**Problem 4.59.** Recall that the concepts of strong and uniform convergence coincide in a finite-dimensional space (Proposition 4.46). Consider the Banach space $\ell_+^p$ for any $p \geq 1$ (which has a Schauder basis — Problem 4.12). Exhibit a sequence of finite-rank (compact) operators on $\ell_+^p$ that converges strongly to a finite-rank (compact) operator but is not uniformly convergent.

*Hint*: Let $P_k$ be the diagonal operator defined in Problem 4.52.

**Problem 4.60.** Let $\mathcal{M}$ be a subspace of an infinite-dimensional Banach space $\mathcal{X}$. Show that an extension over $\mathcal{X}$ of a compact operator on $\mathcal{M}$ may not be compact.

*Hint*: Let $\mathcal{M}$ and $\mathcal{N}$ be Banach spaces over the same scalar field. Suppose $\mathcal{N}$ is infinite-dimensional. Set $\mathcal{X} = \mathcal{M} \oplus \mathcal{N}$ and consider the direct sum $T = K \oplus I$ in $\mathcal{B}[\mathcal{X}]$, where $K$ is a compact operator in $\mathcal{B}_\infty[\mathcal{M}]$ and $I$ is the identity operator in $\mathcal{B}[\mathcal{N}]$, as in Problem 4.16.

**Problem 4.61.** Let $\mathcal{X} \neq \{0\}$ and $\mathcal{Y} \neq \{0\}$ be normed spaces over the same scalar field and let $\mathcal{M}$ be a proper subspace of $\mathcal{X}$. Show that there exists $O \neq T$ in $\mathcal{B}[\mathcal{X}, \mathcal{Y}]$ such that $\mathcal{M} \subseteq \mathcal{N}(T)$ (i.e., such that $T(\mathcal{M}) = \{0\}$). Hint: Corollary 4.63.

**Problem 4.62.** Let $\mathcal{X}$ be a normed space. Since $\big|\,\|x\| - \|y\|\,\big| \leq \|x - y\|$ for every $x, y \in \mathcal{X}$, it follows that the norm on $\mathcal{X}$ is a real-valued contraction that takes each vector of $\mathcal{X}$ to its norm. Show that for each vector in $\mathcal{X}$ there exists a real-valued *linear* contraction on $\mathcal{X}$ that takes that vector to its norm.

**Problem 4.63.** The *annihilator* of a subset $A$ of a normed space $\mathcal{X}$ is the following subset of the dual space $\mathcal{X}^*$:

$$A^\perp = \{ f \in \mathcal{X}^* : \ A \subseteq N(f) \}.$$

(a) Verify that $\varnothing^\perp = \{0\}^\perp = \mathcal{X}$, $\mathcal{X}^\perp = \{0\}$, $B^\perp \subseteq A^\perp$ whenever $A \subseteq B$ and, if $A \neq \varnothing$, then

$$A^\perp = \{ f \in \mathcal{X}^* : \ f(A) = \{0\} \}.$$

(b) Show that $A^\perp$ is a subspace of $\mathcal{X}^*$ for every subset $A$ of $\mathcal{X}$.

(c) Take an arbitrary subset $A$ of $\mathcal{X}$ and show that

$$A^- \subseteq \bigcap_{f \in A^\perp} N(f).$$

*Hint:* If $f \in A^\perp$, then $A \subseteq N(f)$. Thus conclude that $A^- \subseteq N(f)$ for every $f \in A^\perp$ (Proposition 4.13).

Now let $\mathcal{M}$ be a linear manifold of $\mathcal{X}$ and prove the following assertions.

(d) $\bigcap_{f \in \mathcal{M}^\perp} N(f) \subseteq \mathcal{M}^-$.

*Hint:* if $x_0 \in \mathcal{X} \backslash \mathcal{M}^-$, then there exists $f \in \mathcal{M}^\perp$ such that $f(x_0) = 1$ (Corollary 4.63), and hence $x_0 \notin \bigcap_{f \in \mathcal{M}^\perp} N(f)$.

(e) $\mathcal{M}^- = \bigcap_{f \in \mathcal{M}^\perp} N(f)$.

(f) $\mathcal{M}^- = \mathcal{X}$ if and only if $\mathcal{M}^\perp = \{0\}$.

**Problem 4.64.** Let $\{e_i\}_{i=1}^n$ be a Hamel basis for a finite-dimensional normed space $\mathcal{X}$. Verify the following propositions.

(a) For each $i = 1, \ldots, n$ there exists $f_i \in \mathcal{X}^*$ such that $f_i(e_j) = \delta_{ij}$ for every $j = 1, \ldots, n$.

*Hint:* Set $f_i(x) = \xi_i$ for every $x = \sum_{i=1}^n \xi_i e_i \in \mathcal{X}$.

(b) $\{f_i\}_{i=1}^n$ is a Hamel basis for $\mathcal{X}^*$.

*Hint:* If $f \in \mathcal{X}^*$, then $f = \sum_{i=1}^n f(e_i) f_i$.

Now conclude that $\dim \mathcal{X} = \dim \mathcal{X}^*$ whenever $\dim \mathcal{X} < \infty$.

**Problem 4.65.** Let $J: \mathcal{Y} \to \mathcal{X}$ be an isometric isomorphism of a normed space $\mathcal{Y}$ onto a normed space $\mathcal{X}$, and consider the mapping $J^*: \mathcal{X}^* \to \mathcal{Y}^{\mathbb{F}}$ defined by $J^*f = f \circ J$ for every $f \in \mathcal{X}^*$. Show that

(a) $J^*(\mathcal{X}^*) = \mathcal{Y}^*$ so that $J^*: \mathcal{X}^* \to \mathcal{Y}^*$ is surjective,

(b) $J^*: \mathcal{X}^* \to \mathcal{Y}^*$ is linear, and

(c) $\|J^*f\| = \|f\|$ for every $f \in \mathcal{X}^*$. (*Hint*: Problem 4.41.)

Conclude: If $\mathcal{X}$ and $\mathcal{Y}$ are isometrically isomorphic normed spaces, then their duals $\mathcal{X}^*$ and $\mathcal{Y}^*$ are isometric isomorphic too. That is,

$$\mathcal{X} \cong \mathcal{Y} \quad \text{implies} \quad \mathcal{X}^* \cong \mathcal{Y}^*.$$

**Problem 4.66.** Consider the normed space $\ell_+^\infty$ equipped with its usual sup-norm and recall that $\ell_+^c \subseteq \ell_+^\infty$ (Problem 3.59). Let $S_- \in \mathcal{B}[\ell_+^\infty]$ be the backward unilateral shift on $\ell_+^\infty$ (Example 4L and Problem 4.39). A bounded linear functional $f: \ell_+^\infty \to \mathbb{F}$ is called a *Banach limit* if it satisfies the following conditions.

(i)    $\|f\| = 1$,

(ii)    $f(x) = f(S_-x)$ for every $x \in \ell_+^\infty$,

(iii)    If $x = \{\xi_k\}$ lies in $\ell_+^c$, then $f(x) = \lim_k \xi_k$,

(iv)    If $x = \{\xi_k\}$ in $\ell_+^\infty$ is such that $\xi_k \geq 0$ for all $k$, then $f(x) \geq 0$.

Condition (iii) says that Banach limits extend to $\ell_+^\infty$ the limit function on $\ell_+^c$ (i.e., $f$ is defined on $\ell_+^\infty$ and its restriction to $\ell_+^c$, $f|_{\ell_+^c}$, assigns to each convergent sequence its own limit). The remaining conditions represent fundamental properties of a limit function. Conditions (i) and (ii) say that $\lim_k |\xi_k| \leq \sup_k |\xi_k|$ and $\lim_k \xi_k = \lim_k \xi_{k+n}$ for every positive integer $n$, respectively, whenever $\{\xi_k\} \in \ell_+^c$. Condition (iv) says that $f$ is order-preserving for real-valued sequences in $\ell_+^\infty$ (i.e., if $x = \{\xi_k\}$ and $y = \{v_k\}$ are real-valued sequences in $\ell_+^\infty$, then $f(x), f(y) \in \mathbb{R}$ — why? — and $f(x) \leq f(y)$ whenever $\xi_k \leq v_k$ for every $k$). The purpose of this problem is to show how the Hahn–Banach Theorem ensures the existence of Banach limits on $\ell_+^\infty$.

(a) Suppose $\mathbb{F} = \mathbb{R}$ (so that the sequences in $\ell_+^\infty$ are all real-valued). Let $e$ be the constant sequence in $\ell_+^\infty$ whose entries are all ones (i.e., $e = (1, 1, 1, \ldots)$) and set $\mathcal{M} = \mathcal{R}(I - S_-)$. Show that $d(e, \mathcal{M}) = 1$.

*Hint*: Verify that $d(e, \mathcal{M}) \leq 1$ (for $\|e\|_\infty = 1$ and $0 \in \mathcal{M}$). Now take an arbitrary $u = \{v_k\}$ in $\mathcal{M}$. If $v_{k_0} \leq 0$ for some integer $k_0$, then show that $1 \leq \|e - u\|_\infty$. But $u \in \mathcal{R}(I - S_-)$, and hence $v_k = \xi_k - \xi_{k+1}$ for some

$x = \{\xi_k\}$ in $\ell_+^\infty$. If $\upsilon_k \geq 0$ for all $k$, then $\{\xi_k\}$ is decreasing and bounded. Check that $\{\xi_k\}$ converges in $\mathbb{R}$ (Problem 3.10), show that $\upsilon_k \to 0$ (Problem 3.51), and conclude that $1 \leq \|e - u\|_\infty$ whenever $\upsilon_k \geq 0$ for all $k$. Hence $1 \leq \|e - u\|_\infty$ for every $u \in \mathcal{M}$ so that $d(e, \mathcal{M}) \geq 1$.

Therefore (it does not matter whether $\mathcal{M}$ is closed or not), $\mathcal{M}^-$ is a subspace of $\ell_+^\infty$ (Proposition 4.9(a)) and $d(e, \mathcal{M}^-) = 1$ (Problem 3.43(b)). Then, according to Corollary 4.63, there exists a bounded linear functional $\varphi \colon \ell_+^\infty \to \mathbb{R}$ such that

$$\varphi(e) = 1, \quad \varphi(\mathcal{M}^-) = \{0\} \quad \text{and} \quad \|\varphi\| = 1.$$

(b)  Show that $\varphi(x) = \varphi(S_-^n x)$ for every $x \in \ell_+^\infty$ and all $n \geq 1$.

   *Hint*: $\varphi((I - S_-)x) = 0$ because $\varphi(\mathcal{M}) = \{0\}$. This leads to $\varphi(x) = \varphi(S_- x)$ for every $x$ in $\ell_+^\infty$. Conclude the proof by induction.

(c)  Show that $\varphi$ satisfies condition (iii).

   *Hint*: Take an arbitrary $x = \{\xi_k\}$ in $\ell_+^c$ so that $\xi_k \to \xi$ in $\mathbb{R}$ for some $\xi \in \mathbb{R}$. Observe that $|\xi_{k+n} - \xi| \leq |\xi_{k+n} - \xi_n| + |\xi_n - \xi|$ for every pair of positive integers $n, k$. Now use Problem 3.51(a) to show that $\|S_-^n x - \xi e\|_\infty \to 0$. That is, $S_-^n x \to \xi e$ in $\ell_+^\infty$. Next verify that $\varphi(x) = \xi \varphi(e)$.

(d)  Show that $\varphi$ satisfies condition (iv).

   *Hint*: Take any nonzero $x = \{\xi_k\}$ in $\ell_+^c$ such that $\xi_k \geq 0$ for all $k$ and set $x' = (\|x\|_\infty^{-1})x = \{\xi_k'\}$. Verify that $0 \leq \xi_k' \leq 1$ for all $k$, and hence $\|e - x'\|_\infty \leq 1$. Finally, show that $1 - \varphi(x') = \varphi(e - x') \leq 1$, and conclude: $\varphi(x) \geq 0$.

Thus, in the real case, $\varphi \colon \ell_+^\infty \to \mathbb{R}$ is a Banach limit on $\ell_+^\infty$.

(e)  Now suppose $\mathbb{F} = \mathbb{C}$ (so that complex-valued sequences are allowed in $\ell_+^\infty$). For each $x = x_1 + i x_2$ in $\ell_+^\infty$, where $x_1$ and $x_2$ are real-valued sequences in $\ell_+^\infty$, set

$$f(x) = \varphi(x_1) + i \varphi(x_2).$$

Show that this defines a bounded linear functional $f \colon \ell_+^\infty \to \mathbb{C}$.

   *Hint*: $\|f\| \leq 2$.

(f)  Verify that $f$ satisfies conditions (ii), (iii) and (iv).

(g)  Prove that $\|f\| = 1$.

   Hint: Let $\ell_+^\#$ be the set of all scalar-valued sequences that take on only a finite number of values (i.e., that have a finite range). Clearly, $\ell_+^\# \subset \ell_+^\infty$. If $y \in \ell_+^\#$ with $\|y\|_\infty \leq 1$, then there exists a finite partition of $\mathbb{N}$, say $\{N_j\}_{j=1}^m$, and a finite set of scalars $\{\alpha_j\}_{j=1}^m$ with $|\alpha_j| \leq 1$ for all $j$ such that $y = \sum_{j=1}^m \alpha_j \chi_{N_j}$.

Here $\chi_{N_j}$ is the characteristic function of $N_j$ which, in fact, is an element of $\ell_+^{\#}$ (i.e., $\chi_{N_j} = \{\chi_{N_j}(k)\}_{k\in\mathbb{N}}$, where $\chi_{N_j}(k) = 1$ if $k \in N_j$ and $\chi_{N_j}(k) = 0$ if $k \in \mathbb{N}\backslash N_j$). Verify that $f(y) = \sum_{j=1}^{m}\alpha_j f(\chi_{N_j}) = \sum_{j=1}^{m}\alpha_j \varphi(\chi_{N_j})$ and $\sum_{j=1}^{m}\varphi(\chi_{N_j}) = \varphi\left(\sum_{j=1}^{m}\chi_{N_j}\right) = \varphi(\chi_{\mathbb{N}}) = \varphi(e)$. Recall that $\varphi$ satisfies condition (iv) and show that $|f(y)| \leq (\sup_j |\alpha_j|)\varphi(e)$.

Conclusion 1. If $y \in \ell_+^{\#}$ and $\|y\|_{\infty} \leq 1$, then $|f(y)| \leq 1$.

The closed unit ball $B$ with center at the origin of $\mathbb{C}$ is compact. For each positive integer $n$ take a finite $\frac{1}{n}$-net for $B$, say $B_n \subset B$. If $x = \{\xi_k\} \in \ell_+^{\infty}$ is such that $\|x\|_{\infty} \leq 1$, then $\xi_k \in B$ for all $k$. Thus for each $k$ there exists $\upsilon_n(k) \in B_n$ such that $|\upsilon_n(k) - \xi_k| < \frac{1}{n}$. This defines, for each $n$, a $B_n$-valued sequence $y_n = \{\upsilon_n(k)\}$ such that $\|y_n - x\|_{\infty} < \frac{1}{n}$, which in turn defines an $\ell_+^{\#}$-valued sequence $\{y_n\}$ with $\|y_n\|_{\infty} \leq 1$ for all $n$ that converges in $\ell_+^{\infty}$ to $x$.

Conclusion 2. Every $x \in \ell_+^{\infty}$ with $\|x\|_{\infty} \leq 1$ is the limit in $\ell_+^{\infty}$ of an $\ell_+^{\#}$-valued sequence $\{y_n\}$ with $\sup_n \|y_n\|_{\infty} \leq 1$.

Recall that $f\colon \ell_+^{\infty} \to \mathbb{C}$ is continuous. Apply Conclusion 2 to show that $f(y_n) \to f(x)$, and hence $|f(y_n)| \to |f(x)|$. Since $|f(y_n)| \leq 1$ for every $n$ (by Conclusion 1), it follows that $|f(x)| \leq 1$ for every $x \in \ell_+^{\infty}$ with $\|x\|_{\infty} \leq 1$. Therefore $\|f\| \leq 1$. Finally, verify that $\|f\| \geq 1$ (for $f(e) = \varphi(e)$ and $\|e\|_{\infty} = 1$).

Thus, in the complex case, $f\colon \ell_+^{\infty} \to \mathbb{C}$ is a Banach limit on $\ell_+^{\infty}$.

**Problem 4.67.** Let $\mathcal{X}$ be a normed space. An $\mathcal{X}$-valued sequence $\{x_n\}$ is said to be *weakly convergent* if there exists $x \in \mathcal{X}$ such that $\{f(x_n)\}$ converges in $\mathbb{F}$ to $f(x)$ for every $f \in \mathcal{X}^*$. In this case we say that $\{x_n\}$ *converges weakly to* $x \in \mathcal{X}$ (notation: $x_n \overset{w}{\longrightarrow} x$) and $x$ is said to be the *weak limit* of $\{x_n\}$. Prove the following assertions.

(a) The weak limit of a weakly convergent sequence is unique.

  *Hint*: $f(x) = 0$ for all $f \in \mathcal{X}^*$ implies $x = 0$.

(b) $\{x_n\}$ converges weakly to $x$ if and only if every subsequence of $\{x_n\}$ converges weakly to $x$.

  *Hint*: Proposition 3.5.

(c) If $\{x_n\}$ converges in the norm topology to $x$, then it converges weakly to $x$ (i.e., $x_n \to x \implies x_n \overset{w}{\longrightarrow} x$).

  *Hint*: $|f(x_n - x)| \leq \|f\|\|x_n - x\|$.

(d) If $\{x_n\}$ converges weakly, then it is bounded in the norm topology (i.e., $x_n \xrightarrow{w} x \implies \sup_n \|x_n\| < \infty$).

*Hint*: For each $x \in \mathcal{X}$ there exists $\varphi_x \in \mathcal{X}^{**}$ such that $\varphi_x(f) = f(x)$ for every $f \in \mathcal{X}^*$ and $\|\varphi_x\| = \|x\|$. This is the natural embedding of $\mathcal{X}$ into $\mathcal{X}^{**}$ (Theorem 4.66). Verify that $\sup_n |f(x_n)| < \infty$ for every $f \in \mathcal{X}^*$ whenever $x_n \xrightarrow{w} x$, and show that $\sup_n |\varphi_{x_n}(f)| < \infty$ for every $f \in \mathcal{X}^*$. Now use the Banach–Steinhaus Theorem (recall: $\mathcal{X}^*$ is a Banach space).

**Problem 4.68.** Let $\mathcal{X}$ and $\mathcal{Y}$ be normed spaces. A $\mathcal{B}[\mathcal{X}, \mathcal{Y}]$-valued sequence $\{T_n\}$ *converges weakly* in $\mathcal{B}[\mathcal{X}, \mathcal{Y}]$ if $\{T_n x\}$ converges weakly in $\mathcal{Y}$ for every $x \in \mathcal{X}$. Equivalently, if $\{f(T_n x)\}$ converges in $\mathbb{F}$ for every $f \in \mathcal{Y}^*$ and every $x \in \mathcal{X}$.

(a) If $\{T_n\}$ converges weakly in $\mathcal{B}[\mathcal{X}, \mathcal{Y}]$, then there exists a unique $T \in \mathcal{L}[\mathcal{X}, \mathcal{Y}]$ (called the *weak limit* of $\{T_n\}$) such that $T_n x \xrightarrow{w} Tx$ in $\mathcal{Y}$ for every $x \in \mathcal{X}$. Prove.

*Hint*: Problem 4.67(a) ensures the existence of a unique mapping $T : \mathcal{X} \to \mathcal{Y}$, which is linear because every $f$ in $\mathcal{Y}^*$ is linear.

Notation: $T_n \xrightarrow{w} T$ (or $T_n - T \xrightarrow{w} O$). If $\{T_n\}$ does not converge weakly to $T$, then we write $T_n \xrightarrow{w}\!\!\!\!/\;\, T$.

(b) If $\mathcal{X}$ is a Banach space and $T_n \xrightarrow{w} T$, then $\sup_n \|T_n\| < \infty$ and $T \in \mathcal{B}[\mathcal{X}, \mathcal{Y}]$. Prove.

*Hint*: Apply the Banach–Steinhaus Theorem and Problem 4.67(d) to prove boundedness for $\{T_n\}$. Show that $|f(Tx)| \leq \|f\|(\sup_n \|T_n\|)\|x\|$ for every $x \in \mathcal{X}$ and $f \in \mathcal{Y}^*$. Conclude: $\|Tx\| = \sup_{f \in \mathcal{Y}^*, \|f\|=1} |f(Tx)| \leq \sup_n \|T_n\| \|x\|$ for every $x \in \mathcal{X}$.

(c) Show that $T_n \xrightarrow{s} T$ implies $T_n \xrightarrow{w} T$.

*Hint*: $|f((T_n - T)x)| \leq \|f\|\|(T_n - T)x\|$.

Take $T \in \mathcal{B}[\mathcal{X}]$ and consider the power sequence $\{T^n\}$ in the normed algebra $\mathcal{B}[\mathcal{X}]$. The operator $T$ is *weakly stable* if $T^n \xrightarrow{w} O$.

(d) Verify that strong stability implies weak stability,

$$T^n \xrightarrow{s} O \implies T^n \xrightarrow{w} O,$$

which in turn implies power boundedness if $\mathcal{X}$ is a Banach space:

$$T^n \xrightarrow{w} O \implies \sup_n \|T^n\| < \infty \text{ if } \mathcal{X} \text{ is Banach}.$$

**Problem 4.69.** Let $\mathcal{X}$ and $\mathcal{Y}$ be normed spaces. Prove the following propositions.

(a) If $T \in \mathcal{B}[\mathcal{X}, \mathcal{Y}]$, then $Tx_n \xrightarrow{w} Tx$ in $\mathcal{Y}$ whenever $x_n \xrightarrow{w} x$ in $\mathcal{X}$. That is, a continuous linear transformation of $\mathcal{X}$ into $\mathcal{Y}$ takes weakly convergent sequences in $\mathcal{X}$ into weakly convergent sequences in $\mathcal{Y}$.

*Hint*: If $f \in \mathcal{Y}^*$, then $f \circ T \in \mathcal{X}^*$.

(b) If $T \in \mathcal{B}_\infty[\mathcal{X}, \mathcal{Y}]$, then $\|Tx_n - Tx\| \to 0$ whenever $x_n \xrightarrow{w} x$ in $\mathcal{X}$. That is, a compact linear transformation of $\mathcal{X}$ into $\mathcal{Y}$ takes weakly convergent sequences in $\mathcal{X}$ into convergent sequences in $\mathcal{Y}$.

*Hint*: Let $x_n \xrightarrow{w} x$ in $\mathcal{X}$ and take $T \in \mathcal{B}_\infty[\mathcal{X}, \mathcal{Y}]$. Use Theorem 4.49, part (a) of this problem, and Problem 4.67(d) to show that

(b$_1$) $$Tx_n \xrightarrow{w} Tx \text{ in } \mathcal{Y} \quad \text{and} \quad \sup_n \|x_n\| < \infty.$$

Suppose $\{Tx_n\}$ does not converge (in the norm topology of $\mathcal{Y}$) to $Tx$. Use Proposition 3.5 to show that $\{Tx_n\}$ has a subsequence, say $\{Tx_{n_k}\}$, that does not converge to $Tx$. Thus conclude: there exists $\varepsilon_0 > 0$ and a positive integer $k_{\varepsilon_0}$ such that

(b$_2$) $$\|T(x_{n_k} - x)\| > \varepsilon_0 \quad \text{for every} \quad k \geq k_{\varepsilon_0}.$$

Verify from (b$_1$) that $\sup_k \|x_{n_k}\| < \infty$. Apply Theorem 4.52 to show that $\{Tx_{n_k}\}$ has a subsequence, say $\{Tx_{n_{k_j}}\}$, that converges in the norm topology of $\mathcal{Y}$. Now use the weak convergence in (b$_1$) and Problem 4.67(b) to show that $\{Tx_{n_{k_j}}\}$ in fact converges to $Tx$ (i.e., $Tx_{n_{k_j}} \to Tx$ in $\mathcal{Y}$). Therefore, for each $\varepsilon > 0$ there exists a positive integer $j_\varepsilon$ such that

(b$_3$) $$\|T(x_{n_{k_j}} - x)\| < \varepsilon \quad \text{for every} \quad j \geq j_\varepsilon.$$

Finally, verify that (b$_3$) contradicts (b$_2$) and conclude that $\{Tx_n\}$ must converge in $\mathcal{Y}$ to $Tx$.

**Problem 4.70.** Let $\mathcal{X}$ be a normed space. An $\mathcal{X}^*$-valued sequence $\{f_n\}$ is *weakly convergent* if there exists $f \in \mathcal{X}^*$ such that $\{\varphi(f_n)\}$ converges in $\mathbb{F}$ to $\varphi(f)$ for every $\varphi \in \mathcal{X}^{**}$ (cf. Problem 4.67). In this case we write $f_n \xrightarrow{w} f$ in $\mathcal{X}^*$. An $\mathcal{X}^*$-valued sequence $\{f_n\}$ is *weakly* convergent* if there exists $f \in \mathcal{X}^*$ such that $\{f_n(x)\}$ converges in $\mathbb{F}$ to $f(x)$ for every $x \in \mathcal{X}$ (notation: $f_n \xrightarrow{w*} f$). Thus *weak* convergence* in $\mathcal{X}^*$ means pointwise convergence of $\mathcal{B}[\mathcal{X}, \mathbb{F}]$-valued sequences to an element of $\mathcal{B}[\mathcal{X}, \mathbb{F}]$.

(a) Show that weak convergence in $\mathcal{X}^*$ implies weak* convergence in $\mathcal{X}^*$ (i.e., $f_n \xrightarrow{w} f \implies f_n \xrightarrow{w*} f$).

*Hint*: According to the natural embedding of $\mathcal{X}$ into $\mathcal{X}^{**}$ (Theorem 4.66) for each $x \in \mathcal{X}$ there exists $\varphi_x \in \mathcal{X}^{**}$ such that $\varphi_x(f) = f(x)$ for every $f \in \mathcal{X}^*$. Now verify that, for each $x \in \mathcal{X}$, $f_n(x) \to f(x)$ whenever $\varphi_x(f_n) \to \varphi_x(f)$.

(b)  If $\mathcal{X}$ is reflexive, then the concepts of weak convergence in $\mathcal{X}^*$ and weak* convergence in $\mathcal{X}^*$ coincide. Prove.

# 5
# Hilbert Spaces

What is missing? The algebraic structure of a normed space allowed us to operate with vectors (addition and scalar multiplication), and its topological structure (the one endowed by the norm) gave us a notion of closeness (by means of the metric generated by the norm), which interacts harmoniously with the algebraic operations. In particular, it provided the notion of the length of a vector. So what is missing if algebra and topology have already been properly laid on the same underlying set? A full geometric structure is still missing. Just algebra and topology are not enough to extend to abstract spaces the geometric concept of relative direction (or angle) between vectors that is familiar in Euclidean geometry. The keyword here is *orthogonality*, a concept that emerges when we equip a linear space with an *inner product*. This supplies a tremendously rich structure that leads to remarkable simplifications.

## 5.1 Inner Product Spaces

We shall assume throughout this chapter (as we did in Chapter 4) that $\mathbb{F}$ denotes either the real field $\mathbb{R}$ or the complex field $\mathbb{C}$, both equipped with their usual topologies induced by their usual metrics. Recall that an upper bar stands for complex conjugate in $\mathbb{C}$. That is, for each complex number $\lambda = \alpha + i\beta$ in standard form the real numbers $\alpha = \mathrm{Re}\,\lambda$ and $\beta = \mathrm{Im}\,\lambda$ are the *real* and *imaginary parts* of $\lambda$, respectively, and the complex number $\bar{\lambda} = \alpha - i\beta$ is the *conjugate* of $\lambda$. The following are basic properties of conjugates: for every $\lambda, \mu \in \mathbb{C}$, $\overline{(\bar{\mu})} = \mu$, $\overline{(\lambda + \mu)} = \bar{\lambda} + \bar{\mu}$,

$\overline{(\lambda\mu)} = \overline{\lambda}\overline{\mu}$, $\lambda + \overline{\lambda} = 2\operatorname{Re}\lambda$, $\lambda - \overline{\lambda} = 2i\operatorname{Im}\lambda$, $\overline{\lambda}\lambda = |\lambda|^2 = (\operatorname{Re}\lambda)^2 + (\operatorname{Im}\lambda)^2$, and $\lambda = \overline{\lambda}$ if and only if $\lambda \in \mathbb{R}$.

Let $\mathcal{X}$ be a linear space over $\mathbb{F}$. A functional $\sigma: \mathcal{X} \times \mathcal{X} \to \mathbb{F}$, defined on the Cartesian product of a linear space $\mathcal{X}$ with itself, is *symmetric* if $\sigma(x, y) = \sigma(y, x)$ and *Hermitian symmetric* if

$$\sigma(x, y) = \overline{\sigma(y, x)},$$

for every $x, y \in \mathcal{X}$. If $\sigma(\cdot, v): \mathcal{X} \to \mathbb{F}$ and $\sigma(u, \cdot): \mathcal{X} \to \mathbb{F}$ are linear functionals on $\mathcal{X}$ for every $u, v \in \mathcal{X}$ (i.e., if $\sigma$ is linear in both arguments), then it is called a *bilinear form* (or a *bilinear functional*) on $\mathcal{X}$. If $\sigma(\cdot, v): \mathcal{X} \to \mathbb{F}$ is a linear functional on $\mathcal{X}$ for each $v \in \mathcal{X}$, and if $\overline{\sigma(u, \cdot)}: \mathcal{X} \to \mathbb{F}$ is a linear functional on $\mathcal{X}$ for each $u \in \mathcal{X}$, then $\sigma$ is said to be a *sesquilinear form* (or a *sesquilinear functional*) on $\mathcal{X}$. Equivalently, $\sigma$ is a sesquilinear form on $\mathcal{X}$ if

$$\begin{aligned}
\sigma(\alpha u + \beta x, v) &= \alpha\sigma(u, v) + \beta\sigma(x, v), \\
\sigma(u, \alpha v + \beta y) &= \overline{\alpha}\sigma(u, v) + \overline{\beta}\sigma(u, y),
\end{aligned}$$

for every $u, v, x, y \in \mathcal{X}$ and every $\alpha, \beta \in \mathbb{F}$ ("sesqui" means "one-and-a-half"). If $\mathbb{F} = \mathbb{R}$, then it is clear that $\sigma$ is symmetric if and only if it is Hermitian symmetric, and the notions of bilinear and sesquilinear forms coincide as well. It is readily verified that a functional $\sigma: \mathcal{X} \times \mathcal{X} \to \mathbb{F}$ that is linear in the first argument (i.e., $\sigma(\cdot, v): \mathcal{X} \to \mathbb{F}$ is linear for every $v \in \mathbb{F}$) and Hermitian symmetric is a sesquilinear form. Therefore, a Hermitian symmetric sesquilinear form is precisely a Hermitian symmetric functional on $\mathcal{X} \times \mathcal{X}$ that is linear in the first argument. If $\sigma$ is a sesquilinear form on $\mathcal{X}$, then the functional $\phi: \mathcal{X} \to \mathbb{F}$ defined by $\phi(x) = \sigma(x, x)$ for every $x \in \mathcal{X}$ is called a *quadratic form* on $\mathcal{X}$ induced (or generated) by $\sigma$. If $\sigma$ is Hermitian symmetric, then the induced quadratic form $\phi$ is real-valued (i.e., $\sigma(x, x) \in \mathbb{R}$ for every $x \in \mathcal{X}$ whenever $\sigma$ is Hermitian symmetric). Also note that if $\sigma$ is a sesquilinear form, then $\sigma(0, v) = \sigma(u, 0) = 0$ for every $u, v \in \mathcal{X}$ so that $\sigma(0, 0) = 0$. The quadratic form $\phi$ induced by a Hermitian symmetric sesquilinear form $\sigma$ is *nonnegative* or *positive* if

$$\sigma(x, x) \geq 0 \quad \text{for every } x \in \mathcal{X}, \quad \text{or}$$

$$\sigma(x, x) > 0 \quad \text{for every nonzero } x \in \mathcal{X},$$

respectively. Equivalently, $\phi$ is positive if it is nonnegative and $\sigma(x, x) = 0$ only if $x = 0$. An *inner product* (or a *scalar product*) on a linear space $\mathcal{X}$ *is a Hermitian symmetric sesquilinear form that induces a positive quadratic form*. In other words, an inner product on a linear space $\mathcal{X}$ is a functional on the Cartesian product $\mathcal{X} \times \mathcal{X}$ that satisfies the following properties, called the *inner product axioms*.

**Definition 5.1.** Let $\mathcal{X}$ be a linear space over $\mathbb{F}$. A functional

$$\langle\,;\,\rangle: \mathcal{X} \times \mathcal{X} \to \mathbb{F}$$

is an *inner product* on $\mathcal{X}$ if the following conditions are satisfied for all vectors $x$, $y$ and $z$ in $\mathcal{X}$ and all scalars $\alpha$ in $\mathbb{F}$.

$$
\begin{aligned}
&\text{(i)} && \langle x + y \,;\, z \rangle = \langle x \,;\, z \rangle + \langle y \,;\, z \rangle && (\textit{additivity}),\\
&\text{(ii)} && \langle \alpha x \,;\, y \rangle = \alpha \langle x \,;\, y \rangle && (\textit{homogeneity}),\\
&\text{(iii)} && \langle x \,;\, y \rangle = \overline{\langle y \,;\, x \rangle} && (\textit{Hermitian symmetry}),\\
&\text{(iv)} && \langle x \,;\, x \rangle \geq 0 && (\textit{nonnegativeness}),\\
&\text{(v)} && \langle x \,;\, x \rangle = 0 \quad \text{only if} \quad x = 0 && (\textit{positiveness}).
\end{aligned}
$$

(Observe that homogeneity in (ii) really means homogeneity in the first argument.) A linear space $\mathcal{X}$ equipped with an inner product on it is an *inner product space* (or a *pre-Hilbert space*). If $\mathcal{X}$ is a real or complex linear space (so that $\mathbb{F} = \mathbb{R}$ or $\mathbb{F} = \mathbb{C}$) equipped with an inner product on it, then it is referred to as a *real* or *complex inner product space*, respectively.

Axioms (i), (ii) and (iii) are enough by themselves to ensure that, for all vectors $x$, $y$, $z$ in $\mathcal{X}$ and all scalars $\alpha$ in $\mathbb{F}$,

$$
\langle x \,;\, y + z \rangle = \langle x \,;\, y \rangle + \langle x \,;\, z \rangle, \qquad \langle x \,;\, \alpha y \rangle = \overline{\alpha} \langle x \,;\, y \rangle,
$$

$$
\text{and} \quad \langle x \,;\, 0 \rangle = \langle 0 \,;\, x \rangle = \langle 0 \,;\, 0 \rangle = 0.
$$

As a matter of fact, just these three axioms imply (by induction) that

$$
\left\langle \sum_{i=1}^{n} \alpha_i x_i \,;\, \beta_0 y_0 \right\rangle = \sum_{i=1}^{n} \alpha_i \overline{\beta_0} \langle x_i \,;\, y_0 \rangle,
$$

$$
\left\langle \alpha_0 x_0 \,;\, \sum_{i=1}^{n} \beta_i y_i \right\rangle = \sum_{i=1}^{n} \alpha_0 \overline{\beta_i} \langle x_0 \,;\, y_i \rangle,
$$

and hence

$$
\left\langle \sum_{i=0}^{n} \alpha_i x_i \,;\, \sum_{j=0}^{n} \beta_j y_j \right\rangle = \sum_{i,j=0}^{n} \alpha_i \overline{\beta}_j \langle x_i \,;\, y_j \rangle,
$$

for each integer $n \geq 1$ whenever $\alpha_i$ and $\beta_i$ lie in $\mathbb{F}$ and $x_i$ and $y_i$ lie in $\mathcal{X}$, for every $i = 0, \ldots, n$. In particular, for every $x, y \in \mathcal{X}$,

$$
\langle x + y \,;\, x + y \rangle = \langle x \,;\, x \rangle + 2 \operatorname{Re} \langle x \,;\, y \rangle + \langle y \,;\, y \rangle
$$

(because $\langle x \,;\, y \rangle + \langle y \,;\, x \rangle = \langle x \,;\, y \rangle + \overline{\langle x \,;\, y \rangle} = 2 \operatorname{Re} \langle x \,;\, y \rangle$). Moreover, by using axioms (ii) and (v) we get

$$
\langle x \,;\, y \rangle = 0 \quad \text{for all} \quad y \in \mathcal{X} \quad \text{if and only if} \quad x = 0.
$$

Let $\| \ \|^2 : \mathcal{X} \to \mathbb{F}$ be the quadratic form induced by an inner product $\langle \ ; \ \rangle$ on $\mathcal{X}$ (i.e., $\|x\|^2 = \langle x \,;\, x \rangle \geq 0$ for every $x \in \mathcal{X}$ — the notation $\| \ \|^2$ for the quadratic form induced by an inner product is certainly not a mere coincidence as we shall see shortly). Note that

$$\|x + y\|^2 = \|x\|^2 + 2\operatorname{Re}\langle x\,;y\rangle + \|y\|^2$$

for every $x, y \in \mathcal{X}$ (by axioms (i) and (iii)). The next result is of fundamental importance. It is referred to as the *Schwarz* (or *Cauchy–Schwarz*, or even *Cauchy–Bunyakowski–Schwarz*) *inequality*.

**Lemma 5.2.** *Let* $\langle\,;\,\rangle\colon \mathcal{X}\times\mathcal{X} \to \mathbb{F}$ *be an inner product on a linear space* $\mathcal{X}$. *Set* $\|x\| = \langle x\,;x\rangle^{\frac{1}{2}}$ *for each* $x \in \mathcal{X}$. *If* $x, y \in \mathcal{X}$, *then*

$$|\langle x\,;y\rangle| \le \|x\|\|y\|.$$

*Proof.* Take an arbitrary pair of vectors $x$ and $y$ in $\mathcal{X}$. According to axioms (i), (ii), (iii) and (iv) in Definition 5.1,

$$0 \le \langle x - \alpha y\,;x - \alpha y\rangle = \langle x\,;x\rangle - \overline{\alpha}\langle x\,;y\rangle - \alpha\overline{\langle x\,;y\rangle} + |\alpha|^2\langle y\,;y\rangle$$

for every $\alpha \in \mathbb{F}$. In particular, set $\alpha = \frac{\langle x\,;y\rangle}{\beta}$ for any $\beta > 0$ so that

$$0 \le \|x\|^2 - \frac{1}{\beta}\left(2 - \frac{\|y\|^2}{\beta}\right)|\langle x\,;y\rangle|^2.$$

If $\|y\| \neq 0$, then put $\beta = \|y\|^2$ to get the Schwarz inequality. If $\|y\| = 0$, then $0 \le 2|\langle x\,;y\rangle|^2 \le \beta\|x\|^2$ for all $\beta > 0$, and hence $|\langle x\,;y\rangle| = 0$ (which trivially satisfies the Schwarz inequality). $\qquad\Box$

**Proposition 5.3.** *If* $\langle\,;\,\rangle\colon \mathcal{X}\times\mathcal{X} \to \mathbb{F}$ *is an inner product on a linear space* $\mathcal{X}$, *then the function* $\|\,\|\colon \mathcal{X} \to \mathbb{R}$, *defined by*

$$\|x\| = \langle x\,;x\rangle^{\frac{1}{2}}$$

*for each* $x \in \mathcal{X}$, *is a norm on* $\mathcal{X}$.

*Proof.* Axioms (ii), (iii), (iv) and (v) in Definition 5.1 imply the norm axioms (i), (ii) and (iii) of Definition 4.1. The triangle inequality (axiom (iv) of Definition 4.1) is a consequence of the Schwarz inequality:

$$0 \le \|x + y\|^2 = \|x\|^2 + 2\operatorname{Re}\langle x\,;y\rangle + \|y\|^2 \le (\|x\| + \|y\|)^2$$

for every $x$ and $y$ in $\mathcal{X}$. (Reason: $\operatorname{Re}\langle x\,;y\rangle \le |\langle x\,;y\rangle| \le \|x\|\|y\|$.) $\qquad\Box$

A word on notation and terminology. An inner product space in fact is an ordered pair $(\mathcal{X}, \langle\,;\,\rangle)$ where $\mathcal{X}$ is a linear space and $\langle\,;\,\rangle$ is an inner product on $\mathcal{X}$. We shall often refer to an inner space by simply saying that "$\mathcal{X}$ is an inner product space" without explicitly mentioning the inner product $\langle\,;\,\rangle$ that equips the linear space $\mathcal{X}$.

However, there may be occasions when the role played by different inner products should be emphasized and, in these cases, we shall insert a subscript on the inner products (e.g., $(\mathcal{X}, \langle \ ; \ \rangle_X)$ and $(\mathcal{Y}, \langle \ ; \ \rangle_Y)$). If a linear space $\mathcal{X}$ can be equipped with more than one inner product, say $\langle \ ; \ \rangle_1$ and $\langle \ ; \ \rangle_2$, then $(\mathcal{X}, \langle \ ; \ \rangle_1)$ and $(\mathcal{X}, \langle \ ; \ \rangle_2)$ will represent *different* inner product spaces with the same linear space $\mathcal{X}$. The norm $\| \ \|$ of Proposition 5.3 is the *norm induced* (or defined, or generated) by the inner product $\langle \ ; \ \rangle$, so that every inner product space is a special kind of normed space (and hence a very special kind of linear metric space). Whenever we refer to the topological structure of an inner product space $(\mathcal{X}, \langle \ ; \ \rangle)$ it will always be understood that such a topology on $\mathcal{X}$ is that defined by the metric $d$ that is generated by the norm $\| \ \|$, which in turn is the one induced by the inner product $\langle \ ; \ \rangle$. That is,

$$d(x, y) \ = \ \|x - y\| \ = \ \langle x - y \, ; x - y \rangle^{\frac{1}{2}}$$

for every $x, y \in \mathcal{X}$ (see Propositions 4.2 and 5.3). This is the *norm topology* on $\mathcal{X}$ *induced by the inner product*. Since every inner product on a linear space induces a norm, it follows that every inner product space is a normed space (equipped with the induced norm). However, an arbitrary norm on a linear space may not be induced by any inner product on it (so that an arbitrary normed space may not be an inner product space). The next proposition leads to a necessary and sufficient condition that a norm be induced by an inner product.

**Proposition 5.4.** *Let $\langle \ ; \ \rangle$ be an inner product on a linear space $\mathcal{X}$ and let $\| \ \|$ be the induced norm on $\mathcal{X}$. Then*

$$\|x + y\|^2 + \|x - y\|^2 \ = \ 2\big(\|x\|^2 + \|y\|^2\big)$$

*for every $x, y \in \mathcal{X}$. This is called the parallelogram law. If $(\mathcal{X}, \langle \ ; \ \rangle)$ is a complex inner product space, then*

$$\langle x \, ; y \rangle \ = \ \tfrac{1}{4}\big(\|x + y\|^2 - \|x - y\|^2 + i\|x + iy\|^2 - i\|x - iy\|^2\big)$$

*for every $x, y \in \mathcal{X}$. If $(\mathcal{X}, \langle \ ; \ \rangle)$ is a real inner product space, then*

$$\langle x \, ; y \rangle \ = \ \tfrac{1}{4}\big(\|x + y\|^2 - \|x - y\|^2\big)$$

*for every $x, y \in \mathcal{X}$. The above two expressions are referred to as the complex and real polarization identities, respectively.*

*Proof.* According to axioms (i), (ii) and (iii) in Definition 5.1 (also see the displayed identity that precedes Lemma 5.2) we get

$$
\begin{aligned}
\|x + \alpha y\|^2 \ &= \ \|x\|^2 + 2\operatorname{Re}\big(\overline{\alpha}\langle x \, ; y \rangle\big) + |\alpha|^2\|y\|^2 \\
&= \ \|x\|^2 + 2\big[\operatorname{Re}\alpha \operatorname{Re}\langle x \, ; y \rangle + \operatorname{Im}\alpha \operatorname{Im}\langle x \, ; y \rangle\big] + |\alpha|^2\|y\|^2
\end{aligned}
$$

for every $x, y \in \mathcal{X}$ and every $\alpha \in \mathbb{F}$ (recall that $\operatorname{Re}(\lambda \mu) = \operatorname{Re}\lambda \operatorname{Re}\mu - \operatorname{Im}\lambda \operatorname{Im}\mu$ for every $\lambda, \mu \in \mathbb{C}$). The parallelogram law and the (real) polarization identity in a real inner product space follow by setting $\alpha = 1$ and $\alpha = -1$. To get the (complex) polarization identity in a complex inner product space also set $\alpha = i$ and $\alpha = -i$.    □

**Theorem 5.5.** (von Neumann). *Let $\mathcal{X}$ be a linear space. A norm on $\mathcal{X}$ is induced by an inner product on $\mathcal{X}$ if and only if it satisfies the parallelogram law. Moreover, if a norm on $\mathcal{X}$ satisfies the parallelogram law, then the unique inner product that induces it is given by the polarization identity.*

*Proof.* Proposition 5.4 ensures that if a norm on $\mathcal{X}$ is induced by an inner product, then it satisfies the parallelogram law and the inner product on $\mathcal{X}$ can be written in terms of this norm according to the polarization identity. Conversely, suppose a norm $\| \ \|$ on $\mathcal{X}$ satisfies the parallelogram law and consider the mapping $\langle \ ; \ \rangle : \mathcal{X} \times \mathcal{X} \to \mathbb{F}$ defined by the polarization identity. Take $x$, $y$ and $z$ arbitrary in $\mathcal{X}$. Note that

$$x + z = \left( \frac{x+y}{2} + z \right) + \frac{x-y}{2} \quad \text{and} \quad y + z = \left( \frac{x+y}{2} + z \right) - \frac{x-y}{2}.$$

Thus, by the parallelogram law,

$$\|x + z\|^2 + \|y + z\|^2 = 2 \left( \left\| \tfrac{x+y}{2} + z \right\|^2 + \left\| \tfrac{x-y}{2} \right\|^2 \right).$$

Suppose $\mathbb{F} = \mathbb{R}$ so that $\langle \ ; \ \rangle : \mathcal{X} \times \mathcal{X} \to \mathbb{R}$ is the mapping defined by the real polarization identity (on the real normed space $\mathcal{X}$). Hence

$$
\begin{aligned}
\langle x \,;\, z \rangle + \langle y \,;\, z \rangle &= \tfrac{1}{4} \left( \|x + z\|^2 - \|x - z\|^2 + \|y + z\|^2 - \|y - z\|^2 \right) \\
&= \tfrac{1}{4} \left[ \left( \|x + z\|^2 + \|y + z\|^2 \right) - \left( \|x - z\|^2 + \|y - z\|^2 \right) \right] \\
&= \tfrac{1}{2} \left[ \left( \left\| \tfrac{x+y}{2} + z \right\|^2 + \left\| \tfrac{x-y}{2} \right\|^2 \right) - \left( \left\| \tfrac{x+y}{2} - z \right\|^2 + \left\| \tfrac{x-y}{2} \right\|^2 \right) \right] \\
&= \tfrac{1}{2} \left( \left\| \tfrac{x+y}{2} + z \right\|^2 - \left\| \tfrac{x+y}{2} - z \right\|^2 \right) = 2 \langle \tfrac{x+y}{2} \,;\, z \rangle.
\end{aligned}
$$

The above identity holds for arbitrary $x, y, z \in \mathcal{X}$, and so it holds for $y = 0$. Moreover, the polarization identity ensures that $\langle 0 \,;\, z \rangle = 0$ for every $z \in \mathcal{X}$. Therefore, by setting $y = 0$ above we get $\langle x \,;\, z \rangle = 2 \langle \tfrac{x}{2} \,;\, z \rangle$ for every $x, z \in \mathcal{X}$. Then

(i) $$\langle x \,;\, z \rangle + \langle y \,;\, z \rangle = \langle x + y \,;\, z \rangle$$

for arbitrary $x$, $y$ and $z$ in $\mathcal{X}$. It is readily verified (exactly the same argument) that such an identity still holds if $\mathbb{F} = \mathbb{C}$, where the mapping $\langle \ ; \ \rangle : \mathcal{X} \times \mathcal{X} \to \mathbb{C}$ now satisfies the complex polarization identity (on the complex normed space $\mathcal{X}$). This is the axiom (i) of Definition 5.1 (additivity). To verify the axiom (ii) of Definition 5.1 (homogeneity in the first argument) proceed as follows. Take $x$ and $y$ arbitrary in $\mathcal{X}$. The polarization identity ensures that

$$\langle -x \,;\, y \rangle = -\langle x \,;\, y \rangle.$$

Since (i) holds true, it follows by a trivial induction that

$$\langle nx \, ; y \rangle \; = \; n \langle x \, ; y \rangle,$$

and hence $\langle x \, ; y \rangle = \langle n \frac{x}{n} \, ; y \rangle = n \langle \frac{x}{n} \, ; y \rangle$ so that

$$\langle \tfrac{x}{n} \, ; y \rangle \; = \; \tfrac{1}{n} \langle x \, ; y \rangle,$$

for every positive integer $n$. The above three expressions imply that

$$\langle qx \, ; y \rangle = q \langle x \, ; y \rangle$$

for every rational number $q$ (for $\langle 0 \, ; y \rangle = 0$ by the polarization identity). Take an arbitrary $\alpha \in \mathbb{R}$ and recall that $\mathbb{Q}$ is dense in $\mathbb{R}$. Thus there exists a rational-valued sequence $\{q_n\}$ that converges in $\mathbb{R}$ to $\alpha$. Moreover, according to (i) and recalling that $-\langle \alpha x \, ; y \rangle = \langle -\alpha x \, ; y \rangle$,

$$|\langle q_n x \, ; y \rangle - \langle \alpha x \, ; y \rangle| \; = \; |\langle (q_n - \alpha)x \, ; y \rangle|.$$

The polarization identity ensures that $|\langle \alpha_n x \, ; y \rangle| \to 0$ whenever $\alpha_n \to 0$ in $\mathbb{R}$ (because the norm is continuous). Hence $|\langle (q_n - \alpha)x \, ; y \rangle| \to 0$ and so $|\langle q_n x \, ; y \rangle - \langle \alpha x \, ; y \rangle| \to 0$, which means that $\langle q_n x \, ; y \rangle \to \langle \alpha x \, ; y \rangle$. Therefore, $\langle \alpha x \, ; y \rangle = \lim_n \langle q_n x \, ; y \rangle = \lim_n q_n \langle x \, ; y \rangle = \alpha \langle x \, ; y \rangle$. Outcome:

(ii(a)) $$\langle \alpha x \, ; y \rangle \; = \; \alpha \langle x \, ; y \rangle$$

for every $\alpha \in \mathbb{R}$. If $\mathbb{F} = \mathbb{C}$, then the complex polarization identity (on the complex space $\mathcal{X}$) ensures that

$$\langle ix \, ; y \rangle \; = \; i \langle x \, ; y \rangle.$$

Take an arbitrary $\lambda = \alpha + i\beta$ in $\mathbb{C}$ and observe by (i) and (ii(a)) that $\langle \lambda x \, ; y \rangle = \langle (\alpha + i\beta)x \, ; y \rangle = \langle \alpha x \, ; y \rangle + \langle i\beta x \, ; y \rangle = (\alpha + i\beta)\langle x \, ; y \rangle = \lambda \langle x \, ; y \rangle$. Conclusion:

(ii(b)) $$\langle \lambda x \, ; y \rangle \; = \; \lambda \langle x \, ; y \rangle$$

for every $\lambda \in \mathbb{C}$. Axioms (iii), (iv) and (v) of Definition 5.1 (Hermitian symmetry and positiveness) emerge as immediate consequences of the polarization identity. Thus the mapping $\langle \; ; \; \rangle \colon \mathcal{X} \times \mathcal{X} \to \mathbb{F}$ defined by the polarization identity is, in fact, an inner product on $\mathcal{X}$. Moreover, this inner product induces the norm $\| \; \|$; that is, $\langle x \, ; x \rangle = \|x\|^2$ for every $x \in \mathcal{X}$ (polarization identity again). Finally, if $\langle \; ; \; \rangle_0 \colon \mathcal{X} \times \mathcal{X} \to \mathbb{F}$ is an inner product on $\mathcal{X}$ that induces the same norm $\| \; \|$ on $\mathcal{X}$, then it must coincide with $\langle \; ; \; \rangle$. That is, $\langle x \, ; y \rangle_0 = \langle x \, ; y \rangle$ for every $x, y \in \mathcal{X}$ (polarization identity once again). $\qquad\square$

A *Hilbert space* is a complete inner product space. In other words, a Hilbert space is an inner product space that is complete as a metric space with respect to the metric generated by the norm induced by the inner product. In fact, every Hilbert space is a special kind of Banach space: a Hilbert space is a Banach space whose norm is induced by an inner product. According to Theorem 5.5, *a Hilbert space is a Banach space whose norm satisfies the parallelogram law.*

## 5.2   Examples

Theorem 5.5 may suggest that just a few of the classical examples of Section 4.2 survive as inner product spaces. This indeed is the case.

**Example 5A.** Consider the linear space $\mathbb{F}^n$ over $\mathbb{F}$ (with either $\mathbb{F} = \mathbb{R}$ or $\mathbb{F} = \mathbb{C}$) and set

$$\langle x \, ; \, y \rangle = \sum_{i=1}^{n} \xi_i \, \overline{\upsilon_i}$$

for every $x = (\xi_1, \dots, \xi_n)$ and $y = (\upsilon_1, \dots, \upsilon_n)$ in $\mathbb{F}^n$. It is readily verified that this defines an inner product on $\mathbb{F}^n$ (check the axioms in Definition 5.1), which induces the norm $\| \; \|_2$ on $\mathbb{F}^n$. In particular, it induces the *Euclidean norm* on $\mathbb{R}^n$ so that $(\mathbb{R}^n, \langle \; ; \; \rangle)$ is the $n$-dimensional *Euclidean space* (see Example 4A). Since $(\mathbb{F}^n, \| \; \|_2)$ is a Banach space,

$$(\mathbb{F}^n, \langle \; ; \; \rangle) \text{ is a Hilbert space.}$$

Now consider the norms $\| \; \|_p$ (for $p \geq 1$) and $\| \; \|_\infty$ on $\mathbb{F}^n$ defined in Example 4A. If $n > 1$, then all of them, except the norm $\| \; \|_2$, are not induced by any inner product on $\mathbb{F}^n$. Indeed, set $x = (1, 0, \dots, 0)$ and $y = (0, 1, 0, \dots, 0)$ in $\mathbb{F}^n$ and verify that the parallelogram law fails for every norm $\| \; \|_p$ with $p \neq 2$, as it also fails for the sup-norm $\| \; \|_\infty$. Therefore, if $n > 1$, then $(\mathbb{F}^n, \| \; \|_2)$ is the only Hilbert space among the Banach spaces of Example 4A.

**Example 5B.** Consider the Banach spaces $(\ell_+^p, \| \; \|_p)$ for each $p \geq 1$ and $(\ell_+^\infty, \| \; \|_\infty)$ of Example 4B. It is easy to show that, except for $(\ell_+^2, \| \; \|_2)$, these are not Hilbert spaces: the norms $\| \; \|_p$ for every $p \neq 2$ and $\| \; \|_\infty$ do not pass the parallelogram-law test of Theorem 5.5, and hence are not induced by any possible inner product on $\ell_+^p$ ($p \neq 2$) or on $\ell_+^\infty$ (e.g., take $x = e_1 = (1, 0, 0, 0, \dots)$ and $y = e_2 = (0, 1, 0, 0, 0, \dots)$ in $\ell_+^p \cap \ell_+^\infty$). On the other hand, the function $\langle \; ; \; \rangle \colon \ell_+^2 \times \ell_+^2 \to \mathbb{F}$ given by

$$\langle x \, ; \, y \rangle = \sum_{k=1}^{\infty} \xi_k \, \overline{\upsilon_k}$$

for every $x = \{\xi_k\}_{k \in \mathbb{N}}$ and $y = \{\upsilon_k\}_{k \in \mathbb{N}}$ in $\ell_+^2$ is well-defined (i.e., the above infinite series converges in $\mathbb{F}$ for every $x, y \in \ell_+^2$ by the Hölder inequality for $p = q = 2$ and Proposition 4.4). Moreover, it defines an inner product on $\ell_+^2$ (i.e., it satisfies the axioms of Definition 5.1), which induces the norm $\| \; \|_2$ on $\ell_+^2$. Thus, as $(\ell_+^2, \| \; \|_2)$ is a Banach space,

$$(\ell_+^2, \langle \; ; \; \rangle) \text{ is a Hilbert space.}$$

Similarly, the Banach spaces $(\ell^p, \|\ \|_p)$ for any $1 \le p \neq 2$ and $(\ell^\infty, \|\ \|_\infty)$ are not Hilbert spaces. However, the function $\langle\ ;\ \rangle \colon \ell^2 \times \ell^2 \to \mathbb{F}$ given by

$$\langle x\,;\,y \rangle \;=\; \sum_{k=-\infty}^{\infty} \xi_k \overline{\upsilon}_k$$

for every $x = \{\xi_k\}_{k \in \mathbb{Z}}$ and $y = \{\upsilon_k\}_{k \in \mathbb{Z}}$ in $\ell^2$ defines an inner product on $\ell^2$, which induces the norm $\|\ \|_2$ on $\ell^2$. Indeed, the sequence of nonnegative numbers $\{\sum_{k=-n}^{n} |\xi_k \overline{\upsilon}_k|\}_{n \in \mathbb{N}_0}$ converges in $\mathbb{R}$ whenever the sequences $\{\sum_{k=-n}^{n} |\xi_k|^2\}_{n \in \mathbb{N}_0}$ and $\{\sum_{k=-n}^{n} |\upsilon_k|^2\}_{n \in \mathbb{N}_0}$ of nonnegative numbers converge in $\mathbb{R}$ (cf. Hölder inequality for $p = q = 2$), and hence $\{\sum_{k=-n}^{n} \xi_k \overline{\upsilon}_k\}_{n \in \mathbb{N}_0}$ converges in $\mathbb{F}$ (by Proposition 4.4). Therefore, the function $\langle\ ,\ \rangle$ is well-defined and, as it is easy to check, it satisfies all the axioms of Definition 5.1. Since $(\ell^2, \|\ \|_2)$ is a Banach space,

$$(\ell^2, \langle\ ;\ \rangle) \text{ is a Hilbert space.}$$

**Example 5C.** Consider the linear space $C[0, 1]$ equipped with any of the norms $\|\ \|_p$ $(p \ge 1)$ of Example 4D or with the sup-norm $\|\ \|_\infty$ of Example 4G. Among these norms on $C[0, 1]$, the only one that is induced by an inner product on $C[0, 1]$ is the norm $\|\ \|_2$. Indeed, take $x$ and $y$ in $C[0, 1]$ such that $xy = 0$ and $\|x\| = \|y\| \neq 0$, where $\|\ \|$ denotes either $\|\ \|_p$ for some $p \ge 1$ or $\|\ \|_\infty$. That is, suppose $x$ and $y$ are nonzero continuous functions on $[0, 1]$ of equal norms such that their nonzero values are attained on disjoint subsets of $[0, 1]$. For instance,

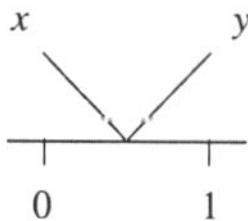


Observe that $\|x + y\|_p^p - \|x - y\|_p^p = 2\|x\|_p^p$ for every $p \ge 1$ and $\|x + y\|_\infty - \|x - y\|_\infty = 2\|x\|_\infty$. Thus $\|\ \|_p$ for $p \neq 2$ and $\|\ \|_\infty$ do not satisfy the parallelogram law, and hence these norms are not induced by any inner product on $C[0, 1]$ (Theorem 5.5). Now consider the function $\langle\ ;\ \rangle \colon C[0, 1] \times C[0, 1] \to \mathbb{F}$ given by

$$\langle x\,;\,y \rangle \;=\; \int_0^1 x(t)\,\overline{y(t)}\,dt$$

for every $x, y \in C[0, 1]$. It is readily verified that $\langle\ ;\ \rangle$ is an inner product on $C[0, 1]$ that induces the norm $\|\ \|_2$. Hence

$$(C[0, 1], \langle\ ;\ \rangle) \text{ is an inner product space}$$

but not a Hilbert space. (Reason: $(C[0, 1], \|\ \|_2)$ is not a Banach space — Example 3D.) As a matter of fact, among the normed spaces $(C[0, 1], \|\ \|_p)$ for each $p \ge 1$

and $(C[0, 1], \| \; \|_\infty)$, the only Banach space is $(C[0, 1], \| \; \|_\infty)$. This leads to a dichotomy: either equip $C[0, 1]$ with $\| \; \|_2$ to get an inner product space that is not a Banach space, or equip it with $\| \; \|_\infty$ to get a Banach space whose norm is not induced by an inner product. In any case, $C[0, 1]$ cannot be made into a Hilbert space. Roughly speaking, the continuous functions on $[0, 1]$ are not large enough a set to be a Hilbert space.

Let $\mathcal{X}$ be a linear space over a field $\mathbb{F}$. A functional $\langle \; ; \; \rangle \colon \mathcal{X} \times \mathcal{X} \to \mathbb{F}$ is a *semi-inner product* on $\mathcal{X}$ if it satisfies the first four axioms of Definition 5.1. The difference between an inner product and a semi-inner product is that a semi-inner product is a Hermitian symmetric sesquilinear form that induces a nonnegative quadratic form which is not necessarily positive (i.e., the axiom (v) of Definition 5.1 may not be satisfied by a semi-inner product). A semi-inner product $\langle \; ; \; \rangle$ on $\mathcal{X}$ induces a seminorm $\| \; \|$, which in turn generates a pseudometric $d$, viz., $d(x, y) = \|x - y\| = \langle x - y \, ; \, x - y \rangle^{\frac{1}{2}}$ for every $x, y \in \mathcal{X}$. A *semi-inner product space* is a linear space equipped with a semi-inner product.

Remark: The identity $\|x + y\|^2 = \|x\|^2 + 2\,\mathrm{Re}\,\langle x \, ; \, y \rangle + \|y\|^2$ for every $x, y \in \mathcal{X}$ still holds for a semi-inner product and its induced seminorm. Indeed, *the Schwarz inequality, the parallelogram law and the polarization identities remain valid in a semi-inner product space* (i.e., they still hold if we replace "inner product" and "norm" with "semi-inner product" and "seminorm", respectively — cf. proofs of Lemma 5.2 and Proposition 5.4). The same happens with respect to Theorem 5.5.

**Proposition 5.6.** *Let $\| \; \|$ be the seminorm induced by a semi-inner product $\langle \; ; \; \rangle$ on a linear space $\mathcal{X}$. Consider the quotient space $\mathcal{X}/\mathcal{N}$, where $\mathcal{N} = \{x \in \mathcal{X} \colon \|x\| = 0\}$ is a linear manifold of $\mathcal{X}$. Set*

$$\langle [x] \, ; \, [y] \rangle_\sim = \langle x \, ; \, y \rangle$$

*for every $[x]$ and $[y]$ in $\mathcal{X}/\mathcal{N}$, where $x$ and $y$ are arbitrary vectors in $[x]$ and $[y]$, respectively. This defines an inner product on $\mathcal{X}/\mathcal{N}$ so that $(\mathcal{X}/\mathcal{N}, \langle \; ; \; \rangle_\sim)$ is an inner product space.*

*Proof.* The seminorm $\| \; \|$ is induced by a semi-inner product so that it satisfies the parallelogram law of Proposition 5.4. Consider the norm $\| \; \|_\sim$ on $\mathcal{X}/\mathcal{N}$ of Proposition 4.5 and note that

$$
\begin{aligned}
\|[x] + [y]\|_\sim^2 + \|[x] - [y]\|_\sim^2 &= \|[x + y]\|_\sim^2 + \|[x - y]\|_\sim^2 \\
&= \|x + y\|^2 + \|x - y\|^2 \\
&= 2\big(\|x\|^2 + \|y\|^2\big) = 2\big(\|[x]\|_\sim^2 + \|[y]\|_\sim^2\big)
\end{aligned}
$$

for every $[x], [y] \in \mathcal{X}/\mathcal{N}$. Thus $\| \; \|_\sim$ satisfies the parallelogram law. This means that it is induced by a (unique) inner product $\langle \; ; \; \rangle_\sim$ on $\mathcal{X}/\mathcal{N}$, which is given in terms of the norm $\| \; \|_\sim$ by the polarization identity (Theorem 5.5). On the other hand, the semi-inner product $\langle\ ;\ \rangle$ on $\mathcal{X}$ also is given in terms of the seminorm $\|\ \|$ through the polarization identity as in Proposition 5.4. Since $\|[x] + \alpha[y]\|_{\sim} = \|x + \alpha y\|$ for every $[x], [y] \in \mathcal{X}/\mathcal{N}$ and every $\alpha \in \mathbb{F}$ (with $x$ and $y$ being arbitrary elements of $[x]$ and $[y]$, respectively), it is readily verified via polarization identity that $\langle[x]\,;[y]\rangle_{\sim} = \langle x\,;y\rangle$. $\qquad\square$

**Example 5D.** For each $p \geq 1$ let $r^p(S)$ be the linear space of all scalar-valued Riemann $p$-integrable functions, on a nondegenerate interval $S$ of the real line, equipped with the seminorm $|\ |_p$ of Example 4C. Again (see Example 5C) it is easy to show that, except for the seminorm $|\ |_2$, these seminorms do not satisfy the parallelogram law. Moreover,

$$\langle x\,;y\rangle = \int_S x(s)\,\overline{y(s)}\,ds$$

since every $x, y \in r^2(S)$ defines a semi-inner product that induces the seminorm $|\ |_2$, viz., $|x|_2 = (\int_S |x(s)|^2\,ds)^{\frac{1}{2}}$ for each $x \in r^2(S)$. Consider the linear manifold $\mathcal{N} = \{x \in r^2(S)\colon\ |x|_2 = 0\}$ and let $R^2(S)$ be the quotient space $r^2(S)/\mathcal{N}$ as in Example 4C. Set

$$\langle[x]\,;[y]\rangle = \langle x\,;y\rangle$$

for every $[x], [y] \in R^2(S)$, where $x$ and $y$ are arbitrary vectors in $[x]$ and $[y]$, respectively. According to Proposition 5.6 this defines an inner product on $R^2(S)$, which is the one that induces the norm $\|\ \|_2$ of Example 4C. Since the normed space $(R^2(S), \|\ \|_2)$ is not a Banach space, it follows that

$$(R^2(S), \langle\ ;\ \rangle)\ \text{is an inner product space}$$

but not a Hilbert space. The completion $(L^2(S), \|\ \|_2)$ of $(R^2(S), \|\ \|_2)$ is a Banach space whose norm is induced by the inner product $\langle\ ;\ \rangle$ so that

$$(L^2(S), \langle\ ;\ \rangle)\ \text{is a Hilbert space.}$$

This, in fact, is the completion of the inner product space $(C[0, 1], \langle\ ;\ \rangle)$ of Example 5C (if $S = [0, 1]$ — see Examples 4C and 4D). We shall discuss the completion of an inner product space in Section 5.6.

**Example 5E.** Let $\{(\mathcal{X}_i, \langle\ ;\ \rangle_i)\}_{i=1}^n$ be a finite collection of inner product spaces, where the linear spaces $\mathcal{X}_i$ are all over the same field $\mathbb{F}$, and let $\bigoplus_{i=1}^n \mathcal{X}_i$ be the direct sum of the family $\{\mathcal{X}_i\}_{i=1}^n$. For each $x = (x_1, \ldots, x_n)$ and $y = (y_1, \ldots, y_n)$ in $\bigoplus_{i=1}^n \mathcal{X}_i$ put

$$\langle x\,;y\rangle = \sum_{i=1}^n \langle x_i\,;y_i\rangle_i.$$

It is easy to check that this defines an inner product on $\bigoplus_{i=1}^n \mathcal{X}_i$ that induces the norm $\|\ \|_2$ of Example 4E. Indeed, if $\|\ \|_i$ is the norm on each $\mathcal{X}_i$ induced by the inner product $\langle \; ; \; \rangle_i$, then $\langle x \; ; \; x \rangle = \sum_{i=1}^{n} \langle x_i \; ; \; x_i \rangle_i = \sum_{i=1}^{n} \|x_i\|_i^2 = \|x\|_2^2$ for every $x = (x_1, \dots, x_n)$ in $\bigoplus_{i=1}^{n} \mathcal{X}_i$. Since $\left( \bigoplus_{i=1}^{n} \mathcal{X}_i, \| \; \|_2 \right)$ is a Banach space if and only if each $(\mathcal{X}_i, \| \; \|_i)$ is a Banach space, it follows that

$$\left( \bigoplus_{i=1}^{n} \mathcal{X}_i, \langle \; ; \; \rangle \right) \text{ is a Hilbert space}$$

whenever each $(\mathcal{X}_i, \langle \; ; \; \rangle_i)$ is a Hilbert space. If the inner product spaces $(\mathcal{X}_i, \langle \; ; \; \rangle_i)$ coincide with a fixed inner product space $(\mathcal{X}, \langle \; ; \; \rangle_X)$, then $\langle x \; ; \; y \rangle = \sum_{i=1}^{n} \langle x_i \; ; \; y_i \rangle_X$ defines an inner product on $\mathcal{X}^n = \bigoplus_{i=1}^{n} \mathcal{X}$ and

$$(\mathcal{X}^n, \langle \; ; \; \rangle) \text{ is a Hilbert space}$$

whenever $(\mathcal{X}, \langle \; ; \; \rangle_X)$ is a Hilbert space. This generalizes Example 5A.

**Example 5F.** Let $\{(\mathcal{X}_k, \langle \; ; \; \rangle_k)\}$ be a countably infinite collection of inner product spaces indexed by $\mathbb{N}$ (or by $\mathbb{N}_0$), where the linear spaces $\mathcal{X}_k$ are all over the same field $\mathbb{F}$. Consider the full direct sum $\bigoplus_{k=1}^{\infty} \mathcal{X}_k$ of $\{\mathcal{X}_k\}_{k=1}^{\infty}$, which is a linear space over $\mathbb{F}$. Let $\left[ \bigoplus_{k=1}^{\infty} \mathcal{X}_k \right]_2$ be the linear manifold of $\bigoplus_{k=1}^{\infty} \mathcal{X}_k$ made up of all *square-summable sequences* $\{x_k\}_{k=1}^{\infty}$ in $\bigoplus_{k=1}^{\infty} \mathcal{X}_k$. That is (see Example 4F),

$$\left[ \bigoplus_{k=1}^{\infty} \mathcal{X}_k \right]_2 = \left\{ \{x_k\}_{k=1}^{\infty} \in \bigoplus_{k=1}^{\infty} \mathcal{X}_k : \; \sum_{k=1}^{\infty} \|x_k\|_k^2 < \infty \right\},$$

where each $\| \; \|_k$ is the norm on $\mathcal{X}_k$ induced by the inner product $\langle \; ; \; \rangle_k$. Take arbitrary sequences $\{x_k\}_{k=1}^{\infty}$ and $\{y_k\}_{k=1}^{\infty}$ in $\left[ \bigoplus_{k=1}^{\infty} \mathcal{X}_k \right]_2$ so that the real-valued sequences $\{\|x_k\|_k\}_{k=1}^{\infty}$ and $\{\|y_k\|_k\}_{k=1}^{\infty}$ lie in $\ell_+^2$. Write $\langle \; ; \; \rangle_{\ell_+^2}$ and $\| \; \|_{\ell_+^2}$ for the inner product and norm on $\ell_+^2$ (as in Example 5B). Use the Schwarz inequality in each inner product space $\mathcal{X}_k$, and also in the Hilbert space $\ell_+^2$, and get

$$\sum_{k=1}^{\infty} |\langle x_k \; ; \; y_k \rangle_k| \leq \sum_{k=1}^{\infty} \|x_k\|_k \|y_k\|_k = \left\langle \{\|x_k\|_k\}_{k=1}^{\infty} \; ; \; \{\|y_k\|_k\}_{k=1}^{\infty} \right\rangle_{\ell_+^2}$$

$$\leq \left\| \{\|x_k\|_k\}_{k=1}^{\infty} \right\|_{\ell_+^2} \left\| \{\|y_k\|_k\}_{k=1}^{\infty} \right\|_{\ell_+^2}.$$

Therefore $\sum_{k=1}^{\infty} |\langle x_k \; ; \; y_k \rangle_k| < \infty$; that is, the infinite series $\sum_{k=1}^{\infty} \langle x_k \; ; \; y_k \rangle_k$ is absolutely convergent in the Banach space $(\mathbb{F}, | \; |)$, and hence it converges in $(\mathbb{F}, | \; |)$ by Proposition 4.4. Set

$$\langle x \; ; \; y \rangle = \sum_{k=1}^{\infty} \langle x_k \; ; \; y_k \rangle_k$$

for each $x = \{x_k\}_{k=1}^{\infty}$ and $y = \{y_k\}_{k=1}^{\infty}$ in $\left[ \bigoplus_{k=1}^{\infty} \mathcal{X}_k \right]_2$. It is easy to show that this defines an inner product on $\left[ \bigoplus_{k=1}^{\infty} \mathcal{X}_k \right]_2$ that induces the norm $\| \; \|_2$ of Example 4F. Moreover, since $\left( \left[ \bigoplus_{k=1}^{\infty} \mathcal{X}_k \right]_2, \| \; \|_2 \right)$ is a Banach space if and only if each $(\mathcal{X}_k, \| \; \|_k)$ is a Banach space, it follows that

$$\left( \left[ \bigoplus_{k=1}^{\infty} \mathcal{X}_k \right]_2, \langle \; ; \; \rangle \right) \text{ is a Hilbert space}$$

whenever each $(\mathcal{X}_k, \langle \ ; \ \rangle_k)$ is a Hilbert space. A similar argument holds if the collection $\{(\mathcal{X}_k, \langle \ ; \ \rangle_k)\}$ is indexed by $\mathbb{Z}$. Indeed, if we set

$$\left[\bigoplus_{k=-\infty}^{\infty} \mathcal{X}_k\right]_2 = \left\{\{x_k\}_{k=-\infty}^{\infty} \in \bigoplus_{k=-\infty}^{\infty} \mathcal{X}_k : \ \sum_{k=-\infty}^{\infty} \|x_k\|_k^2 < \infty\right\},$$

the linear manifold of the full direct sum $\bigoplus_{k=-\infty}^{\infty} \mathcal{X}_k$ of $\{\mathcal{X}_k\}_{k=-\infty}^{\infty}$ made up of all *square-summable nets* $\{x_k\}_{k=-\infty}^{\infty}$ in $\bigoplus_{k=-\infty}^{\infty} \mathcal{X}_k$, then

$$\langle x ; y \rangle = \sum_{k=-\infty}^{\infty} \langle x_k ; y_k \rangle_k$$

for each $x = \{x_k\}_{k=-\infty}^{\infty}$ and $y = \{y_k\}_{k=-\infty}^{\infty}$ in $\left[\bigoplus_{k=-\infty}^{\infty} \mathcal{X}_k\right]_2$ defines the inner product on $\left[\bigoplus_{k=-\infty}^{\infty} \mathcal{X}_k\right]_2$ that induces the norm $\| \ \|_2$ of Example 4F. Again, if each $(\mathcal{X}_k, \langle \ ; \ \rangle_k)$ is a Hilbert space, then

$$\left(\left[\bigoplus_{k=-\infty}^{\infty} \mathcal{X}_k\right]_2, \langle \ ; \ \rangle\right) \text{ is a Hilbert space.}$$

If the inner product spaces $(\mathcal{X}_k, \langle \ ; \ \rangle_k)$ coincide with a fixed inner product space $(\mathcal{X}, \langle \ ; \ \rangle_X)$, then set

$$\ell_+^2(\mathcal{X}) = \left[\bigoplus_{k=1}^{\infty} \mathcal{X}\right]_2 \quad \text{and} \quad \ell^2(\mathcal{X}) = \left[\bigoplus_{k=-\infty}^{\infty} \mathcal{X}\right]_2$$

as in Example 4F. If $(\mathcal{X}, \langle \ ; \ \rangle_X)$ is a Hilbert space, then

$$(\ell_+^2(\mathcal{X}), \langle \ ; \ \rangle) \text{ and } (\ell^2(\mathcal{X}), \langle \ ; \ \rangle) \text{ are Hilbert spaces.}$$

## 5.3 Orthogonality

Let $a$ and $b$ be nonzero vectors in the Euclidean plane $\mathbb{R}^2$, and let $\theta_{ab}$ be the angle between the line segments joining these points to the origin (this is usually called the *angle between $a$ and $b$*). Set $a' = \|a\|^{-1}a = (\alpha_1, \alpha_2)$ and $b' = \|b\|^{-1}b = (\beta_1, \beta_2)$ in the unit circle about the origin. It is a simple exercise of elementary plane geometry to verify that $\cos\theta_{ab} = \alpha_1\beta_1 + \alpha_2\beta_2 = \langle a' ; \beta' \rangle = \|a\|^{-1}\|b\|^{-1}\langle a ; b \rangle$. We shall be particularly concerned with the notion of *orthogonal* (or *perpendicular*) vectors $a$ and $b$. The line segments joining $a$ and $b$ to the origin are perpendicular if $\theta_{ab} = \frac{\pi}{2}$ (equivalently, if $\cos\theta_{ab} = 0$), which means that $\langle a ; b \rangle = 0$. These notions (angle and orthogonality, that is) can be extended from the Euclidean plane to a *real* inner product space $(\mathcal{X}, \langle \ ; \ \rangle)$ by setting

$$\cos\theta_{xy} = \frac{\langle x ; y \rangle}{\|x\| \|y\|}$$

whenever $x$ and $y$ are nonzero vectors in $\mathcal{X} \neq \{0\}$. Observe that $-1 \leq \cos\theta_{xy} \leq 1$ by the Schwarz inequality, and also that $\cos\theta_{xy} = 0$ if and only if $\langle x ; y \rangle = 0$.

**Definition 5.7.** Two vectors $x$ and $y$ in any (real or complex) inner product space $(\mathcal{X}, \langle\ ;\ \rangle)$ are said to be *orthogonal* (notation: $x \perp y$) if $\langle x ; y \rangle = 0$. A vector $x$ in $\mathcal{X}$ is orthogonal to a subset $A$ of $\mathcal{X}$ (notation: $x \perp A$) if it is orthogonal to every vector in $A$ (i.e., if $\langle x ; y \rangle = 0$ for every $y \in A$). Two subsets $A$ and $B$ of $\mathcal{X}$ are orthogonal (notation: $A \perp B$) if every vector in $A$ is orthogonal to every vector in $B$ (i.e., if $\langle x ; y \rangle = 0$ for every $x \in A$ and every $y \in B$).

Thus $A$ and $B$ are orthogonal if there is no $x$ in $A$ and no $y$ in $B$ such that $\langle x ; y \rangle \neq 0$. In this sense the empty set $\varnothing$ is orthogonal to every subset of $\mathcal{X}$. Clearly, $x \perp y$ if and only if $y \perp x$, and hence $A \perp B$ if and only if $B \perp A$, so that $\perp$ is a symmetric relation both on $\mathcal{X}$ and on the power set $\wp(\mathcal{X})$. We write $x \not\perp y$ if $x \in \mathcal{X}$ and $y \in \mathcal{X}$ are not orthogonal. Similarly, $A \not\perp B$ means that $A \subseteq \mathcal{X}$ and $B \subseteq \mathcal{X}$ are not orthogonal. Note that if there exists a nonzero vector $x$ in $A \cap B$, then $\langle x ; x \rangle = \|x\|^2 \neq 0$, and hence $A \not\perp B$. Therefore,

$$A \perp B \quad \text{implies} \quad A \cap B = \{0\}.$$

We shall say that a subset $A$ of an inner product space $\mathcal{X}$ is an *orthogonal set* (or a *set of pairwise orthogonal vectors*) if $x \perp y$ for every pair $\{x, y\}$ of *distinct* vectors in $A$. Similarly, an $\mathcal{X}$-valued sequence $\{x_k\}$ is an *orthogonal sequence* (or a *sequence of pairwise orthogonal vectors*) if $x_k \perp x_j$ whenever $k \neq j$. Since $\|x + y\|^2 = \|x\|^2 + 2\operatorname{Re}\langle x ; y \rangle + \|y\|^2$ for every $x$ and $y$ in $\mathcal{X}$, it follows as an immediate consequence of the definition of orthogonality that

$$x \perp y \quad \text{implies} \quad \|x + y\|^2 = \|x\|^2 + \|y\|^2.$$

This is the *Pythagorean Theorem*. The next result is a generalization of it for a finite orthogonal set.

**Proposition 5.8.** *If $\{x_i\}_{i=0}^{n}$ is a finite set of pairwise orthogonal vectors in an inner product space, then*

$$\left\| \sum_{i=0}^{n} x_i \right\|^2 = \sum_{i=0}^{n} \|x_i\|^2.$$

*Proof.* We have already seen that the result holds for $n = 1$ (i.e., it holds for *every* pair of distinct orthogonal vectors). Suppose it holds for some $n \geq 1$ (i.e., suppose $\|\sum_{i=0}^{n} x_i\|^2 = \sum_{i=0}^{n}\|x_i\|^2$ for *every* orthogonal set $\{x_i\}_{i=0}^{n}$ containing $n + 1$ elements). Let $\{x_i\}_{i=0}^{n+1}$ be an *arbitrary* orthogonal set with $n + 2$ elements. Since $x_{n+1} \perp \{x_i\}_{i=0}^{n}$, it follows that $x_{n+1} \perp \sum_{i=0}^{n} x_i$ (since $\langle x_{n+1} ; \sum_{i=0}^{n} x_i \rangle = \sum_{i=0}^{n}\langle x_{n+1} ; x_i \rangle$). Hence

$$\left\| \sum_{i=0}^{n+1} x_i \right\|^2 = \left\| \sum_{i=0}^{n} x_i + x_{n+1} \right\|^2 = \left\| \sum_{i=0}^{n} x_i \right\|^2 + \|x_{n+1}\|^2 = \sum_{i=0}^{n+1} \|x_i\|^2,$$

so that the result holds for $n + 1$ (i.e., it holds for every orthogonal set with $n + 2$ elements whenever it holds for every orthogonal set with $n + 1$ elements), which completes the proof by induction. $\qquad\square$

Recall that an $\mathcal{X}$-valued sequence $\{x_k\}_{k=1}^{\infty}$ (where $\mathcal{X}$ is any normed space) is *square-summable* if $\sum_{k=1}^{\infty} \|x_k\|^2 < \infty$. Here is a countably infinite version of the Pythagorean Theorem.

**Corollary 5.9.** *Let $\{x_k\}_{k=1}^{\infty}$ be a sequence of pairwise orthogonal vectors in a inner product space $\mathcal{X}$.*

(a) *If the infinite series $\sum_{k=1}^{\infty} x_k$ converges in $\mathcal{X}$, then $\{x_k\}_{k=1}^{\infty}$ is a square-summable sequence and $\|\sum_{k=1}^{\infty} x_k\|^2 = \sum_{k=1}^{\infty} \|x_k\|^2$.*

(b) *If $\mathcal{X}$ is a Hilbert space and $\{x_k\}_{k=1}^{\infty}$ is a square-summable sequence, then the infinite series $\sum_{k=1}^{\infty} x_k$ converges in $\mathcal{X}$.*

*Proof.* Let $\{x_k\}_{k=1}^{\infty}$ be an orthogonal sequence in $\mathcal{X}$.

(a) If the series $\sum_{k=1}^{\infty} x_k$ converges in $\mathcal{X}$; that is, if $\sum_{k=1}^{n} x_k \to \sum_{k=1}^{\infty} x_k \in \mathcal{X}$ as $n \to \infty$, then $\|\sum_{k=1}^{n} x_k\|^2 \to \|\sum_{k=1}^{\infty} x_k\|^2$ as $n \to \infty$. (Reason: norm and squaring are continuous mappings.) But Proposition 5.8 says that $\|\sum_{k=1}^{n} x_k\|^2 = \sum_{k=1}^{n} \|x_k\|^2$ for every $n \geq 1$, and hence $\sum_{k=1}^{n} \|x_k\|^2 \to \|\sum_{k=1}^{\infty} x_k\|^2$ as $n \to \infty$.

(b) Consider the $\mathcal{X}$-valued sequence $\{y_n\}_{n=1}^{\infty}$ of partial sums of $\{x_k\}_{k=1}^{\infty}$; that is, set $y_n = \sum_{k=1}^{n} x_k$ for each integer $n \geq 1$. According to Proposition 5.8 we know that $\|y_{n+m} - y_n\|^2 = \sum_{j=n+1}^{n+m} \|x_{k_j}\|^2$ for every $m, n \geq 1$. If $\sum_{k=1}^{\infty} \|x_k\|^2 < \infty$, then $\sup_{m \geq 1} \|y_{n+m} - y_n\|^2 = \sum_{k=n+1}^{\infty} \|x_k\|^2 \to 0$ as $n \to \infty$ (Problem 3.11), and hence $\{y_n\}_{n=1}^{\infty}$ is a Cauchy sequence in $\mathcal{X}$ (Problem 3.51). If $\mathcal{X}$ is Hilbert, then $\{y_n\}_{n=1}^{\infty}$ converges in $\mathcal{X}$, which means that the series $\sum_{k=1}^{\infty} x_k$ converges in $\mathcal{X}$. $\qquad\square$

Therefore, *if $\{x_k\}_{k=1}^{\infty}$ is an orthogonal sequence in a Hilbert space $\mathcal{H}$, then $\sum_{k=1}^{\infty} \|x_k\|^2 < \infty$ if and only if the infinite series $\sum_{k=1}^{\infty} x_k$ converges in $\mathcal{H}$ and, in this case, $\|\sum_{k=1}^{\infty} x_k\|^2 = \sum_{k=1}^{\infty} \|x_k\|^2$.*

**Example 5G.** Let $\{(\mathcal{X}_k, \langle\ ;\ \rangle_k)\}_{k=1}^{\infty}$ be a sequence of Hilbert spaces. Consider the Hilbert space $\left(\left[\bigoplus_{k=1}^{\infty} \mathcal{X}_k\right]_2, \langle\ ;\ \rangle\right)$, where $\left[\bigoplus_{k=1}^{\infty} \mathcal{X}_k\right]_2$ is the linear space of all square-summable sequences in the full direct sum $\bigoplus_{k=1}^{\infty} \mathcal{X}_k$ and $\langle\ ;\ \rangle$ is the inner product of Example 5F; that is,

$$\langle x\,;y\rangle = \sum_{k=1}^{\infty} \langle x_k\,;y_k\rangle_k$$

for every $x = \{x_k\}_{k=1}^{\infty}$ and $y = \{y_k\}_{k=1}^{\infty}$ in $\left[\bigoplus_{k=1}^{\infty} \mathcal{X}_k\right]_2$. This is referred to as an (external) *orthogonal direct sum*. An *orthogonal* direct sum actually deserves its name. Indeed, if we identify each linear space $\mathcal{X}_i$ with the linear manifold $\bigoplus_{k=1}^{\infty} \mathcal{X}_{i\,(k)}$

of $\left[\bigoplus_{k=1}^{\infty}\mathcal{X}_k\right]_2$ such that $\mathcal{X}_{i\,(k)} = \{0_k\} \subseteq \mathcal{X}_k$ for $k \neq i$ and $\mathcal{X}_{i\,(i)} = \mathcal{X}_i$ (as in Example 4I), then it is clear that

$$\mathcal{X}_i \perp \mathcal{X}_j \quad \text{whenever} \quad i \neq j,$$

where such an orthogonality is interpreted as $\bigoplus_{k=1}^{\infty}\mathcal{X}_{i\,(k)} \perp \bigoplus_{k=1}^{\infty}\mathcal{X}_{j\,(k)}$ with respect to the inner product $\langle\ ;\ \rangle$ on $\left[\bigoplus_{k=1}^{\infty}\mathcal{X}_k\right]_2$. Observe that the norm $\|\ \|$ induced on $\left[\bigoplus_{k=1}^{\infty}\mathcal{X}_k\right]_2$ by this inner product is given by

$$\|x\|^2 = \langle x\,;x\rangle = \sum_{k=1}^{\infty}\langle x_k\,;x_k\rangle_k = \sum_{k=1}^{\infty}\|x_k\|_k^2$$

for every $x = \{x_k\}_{k=1}^{\infty}$ in $\left[\bigoplus_{k=1}^{\infty}\mathcal{X}_k\right]_2$. This can also be verified via Corollary 5.9 as follows. Take an arbitrary vector $x = \{x_k\}_{k=1}^{\infty}$ in $\left[\bigoplus_{k=1}^{\infty}\mathcal{X}_k\right]_2$ (i.e., take an arbitrary square-summable sequence from $\bigoplus_{k=1}^{\infty}\mathcal{X}_k$). Set $x_{i\,(k)} = \delta_{ik}x_k$ in $\mathcal{X}_k$ for every $k, i \geq 1$ (i.e., $x_{i\,(k)} = 0_k$ if $k \neq i$ and $x_{i\,(i)} = x_i$). For each $i \geq 1$ consider the vector $x_i = \{x_{i\,(k)}\}_{k=1}^{\infty}$ so that $x_1 = (x_1, 0_2, 0_3, \dots)$ and, for $i \geq 2$,

$$x_i = \{x_{i\,(k)}\}_{k=1}^{\infty} = (0_1, \dots, 0_{i-1}, x_i, 0_{i+1}, \dots) \quad \text{in} \quad \left[\bigoplus_{k=1}^{\infty}\mathcal{X}_k\right]_2.$$

*Claim*: $\{x_i\}_{i=1}^{\infty}$ is an orthogonal square-summable sequence. (Proof: Just note that $x_i \in \bigoplus_{k=1}^{\infty}\mathcal{X}_{i\,(k)}$, $\|x_i\| = \|x_i\|_i$, and $\{x_i\}_{i=1}^{\infty}$ is square-summable). Thus Corollary 5.9 ensures that (i) the infinite series $\bigoplus_{i=1}^{\infty}x_i$ converges in the Hilbert space $\left(\left[\bigoplus_{k=1}^{\infty}\mathcal{X}_k\right]_2, \langle\ ;\ \rangle\right)$, and (ii) $\|\bigoplus_{i=1}^{\infty}x_i\|^2 = \sum_{i=1}^{\infty}\|x_i\|^2$. Notational warning: We are denoting vector addition in the linear space $\bigoplus_{k=1}^{\infty}\mathcal{X}_k$ by $\oplus$ and vector subtraction by $\ominus$, as usual. From (i) we get $x = \bigoplus_{i=1}^{\infty}x_i$. (Reason: $x \ominus (\bigoplus_{i=1}^{n}x_i) = \bigoplus_{i=n+1}^{\infty}x_i$ for each $n \geq 1$.) and from (ii) we get $\|x\|^2 = \sum_{i=1}^{\infty}\|x_i\|^2$. Therefore,

$$\|x\|^2 = \sum_{i=1}^{\infty}\|x_i\|^2 = \sum_{i=1}^{\infty}\|x_i\|_i^2.$$

If $(\mathcal{X}, \langle\ ;\ \rangle)$ is an inner product space, and if $\mathcal{M}$ is a linear manifold of the linear space $\mathcal{X}$, then it is easy to show that the restriction $\langle\ ;\ \rangle_\mathrm{M} : \mathcal{M}\times\mathcal{M} \to \mathbb{F}$ of the inner product $\langle\ ;\ \rangle : \mathcal{X}\times\mathcal{X} \to \mathbb{F}$ to $\mathcal{M}\times\mathcal{M}$ is an inner product on $\mathcal{M}$, so that $(\mathcal{M}, \langle\ ;\ \rangle_\mathrm{M})$ is an inner product space. Moreover, the norm $\|\ \|_\mathrm{M} : \mathcal{M} \to \mathbb{R}$ induced by the inner product $\langle\ ;\ \rangle_\mathrm{M}$ on $\mathcal{M}$ coincides with the restriction to $\mathcal{M}$ of the norm $\|\ \| : \mathcal{X} \to \mathbb{R}$ induced by the inner product $\langle\ ;\ \rangle$ on $\mathcal{X}$. Thus $(\mathcal{M}, \|\ \|_\mathrm{M})$ is a linear manifold of the normed space $(\mathcal{X}, \|\ \|)$. Whenever a linear manifold of an inner product space is regarded as an inner product space, it will always be understood that the inner product on it is the restricted inner product $\langle\ ;\ \rangle_\mathrm{M}$. We shall drop the subscript and write $(\mathcal{M}\langle\ ;\ \rangle)$ instead of $(\mathcal{M}\langle\ ;\ \rangle_\mathrm{M})$, and refer to the inner product space $(\mathcal{M}\langle\ ;\ \rangle)$ by simply saying that "$\mathcal{M}$ is a linear manifold of $\mathcal{X}$". Recall that a subspace of a normed space is a *closed* linear manifold of it. Hence a subspace of an inner product space $\mathcal{X}$ is a linear manifold of the linear space $\mathcal{X}$ that is closed in the inner product topology. That is, a subspace $\mathcal{M}$ of an inner product space $\mathcal{X}$ is a linear manifold of $\mathcal{X}$ that is closed as a subset of $\mathcal{X}$ when $\mathcal{X}$ is regarded as a metric space whose metric is that generated by the norm that is induced by the inner product. According to Proposition 4.4 a linear manifold of a Hilbert space $\mathcal{H}$ is a Hilbert space if and only if it is a subspace of $\mathcal{H}$. We observed in Section 4.3 that the sum of subspaces is not necessarily a subspace (it is a linear manifold but may not be closed). An extremely important consequence of the Pythagorean Theorem is that *the sum of orthogonal subspaces of a Hilbert space is again a subspace.*

**Theorem 5.10.** (a) *If $\mathcal{M}$ and $\mathcal{N}$ are complete orthogonal linear manifolds of an inner product space $\mathcal{X}$, then the sum $\mathcal{M} + \mathcal{N}$ is a complete linear manifold of $\mathcal{X}$.*

(b) *If $\mathcal{M}$ and $\mathcal{N}$ are orthogonal subspaces of a Hilbert space $\mathcal{H}$, then the sum $\mathcal{M} + \mathcal{N}$ is a subspace of $\mathcal{H}$.*

*Proof.* Let $\mathcal{M}$ and $\mathcal{N}$ be orthogonal linear manifolds of an inner product space $\mathcal{X}$. Take an arbitrary Cauchy sequence $\{x_n\}$ in $\mathcal{M} + \mathcal{N}$ so that $x_n = u_n + v_n$ with $u_n$ in $\mathcal{M}$ and $v_n$ in $\mathcal{N}$ for each $n$. Since $\mathcal{M}$ and $\mathcal{N}$ are linear manifolds of $\mathcal{X}$, it follows that $u_m - u_n$ lies in $\mathcal{M}$ and $v_m - v_n$ lies in $\mathcal{N}$, and hence $u_m - u_n \perp v_m - v_n$, for every pair of integers $m$ and $n$ (because $\mathcal{M} \perp \mathcal{N}$). Writing $x_m - x_n = (u_m - u_n) + (v_m - v_n)$ the Pythagorean Theorem ensures that

$$\|x_m - x_n\|^2 \; = \; \|u_m - u_n\|^2 + \|v_m - v_n\|^2$$

for every $m$ and $n$. This implies that $\{u_n\}$ and $\{v_n\}$ are Cauchy sequences in $\mathcal{M}$ and $\mathcal{N}$, respectively (since $\{x_n\}$ is a Cauchy sequence).

(a) If $\mathcal{M}$ and $\mathcal{N}$ are complete, then $\{u_n\}$ converges in $\mathcal{M}$ and $\{v_n\}$ converges in $\mathcal{N}$. Recalling that addition is a continuous operation (Problem 4.1) we get from Corollary 3.8 that $\{x_n\}$ converges in $\mathcal{M} + \mathcal{N}$. Conclusion: Every Cauchy sequence in $\mathcal{M} + \mathcal{N}$ converges in $\mathcal{M} + \mathcal{N}$. Thus $\mathcal{M} + \mathcal{N}$ is complete (and hence closed in $\mathcal{X}$ by Theorem 3.40(a)).

(b) If $\mathcal{M}$ and $\mathcal{N}$ are closed linear manifolds of a complete inner product space $\mathcal{H}$, then they are complete (by Theorem 3.40(b)). Thus the linear manifold $\mathcal{M} + \mathcal{N}$ of $\mathcal{H}$ is complete (and therefore closed in $\mathcal{H}$) according to item (a).     $\square$

Theorem 5.10(a) fails if $\mathcal{M}$ and $\mathcal{N}$ are either (1) orthogonal but not complete or (2) complete but not orthogonal. Equivalently, Theorem 5.10(b) fails if $\mathcal{M}$ and $\mathcal{N}$ are either (1) orthogonal subspaces of an incomplete inner product space or (2) not orthogonal subspaces of a Hilbert space. That is, completeness and orthogonality are both crucial assumptions in the statement of Theorem 5.10. This will be verified in Problems 5.12 and 5.13.

Let $\{\mathcal{M}_\gamma\}_{\gamma \in \Gamma}$ be an arbitrary nonempty indexed family of subspaces of a Hilbert space $\mathcal{H}$ (i.e., an arbitrary nonempty subcollection of $\mathrm{Lat}(\mathcal{H})$). Recall that the sum $\sum_{\gamma \in \Gamma} \mathcal{M}_\gamma$ of $\{\mathcal{M}_\gamma\}_{\gamma \in \Gamma}$ is the linear manifold of $\mathcal{H}$ consisting of *all finite sums* of vectors in $\mathcal{H}$ with each summand being a vector in one of the subspaces $\mathcal{M}_\gamma$. That is,

$$\sum_{\gamma \in \Gamma} \mathcal{M}_\gamma = \text{span}\left(\bigcup_{\gamma \in \Gamma} \mathcal{M}_\gamma\right).$$

**Corollary 5.11.** *Every finite sum of pairwise orthogonal subspaces of a Hilbert space $\mathcal{H}$ is itself a subspace of $\mathcal{H}$.*

*Proof.* Let $\mathcal{H}$ be a Hilbert space. Theorem 5.10(b) says that the sum of every pair of orthogonal subspaces of $\mathcal{H}$ is again a subspace of $\mathcal{H}$. Take an arbitrary integer $n \geq 2$. Suppose the sum of every set of $n$ pairwise orthogonal subspaces of $\mathcal{H}$ is a subspace of $\mathcal{H}$. Now take an arbitrary collection of $n + 1$ pairwise orthogonal subspaces of $\mathcal{H}$, say, $\{\mathcal{M}_i\}_{i=1}^{n+1}$. If $x \in \sum_{i=1}^{n} \mathcal{M}_i$, then $x = \sum_{i=1}^{n} x_i$ with each $x_i$ in $\mathcal{M}_i$, and hence $\langle x ; x_{n+1} \rangle = \sum_{i=1}^{n} \langle x_i ; x_{n+1} \rangle = 0$ whenever $x_{n+1}$ lies in $\mathcal{M}_{n+1}$ (because $\mathcal{M}_i \perp \mathcal{M}_{n+1}$ for every $i = 1, \ldots, n$). Thus $\sum_{i=1}^{n} \mathcal{M}_i \perp \mathcal{M}_{n+1}$. Since $\sum_{i=1}^{n} \mathcal{M}_i$ was assumed to be a subspace of $\mathcal{H}$, and since

$$\sum_{i=1}^{n+1} \mathcal{M}_i = \text{span}\left(\bigcup_{i=1}^{n+1} \mathcal{M}_i\right) = \text{span}\left(\bigcup_{i=1}^{n} \mathcal{M}_i \cup \mathcal{M}_{n+1}\right) = \sum_{i=1}^{n} \mathcal{M}_i + \mathcal{M}_{n+1},$$

it follows by Theorem 5.10(b) that $\sum_{i=1}^{n+1} \mathcal{M}_i$ is a subspace of $\mathcal{H}$. This completes the proof by induction. $\qquad\square$

## 5.4   Orthogonal Complement

If $A$ is a subset of an inner product space $\mathcal{X}$, then the *orthogonal complement* of $A$ is the set

$$A^\perp = \left\{x \in \mathcal{X} : x \perp A\right\} = \left\{x \in \mathcal{X} : \langle x ; y \rangle = 0 \text{ for every } y \in A\right\}$$

consisting of all vectors in $\mathcal{X}$ that are orthogonal to every vector in $A$. If $A$ is the empty set $\varnothing$, then for every $x$ in $\mathcal{X}$ there is no vector $y$ in $A$ for which $\langle x ; y \rangle \neq 0$, and hence $\varnothing^\perp = \mathcal{X}$. Clearly, $x \perp \{0\}$ for every $x \in \mathcal{X}$, and $x \perp \mathcal{X}$ if and only if $x = 0$. Hence

$$\{0\}^\perp = \mathcal{X} \quad \text{and} \quad \mathcal{X}^\perp = \{0\}.$$

Let $A$ and $B$ be nonempty subsets of $\mathcal{X}$. The next results are immediate consequences of the definition of orthogonal complement.

$$A \perp A^\perp, \quad A \cap A^\perp \subseteq \{0\} \quad \text{and} \quad A \cap A^\perp = \{0\} \text{ whenever } 0 \in A$$

(reason: if there exists $x \in A \cap A^\perp$, then $\langle x ; x \rangle = 0$), and

$$A \perp B \quad \text{if and only if} \quad A \subseteq B^\perp.$$

Since $\perp$ is a symmetric relation (i.e., $A \perp B$ if and only if $B \perp A$), the above equivalent assertions also are equivalent to $B \subseteq A^\perp$. Moreover,

$$A \perp B \quad \text{implies} \quad A \cap B \subseteq \{0\}$$

(if $A \subseteq B^\perp$, then $A \cap B \subseteq B^\perp \cap B \subseteq \{0\}$). It is readily verified that

$$A \subseteq B \quad \text{implies} \quad B^\perp \subseteq A^\perp \quad \text{and so} \quad A^{\perp\perp} \subseteq B^{\perp\perp},$$

where $A^{\perp\perp} = (A^\perp)^\perp$. Since $A \perp A^\perp$ and $A^\perp \perp A^{\perp\perp}$, we get $A \subseteq A^{\perp\perp}$ (so that $A^{\perp\perp\perp} \subseteq A^\perp$) and $A^\perp \subseteq A^{\perp\perp\perp}$, where $A^{\perp\perp\perp} = (A^{\perp\perp})^\perp$. Therefore,

$$A \subseteq A^{\perp\perp} \quad \text{and} \quad A^\perp = A^{\perp\perp\perp}.$$

**Proposition 5.12.** *The orthogonal complement* $A^\perp$ *of every subset $A$ of any inner product space* $\mathcal{X}$ *is a subspace of* $\mathcal{X}$. *Moreover,*

$$A^\perp = (A^\perp)^- = (A^-)^\perp = (\operatorname{span} A)^\perp = \left(\bigvee A\right)^\perp.$$

*The orthogonal complement of every dense subset of* $\mathcal{X}$ *is the zero space:*

$$A^\perp = \{0\} \quad \text{whenever} \quad A^- = \mathcal{X}.$$

*Proof.* Suppose $A \neq \varnothing$ (otherwise the results are trivially verified). Since the inner product is linear in the first argument, it follows at once that $A^\perp$ is a linear manifold of the linear space $\mathcal{X}$. If $x \perp A$, then $x \perp \sum_{i=1}^n \alpha_i y_i$ for every integer $n \geq 1$ whenever $y_i \in A$ and $\alpha_i \in \mathbb{F}$ for each $i = 1, \dots, n$, and hence $A^\perp \subseteq (\operatorname{span} A)^\perp$. On the other hand, $A \subset \operatorname{span} A$ so that $(\operatorname{span} A)^\perp \subseteq A^\perp$. Then

$$A^\perp = (\operatorname{span} A)^\perp.$$

That $A^\perp$ is closed in $\mathcal{X}$ is a consequence of the continuity of the inner product (Problem 5.6). Actually, if $\{x_n\}$ is an $A^\perp$-valued sequence that converges in $\mathcal{X}$ to $x \in \mathcal{X}$, then (cf. Corollary 3.8) $\langle x \, ; y \rangle = \langle \lim x_n \, ; y \rangle = \lim \langle x_n \, ; y \rangle = 0$ for every $y \in A$, which implies that $x \in A^\perp$. Therefore, $A^\perp$ is closed in $\mathcal{X}$ by the Closed Set Theorem (Theorem 3.30); that is,

$$A^\perp = (A^\perp)^-,$$

and so $A^\perp$ is a subspace (i.e., a closed linear manifold) of $\mathcal{X}$. Now take an arbitrary $x$ in $A^\perp$ and an arbitrary $y$ in $A^-$. By Proposition 3.27 there exists an $A$-valued sequence $\{y_n\}$ that converges in $\mathcal{X}$ to $y$. Using Corollary 3.8 again, and recalling that the inner product is continuous, we get $\langle y \, ; x \rangle = \langle \lim y_n \, ; x \rangle = \lim \langle y_n \, ; x \rangle = 0$. Thus $A^\perp \perp A^-$ so that $A^\perp \subseteq (A^-)^\perp$. But $(A^-)^\perp \subseteq A^\perp$ because $A \subseteq A^-$. Hence

$$A^\perp = (A^-)^\perp.$$

Since $A^\perp = (\operatorname{span} A)^\perp$ and $A^\perp = (A^-)^\perp$ for every subset $A$ of $\mathcal{X}$,

$$A^\perp = (\operatorname{span} A)^\perp = \big[(\operatorname{span} A)^-\big]^\perp = (\bigvee A)^\perp.$$

Finally, if $A^- = \mathcal{X}$, then $A^\perp = (A^-)^\perp = \mathcal{X}^\perp = \{0\}$. $\qquad\qquad\square$

Remark: If $L \in \mathcal{L}[\mathcal{X}, \mathcal{Y}]$, where $\mathcal{X}$ is an inner product space, then *the linear transformation* $L|_{\mathcal{N}(L)^\perp}$ *is injective* (i.e., $\mathcal{N}(L|_{\mathcal{N}(L)^\perp}) = \{0\}$). In fact, if $v \in \mathcal{N}(L)^\perp$ lies in $\mathcal{N}(L|_{\mathcal{N}(L)^\perp})$, then $v \in \mathcal{N}(L) \cap \mathcal{N}(L)^\perp = \{0\}$.

The next theorem is of critical importance; it may be thought of as a pivotal result in the theory of Hilbert spaces. Recall that the distance $d(x, M)$ of a point $x$ in a normed space $\mathcal{X}$ to a *nonempty* subset $M$ of $\mathcal{X}$ is the real (nonnegative) number

$$d(x, M) = \inf_{u \in M} \|x - u\|.$$

**Theorem 5.13.** *Let $x$ be an arbitrary vector in a Hilbert space $\mathcal{H}$.*

(a) *If $M$ is a closed convex nonempty subset of $\mathcal{H}$, then there exists a unique vector $u_x$ in $M$ such that*

$$\|x - u_x\| = d(x, M).$$

(b) *Moreover, if $M$ is a subspace of $\mathcal{H}$, then the unique vector in $M$ for which $\|x - u_x\| = d(x, M)$ is the unique vector in $M$ such that the difference $x - u_x$ is orthogonal to $M$; that is, such that*

$$x - u_x \in M^\perp.$$

*Proof.* (a) Let $x$ be an arbitrary vector in $\mathcal{H}$ and let $M$ be a nonempty subset of $\mathcal{H}$ so that $d(x, M) = \inf_{u \in M} \|x - u\|$ exists in $\mathbb{R}$. Therefore, for each integer $n \geq 1$ there exists $u_n \in M$ such that

$$d(x, M) \leq \|x - u_n\| < d(x, M) + \tfrac{1}{n}.$$

Consider the $M$-valued sequence $\{u_n\}$. $\mathcal{H}$ is an inner product space and so the parallelogram law ensures that

$$\|2x - u_m - u_n\|^2 + \|u_n - u_m\|^2 = 2\big(\|x - u_m\|^2 + \|x - u_n\|^2\big)$$

for each $m, n \geq 1$. Since $M$ is convex, it follows that $\tfrac{1}{2}(u_m + u_n) \in M$, and hence $2d(x, M) \leq 2\|\tfrac{1}{2}(u_m + u_n) - x\| = \|2x - u_m - u_n\|$ so that

$$0 \leq \|u_m - u_n\|^2 \leq 2\big(\|x - u_m\|^2 + \|x - u_n\|^2 - 2d(x, M)^2\big)$$

for every $m, n \geq 1$. This inequality and the fact that $\|x - u_n\| \to d(x, M)$ as $n \to \infty$ are enough to ensure that $\{u_n\}$ is a Cauchy sequence in $\mathcal{H}$, and therefore it converges in the Hilbert space $\mathcal{H}$ to, say, $u_x \in \mathcal{H}$. But the norm is a continuous function so that (Corollary 3.8)

$$\|x - u_x\| = \|x - \lim u_n\| = \lim \|x - u_n\| = d(x, M).$$

Moreover, since $M$ is closed in $\mathcal{H}$ and $\{u_n\}$ is an $M$-valued sequence that converges to $u_x$ in $\mathcal{H}$, it follows by the Closed Set Theorem (Theorem 3.30) that $u_x \in M$. Conclusion: There exists $u_x$ in $M$ such that

$$\|x - u_x\| = d(x, M).$$

To prove uniqueness, take any $u$ in $M$ such that $\|x - u\| = d(x, M)$. Observe that $\frac{1}{2}(u_x + u)$ lies in $M$ because $M$ is convex, and hence $d(x, M) \leq \left\| \frac{1}{2}(u_x + u) - x \right\|$. Thus $4d(x, M)^2 \leq \|u_x + u - 2x\|^2$. This inequality and the parallelogram law imply that

$$\begin{aligned} 4d(x, M)^2 + \|u_x - u\|^2 &\leq \|u_x + u - 2x\|^2 + \|u_x - u\|^2 \\ &= 2\big(\|u_x - x\|^2 + \|u - x\|^2\big) = 4d(x, M)^2. \end{aligned}$$

Outcome: $\|u_x - u\|^2 = 0$; that is, $u = u_x$.

(b) Now let $x$ be an arbitrary vector in $\mathcal{H}$ and suppose $M$ is a subspace of $\mathcal{H}$, which obviously implies that $M$ is a closed convex nonempty subset of $\mathcal{H}$. According to item (a) there exists a unique $u_x \in M$ such that $\|x - u_x\| = d(x, M)$. Take an arbitrary nonzero $u \in M$. Since $(u_x + \alpha u) \in M$ for every scalar $\alpha$, it follows that

$$d(x, M)^2 \leq \|x - u_x - \alpha u\|^2 = \|x - u_x\|^2 + |\alpha|^2 \|u\|^2 - 2\operatorname{Re}(\alpha \langle x - u_x \,;\, u \rangle).$$

Setting $\alpha = \|u\|^{-2} \langle x - u_x \,;\, u \rangle$ in the above inequality and recalling that $\|x - u_x\|^2 = d(x, M)^2$, we get $2|\langle x - u_x \,;\, u \rangle|^2 \leq |\langle x - u_x \,;\, u \rangle|^2$, and hence $|\langle x - u_x \,;\, u \rangle| = 0$. Conclusion: $x - u_x \perp u$ for every nonzero $u$ in $M$, which implies

$$x - u_x \perp M.$$

Finally, we show that this $u_x$ is the unique vector in $M$ with the above property. Indeed, if $v \in M$ is such that $x - v \perp M$, then $\langle x - v \,;\, v - u \rangle = \langle x - v \,;\, v \rangle - \langle x - v \,;\, u \rangle = 0$ whenever $u \in M$ so that $x - v \perp v - u$ for every $u \in M$. Thus, by the Pythagorean Theorem,

$$\|x - v\|^2 \leq \|x - v\|^2 + \|v - u\|^2 = \|x - v + v - u\|^2 = \|x - u\|^2$$

for all $u \in M$. In particular, for $u = u_x$,

$$d(x, M) \leq \|x - v\| \leq \|x - u_x\| = d(x, M)$$

so that $d(x, \mathcal{M}) = \|x - v\|$, and hence $v = u_x$ because $u_x$ is the unique vector in $\mathcal{M}$ for which $d(x, \mathcal{M}) = \|x - u_x\|$. $\qquad\square$

Let $A$ be any nonempty subset of a Hilbert space $\mathcal{H}$. Recall that the subspace $\bigvee A = (\text{span } A)^-$ of $\mathcal{H}$ is the closure in $\mathcal{H}$ of the set of all (finite) linear combinations of vectors in $A$. Take an arbitrary vector in $x$ in $\mathcal{H}$. The vector $u_x$ in $\bigvee A$ that minimizes the distance of $x$ to $\bigvee A$ is called the *best linear approximation* of $x$ in terms of $A$, and the difference $x - u_x$ in $\mathcal{H}$ is called the *error* of the approximation of $x$ in $\mathcal{H}$ by $u_x$ in $\bigvee A$. The next result is a straightforward consequence of Theorem 5.13, which is illustrated in the figure below.

**Corollary 5.14.** *The best linear approximation of any vector $x$ in a Hilbert space $\mathcal{H}$ in terms of a nonempty subset $A$ of $\mathcal{H}$ is the vector $u_x$ in $\bigvee A$ for which the error $x - u_x$ is orthogonal to $\bigvee A$.*

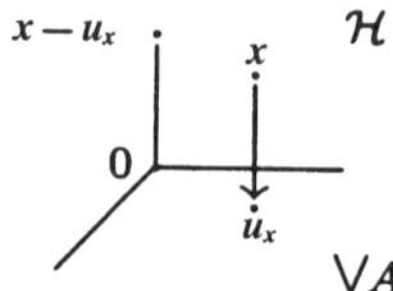


**Proposition 5.15.** *Let $\mathcal{M}$ be a linear manifold of a Hilbert space $\mathcal{H}$.*

$$\mathcal{M}^{\perp\perp} = \mathcal{M}^- \quad \text{and} \quad \mathcal{M}^\perp = \{0\} \text{ if and only if } \mathcal{M}^- = \mathcal{H}.$$

*In particular, if $A$ is any subset of a Hilbert space $\mathcal{H}$, then*

$$A^{\perp\perp} = \bigvee A \quad \text{and} \quad A^\perp = \{0\} \text{ if and only if } \bigvee A = \mathcal{H}.$$

*Proof.* Recall that $\mathcal{M} \subseteq \mathcal{M}^{\perp\perp}$ and $\mathcal{M}^{\perp\perp}$ is closed in $\mathcal{H}$ according to Proposition 5.12. Then

$$\mathcal{M}^- \subseteq \mathcal{M}^{\perp\perp}.$$

Since $\mathcal{M}^{\perp\perp}$ is a subspace (i.e., a closed linear manifold) of a Hilbert space, it follows that it is itself a Hilbert space, and hence $\mathcal{M}^-$ is a subspace of the Hilbert space $\mathcal{M}^{\perp\perp}$. Take an arbitrary $x$ in $\mathcal{M}^{\perp\perp}$. Theorem 5.13 ensures that there exists $u_x \in \mathcal{M}^-$ such that $x - u_x \in (\mathcal{M}^-)^\perp = \mathcal{M}^\perp$ (see Proposition 5.12). But $x - u_x \in \mathcal{M}^{\perp\perp}$ because $u_x \in \mathcal{M}^- \subseteq \mathcal{M}^{\perp\perp}$. Thus $x - u_x \in \mathcal{M}^\perp \cap \mathcal{M}^{\perp\perp}$, and hence $x = u_x \in \mathcal{M}^-$. Conclusion:

$$\mathcal{M}^{\perp\perp} \subseteq \mathcal{M}^-.$$

Therefore $\mathcal{M}^- = \mathcal{M}^{\perp\perp}$. If $\mathcal{M}^\perp = \{0\}$, then $\mathcal{M}^{\perp\perp} = \{0\}^\perp = \mathcal{H}$, and so

$$\mathcal{M}^- = \mathcal{H} \quad \text{whenever} \quad \mathcal{M}^\perp = \{0\}.$$

On the other hand, Proposition 5.12 says that $\mathcal{M}^\perp = \{0\}$ whenever $\mathcal{M}^- = \mathcal{H}$. Finally, if $A$ is any subset of $\mathcal{H}$, then $A^\perp = (\bigvee A)^\perp$ so that $A^{\perp\perp} = (\bigvee A)^{\perp\perp}$

(Proposition 5.12 again). Since $\bigvee A$ is a subspace of $\mathcal{H}$, it follows that $A^{\perp\perp} = \bigvee A$, and $A^{\perp} = \{0\}$ if and only if $\bigvee A = \mathcal{H}$. $\qquad\qquad\square$

It is worth noticing that the inner product space $\mathcal{H}$ in Theorem 5.13 was supposed to be complete only to ensure that the *closed* (convex and nonempty) subset $M$ and the *closed* linear manifold $\mathcal{M}$ are *complete* (see Theorem 3.40(b)). In fact, Theorem 5.13 can be formulated in an inner product space setting by assuming that $M$ is *complete* (as a metric space in the inner-product topology) and $\mathcal{M}$ is a *complete* linear manifold, instead of assuming that $M$ and $\mathcal{M}$ are closed in a inner product space $\mathcal{H}$ that is itself complete. We shall see next that Theorem 5.13 does not hold without the completeness assumption.

**Example 5H.** Let $\mathcal{X}$ be a proper dense linear manifold of a Hilbert space $\mathcal{H}$ so that $\mathcal{X}$ is an inner product space that is not complete (Proposition 4.7). Take $z \in \mathcal{H}\backslash\mathcal{X}$ and set

$$\mathcal{M} = \{z\}^{\perp} \cap \mathcal{X},$$

where $\{z\}^{\perp}$ is the orthogonal complement of $\{z\}$ in $\mathcal{H}$. Since $\{z\}^{\perp}$ is a closed linear manifold of $\mathcal{H}$ (Proposition 5.12), it follows by Problem 3.38(d) that $\{x\}^{\perp} \cap \mathcal{X}$ is closed in $\mathcal{X}$. Thus the intersection $\mathcal{M}$ of the linear manifolds $\{z\}^{\perp}$ and $\mathcal{X}$ of $\mathcal{H}$ is a linear manifold of $\mathcal{X}$ that is closed in $\mathcal{X}$ (i.e., $\mathcal{M}$ is a subspace of $\mathcal{X}$). Moreover, if $\mathcal{M} = \mathcal{X}$, then $\mathcal{X} \subseteq \{z\}^{\perp}$, which implies $\{z\}^{\perp\perp} \subseteq \mathcal{X}^{\perp}$. But $\mathcal{X}^{\perp} = \{0\}$ because $\mathcal{X}^{-} = \mathcal{H}$ (Proposition 5.12 again), and hence $\{z\} \subseteq \{z\}^{\perp\perp} = \{0\}$ so that $z = 0$, which contradicts the fact that $z \notin \mathcal{X}$. Therefore,

$$\mathcal{M} \text{ is a proper subspace of } \mathcal{X}.$$

Take $x \in \mathcal{X}\backslash\mathcal{M}$. Since $x \in \mathcal{H}$, and since $\{z\}^{\perp}$ is a subspace of $\mathcal{H}$, it follows by Theorem 5.13 that there exists a unique $u_x \in \{z\}^{\perp}$ such that $x - u_x \in \{z\}^{\perp\perp} = \bigvee\{z\}$ (see Proposition 5.15), and so $u_x = x + \alpha z$ for some scalar $\alpha$ (recall: $\bigvee\{z\} =$ span $\{z\}$). Hence,

$$u_x \notin \mathcal{X}$$

because $z \notin \mathcal{X}$ and $x \in \mathcal{X}$. Theorem 5.13 also says that $u_x$ is the *unique* vector in $\{z\}^{\perp}$ such that

$$\|x - u_x\| = d(x, \{z\}^{\perp}).$$

If $\mathcal{M}$ is dense in $\{z\}^{\perp}$ (i.e., if $\mathcal{M}^{-} = \{z\}^{\perp}$, where $\mathcal{M}^{-}$ is the closure of $\mathcal{M}$ in $\mathcal{H}$), then $d(x, \mathcal{M}) = d(x, \{z\}^{\perp})$ according to Problem 3.43(b). In this case we may conclude that there is no vector $u$ in $\mathcal{M} = \{z\}^{\perp} \cap \mathcal{X}$ such that $\|x - u\| = d(x, \mathcal{M})$, which shows that Theorem 5.13 does not hold in the incomplete inner product space $\mathcal{X}$. We shall now exhibit a Hilbert space $\mathcal{H}$, a dense linear manifold $\mathcal{X}$ of $\mathcal{H}$, and a vector $z$ in $\mathcal{H}\backslash\mathcal{X}$ for which $\mathcal{M} = \{z\}^{\perp} \cap \mathcal{X}$ is dense in $\{z\}^{\perp}$. Set

$$\mathcal{H} = \ell_+^2 \quad \text{and} \quad \mathcal{X} = \ell_+^0.$$

Recall that $\ell_+^0$ is dense in $\ell_+^2$ (Problem 3.44). Set $z = \{(\frac{1}{2})^k\}_{k=1}^\infty \in \ell_+^2 \backslash \ell_+^0$ and take an arbitrary $y \in \{z\}^\perp$. That is, $y = \{v_k\}_{k=1}^\infty \in \ell_+^2$ is such that

$$\sum_{k=1}^\infty (\tfrac{1}{2})^k v_k = 0.$$

For each integer $n \geq 1$ consider the sequence

$$y_n = \left(v_1, \ldots, v_n, -2^{n+1}\sum_{k=1}^n (\tfrac{1}{2})^k v_k, 0, 0, 0, \ldots, \right)$$

in $\ell_+^0$. It is clear that $\langle y_n ; z \rangle = 0$, and hence $y_n \in \mathcal{M} = \{z\}^\perp \cap \ell_+^0$, for every $n \geq 1$. Moreover,

$$\|y_n - y\|^2 = \left|v_{n+1} + 2^{n+1}\sum_{k=1}^n (\tfrac{1}{2})^k v_k\right|^2 + \sum_{k=n+2}^\infty |v_k|^2$$

for each $n \geq 1$. Since $y \in \{z\}^\perp \subset \ell_+^2$, it follows that $\sum_{k=n+2}^\infty |v_k|^2 \to 0$ as $n \to \infty$ (Problem 3.11) and $\sum_{k=1}^n (\tfrac{1}{2})^k v_k = -\sum_{k=n+1}^\infty (\tfrac{1}{2})^k v_k$ for every $n \geq 1$. Recalling that $\sum_{k=n+1}^\infty (\tfrac{1}{2})^k = (\tfrac{1}{2})^n$ for each $n \geq 0$ we get

$$0 \leq 2^{n+1}\left|\sum_{k=1}^n (\tfrac{1}{2})^k v_k\right| = 2^{n+1}\left|\sum_{k=n+1}^\infty (\tfrac{1}{2})^k v_k\right|$$

$$\leq \left(\sup_{k \geq n+1} |v_k|\right) 2^{n+1} \sum_{k=n+1}^\infty (\tfrac{1}{2})^k$$

$$= \left(\sup_{k \geq n+1} |v_k|\right)\frac{2^{n+1}}{2^n} = 2 \sup_{k \geq n+1} |v_k| \to 0$$

as $n \to \infty$ (for $\lim_n \sup_{k \geq n} |v_k| = \lim \sup_n |v_k| = \lim_n |v_n| = 0$). Thus

$$y_n \to y \quad \text{in} \quad \ell_+^2.$$

Conclusion: For every $y \in \{z\}^\perp$ there exists an $\mathcal{M}$-valued sequence $\{y_n\}_{n=1}^\infty$ that converges in $\ell_+^2$ to $y$, which means that $\mathcal{M}$ is dense in $\{z\}^\perp$ (Proposition 3.32(d)). That is,

$$\mathcal{M}^- = \{z\}^\perp.$$

Note that $\mathcal{M} \neq \{z\}^\perp$ (e.g., the sequence $(\zeta_1 - \|z\|^2(\overline{\zeta}_1)^{-1}, \zeta_2, \zeta_3, \ldots)$ lies in $\{z\}^\perp \backslash \ell_+^0$ for every $z = \{\zeta_k\}_{k=1}^\infty$ in $\ell_+^2 \backslash \ell_+^0$ with $\zeta_1 \neq 0$), which implies that $\mathcal{M}$ is not closed in the Hilbert space $\mathcal{H} = \ell_+^2$, and hence $\mathcal{M}$ is not complete (Corollary 3.41).

The above example shows that completeness of $\mathcal{H}$ was crucial in the proof of Theorem 5.13. But it is not enough. Indeed, Theorem 5.13 does not necessarily hold in a Banach space that is not Hilbert (i.e., in a Banach space whose norm does not satisfy the parallelogram law). This is closely related to Lemma 4.33. In fact, what is behind that lemma is that there may exist a proper subspace $\mathcal{M}$ of a Banach space $\mathcal{X}$ for which $d(x, \mathcal{M}) < 1$ for all $x \in \mathcal{X}$ with $\|x\| = 1$ or, equivalently, $d(x, \mathcal{M}) < \|x\|$ for every nonzero $x$ in $\mathcal{X}$. Suppose this is the case, and take an arbitrary $x \in \mathcal{X} \backslash \mathcal{M}$. Since $d(x, \mathcal{M}) = d(x - u, \mathcal{M})$ for every $x \in \mathcal{X}$ and every $u \in \mathcal{M}$, it follows that $d(x, \mathcal{M}) < \|x - u\|$ whenever $u$ lies in $\mathcal{M}$ and $x$ lies in $\mathcal{X} \backslash \mathcal{M}$. Therefore, if $x$ is a vector in $\mathcal{X} \backslash \mathcal{M}$, then there is no vector $u$ in $\mathcal{M}$ for which $\|x - u\| = d(x, \mathcal{M})$.

## 5.5   Orthogonal Structure

Let $\{\mathcal{M}_\gamma\}_{\gamma \in \Gamma}$ be an arbitrary nonempty indexed family of subspaces of a Hilbert space $\mathcal{H}$ and consider their topological sum

$$\left(\sum_{\gamma \in \Gamma} \mathcal{M}_\gamma\right)^- = \left[\operatorname{span}\left(\bigcup_{\gamma \in \Gamma} \mathcal{M}_\gamma\right)\right]^- = \bigvee\left(\bigcup_{\gamma \in \Gamma} \mathcal{M}_\gamma\right) = \bigvee_{\gamma \in \Gamma} \mathcal{M}_\gamma.$$

If $\{\mathcal{M}_i\}_{i=1}^n$ is a *finite* (nonempty) family of pairwise *orthogonal* (that is, $\mathcal{M}_i \perp \mathcal{M}_j$ whenever $i \neq j$) *subspaces* of a *Hilbert* space, then Corollary 5.11 says that $\sum_{i=1}^n \mathcal{M}_i = \left(\sum_{i=1}^n \mathcal{M}_i\right)^-$. That is, *the topological sum and the ordinary* (algebraic) *sum of a finite family of pairwise orthogonal subspaces of a Hilbert space coincide.* The next theorem is a countably infinite counterpart of the above italicized result, which emerges as an important consequence of Theorem 5.13.

**Theorem 5.16.** (The Orthogonal Structure Theorem). *Let $\mathcal{H}$ be a Hilbert space and suppose $\{\mathcal{M}_k\}_{k \in \mathbb{N}}$ is a countably infinite family of pairwise orthogonal subspaces of $\mathcal{H}$. If $x \in \left(\sum_{k \in \mathbb{N}} \mathcal{M}_k\right)^-$, then there exists a unique $\mathcal{H}$-valued sequence $\{u_k\}_{k=1}^\infty$ with $u_k \in \mathcal{M}_k$ for each $k$ such that*

$$x = \sum_{k=1}^\infty u_k.$$

*Moreover, $\|x\|^2 = \sum_{k=1}^\infty \|u_k\|^2$. Conversely, if $\{u_k\}_{k=1}^\infty$ is an $\mathcal{H}$-valued sequence such that $u_k \in \mathcal{M}_k$ for each $k$ and $\sum_{k=1}^\infty \|u_k\|^2 < \infty$, then the infinite series $\sum_{k=1}^\infty u_k$ converges in $\mathcal{H}$ and $\sum_{k=1}^\infty u_k \in \left(\sum_{k \in \mathbb{N}} \mathcal{M}_k\right)^-$.*

*Proof.* If $x \in \left(\sum_{k \in \mathbb{N}} \mathcal{M}_k\right)^-$, then there exists a sequence $\{x_{(n)}\}_{n=1}^\infty$ of vectors in $\sum_{k \in \mathbb{N}} \mathcal{M}_k$ that converges to $x$ in $\mathcal{H}$ (Proposition 3.27). Take an arbitrary integer $n \geq 1$. Since $x_{(n)} \in \sum_{k \in \mathbb{N}} \mathcal{M}_k$, it follows that $x_{(n)}$ is a *finite* sum where each summand lies in one of the subspaces $\mathcal{M}_k$. Thus $x_{(n)}$ can be written as

$$x_{(n)} = \sum_{k=1}^{m_n} x_{(n)k} \in \sum_{k=1}^{m_n} \mathcal{M}_k$$

with $x(n)_k \in \mathcal{M}_k$ for each $k = 1, \dots, m_n$, where $m_n$ is an integer that depends on $n$. Clearly, we may take $m_{n+1} \geq m_n \geq n$. Observe that the above finite sum may contain finitely many zero summands, and $\{\mathcal{M}_k\}_{k=1}^{m_n} \subseteq \{\mathcal{M}_k\}_{k=1}^{m_{n+1}}$.

*Existence*. Note that $\sum_{k=1}^{m_n} \mathcal{M}_k$ is a subspace of the Hilbert space $\mathcal{H}$ (Corollary 5.10). According to Theorem 5.13 there exists a unique vector

$$u_x(n) = \sum_{k=1}^{m_n} u_k \in \sum_{k=1}^{m_n} \mathcal{M}_k,$$

with each $u_k$ in $\mathcal{M}_k$, such that $\|x - u_x(n)\| \leq \|x - u\|$ for all vectors $u$ in $\sum_{k=1}^{m_n} \mathcal{M}_k$ and $x - u_x(n) \perp \sum_{k=1}^{m_n} \mathcal{M}_k$. In particular,

$$\left\| x - \sum_{k=1}^{m_n} u_k \right\| \leq \|x - x(n)\|.$$

*Claim.* The vectors $u_k$ in the expansion of $u_x(n)$ do not depend on $n$.

*Proof.* Since $\sum_{k=1}^{m_{n+1}} \mathcal{M}_k = \sum_{k=1}^{m_n} \mathcal{M}_k + \sum_{k=m_n+1}^{m_{n+1}} \mathcal{M}_k$ is a subspace of $\mathcal{H}$, it follows by Theorem 5.13 that there exists a unique

$$u_x(n+1) = v + w \in \sum_{k=1}^{m_{n+1}} \mathcal{M}_k,$$

with $v \in \sum_{k=1}^{m_n} \mathcal{M}_k$ and $w \in \sum_{k=m_n+1}^{m_{n+1}} \mathcal{M}_k$, such that $\|x - u_x(n+1)\| \leq \|x - u\|$ for all vectors $u$ in $\sum_{k=1}^{m_{n+1}} \mathcal{M}_k$ and $x - u_x(n+1) \perp \sum_{k=1}^{m_{n+1}} \mathcal{M}_k$. Take an arbitrary $z \in \sum_{k=1}^{m_n} \mathcal{M}_k \subseteq \sum_{k=1}^{m_{n+1}} \mathcal{M}_k$ and note that

$$0 = \langle x - u_x(n+1) ; z \rangle = \langle x - v - w ; z \rangle = \langle x - v ; z \rangle - \langle w ; z \rangle = \langle x - v ; z \rangle$$

(for $\mathcal{M}_j \perp \mathcal{M}_k$ whenever $j \neq k$ so that $\sum_{k=m_n+1}^{m_{n+1}} \mathcal{M}_k \perp \sum_{k=1}^{m_n} \mathcal{M}_k$, and hence $\langle w ; z \rangle = 0$). Thus $x - v \perp \sum_{k=1}^{m_n} \mathcal{M}_k$. But $u_x(n)$ is the only vector in $\sum_{k=1}^{m_n} \mathcal{M}_k$ for which $x - u_x(n) \perp \sum_{k=1}^{m_n} \mathcal{M}_k$. Therefore $v = u_x(n)$. Outcome: $u_x(n+1) = u_x(n) + w = \sum_{k=1}^{m_{n+1}} u_k$ with $u_k \in \mathcal{M}_k$. $\square$

Set $\{u_k\}_{k=1}^{\infty} = \bigcup_{n \geq 1} \{u_k\}_{k=1}^{n}$, which is a sequence of pairwise orthogonal vectors in $\mathcal{H}$ because $\{\mathcal{M}_k\}_{k \in \mathbb{N}}$ is an orthogonal family. Since $n \leq m_n$, it follows by Proposition 5.8 that

$$\sum_{k=1}^{n} \|u_k\|^2 \leq \sum_{k=1}^{m_n} \|u_k\|^2 = \left\| \sum_{k=1}^{m_n} u_k \right\|^2 \leq (\|x - x(n)\| + \|x\|)^2.$$

But $x(n) \to x$ in $\mathcal{H}$ as $n \to \infty$ so that $\sum_{k=1}^{\infty} \|u_k\|^2 \leq \|x\|^2$. Hence (cf. Corollary 5.9) the infinite series $\sum_{k=1}^{\infty} u_k$ converges in the Hilbert space $\mathcal{H}$ and $\|\sum_{k=1}^{\infty} u_k\|^2 =$

$\sum_{k=1}^{\infty} \|u_k\|^2$. Moreover, it in fact converges to $x$ because the sequence of partial sums $\{\sum_{k=1}^{n} u_k\}_{n=1}^{\infty}$ has a subsequence, namely $\{\sum_{k=1}^{m_n} u_k\}_{n=1}^{\infty}$, that converges to $x$ (see Proposition 3.5), and so $\|x\|^2 = \sum_{k=1}^{\infty} \|u_k\|^2$.

*Uniqueness.* Suppose $x = \sum_{k=1}^{\infty} u_k = \sum_{k=1}^{\infty} v_k$, where $u_k$ and $v_k$ lie in $\mathcal{M}_k$ for each $k$. Thus $\sum_{k=1}^{\infty} (u_k - v_k) = 0$ (see Problem 4.9(a)). But $\{(u_k - v_k)\}_{k=1}^{\infty}$ is a sequence of pairwise orthogonal vectors in the inner product space $\mathcal{H}$ (because $u_k - v_k \in \mathcal{M}_k$ for each $k$ and $\{\mathcal{M}_k\}_{k\in\mathbb{N}}$ is an orthogonal family) so that $\sum_{k=1}^{\infty} \|u_k - v_k\|^2 = \|\sum_{k=1}^{\infty} (u_k - v_k)\|^2 = 0$ by Corollary 5.9(a). Hence $u_k = v_k$ for every $k \geq 1$.

*Converse.* If $\{u_k\}_{k=1}^{\infty}$ is an $\mathcal{H}$-valued sequence with $u_k \in \mathcal{M}_k$ for each $k$, then it is a sequence of pairwise orthogonal vectors in $\mathcal{H}$ (since $\mathcal{M}_j \perp \mathcal{M}_k$ whenever $j \neq k$). If $\sum_{k=1}^{\infty} \|u_k\|^2 < \infty$, then the infinite series $\sum_{k=1}^{\infty} u_k$ converges in $\mathcal{H}$ according to Corollary 5.9(b); that is, $\sum_{k=1}^{n} u_k \to \sum_{k=1}^{\infty} u_k$ in $\mathcal{H}$ as $n \to \infty$. Since $\sum_{k=1}^{n} u_k$ lies in $\sum_{k\in\mathbb{N}} \mathcal{M}_k$ for each integer $n \geq 1$, it follows by Proposition 3.27 that the limit $\sum_{k=1}^{\infty} u_k$ lies in $(\sum_{k\in\mathbb{N}} \mathcal{M}_k)^-$.  $\square$

Two immediate consequences of the Orthogonal Structure Theorem:

**Corollary 5.17.** *If $\{\mathcal{M}_k\}_{k\in\mathbb{N}}$ is a countably infinite family of pairwise orthogonal subspaces of a Hilbert space $\mathcal{H}$, then*

$$\left(\sum_{k\in\mathbb{N}} \mathcal{M}_k\right)^- = \left\{\sum_{k=1}^{\infty} u_k \in \mathcal{H}: \ u_k \in \mathcal{M}_k \text{ and } \sum_{k=1}^{\infty} \|u_k\|^2 < \infty\right\}.$$

**Corollary 5.18.** *Let $\{\mathcal{M}_k\}_{k\in\mathbb{N}}$ be a countably infinite orthogonal family of subspaces of a Hilbert space $\mathcal{H}$. If it spans $\mathcal{H}$; that is, if*

$$\bigvee_{k\in\mathbb{N}} \mathcal{M}_k = \mathcal{H},$$

*then every vector $x$ in $\mathcal{H}$ is uniquely expressed as*

$$x = \sum_{k=1}^{\infty} u_k$$

*in terms of an orthogonal sequence $\{u_k\}_{k=1}^{\infty}$ with each $u_k$ in each $\mathcal{M}_k$.*

**Example 5I.** Let $\{\mathcal{M}_k\}$ be a countable collection of subspaces of a Hilbert space $(\mathcal{H}, \langle\ ;\ \rangle)$ so that each $(\mathcal{M}_k, \langle\ ;\ \rangle)$ is itself a Hilbert space. If $\{\mathcal{M}_k\}$ is an infinite collection, then let it be indexed by $\mathbb{N}$. Consider the direct sum $\bigoplus_k \mathcal{M}_k$. Recall that $\bigoplus_k \mathcal{M}_k$ is itself a linear space (but it is not a subset of $\mathcal{H}$). Let $[\bigoplus_k \mathcal{M}_k]_2$ denote the linear manifold of the full direct sum $\bigoplus_{k=1}^{\infty} \mathcal{M}_k$ made up of all square-summable sequences $\{x_k\}$ in $\bigoplus_{k=1}^{\infty} \mathcal{M}_k$. That is,

$$\left[\bigoplus_k \mathcal{M}_k\right]_2 = \left\{\{x_k\} \in \bigoplus_k \mathcal{M}_k: \ \sum_k \|x_k\|^2 < \infty\right\},$$

where $\| \ \|$ is the norm induced on $\mathcal{H}$ by the inner product $\langle \ ; \ \rangle$. (It is clear that $\left[\bigoplus_k \mathcal{M}_k\right]_2$ coincides with $\bigoplus_k \mathcal{M}_k$ if the collection $\{\mathcal{M}_k\}$ is finite.) The function $\langle \ ; \ \rangle_\oplus : \left[\bigoplus_k \mathcal{M}_k\right]_2 \times \left[\bigoplus_k \mathcal{M}_k\right]_2 \to \mathbb{F}$, given by

$$\langle x \, ; y \rangle_\oplus \ = \ \sum_k \langle x_k \, ; y_k \rangle$$

for every $x = \{x_k\} \in \left[\bigoplus_k \mathcal{M}_k\right]_2$ and $y = \{y_k\} \in \left[\bigoplus_k \mathcal{M}_k\right]_2$, is an inner product on $\left[\bigoplus_k \mathcal{M}_k\right]_2$ that makes it into a Hilbert space (Examples 5E and 5F). If the collection $\{\mathcal{M}_k\}$ consists of pairwise *orthogonal* subspaces of $\mathcal{H}$, then the Hilbert space $\left(\left[\bigoplus_k \mathcal{M}_k\right]_2, \langle \ ; \ \rangle_\oplus\right)$ is referred to as an (internal) *orthogonal direct sum*. In this case (i.e., if $\{\mathcal{M}_k\}$ is a collection of orthogonal subspaces), then Corollary 5.17 (or Corollary 5.11, if the collection is finite) says that

$$\left(\textstyle\sum_k \mathcal{M}_k\right)^- \ = \ \left\{\textstyle\sum_k x_k \in \mathcal{H} : \ \{x_k\} \in \left[\bigoplus_k \mathcal{M}_k\right]_2\right\},$$

and

$$\left[\bigoplus_k \mathcal{M}_k\right]_2 \ = \ \left\{\{x_k\} \in \bigoplus_k \mathcal{M}_k : \ \textstyle\sum_k x_k \in \left(\sum_k \mathcal{M}_k\right)^-\right\}.$$

This establishes a natural mapping $\Phi : \left[\bigoplus_k \mathcal{M}_k\right]_2 \to \left(\sum_k \mathcal{M}_k\right)^-$,

$$\Phi(\{x_k\}) \ = \ \sum_k x_k$$

for every $\{x_k\} \in \left[\bigoplus_k \mathcal{M}_k\right]_2$, which is clearly injective and surjective (by the Orthogonal Structure Theorem) and linear as well (addition and scalar multiplication of summable sequences is again a summable sequence — see Problem 4.9). Conclusion: $\Phi$ is an isomorphism of the linear space $\left[\bigoplus_k \mathcal{M}_k\right]_2$ onto the linear manifold $\left(\sum_k \mathcal{M}_k\right)^-$ of the linear space $\mathcal{H}$. Therefore, the orthogonal direct sum $\left[\bigoplus_k \mathcal{M}_k\right]_2$ and the topological sum $\left(\sum_k \mathcal{M}_k\right)^-$ of pairwise orthogonal subspaces of a Hilbert space are isomorphic linear spaces. They are, in fact, isometrically isomorphic Hilbert spaces: the natural mapping $\Phi$ actually is an isometric isomorphism (next section), which provides a natural identification between them. It is customary to use the same notation of the full direct sum to denote $\left[\bigoplus_k \mathcal{M}_k\right]_2$ when it is equipped with the above inner product $\langle \ ; \ \rangle_\oplus$.

*Notation:*
$$\bigoplus_k \mathcal{M}_k \ = \ \left(\left[\bigoplus_k \mathcal{M}_k\right]_2, \langle \ ; \ \rangle_\oplus\right).$$

We shall follow the common usage. From now on $\bigoplus_k \mathcal{M}_k$ will denote the *Hilbert space* whose linear space is the set of all *square-summable* sequences in the full direct sum of a collection $\{\mathcal{M}_k\}$ of pairwise *orthogonal subspaces* of a *Hilbert space* $\mathcal{H}$.

According to Problem 4.34 a pair of subspaces $\mathcal{M}$ and $\mathcal{N}$ of an inner product space $\mathcal{X}$ are *complementary* in $\mathcal{X}$ if they are algebraic complements of each other (i.e., $\mathcal{M} + \mathcal{N} = \mathcal{X}$ and $\mathcal{M} \cap \mathcal{N} = \{0\}$). Recall that $\mathcal{M} \perp \mathcal{N}$ implies $\mathcal{M} \cap \mathcal{N} = \{0\}$.

**Proposition 5.19.** *Orthogonal complementary subspaces in an inner product space are orthogonal complements of each other.*

*Proof.* Let $\mathcal{M}$ and $\mathcal{N}$ be orthogonal complementary subspaces in an inner product space $\mathcal{X}$. Take an arbitrary $x$ in $\mathcal{M}^\perp \subseteq \mathcal{X}$. Since $\mathcal{M} + \mathcal{N} = \mathcal{X}$, it follows that $x = u + v$ with $u$ in $\mathcal{M}$ and $v$ is $\mathcal{N}$, and so

$$\langle x \,;\, u \rangle \;=\; \langle u \,;\, u \rangle + \langle v \,;\, u \rangle.$$

But $\langle x \,;\, u \rangle = \langle v \,;\, u \rangle = 0$ (for $\mathcal{M} \perp \mathcal{M}^\perp$ and $\mathcal{M} \perp \mathcal{N}$). Hence $\|u\|^2 = 0$, which means that $u = 0$. Therefore $x = v \in \mathcal{N}$. Conclusion: $\mathcal{M}^\perp \subseteq \mathcal{N}$. But $\mathcal{N} \subseteq \mathcal{M}^\perp$ because $\mathcal{M} \perp \mathcal{N}$. Outcome: $\mathcal{M}^\perp = \mathcal{N}$. $\qquad\square$

The next result is the central theorem of Hilbert space geometry.

**Theorem 5.20.** (Projection Theorem – First version). *Every Hilbert space $\mathcal{H}$ can be decomposed as*

$$\mathcal{H} = \mathcal{M} + \mathcal{M}^\perp$$

*where $\mathcal{M}$ is any subspace of $\mathcal{H}$.*

*Proof.* Let $\mathcal{M}$ be an arbitrary subspace of a Hilbert space $\mathcal{H}$. Since $\mathcal{M}^\perp$ is a subspace of $\mathcal{H}$ (Proposition 5.12) which is orthogonal to $\mathcal{M}$ (by its very definition), it follows by Theorem 5.10 that $\mathcal{M} + \mathcal{M}^\perp$ is a subspace of $\mathcal{H}$. Moreover, $\mathcal{M} \subseteq \mathcal{M} + \mathcal{M}^\perp$ and $\mathcal{M}^\perp \subseteq \mathcal{M} + \mathcal{M}^\perp$. Thus $(\mathcal{M} + \mathcal{M}^\perp)^\perp \subseteq \mathcal{M}^\perp \cap \mathcal{M}^{\perp\perp} = \mathcal{M}^\perp \cap \mathcal{M}^- = \mathcal{M}^\perp \cap \mathcal{M} = 0$, and hence $\mathcal{M} + \mathcal{M}^\perp = (\mathcal{M} + \mathcal{M}^\perp)^- = \mathcal{H}$ (Proposition 5.15). $\quad\square$

Let $\mathcal{M}$ be an arbitrary subspace of a Hilbert space $\mathcal{H}$. Since $\mathcal{M}^\perp$ is again a subspace of $\mathcal{H}$, and since $\mathcal{M} \cap \mathcal{M}^\perp = \{0\}$, what Theorem 5.20 says is: *If $\mathcal{M}$ is any subspace of a Hilbert space $\mathcal{H}$, then $\mathcal{M}$ and $\mathcal{M}^\perp$ are complementary subspaces of $\mathcal{H}$.* This in fact is the converse of Proposition 5.19 in a Hilbert space setting. Moreover, according to Theorem 2.14, *for each $x \in \mathcal{H} = \mathcal{M} + \mathcal{M}^\perp$ there exists a unique $u$ in $\mathcal{M}$ and a unique $v$ in $\mathcal{M}^\perp$ such that $x = u + v$* and, by the Pythagorean Theorem, $\|x\|^2 = \|u\|^2 + \|v\|^2$.

## 5.6 Unitary Equivalence

Recall that an *isometry* between metric spaces is a map that preserves distance, which is obviously continuous. Proposition 4.37 placed linear isometries in a normed-space setting. The next one places it in an inner-product-space setting.

**Proposition 5.21.** *Let $\mathcal{X}$ and $\mathcal{Y}$ be inner product spaces. A linear transformation $V \in \mathcal{L}[\mathcal{X}, \mathcal{Y}]$ is an isometry if and only if*

$$\langle V x_1 \,;\, V x_2 \rangle \;=\; \langle x_1 \,;\, x_2 \rangle$$

*for every $x_1, x_2 \in \mathcal{X}$.*

*Proof.* First note that the inner product on the left-hand side is the inner product on $\mathcal{Y}$, and that on the right-hand side is the inner product on $\mathcal{X}$. If the above identity holds, then $\|Vx\|^2 = \langle Vx \, ; Vx \rangle = \langle x \, ; x \rangle = \|x\|^2$ for every $x \in \mathcal{X}$. Conversely, if $\|Vx\| = \|x\|$ for every $x \in \mathcal{X}$, then $\langle Vx_1 \, ; Vx_2 \rangle = \langle x_1 \, ; x_2 \rangle$ for every $x_1, x_2 \in \mathcal{X}$ by the polarization identity of Proposition 5.4. Therefore, $\langle Vx_1 \, ; Vx_2 \rangle = \langle x_1 \, ; x_2 \rangle$ for every $x_1, x_2 \in \mathcal{X}$ if and only if $\|Vx\| = \|x\|$ for every $x \in \mathcal{X}$. Equivalently, if and only if $V \in \mathcal{L}[\mathcal{X}, \mathcal{Y}]$ is an isometry (Proposition 4.37).    $\Box$

In other words, *a linear isometry between inner product spaces is a linear transformation that preserves the inner product.* Now recall that an *isomorphism* is an invertible linear transformation between linear spaces. These concepts (isometry and isomorphism) were combined in Section 4.7 to yield the notion of an *isometric isomorphism* between normed spaces, which is precisely a *linear surjective isometry*. Between inner product spaces it has a name of its own: an isometric isomorphism between inner product spaces is called a *unitary transformation*. That is, a unitary transformation of an inner product space *onto* an inner product space is a linear surjective isometry or, equivalently, an invertible linear isometry. According to Proposition 5.21, a *unitary transformation between inner product spaces is a linear-space isomorphism that preserves the inner product.* Thus a unitary transformation preserves the algebraic structure, the topological structure, and also the geometric structure between inner product spaces. In particular, it preserves convergence and Cauchy sequences (see Section 4.6), and so it also preserves separability and completeness (cf. Problem 3.48(b) and Theorem 3.44). Two inner product spaces, say $\mathcal{X}_1$ and $\mathcal{X}_2$, are called *unitarily equivalent* if there exists a unitary transformation between them (equivalently, if they are *isometrically isomorphic* — Notation: $\mathcal{X}_1 \cong \mathcal{X}_2$). Unitarily equivalent inner product spaces are regarded as essentially the same inner product space (i.e., they are indistinguishable, except perhaps by the nature of their points).

The continuous linear extension results of Section 4.7 (viz., Theorem 4.35 and Corollaries 4.36 and 4.38) are trivially extended to inner product spaces (simply replace "normed space", "Banach space" and "isometric isomorphism" with "inner product space", "Hilbert space" and "unitary transformation", respectively). The completion results of Section 4.7 are (almost) immediately translated into the inner product space language by just recalling that in an inner-product-space setting two spaces are unitarily equivalent if and only if they are isometrically isomorphic.

**Definition 5.22.** If the image of a linear isometry on an inner product space $\mathcal{X}$ is a dense linear manifold of a Hilbert $\mathcal{H}$, then $\mathcal{H}$ is a *completion* of $\mathcal{X}$. Equivalently, if an inner product space $\mathcal{X}$ is unitarily equivalent to a dense linear manifold of a Hilbert space $\mathcal{H}$, then $\mathcal{H}$ is a completion of $\mathcal{X}$.

**Theorem 5.23.** *Every inner product space has a completion. Any two completions of an inner product space are unitarily equivalent. If $\mathcal{H}$ and $\mathcal{K}$ are completions of* *two inner product spaces $\mathcal{X}$ and $\mathcal{Y}$, respectively, then every operator $T \in \mathcal{B}[\mathcal{X}, \mathcal{Y}]$ has an extension $\widehat{T} \in \mathcal{B}[\mathcal{H}, \mathcal{K}]$ over the completion $\mathcal{H}$ of $\mathcal{X}$ into the completion $\mathcal{K}$ of $\mathcal{Y}$. Moreover, $\widehat{T}$ is unique up to unitary transformations and $\|\widehat{T}\| = \|T\|$.*

*Proof.* According to Theorem 4.40, every inner product space has a completion (as a normed space). The only question that has not been answered in Theorems 4.40 to 4.42 is whether this completion (which certainly is a Banach space) is a Hilbert space. In other words, whether the norm $\| \ \|_{\widehat{X}}$ in the proof of Theorem 4.40 satisfies the parallelogram law whenever the norm $\| \ \|_X$ does. Consider the setup of the proof of Theorem 4.40 and suppose $\mathcal{X}$ is an inner product space. Recall that

$$\|[x]\|_{\widehat{X}} = \lim \|x_n\|_X,$$

where $x = \{x_n\}$ is an arbitrary element of an arbitrary coset $[x]$ in the quotient space $\widehat{\mathcal{X}}$. Take any pair of cosets $[x]$ and $[y]$ in $\widehat{\mathcal{X}}$. Since the norm $\| \ \|_X$ on the inner product space $\mathcal{X}$ satisfies the parallelogram law,

$$
\begin{aligned}
\|[x] + [y]\|_{\widehat{X}}^2 + \|[x] - [y]\|_{\widehat{X}}^2 &= \|[x + y]\|_{\widehat{X}}^2 + \|[x - y]\|_{\widehat{X}}^2 \\
&= \lim \|x_n + y_n\|_X^2 + \lim \|x_n - y_n\|_X^2 \\
&= \lim \left( \|x_n + y_n\|_X^2 + \|x_n - y_n\|_X^2 \right) \\
&= \lim 2\left( \|x_n\|_X^2 + \|y_n\|_X^2 \right) \\
&= 2\left( \lim \|x_n\|_X^2 + \lim \|y_n\|_X^2 \right) \\
&= 2\left( \|[x]\|_{\widehat{X}}^2 + \|[y]\|_{\widehat{X}}^2 \right)
\end{aligned}
$$

by continuity (apply Corollary 3.8 recalling that squaring, addition and scalar multiplication are continuous). Therefore, the norm $\| \ \|_{\widehat{X}}$ on $\widehat{\mathcal{X}}$ also satisfies the parallelogram law so that this Banach space is, in fact, a Hilbert space. $\qquad\square$

We shall now return to orthogonal complements. Let $\mathcal{M}$ and $\mathcal{N}$ be linear manifolds of an inner product space $\mathcal{X}$. If $\mathcal{M}$ and $\mathcal{N}$ are algebraic complements of each other (i.e., if $\mathcal{M} + \mathcal{N} = \mathcal{X}$ and $\mathcal{M} \cap \mathcal{N} - \{0\}$), then they are said to be *complementary linear manifolds* in $\mathcal{X}$. Consider the *natural mapping* $\Phi$ of the linear space $\mathcal{M} \oplus \mathcal{N}$ into the linear space $\mathcal{M} + \mathcal{N}$, which was defined in Section 2.8 by the formula

$$\Phi((u, v)) = u + v$$

for each $(u, v) \in \mathcal{M} \oplus \mathcal{N}$. Let $\langle \ ; \ \rangle$ be the inner product on $\mathcal{X}$ and equip the direct sum $\mathcal{M} \oplus \mathcal{N}$ with the inner product $\langle \ ; \ \rangle_\oplus$, viz.,

$$\langle (u_1, v_1) ; (u_2 ; v_2) \rangle_\oplus = \langle u_1 ; u_2 \rangle + \langle v_1 ; v_2 \rangle$$

for every $(u_1, v_1)$ and $(u_2, v_2)$ in $\mathcal{M} \oplus \mathcal{N}$, as in Example 5E.

**Proposition 5.24.** *If $\mathcal{M}$ and $\mathcal{N}$ are orthogonal complementary linear manifolds in an inner product space $\mathcal{X}$, then the natural mapping $\Phi: \mathcal{M} \oplus \mathcal{N} \to \mathcal{M} + \mathcal{N} = \mathcal{X}$*

*is a unitary transformation so that $\mathcal{M} \oplus \mathcal{N}$ and $\mathcal{X}$ are unitarily equivalent.* (i.e., $\mathcal{X} \cong \mathcal{M} \oplus \mathcal{N}$).

*Proof.* Since $\mathcal{M}$ and $\mathcal{N}$ are complementary linear manifolds in $\mathcal{X}$, it follows by Theorem 2.14 that $\Phi$ is an isomorphism of the linear space $\mathcal{M} \oplus \mathcal{N}$ onto the linear space $\mathcal{M} + \mathcal{N} = \mathcal{X}$. If $\mathcal{M} \perp \mathcal{N}$, then

$$
\begin{aligned}
\langle \Phi((u_1, v_1)) \,;\, \Phi((u_2, v_2)) \rangle &= \langle u_1 + v_1 \,;\, u_2 + v_2 \rangle \\
&= \langle u_1 \,;\, u_2 \rangle + \langle v_1 \,;\, v_2 \rangle \\
&= \langle (u_1, v_1) \,;\, (u_2 \,;\, v_2) \rangle_{\oplus}
\end{aligned}
$$

for every $(u_1, v_1)$ and $(u_2, v_2)$ in $\mathcal{M} \oplus \mathcal{N}$. Thus the natural mapping $\Phi$ is a linear-space isomorphism that preserves the inner product; that is, $\Phi$ is a unitary transformation. $\qquad\square$

In light of Proposition 5.24 we may identify the inner product space $\mathcal{X}$ with the orthogonal direct sum $\mathcal{M} \oplus \mathcal{N}$ (equipped with the above inner product $\langle \;;\; \rangle_{\oplus}$) through the natural mapping $\Phi$, whenever $\mathcal{M}$ and $\mathcal{N}$ are orthogonal complementary linear manifolds in $\mathcal{X}$. Next suppose $\mathcal{M}$ and $\mathcal{N}$ are *orthogonal complementary subspaces* in a *Hilbert space* $\mathcal{H}$. That is, suppose $\mathcal{M}$ and $\mathcal{N}$ are closed linear manifolds of a Hilbert space $\mathcal{H}$ such that $\mathcal{M} + \mathcal{N} = \mathcal{H}$ and $\mathcal{M} \perp \mathcal{N}$ (and so $\mathcal{M} \cap \mathcal{N} = \{0\}$). According to Proposition 5.19 $\mathcal{N} = \mathcal{M}^{\perp}$ so that $\mathcal{M} + \mathcal{M}^{\perp} = \mathcal{H}$, and hence $\mathcal{M} \oplus \mathcal{M}^{\perp} \cong \mathcal{H}$ by Proposition 5.24. In this case, it is usual to identify the orthogonal direct sum $\mathcal{M} \oplus \mathcal{M}^{\perp}$ with its unitarily equivalent image, $\Phi(\mathcal{M} \oplus \mathcal{M}^{\perp}) = \mathcal{M} + \mathcal{M}^{\perp} = \mathcal{H}$, and write $\mathcal{M} \oplus \mathcal{M}^{\perp} = \mathcal{H}$. Therefore, under such an identification, the central Theorem 5.20 can be restated as follows.

**Theorem 5.25.** (Projection Theorem – Second version). *Every Hilbert space $\mathcal{H}$ has an orthogonal direct sum decomposition*

$$
\mathcal{H} = \mathcal{M} \oplus \mathcal{M}^{\perp}
$$

*where $\mathcal{M}$ is any subspace of $\mathcal{H}$.*

This leads to a useful notation for the orthogonal complement of a *subspace* $\mathcal{M}$ of a *Hilbert space* $\mathcal{H}$, namely,

$$
\mathcal{M}^{\perp} = \mathcal{H} \ominus \mathcal{M}.
$$

Observe that both $\mathcal{M}$ and $\mathcal{M}^{\perp}$ are themselves Hilbert spaces so that each of them can be further decomposed as direct sums of orthogonal complementary subspaces.

**Example 5J.** Let $\{\mathcal{M}_k\}$ be a countable collection of orthogonal subspaces of a Hilbert space $(\mathcal{H}, \langle \;;\; \rangle)$. If $\{\mathcal{M}_k\}$ is countably infinite, then assume that it is indexed by $\mathbb{N}$. Consider the natural isomorphism $\Phi$ of $\bigoplus_k \mathcal{M}_k$ onto $\left(\sum_k \mathcal{M}_k\right)^{-}$ that was defined in Example 5I:

$$
\Phi(\{x_k\}) = \sum_k x_k \quad \text{for every } \{x_k\} \in \bigoplus_k \mathcal{M}_k.
$$

Recall that $\bigoplus_k \mathcal{M}_k$ denotes the Hilbert space of all square-summable sequences in the full direct sum of $\{\mathcal{M}_k\}$, and that the inner product on $\bigoplus_k \mathcal{M}_k$ is given by

$$\langle \{x_k\} ; \{y_k\}\rangle_{\oplus} = \sum_k \langle x_k ; y_k \rangle$$

for every $\{x_k\}$ and $\{y_k\}$ in $\bigoplus_k \mathcal{M}_k$ (see Example 5I). Since $\mathcal{M}_j \perp \mathcal{M}_k$ whenever $j \neq k$, and since the inner product is continuous (in both arguments — Problem 5.6), it follows that

$$\langle \Phi(\{x_k\}) ; \Phi(\{y_k\}) \rangle = \Big\langle \sum_j x_j ; \sum_k y_k \Big\rangle = \sum_j \sum_k \langle x_j ; y_k \rangle$$

$$= \sum_k \langle x_k ; y_k \rangle = \langle \{x_k\} ; \{y_k\} \rangle_{\oplus}$$

for every pair of square-summable sequences $\{x_k\}$ and $\{y_k\}$ in $\bigoplus_k \mathcal{M}_k$. Thus the linear-space isomorphism $\Phi$ preserves the inner product, which means that it is a unitary transformation. Conclusion: The orthogonal direct sum $\bigoplus_k \mathcal{M}_k$ and the topological sum $\left(\sum_k \mathcal{M}_k\right)^-$ of pairwise orthogonal subspaces of a Hilbert space are unitarily equivalent Hilbert spaces. That is,

$$\bigoplus_k \mathcal{M}_k \cong \left(\sum_k \mathcal{M}_k\right)^-$$

whenever $\{\mathcal{M}_k\}$ is a collection of pairwise orthogonal subspaces of a Hilbert space. This, in fact, is a restatement of the Orthogonal Structure Theorem (Theorem 5.16). (Recall: for a finite collection, $\left(\sum_k \mathcal{M}_k\right)^- = \sum_k \mathcal{M}_k$ by Corollary 5.11 so that $\bigoplus_k \mathcal{M}_k \cong \sum_k \mathcal{M}_k$.) If, in addition, the orthogonal collection $\{\mathcal{M}_k\}$ spans $\mathcal{H}$ (i.e., if $\bigvee_k \mathcal{M}_k = \mathcal{H}$), then it is usual to identify the orthogonal direct sum $\bigoplus_k \mathcal{M}_k$ with its unitarily equivalent image $\Phi(\bigoplus_k \mathcal{M}_k) = \left(\sum_k \mathcal{M}_k\right)^- = \bigvee_k \mathcal{M}_k = \mathcal{H}$. Therefore, under such an identification, Corollary 5.18 is restated as follows. *If the orthogonal collection $\{\mathcal{M}_k\}$ spans $\mathcal{H}$, then $\mathcal{H} = \bigoplus_k \mathcal{M}_k$.*

## 5.7  Summability

The first half of this section pertains to Banach spaces and, as such, could have been introduced in Chapter 4. The final part, however, is a genuine Hilbert space subject.

**Definition 5.26.** Let $\Gamma$ be any index set and let $\{x_\gamma\}_{\gamma \in \Gamma}$ be an indexed family of vectors in a normed space $\mathcal{X}$. $\{x_\gamma\}_{\gamma \in \Gamma}$ is a *summable family* with *sum* $x \in \mathcal{X}$ (notation: $x = \sum_{\gamma \in \Gamma} x_\gamma$) if for each $\varepsilon > 0$ there exists a finite set of indices $N_\varepsilon \subseteq \Gamma$ such that

$$\Big\| \sum_{k \in N} x_k - \sum_{\gamma \in \Gamma} x_\gamma \Big\| < \varepsilon$$

for every finite subset $N$ of $\Gamma$ that includes $N_\varepsilon$ (i.e., for every finite $N$ such that $N_\varepsilon \subseteq N \subseteq \Gamma$). It is called a *p-summable family* for some $p \geq 1$ if $\{\|x_\gamma\|^p\}_{\gamma \in \Gamma}$ is a summable family of positive numbers. In particular, $\{x_\gamma\}_{\gamma \in \Gamma}$ is an *absolutely summable family* if $\{\|x_\gamma\|\}_{\gamma \in \Gamma}$ is a summable family, and a *square-summable family* if $\{\|x_\gamma\|^2\}_{\gamma \in \Gamma}$ is a summable family.

It is readily verified from Definition 5.26 that if $\{x_\gamma\}_{\gamma \in \Gamma}$ and $\{y_\gamma\}_{\gamma \in \Gamma}$ are (similarly indexed) summable families of vectors in a normed space $\mathcal{X}$ with sums $\sum_{\gamma \in \Gamma} x_\gamma$ and $\sum_{\gamma \in \Gamma} y_\gamma$ in $\mathcal{X}$, respectively, then $\{\alpha x_\gamma + \beta y_\gamma\}_{\gamma \in \Gamma} = \alpha \{x_\gamma\}_{\gamma \in \Gamma} + \beta \{y_\gamma\}_{\gamma \in \Gamma}$ is again a summable family of vectors in $\mathcal{X}$ with sum

$$\sum_{\gamma \in \Gamma} (\alpha x_\gamma + \beta y_\gamma) = \alpha \sum_{\gamma \in \Gamma} x_\gamma + \beta \sum_{\gamma \in \Gamma} y_\gamma$$

for every pair of scalars $\alpha$ and $\beta$. This shows that the collection of all summable families of vectors in $\mathcal{X}$ is a linear manifold of the linear space $\mathcal{X}^\Gamma$ (see Example 2F), and hence is a linear space itself. It is also easy to verify that if $\{x_\gamma\}_{\gamma \in \Gamma}$ is a summable family of vectors in $\mathcal{X}$ with sum $\sum_{\gamma \in \Gamma} x_\gamma$ in $\mathcal{X}$, and if $T \in \mathcal{B}[\mathcal{X}, \mathcal{Y}]$ for some normed space $\mathcal{Y}$, then $\{Tx_\gamma\}_{\gamma \in \Gamma}$ is a summable family of vectors in $\mathcal{Y}$ with sum

$$\sum_{\gamma \in \Gamma} Tx_\gamma = T \sum_{\gamma \in \Gamma} x_\gamma.$$

**Theorem 5.27.** (Cauchy Criterion). *Let $\{x_\gamma\}_{\gamma \in \Gamma}$ be an indexed family of vectors in a normed space $\mathcal{X}$ and consider the following assertions.*

(a) $\{x_\gamma\}_{\gamma \in \Gamma}$ *is a summable family.*

(b) *For every $\varepsilon > 0$ there exists a finite set of indices $N_\varepsilon \subseteq \Gamma$ such that*

$$\left\| \sum_{k \in N} x_k \right\| < \varepsilon$$

*for every finite subset $N$ of $\Gamma$ such that $N$ is disjoint with $N_\varepsilon$ (i.e., whenever $N \subseteq \Gamma$ is finite and $N \cap N_\varepsilon = \varnothing$).*

*Claim*: (a) *implies* (b), *and* (b) *implies* (a) *if $\mathcal{X}$ is a Banach space.*

*Proof.* Suppose (a) holds true and take an arbitrary $\varepsilon > 0$. According to Definition 5.26 there exists a finite $N_\varepsilon \subseteq \Gamma$ such that

$$\left\| \sum_{k \in N'} x_k - \sum_{\gamma \in \Gamma} x_\gamma \right\| < \tfrac{\varepsilon}{2}$$

whenever $N'$ is finite and $N_\varepsilon \subseteq N' \subseteq \Gamma$. If $N$ is a finite subset of $\Gamma$ and $N \cap N_\varepsilon = \varnothing$, then $\sum_{k \in N} x_k = \sum_{k \in N \cup N_\varepsilon} x_k - \sum_{k \in N_\varepsilon} x_k$. Since the set $N \cup N_\varepsilon$ is finite and

$N_\varepsilon \subseteq N \cup N_\varepsilon \subseteq \Gamma$, it follows that

$$\left\| \sum_{k\in N} x_k \right\| \leq \left\| \sum_{k\in N\cup N_\varepsilon} x_k - \sum_{\gamma\in\Gamma} x_\gamma \right\| + \left\| \sum_{k\in N_\varepsilon} x_k - \sum_{\gamma\in\Gamma} x_\gamma \right\| < \varepsilon.$$

Thus (a)$\Rightarrow$(b). Conversely, if (b) holds true, then there exists a sequence $\{N_i'\}_{i=1}^\infty$ of finite subsets of $\Gamma$ such that, for each $i \geq 1$, $\|\sum_{k\in N'} x_k\| < \frac{1}{i}$ whenever $N' \subseteq \Gamma$ is finite and $N' \cap N_i' = \varnothing$. Set $N_i = \bigcup_{j=1}^i N_j'$, which is a finite subset of $\Gamma$ such that $N_i \subseteq N_{i+1}$ for each $i \geq 1$. Take any $i \geq 1$ and an arbitrary finite $N \subseteq \Gamma$ such that $N \cap N_i = \varnothing$. Since $N \cap N_i' = \varnothing$, it follows that $\|\sum_{k\in N} x_k\| < \frac{1}{i}$. Conclusion: $\{N_i\}_{i=1}^\infty$ is an *increasing* sequence of finite subsets of $\Gamma$ such that, for each $i \geq 1$,

$$\left\| \sum_{k\in N} x_k \right\| < \frac{1}{i}$$

whenever $N \subseteq \Gamma$ is finite and $N \cap N_i = \varnothing$. Now set $y_i = \sum_{k\in N_i} x_k$ in $\mathcal{X}$ for each $i \geq 1$. Since $\{N_i\}_{i=0}^\infty$ is an increasing sequence we get

$$y_{i+j} - y_i = \sum_{k\in N_{i+j}} x_k - \sum_{k\in N_i} x_k = \sum_{k\in N_{i+j}\setminus N_i} x_k$$

for every $i$, $j \geq 1$, which implies

$$\|y_{i+j} - y_i\| \leq \left\| \sum_{k\in N_{i+j}\setminus N_i} x_k \right\| < \frac{1}{i}$$

for every $i \geq 1$ and all $j \geq 1$ (reason: for every $i$, $j \geq 1$, $N_{i+j}\setminus N_i$ is a finite subset of $\Gamma$ and $(N_{i+j}\setminus N_i) \cap N_i = \varnothing$). Hence (see Problem 3.51) $\{y_i\}_{i=1}^\infty$ is a Cauchy sequence in $\mathcal{X}$. If $\mathcal{X}$ is a Banach space, then $\{y_i\}_{i=1}^\infty$ converges in $\mathcal{X}$ to, say, $x \in \mathcal{X}$. Therefore, for each $\varepsilon > 0$ there exists

(i)   an integer $i_\varepsilon \geq 1$ such that $\|\sum_{k\in N_i} x_k - x\| < \frac{\varepsilon}{2}$ whenever $i \geq i_\varepsilon$

and, as (b) holds true,

(ii)   a finite $N_\varepsilon \subseteq \Gamma$ such that $\|\sum_{k\in N'} x_k\| < \frac{\varepsilon}{2}$ whenever $N' \subseteq \Gamma$ is finite and $N' \cap N_\varepsilon = \varnothing$.

Since $N_\varepsilon \cup N_{i_\varepsilon}$ is finite and $\{N_i\}_{i=1}^\infty$ is increasing, it follows that there exists an integer $i_\varepsilon^* \geq i_\varepsilon$, and consequently a finite subset $N_{i_\varepsilon^*}$ of $\Gamma$, such that $N_\varepsilon \cup N_{i_\varepsilon} \subseteq N_{i_\varepsilon^*}$. If $N$ is finite and $N_{i_\varepsilon^*} \subseteq N \subseteq \Gamma$, then $N\setminus N_{i_\varepsilon^*} \subseteq \Gamma$ is finite and $(N\setminus N_{i_\varepsilon^*}) \cap N_\varepsilon = \varnothing$. Hence

$$\left\| \sum_{k\in N} x_k - x \right\| = \left\| \sum_{k\in N\setminus N_{i_\varepsilon^*}} x_k + \sum_{k\in N_{i_\varepsilon^*}} x_k - x \right\|$$

$$\leq \left\| \sum_{k\in N\setminus N_{i_\varepsilon^*}} x_k \right\| + \left\| \sum_{k\in N_{i_\varepsilon^*}} x_k - x \right\| < \varepsilon,$$

and therefore (b)$\Rightarrow$(a).                                     $\square$

**Corollary 5.28.** *If $\{x_\gamma\}_{\gamma \in \Gamma}$ is a summable family of vectors in a normed space $\mathcal{X}$, then the set $\{\gamma \in \Gamma: \ x_\gamma \neq 0\}$ is countable.*

*Proof.* Let $\{x_\gamma\}_{\gamma \in \Gamma}$ be a summable family of vectors in a normed space $\mathcal{X}$. According to the previous theorem, for every integer $n \geq 1$ there exists a finite subset $N_n$ of $\Gamma$ such that $\|\sum_{k \in N} x_k\| < \frac{1}{n}$ whenever $N \subseteq \Gamma$ is finite and $N \cap N_n = \varnothing$. Put $S = \bigcup_{n=1}^{\infty} N_n \subseteq \Gamma$ and recall from Corollary 1.11 that $S$ is a countable set. If $\gamma \in \Gamma \backslash S$, then $\{\gamma\} \cup N_n = \varnothing$, and hence $\|x_\gamma\| < \frac{1}{n}$, for every $n \geq 1$, which implies that $x_\gamma = 0$. Thus $x_\gamma$ is a nonzero vector in $\mathcal{X}$ only if $\gamma$ lies in the countable set $S$.                                     $\square$

What Corollary 5.28 says is that an uncountable indexed family of vectors in a normed space may be summable but, in this case, it has only a countable number of nonzero vectors.

**Corollary 5.29.** *Every absolutely summable family of vectors in a Banach space is a summable family.*

*Proof.* Let $\{x_\gamma\}_{\gamma \in \Gamma}$ be an absolutely summable family of vectors in a normed space $\mathcal{X}$ so that $\{\|x_\gamma\|\}_{\gamma \in \Gamma}$ is a summable family of nonnegative numbers, and hence (cf. Theorem 5.27) for every $\varepsilon > 0$ there exists a finite $N_\varepsilon \subseteq \Gamma$ such that

$$\left\| \sum_{k \in N} x_k \right\| \leq \sum_{k \in N} \|x_k\| < \varepsilon$$

whenever $N \subseteq \Gamma$ is finite and $N \cap N_\varepsilon = \varnothing$. Another application of Theorem 5.27 ensures that $\{x_\gamma\}_{\gamma \in \Gamma}$ is a summable family if $\mathcal{X}$ is a Banach space.                                     $\square$

The converse of Corollary 5.29 holds for finite-dimensional Banach spaces. That is, *if $\mathcal{X}$ is a finite-dimensional normed space, then every summable family of vectors in $\mathcal{X}$ is absolutely summable*. But it fails in general. In fact, Dvoretzky and Rogers proved in 1950 that there exist summable families of vectors in infinite-dimensional Banach spaces that are not absolutely summable.

**Proposition 5.30.** *If $\mathcal{X}$ is a finite-dimensional normed space, then $\{x_\gamma\}_{\gamma \in \Gamma}$ is a summable family of vectors in $\mathcal{X}$ if and only if it is absolutely summable.*

*Proof.* Recall that a finite-dimensional normed space is a Banach space (Corollary 4.28). Then, according to Corollary 5.29, it remains to show that every summable family of vectors in a finite-dimensional normed space is absolutely summable. Consider a normed space $(\mathcal{X}, \| \ \|)$ with $\dim \mathcal{X} = n$ for some positive integer $n$ and let $B = \{e_j\}_{j=1}^{n}$ be a Hamel basis for $\mathcal{X}$. Take an arbitrary $x \in \mathcal{X}$ and consider its *unique* expansion on $B$,

$$x = \sum_{j=1}^{n} \xi_j e_j,$$

where $\{\xi_j\}_{j=1}^n$ is a family of scalars consisting of the coordinates of $x$ with respect to the basis $B$. It is readily verified that the function $\|\ \|_1 \colon \mathcal{X} \to \mathbb{R}$, defined by

$$\|x\|_1 = \sum_{j=1}^n |\xi_j|$$

for every $x \in \mathcal{X}$, is a norm on $\mathcal{X}$. Since any two norms on $\mathcal{X}$ are equivalent, it follows that there exist real constants $\alpha > 0$ and $\beta > 0$ such that

$$\|x\| \le \alpha \|x\|_1 \quad \text{and} \quad \|x\|_1 \le \beta \|x\|$$

for every $x \in \mathcal{X}$ (Proposition 4.26 and Theorem 4.27). Now suppose $\{x_\gamma\}_{\gamma \in \Gamma}$ is a summable family of vectors in $\mathcal{X}$ and take an arbitrary $\varepsilon > 0$. Theorem 5.27 ensures the existence of a finite $N_\varepsilon \subseteq \Gamma$ such that

$$\sum_{j=1}^n \left| \sum_{k \in N} \xi_j(k) \right| = \left\| \sum_{k \in N} x_k \right\|_1 \le \beta \left\| \sum_{k \in N} x_k \right\| < \beta\varepsilon,$$

and hence

$$\left| \sum_{k \in N} \xi_j(k) \right| < \beta\varepsilon \quad \text{for every} \quad j = 1, \dots, n,$$

whenever $N \subseteq \Gamma$ is finite and $N \cap N_\varepsilon = \varnothing$, where $\{\xi_j(k)\}_{j=1}^n$ are the coordinates of each $x_k$ with respect to $B$. That is, $x_k = \sum_{j=1}^n \xi_j(k) e_j$ with $\|x_k\|_1 = \sum_{j=1}^n |\xi_j(k)|$, and therefore $\sum_{k \in N} x_k = \sum_{k \in N} \sum_{j=1}^n \xi_j(k) e_j = \sum_{j=1}^n \sum_{k \in N} \xi_j(k) e_j$ is the unique expansion of $\sum_{k \in N} x_k$ on $B$ so that $\|\sum_{k \in N} x_k\|_1 = \sum_{j=1}^n |\sum_{k \in N} \xi_j(k)|$. Take any finite subset $N$ of $\Gamma$ for which $N \cap N_\varepsilon = \varnothing$ and observe that

$$\sum_{k \in N} \|x_k\| \le \alpha \sum_{k \in N} \|x_k\|_1 = \alpha \sum_{k \in N} \sum_{j=1}^n |\xi_j(k)|$$

$$= \alpha \sum_{j=1}^n \sum_{k \in N} |\mathrm{Re}\, \xi_j(k)| + \alpha \sum_{j=1}^n \sum_{k \in N} |\mathrm{Im}\, \xi_j(k)|.$$

Set $N_j^+ = \{k \in N \colon \mathrm{Re}\, \xi_j(k) \ge 0\}$ and $N_j^- = \{k \in N \colon \mathrm{Re}\, \xi_j(k) < 0\}$ for each $j = 1, \dots, n$, which (as subsets of $N$) are finite subsets of $\Gamma$ such that $N_j^+ \cap N_\varepsilon = \varnothing$ and $N_j^- \cap N_\varepsilon = \varnothing$ for every $j = 1, \dots, n$. Thus $|\sum_{k \in N_j^+} \xi_j(k)| < \beta\varepsilon$ and

$|\sum_{k\in N_j^-}\xi_j(k)| < \beta\varepsilon$ for all $j = 1, \dots, n$. Hence,

$$
\begin{aligned}
\sum_{j=1}^n \sum_{k\in N} |\mathrm{Re}\,\xi_j(k)| &= \sum_{j=1}^n \left|\sum_{k\in N_j^+} \mathrm{Re}\,\xi_j(k)\right| + \sum_{j=1}^n \left|\sum_{k\in N_j^-} \mathrm{Re}\,\xi_j(k)\right| \\
&= \sum_{j=1}^n \left|\mathrm{Re}\left(\sum_{k\in N_j^+}\xi_j(k)\right)\right| + \sum_{j=1}^n \left|\mathrm{Re}\left(\sum_{k\in N_j^-}\xi_j(k)\right)\right| \\
&\leq \sum_{j=1}^n \left|\sum_{k\in N_j^+}\xi_j(k)\right| + \sum_{j=1}^n \left|\sum_{k\in N_j^-}\xi_j(k)\right| < 2n\beta\varepsilon.
\end{aligned}
$$

Similarly, $\sum_{j=1}^n \sum_{k\in N} |\mathrm{Im}\,\xi_j(k)| < 2n\beta\varepsilon$, and so

$$
\sum_{k\in N} \|x_k\| < 4n\alpha\beta\varepsilon.
$$

Conclusion: $\{x_\gamma\}_{\gamma\in\Gamma}$ is an absolutely summable family. $\qquad\square$

Remark: If a family of vectors in a normed space $\mathcal{X}$ is indexed by $\mathbb{N}$ (or by $\mathbb{N}_0$), then it can be viewed as an $\mathcal{X}$-valued sequence (the very indexing process establishes a function from $\mathbb{N}$ to $\mathcal{X}$). If $\{x_k\}_{k\in\mathbb{N}}$ is a summable family, then $\{x_k\}_{k=1}^\infty$ is a summable sequence (or, equivalently, the infinite series $\sum_{k=1}^\infty x_k$ converges). Indeed, if for each $\varepsilon > 0$ there exists a finite $N_\varepsilon \subset \mathbb{N}$ such that $\|\sum_{k\in N} x_k - x\| < \varepsilon$ for some $x \in \mathcal{X}$ whenever $N$ is finite and $N_\varepsilon \subseteq N \subset \mathbb{N}$, then by setting $n_\varepsilon = \#N_\varepsilon$ it follows that $\|\sum_{k=1}^n x_k - x\| < \varepsilon$ whenever $n \geq n_\varepsilon$. However, the converse fails even for scalar-valued sequences. For instance, consider the sequence $\{x_k\}_{k=1}^\infty$ with $x_{2k-1} = -x_{2k} = \frac{1}{k}$. It is clear that the infinite series $\sum_{k=1}^\infty x_k$ converges (since $|\sum_{k=1}^n x_k| \leq \frac{2}{n+1}$ for every $n \geq 1$) but is not absolutely convergent (since $\sum_{k=1}^{2n} |x_k| = 2\sum_{k=1}^n \frac{1}{k}$ for every $n \geq 1$). Thus $\{x_k\}_{k=1}^\infty$ is a summable sequence but not an absolutely summable sequence, and hence $\{x_k\}_{k\in\mathbb{N}}$ is not an absolutely summable family of vectors in the one-dimensional normed space $\mathbb{R}$, so that $\{x_k\}_{k\in\mathbb{N}}$ is not a summable family by Proposition 5.30. An $\mathcal{X}$-valued sequence $\{x_k\}_{k=1}^\infty$ for which $\{x_k\}_{k\in\mathbb{N}}$ is a summable family of vectors in $\mathcal{X}$ is referred to as an *unconditionally summable sequence*. In this case it is also common to say that the infinite series $\sum_{k=1}^\infty x_k$ is *unconditionally convergent*.

**Proposition 5.31.** *Let $\{x_\gamma\}_{\gamma\in\Gamma}$ be a family of vectors in a normed space $\mathcal{X}$ and take any $p \geq 1$. The following assertions are equivalent.*

(a) *$\{x_\gamma\}_{\gamma\in\Gamma}$ is a p-summable family.*

(b) *$\sup_N \sum_{k\in N} \|x_k\|^p < \infty$, where the supremum is taken over all finite subsets of $\Gamma$, which is expressed by writing $\sum_{\gamma\in\Gamma} \|x_\gamma\|^p < \infty$.*

*Moreover, if any of the above equivalent assertions holds, then*

$$\sum_{\gamma\in\Gamma} \|x_\gamma\|^p = \sup_N \sum_{k\in N} \|x_k\|^p.$$

*Proof.* Suppose (a) holds true. Theorem 5.27 ensures that for each $\varepsilon > 0$ there exists a finite subset $N_\varepsilon$ of $\Gamma$ such that $\sum_{k\in N} \|x_k\|^p < \varepsilon$ whenever $N$ is a finite subset of $\Gamma$ such that $N \cap N_\varepsilon = \varnothing$. If (b) fails, then for every integer $n \geq 1$ there exists a finite subset $N_n$ of $\Gamma$ such that $\sum_{k\in N_n} \|x_k\|^p > n$. Therefore, since $(N_n \backslash N_\varepsilon) \cap N_\varepsilon = \varnothing$, it follows that

$$n < \sum_{k\in N_n} \|x_k\|^p = \sum_{k\in N_n\cap N_\varepsilon} \|x_k\|^p + \sum_{k\in N_n\backslash N_\varepsilon} \|x_k\|^p < \sum_{k\in N_\varepsilon} \|x_k\|^p + \varepsilon$$

for every integer $n \geq 1$, which is a contradiction. Hence (a) implies (b). Conversely, Suppose (b) holds and set $\alpha = \sup_N \sum_{N\in\Gamma} \|x_k\|^p$. Thus for every $\varepsilon > 0$ there exists a finite subset $N_\varepsilon$ of $\Gamma$ for which $\sum_{k\in N_\varepsilon} \|x_k\|^p < \alpha - \varepsilon$. If $N$ is any finite subset of $\Gamma$ such that $N \cap N_\varepsilon = \varnothing$, then

$$\sum_{k\in N} \|x_k\|^p = \sum_{k\in N\cup N_\varepsilon} \|x_k\|^p - \sum_{k\in N_\varepsilon} \|x_k\|^p < \alpha - (\alpha - \varepsilon) = \varepsilon.$$

Hence (b) implies (a) by Theorem 5.27 (in the Banach space $\mathbb{R}$). Finally, if any of the above equivalent assertions holds, then for every $\varepsilon > 0$ there exists a finite subset $N_\varepsilon$ of $\Gamma$ such that

$$\left| \sum_{k\in N} \|x_k\|^p - \sum_{\gamma\in\Gamma} \|x_\gamma\|^p \right| < \varepsilon$$

or, equivalently, such that

$$\sum_{k\in N} \|x_k\|^p < \sum_{\gamma\in\Gamma} \|x_\gamma\|^p + \varepsilon \quad \text{and} \quad \sum_{\gamma\in\Gamma} \|x_\gamma\|^p < \sum_{k\in N} \|x_k\|^p + \varepsilon,$$

whenever $N$ is finite and $N_\varepsilon \subseteq N \subseteq \Gamma$ (Definition 5.26). Hence

$$\sup_N \sum_{k\in N} \|x_k\|^p \leq \sum_{\gamma\in\Gamma} \|x_\gamma\|^p + \varepsilon \leq \sup_N \sum_{k\in N} \|x_k\|^p + 2\varepsilon$$

for every $\varepsilon > 0$, where the supremum is taken over all finite subsets of $\Gamma$ (recall: $\sum_{k\in N_1} \|x_k\|^p \leq \sum_{k\in N_2} \|x_k\|^p$ whenever $N_1$ and $N_2$ are finite subsets of $\Gamma$ such that $N_1 \subseteq N_2$). By taking the infimum over all $\varepsilon > 0$ in the above inequalities,

$$\sum_{\gamma\in\Gamma} \|x_\gamma\|^p = \sup_N \sum_{k\in N} \|x_k\|^p. \qquad \square$$

Let $\{x_\gamma\}_{\gamma\in\Gamma}$ be a family of vectors in a normed space $\mathcal{X}$. It is obvious that $\sum_{k\in N} \|x_k\|^p < \infty$ for every finite subset $N$ of $\Gamma$. By convention, the *empty sum* is null (i.e., $\sum_{k\in\varnothing}\|x_k\|^p = 0$ — in general, $\sum_{k\in\varnothing}x_k = 0 \in \mathcal{X}$). If $\{x_\gamma\}_{\gamma\in\Gamma}$ is a summable family, then it has only a countable set of nonzero vectors (Corollary 5.28). If this set is infinite (countably infinite, that is), then the family of all nonzero vectors from $\{x_\gamma\}_{\gamma\in\Gamma}$ can be indexed by $\mathbb{N}$, say $\{x_k\}_{k\in\mathbb{N}}$ (or by any other countably infinite index set). Clearly, $\{x_\gamma\}_{\gamma\in\Gamma}$ is a summable (a $p$-summable) family if and only if $\{x_k\}_{k\in\mathbb{N}}$ is. If $\{x_k\}_{k=1}^{\infty}$ is *any* sequence containing all vectors from $\{x_k\}_{k\in\mathbb{N}}$, then it follows by the remark that precedes Proposition 5.31 that $\{x_\gamma\}_{\gamma\in\Gamma}$ is a summable family if and only if $\{x_k\}_{k=1}^{\infty}$ is an unconditionally summable sequence. In particular, $\{x_\gamma\}_{\gamma\in\Gamma}$ is a $p$-summable family if and only if $\{\|x_k\|^p\}_{k=1}^{\infty}$ is an unconditionally summable sequence of positive numbers.

**Example 5K.** Let $\Gamma$ be any index set and let $\ell_\Gamma^2$ denote the collection of all square-summable families $\{\xi_\gamma\}_{\gamma\in\Gamma}$ of scalars in $\mathbb{F}$ (as usual, $\mathbb{F}$ stands either for the real field $\mathbb{R}$ or for the complex field $\mathbb{C}$). It is easy to check that $\ell_\Gamma^2$ is a linear space over $\mathbb{F}$. Observe that $\{\xi_\gamma \overline{\upsilon}_\gamma\}_{\gamma\in\Gamma}$ is a summable family whenever $x = \{\xi_\gamma\}_{\gamma\in\Gamma}$ and $y = \{\upsilon_\gamma\}_{\gamma\in\Gamma}$ are square-summable families. Indeed, by the Hölder inequality for finite sums,

$$\sum_{k\in N} |\xi_k \overline{\upsilon}_k| \le \left( \sum_{k\in N} |\xi_k|^2 \right)^{\frac{1}{2}} \left( \sum_{k\in N} |\upsilon_k|^2 \right)^{\frac{1}{2}}$$

for every finite subset $N$ of $\Gamma$. Therefore,

$$\sum_{\gamma\in\Gamma} |\xi_\gamma \overline{\upsilon}_\gamma| \le \left( \sum_{\gamma\in\Gamma} |\xi_\gamma|^2 \right)^{\frac{1}{2}} \left( \sum_{\gamma\in\Gamma} |\upsilon_\gamma|^2 \right)^{\frac{1}{2}}$$

by Proposition 5.31, which implies that $\{\xi_\gamma \overline{\upsilon}_\gamma\}_{\gamma\in\Gamma}$ is an absolutely summable family of scalars, and hence a summable family in the Banach space $\mathbb{F}$ (Corollary 5.29). Thus the function $\langle\ ;\ \rangle: \ell_\Gamma^2 \times \ell_\Gamma^2 \to \mathbb{F}$, given by

$$\langle x\,;\,y \rangle = \sum_{\gamma\in\Gamma} \xi_\gamma \overline{\upsilon}_\gamma$$

for every $x = \{\xi_\gamma\}_{\gamma\in\Gamma}$ and $y = \{\upsilon_\gamma\}_{\gamma\in\Gamma}$ in $\ell_\Gamma^2$, is well-defined. Moreover, it is readily verified that it defines an inner product on $\ell_\Gamma^2$. In fact, it is not difficult to show that

$$(\ell_\Gamma^2, \langle\ ;\ \rangle)\ \text{is a Hilbert space.}$$

In particular, if $\Gamma = \mathbb{N}$, then $(\ell_\Gamma^2, \langle\ ;\ \rangle)$ is reduced to the Hilbert space $(\ell_+^2, \langle\ ;\ \rangle)$ of Example 5B. The natural generalization is easy. Let $\ell_\Gamma^2(\mathcal{H})$ denote the linear space of all square-summable families $\{x_\gamma\}_{\gamma\in\Gamma}$ of vectors in a Hilbert space $(\mathcal{H}, \langle\ ;\ \rangle_{\mathrm{H}})$. Proceeding as above, it can be shown that the function $\langle\ ;\ \rangle: \ell_\Gamma^2(\mathcal{H}) \times \ell_\Gamma^2(\mathcal{H}) \to \mathbb{F}$, given by

$$\langle \{x_\gamma\}_{\gamma\in\Gamma}\,;\,\{y_\gamma\}_{\gamma\in\Gamma} \rangle = \sum_{\gamma\in\Gamma} \langle x_\gamma\,;\,y_\gamma \rangle_{\mathrm{H}}$$

for every $\{x_\gamma\}_{\gamma \in \Gamma}$ and $\{y_\gamma\}_{\gamma \in \Gamma}$ in $\ell^2_\Gamma(\mathcal{H})$, defines an inner product on $\ell^2_\Gamma(\mathcal{H})$, and

$$(\ell^2_\Gamma(\mathcal{H}), \langle\ ;\ \rangle) \text{ is a Hilbert space.}$$

Again, if $\Gamma = \mathbb{N}$, then $(\ell^2_\Gamma(\mathcal{H}), \langle\ ;\ \rangle)$ is reduced to the Hilbert space $(\ell^2_+(\mathcal{H}), \langle\ ;\ \rangle)$ of Example 5F.

From now on suppose $\mathcal{X}$ is an inner product space. It is easy to show from Definition 5.26 that if $\{x_\gamma\}_{\gamma \in \Gamma}$ and $\{y_\gamma\}_{\gamma \in \Gamma}$ are (similarly indexed) summable families of vectors in $\mathcal{X}$ with sums $\sum_{\gamma \in \Gamma} x_\gamma$ and $\sum_{\gamma \in \Gamma} y_\gamma$ in $\mathcal{X}$, respectively, and if $x$ and $y$ are arbitrary vectors in $\mathcal{X}$, then $\{\langle x_\gamma\ ;\ y\rangle\}_{\gamma \in \Gamma}$ and $\{\langle x\ ;\ y_\gamma\rangle\}_{\gamma \in \Gamma}$ are summable families of scalars with sums

$$\sum_{\gamma \in \Gamma} \langle x_\gamma\ ;\ y\rangle = \Big\langle \sum_{\gamma \in \Gamma} x_\gamma\ ;\ y \Big\rangle \quad \text{and} \quad \sum_{\gamma \in \Gamma} \langle x\ ;\ y_\gamma\rangle = \Big\langle x\ ;\ \sum_{\gamma \in \Gamma} y_\gamma \Big\rangle.$$

The next result is the general version of the Pythagorean Theorem. It extends Corollary 5.9 to any infinite orthogonal family of vectors in an inner product space.

**Theorem 5.32.** *Let $\{x_\gamma\}_{\gamma \in \Gamma}$ be a family of pairwise orthogonal vectors in an inner product space $\mathcal{X}$.*

(a) *If $\{x_\gamma\}_{\gamma \in \Gamma}$ is a summable family, then it is a square-summable family and $\|\sum_{\gamma \in \Gamma} x_\gamma\|^2 = \sum_{\gamma \in \Gamma} \|x_\gamma\|^2$.*

(b) *If $\mathcal{X}$ is a Hilbert space and $\{x_\gamma\}_{\gamma \in \Gamma}$ is a square-summable family, then $\{x_\gamma\}_{\gamma \in \Gamma}$ is a summable family.*

*Proof.* Let $\{x_\gamma\}_{\gamma \in \Gamma}$ be an orthogonal family of vectors in $\mathcal{X}$.

(a)  If $\{x_\gamma\}_{\gamma \in \Gamma}$ is a summable family, then for every $\varepsilon > 0$ there exists a finite $N_\varepsilon \subseteq \Gamma$ such that

$$\sum_{k \in N} \|x_k\|^2 = \Big\| \sum_{k \in N} x_k \Big\|^2 < \varepsilon^2$$

whenever $N \subseteq \Gamma$ is finite and $N \cap N_\varepsilon = \varnothing$ (by Proposition 5.8 and Theorem 5.27). Another application of Theorem 5.27 ensures that $\{\|x_\gamma\|^2\}_{\gamma \in \Gamma}$ is a summable family in the Banach space $\mathbb{R}$. Moreover, since $x_\alpha \perp x_\beta$ whenever $x_\alpha$ and $x_\beta$ are distinct vectors from $\{x_\gamma\}_{\gamma \in \Gamma}$ (i.e., $\langle x_\alpha\ ;\ x_\beta\rangle = 0$ for every $\alpha, \beta \in \Gamma$ such that $\alpha \neq \beta$), it follows that $\{\langle x_\gamma\ ;\ x_\gamma\rangle\}_{\gamma \in \Gamma}$ includes the family of all nonzero scalars from the family $\{\langle x_\alpha\ ;\ x_\beta\rangle\}_{(\alpha,\beta) \in \Gamma \times \Gamma}$. Therefore,

$$\Big\| \sum_{\gamma \in \Gamma} x_\gamma \Big\|^2 = \Big\langle \sum_{\gamma \in \Gamma} x_\gamma\ ;\ \sum_{\gamma \in \Gamma} x_\gamma \Big\rangle = \sum_{\alpha \in \Gamma} \Big\langle x_\alpha\ ;\ \sum_{\beta \in \Gamma} x_\beta \Big\rangle$$

$$= \sum_{\alpha \in \Gamma} \sum_{\beta \in \Gamma} \langle x_\alpha\ ;\ x_\beta\rangle = \sum_{\gamma \in \Gamma} \langle x_\gamma\ ;\ x_\gamma\rangle = \sum_{\gamma \in \Gamma} \|x_\gamma\|^2.$$

(b)  If $\{\|x_\gamma\|^2\}_{\gamma\in\Gamma}$ is a summable family of nonnegative numbers, then for every $\varepsilon > 0$ there exists a finite $N_\varepsilon \subseteq \Gamma$ such that

$$\left\|\sum_{k\in N} x_k\right\| = \left(\sum_{k\in N}\|x_k\|^2\right)^{\frac{1}{2}} < \varepsilon^{\frac{1}{2}}$$

whenever $N \subseteq \Gamma$ is finite and $N \cap N_\varepsilon = \varnothing$ (by Proposition 5.8 and Theorem 5.27 again). Another application of Theorem 5.27 ensures that $\{x_\gamma\}_{\gamma\in\Gamma}$ is a summable family in the Hilbert space $\mathcal{X}$. $\qquad\square$


## 5.8   Orthonormal Basis

If $A$ is an orthogonal set in an inner product space $\mathcal{X}$, then $A$ may contain the origin of $\mathcal{X}$, which is orthogonal to every vector in $\mathcal{X}$. The next proposition presents the main algebraic property of orthogonal sets that do not contain the origin.

**Proposition 5.33.** *If $A$ is an orthogonal set consisting of nonzero vectors in an inner product space, then $A$ is linearly independent.*

*Proof.*  Let $A$ be a set of pairwise orthogonal nonzero vectors in an inner product space $\mathcal{X}$. Suppose there exists a finite subset of $A$ containing more than one vector that is not linearly independent. For instance, $\{x_i\}_{i=0}^n \subseteq A$ such that $x_0 = \sum_{i=1}^n \alpha_i x_i$ for some integer $n \geq 1$ and some set of scalars $\{\alpha_i\}_{i=1}^n$. Since $x_0 \perp x_i$ for every $i = 1,\dots,n$ we get $\|x_0\|^2 = \langle x_0 ; x_0 \rangle = \sum_{i=1}^n \alpha_i \langle x_0 ; x_i \rangle = 0$, which is a contradiction (because $A$ does not contain the origin). Conclusion: Every finite subset of $A$ containing more than one vector is linearly independent. Recall that every singleton $\{x\}$ in a linear space $\mathcal{X}$ such that $x \neq 0$ is linearly independent. Therefore, every finite subset of $A$ is linearly independent, which means that $A$ is itself linearly independent (Proposition 2.3). $\qquad\square$

A *unit vector* in a normed space is a vector with norm equal to one. An *orthonormal set* in an inner product space $\mathcal{X}$ is an orthogonal set consisting of unit vectors. That is, a subset $A$ of $\mathcal{X}$ is an orthonormal set if $x \perp y$ for every pair $\{x, y\}$ of distinct vectors in $A$ and $\|x\| = 1$ for every $x \in A$. Equivalently, $\{e_\gamma\}_{\gamma\in\Gamma}$ is an orthonormal family of vectors in $\mathcal{X}$ if $\langle e_\alpha ; e_\beta \rangle = \delta_{\alpha\beta}$ for every $\alpha, \beta \in \Gamma$, where $\delta_{\alpha\beta}$ is the Kronecker delta (i.e., $\langle e_\alpha ; e_\beta \rangle = 0$ for every $\alpha, \beta \in \Gamma$ such that $\alpha \neq \beta$ and $\|e_\gamma\|^2 = \langle e_\gamma ; e_\gamma \rangle = 1$ for every $\gamma \in \Gamma$). Each orthogonal set consisting of nonzero vectors can be normalized. In fact, if $A$ is an orthogonal subset of $\mathcal{X}$ such that $x \neq 0$ for every $x \in A$, then the set $\{\|x\|^{-1}x \in \mathcal{X}: \ x \in A\}$ is an orthonormal subset of $\mathcal{X}$. The next two results are immediate consequences of the definition of orthonormal sets (Proposition 5.34 below is just a particular case of Proposition 5.33).

**Proposition 5.34.** *Every orthonormal set in any inner product space is linearly independent.*

**Proposition 5.35.** *If $A$ is an orthonormal set in an inner product space $\mathcal{X}$, and if there exists $x \in \mathcal{X}$ such that $x \perp A$ and $\|x\| = 1$, then $A \cup \{x\}$ is an orthonormal set in $\mathcal{X}$ (that properly includes $A$).*

Let $\mathcal{O}$ be the collection of all orthonormal sets in an inner product space $\mathcal{X}$. Note that we may argue by contradiction that *any singleton $\{x\}$ with $\|x\| = 1$ is an orthonormal set in $\mathcal{X}$* (although the expression "pairwise orthonormal" is meaningless in this case). As a subcollection of the power set $\wp(\mathcal{X})$, $\mathcal{O}$ is partially ordered in the inclusion ordering of $\wp(\mathcal{X})$. Recall from Section 1.5 that a set $A \in \mathcal{O}$ (i.e., an orthonormal set $A$ in an inner product space $\mathcal{X}$) is maximal in $\mathcal{O}$ if there is no set $A' \in \mathcal{O}$ such that $A \subset A'$ (i.e., if there is no orthonormal set $A'$ in $\mathcal{X}$ that properly includes $A$). If $A$ is maximal in $\mathcal{O}$, then we say that $A$ is a *maximal orthonormal set* in the inner product space $\mathcal{X}$.

**Proposition 5.36.** *Let $\mathcal{X}$ be an inner product space and let $A$ be an orthonormal set in $\mathcal{X}$. The following assertions are pairwise equivalent.*

(a) *$A$ is a maximal orthonormal set in $\mathcal{X}$.*

(b) *There is no unit vector $x$ for which $A \cup \{x\}$ is an orthonormal set.*

(c) *If $x \perp A$, then $x = 0$   (i.e., $A^{\perp} = \{0\}$).*

*Proof.* Let $A$ be an orthonormal set in an inner product space $\mathcal{X}$.

*Proof of* (a)$\Rightarrow$(b). If there exists a unit vector $x$ in $\mathcal{X}$ for which $A \cup \{x\}$ is an orthonormal set, then $A \cup \{x\}$ is an orthonormal set that properly includes $A$ (since $x \perp A$). Thus $A$ is not a maximal orthonormal set in $\mathcal{X}$.

*Proof of* (b)$\Rightarrow$(c). If there exists a nonzero vector $x$ in $\mathcal{X}$ such that $x \perp A$, then there exists a unit vector $x' = \|x\|^{-1}x$ in $\mathcal{X}$ such that $A \cup \{x'\}$ is an orthonormal set.

*Proof of* (c)$\Rightarrow$(a). If (a) fails, then there exists an orthonormal set $A'$ in $\mathcal{X}$ that properly includes $A$ so that $A'\backslash A \neq \varnothing$. Take any $x$ in $A'\backslash A$, which is a nonzero vector (actually, $x$ is a unit vector) orthogonal to $A$ (for $x \in A'$, $A \subset A'$, and $A'$ is an orthonormal set). Thus (c) fails. $\qquad\qquad\square$

**Proposition 5.37.** *If $A$ is an orthonormal set in an inner product space $\mathcal{X}$, then there exists a maximal orthonormal set $B$ in $\mathcal{X}$ such that $A \subseteq B$.*

*Proof.* Let $A$ be an orthonormal set in an inner product space $\mathcal{X}$. Put

$$\mathcal{O}_A = \big\{ S \in \wp(\mathcal{X}) : \ S \text{ is an orthonormal set in } \mathcal{X} \text{ and } A \subseteq S \big\},$$

the collection of all orthonormal sets in $\mathcal{X}$ that include $A$. Recall that, as a nonempty subcollection ($A \in \mathcal{O}_A$) of the power set $\wp(\mathcal{X})$, $\mathcal{O}_A$ is partially ordered in the inclusion ordering. Take an arbitrary chain $\mathcal{C}$ in $\mathcal{O}_A$ and consider the union $\bigcup \mathcal{C}$ of all sets in $\mathcal{C}$. If $x$ and $y$ are distinct vectors in $\bigcup\mathcal{C}$, then $x \in C$ and $y \in D$, where $C, D \in \mathcal{C} \subseteq \mathcal{O}_A$. Since $\mathcal{C}$ is a chain, it follows that either $C \subseteq D$ or $D \subseteq C$. Suppose (without loss of generality) that $C \subseteq D$. Thus $x, y \in D$, and so $\bigcup\mathcal{C}$ is an orthonormal set (for $D \in \mathcal{O}_A$). Moreover, $A \subseteq \bigcup\mathcal{C}$ (reason: every set in $\mathcal{C}$ includes $A$). Outcome: $\bigcup\mathcal{C} \in \mathcal{O}_A$. Since $\bigcup\mathcal{C}$ is an upper bound for $\mathcal{C}$ we may conclude: Every chain in $\mathcal{O}_A$ has an upper bound in $\mathcal{O}_A$. Hence $\mathcal{O}_A$ has a maximal element by Zorn's Lemma. Let $B$ be a maximal element of $\mathcal{O}_A$, which clearly is an orthonormal set in $\mathcal{X}$ that includes $A$. If there exists a unit vector $x$ in $\mathcal{X}$ such that $B \cup \{x\}$ is an orthonormal set, then $B \cup \{x\}$ lies in $\mathcal{O}_A$ and properly includes $B$, which contradicts the fact that $B$ is a maximal element of $\mathcal{O}_A$. Therefore, there is no unit vector $x$ in $\mathcal{X}$ such that $B \cup \{x\}$ is an orthonormal set, and hence $B$ is a maximal orthonormal set in $\mathcal{X}$ (Proposition 5.36). $\qquad\square$

Proposition 5.37 says that there are plenty of maximal orthonormal sets in any inner product space of dimension greater than 1. The next proposition says that the maximal orthonormal sets in a Hilbert space $\mathcal{H}$ are precisely those orthonormal sets that span $\mathcal{H}$.

**Proposition 5.38.** *Let $A$ be an orthonormal set in an inner product space $\mathcal{X}$.*

(a) *If $\bigvee A = \mathcal{X}$, then $A$ is a maximal orthonormal set.*

(b) *If $\mathcal{X}$ is a Hilbert space and $A$ is a maximal orthonormal set, then $\bigvee A = \mathcal{X}$.*

*Proof.* Take any orthonormal set $A$ in $\mathcal{X}$. According to Proposition 5.36, $A$ is a maximal orthonormal set if and only if $A^{\perp} = \{0\}$. If $\bigvee A = \mathcal{X}$, then $(\bigvee A)^{\perp} = \mathcal{X}^{\perp} = \{0\}$. But Proposition 5.12 says that $A^{\perp} = (\bigvee A)^{\perp}$, and hence $A^{\perp} = \{0\}$. The converse is an immediate consequence of Proposition 5.15. If $\mathcal{X}$ is a Hilbert space and $A^{\perp} = \{0\}$, then $\bigvee A = \mathcal{X}$. $\qquad\square$

An orthonormal set in an inner product space $\mathcal{X}$ that spans $\mathcal{X}$ is called an *orthonormal basis* for $\mathcal{X}$. In other words, a subset $B$ of an inner product space $\mathcal{X}$ is an orthonormal basis if

$\quad$(i)$\quad$ $B$ is an orthonormal set, and

$\quad$(ii)$\quad$ $\bigvee B = \mathcal{X}$.

This is a combined topological and algebraic concept, while the Hamel basis of Section 2.4 is a purely algebraic concept. However, *every orthonormal basis for a given inner product space $\mathcal{X}$ is included in a Hamel basis for the linear space $\mathcal{X}$* (Proposition 5.34 and Theorem 2.5). Recall from Proposition 5.37 that *every nonzero inner product space has a maximal orthonormal set*. Using the above terminology, Proposition 5.38(a) says that *if $B$ is an orthonormal basis for an inner product space $\mathcal{X}$, then $B$ is a maximal orthonormal set in $\mathcal{X}$*. Note that this does not ensure the existence of an orthonormal basis in incomplete inner product spaces, but in nonzero Hilbert spaces they do exist. In fact, in a Hilbert-space setting the concepts of maximal orthonormal set and orthonormal basis coincide (Proposition 5.38). That is, *B is an orthonormal basis for a Hilbert space $\mathcal{H}$ if and only if B is a maximal orthonormal set in $\mathcal{H}$.* Therefore (cf. Proposition 5.37 again), *every nonzero Hilbert space has an orthonormal basis.* As we could expect (suggested perhaps by Section 2.4), *the cardinality of all maximal orthonormal sets in an inner product space $\mathcal{X}$ is an invariant for $\mathcal{X}$.* First we prove this statement for finite-dimensional spaces.

**Proposition 5.39.** *Let $\mathcal{X}$ be an inner product space.*

(a) *If $\mathcal{X}$ is finite-dimensional, then every orthonormal basis for $\mathcal{X}$ is a Hamel basis for $\mathcal{X}$.*

(b) *If there exists an orthonormal basis for $\mathcal{X}$ with a finite number of vectors, then every orthonormal basis for $\mathcal{X}$ is a Hamel basis for $\mathcal{X}$.*

*Consequently, in both cases, every orthonormal basis for $\mathcal{X}$ has the same finite cardinality, which coincides with the linear dimension of $\mathcal{X}$.*

*Proof.* Let $\mathcal{X}$ be an inner product space.

(a) If $\mathcal{X}$ is finite-dimensional, then $\mathcal{M} = \mathcal{M}^-$ for every linear manifold $\mathcal{M}$ of $\mathcal{X}$. In particular, span $B = \bigvee B$ for every orthonormal basis $B$ for $\mathcal{X}$. Recall that every orthonormal basis for $\mathcal{X}$ is linearly independent (Proposition 5.34). Therefore, if $B$ is an orthonormal basis for $\mathcal{X}$, then $B$ is a linearly independent subset of $\mathcal{X}$ such that span $B = \mathcal{X}$; that is, $B$ is a Hamel basis for $\mathcal{X}$.

(b) If $B = \{e_i\}_{i=1}^n$ is an orthonormal basis for $\mathcal{X}$, then span $\{e_i\}_{i=1}^n$ is an $n$-dimensional linear manifold of $\mathcal{X}$, which in fact is a subspace of $\mathcal{X}$ (Corollary 4.29). Hence span $B = \bigvee B$. But $\bigvee B = \mathcal{X}$ so that $B$ is a linearly independent subset of $\mathcal{X}$ (Proposition 5.34) such that span $B = \mathcal{X}$. In other words, $B$ is a Hamel basis for $\mathcal{X}$.

Finally, recall that the cardinality of any Hamel basis for $\mathcal{X}$ is an invariant for $\mathcal{X}$: the linear dimension of $\mathcal{X}$ (Theorem 2.7). Thus, in both cases, $\#B = \dim \mathcal{X} \in \mathbb{N}$ for every orthonormal basis $B$ for $\mathcal{X}$. $\qquad\square$

To verify such an invariance for infinite-dimensional spaces we shall use the following fundamental inequality.

**Lemma 5.40.** (Bessel Inequality). *Let $\{e_\gamma\}_{\gamma \in \Gamma}$ be a family of vectors in an inner product space $\mathcal{X}$ and let $x$ be any vector in $\mathcal{X}$. If $\{e_\gamma\}_{\gamma \in \Gamma}$ is an orthonormal family, then $\{\langle x \,;\, e_\gamma \rangle\}_{\gamma \in \Gamma}$ is a square-summable family of scalars and*

$$\sum_{\gamma \in \Gamma} |\langle x \,;\, e_\gamma \rangle|^2 \leq \|x\|^2.$$

*Proof.* Take an arbitrary $x \in \mathcal{X}$ and an arbitrary finite set of indices $N \subseteq \Gamma$. Since $\{e_\gamma\}_{\gamma \in \Gamma}$ is an orthonormal family, it follows by the Pythagorean Theorem

(Proposition 5.8) that

$$
\begin{aligned}
0 \;\leq\; & \left\| \sum_{k\in N} \langle x\,;e_k\rangle e_k - x \right\|^2 \\
=\; & \left\| \sum_{k\in N} \langle x\,;e_k\rangle e_k \right\|^2 - 2\mathrm{Re}\sum_{k\in N} \langle x\,;e_k\rangle\langle e_k\,;x\rangle + \|x\|^2 \\
=\; & \sum_{k\in N} |\langle x\,;e_k\rangle|^2 - 2\sum_{k\in N} |\langle x\,;e_k\rangle|^2 + \|x\|^2,
\end{aligned}
$$

and hence

$$
\sum_{k\in N} |\langle x\,;e_k\rangle|^2 \;\leq\; \|x\|^2.
$$

Therefore, as $N$ is an arbitrary finite subset of $\Gamma$,

$$
\sum_{\gamma\in\Gamma} |\langle x\,;e_\gamma\rangle|^2 \;=\; \sup_N \sum_{k\in N} |\langle x\,;e_k\rangle|^2 \;\leq\; \|x\|^2,
$$

where the supremum is taken over all finite subsets of $\Gamma$. The family of scalars $\{\langle x\,;e_\gamma\rangle\}_{\gamma\in\Gamma}$ is then square-summable by Proposition 5.31. $\qquad\square$

**Corollary 5.41.** *Let $\{e_\gamma\}_{\gamma\in\Gamma}$ be any orthonormal family of vectors in an inner product space $X$. For each $x \in X$ the set $\{\gamma \in \Gamma\colon \langle x\,;e_\gamma\rangle \neq 0\}$ is countable.*

*Proof.* According to Lemma 5.40 $\{|\langle x\,;e_\gamma\rangle|^2\}_{\gamma\in\Gamma}$ is a summable family of non-negative numbers. Thus Corollary 5.28 ensures that $\{\gamma \in \Gamma\colon |\langle x\,;e_\gamma\rangle|^2 \neq 0\}$ is a countable set. That is, $\{\gamma \in \Gamma\colon \langle x\,;e_\gamma\rangle \neq 0\}$ is a countable set (since $\langle x\,;e_\gamma\rangle \neq 0$ if and only if $|\langle x\,;e_\gamma\rangle|^2 \neq 0$). $\qquad\square$

**Theorem 5.42.** *Every orthonormal basis for a given Hilbert space $\mathcal{H}$ has the same cardinality.*

*Proof.* If $\mathcal{H}$ is finite-dimensional, then the above result follows by Proposition 5.39. Suppose $\mathcal{H}$ is infinite-dimensional and let $B$ and $C$ be an arbitrary orthonormal basis for $\mathcal{H}$. Proposition 5.39(b) ensures that $B$ and $C$ are infinite, and so $\#\mathbb{N} \leq \#B$. For each $b \in B$ set $C_b = \{c \in C\colon \langle c\,;b\rangle \neq 0\}$. According to Corollary 5.41 $\#C_b \leq \#\mathbb{N}$, and hence $\#C_b \leq \#B$ for all $b \in B$. Then (cf. Theorems 1.10 and 1.9)

$$
\#\!\left( \bigcup_{b\in B} C_b \right) \;\leq\; \#(B\times B) \;=\; \#B
$$

because $B$ is infinite. Now observe that if $c \in C$, then $c \in C_b$ for some $b \in B$. (Reason: since $B$ is a maximal orthonormal set, it follows that if $c \in C$ and $c \perp B$, then $c = 0$, which contradicts the fact that $\|c\| = 1$; hence $\langle c\,;b\rangle \neq 0$ for some $b \in B$.) Therefore $C \subseteq \bigcup_{b\in B} C_b$. But $\bigcup_{b\in B} C_b \subseteq C$ (each $C_b$ is a subset of $C$). Thus $C = \bigcup_{b\in B} C_b$ and, consequently, $\#C \leq \#B$. Since $C$ also is infinite, we may swap $B$ and $C$, apply the same argument, and get $\#B \leq \#C$. Hence $\#C = \#B$ by the Cantor–Bernstein Theorem (Theorem 1.6). $\qquad\square$

It is worth noticing that the above theorem might be restated as "*every maximal orthonormal set in a given inner product space $\mathcal{X}$ has the same cardinality*". The proof remains essentially the same. Such an invariant (i.e., the cardinality of every orthonormal basis for a given Hilbert space) is called the *orthogonal dimension* of the Hilbert space $\mathcal{H}$. According to Proposition 5.39 the orthogonal dimension of $\mathcal{H}$ is finite if and only if its linear dimension is finite and, in this case, these two dimensions coincide. Therefore, the concepts of "infinite-dimensional" and "finite-dimensional" Hilbert spaces are unambiguously defined.

**Proposition 5.43.** *If $\mathcal{X}$ is a separable inner product space, then every orthonormal set in $\mathcal{X}$ is countable.*

*Proof.* Let $B = \{e_\gamma\}_{\gamma \in \Gamma}$ be any family of orthonormal vectors in the inner product space $\mathcal{X}$. If $\mathcal{X}$ is separable (as a metric space), then there exists a countable dense subset $A = \{a_k\}_{k \in \mathbb{N}}$ of $\mathcal{X}$. Since $A$ is dense in $\mathcal{X}$, it follows by Proposition 3.32 that every nonempty open ball centered at any point of $\mathcal{X}$ meets $A$. In particular, every open ball of radius, say $\frac{1}{2}$, centered at any point of $B$ meets $A$. Then for each $\gamma \in \Gamma$ there exists an integer $k_\gamma \in \mathbb{N}$ such that $\|e_\gamma - a_{k_\gamma}\| < \frac{1}{2}$. This establishes a function $F \colon \Gamma \to \mathbb{N}$ that assigns to each $\gamma \in \Gamma$ the integer $k_\gamma \in \mathbb{N}$. We shall show that this function is injective. Indeed, consider the family $\{a_{k_\gamma}\}_{\gamma \in \Gamma}$ and recall that $B$ is an orthonormal set. Thus

$$\begin{aligned}
\sqrt{2} = \|e_\alpha - e_\beta\| &= \|e_\alpha - a_{k_\alpha} - e_\beta + a_{k_\beta} + a_{k_\alpha} - a_{k_\beta}\| \\
&\leq \|e_\alpha - a_{k_\alpha}\| + \|e_\beta - a_{k_\beta}\| + \|a_{k_\alpha} - a_{k_\beta}\| \\
&\leq 1 + \|a_{k_\alpha} - a_{k_\beta}\|,
\end{aligned}$$

and hence $\|a_{k_\alpha} - a_{k_\beta}\| \geq \sqrt{2} - 1 > 0$, for every pair of distinct indices $\alpha, \beta \in \Gamma$. This implies that $k_\alpha \neq k_\beta$ whenever $\alpha \neq \beta$, which means that $F \colon \Gamma \to \mathbb{N}$ is injective. Therefore $\#\Gamma \leq \#\mathbb{N}$, and so $B$ is countable. $\qquad\square$

Recall that every Hilbert space has an orthonormal basis. The next theorem characterizes the separable Hilbert spaces in terms of their orthogonal dimension.

**Theorem 5.44.** *A Hilbert space is separable if and only if it has a countable orthonormal basis.*

*Proof.* Propositions 4.9(b) and 5.43. $\qquad\square$

We close this section with a useful result for constructing countable orthonormal sets, which is known as the *Gram–Schmidt orthogonalization process*.

**Proposition 5.45.** *Let $\mathcal{X}$ be an inner product space. If $A = \{a_k\}$ is a countable linearly independent nonempty subset of $\mathcal{X}$, then there exists a countable orthonormal* *subset $B = \{e_k\}$ of $\mathcal{X}$ with the following property:* span $\{e_k\}_{k=1}^n =$ span $\{a_k\}_{k=1}^n$ *for every integer $n$ such that $1 \leq n \leq \#A$, and hence* span $B =$ span $A$.

*Proof.* Let $A = \{a_k\}$ be a countable (either a finite and nonempty or a countably infinite) linearly independent subset of an inner product space $\mathcal{X}$.

*Claim.* For every integer $n \geq 1$ ($n \leq \#A$) there exists an orthonormal subset $\{e_k\}_{k=1}^n$ of $\mathcal{X}$ such that span $\{e_k\}_{k=1}^n =$ span $\{a_k\}_{k=1}^n$.

*Proof.* $a_1 \neq 0$ (for $A = \{a_k\}$ is a linearly independent subset of $\mathcal{X}$). Set $e_1 = \|a_1\|^{-1} a_1$ in $\mathcal{X}$ so that the result holds for $n = 1$. (Recall that any singleton $\{x\}$ such that $\|x\| = 1$ is an orthonormal set in $\mathcal{X}$.) If $\#A = 1$ the proposition is proved. Thus assume that $1 < \#A \leq \aleph_0$. Suppose the result holds for some integer $n$ such that $1 \leq n < \#A$. Observe that

$$a_{n+1} - \sum_{k=1}^n \langle a_{n+1} ; e_k \rangle e_k \neq 0$$

(otherwise $a_{n+1} \in$ span $\{e_k\}_{k=1}^n =$ span $\{a_k\}_{k=1}^n$, which contradicts the fact that $A$ is linearly independent). Set

$$e_{n+1} = \beta_{n+1} \left( a_{n+1} - \sum_{k=1}^n \langle a_{n+1} ; e_k \rangle e_k \right),$$

where $\beta_{n+1} = \|a_{n+1} - \sum_{k=1}^n \langle a_{n+1} ; e_k \rangle e_k \|^{-1}$ so that $\|e_{n+1}\| = 1$. Take any integer $j = 1, \dots, n$ and note that

$$\begin{aligned}
\langle e_{n+1} ; e_j \rangle &= \beta_{n+1} \left( \langle a_{n+1} ; e_j \rangle - \sum_{k=1}^n \langle a_{n+1} ; e_k \rangle \langle e_k ; e_j \rangle \right) \\
&= \beta_{n+1} \left( \langle a_{n+1} ; e_j \rangle - \langle a_{n+1} ; e_j \rangle \right) = 0
\end{aligned}$$

because $\{e_k\}_{k=1}^n$ is an orthonormal set. Then $e_{n+1} \perp \{e_k\}_{k=1}^n$, and hence $\{e_k\}_{k=1}^{n+1}$ is an orthonormal set. But $e_{n+1} \in$ span $(\{e_k\}_{k=1}^n \cup \{a_{n+1}\}) =$ span $\{a_k\}_{k=1}^{n+1}$. Therefore, span $\{e_k\}_{k=1}^{n+1} =$ span $(\{e_k\}_{k=1}^n \cup \{e_{n+1}\}) =$ span $(\{a_k\}_{k=1}^n \cup \{e_{n+1}\}) =$ span $\{a_k\}_{k=1}^{n+1}$, which completes the proof by induction. $\square$

Finally, put $B = \bigcup_{n=1}^{\#A} \{e_k\}_{k=1}^n$. Since $\{e_k\}_{k=1}^n \subseteq \{e_k\}_{k=1}^{n+1}$ for each integer $n$ such that $1 \leq n \leq \#A$, every pair of distinct vectors in $B$, say $e$ and $e'$, lies in $\{e_k\}_{k=1}^m$ for some integer $m$ such that $1 \leq m \leq \#A$. Since $\{e_k\}_{k=1}^m$ is an orthonormal set, $e \perp e'$ and $\|e\| = \|e'\| = 1$. Thus $B$ is an orthonormal set. Moreover, if span $\{e_k\}_{k=1}^n =$ span $\{a_k\}_{k=1}^n$ for every integer $n$ such that $1 \leq n \leq \#A$, then span $B =$ span $A$. $\square$

**Corollary 5.46.** *There is no Hilbert space with a countably infinite Hamel basis. In other words, a Hamel basis for a Hilbert space is either finite or uncountable.*

*Proof.* Let $\mathcal{H}$ be a Hilbert space and suppose there exists a countably infinite Hamel basis for the linear space $\mathcal{H}$, say $\{f_k\}_{k=1}^{\infty}$. The Gram–Schmidt orthogonalization process ensures the existence of a countably infinite orthonormal set, say $\{e_k\}_{k=1}^{\infty}$, such that span $\{e_k\}_{k=1}^{n} = $ span $\{f_k\}_{k=1}^{n}$ for every $n \geq 1$. If $\{\alpha_k\}_{k=1}^{\infty}$ is any square-summable sequence of scalars, then $\{\alpha_k e_k\}_{k=1}^{\infty}$ is a square-summable sequence of pairwise orthogonal vectors in $\mathcal{H}$, and hence the infinite series $\sum_{k=1}^{\infty} \alpha_k e_k$ converges in the Hilbert space $\mathcal{H}$ by Corollary 5.9(b). Take an arbitrary square-summable sequence $\{\alpha_k\}_{k=1}^{\infty}$ of nonzero scalars (e.g., $\alpha_k = \frac{1}{k}$ for each $k \geq 1$) and set $x = \sum_{k=1}^{\infty} \alpha_k e_k$ in $\mathcal{H}$. Since $x \notin $ span $\{e_k\}_{k=1}^{n}$ and span $\{e_k\}_{k=1}^{n} = $ span $\{f_k\}_{k=1}^{n}$ for every $n \geq 1$, it follows that $x \notin $ span $\{f_k\}_{k=1}^{\infty}$ (i.e., $x$ is not a finite linear combination of vectors from $\{f_k\}_{k=1}^{\infty}$), which contradicts the assumption that $\{f_k\}_{k=1}^{\infty}$ is a Hamel basis for $\mathcal{H}$. Conclusion: There is no countably infinite Hamel basis for $\mathcal{H}$. $\qquad\square$

This result is the main reason why the concept of linear dimension is of no interest in Hilbert space theory. Here the useful notion is orthogonal dimension and, from now on, $\dim \mathcal{H}$ will always mean the orthogonal dimension of the Hilbert space $\mathcal{H}$. Observe that the notions of finite-dimensional and infinite-dimensional Hilbert spaces remain unchanged.

## 5.9 The Fourier Series Theorem

The Fourier Series Theorem states the fundamental properties of an orthonormal basis in a Hilbert space. Precisely, it exhibits a collection of necessary and sufficient conditions for an orthonormal set be an orthonormal basis in a Hilbert space. Before stating and proving it, we need the following auxiliary result.

**Proposition 5.47.** *Let $\{e_k\}_{k\in N}$ be a finite orthonormal set in a Hilbert space $\mathcal{H}$ and let $x$ be an arbitrary vector in $\mathcal{H}$.*

$$u_x = \sum_{k \in N} \langle x \, ; e_k \rangle e_k$$

*is the unique vector in* span $\{e_k\}_{k \in N}$ *such that*

$$\|x - u_x\| \leq \|x - u\| \quad \text{for every} \quad u \in \text{span} \, \{e_k\}_{k \in N}.$$

*Proof.* Since span $\{e_k\}_{k \in N}$ is a finite-dimensional linear manifold of $\mathcal{H}$ (Theorem 2.6), it follows by Corollary 4.29 that it is a subspace of $\mathcal{H}$ so that span $\{e_k\}_{k \in N} = \left(\text{span} \, \{e_k\}_{k \in N}\right)^{-} = \bigvee \{e_k\}_{k \in N}$. Theorem 5.13 says that there exists a unique vector $u_x$ in $\bigvee \{e_k\}_{k \in N}$ such that

$$\|x - u_x\| \leq \|x - u\| \quad \text{for every} \quad u \in \bigvee \{e_k\}_{k \in N}.$$

Moreover, Theorem 5.13 also says that this $u_x$ is the unique vector in $\bigvee \{e_k\}_{k \in N}$ such that

$$x - u_x \perp \bigvee \{e_k\}_{k \in N}.$$

Since $u_x \in \text{span }\{e_k\}_{k \in N}, u_x = \sum_{k \in N} \alpha_k e_k$ for some finite family of scalars $\{\alpha_k\}_{k \in N}$. Since $x - u_x \perp \bigvee \{e_k\}_{k \in N}, x - u_x \perp e_j$ for every $j \in N$. Thus, recalling that $\{e_k\}_{k \in N}$ is an orthonormal set,

$$0 = \langle x - u_x \, ; e_j \rangle = \langle x \, ; e_j \rangle - \sum_{k \in N} \alpha_k \langle e_k \, ; e_j \rangle = \langle x \, ; e_j \rangle - \alpha_j$$

for every $j \in N$, and hence $u_x = \sum_{k \in N} \langle x \, ; e_k \rangle e_k$.    $\square$

**Theorem 5.48.** (The Fourier Series Theorem). *Let* $B = \{e_\gamma\}_{\gamma \in \Gamma}$ *be an orthonormal set in a Hilbert space* $\mathcal{H}$. *The following assertions are pairwise equivalent.*

(a)  *$B$ is an orthonormal basis for* $\mathcal{H}$.

(b)  *Every $x \in \mathcal{H}$ has a unique expansion on $B$, namely*

$$x = \sum_{\gamma \in \Gamma} \langle x \, ; e_\gamma \rangle e_\gamma.$$

*This is referred to as the Fourier series expansion of $x$. The elements of the family of scalars $\{\langle x \, ; e_\gamma \rangle\}_{\gamma \in \Gamma}$ are called the Fourier coefficients of $x$ with respect to $B$.*

(c)  *For every pair of vectors $x$, $y$ in $\mathcal{H}$,*

$$\langle x \, ; y \rangle = \sum_{\gamma \in \Gamma} \langle x \, ; e_\gamma \rangle \overline{\langle y \, ; e_\gamma \rangle}.$$

(d)  *For every $x \in \mathcal{H}$,*

$$\|x\|^2 = \sum_{\gamma \in \Gamma} |\langle x \, ; e_\gamma \rangle|^2.$$

*This is the Parseval identity.*

(e)  *Every linear manifold of $\mathcal{H}$ that includes $B$ is dense in $\mathcal{H}$.*

*Proof.* We shall verify that (e)$\Leftrightarrow$(a)$\Leftrightarrow$(d) and (b)$\Rightarrow$(c)$\Rightarrow$(d)$\Rightarrow$(b).

*Proof of* (a)$\Leftrightarrow$(d). Take any $x$ in $\mathcal{H}$. If $(\text{span } B)^- = \mathcal{H}$, then every nonempty open ball centered at $x$ meets span $B$ (Proposition 3.32). That is, for every $\varepsilon > 0$ there exists a finite set $N_\varepsilon \subseteq \Gamma$ and a vector $u \in \text{span }\{e_k\}_{k \in N_\varepsilon}$ such that $\|x - u\| < \varepsilon$. Set $u_x = \sum_{k \in N_\varepsilon} \langle x \, ; e_k \rangle e_k$ in span $\{e_k\}_{k \in N_\varepsilon}$. Proposition 5.47 ensures that $\|x - u_x\| \le \|x - u\|$, and hence

$$\left\| x - \sum_{k \in N_\varepsilon} \langle x \, ; e_k \rangle e_k \right\|^2 < \varepsilon.$$

Since $\{e_k\}_{k\in N_\varepsilon}$ is an orthonormal set, it follows by the Pythagorean Theorem (Proposition 5.8) that

$$
\left\| \sum_{k\in N_\varepsilon} \langle x \,;\, e_k \rangle e_k - x \right\|^2 \;=\; \left\| \sum_{k\in N_\varepsilon} \langle x \,;\, e_k \rangle e_k \right\|^2 - 2\mathrm{Re}\sum_{k\in N_\varepsilon} \langle x \,;\, e_k \rangle \langle e_k \,;\, x \rangle + \|x\|^2
$$

$$
=\; \sum_{k\in N_\varepsilon} |\langle x \,;\, e_k \rangle|^2 - 2\sum_{k\in N_\varepsilon} |\langle x \,;\, e_k \rangle|^2 + \|x\|^2
$$

$$
=\; \|x\|^2 - \sum_{k\in N_\varepsilon} |\langle x \,;\, e_k \rangle|^2.
$$

Therefore, by the Bessel inequality (Lemma 5.40) and since $\sum_{k\in N_\varepsilon} |\langle x \,;\, e_k \rangle|^2 \le \sum_{\gamma\in\Gamma} |\langle x \,;\, e_\gamma \rangle|^2$ (Proposition 5.31), we get

$$
\left| \|x\|^2 - \sum_{\gamma\in\Gamma} |\langle x \,;\, e_\gamma \rangle|^2 \right| \;=\; \|x\|^2 - \sum_{\gamma\in\Gamma} |\langle x \,;\, e_\gamma \rangle|^2 \;\le\; \|x\|^2 - \sum_{k\in N_\varepsilon} |\langle x \,;\, e_k \rangle|^2
$$

$$
=\; \left\| x - \sum_{k\in N_\varepsilon} \langle x \,;\, e_k \rangle e_k \right\|^2 \;<\; \varepsilon^2
$$

for all $\varepsilon > 0$. Hence $\|x\|^2 = \sum_{\gamma\in\Gamma} |\langle x \,;\, e_\gamma \rangle|^2$. Outcome: (a) implies (d). Conversely, If the orthonormal set $B$ is not an orthonormal basis for $\mathcal{H}$, then $B$ is not a maximal orthonormal set in the Hilbert space $\mathcal{H}$ (Proposition 5.38(b)). Thus there exists a unit vector $e$ in $\mathcal{H}$ such that $B \cup \{e\}$ is an orthonormal set (Proposition 5.36). Therefore, $\langle e \,;\, e_\gamma \rangle = 0$ for all $\gamma \in \Gamma$, and hence $1 = \|e\| \neq \sum_{\gamma\in\Gamma} |\langle e \,;\, e_\gamma \rangle|^2 = 0$. Conclusion: If (a) fails then (d) fails. Equivalently, (d) implies (a).

Proposition 4.9(a) ensures that (a)$\Leftrightarrow$(e). It is readily verified that (b)$\Rightarrow$(c). Indeed, if (b) holds, then

$$
\langle x \,;\, y \rangle \;=\; \left\langle \sum_{\alpha\in\Gamma} \langle x \,;\, e_\alpha \rangle e_\alpha \,;\, \sum_{\beta\in\Gamma} \langle y \,;\, e_\beta \rangle e_\beta \right\rangle = \sum_{\alpha\in\Gamma} \langle x \,;\, e_\alpha \rangle \left\langle e_\alpha \,;\, \sum_{\beta\in\Gamma} \langle y \,;\, e_\beta \rangle e_\beta \right\rangle
$$

$$
=\; \sum_{\alpha\in\Gamma} \langle x \,;\, e_\alpha \rangle \sum_{\beta\in\Gamma} \overline{\langle y \,;\, e_\beta \rangle} \langle e_\alpha \,;\, e_\beta \rangle \;=\; \sum_{\alpha\in\Gamma} \langle x \,;\, e_\alpha \rangle \overline{\langle y \,;\, e_\alpha \rangle}
$$

for every $x, y \in \mathcal{H}$ (because $\{e_\gamma\}_{\gamma\in\Gamma}$ is an orthonormal set). Hence (b) implies (c). Moreover, (c)$\Rightarrow$(d) trivially.

*Proof of* (d)$\Rightarrow$(b). Take any $x$ in $\mathcal{H}$. Assertion (d) implies that $\{\langle x \,;\, e_\gamma \rangle e_\gamma\}_{\gamma\in\Gamma}$ is a square-summable family of orthogonal vectors in $\mathcal{H}$. Thus $\{\langle x \,;\, e_\gamma \rangle e_\gamma\}_{\gamma\in\Gamma}$ is a summable family of orthogonal vectors in the Hilbert space $\mathcal{H}$ by Theorem 5.32(b). Let $x'$ be the sum of $\{\langle x \,;\, e_\gamma \rangle e_\gamma\}_{\gamma\in\Gamma}$. That is, $x' = \sum_{\gamma\in\Gamma} \langle x \,;\, e_\gamma \rangle e_\gamma \in \mathcal{H}$. Since

$\{e_\gamma\}_{\gamma \in \Gamma}$ is an orthonormal set, it follows by Theorem 5.32(a) that

$$
\begin{aligned}
\|x' - x\|^2 &= \left\| \sum_{\gamma \in \Gamma} \langle x \,;\, e_\gamma \rangle e_\gamma - x \right\|^2 \\[2mm]
&= \left\| \sum_{\gamma \in \Gamma} \langle x \,;\, e_\gamma \rangle e_\gamma \right\|^2 - 2\,\mathrm{Re} \sum_{\gamma \in \Gamma} \langle x \,;\, e_\gamma \rangle \langle e_\gamma \,;\, x \rangle + \|x\|^2 \\[2mm]
&= \sum_{\gamma \in \Gamma} |\langle x \,;\, e_\gamma \rangle|^2 - 2 \sum_{\gamma \in \Gamma} |\langle x \,;\, e_\gamma \rangle|^2 + \|x\|^2 = 0
\end{aligned}
$$

because assertion (d) holds true (i.e., because $\|x\|^2 = \sum_{\gamma \in \Gamma} |\langle x \,;\, e_\gamma \rangle|^2$). Therefore, $x' = x$ so that

$$
x = \sum_{\gamma \in \Gamma} \langle x \,;\, e_\gamma \rangle e_\gamma .
$$

Finally, if $x = \sum_{\gamma \in \Gamma} \alpha_\gamma(x) e_\gamma$ for some family of scalars $\{\alpha_\gamma(x)\}_{\gamma \in \Gamma}$, then $\sum_{\gamma \in \Gamma} \big(\alpha_\gamma(x) - \langle x \,;\, e_\gamma \rangle\big) e_\gamma = 0$ so that $\sum_{\gamma \in \Gamma} |\alpha_\gamma(x) - \langle x \,;\, e_\gamma \rangle|^2 = 0$ by Theorem 5.32(a). Thus $\alpha_\gamma(x) = \langle x \,;\, e_\gamma \rangle$ for every $\gamma \in \Gamma$, which proves that the Fourier series expansion of $x$ is unique. $\qquad\square$

Remark: Consider the sums in (b), (c) and (d) — Theorem 5.48. If $\mathcal{H}$ is a finite-dimensional Hilbert space, then any orthonormal basis for $\mathcal{H}$ is finite (Proposition 5.39), and hence these are finite sums. If $\mathcal{H}$ is infinite-dimensional and separable, then any orthonormal basis $B$ for $\mathcal{H}$ is countably infinite (Proposition 5.43) so that $B$ can be indexed by $\mathbb{N}$ (or by $\mathbb{N}_0$, or by $\mathbb{Z}$). For instance, suppose $B = \{e_k\}_{k \in \mathbb{N}}$ is an orthonormal basis for an infinite-dimensional separable Hilbert space $\mathcal{H}$. In this case we have a countable summable family of vectors $\{\langle x \,;\, e_k \rangle e_k\}_{k \in \mathbb{N}}$ in (b), a countable summable family of scalars $\{\langle x \,;\, e_k \rangle \overline{\langle y \,;\, e_k \rangle}\}_{k \in \mathbb{N}}$ in (c), and a countable summable family of nonnegative numbers $\{|\langle x \,;\, e_k \rangle|^2\}_{k \in \mathbb{N}}$ in (d). If $\{e_k\}_{k=1}^{\infty}$ is *any* $\mathcal{H}$-valued orthonormal sequence containing all vectors from the orthonormal basis $\{e_k\}_{k \in \mathbb{N}}$ for $\mathcal{H}$ (which is itself an orthonormal basis for $\mathcal{H}$), then the Fourier series expansion for any $x \in \mathcal{H}$ can be written as

$$
x = \sum_{k=1}^{\infty} \langle x \,;\, e_k \rangle e_k .
$$

Similarly,

$$
\langle x \,;\, y \rangle = \sum_{k=1}^{\infty} \langle x \,;\, e_k \rangle \overline{\langle y \,;\, e_k \rangle}
$$

for every $x, y$ in $\mathcal{H}$, and

$$
\|x\|^2 = \sum_{k=1}^{\infty} |\langle x \,;\, e_k \rangle|^2
$$

for every $x \in \mathcal{H}$, where all the above infinite series are unconditionally convergent. If $\mathcal{H}$ is a nonseparable Hilbert space, then any orthonormal basis for $\mathcal{H}$ is uncountable (Theorem 5.44). However, Corollary 5.41 says that, even in this case, the sums in (b), (c) and (d) have only a countable number of nonzero summands for each $x, y \in \mathcal{H}$.

**Example 5L.** In this example we shall exhibit an orthonormal basis for some classical separable Hilbert spaces.

(a) First recall that, for any integer $n \geq 1$, $\mathbb{F}^n$ is a Hilbert space (see Example 5A). Moreover, the finite set $\{e_k\}_{k=1}^n$, where each $e_k = (\delta_{k1}, \dots, \delta_{kn}) \in \mathbb{F}^n$ has 1 at the $k$th position and zeros elsewhere, clearly is an orthonormal set in $\mathbb{F}^n$ and also a Hamel basis for the linear space $\mathbb{F}^n$ (Example 2I). Then $\{e_k\}_{k=1}^n$ is an orthonormal basis for the finite-dimensional Hilbert space $\mathbb{F}^n$, which is called the *canonical orthonormal basis* for $\mathbb{F}^n$.

(b) Now let $\ell_+^2$ be the Hilbert space of Example 5B. Consider the $\ell_+^2$-valued sequence $\{e_k\}_{k\in\mathbb{N}}$, where each $e_k$ is an scalar-valued sequence in $\ell_+^2$ with just one nonzero entry (equal to 1) at the $k$th position. That is, $e_k = \{\delta_{kj}\}_{j\in\mathbb{N}} \in \ell_+^2$ for every $k \in \mathbb{N}$ with $\delta_{kj} = 1$ if $j = k$ and $\delta_{kj} = 0$ if $j \neq k$. Again, it is clear that $\{e_k\}_{k\in\mathbb{N}}$ is an orthonormal sequence of vectors in $\ell_+^2$. Moreover, recall that $x = \{\xi_j\}_{j\in\mathbb{N}}$ lies in $\ell_+^2$ if and only if $\sum_{j=1}^\infty |\xi_j|^2 < \infty$. Therefore, if $x = \{\xi_j\}_{l\in\mathbb{N}} \in \ell_+^2$, then $x = \lim x_n$, where $x_n = (\xi_1, \dots, \xi_n, 0, 0, 0, \dots) \in \ell_+^2$ for every $n \in \mathbb{N}$. Since each $x_n \in \mathrm{span}\,\{e_k\}_{k\in\mathbb{N}}$ (for $x_n = \sum_{k=1}^n \xi_k e_k$), it follows by the Closed Set Theorem that $x \in \left(\mathrm{span}\,\{e_k\}_{k\in\mathbb{N}}\right)^- = \bigvee\{e_k\}_{k\in\mathbb{N}}$. Hence $\ell_+^2 \subseteq \bigvee\{e_k\}_{k\in\mathbb{N}} \subseteq \ell_+^2$, which means that $\bigvee\{e_k\}_{k\in\mathbb{N}} = \ell_+^2$. Conclusion: The orthonormal sequence $\{e_k\}_{k\in\mathbb{N}}$ is an orthonormal basis for $\ell_+^2$. It can be similarly shown that $\{e_k\}_{k\in\mathbb{Z}}$, where each $e_k$ is a scalar-valued net in $\ell^2$ with just one nonzero entry (equal to 1) at the $k$th position (i.e., $e_k = \{\delta_{kj}\}_{j\in\mathbb{Z}} \in \ell^2$ for every $k \in \mathbb{Z}$ with $\delta_{kj} = 1$ if $j = k$ and $\delta_{kj} = 0$ if $j \neq k$) is an orthonormal basis for the Hilbert space $\ell^2$ of Example 5B. These are referred to as the *canonical orthonormal basis* for $\ell_+^2$ and $\ell^2$, respectively.

Next consider the complex Hilbert space $L^2(S)$ for some nondegenerate interval $S$ of the real line (see Example 5D). Recall that $L^2(S)$ is the completion of $R^2(S)$, which in turn is the quotient space $r^2(S)/\!\sim$ of Example 3E. Also recall that, according to the usual convention, we write $x \in L^2(S)$ instead of $[x] \in L^2(S)$, where $x$ is any representative of the equivalence class $[x]$.

(c) Suppose $S = [a, b]$ for some pair of real numbers $a < b$ so that $L^2(S) = L^2[a, b]$. It is not difficult to verify that the countable set $\{e_k\}_{k\in\mathbb{Z}}$, with each $e_k$ in $L^2[a, b]$ given by

$$e_k(t) = (b - a)^{-\frac{1}{2}} \exp\left(2\pi i k\, \tfrac{t-a}{b-a}\right) \quad \text{for every} \quad t \in [a, b],$$

is a maximal orthonormal set in $L^2[a, b]$, and hence an orthonormal basis for the Hilbert space $L^2[a, b]$. In particular, for $S = [0, 2\pi]$, the countable set $\{e_k\}_{k\in\mathbb{Z}}$, with each $e_k$ in $L^2[0, 2\pi]$ given by

$$e_k(t) = \frac{1}{\sqrt{2\pi}} e^{ikt} = \frac{1}{\sqrt{2\pi}}\left(\cos kt + i \sin kt\right) \quad \text{for every} \quad t \in [0, 2\pi],$$

is an orthonormal basis for the Hilbert space $L^2[0, 2\pi]$. Similarly, let $\Delta$ be the open unit disk (about the origin) in the complex plane $\mathbb{C}$, and let $\Gamma = \partial\Delta$ denote the unit circle in $\mathbb{C}$. Suppose the length of $\Gamma$ is normalized (i.e., suppose $L^2(\Gamma)$ is the Hilbert space of all equivalence classes of square-integrable complex functions defined on $\Gamma$ with respect to normalized Lebesgue measure $\mu$ on the unit circle so that $\mu(\Gamma) = 1$). The countable set $\{e_k\}_{k\in\mathbb{Z}}$, with each $e_k$ in $L^2(\Gamma)$ given by

$$e_k(z) = z^k \quad \text{for every} \quad z = e^{i\theta} \in \Gamma \quad (0 \le \theta < 2\pi),$$

is an orthonormal basis for the Hilbert space $L^2(\Gamma)$.

(d) If $S = [0, \infty)$, then it can be shown that the sequence $\{e_n\}_{n=0}^{\infty}$ with each $e_n$ in $L^2[0, \infty)$ given by

$$e_n(t) = \tfrac{1}{n!} e^{-\frac{t}{2}} L_n(t) \quad \text{for every} \quad t \ge 0,$$

where each $L_n \in L^2[0, \infty)$ is defined by

$$L_n(t) = e^t \frac{d^n}{dt^n}\left(t^n e^{-t}\right) = \textstyle\sum_{k=0}^{n}(-1)^k \binom{n}{k} \frac{n!}{k!} t^k \quad \text{for every} \quad t \ge 0,$$

is an orthonormal basis for the Hilbert space $L^2[0, \infty)$. (Note: the elements of $\{e_n\}_{n=0}^{\infty}$ and $\{L_n\}_{n=0}^{\infty}$ are referred to as *Chebyshev–Laguerre functions* and *Chebyshev–Laguerre polynomials*, respectively.) If $S = \mathbb{R}$, then it can also be shown that the sequence $\{e_n\}_{n=0}^{\infty}$ with each $e_n$ in $L^2(-\infty, \infty)$ given by

$$e_n(t) = \left(2^n n! \sqrt{\pi}\right)^{-\frac{1}{2}} e^{-\frac{t^2}{2}} H_n(t) \quad \text{for every} \quad t \in \mathbb{R},$$

where each $H_n \in L^2(-\infty, \infty)$ is defined by

$$H_n(t) = (-1)^n e^{t^2} \frac{d^n}{dt^n} e^{-t^2} \quad \text{for every} \quad t \in \mathbb{R},$$

is an orthonormal basis for the Hilbert space $L^2(-\infty, \infty)$. (Note: the elements of $\{e_n\}_{n=0}^{\infty}$ and $\{H_n\}_{n=0}^{\infty}$ are called *Chebyshev–Hermite functions* and *Chebyshev–Hermite polynomials*, respectively.)

**Example 5M.** All the orthonormal bases in the previous example are countable, and so those Hilbert spaces are all separable Hilbert spaces. However, there exist nonseparable Hilbert spaces. Actually, *there exist Hilbert spaces of arbitrary orthogonal dimension*. Indeed, let $\Gamma$ be any index set, and let $\ell_\Gamma^2$ be the Hilbert space of Example 5K. Consider the family $\{e_\gamma\}_{\gamma\in\Gamma}$ of vectors in $\ell_\Gamma^2$, where each $e_\gamma = \{\delta_{\gamma\alpha}\}_{\alpha\in\Gamma}$ is a family of scalars such that $\delta_{\gamma\alpha} = 1$ if $\alpha = \gamma$ and $\delta_{\gamma\alpha} = 0$

if $\alpha \neq \gamma$. Note that each $e_\gamma$ is a square-summable family of scalars (with just one nonzero element equal to 1) so that each $e_\gamma$ in fact lies in $\ell_\Gamma^2$. It is also clear that $\langle e_\alpha ; e_\beta \rangle = \delta_{\alpha\beta}$ for every $\alpha, \beta \in \Gamma$, and hence $\{e_\gamma\}_{\gamma \in \Gamma}$ is an orthonormal family of vectors in $\ell_\Gamma^2$. Let $x$ be any unit vector in $\ell_\Gamma^2$. That is, $x = \{\xi_\gamma\}_{\gamma \in \Gamma}$ is a square-summable family of scalars such that $\|x\|^2 = \sum_{\gamma \in \Gamma} |\xi_\gamma|^2 = 1$. If $x \perp e_\gamma$ for all $\gamma \in \Gamma$, then $0 = \langle x ; e_\gamma \rangle = \sum_{\alpha \in \Gamma} \xi_\alpha \delta_{\gamma\alpha} = \xi_\gamma$ for all $\gamma \in \Gamma$, and hence $x = 0$. But this contradicts the fact that $\|x\| = 1$. Therefore, there is no unit vector $x$ in $\ell_\Gamma^2$ for which $\{e_\gamma\}_{\gamma \in \Gamma} \cup \{x\}$ is an orthonormal set. That is (Proposition 5.36), $\{e_\gamma\}_{\gamma \in \Gamma}$ is a maximal orthonormal set in the Hilbert space $\ell_\Gamma^2$ so that $\{e_\gamma\}_{\gamma \in \Gamma}$ is an orthonormal basis for $\ell_\Gamma^2$ (Proposition 5.38). Hence

$$\dim \ell_\Gamma^2 = \#\Gamma.$$

If the index set $\Gamma$ is uncountable, then $\{e_\gamma\}_{\gamma \in \Gamma}$ is an uncountable orthonormal basis for $\ell_\Gamma^2$ so that, in this case, the Hilbert space $\ell_\Gamma^2$ is not separable (Theorem 5.44).

**Theorem 5.49.** *Two Hilbert spaces are unitarily equivalent if and only if they have the same orthogonal dimension.*

*Proof.* Let $\mathcal{H}$ and $\mathcal{K}$ be Hilbert spaces (over the same field $\mathbb{F}$) and let $B = \{e_\gamma\}_{\gamma \in \Gamma}$ be an orthonormal basis for $\mathcal{H}$.

(a) If $\mathcal{H}$ and $\mathcal{K}$ are unitarily equivalent, then there exists a unitary transformation $U \in \mathcal{B}[\mathcal{H}, \mathcal{K}]$ so that $U(B)$ is an orthonormal basis for $\mathcal{K}$. Indeed, since $B$ is an orthonormal set in $\mathcal{H}$, and since $U$ preserves the inner product, it follows that $U(B)$ is an orthonormal set in $\mathcal{K}$. Moreover, as $U$ is surjective, for every $y \in \mathcal{K}$ there exists $x \in \mathcal{H}$ such that $y = Ux$, and hence (Theorem 5.48)

$$y = U \sum_{\gamma \in \Gamma} \langle x ; e_\gamma \rangle e_\gamma = \sum_{\gamma \in \Gamma} \langle Ux ; Ue_\gamma \rangle Ue_\gamma = \sum_{\gamma \in \Gamma} \langle y ; Ue_\gamma \rangle Ue_\gamma$$

because $U$ is a bounded linear transformation that preserves the inner product. Thus the orthogonal set $U(B) = \{Ue_\gamma\}_{\gamma \in \Gamma}$ is an orthonormal basis for $\mathcal{K}$ according to Theorem 5.48. Since $U$ is invertible, it establishes a one-to-one correspondence between $B$ and $U(B)$, and so $\#B = \#U(B)$. Therefore, $\dim \mathcal{H} = \dim \mathcal{K}$.

(b) Conversely, suppose $\dim \mathcal{H} = \dim \mathcal{K}$. Then $\#B = \#C$, where $C$ is an arbitrary orthonormal basis for $\mathcal{K}$, so that $C$ and $B$ can be similarly indexed, say $C = \{f_\gamma\}_{\gamma \in \Gamma}$. Now consider the Hilbert space $\ell_\Gamma^2$ of Example 5K (over the field $\mathbb{F}$) and take an arbitrary $x \in \mathcal{H}$. The Parseval identity (Theorem 5.48) says that $\|x\|^2 = \sum_{\gamma \in \Gamma} |\langle x ; e_\gamma \rangle|^2$, and hence $\{\langle x ; e_\gamma \rangle\}_{\gamma \in \Gamma}$ lies in $\ell_\Gamma^2$. Consider the mapping $U : \mathcal{H} \to \ell_\Gamma^2$ defined by $Ux = \{\langle x ; e_\gamma \rangle\}_{\gamma \in \Gamma}$ for every $x \in \mathcal{H}$. It is readily verified that $U$ is a linear transformation (by the linearity of the inner product in the first argument), and

$$\|Ux\|^2 = \sum_{\gamma \in \Gamma} |\langle x ; e_\gamma \rangle|^2 = \|x\|^2$$

for every $x \in \mathcal{H}$ (Parseval identity again). Thus $U$ is a linear isometry. Next we verify that $U$ is surjective. If $\{\alpha_\gamma\}_{\gamma \in \Gamma}$ is any family of scalars in $\ell_\Gamma^2$ (i.e., if $\sum_{\gamma \in \Gamma} |\alpha_\gamma|^2 < \infty$ — see Proposition 5.31), then $\{\alpha_\gamma e_\gamma\}_{\gamma \in \Gamma}$ is a summable family of vectors in $\mathcal{H}$. Indeed, as $\{e_\gamma\}_{\gamma \in \Gamma}$ is an orthonormal set, $\sum_{\gamma \in \Gamma} \|\alpha_\gamma e_\gamma\|^2 = \sum_{\gamma \in \Gamma} |\alpha_\gamma|^2 < \infty$ so that $\{\alpha_\gamma e_\gamma\}_{\gamma \in \Gamma}$ is a square-summable family, and hence a summable family of vectors in the Hilbert space $\mathcal{H}$ (Proposition 5.31 and Theorem 5.32). Therefore, for every $\{\alpha_\gamma\}_{\gamma \in \Gamma} \in \ell_\Gamma^2$ there exists an $x \in \mathcal{H}$ such that $x = \sum_{\gamma \in \Gamma} \alpha_\gamma e_\gamma$. But the Fourier series expansion of $x$ is unique, which implies that $\alpha_\gamma = \langle x \, ; e_\gamma \rangle$ for every $\gamma \in \Gamma$. Then $Ux = \{\langle x \, ; e_\gamma \rangle\}_{\gamma \in \Gamma} = \{\alpha_\gamma\}_{\gamma \in \Gamma}$ so that $\{\alpha_\gamma\}_{\gamma \in \Gamma} \in \mathcal{R}(U)$. That is, $\ell_\Gamma^2 \subseteq \mathcal{R}(U)$. Since $\mathcal{R}(U) \subseteq \ell_\Gamma^2$ trivially, it follows that $\mathcal{R}(U) = \ell_\Gamma^2$. Conclusion: $U$ is a linear surjective isometry, which means that $U$ is a unitary transformation. Thus $\mathcal{H}$ and $\ell_\Gamma^2$ are unitarily equivalent. As $B$ and $C$ are indexed by a common index set $\Gamma$, the same argument shows that $\mathcal{K}$ and $\ell_\Gamma^2$ also are unitarily equivalent. Therefore, $\mathcal{H}$ and $\mathcal{K}$ are unitarily equivalent (composition of isometric isomorphism is again an isometric isomorphism). $\qquad\qquad\square$

According to Theorem 5.44 and Examples 5L and 5M, the next result is an immediate consequence of Theorem 5.49.

**Corollary 5.50.** *Let $\Gamma$ be an arbitrary index set. Every Hilbert space $\mathcal{H}$ with* $\dim \mathcal{H} = \#\Gamma$ *is unitarily equivalent to $\ell_\Gamma^2$. In particular, every $n$-dimensional Hilbert space (for any integer $n \in \mathbb{N}$) is unitarily equivalent to $\mathbb{F}^n$, and every infinite-dimensional separable Hilbert space is unitarily equivalent to $\ell_+^2$.*

Our first example of an unbounded linear transformation defined on a Banach space appeared in the proof of Corollary 4.30 part (b). It is easy to exhibit unbounded linear transformations defined on linear manifolds of a Hilbert space. (Hint: see Problem 4.33(b).) Now we shall apply the Fourier Series Theorem to exhibit an unbounded linear transformation defined on a whole Hilbert space; precisely, defined on an infinite-dimensional separable Hilbert space.

**Example 5N.** Let $\mathcal{H}$ be an infinite-dimensional separable Hilbert space and let $\{e_k\}_{k=1}^\infty$ be an orthonormal basis for $\mathcal{H}$. Let $B = \{f_\gamma\}_{\gamma \in \Gamma}$ be a Hamel basis for $\mathcal{H}$ that properly includes $\{e_k\}_{k=1}^\infty$ (see Theorem 2.5 and Corollary 5.46). Take any $f_0 \in B \backslash \{e_k\}_{k=1}^\infty$ (obviously, $f_0 \neq 0$) and consider the mapping $F \colon B \to \mathcal{H}$ such that $Ff_0 = f_0$ and $Ff = 0$ for all $f \neq f_0$ in $B$. Now consider the transformation $L \colon \mathcal{H} \to \mathcal{H}$ defined as follows. For each $x \in \mathcal{H}$ take its unique representation as a (finite) linear combination of vectors in the Hamel basis $B$, say

$$x = \sum_{k=1}^{n(x)} \alpha_k(x) f_k.$$

Here $n(x)$ is a positive integer and $\{\alpha_k(x)\}_{k=1}^{n(x)}$ is a finite family of scalars containing all nonzero coordinates of $x$ with respect to the Hamel basis $B$. Set

$$Lx = \sum_{k=1}^{n(x)} \alpha_k(x) F f_k.$$

It is not difficult to verify that $L$ is linear (i.e., $L \in \mathcal{L}[\mathcal{H}]$). Moreover, $Lf = 0$ for every $f \in B$ such that $f \neq f_0$, and $Lf_0 = f_0$ (so that $L \neq O$). In particular, $Le_k = 0$ for all $k \geq 1$. Take an arbitrary $x \in \mathcal{H}$ and consider its Fourier series expansion, viz., $x = \sum_{k=1}^{\infty} \langle x \, ; e_k \rangle e_k$. If $L$ is bounded (i.e., if $L$ is continuous), then it follows by Corollary 3.8 that $Lx = \sum_{k=1}^{\infty} \langle x \, ; e_k \rangle Le_k = 0$. But this implies that $Lx = 0$ for all $x \in \mathcal{H}$ (i.e., $L = O$), which is a contradiction. Thus $L$ is unbounded; that is $L \in \mathcal{L}[\mathcal{H}] \backslash \mathcal{B}[\mathcal{H}]$.

## 5.10   Orthogonal Projection

A projection is an idempotent linear transformation $P \colon \mathcal{X} \to \mathcal{X}$ of a linear space $\mathcal{X}$ into itself (Section 2.9). An *orthogonal projection* on an inner product space $\mathcal{X}$ is a projection $P \colon \mathcal{X} \to \mathcal{X}$ such that $\mathcal{R}(P) \perp \mathcal{N}(P)$. Since orthogonal projections are projections, all the results of Section 2.9 apply to orthogonal projections in particular. If $P$ is an orthogonal projection on $\mathcal{X}$, then so is the complementary projection $E = (I - P) \colon \mathcal{X} \to \mathcal{X}$. (Reason: $E = (I - P)$ is a projection with $\mathcal{R}(E) = \mathcal{N}(P)$ and $\mathcal{N}(E) = \mathcal{R}(P)$.)

**Proposition 5.51.** *If $P$ is an orthogonal projection on an inner product space $\mathcal{X}$, then*

(a)  $P \in \mathcal{B}[\mathcal{H}]$ *and* $\|P\| = 1$ *whenever* $P \neq O$,

(b)  $\mathcal{N}(P)$ *and* $\mathcal{R}(P)$ *are subspaces of* $\mathcal{X}$,

(c)  $\mathcal{N}(P) = \mathcal{R}(P)^{\perp}$ *and* $\mathcal{R}(P) = \mathcal{N}(P)^{\perp}$,

(d)  $\mathcal{R}(P)$ *and* $\mathcal{N}(P)$ *are orthogonal complementary subspaces in* $\mathcal{X}$. *That is, besides being orthogonal subspaces of* $\mathcal{X}$, $\mathcal{R}(P)$ *and* $\mathcal{N}(P)$ *are such that*

$$\mathcal{X} = \mathcal{R}(P) + \mathcal{N}(P).$$

*Thus $\mathcal{X}$ can be decomposed as an orthogonal direct sum*

$$\mathcal{X} = \mathcal{R}(P) \oplus \mathcal{N}(P).$$

*Proof.* Let $P \colon \mathcal{X} \to \mathcal{X}$ be an orthogonal projection on an inner product space $\mathcal{X}$.

(a)  Take an arbitrary $x \in \mathcal{X}$. Since $\mathcal{R}(P)$ and $\mathcal{N}(P)$ are algebraic complements of each other (Theorem 2.19), we can write $x = u + v$ with $u$ in $\mathcal{R}(P)$ and $v$ in $\mathcal{N}(P)$. Moreover, $u \perp v$ because $\mathcal{R}(P) \perp \mathcal{N}(P)$). Then $\|x\|^2 = \|u\|^2 + \|v\|^2$

by the Pythagorean Theorem. Recalling that $\mathcal{R}(P) = \{u \in \mathcal{X}: Pu = u\}$, we get $Px = Pu + Pv = u$ so that $\|Px\|^2 = \|u\|^2 \leq \|x\|^2$. Hence $\|P\| \leq 1$. That is, $P$ is a contraction. If $P \neq O$, then $\mathcal{R}(P) \neq \{0\}$, and so $\|Pu\| = \|u\| \neq 0$ (because $Pu = u$) for all nonzero $u$ in $\mathcal{R}(P)$. Therefore $\|P\| \geq 1$. Outcome: $\|P\| = 1$.

(b) According to item (a), $P \in \mathcal{B}[\mathcal{X}]$. Thus Proposition 4.13 ensures that $\mathcal{N}(P)$ is a subspace of $\mathcal{X}$. Since $E = I - P$ is an orthogonal projection on $\mathcal{X}$, it follows that $E \in \mathcal{B}[\mathcal{X}]$ and, by Proposition 4.13 again, that $\mathcal{R}(P) = \mathcal{N}(E)$ is a subspace of $\mathcal{X}$.

(c) Recall from the definition of orthogonal complement that if $A$ and $B$ are subsets of $\mathcal{X}$ such that $A \perp B$, then $A \subseteq B^\perp$. Therefore, as $\mathcal{N}(P) \perp \mathcal{R}(P)$, we get $\mathcal{N}(P) \subseteq \mathcal{R}(P)^\perp$. Now take an arbitrary $x \in \mathcal{R}(P)^\perp$ so that $x \perp \mathcal{R}(P)$. Theorem 2.19 says that $x = u + v$ with $u \in \mathcal{R}(P)$ and $v \in \mathcal{N}(P)$. Hence $0 = \langle x \,;\, u \rangle = \langle u \,;\, u \rangle + \langle v \,;\, u \rangle = \|u\|^2$ (for $\mathcal{R}(P) \perp \mathcal{N}(P)$) and so $u = 0$, which implies that $x = v \in \mathcal{N}(P)$. Therefore $\mathcal{R}(P)^\perp \subseteq \mathcal{N}(P)$. Outcome: $\mathcal{N}(P) = \mathcal{R}(P)^\perp$. Considering the complementary orthogonal projection $E = I - P$ we conclude that $\mathcal{R}(P) = \mathcal{N}(E) = \mathcal{R}(E)^\perp = \mathcal{N}(P)^\perp$.

(d) Theorem 2.19 says that $\mathcal{N}(P)$ and $\mathcal{R}(P)$ are algebraic complements of each other so that $\mathcal{X} = \mathcal{R}(P) + \mathcal{N}(P)$. Since $\mathcal{R}(P) \perp \mathcal{N}(P)$, it follows by Proposition 5.24 that $\mathcal{X} \cong \mathcal{R}(P) \oplus \mathcal{N}(P)$. As usual, we identify the orthogonal direct sum $\mathcal{R}(P) \oplus \mathcal{N}(P)$ with its unitarily equivalent image $\mathcal{R}(P) + \mathcal{N}(P) = \mathcal{X}$, and write $\mathcal{X} = \mathcal{R}(P) \oplus \mathcal{N}(P)$. $\qquad\qquad\square$

**Theorem 5.52.** (Projection Theorem – Third version). *For every subspace $\mathcal{M}$ of a Hilbert space $\mathcal{H}$ there exists a unique orthogonal projection $P \in \mathcal{B}[\mathcal{H}]$ with $\mathcal{R}(P) = \mathcal{M}$.*

*Proof. Existence.* Theorem 5.20 says that $\mathcal{H}$ can be decomposed as

$$\mathcal{H} = \mathcal{M} + \mathcal{M}^\perp.$$

Since $\mathcal{M}$ and $\mathcal{M}^\perp$ are algebraic complements of each other, for every $x \in \mathcal{H}$ there exists a unique $u \in \mathcal{M}$ and a unique $v \in \mathcal{M}^\perp$ such that $x = u + v$ (Theorem 2.14). For each $x \in \mathcal{H}$ set $Px = P(u + v) = u$ in $\mathcal{M} \subseteq \mathcal{H}$. It is easy to check that this defines a linear transformation $P \colon \mathcal{H} \to \mathcal{H}$ with $\mathcal{R}(P) = \mathcal{M}$. Moreover, $P^2 x = Pu = u = Px$ (because $u = u + 0$ is the unique decomposition of $u \in \mathcal{M}$ in $\mathcal{M} + \mathcal{M}^\perp$) so that $P$ is idempotent. Furthermore, $Px = 0$ if and only if $x = 0 + v$ for some $v \in \mathcal{M}^\perp$, and hence $\mathcal{N}(P) = \mathcal{M}^\perp$. Conclusion: $P$ is an orthogonal projection on $\mathcal{H}$ with $\mathcal{R}(P) = \mathcal{M}$.

*Uniqueness.* Take an arbitrary vector $x \in \mathcal{H}$ and consider its unique decomposition $x = u + v \in \mathcal{M} + \mathcal{M}^\perp = \mathcal{H}$ with $u \in \mathcal{M}$ and $v \in \mathcal{M}^\perp$. Suppose $P'$ is an orthonormal projection on $\mathcal{H}$ such that $\mathcal{R}(P') = \mathcal{M}$. Thus $P'u = u = Pu$ for every $u \in \mathcal{M}$ (since the range of any projection is the set of all its fixed points — see proof of Theorem 2.19), and $P'v = 0 = Pv$ for every $v \in \mathcal{M}^\perp$. (Reason:

if $\mathcal{M} = \mathcal{R}(P') = \mathcal{R}(P)$, then $\mathcal{M}^\perp = \mathcal{R}(P')^\perp = \mathcal{R}(P)^\perp = \mathcal{N}(P') = \mathcal{N}(P)$.) Therefore, $P'x = P'(u + v) = P(u + v) = Px$, and hence $P' = P$. $\qquad\square$

Let $\mathcal{M}$ be any subspace of a Hilbert space $\mathcal{H}$. The unique orthogonal projection $P \colon \mathcal{H} \to \mathcal{H}$ for which $\mathcal{R}(P) = \mathcal{M}$ is called *the orthogonal projection onto* $\mathcal{M}$. The above proof shows that Theorem 5.20 implies Theorem 5.52. In fact, all the versions of the Projection Theorem, viz., Theorems 5.20, 5.25 and 5.52, are pairwise equivalent. That Theorems 5.20 and 5.25 are equivalent is an immediate consequence of Proposition 5.24. We shall verify that Theorem 5.52 implies Theorem 5.20. Indeed, if $P \colon \mathcal{H} \to \mathcal{H}$ is any orthogonal projection on a Hilbert space $\mathcal{H}$, then $\mathcal{H} = \mathcal{R}(P) + \mathcal{N}(P)$ by Theorem 2.19, and hence $\mathcal{H} = \mathcal{R}(P) + \mathcal{R}(P)^\perp$, where $\mathcal{R}(P)$ is a subspace of $\mathcal{H}$ (Proposition 5.51). But Theorem 5.52 says that for every subspace $\mathcal{M}$ of $\mathcal{H}$ there exists an orthogonal projection $P \colon \mathcal{H} \to \mathcal{H}$ such $\mathcal{R}(P) = \mathcal{M}$. Therefore, for every subspace $\mathcal{M}$ of $\mathcal{H}$ we get $\mathcal{H} = \mathcal{M} + \mathcal{M}^\perp$. Outcome: Theorem 5.52 implies Theorem 5.20. Such an equivalence justifies the terminology "Projection Theorem" for Theorems 5.20 and 5.25. The pivotal result of Theorem 5.13 can also be translated into the orthogonal projection language. Actually, the unique vector $u_x \in \mathcal{M}$ of Theorem 5.13 is given by $u_x = Px$, where $P \colon \mathcal{H} \to \mathcal{H}$ is the orthogonal projection onto $\mathcal{M}$.

**Theorem 5.53.** *Let $\mathcal{M}$ be a subspace of a Hilbert space $\mathcal{H}$ and let $P \in \mathcal{B}[\mathcal{H}]$ be the orthogonal projection onto $\mathcal{M}$. Take any $x$ in $\mathcal{H}$. $Px$ is the unique vector in $\mathcal{M}$ such that*

$$\|x - Px\| = d(x, \mathcal{M}).$$

*Moreover, $Px$ also is the unique vector in $\mathcal{M}$ such that*

$$x - Px \perp \mathcal{M}.$$

*Proof.* Let $P$ be *the* orthogonal projection onto $\mathcal{M}$. $P(x - Px) = Px - P^2x = 0$ so that $x - Px \in \mathcal{N}(P) = \mathcal{R}(P)^\perp = \mathcal{M}^\perp$. Thus $Px$ is the unique vector in $\mathcal{M}$ such that $x - Px \in \mathcal{M}^\perp$, which in turn is the unique vector in $\mathcal{M}$ such that $\|x - Px\| = d(x, \mathcal{M})$ (Theorem 5.13). $\qquad\square$

The next two results connect the notion of orthogonal projection with the Fourier Series Theorem.

**Theorem 5.54.** *Let $\{e_\gamma\}_{\gamma \in \Gamma}$ be an orthonormal family of vectors in a Hilbert space $\mathcal{H}$ and set $\mathcal{M} = \bigvee \{e_\gamma\}_{\gamma \in \Gamma}$. For every $x \in \mathcal{H}$, $\{\langle x\,; e_\gamma \rangle e_\gamma\}_{\gamma \in \Gamma}$ is a summable family of vectors in $\mathcal{M}$ and the mapping $P \colon \mathcal{H} \to \mathcal{H}$, defined by*

$$Px = \sum_{\gamma \in \Gamma} \langle x\,; e_\gamma \rangle e_\gamma,$$

*is the orthogonal projection onto $\mathcal{M}$.*

*Proof.* Take an arbitrary $x \in \mathcal{H}$. The Bessel inequality (Lemma 5.40) says that $\sum_{\gamma \in \Gamma} |\langle x ; e_\gamma \rangle|^2 \le \|x\|^2$, and hence $\{\langle x ; e_\gamma \rangle e_\gamma\}_{\gamma \in \Gamma}$ is a square-summable family (cf. Proposition 5.31) of vectors in the Hilbert space $\mathcal{M}$ (see Proposition 4.7). Then $\{\langle x ; e_\gamma \rangle e_\gamma\}_{\gamma \in \Gamma}$ is a summable family of vectors in $\mathcal{M}$ by Theorem 5.32. Set $Px = \sum_{\gamma \in \Gamma} \langle x ; e_\gamma \rangle e_\gamma \in \mathcal{M}$. It is readily verified that this defines a linear transformation $P \colon \mathcal{H} \to \mathcal{H}$ such that $\mathcal{R}(P) \subseteq \mathcal{M}$. Moreover, $\|Px\|^2 = \sum_{\gamma \in \Gamma} |\langle x ; e_\gamma \rangle|^2$ (Theorem 5.32 again), and hence $\|Px\|^2 \le \|x\|^2$. Thus $P \in \mathcal{B}[\mathcal{H}]$ (i.e., $P$ is continuous — a contraction, actually), which implies that

$$P^2 x = P\Big(\sum_{\gamma \in \Gamma} \langle x ; e_\gamma \rangle e_\gamma\Big) = \sum_{\gamma \in \Gamma} \langle x ; e_\gamma \rangle P e_\gamma.$$

But the very definition of $P$ ensures that $P e_\alpha = \sum_{\gamma \in \Gamma} \langle e_\alpha ; e_\gamma \rangle e_\gamma = e_\alpha$ for every $\alpha \in \Gamma$. Therefore,

$$P^2 x = \sum_{\gamma \in \Gamma} \langle x ; e_\gamma \rangle e_\gamma = Px$$

and so $P$ is a projection. Since $\{e_\gamma\}_{\gamma \in \Gamma}$ is an orthonormal basis for the Hilbert space $\mathcal{M}$, the Fourier Series Theorem ensures that every $u \in \mathcal{M}$ has a unique Fourier series expansion

$$u = \sum_{\gamma \in \Gamma} \langle u ; e_\gamma \rangle e_\gamma.$$

Hence $Pu = \sum_{\gamma \in \Gamma} \langle u ; e_\gamma \rangle P e_\gamma = \sum_{\gamma \in \Gamma} \langle u ; e_\gamma \rangle e_\gamma = u$ so that $u \in \mathcal{R}(P)$. Therefore, $\mathcal{M} \subseteq \mathcal{R}(P)$ and so $\mathcal{R}(P) = \mathcal{M}$. Note that $v \in \mathcal{N}(P)$ (i.e., $Pv = 0$) if and only if $\|Pv\|^2 = \sum_{\gamma \in \Gamma} |\langle v ; e_\gamma \rangle|^2 = 0$. Equivalently, if and only if $\langle v ; e_\gamma \rangle = 0$ for all $\gamma \in \Gamma$, which means that $v \in \big(\{e_\gamma\}_{\gamma \in \Gamma}\big)^\perp = \big(\bigvee \{e_\gamma\}_{\gamma \in \Gamma}\big)^\perp = \mathcal{M}^\perp = \mathcal{R}(P)^\perp$. Thus $\mathcal{N}(P) = \mathcal{R}(P)^\perp$, which implies that $\mathcal{R}(P) \perp \mathcal{N}(P)$. Conclusion: $P \colon \mathcal{H} \to \mathcal{H}$ is an orthogonal projection with $\mathcal{R}(P) = \mathcal{M}$; that is, *the* projection onto $\mathcal{M}$ (Theorem 5.52). $\qquad\square$

**Corollary 5.55.** *Let $\{e_\gamma\}_{\gamma \in \Gamma}$ be any orthonormal basis for a subspace $\mathcal{M}$ of a Hilbert space $\mathcal{H}$. The orthogonal projection $P \in \mathcal{B}[\mathcal{H}]$ onto $\mathcal{M}$ is given by*

$$Px = \sum_{\gamma \in \Gamma} \langle x ; e_\gamma \rangle e_\gamma \quad \text{for every} \quad x \in \mathcal{H}.$$

Now we consider "orthogonal families of orthogonal projections". This is a rather important notion upon which the Spectral Theorem of next chapter will be built.

**Proposition 5.56.** *Let $P_1 \in \mathcal{B}[\mathcal{X}]$ and $P_2 \in \mathcal{B}[\mathcal{X}]$ be orthogonal projections on an inner product space $\mathcal{X}$. The following assertions are pairwise equivalent.*

(a) $\mathcal{R}(P_1) \perp \mathcal{R}(P_2)$.

(b) $P_1 P_2 = O$.

(c) $P_2 P_1 = O$.

(d) $\mathcal{R}(P_2) \subseteq \mathcal{N}(P_1)$.

(e) $\mathcal{R}(P_1) \subseteq \mathcal{N}(P_2)$.

*Proof.* Clearly, $\mathcal{R}(P_1) \perp \mathcal{R}(P_2)$ implies $\mathcal{R}(P_2) \subseteq \mathcal{R}(P_1)^{\perp} = \mathcal{N}(P_1)$. Conversely, if $\mathcal{R}(P_2) \subseteq \mathcal{N}(P_1) = \mathcal{R}(P_1)^{\perp}$, then $\mathcal{R}(P_2) \perp \mathcal{R}(P_1)$. Hence (a)$\Leftrightarrow$(d). Similarly (orthogonality is symmetric so that $\mathcal{R}(P_1) \perp \mathcal{R}(P_2)$ if and only if $\mathcal{R}(P_2) \perp \mathcal{R}(P_1)$) we get (a)$\Leftrightarrow$(e). Since $\mathcal{R}(P_2) \subseteq \mathcal{N}(P_1)$ if and only if $P_1 P_2 = O$, it follows that (d)$\Leftrightarrow$(b). Swap $P_1$ and $P_2$ to get (e)$\Leftrightarrow$(c). $\qquad\square$

If two orthogonal projections $P_1$ and $P_2$ on an inner product space $\mathcal{X}$ satisfy any of the equivalent assertions of Proposition 5.56, then they are said to be *orthogonal to each other* (or *mutually orthogonal*). If $\{P_\gamma\}_{\gamma \in \Gamma}$ is a family of mutually orthogonal projections on an inner product space $\mathcal{X}$ (i.e., $\mathcal{R}(P_\alpha) \perp \mathcal{R}(P_\beta)$ whenever $\alpha \neq \beta$), then we say that $\{P_\gamma\}_{\gamma \in \Gamma}$ is an *orthogonal family of orthogonal projections* on $\mathcal{X}$. An *orthogonal sequence of orthogonal projections* $\{P_k\}_{k=0}^{\infty}$ is similarly defined. If $\{P_\gamma\}_{\gamma \in \Gamma}$ is an orthogonal family of orthogonal projections and

$$\sum_{\gamma \in \Gamma} P_\gamma x = x \quad \text{for every} \quad x \in \mathcal{X},$$

then $\{P_\gamma\}_{\gamma \in \Gamma}$ is called a *resolution of the identity* on $\mathcal{X}$. (For each $x \in \mathcal{X}$ the sum $x = \sum_{\gamma \in \Gamma} P_\gamma x$ has only a countable number of nonzero vectors — why?) If $\{P_k\}_{k=0}^{\infty}$ is an infinite sequence, then the above identity in $\mathcal{X}$ means convergence in the strong operator topology:

$$\sum_{k=0}^{n} P_k \overset{s}{\longrightarrow} I.$$

If $\{P_k\}_{k=0}^{n}$ is a finite family, then the above identity in $\mathcal{X}$ obviously coincides with the identity $\sum_{k=0}^{n} P_k = I$ in $\mathcal{B}[\mathcal{X}]$. For instance, if $P$ is an orthogonal projection on $\mathcal{X}$ and $E = I - P$ is the complementary projection on $\mathcal{X}$, then $P$ and $E$ are orthogonal projections orthogonal to each other (for $PE = P - P^2 = O$). Moreover, $\{P, E\}$ clearly is a resolution of the identity on $\mathcal{X}$ for $P + E = I$.

**Proposition 5.57.** *Let $\{e_\gamma\}_{\gamma \in \Gamma}$ be an orthonormal basis for a Hilbert space $\mathcal{H}$. For each $\gamma \in \Gamma$ define the mapping $P_\gamma : \mathcal{H} \to \mathcal{H}$ by*

$$P_\gamma x = \langle x \,; e_\gamma \rangle e_\gamma \quad \text{for every} \quad x \in \mathcal{H}.$$

*Claim:* $\{P_\gamma\}_{\gamma \in \Gamma}$ *is a resolution of the identity on* $\mathcal{H}$.

*Proof.* Each $P_\gamma$ is the orthogonal projection onto the one-dimensional space $\mathcal{M} = \text{span}\{e_\gamma\}$. This is a particular case of Theorem 5.54. It is clear that $P_\alpha P_\beta x = \langle x ; e_\beta\rangle\langle e_\beta ; e_\alpha\rangle e_\alpha$, and so $P_\alpha P_\beta x = 0$ whenever $\alpha \neq \beta$, for every $x \in \mathcal{H}$. Thus $P_\alpha P_\beta = O$ for every $\alpha, \beta \in \Gamma$ such that $\alpha \neq \beta$. Hence $\{P_\gamma\}_{\gamma\in\Gamma}$ is an orthogonal family of orthogonal projections on $\mathcal{H}$. The Fourier Series Theorem ensures that $\{\langle x ; e_\gamma\rangle e_\gamma\}_{\gamma\in\Gamma}$ is a summable family of vectors in $\mathcal{H}$ and, for every $x \in \mathcal{H}$,

$$x = \sum_{\gamma\in\Gamma}\langle x ; e_\gamma\rangle e_\gamma = \sum_{\gamma\in\Gamma} P_\gamma x.$$

Conclusion: $\{P_\gamma\}_{\gamma\in\Gamma}$ is a resolution of the identity on $\mathcal{H}$.    $\square$

**Example 5O.** Let $\{e_k\}_{k=1}^\infty$ be an orthonormal basis for an infinite-dimensional separable Hilbert space $\mathcal{H}$. According to Theorem 5.54 and Proposition 5.57, for each $k \geq 1$ the mapping $P_k : \mathcal{H} \to \mathcal{H}$ defined by

$$P_k x = \langle x ; e_k\rangle e_k \quad \text{for every} \quad x \in \mathcal{H}$$

is an orthonormal projection, and $\{P_k\}_{k=1}^\infty$ is a resolution of the identity on $\mathcal{H}$. Therefore, the sequence of operators $\{\sum_{k=1}^n P_k\}_{n=1}^\infty$ converges strongly to the identity $I$ on $\mathcal{H}$. In fact, take an arbitrary $x \in \mathcal{H}$. By the Fourier Series Theorem,

$$\sum_{k=1}^n P_k x - x = \sum_{k=1}^n \langle x ; e_k\rangle e_k - \sum_{k=1}^\infty \langle x ; e_k\rangle e_k = -\sum_{k=n+1}^\infty \langle x ; e_k\rangle e_k \to 0$$

as $n \to \infty$ because the infinite series $\sum_{k=1}^\infty \langle x ; e_k\rangle e_k$ converges in $\mathcal{H}$ (see Problem 4.9(b)). This confirms that

$$\sum_{k=1}^n P_k \xrightarrow{s} I.$$

We shall see now that the sequence of operators $\{\sum_{k=1}^n P_k\}_{n=1}^\infty$, which converges strongly to the identity $I$ on $\mathcal{H}$, does not converge uniformly. Indeed, if $\{\sum_{k=1}^n P_k\}_{n=1}^\infty$ converges uniformly, then the identity $I$ must be its uniform limit. (Reason: $\sum_{k=1}^n P_k \xrightarrow{s} I$, and uniform convergence implies strong convergence to the same limit.) However, for each $n \geq 1$,

$$\left\|\left(I - \sum_{k=1}^n P_k\right)e_{n+1}\right\| = \left\|\sum_{k=n+1}^\infty \langle e_{n+1} ; e_k\rangle e_k\right\| = \|e_{n+1}\| = 1.$$

Thus $\left\|I - \sum_{k=1}^n P_k\right\| = \sup_{\|x\|=1}\left\|(I - \sum_{k=1}^n P_k)x\right\| \geq 1$ for every $n \geq 1$, and hence $\sum_{k=1}^n P_k \xrightarrow{u}\!\!\!\!\!/\ \ I$. Conclusion: $\{\sum_{k=1}^n P_k\}_{n=1}^\infty$ does not converge uniformly.

**Proposition 5.58.** *If* $\{P_k\}_{k\in\mathbb{N}}$ *is an orthogonal sequence of orthogonal projections on a Hilbert space* $\mathcal{H}$, *then* $\{\sum_{k=1}^n P_k\}_{n\in\mathbb{N}}$ *converges strongly to the orthogonal* *projection* $P \colon \mathcal{H} \to \mathcal{H}$ *onto* $\left(\sum_{k\in\mathbb{N}}\mathcal{R}(P_k)\right)^-$. *In other words,* $\sum_{k=1}^{n} P_k \xrightarrow{s} P$, *where* $P \in \mathcal{B}[\mathcal{H}]$ *is the orthogonal projection with* $\mathcal{R}(P) = \left(\sum_{k\in\mathbb{N}}\mathcal{R}(P_k)\right)^-$.

*Proof.* Set $\mathcal{M} = \left(\sum_{k\in\mathbb{N}}\mathcal{R}(P_k)\right)^-$, which is a subspace of the Hilbert space $\mathcal{H}$, and consider the decomposition

$$\mathcal{H} = \mathcal{M} + \mathcal{M}^\perp$$

of Theorem 5.20. Take any $x \in \mathcal{H}$ so that $x = u + v$ with $u \in \mathcal{M}$ and $v \in \mathcal{M}^\perp$. Observe that $v \in \left(\sum_{k\in\mathbb{N}}\mathcal{R}(P_k)\right)^\perp$ by Proposition 5.12 so that $v \perp \mathcal{R}(P_k)$ for every $k \in \mathbb{N}$. Thus $v \in \mathcal{R}(P_k)^\perp = \mathcal{N}(P_k)$, and hence $P_k v = 0$, for every $k \in \mathbb{N}$, which implies that $\sum_{k=1}^{\infty} P_k v = 0$. Since $u \in \left(\sum_{k\in\mathbb{N}}\mathcal{R}(P_k)\right)^-$, where $\{\mathcal{R}(P_k)\}_{k\in\mathbb{N}}$ is a countably infinite family of pairwise orthogonal subspaces of $\mathcal{H}$, it follows by the Orthogonal Structure Theorem (Theorem 5.16) that $u = \sum_{k=1}^{\infty} u_k$ with $u_k \in \mathcal{R}(P_k)$ for each $k$. Since $P_j$ is continuous, we get $P_j u = \sum_{k=1}^{\infty} P_j u_k = u_j$ (reason: $u_k \in \mathcal{R}(P_k)$, $P_j P_k = O$ whenever $j \neq k$, and $P_j u_j = u_j$) for each $j \in \mathbb{N}$. Hence $u = \sum_{k=1}^{\infty} P_k u$. Therefore, for every $x \in \mathcal{H}$,

$$\sum_{k=1}^{\infty} P_k x = \sum_{k=1}^{\infty} P_k u + \sum_{k=1}^{\infty} P_k v = u$$

so that the $\mathcal{B}[\mathcal{H}]$-valued sequence $\{\sum_{k=1}^{n} P_k\}_{n\in\mathbb{N}}$ is strongly convergent (Proposition 4.44). Let $P \in \mathcal{B}[\mathcal{H}]$ be the strong limit of $\{\sum_{k=1}^{n} P_k\}_{n\in\mathbb{N}}$ so that

$$P x = \sum_{k=1}^{\infty} P_k x \quad \text{for every} \quad x \in \mathcal{H}.$$

Then $P^2 x = \sum_{k=1}^{\infty} P P_k x = \sum_{k=1}^{\infty} \sum_{j=1}^{\infty} P_j P_k x = \sum_{k=1}^{\infty} P_k x = P x$ for every $x$ in $\mathcal{H}$ (because $P$ is continuous and $P_j P_k = O$ whenever $j \neq k$), and hence $P$ is idempotent. Moreover, $\mathcal{R}(P) = \mathcal{M}$ and $\mathcal{N}(P) = \mathcal{M}^\perp$. Indeed, $P x = P(u+v) = u$ for every $x \in \mathcal{H} = \mathcal{M} + \mathcal{M}^\perp$, where $u$ is the unique vector in $\mathcal{M}$ and $v$ is the unique vector in $\mathcal{M}^\perp$ such that $x = u + v$. Therefore, $\mathcal{R}(P) \perp \mathcal{N}(P)$. Outcome: $P \in \mathcal{B}[\mathcal{H}]$ is the unique orthogonal projection onto $\mathcal{M}$ (cf. Theorem 5.52).    $\square$

The above proposition is a consequence of the Projection Theorem (i.e., Theorems 5.20 and 5.52) and the Orthogonal Structure Theorem (Theorem 5.16). Here is the full version of the Orthogonal Structure Theorem in terms of orthogonal projections.

**Theorem 5.59.** *Let $\mathcal{H}$ be a Hilbert space. If $\{P_k\}_{k\in\mathbb{N}}$ is a resolution of the identity on $\mathcal{H}$, then*

$$\mathcal{H} = \left(\sum_{k\in\mathbb{N}} \mathcal{R}(P_k)\right)^-.$$

*Conversely, if $\{\mathcal{M}_k\}_{k\in\mathbb{N}}$ is a sequence of pairwise orthogonal subspaces of $\mathcal{H}$ such that $\mathcal{H} = \left(\sum_{k\in\mathbb{N}}\mathcal{M}_k\right)^-$, then the sequence $\{P_k\}_{k\in\mathbb{N}}$ consisting of the orthogonal projections $P_k \in \mathcal{B}[\mathcal{H}]$ onto each $\mathcal{M}_k$ is a resolution of the identity on $\mathcal{H}$.*

*Proof.* Set $\mathcal{M} = \left(\sum_{k\in\mathbb{N}}\mathcal{R}(P_k)\right)^-$, which is a subspace of the Hilbert space $\mathcal{H}$. If $x \in \mathcal{M}^\perp = \left(\sum_{k\in\mathbb{N}}\mathcal{R}(P_k)\right)^\perp$, then $x \perp \mathcal{R}(P_k)$ for every $k \in \mathbb{N}$. Hence $x$ lies in $\mathcal{R}(P_k)^\perp = \mathcal{N}(P_k)$, so that $P_k x = 0$, for every $k \in \mathbb{N}$. Then $\sum_{k\in\mathbb{N}} P_k x = 0$. But $\sum_{k\in\mathbb{N}} P_k x = x$ for every $x \in \mathcal{H}$ because $\{P_k\}_{k\in\mathbb{N}}$ is a resolution of the identity on $\mathcal{H}$. Therefore, $x \in \mathcal{M}^\perp$ implies $x = 0$ so that $\mathcal{M}^\perp = \{0\}$, and hence

$$\mathcal{M} = \mathcal{H}$$

by Proposition 5.15. Conversely, let $\{\mathcal{M}_k\}_{k\in\mathbb{N}}$ be a sequence of pairwise orthogonal subspaces of $\mathcal{H}$ such that $\mathcal{H} = \left(\sum_{k\in\mathbb{N}}\mathcal{M}_k\right)^-$. For each $k \in \mathbb{N}$ let $P_k \in \mathcal{B}[\mathcal{H}]$ be the orthogonal projection onto $\mathcal{M}_k$ (Theorem 5.52), and note that $\mathcal{R}(P_j) = \mathcal{M}_j \perp \mathcal{M}_k = \mathcal{R}(P_k)$ whenever $j \neq k$. Thus $\{P_k\}_{k\in\mathbb{N}}$ is an orthogonal sequence of orthogonal projections on $\mathcal{H}$. Therefore, according to Proposition 5.58,

$$\sum_{k=1}^{n} P_k \overset{s}{\longrightarrow} I,$$

where the identity $I \in \mathcal{B}[\mathcal{H}]$ is the unique orthogonal projection on $\mathcal{H}$ with $\mathcal{R}(I) = \mathcal{H} = \left(\sum_{k\in\mathbb{N}}\mathcal{M}_k\right)^- = \left(\sum_{k\in\mathbb{N}}\mathcal{R}(P_k)\right)^-$. Outcome: $\{P_k\}_{k\in\mathbb{N}}$ is a resolution of the identity on $\mathcal{H}$. $\qquad\square$

It is worth noticing that, as $\{\mathcal{R}(P_k)\}_{k\in\mathbb{N}}$ is a sequence of pairwise orthogonal subspaces of a Hilbert space $\mathcal{H}$, then $\bigoplus_{k\in\mathbb{N}}\mathcal{R}(P_k) \cong \left(\sum_{k\in\mathbb{N}}\mathcal{R}(P_k)\right)^-$ (see Example 5J). Therefore, under the usual identification, Proposition 5.58 says that $\sum_{k=1}^{n} P_k \overset{s}{\longrightarrow} P$, *where $P$ in $\mathcal{B}[\mathcal{H}]$ is the orthogonal projection with $\mathcal{R}(P) = \bigoplus_{k\in\mathbb{N}}\mathcal{R}(P_k)$*, and Theorem 5.59 says that $\mathcal{H} = \bigoplus_{k\in\mathbb{N}}\mathcal{R}(P_k)$ *whenever $\{\mathcal{R}(P_k)\}_{k\in\mathbb{N}}$ is a resolution of the identity on $\mathcal{H}$.*

**Definition 5.60.** Let $\{P_\gamma\}_{\gamma\in\Gamma}$ be a resolution of the identity on a Hilbert space $\mathcal{H} \neq \{0\}$, where $P_\gamma \neq O$ for every $\gamma \in \Gamma$. Let $\{\lambda_\gamma\}_{\gamma\in\Gamma}$ be a similarly indexed family of scalars. Set

$$\mathcal{D}(T) = \left\{x \in \mathcal{H} \colon \{\lambda_\gamma P_\gamma x\}_{\gamma\in\Gamma} \text{ is a summable family of vectors in } \mathcal{H}\right\}.$$

The mapping $T \colon \mathcal{D}(T) \to \mathcal{H}$, defined by

$$Tx = \sum_{\gamma\in\Gamma} \lambda_\gamma P_\gamma x \quad \text{for every} \quad x \in \mathcal{D}(T),$$

is said to be a *weighted sum of projections*.

**Proposition 5.61.** *Every weighted sum of projections is a linear transformation. Moreover, if $T \in \mathcal{L}[\mathcal{D}(T), \mathcal{H}]$ is a weighted sum of projections, then the following assertions are pairwise equivalent.*

(a)  $\mathcal{D}(T) = \mathcal{H}.$

(b) $\{\lambda_\gamma\}_{\gamma\in\Gamma}$ *is a bounded family of scalars.*

(c) *$T$ is bounded.*

*If any of the above equivalent assertions holds true, then $T \in \mathcal{B}[\mathcal{H}]$ is such that $\|T\| = \sup_{\gamma\in\Gamma}|\lambda_\gamma|$.*

*Proof.* It is readily verified that the domain $\mathcal{D}(T)$ of a weighted sum of projections is a linear manifold of $\mathcal{H}$, and also that $T: \mathcal{D}(T) \to \mathcal{H}$ is linear (see the remark that follows Definition 5.26).

*Proof of* (a)$\Rightarrow$(b). If $\{\lambda_\gamma\}_{\gamma\in\Gamma}$ is not bounded, then for each integer $n \geq 1$ there exists a $\gamma_n \in \Gamma$ such that $|\lambda_{\gamma_n}| \geq n$. Consider the scalar-valued sequence $\{\lambda_{\gamma_n}\}_{n=1}^\infty$, which is clearly unbounded. Consider the sequence $\{P_{\gamma_n}\}_{n=1}^\infty$ from $\{P_\gamma\}_{\gamma\in\Gamma}$ and, for each $n \geq 1$, take $e_{\gamma_n} \in \mathcal{R}(P_{\gamma_n})$ such that $\|e_{\gamma_n}\| = 1$. (Recall: $\mathcal{R}(P_{\gamma_n}) \neq \{0\}$ because $P_\gamma \neq O$ for every $\gamma \in \Gamma$.) Observe that the orthogonal sequence $\{\lambda_{\gamma_n}^{-1} e_{\gamma_n}\}_{n=1}^\infty$ is square-summable (for $\sum_{n=1}^\infty \|\lambda_{\gamma_n}^{-1} e_{\gamma_n}\|^2 = \sum_{n=1}^\infty |\lambda_{\gamma_n}|^{-2} \leq \sum_{n=1}^\infty \frac{1}{n^2} < \infty$), and hence it is a summable sequence in the Hilbert space $\mathcal{H}$ (Corollary 5.9(b)). Set $x = \sum_{n=1}^\infty \lambda_{\gamma_n}^{-1} e_{\gamma_n}$ in $\mathcal{H}$ and note that (since each $P_{\gamma_n}$ is continuous)

$$\lambda_{\gamma_n} P_{\gamma_n} x = \sum_{k=1}^\infty \lambda_{\gamma_n} \lambda_{\gamma_k}^{-1} P_{\gamma_n} e_{\gamma_k} = e_{\gamma_n} \qquad \text{for each} \quad n \geq 1$$

(because $P_{\gamma_n} e_{\gamma_k} = 0$ if $k \neq n$ and $P_{\gamma_n} e_{\gamma_n} = e_{\gamma_n}$, once $e_{\gamma_k} \in \mathcal{R}(P_{\gamma_k})$ for each $k \geq 1$ and $P_{\gamma_n} P_{\gamma_k} = O$ whenever $k \neq n$). But $\{e_{\gamma_n}\}_{n=1}^\infty$ is not a square-summable sequence (for it is an orthogonal sequence of unit vectors). Thus $\{\lambda_{\gamma_n} P_{\gamma_n} x\}_{n=1}^\infty$ is not a square-summable sequence of vectors in $\mathcal{H}$, so that $\sup_{n\geq 1}\sum_{k=1}^n \|\lambda_{\gamma_k} P_{\gamma_k} x\|^2 = \infty$, and hence the orthogonal family $\{\lambda_\gamma P_\gamma x\}_{\gamma\in\Gamma}$ of vectors in $\mathcal{H}$ is not square-summable (Proposition 5.31). Therefore, $\{\lambda_\gamma P_\gamma x\}_{\gamma\in\Gamma}$ is not a summable family of vectors in $\mathcal{H}$ by Theorem 5.32, which means that $x \notin \mathcal{D}(T)$. Conclusion: If $\{\lambda_\gamma\}_{\gamma\in\Gamma}$ is not bounded, then $\mathcal{D}(T) \neq \mathcal{H}$. Equivalently, (a) implies (b).

*Proof of* (b)$\Rightarrow$(a,c). Let $x$ be an arbitrary vector in $\mathcal{H}$. Since $\{P_\gamma\}_{\gamma\in\Gamma}$ is a resolution of the identity on $\mathcal{H}$, it follows that $\{P_\gamma x\}_{\gamma\in\Gamma}$ is a summable family (for $\sum_{\gamma\in\Gamma} P_\gamma x = x$) of orthogonal vectors in $\mathcal{H}$, and hence a square-summable family by Theorem 5.32(a). Suppose $\{\lambda_\gamma\}_{\gamma\in\Gamma}$ is bounded and set $\beta = \sup_{\gamma\in\Gamma}|\lambda_\gamma|$ (which is a nonnegative real number). Then, for any finite $N \subseteq \Gamma$,

$$\sum_{k\in N} \|\lambda_k P_k x\|^2 \leq \beta^2 \sum_{k\in N} \|P_k x\|^2 \leq \beta^2 \sum_{\gamma\in\Gamma} \|P_\gamma x\|^2 < \infty$$

so that $\{\lambda_\gamma P_\gamma x\}_{\gamma\in\Gamma}$ is a square-summable family of orthogonal vectors in $\mathcal{H}$ (Proposition 5.31), and hence a summable family of vectors in the Hilbert space $\mathcal{H}$ by Theorem 5.32(b). Thus $x \in \mathcal{D}(T)$, and therefore (b) implies (a). Moreover, since $\{\lambda_\gamma P_\gamma x\}_{\gamma\in\Gamma}$ is an orthogonal family of vectors in $\mathcal{H}$, and $\{P_\gamma\}_{\gamma\in\Gamma}$ is a resolution of the identity on $\mathcal{H}$, we get

$$
\begin{aligned}
\|Tx\|^2 &= \left\|\sum_{\gamma\in\Gamma}\lambda_\gamma P_\gamma x\right\|^2 = \sum_{\gamma\in\Gamma}|\lambda_\gamma|^2\|P_\gamma x\|^2 \\
&\le \beta^2\sum_{\gamma\in\Gamma}\|P_\gamma x\|^2 = \beta^2\left\|\sum_{\gamma\in\Gamma}P_\gamma x\right\|^2 = \beta^2\|x\|^2
\end{aligned}
$$

(see Theorem 5.32 again). Hence (b) implies (c).

*Proof of* (c)$\Rightarrow$(b). For each $\gamma\in\Gamma$ take a unit vector $e_\gamma$ in $\mathcal{R}(P_\gamma)$. Thus

$$
Te_\gamma = \sum_{\gamma\in\Gamma}\lambda_\gamma P_\gamma e_\gamma = \lambda_\gamma e_\gamma.
$$

If $T$ is bounded, then

$$
|\lambda_\gamma| = \|\lambda_\gamma e_\gamma\| = \|Te_\gamma\| \le \|T\|\,\|e_\gamma\| = \|T\|
$$

for all $\gamma\in\Gamma$ so that (c) implies (d).

Finally, the above inequality says that $\sup_{\gamma\in\Gamma}|\lambda_\gamma| \le \|T\|$. On the other hand, we have already seen that $\|Tx\| \le \beta\|x\|$ for every $x\in\mathcal{H}$, where $\beta = \sup_{\gamma\in\Gamma}|\lambda_\gamma|$, which implies that $\|T\| \le \sup_{\gamma\in\Gamma}|\lambda_\gamma|$. Hence $\|T\| = \sup_{\gamma\in\Gamma}|\lambda_\gamma|$. $\qquad\square$

Let the infinite sequence $\{P_k\}_{k=1}^\infty$ be a resolution of the identity on $\mathcal{H}$, where $P_k \ne O$ for every $k \ge 1$, and let $\{\lambda_k\}_{k=1}^\infty$ be a bounded sequence of scalars. Observe that the identity in $\mathcal{H}$

$$
Tx = \sum_{k=1}^\infty \lambda_k P_k x \quad\text{for every}\quad x\in\mathcal{H}
$$

that defines the weighted sum of projections $T\in\mathcal{B}[\mathcal{H}]$ actually means convergence in the strong topology; that is,

$$
\sum_{k=1}^n \lambda_k P_k \xrightarrow{\ s\ } T.
$$

## 5.11  The Riesz Representation Theorem and Weak Convergence

Let $y$ be an arbitrary vector in an inner product space $\mathcal{X}$ and consider the functional $f:\mathcal{X}\to\mathbb{F}$ defined by

$$
f(x) = \langle x\,;\,y\rangle \quad\text{for every}\quad x\in\mathcal{X}.
$$

This is a linear and bounded functional. Indeed, $f$ is linear because the inner product is linear in the first argument; and bounded by the Schwarz inequality: $|f(x)| = |\langle x\,;\,y\rangle| \leq \|y\|\,\|x\|$ for every $x \in \mathcal{X}$. Then $f \in \mathcal{B}[\mathcal{X}, \mathbb{F}]$ and $\|f\| = \sup_{\|x\|=1}|f(x)| \leq \|y\|$. On the other hand, $\|y\|\,\|y\| = |\langle y\,;\,y\rangle| = |f(y)| \leq \|f\|\,\|y\|$ so that $\|y\| \leq \|f\|$. Therefore,

$$\|f\| = \|y\|.$$

Outcome: A bounded (i.e., continuous) linear functional $f$ is naturally associated with each vector $y$ in an inner product space $\mathcal{X}$. The remarkable fact is that the converse holds true in a Hilbert space.

**Theorem 5.62.** (The Riesz Representation Theorem). *For every bounded linear functional $f$ on a Hilbert space $\mathcal{H}$ there exists a unique vector $y \in \mathcal{H}$ such that*

$$f(x) = \langle x\,;\,y\rangle \quad \text{for every} \quad x \in \mathcal{H}.$$

*Moreover, $\|f\| = \|y\|$. Such a unique vector $y$ in $\mathcal{H}$ is called the Riesz representation of the functional $f$ in $\mathcal{B}[\mathcal{H}, \mathbb{F}]$.*

*Proof.* If $f = 0$, then it is clear that $y = 0$ is the unique vector in $\mathcal{H}$ for which $f(x) = \langle x\,;\,y\rangle$ for every $x \in \mathcal{H}$. Thus suppose $f \neq 0$.

*Existence.* Consider the null space $\mathcal{N}(f)$ of $f \in \mathcal{B}[\mathcal{H}, \mathbb{F}]$, which is a proper subspace of $\mathcal{H}$ (i.e., $\mathcal{N}(f)$ is a subspace of $\mathcal{H}$ by Proposition 4.13 and $\mathcal{N}(f) \neq \mathcal{H}$ because $f \neq 0$). Hence, as $\mathcal{H}$ is a Hilbert space, $\mathcal{N}(f)^{\perp} \neq \{0\}$ according to Proposition 5.15. Let $z$ be any nonzero vector in $\mathcal{N}(f)^{\perp}$. Since $z \notin \mathcal{N}(f)$ (for $\mathcal{N}(f) \cap \mathcal{N}(f)^{\perp} = \{0\}$), it follows that $f(z) \neq 0$. Now take an arbitrary $x$ in $\mathcal{H}$ and note that

$$f\left(x - \tfrac{f(x)}{f(z)}z\right) - f(x) - f(x)\tfrac{f(z)}{f(z)} = 0.$$

Thus $x - \tfrac{f(x)}{f(z)}z \in \mathcal{N}(f)$. Since $z \in \mathcal{N}(f)^{\perp}$ we get

$$0 = \left\langle x - \tfrac{f(x)}{f(z)}z\,;\,z\right\rangle = \langle x\,;\,z\rangle - \tfrac{f(x)}{f(z)}\|z\|^2,$$

and hence $f(x) = \left\langle x\,;\,\tfrac{\overline{f(z)}}{\|z\|^2}z\right\rangle$. Then there exists a vector $y = \tfrac{\overline{f(z)}}{\|z\|^2}z$ in $\mathcal{H}$, which does not depend on $x$, such that $f(x) = \langle x\,;\,y\rangle$.

*Uniqueness.* If $y' \in \mathcal{H}$ is such that $f(x) = \langle x\,;\,y'\rangle$ for every $x \in \mathcal{H}$, then $\langle x\,;\,y\rangle = \langle x\,;\,y'\rangle$ so that $\langle x\,;\,y - y'\rangle = 0$ for every $x \in \mathcal{H}$. Therefore, $y - y' \in \mathcal{H}^{\perp} = \{0\}$.

Finally, as we have already seen in the introduction of this section, if $y \in \mathcal{H}$ is such that $f(x) = \langle x\,;\,y\rangle$ for every $x \in \mathcal{H}$, then $\|f\| = \|y\|$. $\qquad\square$

**Corollary 5.63.** *For every Hilbert space $\mathcal{H}$ there exists a surjective isometry $\Psi\colon \mathcal{H}^* \to \mathcal{H}$ of the dual $\mathcal{H}^*$ of $\mathcal{H}$ onto $\mathcal{H}$, which is additive and conjugate homogeneous (i.e., $\Psi(\alpha f) = \overline{\alpha}\,\Psi(f)$ for every $f \in \mathcal{H}^*$ and every $\alpha \in \mathbb{F}$).*

*Proof.* Let $\mathcal{H}$ be a Hilbert space and let $\mathcal{H}^* = \mathcal{B}[\mathcal{H}, \mathbb{F}]$ be the dual of $\mathcal{H}$. According to the Riesz Representation Theorem, for each $f \in \mathcal{H}^*$ there exists a unique $y \in \mathcal{H}$ such that $f(x) = \langle x\,;y \rangle$ for every $x \in \mathcal{H}$ and $\|f\| = \|y\|$. Conversely, for each $y \in \mathcal{H}$ the functional $f\colon \mathcal{H} \to \mathbb{F}$ given by $f(x) = \langle x\,;y \rangle$ for every $x \in \mathcal{H}$ is linear and bounded, which means that $f \in \mathcal{H}^*$. This establishes a surjective isometry (i.e., an invertible isometry) $\Psi\colon \mathcal{H}^* \to \mathcal{H}$ of the dual $\mathcal{H}^*$ of $\mathcal{H}$ onto $\mathcal{H}$:

$$\Psi(f) = y \quad \text{for every} \quad f \in \mathcal{H}^*,$$

where $y \in \mathcal{H}$ is the (unique) Riesz representation of $f \in \mathcal{H}^*$. Therefore, every $f$ in $\mathcal{H}^*$ is such that

$$f(x) = \langle x\,;\Psi(f) \rangle \quad \text{for every} \quad x \in \mathcal{H}.$$

Observe that $\Psi$ is additive. Indeed, if $f, g \in \mathcal{H}^*$, then

$$\begin{aligned}
\langle x\,;\Psi(f+g) \rangle &= (f+g)(x) = f(x) + g(x) \\
&= \langle x\,;\Psi(f) \rangle + \langle x\,;\Psi(g) \rangle = \langle x\,;\Psi(f) + \Psi(g) \rangle
\end{aligned}$$

for every $x \in \mathcal{H}$, so that $\Psi(f+g) = \Psi(f) + \Psi(g)$. Moreover, if $f \in \mathcal{H}^*$ and $\alpha \in \mathbb{F}$, then

$$\langle x\,;\ \Psi(\alpha f) \rangle = \alpha f(x) = f(\alpha x) = \langle \alpha x\,;\Psi(f) \rangle = \langle x\,;\overline{\alpha}\,\Psi(f) \rangle$$

for every $x \in \mathcal{H}$, and hence $\Psi(\alpha f) = \overline{\alpha}\,\Psi(f)$. $\qquad\qquad\square$

From the above corollary we may conclude: *Every Hilbert space is isometrically equivalent to its dual.* In particular, *every real Hilbert space is isometrically isomorphic to its dual.*

**Corollary 5.64.** *Every Hilbert space is reflexive.*

*Proof.* Let $\Psi\colon \mathcal{H}^* \to \mathcal{H}$ be the surjective isometry of Corollary 5.63, which is additive and conjugate homogeneous. Consider the function $\langle\ \ \rangle_*\colon \mathcal{H}^* \times \mathcal{H}^* \to \mathbb{F}$ given by

$$\langle f\,;g \rangle_* = \langle \Psi(g)\,;\Psi(f) \rangle$$

for every $f, g \in \mathcal{H}^*$, where $\langle\ ;\ \rangle$ stands for the inner product on $\mathcal{H}$. This defines an inner product on $\mathcal{H}^*$. Indeed, $\langle\ ;\ \rangle_*$ is additive because $\Psi$ is additive. Since $\Psi$ is conjugate homogeneous,

$$\langle \alpha f\,;g \rangle_* = \langle \Psi(g)\,;\Psi(\alpha f) \rangle = \langle \Psi(g)\,;\overline{\alpha}\,\Psi(f) \rangle = \alpha\langle \Psi(g)\,;\Psi(f) \rangle = \alpha\langle f\,;g \rangle_*$$

for every $f, g \in \mathcal{H}^*$ and every $\alpha \in \mathbb{F}$, and hence $\langle\ ;\ \rangle_*$ is homogeneous in the first argument. It is clear that $\langle\ ;\ \rangle_*$ is Hermitian symmetric and positive. Actually,

$$\|f\|_* = \|\Psi(f)\| = \|f\|$$

for every $f \in \mathcal{H}^*$ so that the norm $\| \ \|_*$ induced on $\mathcal{H}^*$ by the inner product $\langle \ ; \ \rangle_*$ coincides with the usual (induced uniform) norm on $\mathcal{H}^* = \mathcal{B}[\mathcal{H}, \mathbb{F}]$. Since the dual space of every normed space is a Banach space, it follows that $(\mathcal{H}^*, \| \ \|)$ is a Banach space, and hence $(\mathcal{H}^*, \| \ \|_*)$ is a Hilbert space. We shall now apply the Riesz Representation Theorem to the Hilbert space $\mathcal{H}^*$. Take an arbitrary $\varphi \in \mathcal{H}^{**}$. Theorem 5.62 ensures that there exists a unique $g \in \mathcal{H}^*$ such that

$$\varphi(f) = \langle f ; g \rangle_* = \langle \Psi(g) ; \Psi(f) \rangle$$

for every $f \in \mathcal{H}^*$. According to Theorem 5.62 every $f \in \mathcal{H}^*$ is given by $f(x) = \langle x ; y \rangle$ for every $x \in \mathcal{H}$, where $y = \Psi(f) \in \mathcal{H}$. Set $z = \Psi(g) \in \mathcal{H}$ so that

$$f(z) = \langle z ; y \rangle = \langle \Psi(g) ; \Psi(f) \rangle.$$

Therefore, there exists $z \in \mathcal{H}$ such that

$$\varphi(f) = f(z) \quad \text{for every} \quad f \in \mathcal{H}^*.$$

Hence $\mathcal{H}$ is reflexive by Proposition 4.67. $\qquad\qquad\qquad\qquad\qquad\qquad\square$

Let $T \in \mathcal{B}[\mathcal{H}, \mathcal{Y}]$ be a bounded linear transformation of a Hilbert space $\mathcal{H}$ to an inner product space $\mathcal{Y}$. Take an arbitrary $y \in \mathcal{Y}$ and consider the functional $f_y \colon \mathcal{H} \to \mathbb{F}$ defined by

$$f_y(x) = \langle Tx ; y \rangle$$

for every $x \in \mathcal{H}$, where the above inner product is that on $\mathcal{Y}$. It is easy to show that $f_y \colon \mathcal{H} \to \mathbb{F}$ is a bounded linear functional (i.e., $f_y \in \mathcal{H}^*$). Indeed, $f_y$ is linear because $T$ is linear and the inner product is linear in the first argument; and bounded by the Schwarz inequality: $|f(x)| \leq \|Tx\|\|y\| \leq \|T\|\|y\|\|x\|$ for every $x \in \mathcal{H}$. The Riesz Representation Theorem says that there exists a unique $z_y$ in $\mathcal{H}$ such that

$$\langle Tx ; y \rangle = f_y(x) = \langle x ; z_y \rangle$$

for every $x \in \mathcal{H}$, where the right-hand side inner product is that on $\mathcal{H}$. This establishes a mapping $T^* \colon \mathcal{Y} \to \mathcal{H}$ that assigns to each $y$ in $\mathcal{Y}$ this unique $z_y$ in $\mathcal{H}$ (i.e., $T^*y = z_y$ for every $y \in \mathcal{Y}$), and therefore satisfies the following identity for every $x$ in $\mathcal{H}$ and every $y$ in $\mathcal{Y}$:

$$\langle Tx ; y \rangle = \langle x ; T^*y \rangle.$$

The mapping $T^* \colon \mathcal{Y} \to \mathcal{H}$ is referred to as *the adjoint* of $T \in \mathcal{B}[\mathcal{H}, \mathcal{Y}]$. In fact, as we shall see below, the adjoint $T^*$ of $T$ is unique.

**Proposition 5.65.** *Take any $T \in \mathcal{B}[\mathcal{H}, \mathcal{Y}]$, where $\mathcal{H}$ is a Hilbert space and $\mathcal{Y}$ is an inner product space.*

(a) *The adjoint $T^*$ of $T$ is the unique mapping of $\mathcal{Y}$ into $\mathcal{H}$ such that $\langle Tx ; y \rangle = \langle x ; T^*y \rangle$ for every $x \in \mathcal{H}$ and every $y \in \mathcal{Y}$.*

(b)   $T^*$ *is a bounded linear transformation*   (i.e., $T^* \in \mathcal{B}[\mathcal{Y}, \mathcal{H}]$).

*Moreover, if $\mathcal{Y}$ also is a Hilbert space, then*

(c)   $T^{**} = T$,    *and*

(d)   $\|T^*\|^2 = \|T^*T\| = \|T\,T^*\| = \|T\|^2$.

*Proof.* Take $T \in \mathcal{B}[\mathcal{H}, \mathcal{Y}]$ and let $T^*\colon \mathcal{Y} \to \mathcal{H}$ be a mapping such that $\langle Tx \,;\, y \rangle = \langle x \,;\, T^*y \rangle$ for every $x \in \mathcal{H}$ and every $y \in \mathcal{Y}$.

*Proof of* (a). If $T^{\#}\colon \mathcal{Y} \to \mathcal{H}$ satisfies the identity $\langle Tx \,;\, y \rangle = \langle x \,;\, T^{\#}y \rangle$ for every $x \in \mathcal{H}$ and every $y \in \mathcal{Y}$, then for each $y$ in $\mathcal{Y}$ $\langle x \,;\, T^{\#}y \rangle = \langle x \,;\, T^*y \rangle$ for every $x \in \mathcal{H}$, and hence $T^{\#}y = T^*y$ for every $y \in \mathcal{Y}$.

*Proof of* (b). Take $y_1, y_2 \in \mathcal{Y}$ and $\alpha_1, \alpha_2 \in \mathbb{F}$ arbitrary. Note that

$$
\begin{aligned}
\langle x \,;\, T^*(\alpha_1 y_1 + \alpha_2 y_2) \rangle
&= \langle Tx \,;\, \alpha_1 y_1 + \alpha_2 y_2 \rangle \\
&= \overline{\alpha}_1 \langle Tx \,;\, y_1 \rangle + \overline{\alpha}_2 \langle Tx \,;\, y_2 \rangle \\
&= \overline{\alpha}_1 \langle x \,;\, T^*y_1 \rangle + \overline{\alpha}_2 \langle x \,;\, T^*y_2 \rangle \\
&= \langle x \,;\, \alpha_1 T^*y_1 + \alpha_2 T^*y_2 \rangle
\end{aligned}
$$

for every $x \in \mathcal{H}$, and hence $T^*(\alpha_1 y_1 + \alpha_2 y_2) = \alpha_1 T^*y_1 + \alpha_2 T^*y_2$ so that $T^*$ is linear. Moreover, by the Schwarz inequality,

$$
\|T^*y\|^2 = \langle T^*y \,;\, T^*y \rangle = \langle T\,T^*y \,;\, y \rangle \le \|T\,T^*y\| \, \|y\| \le \|T\| \, \|T^*y\| \, \|y\|
$$

for every $y \in \mathcal{Y}$. This implies that $\|T^*y\| \le \|T\| \, \|y\|$ for every $y \in \mathcal{Y}$. Thus $T^*$ is bounded and

$$
\|T^*\| \le \|T\|.
$$

Now suppose $\mathcal{Y}$ is a Hilbert space so that $T^* \in \mathcal{B}[\mathcal{Y}, \mathcal{H}]$ has a unique adjoint $(T^*)^* \in \mathcal{B}[\mathcal{H}, \mathcal{Y}]$ (notation: $T^{**} = (T^*)^*$) such that

$$
\langle T^*y \,;\, x \rangle = \langle y \,;\, T^{**}x \rangle
$$

for every $x \in \mathcal{H}$ and every $y \in \mathcal{Y}$.

*Proof of* (c). Since $\langle y \,;\, Tx \rangle = \overline{\langle Tx \,;\, y \rangle} = \overline{\langle x \,;\, T^*y \rangle} = \langle T^*y \,;\, x \rangle$, for every $x \in \mathcal{H}$ and every $y \in \mathcal{Y}$, it follows by the above identity that

$$
\langle y \,;\, Tx \rangle = \langle y \,;\, T^{**}x \rangle
$$

for every $x \in \mathcal{H}$ and every $y \in \mathcal{Y}$, and hence $Tx = T^{**}x$ for every $x \in \mathcal{H}$.

*Proof of* (d). We have already seen that $\|T^*\| \le \|T\|$. Then, as $T = T^{**}$, we get $\|T\| = \|T^{**}\| \le \|T^*\|$. Hence $\|T^*\| = \|T\|$. Therefore, $\|T^*T\| \le \|T^*\| \, \|T\| = \|T\|^2$. However, the Schwarz inequality ensures that $\|Tx\|^2 = \langle Tx \,;\, Tx \rangle =$

$\langle T^*Tx \, ; x \rangle \leq \|T^*Tx\| \, \|x\| \leq \|T^*T\| \, \|x\|^2$ for every $x \in \mathcal{H}$ so that $\|T\|^2 \leq \|T^*T\|$. Thus $\|T\|^2 = \|T^*T\|$. Again, as $T^{**} = T$, we get $\|T^*\|^2 = \|T^{**}T^*\| = \|T\,T^*\|$. $\square$

Let $\mathcal{X}$ be an inner product space and consider the definition of weak convergence (cf. Problem 4.67): An $\mathcal{X}$-valued sequence $\{x_n\}$ converges weakly to $x \in \mathcal{X}$ (notation: $x_n \xrightarrow{w} x$) if the scalar-valued sequence $\{f(x_n)\}$ converges in $\mathbb{F}$ to $f(x)$ for every $f \in \mathcal{X}^*$ (i.e., if $f(x_n - x) \to 0$ for every $f \in \mathcal{X}^*$). Recall from Problem 4.67 that convergence in the norm topology implies weak convergence (to the same and unique limit):

$$x_n \to x \quad \text{implies} \quad x_n \xrightarrow{w} x.$$

Since $\langle \cdot \, ; y \rangle \colon \mathcal{X} \to \mathbb{F}$ lies in $\mathcal{X}^*$ for every $y \in \mathcal{X}$, it follows that

$$x_n \xrightarrow{w} x \quad \text{implies} \quad \langle x_n \, ; y \rangle \to \langle x \, ; y \rangle \quad \text{for every } y \in \mathcal{X}.$$

The Riesz Representation Theorem ensures the converse in a Hilbert space: If $\mathcal{X}$ is a Hilbert space, then

$$x_n \xrightarrow{w} x \quad \text{if and only if} \quad \langle x_n \, ; y \rangle \to \langle x \, ; y \rangle \quad \text{for every } y \in \mathcal{X}.$$

In particular, $x_n \xrightarrow{w} x$ implies $\langle x_n \, ; x \rangle \to \langle x \, ; x \rangle = \|x\|^2$. By recalling that $\|x_n - x\|^2 = \|x_n\|^2 - 2\operatorname{Re}\langle x_n \, ; x \rangle + \|x\|^2$, we may conclude the nontrivial part of the next equivalence, which holds in any inner product space.

$$x_n \to x \quad \text{if and only if} \quad x_n \xrightarrow{w} x \quad \text{and} \quad \|x_n\| \to \|x\|.$$

Now suppose $\mathcal{X}$ and $\mathcal{Y}$ are inner product spaces and let $\{T_n\}$ be a $\mathcal{B}[\mathcal{X}, \mathcal{Y}]$-valued sequence. If the $\mathcal{Y}$-valued sequence $\{T_n x\}$ converges weakly for every $x \in \mathcal{X}$, then we say that $\{T_n\}$ *converges weakly* (or *converges in the weak (operator) topology*). Equivalently, $\{T_n\}$ converges weakly if there exists a unique $T \in \mathcal{B}[\mathcal{X}, \mathcal{Y}]$ such that

$$T_n x \xrightarrow{w} Tx \quad \text{in } \mathcal{Y} \text{ for every } x \in \mathcal{X},$$

which is called the *weak limit* of $\{T_n\}$ (see Problem 4.68). Notation: $T_n \xrightarrow{w} T$. Since $\langle \cdot \, ; y \rangle \colon \mathcal{Y} \to \mathbb{F}$ lies in $\mathcal{Y}^*$ for every $y \in \mathcal{Y}$, it follows that

$$T_n \xrightarrow{w} T \quad \text{implies} \quad \langle (T_n - T)x \, ; y \rangle \to 0 \text{ for every } x \in \mathcal{X} \text{ and every } y \in \mathcal{Y}.$$

We shall see below (Proposition 5.67) that the converse holds if $\mathcal{Y}$ is a Hilbert space. Recall that uniform convergence implies strong convergence, which in turn implies weak convergence (to the same limit — see Definition 4.45 and Problem 4.68):

$$T_n \xrightarrow{u} T \quad \Longrightarrow \quad T_n \xrightarrow{s} T \quad \Longrightarrow \quad T_n \xrightarrow{w} T.$$

Since $y_n \to y$ if and only if $y_n \xrightarrow{w} y$ and $\|y_n\| \to \|y\|$, it follows that

$$T_n \xrightarrow{s} T \quad \text{if and only if} \quad T_n \xrightarrow{w} T \quad \text{and} \quad \|T_n x\| \to \|Tx\| \text{ for every } x \in \mathcal{X}.$$

We have already seen that strong and uniform convergence coincide if $\mathcal{X}$ is finite-dimensional. In fact, *uniform, strong and weak convergence coincide if $\mathcal{X}$ and $\mathcal{Y}$ are finite-dimensional inner product spaces.*

**Proposition 5.66.** (a) *In a finite-dimensional inner product space, weak convergence and convergence in the norm topology coincide.*

(b) *Let $\mathcal{X}$ and $\mathcal{Y}$ be inner product spaces. Consider a $\mathcal{B}[\mathcal{X},\mathcal{Y}]$-valued sequence $\{T_k\}$ and let $T$ be a transformation in $\mathcal{B}[\mathcal{X},\mathcal{Y}]$. If $\mathcal{Y}$ is finite-dimensional, then $T_k \xrightarrow{\;w\;} T$ if and only if $T_k \xrightarrow{\;s\;} T$.*

*Proof.* (a) Suppose $\mathcal{X}$ is a finite-dimensional inner product space so that it is a Hilbert space (Corollary 4.28) and let $B = \{e_i\}_{i=1}^n$ be an orthonormal basis for $\mathcal{X}$. Take an arbitrary weakly convergent sequence $\{x_k\}$ in $\mathcal{X}$ so that $x_k \xrightarrow{\;w\;} x$ for some $x \in \mathcal{X}$, and hence $\langle x_k - x \,;\, e_i \rangle \to 0$ as $k \to \infty$ for each $e_i \in B$. Therefore, for each $i = 1,\dots,n$ and every $\varepsilon > 0$, there exists a positive integer $k_{i,\varepsilon}$ such that $|\langle x_k - x \,;\, e_i \rangle| < \varepsilon$ whenever $k \geq k_{i,\varepsilon}$. Then the Fourier Series Theorem (Theorem 5.48) ensures that

$$\|x_k - x\|^2 = \sum_{i=1}^n |\langle x_k - x \,;\, e_i \rangle|^2 < n\varepsilon^2$$

whenever $k \geq k_\varepsilon = \max\{k_{i,\varepsilon}\}_{i=1}^n$. Hence $\|x_k - x\| \to 0$ as $k \to \infty$. That is, $x_k \to x$. Summing up: $x_k \xrightarrow{\;w\;} x$ implies $x_k \to x$ whenever $\mathcal{X}$ is a finite-dimensional inner product space. This concludes the proof of (a), for norm convergence always implies weak convergence.

(b) $T_k \xrightarrow{\;w\;} T$ means $T_k x \xrightarrow{\;w\;} Tx$ in $\mathcal{Y}$ for every $x \in \mathcal{X}$. But item (a) says that this is equivalent to $T_k x \to Tx$ in $\mathcal{Y}$ for every $x \in \mathcal{X}$, whenever $\mathcal{Y}$ is finite-dimensional, which means $T_k \xrightarrow{\;s\;} T$. $\qquad\qquad\square$

Here are equivalent conditions for weak convergence of bounded linear transformations between Hilbert spaces.

**Proposition 5.67.** *Let $\{T_n\}$ be a sequence of bounded linear transformations of a Hilbert space $\mathcal{H}$ into a Hilbert space $\mathcal{K}$ (i.e., $\{T_n\}$ is a $\mathcal{B}[\mathcal{H},\mathcal{K}]$-valued sequence). The following three assertions are pairwise equivalent.*

(a) *There exists $T \in \mathcal{B}[\mathcal{H},\mathcal{K}]$ such that $\{T_n\}$ converges weakly to $T$ (i.e., $T_n \xrightarrow{\;w\;} T$ or, equivalently, $T_n - T \xrightarrow{\;w\;} O$).*

(b) *There exists $T \in \mathcal{B}[\mathcal{H},\mathcal{K}]$ such that $\langle T_n x \,;\, y \rangle \to \langle Tx \,;\, y \rangle$ as $n \to \infty$ for every $x$ in $\mathcal{H}$ and every $y$ in $\mathcal{K}$.*

(c) *The scalar-valued sequence $\{\langle T_n x \,;\, y \rangle\}$ converges in $\mathbb{F}$ for every $x$ in $\mathcal{H}$ and every $y$ in $\mathcal{K}$.*

*Now set $\mathcal{K} = \mathcal{H}$ and consider the following further assertions.*

(d)  *There exists $T \in \mathcal{B}[\mathcal{H}]$ such that $\langle T_n x \, ; x \rangle \to \langle Tx \, ; x \rangle$ as $n \to \infty$ for every $x \in \mathcal{H}$.*

(e)  *The scalar-valued sequence $\{\langle T_n x \, ; x \rangle\}$ converges in $\mathbb{F}$ for every $x \in \mathcal{H}$.*

*Clearly,* (b) *implies* (d), *which implies* (e). *If $\mathcal{K} = \mathcal{H}$ is a complex* Hilbert space, *then these five assertions are all pairwise equivalent.*

*Proof.* If there exists $T \in \mathcal{B}[\mathcal{H}, \mathcal{K}]$ such that $f(T_n x) \to f(Tx)$ in $\mathbb{F}$ as $n \to \infty$ for every $f \in \mathcal{K}^*$ and every $x \in \mathcal{H}$, then $\langle T_n x \, ; y \rangle \to \langle Tx \, ; y \rangle$ as $n \to \infty$ for every $y \in \mathcal{K}$ and every $x \in \mathcal{H}$ (because $\langle \cdot \, ; y \rangle \colon \mathcal{K} \to \mathbb{F}$ lies in $\mathcal{K}^*$ for every $y \in \mathcal{K}$). Hence (a)$\Rightarrow$(b). Conversely, suppose (b) holds true and take any $f \in \mathcal{K}^*$. Since $\mathcal{K}$ is a Hilbert space, the Riesz Representation Theorem (Theorem 5.62) ensures that $f(T_n x) \to f(Tx)$ in $\mathbb{F}$ as $n \to \infty$ for every $x \in \mathcal{H}$. Thus (b)$\Rightarrow$(a). It is clear that (b)$\Rightarrow$(c). Now suppose assertion (c) holds true. Take an arbitrary $y$ in $\mathcal{K}$ and consider the functional $f_y \colon \mathcal{H} \to \mathbb{F}$ defined by

$$f_y(x) = \lim_n \langle T_n x \, ; y \rangle$$

for every $x \in \mathcal{H}$. Observe that $f_y$ is linear (because $T_n$ is linear for each $n$, the inner product is linear in the first argument, and the linear operations in $\mathbb{F}$ are continuous). Since $\{\langle T_n x \, ; y \rangle\}$ converges in $\mathbb{F}$ for every $x \in \mathcal{H}$, it follows that $\{\langle T_n x \, ; y \rangle\}$ is a bounded sequence for every $x \in \mathcal{H}$ (Proposition 3.39). Then $\sup_n |\langle T_n x \, ; y \rangle| < \infty$ for every $x$ in $\mathcal{H}$ and every $y$ in $\mathcal{K}$, which implies that $\sup_n \|T_n\| < \infty$ (a consequence of the Banach–Steinhaus Theorem — see Problem 5.5). Hence $|f_y(x)| = |\lim_n \langle T_n x \, ; y \rangle| = \lim_n |\langle T_n x \, ; y \rangle| \le \sup_n |\langle T_n x \, ; y \rangle| \le \sup_n \|T_n x\| \|y\| \le \sup_n \|T_n\| \|x\| \|y\|$ for every $x \in \mathcal{H}$ so that

$$\|f_y\| \le \sup_n \|T_n\| \|y\|.$$

That is, $f_y$ is bounded, and therefore $f_y \in \mathcal{H}^*$ ($f_y$ is a bounded linear functional on $\mathcal{H}$). Since $\mathcal{H}$ is a Hilbert space, the Riesz Representation Theorem says that there exists a unique $z_y \in \mathcal{H}$ such that

$$f_y(x) = \langle x \, ; z_y \rangle$$

for every $x \in \mathcal{H}$. Consider the mapping $S \colon \mathcal{K} \to \mathcal{H}$ that assigns to each $y$ in $\mathcal{H}$ this unique $z_y$ in $\mathcal{H}$, $Sy = z_y$ for every $y \in \mathcal{K}$, so that

$$\lim_n \langle T_n x \, ; y \rangle = f_y(x) = \langle x \, ; z_y \rangle = \langle x \, ; Sy \rangle$$

for every $x \in \mathcal{H}$ and every $y \in \mathcal{K}$. Note that $S$ is linear and bounded (i.e., $S$ lies in $\mathcal{B}[\mathcal{K}, \mathcal{H}]$). Indeed, if $y_1, y_2 \in \mathcal{Y}$ and $\alpha_1, \alpha_2 \in \mathbb{F}$, then

$$
\begin{aligned}
\langle x \,;\, S(\alpha_1 y_1 + \alpha_2 y_2)\rangle &= \lim_n \langle T_n x \,;\, \alpha_1 y_1 + \alpha_2 y_2\rangle \\
&= \overline{\alpha}_1 \lim_n \langle T_n x \,;\, y_1\rangle + \overline{\alpha}_2 \lim_n \langle T_n x \,;\, y_2\rangle \\
&= \overline{\alpha}_1 \langle x \,;\, S y_1\rangle + \overline{\alpha}_2 \langle x \,;\, S y_2\rangle \\
&= \langle x \,;\, \alpha_1 S y_1 + \alpha_2 S y_2\rangle
\end{aligned}
$$

for every $x \in \mathcal{H}$. Hence $S(\alpha_1 y_1 + \alpha_2 y_2) = \alpha_1 S y_1 + \alpha_2 S y_2$, which means that $S$ is linear. Moreover,

$$
\|S y\| = \|z_y\| = \|f_y\| \le \sup_n \|T_n\| \, \|y\|
$$

for every $y \in \mathcal{K}$ so that $S$ is bounded with $\|S\| \le \sup_n \|T_n\|$. Setting $T = S^*$ in $\mathcal{B}[\mathcal{H}, \mathcal{K}]$, the adjoint of $S$, we get (cf. Proposition 5.65)

$$
\lim_n \langle T_n x \,;\, y\rangle = \langle x \,;\, S y\rangle = \langle x \,;\, S^{**} y\rangle = \langle S^* x \,;\, y\rangle = \langle T x \,;\, y\rangle
$$

for every $x \in \mathcal{H}$ and every $y \in \mathcal{K}$. Therefore, (c)$\Rightarrow$(b). Finally, set $\mathcal{K} = \mathcal{H}$ so that (b)$\Rightarrow$(d) and (c)$\Rightarrow$(e) trivially. According to Problem 5.3(b) (with $L = I$), it follows that (d)$\Rightarrow$(b) and (e)$\Rightarrow$(c) whenever $\mathcal{H}$ is a complex Hilbert space. $\qquad\square$

Observe from Proposition 5.67 and Problem 5.5 that

$$
T_n \xrightarrow{\ w\ } T \quad \Longrightarrow \quad \sup_n \|T_n\| < \infty \quad \text{whenever } \mathcal{H} \text{ and } \mathcal{K} \text{ are Hilbert.}
$$

Take $T \in \mathcal{B}[\mathcal{X}]$, where $\mathcal{X}$ is an inner product space, and consider the power sequence $\{T^n\}$ so that each $T^n$ lies in $\mathcal{B}[\mathcal{H}]$. The operator $T$ is *weakly stable* if the power sequence $\{T^n\}$ converges weakly to the null operator; that is, $T^n \xrightarrow{\ w\ } O$. If $\mathcal{X}$ is a Hilbert space, then Proposition 5.67 says that this is equivalent to any of those five assertions with $T_n$ replaced with $T^n$ and $T$ replaced with $O$. In particular, if $\mathcal{X}$ is a Hilbert space, then $T^n \xrightarrow{\ w\ } O$ if and only if $\langle T^n x \,;\, x\rangle \to 0$ as $n \to \infty$ for every $x \in \mathcal{X}$. Clearly, uniform stability implies strong stability that implies weak stability,

$$
T^n \xrightarrow{\ u\ } O \quad \Longrightarrow \quad T^n \xrightarrow{\ s\ } O \quad \Longrightarrow \quad T^n \xrightarrow{\ w\ } O,
$$

which in turn implies power boundedness whenever $\mathcal{X}$ is a Hilbert space:

$$
T^n \xrightarrow{\ w\ } O \quad \Longrightarrow \quad \sup_n \|T^n\| < \infty \quad \text{if } \mathcal{X} \text{ is Hilbert.}
$$

The converses, however, fail (see Example 4K and Problem 5.29(c)).

Let $\mathcal{X}$ and $\mathcal{Y}$ be inner product spaces. We say that a subset $\Theta$ of $\mathcal{B}[\mathcal{X}, \mathcal{Y}]$ is *weakly closed* in $\mathcal{B}[\mathcal{X}, \mathcal{Y}]$ if every $\Theta$-valued weakly convergent sequence $\{T_n\}$ has its (weak) limit $T$ in $\Theta$. Recall from Proposition 4.48: If $\Theta$ is strongly closed in $\mathcal{B}[\mathcal{X}, \mathcal{Y}]$, then it is (uniformly) closed in $\mathcal{B}[\mathcal{X}, \mathcal{Y}]$.

**Proposition 5.68.** *If $\Theta \subseteq \mathcal{B}[\mathcal{X}, \mathcal{Y}]$ is weakly closed in $\mathcal{B}[\mathcal{X}, \mathcal{Y}]$, then it is strongly closed in $\mathcal{B}[\mathcal{X}, \mathcal{Y}]$.*

*Proof.* Take an arbitrary $\Theta$-valued strongly convergent sequence, say $\{T_n\}$, and let $T \in \mathcal{B}[\mathcal{X}, \mathcal{Y}]$ be its (strong) limit. Since strong convergence implies weak convergence to the same limit, it follows that $\{T_n\}$ converges weakly to $T$. If every $\Theta$-valued weakly convergent sequence has its (weak) limit in $\Theta$, then $T \in \Theta$. Conclusion: Every $\Theta$-valued strongly convergent sequence has its (strong) limit in $\Theta$. $\qquad\qquad\square$

Remark: If $\mathcal{Y}$ is finite-dimensional, then weak convergence coincides with strong convergence (Proposition 5.66(b)), and hence the concepts of weakly closed and strongly closed in $\mathcal{B}[\mathcal{X}, \mathcal{Y}]$ coincide whenever $\mathcal{Y}$ is a finite-dimensional inner product space. Then (cf. Remark after Proposition 4.48), *if $\mathcal{X}$ and $\mathcal{Y}$ are finite-dimensional inner product spaces, then all the three concepts of weakly, strongly and uniformly closed in $\mathcal{B}[\mathcal{X}, \mathcal{Y}]$ coincide.*

Recall that a subset $A$ of a metric space is compact if and only if it is sequentially compact (Theorem 3.80), which means by Definition 3.76 that every $A$-valued sequence has a convergent subsequence (that converges to a point in $A$). The Heine–Borel Theorem (Theorem 3.83) ensures that every bounded subset $B$ of a finite-dimensional inner product space has a compact closure. Thus every $B$-valued sequence has a convergent subsequence (that converges to a point in $B^-$). Therefore, *every bounded nonempty subset of a finite-dimensional inner product space has a convergent sequence* (whose limit lies in its closure). This no longer holds in an infinite-dimensional inner product space (e.g., if $\{e_n\}$ is an infinite orthonormal sequence, then $\|e_n - e_m\|^2 = 2$ for every $m \neq n$). Now recall that a finite-dimensional inner product space is a Hilbert space where convergence in the norm topology coincides with weak convergence (Proposition 5.66(a)), so that the next lemma actually is an extension to infinite-dimensional Hilbert spaces of the above italicized result.

**Lemma 5.69.** *Every bounded nonempty subset of a Hilbert space has a weakly convergent sequence. In particular, every bounded sequence in a Hilbert space has a weakly convergent subsequence.*

*Proof.* Let $B$ be a bounded nonempty subset of an inner product space $\mathcal{H}$. The result is trivial if $B = \{0\}$. Suppose $B \neq \{0\}$ and let $A \neq \{0\}$ be a nonempty countable subset of $B$. Put $\mathcal{X} = \bigvee A$, which is a subspace of $\mathcal{H}$. Thus $0 < \sup_{x \in A} \|x\| < \infty$ and $\mathcal{X}$ is a separable inner product space by Proposition 4.9(b). Consider the following subset of the dual of $\mathcal{X}$:

$$\Phi = \big\{ \langle \,\cdot\, ; x\rangle \colon \mathcal{X} \to \mathbb{F} \colon \ x \in A \big\} \subset \mathcal{X}^*.$$

Recall the definition of "pointwise total boundedness" and "equicontinuity" of Example 3Z.

*Claim 1.*  $\Phi$ is pointwise totally bounded.

*Proof.* Since total boundedness coincides with plain boundedness in $\mathbb{F}$ (cf. proof of Theorem 3.83), it follows that $\Phi$ is pointwise totally bounded if and only if it is pointwise bounded. But $|\langle w \,;\, x\rangle| \leq \sup_{x \in A} \|x\| \|w\|$ for every $w \in \mathcal{X}$, and so $\Phi$ is pointwise bounded. $\square$

*Claim 2.*  $\Phi$ is (uniformly) equicontinuous.

*Proof.* Indeed, $|\langle u \,;\, x\rangle - \langle v \,;\, x\rangle| = |\langle u - v \,;\, x\rangle| \leq \sup_{x \in A} \|x\| \|u - v\|$ for every $u, v \in \mathcal{X}$ and every $x \in A$. This implies that for every $\varepsilon > 0$ there exists $\delta = (\sup_{x \in A} \|x\|)^{-1}\varepsilon > 0$ such that $|\langle u \,;\, x\rangle - \langle v \,;\, x\rangle| < \varepsilon$ whenever $\|u - v\| < \delta$ for all $u, v \in \mathcal{X}$ and every $x \in A$. $\square$

Then, according to the proof of Example 3Z (set $f_n = \langle \cdot \,;\, x_n\rangle \in \Phi$), we may conclude: $\Phi$ is totally bounded because $\mathcal{X}$ is separable, which means that every $\Phi$-valued sequence has a Cauchy subsequence (Lemma 3.73). Take an arbitrary $A$-valued sequence $\{x_n\}$ so that the $\Phi$-valued sequence $\{\langle \cdot \,;\, x_n\rangle\}$ has a subsequence, say $\{\langle \cdot \,;\, x_{n_k}\rangle\}$, that is Cauchy in $\mathcal{X}^*$. But $\mathcal{X}^*$ is always complete so that $\{\langle \cdot \,;\, x_{n_k}\rangle\}$ converges in $\mathcal{X}^*$. That is, there exists $f \in \mathcal{X}^*$ such that $\langle \cdot \,;\, x_{n_k}\rangle \to f$ in $\mathcal{X}^*$. If $\mathcal{H}$ is a Hilbert space, then the subspace $\mathcal{X}$ is itself a Hilbert space, and the Riesz Representation Theorem says that there exists $x \in \mathcal{X}$ for which $f = \langle \cdot \,;\, x\rangle$. Therefore, $\langle x_{n_k} \,;\, w\rangle \to \langle x \,;\, w\rangle$ for every $w \in \mathcal{X}$. Using the Riesz Representation Theorem again we get $x_{n_k} \xrightarrow{w} x$ in $\mathcal{X}$. Hence $x_{n_k} \xrightarrow{w} x$ in $\mathcal{H}$ (cf. Problem 5.20). $\square$

**Theorem 5.70.** *Let* $\{T_n\}$ *be a* $\mathcal{B}[\mathcal{H}, \mathcal{K}]$*-valued sequence, where* $\mathcal{H}$ *and* $\mathcal{K}$ *are Hilbert spaces. If* $\mathcal{H}$ *is a separable Hilbert space and* $\sup_n \|T_n\| < \infty$*, then* $\{T_n\}$ *has a weakly convergent subsequence.*

*Proof.* Suppose $\mathcal{H} \neq \{0\}$ to avoid trivialities. If $\mathcal{H}$ is separable, then there exists a countably infinite dense subset $A$ of nonzero linear space $\mathcal{H}$. Let $\{a_i\}_{i \geq 1}$ be an $A$-valued sequence consisting of an enumeration of all points of $A$. Since $A^- = \mathcal{H}$, it follows by Proposition 3.32 that

$$\inf_i \|x - a_i\| = 0 \quad \text{for every} \quad x \in \mathcal{H}.$$

If $\sup_n \|T_n\| < \infty$, then $\{T_n a_1\}_{n \geq 1}$ is a bounded sequence in $\mathcal{K}$. Thus Lemma 5.69 ensures the existence of a subsequence of $\{T_n\}_{n \geq 1}$, say $\{T_n^{(1)}\}_{n \geq 1}$, such that $\{T_n^{(1)} a_1\}_{n \geq 1}$ converges weakly in $\mathcal{K}$. Again, since $\sup_n \|T_n\| < \infty$, it follows that $\{T_n a_2\}_{n \geq 1}$ is bounded in $\mathcal{K}$ and, in particular, the subsequence $\{T_n^{(1)} a_2\}_{n \geq 1}$ is bounded in $\mathcal{K}$. Hence another application of Lemma 5.69 ensures that there exists a subsequence of $\{T_n^{(1)}\}_{n \geq 1}$, say $\{T_n^{(2)}\}_{n \geq 1}$, such that $\{T_n^{(2)} a_2\}_{n \geq 1}$ converges weakly in $\mathcal{K}$. Note that $\{T_n^{(2)} a_1\}_{n \geq 1}$ also converges weakly in $\mathcal{K}$. (Reason: $\{T_n^{(2)}\}_{n \geq 1}$ is a subsequence of $\{T_n^{(1)}\}_{n \geq 1}$ and $\{T_n^{(1)} a_1\}_{n \geq 1}$ converges weakly in $\mathcal{K}$ — see Problem 4.67(b).) This leads to an inductive construction of a sequence of $\mathcal{B}[\mathcal{H}, \mathcal{K}]$-valued sequences, $\{\{T_n^{(k)}\}_{n \geq 1}\}_{k \geq 1}$, with the following properties.

**Property (1).** $\{T_n^{(k+1)}\}_{n\geq 1}$ is a subsequence of $\{T_n^{(k)}\}_{n\geq 1}$, which is a subsequence of $\{T_n\}_{n\geq 1}$, for every $k \geq 1$.

**Property(2).** $\{T_n^{(k)} a_i\}_{n\geq 1}$ converges weakly in $\mathcal{K}$ whenever $k \geq i$.

Consider the "diagonal" sequence $\{T_n^{(n)}\}_{n\geq 1}$, which is a subsequence of $\{T_n\}_{n\geq 1}$. If $\{T_n^{(n)}\}_{n\geq 1}$ is weakly convergent, then the theorem is proved.

*Claim.* $\{T_n^{(n)}\}_{n\geq 1}$ is weakly convergent.

*Proof.* Take $x \in \mathcal{H}$, $y \in \mathcal{K}$, and $\varepsilon > 0$ arbitrary. Since $\inf_i \|x - a_i\| = 0$, there exists an integer $i_\varepsilon \geq 1$ such that

$$\|x - a_{i_\varepsilon}\| < \varepsilon.$$

According to Property (1) $\{T_n^{(n)}\}_{n\geq i_\varepsilon}$ is a subsequence of $\{T_n^{(i_\varepsilon)}\}_{n\geq 1}$, and hence $\{T_n^{(n)} a_{i_\varepsilon}\}_{n\geq i_\varepsilon}$ is a subsequence of $\{T_n^{(i_\varepsilon)} a_{i_\varepsilon}\}_{n\geq 1}$. Since $\{T_n^{(i_\varepsilon)} a_{i_\varepsilon}\}_{n\geq 1}$ converges weakly in $\mathcal{K}$ by Property (2), it follows that its subsequence $\{T_n^{(n)} a_{i_\varepsilon}\}_{n\geq i_\varepsilon}$ also converges weakly in $\mathcal{K}$. This implies that $\{T_n^{(n)} a_{i_\varepsilon}\}_{n\geq 1}$ converges weakly in $\mathcal{K}$ so that $\{\langle T_n^{(n)} a_{i_\varepsilon} ; y\rangle\}_{n\geq 1}$ converges in $\mathbb{F}$, and therefore is a Cauchy sequence in $\mathbb{F}$. That is, there exists an integer $n_\varepsilon \geq 1$ such that

$$m, n \geq n_\varepsilon \quad \text{implies} \quad |\langle (T_n^{(n)} - T_m^{(m)}) a_{i_\varepsilon} ; y\rangle| < \varepsilon.$$

Note that $\|T_n^{(n)} - T_m^{(m)}\| \leq 2 \sup_k \|T_k\|$ for all $m, n \geq 1$ because $\{T_n^{(n)}\}_{n\geq 1}$ is a subsequence of $\{T_n\}_{n\geq 1}$. Hence

$$
\begin{aligned}
|\langle (T_n^{(n)} - T_m^{(m)}) x ; y\rangle| &= |\langle (T_n^{(n)} - T_m^{(m)})(a_{i_\varepsilon} + x - a_{i_\varepsilon}); y\rangle| \\
&\leq |\langle (T_n^{(n)} - T_m^{(m)}) a_{i_\varepsilon}; y\rangle| + 2 \sup_k \|T_k\| \|x - a_{i_\varepsilon}\| \|y\| \\
&\leq \left(1 + 2 \sup_k \|T_k\| \|y\|\right) \varepsilon
\end{aligned}
$$

whenever $m, n \geq n_\varepsilon$. Conclusion: $\{\langle T_n^{(n)} x ; y\rangle\}_{n\geq 1}$ is a Cauchy sequence in $\mathbb{F}$ so that it converges in $\mathbb{F}$ (since $\mathbb{F}$ is complete). As $x$ and $y$ are arbitrary vectors in $\mathcal{H}$ and $\mathcal{K}$, respectively, this implies that the scalar-valued sequence $\{\langle T_n^{(n)} x ; y\rangle\}_{n\geq 1}$ converges in $\mathbb{F}$ for every $x \in \mathcal{H}$ and every $y \in \mathcal{K}$, which means that $T_n^{(n)} \xrightarrow{w} T$ for some $T \in \mathcal{B}[\mathcal{H}, \mathcal{K}]$ by Theorem 5.67. $\qquad\square$

## 5.12   The Adjoint Operator

Let $T \in \mathcal{B}[\mathcal{H}, \mathcal{Y}]$ be a bounded linear transformation of a Hilbert space $\mathcal{H}$ into an inner product space $\mathcal{Y}$. The *adjoint* of $T$ was defined in the previous section as the unique mapping $T^* : \mathcal{Y} \to \mathcal{H}$ such that

$$\langle Tx ; y\rangle = \langle x ; T^* y\rangle$$

for every $x \in \mathcal{H}$ and every $y \in \mathcal{Y}$, whose existence was established in Section 5.11 as a consequence of the Riesz Representation Theorem. The basic facts about the adjoint $T^*$ were stated in Proposition 5.65. In particular, it is linear and bounded (i.e., $T^* \in \mathcal{B}[\mathcal{Y}, \mathcal{H}]$). Here is a useful corollary of Proposition 5.65.

**Corollary 5.71.** If $\mathcal{H}$ and $\mathcal{K}$ are Hilbert spaces and $T \in \mathcal{B}[\mathcal{H}, \mathcal{K}]$, then

$$\|T\| = \sup_{\|x\|=\|y\|=1} |\langle Tx ; y \rangle|.$$

*Proof.* Since $\|z\| = \sup_{\|y\|=1} |\langle z ; y \rangle|$ for every $z \in \mathcal{K}$ (see Problem 5.1), it follows that $\|Tx\| = \sup_{\|y\|=1} |\langle Tx ; y \rangle|$ for every $x \in \mathcal{H}$. Hence

$$\|T\| = \sup_{\|x\|=1} \sup_{\|y\|=1} |\langle Tx ; y \rangle|.$$

Recalling that $T^{**} = T$ (see Proposition 5.65), it also follows that $\|T^*y\| = \sup_{\|x\|=1} |\langle T^*y ; x \rangle| = \sup_{\|x\|=1} |\langle y ; Tx \rangle| = \sup_{\|x\|=1} |\langle Tx ; y \rangle|$ for every $y \in \mathcal{K}$. Therefore, as $\|T^*\| = \|T\|$ (cf. Proposition 5.65 again),

$$\|T\| = \|T^*\| = \sup_{\|y\|=1} \sup_{\|x\|=1} |\langle Tx ; y \rangle|. \qquad \square$$

Let us see some further elementary properties of the adjoint. Consider the linear space $\mathcal{B}[\mathcal{H}, \mathcal{Y}]$, where $\mathcal{H}$ is a Hilbert space and $\mathcal{Y}$ is an inner product space. First observe that

$$O^* = O.$$

This means that the adjoint $O^* \in \mathcal{B}[\mathcal{Y}, \mathcal{H}]$ of the null transformation $O \in \mathcal{B}[\mathcal{H}, \mathcal{Y}]$ coincides with the null transformation $O \in \mathcal{B}[\mathcal{Y}, \mathcal{H}]$. In fact, $0 = \langle Ox ; y \rangle = \langle x ; O^*y \rangle$ for every $x \in \mathcal{H}$ and every $y \in \mathcal{Y}$, and hence $O^*y = 0$ for every $y \in \mathcal{Y}$. Now take $S$ and $T$ in $\mathcal{B}[\mathcal{H}, \mathcal{Y}]$ so that $S + T$ and $\alpha T$ lie in $\mathcal{B}[\mathcal{H}, \mathcal{Y}]$, where $\alpha$ is any scalar. Consider their adjoints, which are the unique transformations in $\mathcal{B}[\mathcal{Y}, \mathcal{H}]$ such that $\langle (S + T)x ; y \rangle = \langle x ; (S + T)^*y \rangle$ and $\langle \alpha Tx ; y \rangle = \langle x ; (\alpha T)^*y \rangle$ for every $x \in \mathcal{H}$ and $y \in \mathcal{Y}$, respectively. These two identities imply that $\langle x ; (S + T)^*y \rangle = \langle x ; (S^* + T^*)y \rangle$ and $\langle x ; (\alpha T)^*y \rangle = \langle x ; \overline{\alpha} T^*y \rangle$ for every $x \in \mathcal{H}$ and every $y \in \mathcal{Y}$. Therefore,

$$(S^* + T^*) = S^* + T^* \quad \text{and} \quad (\alpha T)^* = \overline{\alpha} T^*.$$

Next take $T$ in $\mathcal{B}[\mathcal{H}, \mathcal{K}]$ and $S$ in $\mathcal{B}[\mathcal{K}, \mathcal{Y}]$, where $\mathcal{H}$ and $\mathcal{K}$ are Hilbert spaces and $\mathcal{Y}$ is an inner product space, so that $ST$ lies in $\mathcal{B}[\mathcal{H}, \mathcal{Y}]$ by Proposition 4.16. Consider its adjoint, which is the unique transformation in $\mathcal{B}[\mathcal{Y}, \mathcal{H}]$ such that $\langle STx ; y \rangle = \langle x ; (ST)^*y \rangle$ for every $x \in \mathcal{H}$ and $y \in \mathcal{Y}$. This implies that $\langle x ; (ST)^*y \rangle = \langle x ; T^*S^*y \rangle$ for every $x \in \mathcal{H}$ and every $y \in \mathcal{Y}$. Then

$$(ST)^* = T^*S^*.$$

Finally, consider the algebra $\mathcal{B}[\mathcal{H}]$, where $\mathcal{H}$ is a Hilbert space. It is clear by the very definition of adjoint that

$$I = I^*,$$

where $I$ is the identity operator in $\mathcal{B}[\mathcal{H}]$. If $T \in \mathcal{G}[\mathcal{H}, \mathcal{K}]$, where $\mathcal{H}$ and $\mathcal{K}$ are Hilbert spaces (i.e., if $T$ is invertible in $\mathcal{B}[\mathcal{H}, \mathcal{K}]$ so that $T^{-1} \in \mathcal{B}[\mathcal{K}, \mathcal{H}]$ by the Inverse Mapping Theorem), then the above identities ensure that $I = I^* = (T^{-1}T)^* = T^*(T^{-1})^*$, the identity operator in $\mathcal{B}[\mathcal{H}]$, and $I = I^* = (T T^{-1})^* = (T^{-1})^*T^*$, the identity operator in $\mathcal{B}[\mathcal{K}]$. Hence $T^* \in \mathcal{G}[\mathcal{K}, \mathcal{H}]$ and

$$(T^*)^{-1} = (T^{-1})^*.$$

**Example 5P.** Consider the Hilbert spaces $\mathbb{F}^n$ and $\mathbb{F}^m$ (for arbitrary positive integers $m$ and $n$) equipped with their usual inner products as in Example 5A. Take any $A \in \mathcal{B}[\mathbb{F}^n, \mathbb{F}^m]$ and recall that $\mathcal{B}[\mathbb{F}^n, \mathbb{F}^m] = \mathcal{L}[\mathbb{F}^n, \mathbb{F}^m]$ by Corollary 4.30. As usual (cf. Example 2L), we shall identify the linear transformation $A \in \mathcal{L}[\mathbb{F}^n, \mathbb{F}^m]$ with its matrix

$$[A] = [\alpha_{ij}] = \begin{pmatrix} \alpha_{11} & \cdots & \alpha_{1n} \\ \vdots & & \vdots \\ \alpha_{m1} & \cdots & \alpha_{mn} \end{pmatrix} \in \mathbb{F}_{m \times n}$$

relative to the canonical bases for $\mathbb{F}^n$ and $\mathbb{F}^m$. Thus for every $x = (\xi_1, \dots, \xi_n) \in \mathbb{F}^n$ the vector $y = Ax = (\upsilon_1, \dots, \upsilon_n) \in \mathbb{F}^m$ is such that

$$\upsilon_i = \sum_{j=1}^{n} \alpha_{ij} \xi_j$$

for every $i = 1, \dots, m$. In terms of common matrix notation, and according to ordinary matrix operations, the matrix equation

$$[y] = [A][x]$$

represents the identity $y = Ax$. Here

$$[y] = \begin{pmatrix} \upsilon_1 \\ \vdots \\ \upsilon_m \end{pmatrix} \in \mathbb{F}_{m \times 1} \quad \text{and} \quad [x] = \begin{pmatrix} \xi_1 \\ \vdots \\ \xi_n \end{pmatrix} \in \mathbb{F}_{n \times 1}$$

are the matrices of $y$ and $x$ with respect to the canonical bases for $\mathbb{F}^m$ and $\mathbb{F}^n$, respectively. Recall: $\overline{[A]} = [\overline{\alpha}_{ij}] \in \mathbb{F}_{m \times n}$, $[A]^T = [\alpha_{ji}] \in \mathbb{F}_{n \times m}$, and $\overline{[A]}^T = \overline{[A]^T} = [\overline{\alpha}_{ji}] \in \mathbb{F}_{n \times m}$ denote conjugate, transpose, and transpose conjugate of $[A] = [\alpha_{ij}] \in \mathbb{F}_{m \times n}$, respectively. Now observe that the inner product on $\mathbb{F}^m$ can be written as

$$\langle y ; z \rangle = \sum_{i=1}^{m} \upsilon_i \overline{\zeta}_i = [y]^T \overline{[z]}$$

for every $y = (\upsilon_1, \dots, \upsilon_m) \in \mathbb{F}^m$ and for every $z = (\zeta_1, \dots, \zeta_m) \in \mathbb{F}^m$. Using standard matrix algebra we get

$$\langle Ax ; y \rangle = ([A][x])^T \overline{[y]} = [x]^T [A]^T \overline{[y]} = [x]^T \overline{\left( \overline{[A]^T} [y] \right)}$$

for every $x \in \mathbb{F}^n$ and every $y \in \mathbb{F}^m$. Next consider the adjoint $A^* \in \mathcal{B}[\mathbb{F}^m, \mathbb{F}^n]$ of $A \in \mathcal{B}[\mathbb{F}^n, \mathbb{F}^m]$, and let $[A^*] \in \mathbb{F}_{n \times m}$ denote the matrix of $A^*$ relative to the canonical bases for $\mathbb{F}^m$ and $\mathbb{F}^n$. Then

$$\langle Ax ; y \rangle = \langle x ; A^* y \rangle = [x]^T \overline{\left( [A^*][y] \right)}$$

for every $x \in \mathbb{F}^n$ and every $y \in \mathbb{F}^m$. Therefore,

$$[A^*] = \overline{[A]^T}.$$

That is, the matrix of the adjoint $A^*$ of $A$ is the transpose conjugate of the matrix of $A$.

**Example 5Q.** Let $S$ be a nondegenerate interval of the real line $\mathbb{R}$. Consider the Hilbert space $L^2(S)$ equipped with its usual inner product (see Example 5D). As always, we shall write $x \in L^2(S)$ instead of $[x] \in L^2(S)$, where $x$ is any representative of the equivalence class $[x]$. Take an arbitrary $a \in L^2(S \times S)$ so that $a(s, \cdot) \in L^2(S)$ for each $s \in S$, $a(\cdot, t) \in L^2(S)$ for each $t \in S$, and

$$\begin{aligned}
\|a\|^2 &= \int_S \|a(s, \cdot)\|^2 ds = \int_S \|a(\cdot, t)\|^2 dt \\
&= \int_S \int_S |a(s, t)|^2 ds\, dt = \int_S \int_S |a(s, t)|^2 dt\, ds < \infty.
\end{aligned}$$

This is due to a well-known result in integration theory called *Fubini's Theorem*. Now define the integral mapping $A \colon L^2(S) \to L^2(S)$ as follows. For each $x$ in $L^2(S)$ let $z = Ax \in L^2(S)$ be given by

$$z(s) = \int_S a(s, t) x(t)\, dt = \langle a(s, \cdot) ; \overline{x} \rangle = \langle x ; \overline{a(s, \cdot)} \rangle$$

for every $s \in S$. Note that $z = Ax$ actually lies in $L^2(S)$ because $a$ lies in $L^2(S \times S)$. Indeed, by the Schwarz inequality,

$$\|z\|^2 = \int_S |\langle x ; \overline{a(s, \cdot)} \rangle|^2 ds \le \int_S \|x\|^2 \|\overline{a(s, \cdot)}\|^2 ds = \|x\|^2 \|a\|^2,$$

and hence, $\|Ax\| \le \|a\| \|x\|$ for every $x \in L^2(S)$ so that $A$ is bounded. Since $A$ is certainly linear, it follows that $A \in \mathcal{B}[L^2(S)]$. Also note that

$$\begin{aligned}
\langle Ax ; y \rangle &= \int_S \left( \int_S a(s, t) x(t)\, dt \right) \overline{y(s)}\, ds \\
&= \int_S \int_S a(s, t) x(t) \overline{y(s)}\, dt\, ds = \int_S x(t) \left( \overline{\int_S \overline{a(s, t)}\, y(s)\, ds} \right) dt
\end{aligned}$$

for every $x, y \in L^2(S)$. Now consider the adjoint $A^* \in \mathcal{B}[L^2(S)]$ of $A \in \mathcal{B}[L^2(S)]$. For each $y$ in $L^2(S)$ set $w = A^* y \in L^2(S)$ so that

$$\langle Ax ; y \rangle = \langle x ; A^* y \rangle = \langle x ; w \rangle = \int_S x(t) \overline{w(t)}\, dt,$$

for every $x, y \in L^2(S)$. Then $w = A^*y \in L^2(S)$ is given by

$$w(s) = \int_S \overline{a(t, s)}\, y(t)\, dt = \int_S a^*(s, t)\, y(t)\, dt$$

for every $s \in S$. Therefore, the adjoint $A^*$ of the integral operator $A$ is again an integral operator whose kernel $a^* \in L^2(S \times S)$ is related to the kernel $a \in L^2(S \times S)$ of $A$ as follows. For every $(s, t) \in S \times S$,

$$a^*(s, t) = \overline{a(t, s)}.$$

An isometry between metric spaces is a map that preserves distance, and hence every isometry is an injective contraction. A linear isometry between normed spaces is a linear transformation that preserves norm and, between inner product spaces, a linear isometry is a linear transformation that preserves inner product. Propositions 4.37 and 5.21 gave some necessary and sufficient conditions that a linear transformation be an isometry. Here is another one for linear transformations between Hilbert spaces that is stated in terms of the adjoint.

**Proposition 5.72.** *An operator $V \in \mathcal{B}[\mathcal{H}, \mathcal{K}]$ of a Hilbert space $\mathcal{H}$ into a Hilbert space $\mathcal{K}$ is an isometry if and only if $V^*V = I$.*

*Proof.* According to Proposition 5.21, $V$ is an isometry if and only if it preserves inner product; that is, $\langle Vx\,;\,Vy \rangle = \langle x\,;\,y \rangle$ for every $x, y \in \mathcal{H}$. Equivalently, $\langle (V^*V - I)x\,;\,y \rangle = 0$ for every $x, y \in \mathcal{H}$, which means that $(V^*V - I)x = 0$ for every $x \in \mathcal{H}$, where $I$ is the identity on $\mathcal{H}$.                    $\square$

A *coisometry* is a transformation $T \in \mathcal{B}[\mathcal{H}, \mathcal{K}]$ such that its adjoint $T^* \in \mathcal{B}[\mathcal{K}, \mathcal{H}]$ is an isometry. Thus the previous proposition says that $T$ *is a coisometry if and only if $TT^* = I$* (identity on $\mathcal{K}$). Recall that a unitary transformation $U \in \mathcal{B}[\mathcal{H}, \mathcal{K}]$ is an isometric isomorphism between $\mathcal{H}$ and $\mathcal{K}$. Equivalently, a linear surjective isometry, which means an invertible isometry (i.e., an isometry in $\mathcal{G}[\mathcal{H}, \mathcal{K}]$).

**Proposition 5.73.** *Take $U \in \mathcal{B}[\mathcal{H}, \mathcal{K}]$, where $\mathcal{H}$ and $\mathcal{K}$ are Hilbert spaces. The following assertions are pairwise equivalent.*

(a) *$U$ is unitary    (i.e., $U$ is a surjective isometry).*

(b) *$U$ lies in $\mathcal{G}[\mathcal{H}, \mathcal{K}]$ and $U^{-1} = U^*$.*

(c) *$U^*U = I$ (identity on $\mathcal{H}$) and $UU^* = I$ (identity on $\mathcal{K}$).*

(d) *$U$ is an isometry and a coisometry.*

*Proof.* Let $U$ be a transformation in $\mathcal{B}[\mathcal{H}, \mathcal{K}]$. It is trivially verified that (b) $\Leftrightarrow$ (c) by the Open Map Theorem (Theorem 4.22), and (c)$\Leftrightarrow$(d) by Proposition 5.72.

*Proof of* (a)$\Leftrightarrow$(b). If $U$ is unitary, then it is an isometry that lies in $\mathcal{G}[\mathcal{H}, \mathcal{K}]$ so that there exists $U^{-1} \in \mathcal{G}[\mathcal{K}, \mathcal{H}]$ such that $UU^{-1} = I$ (identity on $\mathcal{K}$). Proposition 5.21 ensures that

$$\langle Ux_1 ; Ux_2 \rangle = \langle x_1 ; x_2 \rangle$$

for every $x_1, x_2 \in \mathcal{H}$. Hence

$$\langle x ; U^{-1}y \rangle = \langle Ux ; UU^{-1}y \rangle = \langle Ux ; y \rangle = \langle x ; U^*y \rangle$$

for every $x \in \mathcal{H}$ and every $y \in \mathcal{K}$. Therefore, as the adjoint is unique (Proposition 5.65(a)), $U^{-1} = U^*$. Conversely, if $U$ lies in $\mathcal{G}[\mathcal{H}, \mathcal{K}]$ and $U^{-1} = U^*$, then $U^*U = I$ (identity on $\mathcal{H}$), and hence (cf. Proposition 5.72) $U$ is a surjective isometry. $\quad\square$

Let $T$ be an operator in $\mathcal{B}[\mathcal{X}]$, where $\mathcal{X}$ is an inner product space, and let $\mathcal{M}$ be a subspace of $\mathcal{X}$. Recall: $\mathcal{M}$ is an *invariant subspace for T* (or *invariant under T*, or *T-invariant*) if $T(\mathcal{M}) \subseteq \mathcal{M}$ (i.e., $Tx \in \mathcal{M}$ whenever $x \in \mathcal{M}$). A *nontrivial invariant subspace* for $T$ is an invariant subspace $\mathcal{M}$ for $T$ such that $\{0\} \neq \mathcal{M} \neq \mathcal{X}$ (see Problems 4.18 to 4.20). If $\mathcal{M}$ and its orthogonal complement $\mathcal{M}^{\perp}$ are both invariant for $T$ (i.e., if $T(\mathcal{M}) \subseteq \mathcal{M}$ and $T(\mathcal{M}^{\perp}) \subseteq \mathcal{M}^{\perp}$), then we say that $\mathcal{M}$ *reduces T* (or $\mathcal{M}$ is a *reducing subspace* for $T$). Accordingly, a *nontrivial reducing subspace* for $T$ is a reducing subspace for $T$ such that $\{0\} \neq \mathcal{M} \neq \mathcal{X}$. An operator is *reducible* if it has a nontrivial reducing subspace. Now let $\mathcal{X} = \mathcal{H}$ be a Hilbert space and consider the orthogonal direct sum $\mathcal{H} = \mathcal{M} \oplus \mathcal{M}^{\perp}$ of Theorem 5.25. Observe from Example 2O (also see Problem 4.16) that the following assertions are pairwise equivalent.

(a)  $\mathcal{M}$ reduces $T$.

(b)  $T = T|_{\mathcal{M}} \oplus T|_{\mathcal{M}^{\perp}} = \begin{pmatrix} T|_{\mathcal{M}} & O \\ O & T|_{\mathcal{M}^{\perp}} \end{pmatrix} : \mathcal{H} = \mathcal{M} \oplus \mathcal{M}^{\perp} \to \mathcal{H} = \mathcal{M} \oplus \mathcal{M}^{\perp}$.

(c)  $PT = TP$, where $P = \begin{pmatrix} I & O \\ O & O \end{pmatrix} : \mathcal{H} = \mathcal{M} \oplus \mathcal{M}^{\perp} \to \mathcal{H} = \mathcal{M} \oplus \mathcal{M}^{\perp}$ is the orthogonal projection onto $\mathcal{M}$.

This suggests that, if $\mathcal{M}$ reduces $T$, then the investigation of $T$ is *reduced* to the investigation of smallest operators (viz., $T|_{\mathcal{M}}$ and $T|_{\mathcal{M}^{\perp}}$), which justifies the terminology "reducing subspace".

**Proposition 5.74.** *Let $T$ be any operator on a Hilbert space $\mathcal{H}$. A subspace $\mathcal{M}$ of $\mathcal{H}$ is invariant for $T$ if and only if $\mathcal{M}^{\perp}$ is invariant for $T^*$. Thus $T$ has a nontrivial invariant subspace if and only if $T^*$ has.*

*Proof.* Let $\mathcal{M}$ be a subspace of a Hilbert space $\mathcal{H}$, let $T$ be any operator in $\mathcal{B}[\mathcal{H}]$, and take an arbitrary $y \in \mathcal{M}^{\perp}$. If $Tx \in \mathcal{M}$ whenever $x \in \mathcal{M}$, then $\langle x ; T^*y \rangle = \langle Tx ; y \rangle = 0$ for every $x \in \mathcal{M}$ so that $T^*y \perp \mathcal{M}$. Then $T^*y \in \mathcal{M}^{\perp}$ for every $y \in \mathcal{M}^{\perp}$. Conclusion: $T(\mathcal{M}) \subseteq \mathcal{M}$ implies $T^*(\mathcal{M}^{\perp}) \subseteq \mathcal{M}^{\perp}$. Conversely, since this happens for every $T \in \mathcal{B}[\mathcal{H}]$, it follows that $T^*(\mathcal{M}^{\perp}) \subseteq \mathcal{M}^{\perp}$ implies $T^{**}(\mathcal{M}^{\perp\perp}) \subseteq \mathcal{M}^{\perp\perp}$. But $T^{**} = T$ and $\mathcal{M}^{\perp\perp} = \mathcal{M}^- = \mathcal{M}$ (cf. Propositions 5.15

and 5.65(c)). Therefore, $T^*(\mathcal{M}^\perp) \subseteq \mathcal{M}^\perp$ implies $T(\mathcal{M}) \subseteq \mathcal{M}$. Finally, note that $\{0\} \neq \mathcal{M} \neq \mathcal{H}$ if and only if $\{0\} \neq \mathcal{M}^\perp \neq \mathcal{H}$ (cf. Proposition 5.15).     □

**Corollary 5.75.** *A subspace $\mathcal{M}$ of a Hilbert space $\mathcal{H}$ reduces $T \in \mathcal{B}[\mathcal{H}]$ if and only if it is invariant for both $T$ and $T^*$. In this case $(T|_\mathcal{M})^* = T^*|_\mathcal{M}$.*

*Proof.* Since $\mathcal{M}^{\perp\perp} = \mathcal{M}$, the previous proposition says that $T(\mathcal{M}^\perp) \subseteq \mathcal{M}^\perp$ if and only if $T^*(\mathcal{M}) \subseteq \mathcal{M}$. Therefore, $T(\mathcal{M}) \subseteq \mathcal{M}$ and $T(\mathcal{M}^\perp) \subseteq \mathcal{M}^\perp$ if and only if $T(\mathcal{M}) \subseteq \mathcal{M}$ and $T^*(\mathcal{M}) \subseteq \mathcal{M}$. Moreover, in this case, we get $\langle (T|_\mathcal{M})x\,;\,y\rangle = \langle Tx\,;\,y\rangle = \langle x\,;\,T^*y\rangle = \langle x\,;\,(T^*|_\mathcal{M})y\rangle$ for every $x$ and $y$ in the Hilbert space $\mathcal{M}$.     □

Recall that $\mathcal{N}(T)$ and $\mathcal{R}(T)^-$ are invariant subspaces for $T \in \mathcal{B}[\mathcal{H}]$ (see Problems 4.20 to 4.22). In fact, null spaces of Hilbert space operators constitute an important source of invariant subspaces. The next result shows that $\mathcal{N}(T^*)^\perp$, $\mathcal{R}(T^*)^\perp$, $\mathcal{N}(T^*T)$, and $\mathcal{R}(TT^*)^-$ also are invariant subspaces for $T \in \mathcal{B}[\mathcal{H}]$.

**Proposition 5.76.** *If $T$ is a bounded linear transformation of a Hilbert space $\mathcal{H}$ into a Hilbert space $\mathcal{K}$, then*

(a)
$$\mathcal{N}(T) = \mathcal{R}(T^*)^\perp = \mathcal{N}(T^*T),$$

(b)
$$\mathcal{R}(T)^- = \mathcal{N}(T^*)^\perp = \mathcal{R}(TT^*)^-,$$

(a*)
$$\mathcal{N}(T^*) = \mathcal{R}(T)^\perp = \mathcal{N}(TT^*),$$

(b*)
$$\mathcal{R}(T^*)^- = \mathcal{N}(T)^\perp = \mathcal{R}(T^*T)^-.$$

*Proof.* Note that $x \in \mathcal{R}(T^*)^\perp$ if and only if $\langle x\,;\,T^*y\rangle = 0$ for every $y \in \mathcal{K}$. By the definition of adjoint, this is equivalent to $\langle Tx\,;\,y\rangle = 0$ for every $y \in \mathcal{K}$, which means that $Tx = 0$; that is, $x \in \mathcal{N}(T)$. Hence

$$\mathcal{R}(T^*)^\perp = \mathcal{N}(T).$$

Moreover, since $\|Tx\|^2 = \langle Tx\,;\,Tx\rangle = \langle T^*Tx\,;\,x\rangle$ for every $x \in \mathcal{H}$, it follows that $\mathcal{N}(T^*T) \subseteq \mathcal{N}(T)$. But $\mathcal{N}(T) \subseteq \mathcal{N}(T^*T)$ trivially, and so

$$\mathcal{N}(T) = \mathcal{N}(T^*T),$$

which completes the proof of (a). Since (a) holds true for every $T \in \mathcal{B}[\mathcal{H}, \mathcal{K}]$, it also holds for $T^* \in \mathcal{B}[\mathcal{K}, \mathcal{H}]$ and $TT^* \in \mathcal{B}[\mathcal{K}]$. Therefore (cf. Propositions 5.15 and 5.65(c)),

$$\mathcal{R}(T)^- = \mathcal{R}(T^{**})^{\perp\perp} = \mathcal{N}(T^*)^\perp = \mathcal{N}(T^{**}T^*)^\perp$$
$$= \mathcal{N}(TT^*)^\perp = \mathcal{R}((TT^*)^*)^{\perp\perp} = \mathcal{R}(T^{**}T^*)^{\perp\perp} = \mathcal{R}(TT^*)^-,$$

which proves assertion (b). Since $T^{**} = T$ we get the dual expressions (a*) and (b*).    $\square$

Here is a useful result concerning closed ranges and adjoints.

**Proposition 5.77.** *Let $\mathcal{H}$ and $\mathcal{K}$ be Hilbert spaces and take any $T \in \mathcal{B}[\mathcal{H}, \mathcal{K}]$. The following assertions are pairwise equivalent.*

(a)  $\mathcal{R}(T) = \mathcal{R}(T)^-$.

(b)  $\mathcal{R}(T^*) = \mathcal{R}(T^*)^-$.

(c)  $\mathcal{R}(T^*T) = \mathcal{R}(T^*T)^-$.

(d)  $\mathcal{R}(T T^*) = \mathcal{R}(T T^*)^-$.

*Proof.* Let $T$ be an arbitrary bounded linear transformation of a Hilbert space $\mathcal{H}$ into a Hilbert space $\mathcal{K}$.

*Proof of* (a)$\Leftrightarrow$(b). Set $T_0 = T|_{\mathcal{N}(T)^\perp} \in \mathcal{B}[\mathcal{N}(T)^\perp, \mathcal{K}]$, the restriction of $T$ to $\mathcal{N}(T)^\perp$. Recall that $\mathcal{H} = \mathcal{N}(T) + \mathcal{N}(T)^\perp$ (Proposition 4.13 and Theorem 5.20), and hence every $x \in \mathcal{H}$ can be written as $x = u + v$ with $u \in \mathcal{N}(T)$ and $v \in \mathcal{N}(T)^\perp$. If $y \in \mathcal{R}(T)$, then $y = Tx = Tu + Tv = Tv = T|_{\mathcal{N}(T)^\perp}v$ for some $x \in \mathcal{H}$ so that $y \in \mathcal{R}(T|_{\mathcal{N}(T)^\perp})$. Therefore, $\mathcal{R}(T) \subseteq \mathcal{R}(T|_{\mathcal{N}(T)^\perp})$. Since $\mathcal{R}(T|_{\mathcal{N}(T)^\perp}) \subseteq \mathcal{R}(T)$, it follows that

$$\mathcal{R}(T_0) = \mathcal{R}(T) \quad \text{and} \quad \mathcal{N}(T_0) = \{0\}$$

because $\mathcal{N}(T|_{\mathcal{N}(T)^\perp}) = \{0\}$. If $\mathcal{R}(T) = \mathcal{R}(T)^-$, then Corollary 4.24 ensures the existence of $T_0^{-1} \in \mathcal{B}[\mathcal{R}(T), \mathcal{N}(T)^\perp]$. Now take an arbitrary $w \in \mathcal{N}(T)^\perp$ and consider the functional $f_w \colon \mathcal{R}(T) \to \mathbb{F}$ defined by

$$f_w(y) = \langle T_0^{-1}y \,;\, w \rangle$$

for every $y \in \mathcal{R}(T)$, which is linear (reason: $T_0^{-1}$ is linear and the inner product is linear in the first argument) and bounded (in fact, $|f_w(y)| \leq \|T_0^{-1}\| \|w\| \|y\|$ for every $y \in \mathcal{R}(T)$). The Riesz Representation Theorem (Theorem 5.62) says that there exists $z_w$ in the Hilbert space $\mathcal{R}(T)$ (recall: $\mathcal{R}(T)$ is a subspace of the Hilbert space $\mathcal{K}$, and so a Hilbert space itself, whenever $\mathcal{R}(T) = \mathcal{R}(T)^-$) such that

$$f_w(y) = \langle y \,;\, z_w \rangle$$

for every $y \in \mathcal{R}(T)$. Consider the decomposition $\mathcal{H} = \mathcal{N}(T) + \mathcal{N}(T)^\perp$ (again) and take any $x \in \mathcal{H}$ so that $x = u + v$ with $u \in \mathcal{N}(T)$ and $v \in \mathcal{N}(T)^\perp$. Then

$$
\begin{aligned}
\langle x \,;\, T^*z_w \rangle &= \langle Tx \,;\, z_w \rangle = \langle Tu \,;\, z_w \rangle + \langle Tv \,;\, z_w \rangle = \langle Tv \,;\, z_w \rangle \\
&= f_w(Tv) = \langle T_0^{-1}Tv \,;\, w \rangle = \langle T_0^{-1}T_0 v \,;\, w \rangle \\
&= \langle v \,;\, w \rangle = \langle u \,;\, w \rangle + \langle v \,;\, w \rangle = \langle x \,;\, w \rangle.
\end{aligned}
$$

Hence $\langle x\,;T^*z_w-w\rangle=0$ for every $x\in\mathcal{H}$, which means that $T^*z_w=w$. Therefore, $w\in\mathcal{R}(T^*)$. This shows that $\mathcal{N}(T)^\perp\subseteq\mathcal{R}(T^*)$. On the other hand, $\mathcal{R}(T^*)\subseteq\mathcal{R}(T^*)^-=\mathcal{N}(T)^\perp$ by Proposition 5.73(b$^*$) so that

$$\mathcal{R}(T^*)=\mathcal{N}(T)^\perp.$$

Thus (a) implies (b) by Proposition 5.12 (or by Proposition 5.76(b$^*$)). Since (a) implies (b), it follows that (b) implies (a) because $T^{**}=T$.

*Proof of* (a)$\Rightarrow$(c) *and* (b)$\Rightarrow$(d). Let $T_1\in\mathcal{B}[\mathcal{H},\mathcal{R}(T)]$ be defined by $T_1x=Tx$ for every $x\in\mathcal{H}$ (i.e., $T_1$ is surjective and coincides with $T$ on $\mathcal{H}$). It is clear that

$$\mathcal{R}(T)=\mathcal{R}(T_1).$$

Let $T_1^*\in\mathcal{B}[\mathcal{R}(T),\mathcal{H}]$ be the adjoint of $T_1$. Now consider the restriction $T^*|_{\mathcal{R}(T)}\in\mathcal{B}[\mathcal{R}(T),\mathcal{H}]$ of the adjoint $T^*\in\mathcal{B}[\mathcal{K},\mathcal{H}]$ of $T$ to $\mathcal{R}(T)$, and note that

$$\langle x\,;T_1^*y\rangle=\langle T_1x\,;y\rangle=\langle Tx\,;y\rangle=\langle x\,;T^*y\rangle=\langle x\,;T^*|_{\mathcal{R}(T)}y\rangle$$

for every $x\in\mathcal{H}$ and every $y\in\mathcal{R}(T)$. Then $T_1^*y=T^*|_{\mathcal{R}(T)}y$ for every $y\in\mathcal{R}(T)$ (i.e., $T_1^*=T^*|_{\mathcal{R}(T)}$), and hence

$$\mathcal{R}(T_1^*)=\mathcal{R}(T^*|_{\mathcal{R}(T)}).$$

Observe that $x$ lies in $\mathcal{R}(T^*|_{\mathcal{R}(T)})$ if and only if $x=T^*|_{\mathcal{R}(T)}y=T^*y$ for some $y\in\mathcal{R}(T)$. But this is equivalent to $x=T^*Tu$ for some $u\in\mathcal{H}$, which means $x\in\mathcal{R}(T^*T)$. That is,

$$\mathcal{R}(T^*|_{\mathcal{R}(T)})=\mathcal{R}(T^*T)$$

so that

$$\mathcal{R}(T_1^*)=\mathcal{R}(T^*T).$$

If $\mathcal{R}(T)=\mathcal{R}(T)^-$, then $\mathcal{R}(T_1)=\mathcal{R}(T_1)^-$. Since (a) implies (b), it follows that $\mathcal{R}(T_1^*)=\mathcal{R}(T_1^*)^-$. Therefore,

$$\mathcal{R}(T^*T)=\mathcal{R}(T^*T)^-.$$

Conclusion: (a) implies (c), and hence (b) implies (d) (for $T^{**}=T$).

*Proof of* (d)$\Rightarrow$(a) *and* (c)$\Rightarrow$(b). According to Proposition 5.76(b),

$$\mathcal{R}(TT^*)\subseteq\mathcal{R}(T)\subseteq\mathcal{R}(T)^-=\mathcal{R}(TT^*)^-.$$

Then (d) implies (a), and so (c) implies (b) (because $T^{**}=T$).    $\square$

We close this section by introducing an important notion. Let $\mathcal{A}$ be a unital (complex) Banach algebra (cf. Definition 4.17) and suppose there exists a mapping

$\mathcal{A} \to \mathcal{A}$ defined by $A \mapsto A^*$ that satisfies the following conditions for all $A$, $B \in \mathcal{A}$ and all $\alpha \in \mathbb{C}$.

$$
\begin{aligned}
\text{(i)} \quad & (A^*)^* = A, \\
\text{(ii)} \quad & (AB)^* = B^*A^*, \\
\text{(iii)} \quad & (A + B)^* = A^* + B^*, \\
\text{(iv)} \quad & (\alpha A)^* = \overline{\alpha}A^*.
\end{aligned}
$$

Such a mapping is called an *involution* on $\mathcal{A}$. A *$C^*$-algebra* is a unital Banach algebra with an involution $*\colon \mathcal{A} \to \mathcal{A}$ such that

$$
\text{(v)} \quad \|A^*A\| = \|A\|^2
$$

for every $A \in \mathcal{A}$. It is clear that *if $\mathcal{H}$ is a complex Hilbert space, then $\mathcal{B}[\mathcal{H}]$ is a $C^*$-algebra, where $*$ is the adjoint operation.* Every $T \in \mathcal{B}[\mathcal{H}]$ determines a $C^*$-subalgebra $C^*(T)$ of $\mathcal{B}[\mathcal{H}]$, which is the smallest $C^*$-algebra of operators from $\mathcal{B}[\mathcal{H}]$ containing both $T$ and the identity $I$. It can be shown that $C^*(T) = P(T, T^*)^-$, the closure in $\mathcal{B}[\mathcal{H}]$ of all polynomials in $T$ and $T^*$ with complex coefficients. We mention the *Gelfand–Naimark Theorem* that asserts the converse: *Every $C^*$-algebra is isometrically $*$-isomorphic to a $C^*$-subalgebra of $\mathcal{B}[\mathcal{H}]$.* That is, for every $C^*$-algebra $\mathcal{A}$ there exists an isometric isomorphism of $\mathcal{A}$ onto a $C^*$-subalgebra of $\mathcal{B}[\mathcal{H}]$ that preserves the involution $*$. A great deal of the rest of this book can be posed in an abstract $C^*$-algebra. However, we shall stick to $\mathcal{B}[\mathcal{H}]$.

## 5.13   Self-Adjoint Operators

Throughout this section $\mathcal{H}$ and $\mathcal{K}$ will stand for Hilbert spaces. An operator $T$ in $\mathcal{B}[\mathcal{H}]$ is *self-adjoint* (or *Hermitian*) if $T^* = T$. By the definition of the adjoint operator, $T \in \mathcal{B}[\mathcal{H}]$ is self-adjoint if and only if

$$
\langle Tx\,;\,y \rangle = \langle x\,;\,Ty \rangle \quad \text{for every} \quad x, y \in \mathcal{H}.
$$

**Proposition 5.78.** *If $T \in \mathcal{B}[\mathcal{H}]$ is self-adjoint, then*

$$
\|T\| = \sup_{\|x\|=1} |\langle Tx\,;\,x \rangle|.
$$

*Proof.* Let $T$ be any operator in $\mathcal{B}[\mathcal{H}]$. The Schwarz inequality says that $\langle Tx\,;\,x \rangle \leq \|Tx\|\,\|x\| \leq \|T\|\,\|x\|^2$ for every $x \in \mathcal{H}$. Then

$$
\sup_{\|u\|=1} |\langle Tu\,;\,u \rangle \leq \|T\|.
$$

On the other hand, note that $\langle T(x \pm y) ; x \pm y \rangle = \langle Tx ; x \rangle \pm \langle Tx ; y \rangle \pm \langle Ty ; x \rangle + \langle Ty ; y \rangle$ for every $x, y \in \mathcal{H}$. If $T = T^*$, then $\langle Ty ; x \rangle = \langle y ; Tx \rangle = \overline{\langle Tx ; y \rangle}$, and hence $\langle Tx ; y \rangle + \langle Ty ; x \rangle = 2 \operatorname{Re} \langle Tx ; y \rangle$. Thus

$$\langle T(x \pm y) ; x \pm y \rangle = \langle Tx ; x \rangle \pm 2 \operatorname{Re} \langle Tx ; y \rangle + \langle Ty ; y \rangle$$

so that

$$\langle T(x + y) ; x + y \rangle - \langle T(x - y) ; x - y \rangle = 4 \operatorname{Re} \langle Tx ; y \rangle$$

for every $x, y \in \mathcal{H}$. But $|\langle Tz ; z \rangle| = |\langle T(\|z\|^{-1}z) ; \|z\|^{-1}z \rangle| \, \|z\|^2$ for every nonzero vector $z$ in $\mathcal{H}$, and so

$$|\langle Tz ; z \rangle| \leq \sup_{\|u\|=1} |\langle Tu ; u \rangle| \, \|z\|^2$$

for every $z \in \mathcal{H}$. By the above two relations and the parallelogram law,

$$\begin{aligned}
4 \operatorname{Re} \langle Tx ; y \rangle &\leq |\langle T(x + y) ; x + y \rangle| + |\langle T(x - y) ; x - y \rangle| \\
&\leq \sup_{\|u\|=1} |\langle Tu ; u \rangle| \left( \|x + y\|^2 + \|x - y\|^2 \right) \\
&= 2 \sup_{\|u\|=1} |\langle Tu ; u \rangle| \left( \|x\|^2 + \|y\|^2 \right) = 4 \sup_{\|u\|=1} |\langle Tu ; u \rangle|
\end{aligned}$$

when $\|x\| = \|y\| = 1$. Consider the polar representation $\langle Tx ; y \rangle = |\langle Tx ; y \rangle| e^{i\theta}$. Set $x' = e^{-i\theta}x$ so that $|\langle Tx ; y \rangle| = e^{-i\theta} \langle Tx ; y \rangle = \langle Tx' ; y \rangle$. Since $\|x'\| = \|x\| = 1$ and $|\langle Tx ; y \rangle| = \langle Tx' ; y \rangle = \operatorname{Re} \langle Tx' ; y \rangle$,

$$|\langle Tx ; y \rangle| \leq \sup_{\|u\|=1} |\langle Tu ; u \rangle| \quad \text{whenever} \quad \|x\| = \|y\| = 1.$$

Therefore, according to Corollary 5.71,

$$\|T\| = \sup_{\|x\|=\|y\|=1} |\langle Tx ; y \rangle| \leq \sup_{\|u\|=1} |\langle Tu ; u \rangle|. \qquad \square$$

It is worth noticing now that there are non-self-adjoint operators $T$ in $\mathcal{B}[\mathcal{H}]$ such that $\|T\| = \sup_{\|x\|=1}|\langle Tx ; x \rangle|$. The class of all operators for which this norm identity holds will be characterized in Chapter 6 (these are the *normaloid* operators, which includes the *normal* operators of Section 6.1). The next proposition gives a necessary and sufficient condition that an operator be self-adjoint on a *complex* Hilbert space.

**Proposition 5.79.** *If $\mathcal{H}$ is a complex Hilbert space, then $T \in \mathcal{B}[\mathcal{H}]$ is self-adjoint if and only if $\langle Tx ; x \rangle \in \mathbb{R}$ for every $x \in \mathcal{H}$.*

*Proof.* If $T = T^*$, then $\langle Tx ; x \rangle = \langle x ; Tx \rangle = \overline{\langle Tx ; x \rangle}$, and so $\langle Tx ; x \rangle$ is a real number, for every $x \in \mathcal{H}$. Conversely, if $\langle Tx ; x \rangle \in \mathbb{R}$ for every $x \in \mathcal{H}$, then $\langle Tx ; x \rangle = \overline{\langle Tx ; x \rangle} = \langle x ; Tx \rangle$ for every $x \in \mathcal{H}$, and hence (cf. Problem 5.3(b))

$$\langle Tx ; y \rangle = \langle x ; Ty \rangle$$

for every $x, y \in \mathcal{H}$; that is, $T$ is self-adjoint.    □

Remark: If $\mathcal{H}$ is a real Hilbert space, then $\langle Tx \,;\, x \rangle \in \mathbb{R}$ for every $x \in \mathcal{H}$ for all $T$ in $\mathcal{B}[\mathcal{H}]$. Since there exist non-self-adjoint operators on real Hilbert spaces, the above proposition fails when $\mathcal{H}$ is a real Hilbert space. Sample: If $A = \left(\begin{smallmatrix} 0 & 1 \\ -1 & 0 \end{smallmatrix}\right) \in \mathcal{B}[\mathbb{R}^2]$, then $O \neq A \neq A^* = \left(\begin{smallmatrix} 0 & -1 \\ 1 & 0 \end{smallmatrix}\right)$ and $\langle Ax \,;\, x \rangle = 0$ for all $x \in \mathbb{R}^2$.

Take an arbitrary $z$ in $\mathcal{K}$, where $\mathcal{K}$ is any (real or complex) Hilbert space. Recall that $\langle z \,;\, y \rangle = 0$ for every $y \in \mathcal{K}$ if and only if $z = 0$. Now take $T \in \mathcal{B}[\mathcal{H}, \mathcal{K}]$. Since $T = O$ if and only if $Tx = 0$ in $\mathcal{K}$ for all $x \in \mathcal{H}$, it follows that $T = O$ if and only if $\langle Tx \,;\, y \rangle = 0$ for every $x \in \mathcal{H}$ and every $y \in \mathcal{K}$. Next set $\mathcal{K} = \mathcal{H}$ and take $T \in \mathcal{B}[\mathcal{H}]$. If $\mathcal{H}$ is a *complex* Hilbert space, then $T = O$ if and only if $\langle Tx \,;\, x \rangle = 0$ for all $x \in \mathcal{H}$ (cf. Problem 5.4). This is in general false for a *real* Hilbert space (see the above sample). However, this certainly holds for every operator (on any Hilbert space) that satisfies the norm identity of Proposition 5.78. In particular, it holds for self-adjoint operators on a real Hilbert space.

**Corollary 5.80.** *Let $\mathcal{H}$ be any Hilbert space. If $T \in \mathcal{B}[\mathcal{H}]$ is self-adjoint, then $T = O$ if and only if $\langle Tx \,;\, x \rangle = 0$ for all $x \in \mathcal{H}$.*

Proposition 5.79 leads to a partial ordering of the set of all self-adjoint operators. Let $Q \in \mathcal{B}[\mathcal{H}]$ be a *self-adjoint* operator. We say that $Q$ is *nonnegative* (notation: $Q \geq O$ or $O \leq Q$) if $0 \leq \langle Qx \,;\, x \rangle$ for every $x \in \mathcal{H}$. If $0 < \langle Qx \,;\, x \rangle$ for every nonzero $x$ in $\mathcal{H}$, then $Q$ is called *positive* (notation: $Q > O$ or $O < Q$). If there exists a real number $\alpha > 0$ such that $\alpha \|x\|^2 \leq \langle Qx \,;\, x \rangle$ for every $x \in \mathcal{H}$, then $Q$ is called *strictly positive* (notation: $Q \succ O$ or $O \prec Q$). Trivially,

$$ Q \succ O \quad \Longrightarrow \quad Q > O \quad \Longrightarrow \quad Q \geq O. $$

Observe that $T^*T \in \mathcal{B}[\mathcal{H}]$ and $TT^* \in \mathcal{B}[\mathcal{K}]$ are self-adjoint operators for every $T \in \mathcal{B}[\mathcal{H}, \mathcal{K}]$. In fact, since $0 \leq \|Tx\|^2 = \langle Tx \,;\, Tx \rangle = \langle T^*Tx \,;\, x \rangle$ for every $x$ in $\mathcal{H}$, it follows that $T^*T \geq O$. Dually, since $T^{**} = T$, it also follows that $TT^* \geq O$. The next proposition uses this fact to give necessary and sufficient conditions that a projection be an orthogonal projection.

**Proposition 5.81.** *If $P \in \mathcal{B}[\mathcal{H}]$ is a nonzero projection, then the following assertions are pairwise equivalent.*

(a)  *$P$ is an orthogonal projection.*

(b)  *$P$ is self-adjoint.*

(c)  *$P$ is nonnegative.*

(d)  *$\|P\| = 1$.*

*Proof.* If $P$ is an orthogonal projection, then Proposition 5.51 says that $\mathcal{R}(T) = \mathcal{R}(T)^-$ (so that $\mathcal{R}(T^*) = \mathcal{R}(T^*)^-$ by Proposition 5.77) and $\mathcal{R}(T) = \mathcal{N}(P)^\perp$. But $\mathcal{N}(P)^\perp = \mathcal{R}(P^*)^-$ by Proposition 5.76. Then

$$\mathcal{R}(P) = \mathcal{R}(P^*).$$

Now take an arbitrary $x \in \mathcal{H}$. The above identity ensures that $Px = P^*z$ for some $z \in \mathcal{H}$, and hence $P^*Px = (P^*)^2z = (P^2)^*z = P^*z = Px$. That is, $P^*P = P$, which implies that $P$ is self-adjoint. Therefore (a)$\Rightarrow$(b). If $P = P^*$ (i.e., if $P$ is self-adjoint), then $\langle Px\,;x\rangle = \langle P^2x\,;x\rangle = \langle Px\,;Px\rangle = \|Px\|^2 \geq 0$ for every $x \in \mathcal{H}$. This shows that (b)$\Rightarrow$(c). If $P \geq O$ (i.e., if $P$ is nonnegative), then $P = P^*$ (by definition) so that $\|P\|^2 = \|P^*P\| = \|P^2\| = \|P\|$ (cf. Proposition 5.65), and hence $\|P\| = 1$ (for $P \neq O$). Thus (c)$\Rightarrow$(d). Finally, suppose $\|P\| = 1$ and take an arbitrary $v \in \mathcal{N}(P)^\perp$. Since $\mathcal{R}(I - P) = \mathcal{N}(P)$, it follows that $(I - P)v \in \mathcal{N}(P)$. Then $(I - P)v \perp v$ so that $0 = \langle(I - P)v\,;v\rangle = \|v\|^2 - \langle Pv\,;v\rangle$, and hence $\|v\|^2 = \langle Pv\,;v\rangle \leq \|Pv\|\|v\| \leq \|P\|\|v\|^2 = \|v\|^2$. This implies that $\|Pv\| = \|v\| = \langle Pv\,;v\rangle^{\frac{1}{2}}$. Therefore,

$$\|(I - P)v\|^2 = \|Pv - v\|^2 = \|Pv\|^2 - 2\mathrm{Re}\,\langle Pv\,;v\rangle + \|v\|^2 = 0,$$

and so $v \in \mathcal{N}(I - P) = \mathcal{R}(P)$. Hence $\mathcal{N}(P)^\perp \subseteq \mathcal{R}(P)$. On the other hand, if $y \in \mathcal{R}(P)$, then $y = u + v$, where $u \in \mathcal{N}(P)$ and $v \in \mathcal{N}(P)^\perp$ (reason: $\mathcal{R}(P) \subseteq \mathcal{H} = \mathcal{N}(P) + \mathcal{N}(P)^\perp$). Since $\mathcal{N}(P)^\perp \subseteq \mathcal{R}(P) = \{x \in \mathcal{H}:\ Px = x\}$, it follows that $y = Py = Pu + Pv = Pv = v$. Thus $y \in \mathcal{N}(P)^\perp$, and hence $\mathcal{R}(P) \subseteq \mathcal{N}(P)^\perp$. Then $\mathcal{R}(P) = \mathcal{N}(P)^\perp$ so that $\mathcal{R}(P) \perp \mathcal{N}(P)$. Outcome: (d)$\Rightarrow$(a). $\qquad\square$

Take $S, T \in \mathcal{B}[\mathcal{H}]$. If $T - S$ is self-adjoint (in particular, if $T$ and $S$ are self-adjoint) and $O \leq T - S$, then we write $S \leq T$ so that $S \leq O$ means $O \leq -S$. It is easy to show that $\leq$ defines a reflexive, transitive and antisymmetric relation on the set of all self-adjoint operators on $\mathcal{H}$, and hence a partial ordering of it. Similarly, we write $S < T$ or $S \prec T$ if $O < T - S$ or $O \prec T - S$ (and so $S < O$ or $S \prec O$ mean $O < -S$ or $O \prec -S$), respectively. Note that $O \prec Q$ if and only if $\alpha I \leq Q$ for some $\alpha > 0$. Moreover, $\|T\| \leq 1$ if and only if $T^*T \leq I$, and $\|T\| < 1$ if and only if $T^*T \prec I$ (i.e., $T^*T \leq \beta I$ for some $\beta \in (0, 1)$). Why?

**Proposition 5.82.** *If $Q \in \mathcal{B}[\mathcal{H}]$ is nonnegative, then*

$$|\langle Qx\,;y\rangle|^2 \leq \langle Qx\,;x\rangle\langle Qy\,;y\rangle$$

*for every $x, y \in \mathcal{H}$, and hence*

$$\|Qx\|^2 \leq \|Q\|\langle Qx\,;x\rangle \quad \text{for every} \quad x \in \mathcal{H}.$$

*Proof.* Take a nonnegative operator $Q \in \mathcal{B}[\mathcal{H}]$ and consider the function $\langle\ ;\ \rangle_Q\colon \mathcal{H}\times\mathcal{H} \to \mathbb{F}$ given by

$$\langle x\,;y\rangle_Q = \langle Qx\,;y\rangle$$

for every $x, y \in \mathcal{H}$. It is really easy to verify that $\langle \ ; \ \rangle_Q$ is a semi-inner product on $\mathcal{H}$. Let $\| \ \|_Q$ be the seminorm induced on $\mathcal{H}$ by $\langle \ ; \ \rangle_Q$ so that $\|x\|_Q^2 = \langle Qx ; x \rangle$ for every $x \in \mathcal{H}$. Since the Schwarz inequality holds in a semi-inner product space, it follows that

$$|\langle Qx ; y \rangle|^2 = |\langle x ; y \rangle_Q|^2 \leq \|x\|_Q^2 \|y\|_Q^2 = \langle Qx ; x \rangle \langle Qy ; y \rangle$$

for every $x, y \in \mathcal{H}$. In particular, by setting $y = Qx$ we get

$$\|Qx\|^4 = \langle Qx ; Qx \rangle^2 \leq \langle Qx ; x \rangle \langle Q^2 x ; Qx \rangle$$
$$\leq \langle Qx ; x \rangle \|Q^2 x\| \|Qx\| \leq \langle Qx ; x \rangle \|Q\| \|Qx\|^2,$$

which implies that $\|Qx\|^2 \leq \|Q\| \langle Qx ; x \rangle$, for every $x \in \mathcal{H}$.    $\square$

Positive operators are not necessarily invertible. Indeed,

$$Q > O \iff Q \geq O \text{ and } \mathcal{N}(Q) = \{0\} \iff Q \geq O \text{ and } \mathcal{R}(Q)^- = \mathcal{H}.$$

The nontrivial part of the first equivalence follows from Proposition 5.82, and the second one is a straightforward consequence of Propositions 5.15 and 5.76(b*): $\mathcal{R}(Q)^\perp = \mathcal{N}(Q^*) = \mathcal{N}(Q)$ so that $\mathcal{R}(Q)^- = \mathcal{H}$ if and only if $\mathcal{N}(Q) = \{0\}$. Therefore, $Q > O$ has an inverse on its range, which is not necessarily bounded. However, strictly positive operators are invertible. In fact, if $Q \succ O$, then $Q$ is bounded below (Schwarz inequality). Conversely, if $Q \geq O$ is bounded below, then $Q \succ O$ (cf. Proposition 5.82). Thus, by the above displayed equivalence and Corollary 4.24,

$$Q \succ O \iff O \leq Q \in \mathcal{G}[\mathcal{H}] \iff O \leq Q \text{ and } \mathcal{R}(Q) = \mathcal{H}.$$

Note: If $\mathcal{H}$ is finite-dimensional, then $Q \succ O$ if and only if $Q > O$. Why?

**Example 5R.** Take a sequence $a = \{\alpha_k\}_{k=0}^\infty$ in $\ell_+^\infty$ and consider the diagonal operator $D_a = \mathrm{diag}(\{\alpha_k\}_{k=0}^\infty)$ in $\mathcal{B}[\ell_+^2]$ of Examples 4H and 4J, now acting on the Hilbert space $\ell_+^2$. It is readily verified that

$$
\begin{array}{lll}
D_a = D_a^* & \text{if and only if} & \alpha_k \in \mathbb{R} \text{ for all } k \geq 0, \\
D_a \geq O & \text{if and only if} & \alpha_k \geq 0 \text{ for all } k \geq 0, \\
D_a > O & \text{if and only if} & \alpha_k > 0 \text{ for all } k \geq 0, \\
D_a \succ O & \text{if and only if} & D_a = D_a^* \text{ and } \inf_k \alpha_k > 0.
\end{array}
$$

Let $\mathcal{B}^+[\mathcal{H}]$ denote the weakly closed convex cone of all nonnegative operators in $\mathcal{B}[\mathcal{H}]$ (see Problem 5.49) and set $\mathcal{G}^+[\mathcal{H}] = \mathcal{B}^+[\mathcal{H}] \cap \mathcal{G}[\mathcal{H}]$: the class of all strictly positive operators on $\mathcal{H}$. Here is a rather useful corollary of Proposition 5.82.

**Corollary 5.83.** *If $\{Q_n\}$ is a sequence of nonnegative operators on $\mathcal{H}$ (i.e., if $Q_n \in \mathcal{B}^+[\mathcal{H}]$ for every integer $n$), then*

$$Q_n \xrightarrow{w} O \quad \text{if and only if} \quad Q_n \xrightarrow{s} O.$$

*Proof.* Recall that strong convergence always implies weak convergence (to the same limit). On the other hand, since $\mathcal{H}$ is a Hilbert space, $\sup_n \|Q_n\| < \infty$ whenever $\{Q_n\}$ converges weakly. Therefore, according to Proposition 5.82, $Q_n \xrightarrow{w} O$ implies $Q_n \xrightarrow{s} O$, for

$$\|Q_n x\|^2 \le \sup_k \|Q_k\| \langle Q_n x \, ; x \rangle$$

for every $x \in \mathcal{H}$ and every integer $n$ (see Proposition 5.67). $\qquad\square$

An immediate consequence of the above corollary reads as follows. *If $T_n \in \mathcal{B}[\mathcal{H}]$ is such that, for each $n \ge 1$, either $O \le T - T_n$ or $O \le T_n - T$, then $T_n \xrightarrow{w} T$ if and only if $T_n \xrightarrow{s} T$.* This is what is behind the next proposition. Let $\{T_n\}$ be a $\mathcal{B}[\mathcal{H}]$-valued sequence. We say that $\{T_n\}$ is *increasing* or *decreasing* if $O \le T_{n+1} - T_n$ or $O \le T_n - T_{n+1}$, respectively, for every integer $n$. If it is increasing or decreasing, then we say that $\{T_n\}$ is *monotone*.

**Proposition 5.84.** *A bounded monotone sequence of self-adjoint operators converges strongly.*

*Proof.* Suppose $\{T_n\}$ is a bounded decreasing sequence of self-adjoint operators in $\mathcal{B}[\mathcal{H}]$ and take an arbitrary $x \in \mathcal{H}$. Thus $\{\langle T_n x \, ; x \rangle\}$ is a real-valued bounded decreasing sequence (see Proposition 5.79 and recall that $\sup_n |\langle T_n x \, ; x \rangle| \le \sup_n \|T_n\| \|x\|^2$). Therefore, $\{\langle T_n x \, ; x \rangle\}$ converges in $\mathbb{R}$ (Problem 3.10), and so in $\mathbb{F}$, which implies that $T_n - T \xrightarrow{w} O$ for some $T \in \mathcal{B}[\mathcal{H}]$ (Problem 5.48). Moreover, as $\{T_n\}$ is decreasing, $\langle T_n x \, ; x \rangle \ge \langle T_{n+k} x \, ; x \rangle \to \langle T x \, ; x \rangle$ as $k \to \infty$ so that $T_n - T \in \mathcal{B}^+[\mathcal{H}]$ for every integer $n$, and hence Corollary 5.83 ensures that $T_n - T \xrightarrow{s} O$. If $\{T_n\}$ is a bounded increasing sequence of self-adjoint operators, then it is clear that $\{-T_n\}$ is a bounded decreasing sequence of self-adjoint operators. The above argument ensures that $\{-T_n\}$ also converges strongly, and so does $\{T_n\}$. $\qquad\square$

## 5.14  Square Root and Polar Decomposition

We shall also assume throughout this section that $\mathcal{H}$ and $\mathcal{K}$ are Hilbert spaces. If $T$ is an operator in $\mathcal{B}[\mathcal{H}]$, and if there exists an operator $S$ in $\mathcal{B}[\mathcal{H}]$ such that $S^2 = T$, then $S$ is referred to as a *square root* of $T$. Nonnegative operators have a unique nonnegative square root.

**Theorem 5.85.** *Every operator $Q$ in $\mathcal{B}^+[\mathcal{H}]$ has a unique square root $Q^{\frac{1}{2}}$ in $\mathcal{B}^+[\mathcal{H}]$, which commutes with every operator in $\mathcal{B}[\mathcal{H}]$ that commutes with $Q$.*

*Proof.* (a) First we show that there is no loss of generality in assuming $O \leq Q \leq I$. In fact, if $O \leq Q$ and $O \neq Q$, then $O \leq \|Q\|^{-1} Q \leq I$ (since $\langle \|Q\|^{-1} Qx \,;\, x \rangle = \|Q\|^{-1} \langle Qx \,;\, x \rangle \leq \|Q\|^{-1} \|Q\| \|x\|^2 = \langle x \,;\, x \rangle$ for every $x \in \mathcal{H}$). Hence, if there exists a unique $S \geq O$ such that $S^2 = \|Q\|^{-1} Q$, then set $Q^{\frac{1}{2}} = \|Q\|^{\frac{1}{2}} S$ so that $Q^{\frac{1}{2}} \geq O$ and $(Q^{\frac{1}{2}})^2 = \|Q\| S^2 = Q$. Moreover, if $S$ is the unique nonnegative operator such that $S^2 = \|Q\|^{-1} Q$, then $Q^{\frac{1}{2}}$ is the unique nonnegative operator such that $(Q^{\frac{1}{2}})^2 = Q$. Indeed, if $A \geq O$ is such that $A^2 = Q$, then $\|Q\|^{-\frac{1}{2}} A \geq O$ and $(\|Q\|^{-\frac{1}{2}} A)^2 = \|Q\|^{-1} Q$. If $S$ is the unique nonnegative operator such that $S^2 = \|Q\|^{-1} Q$, then $\|Q\|^{-\frac{1}{2}} A = S$ so that $A = \|Q\|^{\frac{1}{2}} S = Q^{\frac{1}{2}}$. Finally, it is clear that if $\|Q\|^{-1} QT = T\|Q\|^{-1} Q$ implies $ST = TS$, then (since $S = \|Q\|^{-\frac{1}{2}} Q^{\frac{1}{2}}$) $QT = TQ$ implies $Q^{\frac{1}{2}} T = T Q^{\frac{1}{2}}$.

(b) Thus suppose $O \leq Q \leq I$ and set $R = I - Q$ so that $O \leq R \leq I$. Consider the sequence $\{B_n\}_{n=0}^{\infty}$ in $\mathcal{B}[\mathcal{H}]$ recursively defined as follows.

$$B_{n+1} = \tfrac{1}{2}(R + B_n^2) \quad \text{with} \quad B_0 = O.$$

*Claim 1.* $O \leq B_n$ for every integer $n \geq 0$.

*Proof.* The proof goes by induction. First observe that $B_0$ is trivially self-adjoint. If $B_n$ is self-adjoint for some $n \geq 0$, then $B_{n+1}$ is self-adjoint because $(B_n^2)^* = (B_n^*)^2 = B_n^2$ and $R = R^*$. Hence $B_n$ is self-adjoint for every $n \geq 0$. Now note that $B_0 \geq O$ trivially. If $B_n \geq O$ for some $n \geq 0$, then $\langle B_{n+1} x \,;\, x \rangle = \tfrac{1}{2}(\langle Rx \,;\, x \rangle + \|B_n x\|^2) \geq 0$ for every $x \in \mathcal{H}$. $\square$

*Claim 2.* $\|B_n\| \leq 1$ so that $B_n \leq I$ for all $n \geq 0$.

*Proof.* $\|R\| = \sup_{\|x\|=1} \langle Rx \,;\, x \rangle \leq \sup_{\|x\|=1} \langle x \,;\, x \rangle = 1$ by Proposition 5.78. Trivially, $\|B_0\| \leq 1$. If $\|B_n\| \leq 1$ for some $n \geq 0$, then $\|B_{n+1}\| \leq \tfrac{1}{2}(\|R\| + \|B_n\|^2) \leq 1$, which proves by induction that $\|B_n\| \leq 1$ for all $n \geq 0$. Then $O \leq B_n \leq I$ for all $n \geq 0$ (cf. Claim 1 and Problem 5.55). $\square$

*Claim 3.* $\{B_n\}$ is an increasing sequence.

*Proof.* It is readily verified by induction that each $B_n$ is a polynomial in $R$ with positive coefficients. This implies that $B_n B_m = B_m B_n$ for every $m, n \geq 0$. Hence

$$B_{n+2} - B_{n+1} = \tfrac{1}{2}\big(B_{n+1}^2 - B_n^2\big) = \tfrac{1}{2}(B_{n+1} - B_n)(B_{n+1} + B_n)$$

for each $n \geq 0$. Observe that $B_1 - B_0 = \tfrac{1}{2} R$ and $B_2 - B_1 = \tfrac{1}{2} R + \tfrac{1}{8} R^2$ are polynomials in $R$ with positive coefficients. Since $B_{n+1} + B_n$ is a polynomial in $R$ with positive coefficients (because each $B_n$ is), it follows that $B_{n+2} - B_{n+1}$ is a polynomial in $R$ with positive coefficients whenever $B_{n+1} - B_n$ is. This proves by induction that $B_{n+1} - B_n$ is a polynomial in $R$ with positive coefficients for every $n \geq 0$, and so $B_{n+1} - B_n \geq O$ for every $n \geq 0$ because $R \geq O$ (cf. Problem 5.52(e)). $\square$

Since $\{B_n\}$ is a bounded monotone sequence of self-adjoint operators, it converges strongly by Proposition 5.84. That is,

$$B_n \xrightarrow{s} B \quad \text{and hence} \quad I - B_n \xrightarrow{s} I - B$$

for some $B \in \mathcal{B}[\mathcal{H}]$. Since $O \leq B_n \leq I$ (i.e., both $B_n$ and $I - B_n$ lie in $\mathcal{B}^+[\mathcal{H}]$) for each $n$, and since $\mathcal{B}^+[\mathcal{H}]$ is weakly (thus strongly) closed in $\mathcal{B}[\mathcal{H}]$ (cf. Problem 5.49), it follows that $O \leq B \leq I$. Moreover, as $B_n \xrightarrow{s} B$, we get $B_n^2 \xrightarrow{s} B^2$ (Problem 4.46(a)). Therefore,

$$O \leq B = \tfrac{1}{2}(R + B^2) \leq I$$

so that $R = 2B - B^2$, and hence $Q = I - R = B^2 - 2B + I = (I - B)^2$. Recalling that $O \leq I - B \leq I$, set $Q^{\frac{1}{2}} = I - B$. Then $O \leq Q^{\frac{1}{2}} \leq I$ and $(Q^{\frac{1}{2}})^2 = Q$ so that there exists a nonnegative square root $Q^{\frac{1}{2}}$ of $Q$.

(c) Suppose $TQ = QT$ for some $T \in \mathcal{B}[\mathcal{H}]$. Thus $T(I - R) = (I - R)T$ so that $TR = RT$, and hence $Tp(R) = p(R)T$ for every polynomial $p$ in $R$. Then $TB_n = B_nT$ (cf. proof of Claim 3) for each $n$. Since $TB_n \xrightarrow{s} TB$ and $B_nT \xrightarrow{s} BT$, it follows that $TB = BT$ (the strong limit is unique). Therefore, $T(I - B) = (I - B)T$ so that $TQ^{\frac{1}{2}} = Q^{\frac{1}{2}}T$. That is, $Q^{\frac{1}{2}}$ commutes with every operator that commutes with $Q$.

(d) Finally, we show that the nonnegative square root $Q^{\frac{1}{2}}$ of $Q$ is unique. Take any $A \geq O$ such that $A^2 = Q$. Since $AQ = A^3 = QA$, it follows by (c) that $AQ^{\frac{1}{2}} = Q^{\frac{1}{2}}A$. Then

$$(Q^{\frac{1}{2}} - A)Q^{\frac{1}{2}}(Q^{\frac{1}{2}} - A) + (Q^{\frac{1}{2}} - A)A(Q^{\frac{1}{2}} - A)$$
$$= (Q^{\frac{1}{2}} - A)(Q^{\frac{1}{2}} + A)(Q^{\frac{1}{2}} - A) = (Q^{\frac{1}{2}} - A)(Q - A^2) = O.$$

But $(Q^{\frac{1}{2}} - A)Q^{\frac{1}{2}}(Q^{\frac{1}{2}} - A) \geq O$ and $(Q^{\frac{1}{2}} - A)A(Q^{\frac{1}{2}} - A) \geq O$ (for $Q^{\frac{1}{2}}$ and $A$ are nonnegative — cf. Problem 5.51(a)), which implies that these two operators are null, and so is their difference. That is, $(Q^{\frac{1}{2}} - A)^3 = (Q^{\frac{1}{2}} - A)Q^{\frac{1}{2}}(Q^{\frac{1}{2}} - A) - (Q^{\frac{1}{2}} - A)A(Q^{\frac{1}{2}} - A) = O$, and hence $(Q^{\frac{1}{2}} - A)^4 = O$. Since $(Q^{\frac{1}{2}} - A)$ is self-adjoint, $\|(Q^{\frac{1}{2}} - A)\|^4 = \|(Q^{\frac{1}{2}} - A)^4\| = 0$ (see Problem 5.45(e)), so that $A = Q^{\frac{1}{2}}$. $\square$

**Proposition 5.86.** *If $Q \in \mathcal{B}^+[\mathcal{H}]$, then*

(a) $$\|Q^{\frac{1}{2}}\|^2 = \|Q\| = \|Q^2\|^{\frac{1}{2}},$$

(b)   $\mathcal{N}(Q^{\frac{1}{2}}) = \mathcal{N}(Q) = \mathcal{N}(Q^2)$   and   $\mathcal{R}(Q^{\frac{1}{2}})^- = \mathcal{R}(Q)^- = \mathcal{R}(Q^2)^-$.

*Proof.* If $Q$ is nonnegative, then $Q^2$ is nonnegative (Problem 5.52). Hence $Q$ is the unique nonnegative square root of $Q^2$ (Theorem 5.85). Therefore, the identities involving $Q$ and $Q^2$ follow at once if the identities involving $Q^{\frac{1}{2}}$ and $Q$ are established.

(a)  $Q^{\frac{1}{2}} = (Q^{\frac{1}{2}})^*$ implies $\|Q^{\frac{1}{2}}\|^2 = \|(Q^{\frac{1}{2}})^2\| = \|Q\|$ (Proposition 5.65).

(b)  Since $Q = Q^{\frac{1}{2}}Q^{\frac{1}{2}}$, it follows that $\mathcal{N}(Q^{\frac{1}{2}}) \subseteq \mathcal{N}(Q)$. On the other hand, $\|Q^{\frac{1}{2}}x\|^2 = \langle Q^{\frac{1}{2}}x \, ; \, Q^{\frac{1}{2}}x \rangle = \langle Qx \, ; \, x \rangle$ for every $x \in \mathcal{H}$, and hence $\mathcal{N}(Q) \subseteq \mathcal{N}(Q^{\frac{1}{2}})$. Then $\mathcal{N}(Q) = \mathcal{N}(Q^{\frac{1}{2}})$. Therefore, by Proposition 5.76, $\mathcal{R}(Q)^- = \mathcal{N}(Q)^\perp = \mathcal{N}(Q^{\frac{1}{2}})^\perp = \mathcal{R}(Q^{\frac{1}{2}})^-$.   $\square$

A *partial isometry* is a bounded linear transformation that acts isometrically on the orthogonal complement of its null space. That is, $W \in \mathcal{B}[\mathcal{H}, \mathcal{K}]$ is a partial isometry if $W|_{\mathcal{N}(W)^\perp} \colon \mathcal{N}(W)^\perp \to \mathcal{K}$, the restriction of $W$ to $\mathcal{N}(W)^\perp$, is an isometry. Note: In this context, "isometry" means "linear isometry".

**Proposition 5.87.** *If* $W \colon \mathcal{H} \to \mathcal{K}$ *is a partial isometry, then*

$$W = VP,$$

*where* $V \colon \mathcal{N}(W)^\perp \to \mathcal{K}$ *is an isometry and* $P \colon \mathcal{H} \to \mathcal{H}$ *is the orthogonal projection onto* $\mathcal{N}(W)^\perp$. *Conversely, let* $\mathcal{M}$ *be any subspace of* $\mathcal{H}$. *If* $V \colon \mathcal{M} \to \mathcal{H}$ *is an isometry and* $P \colon \mathcal{H} \to \mathcal{H}$ *is the orthogonal projection onto* $\mathcal{M}$, *then* $W = VP \colon \mathcal{H} \to \mathcal{K}$ *is a partial isometry.*

*Proof.* If $W \colon \mathcal{H} \to \mathcal{K}$ is a partial isometry, then let $P \colon \mathcal{H} \to \mathcal{H}$ be the orthogonal projection onto $\mathcal{N}(W)^\perp$ (so that $\mathcal{R}(P) = \mathcal{N}(W)^\perp$, and hence $\mathcal{R}(I - P) = \mathcal{N}(P) = \mathcal{R}(P)^\perp = \mathcal{N}(W)$ — cf. Propositions 4.13, 5.15 and 5.51). Set $V = W|_{\mathcal{N}(W)^\perp} \colon \mathcal{N}(W)^\perp \to \mathcal{K}$, which is an isometry. Thus, for every $x \in \mathcal{H}$,

$$\begin{aligned} Wx &= W(Px + (I - P)x) &= WPx + W(I - P)x \\ &= WPx = W|_{\mathcal{N}(W)^\perp}Px &= VPx. \end{aligned}$$

Conversely, let $V \colon \mathcal{M} \to \mathcal{K}$ be an isometry, and let $P \colon \mathcal{H} \to \mathcal{H}$ be the orthogonal projection onto $\mathcal{M}$, where $\mathcal{M}$ is an arbitrary subspace of $\mathcal{H}$. If $W = VP$, then $W|_{\mathcal{M}}v = Wv = VPv = Vv$ for every $v \in \mathcal{R}(P) = \mathcal{M}$, and hence $W$ acts isometrically on $\mathcal{M}$. Moreover, if $Wu = 0$, then $VPu = 0$ so that $Pu \in \mathcal{N}(V) = \{0\}$ (recall: $V$ is an isometry), which implies that $u \in \mathcal{N}(P)$. Then $\mathcal{N}(W) \subseteq \mathcal{N}(P)$, and so $\mathcal{N}(W) = \mathcal{N}(P)$ (reason: $\mathcal{N}(P) \subseteq \mathcal{N}(W)$ because $W = VP$). Therefore, $\mathcal{M} = \mathcal{R}(P) = \mathcal{N}(P)^\perp = \mathcal{N}(W)^\perp$. Conclusion: $W = VP \colon \mathcal{H} \to \mathcal{K}$ is a bounded linear transformation (composition of bounded linear transformations) that acts isometrically on $\mathcal{N}(W)^\perp$; that is, $W$ is a partial isometry.   $\square$

Since $W = VP$ and $V = W|_{\mathcal{N}(W)^{\perp}}$, $\|V\| \leq \|W\| = \|VP\| \leq \|V\|\|P\|$ and $\mathcal{R}(W) \subseteq \mathcal{R}(V) = \mathcal{R}(W|_{\mathcal{N}(W^{\perp})}) \subseteq \mathcal{R}(W)$. Then $\|W\| = \|V\| = 1$ and $\mathcal{R}(W) = \mathcal{R}(V)$. Recall: *The range of a linear isometry on a Hilbert space is a subspace* (cf. Propositions 4.20 and 4.37). That is, $\mathcal{R}(V) = \mathcal{R}(V)^{-}$, and hence $\mathcal{R}(W) = \mathcal{R}(W)^{-}$. Outcome: *The range of a partial isometry also is a subspace.* $\mathcal{N}(W)^{\perp}$ and $\mathcal{R}(W)$ are called *initial* and *final spaces* of the partial isometry $W$, respectively.

**Proposition 5.88.** *Take $W \in \mathcal{B}[\mathcal{H}, \mathcal{K}]$. The following assertions are pairwise equivalent.*

(a)  *$W$ is a partial isometry with initial space $\mathcal{M}$ and final space $\mathcal{R}$.*

(b)  *$W^*W$ is the orthogonal projection onto $\mathcal{M}$ and $\mathcal{R}(W) = \mathcal{R}$.*

(c)  *$WW^*W = W$*

(d)  *$W^*WW^* = W^*$*

(e)  *$WW^*$ is the orthogonal projection onto $\mathcal{R}$ and $\mathcal{N}(W)^{\perp} = \mathcal{M}$.*

(f)  *$W^*$ is a partial isometry with initial space $\mathcal{R}$ and final space $\mathcal{M}$.*

*Proof.* Let $W$ be a bounded linear transformation of $\mathcal{H}$ into $\mathcal{K}$.

*Proof of* (a)$\Rightarrow$(c). If (a) holds, then $W = VP$, where $V\colon \mathcal{M} \to \mathcal{K}$ is an isometry, $P\colon \mathcal{H} \to \mathcal{H}$ is the orthogonal projection onto $\mathcal{M}$, and $\mathcal{M} = \mathcal{N}(W)^{\perp}$ (Proposition 5.87). Consider the adjoint $V^*\colon \mathcal{K} \to \mathcal{M}$ of $V$, and recall from Proposition 5.72 that $V^*V = I$, the identity on the Hilbert space $\mathcal{M}$ (for $\mathcal{M} = \mathcal{N}(W)^{\perp}$ is a subspace of the Hilbert space $\mathcal{H}$ by Proposition 5.12, and hence a Hilbert space itself). Since $P$ is an orthogonal projection, it follows by Proposition 5.81 that $P = P^*$ and so $W^* = P^*V^* = PV^*$. Therefore, $WW^*W = VPPV^*VP = VP = W$.

*Proof of* (c)$\Rightarrow$(b). If $WW^*W = W$, then $(W^*W)^2 = W^*WW^*W = W^*W$ so that $W^*W$ is a (continuous) projection, and hence it has a closed range. (Reason: if $E = E^2$, then $\mathcal{R}(E) = \mathcal{N}(I - E)$, and $\mathcal{N}(I - E)$ is closed whenever the linear transformation $E$ is bounded — see Proposition 4.13). Thus $\mathcal{R}(W^*W) = \mathcal{R}(W^*W)^{-} = \mathcal{N}((W^*W)^*)^{\perp} = \mathcal{N}(W^*W)^{\perp} = \mathcal{N}(W)^{\perp}$ (Proposition 5.76). Hence $\mathcal{R}(W^*W) = \mathcal{N}(W)^{\perp}$ and $\mathcal{R}(W^*W) \perp \mathcal{N}(W^*W)$, which implies that $W^*W$ is the orthogonal projection onto $\mathcal{M} = \mathcal{N}(W)^{\perp}$.

*Proof of* (b)$\Rightarrow$(a). If (b) holds, then $\mathcal{M} = \mathcal{R}(W^*W) = \mathcal{R}(W^*W)^{-} = \mathcal{N}(W)^{\perp}$ (cf. Propositions 5.51 and 5.76). If $v \in \mathcal{M} = \mathcal{R}(W^*W)$, then $W^*Wv = v$, and hence $\|Wv\|^2 = \langle Wv\,;Wv \rangle = \langle W^*Wv\,;v \rangle = \langle v\,;v \rangle = \|v\|^2$ so that $W|_{\mathcal{M}}$ is an isometry. Since $\mathcal{M} = \mathcal{N}(W)^{\perp}$, it follows that (b) implies (a).

Conclusion: Assertions (a), (b) and (c) are pairwise equivalent. If $W^*W$ is an orthogonal projection, then $\mathcal{R}(W^*W)$ is closed, which implies that $\mathcal{R}(W^*)$ and

$\mathcal{R}(W)$ are closed as well. Similarly, if $W^*$ is a partial isometry, then $\mathcal{R}(W^*)$ is closed, and so is $\mathcal{R}(W)$. Therefore, in both cases, $\mathcal{R}(W^*) = \mathcal{N}(W)^\perp$ and $\mathcal{N}(W^*)^\perp = \mathcal{R}(W)$. Since assertions (a), (b) and (c) are pairwise equivalent, it follows that the dual assertions (f), (e) and (d) are pairwise equivalent too. Finally, observe that (c) and (d) are trivially equivalent.    $\square$

If a transformation $T \in \mathcal{B}[\mathcal{H}, \mathcal{K}]$ is the product of a partial isometry $W \in \mathcal{B}[\mathcal{H}, \mathcal{K}]$ and a nonnegative operator $Q \in \mathcal{B}[\mathcal{H}]$, and if $\mathcal{N}(W) = \mathcal{N}(Q)$, then the representation $T = WQ$ is called the *polar decomposition* of $T$. The next theorem says that every bounded linear transformation in $\mathcal{B}[\mathcal{H}, \mathcal{K}]$ has a unique polar decomposition.

**Theorem 5.89.** *If $T \in \mathcal{B}[\mathcal{H}, \mathcal{K}]$, then there exists a partial isometry $W \in \mathcal{B}[\mathcal{H}, \mathcal{K}]$ with initial space $\mathcal{N}(T)^\perp$ and final space $\mathcal{R}(T)^-$ such that*

$$T = W(T^*T)^{\frac{1}{2}}$$

*and $\mathcal{N}(W) = \mathcal{N}((T^*T)^{\frac{1}{2}})$. Moreover, if $T = ZQ$, where $Q$ is a nonnegative operator in $\mathcal{B}[\mathcal{H}]$ and $Z \in \mathcal{B}[\mathcal{H}, \mathcal{K}]$ is a partial isometry with $\mathcal{N}(Z) = \mathcal{N}(Q)$, then $Q = (T^*T)^{\frac{1}{2}}$ and $Z = W$.*

*Proof.* Take any $T \in \mathcal{B}[\mathcal{H}, \mathcal{K}]$. Recall that $T^*T \in \mathcal{B}^+[\mathcal{H}]$ and $\mathcal{R}((T^*T)^{\frac{1}{2}})^- = \mathcal{R}(T^*T)^- = \mathcal{N}(T^*T)^\perp = \mathcal{N}((T^*T)^{\frac{1}{2}})^\perp = \mathcal{N}(T)^\perp$ (cf. Corollary 5.75 and Proposition 5.86).

*Existence.* Consider the mapping $V_0 \colon \mathcal{R}((T^*T)^{\frac{1}{2}}) \to \mathcal{K}$ defined by $V_0(T^*T)^{\frac{1}{2}}x = Tx$ for every $x \in \mathcal{H}$, so that $\mathcal{R}(V_0) \subseteq \mathcal{R}(T)$. First note that $V_0$ is linear. Indeed, if $y$ and $z$ lie in $\mathcal{R}((T^*T)^{\frac{1}{2}}) \subseteq \mathcal{H}$, then $y = (T^*T)^{\frac{1}{2}}u$ and $z = (T^*T)^{\frac{1}{2}}v$ for some $u$ and $v$ in $\mathcal{H}$, and hence

$$\begin{aligned}
V_0(\alpha y + \beta z) &= V_0(\alpha(T^*T)^{\frac{1}{2}}u + \beta(T^*T)^{\frac{1}{2}}v) \\
&= V_0(T^*T)^{\frac{1}{2}}(\alpha u + \beta v) = T(\alpha u + \beta v) = \alpha Tu + \beta Tv \\
&= \alpha V_0(T^*T)^{\frac{1}{2}}u + \beta V_0(T^*T)^{\frac{1}{2}}v = \alpha V_0 y + \beta V_0 z
\end{aligned}$$

for every $\alpha, \beta \in \mathbb{F}$. Moreover, since

$$\|Tx\|^2 = \langle Tx \,;\, Tx \rangle = \langle T^*Tx \,;\, x \rangle = \langle (T^*T)^{\frac{1}{2}}x \,;\, (T^*T)^{\frac{1}{2}}x \rangle = \|(T^*T)^{\frac{1}{2}}x\|^2,$$

it follows that $\|V_0(T^*T)^{\frac{1}{2}}x\| = \|Tx\| = \|(T^*T)^{\frac{1}{2}}x\|$, for every $x \in \mathcal{H}$. Hence $\|V_0 y\| = \|y\|$ for every $y \in \mathcal{R}((T^*T)^{\frac{1}{2}})$ so that $V_0 \colon \mathcal{R}((T^*T)^{\frac{1}{2}}) \to \mathcal{K}$ is a linear isometry. Extend it to $\mathcal{R}(T^*T)^- = \mathcal{R}((T^*T)^{\frac{1}{2}})^-$ and get the mapping

$$V \colon \mathcal{R}(T^*T)^- \to \mathcal{K}.$$

This is a linear isometry with $\mathcal{R}(V) \subseteq \mathcal{R}(V_0)^- \subseteq \mathcal{R}(T)^-$. Indeed, since the mapping $V_0 \colon \mathcal{R}((T^*T)^{\frac{1}{2}}) \to \mathcal{R}(V_0)$ is an isometric isomorphism, it follows by

Corollary 4.30 that the mapping $V \colon \mathcal{R}((T^*T)^{\frac{1}{2}})^- \to \mathcal{R}(V_0)^-$ is again an isometric isomorphism. Since $V(T^*T)^{\frac{1}{2}}x = V_0(T^*T)^{\frac{1}{2}}x = Tx$ for every $x \in \mathcal{H}$,

$$T = V(T^*T)^{\frac{1}{2}},$$

and hence $\mathcal{R}(T) \subseteq \mathcal{R}(V)$, where

$$V \colon \mathcal{N}(T)^\perp \to \mathcal{K} \ \text{ is a linear isometry with } \ \mathcal{R}(V) = \mathcal{R}(T)^-.$$

(Recall: $\mathcal{N}(T)^\perp$ is a Hilbert space so that $\mathcal{R}(V) = \mathcal{R}(V)^-$.) Let $P \colon \mathcal{H} \to \mathcal{H}$ be the orthogonal projection onto $\mathcal{N}(T)^\perp$. Then $\mathcal{R}(P) = \mathcal{N}(T)^\perp$, and so $VPy = Vy$ for every $y \in \mathcal{N}(T)^\perp = \mathcal{R}((T^*T)^{\frac{1}{2}})$, which implies $VP(T^*T)^{\frac{1}{2}}x = V(T^*T)^{\frac{1}{2}}x$ for every $x \in \mathcal{H}$. That is, $VP(T^*T)^{\frac{1}{2}} = V(T^*T)^{\frac{1}{2}}$. Setting $W = VP \colon \mathcal{H} \to \mathcal{K}$ we get

$$T = W(T^*T)^{\frac{1}{2}}.$$

Since $V = W|_{\mathcal{R}(P)} = W|_{\mathcal{N}(T)^\perp}$ is an isometry and $P$ is an orthogonal projection, it follows that $\mathcal{N}(W) = \mathcal{N}(VP) = \mathcal{N}(P) = \mathcal{R}(P)^\perp = \mathcal{N}(T) = \mathcal{N}((T^*T)^{\frac{1}{2}})$ and $\mathcal{R}(W) = \mathcal{R}(V) = \mathcal{R}(T)^-$. Therefore,

$$W \colon \mathcal{H} \to \mathcal{K} \ \text{ is a partial isometry}$$

with initial space $\mathcal{N}(T)^\perp$, final space $\mathcal{R}(T)^-$, and $\mathcal{N}(W) = \mathcal{N}((T^*T)^{\frac{1}{2}})$.

*Uniqueness.* Let $Z \in \mathcal{B}[\mathcal{H}, \mathcal{K}]$ be a partial isometry with $\mathcal{N}(Z) = \mathcal{N}(Q)$, where $Q$ is a nonnegative operator in $\mathcal{B}[\mathcal{H}]$. Suppose $T = ZQ$. Since $Z^*Z$ is the orthogonal projection onto $\mathcal{N}(Z)^\perp = \mathcal{N}(Q)^\perp = \mathcal{R}(Q)^-$ (Propositions 5.76(b*) and 5.88(b)), it follows that $T^*T = QZ^*ZQ = Q^2$. Thus $Q = (T^*T)^{\frac{1}{2}}$ by uniqueness of the nonnegative square root. Hence $Z(T^*T)^{\frac{1}{2}} = T = W(T^*T)^{\frac{1}{2}}$ so that $Z|_{\mathcal{R}((T^*T)^{1/2})} = W|_{\mathcal{R}((T^*T)^{1/2})}$, and therefore

$$\mathcal{N}(Z) = \mathcal{N}(W) \quad \text{and} \quad Z|_{\mathcal{N}(Z)^\perp} = W|_{\mathcal{N}(W)^\perp}$$

because $\mathcal{R}((T^*T)^{\frac{1}{2}})^- = \mathcal{R}((T^*T))^- = \mathcal{N}(T)^\perp$ and $\mathcal{N}(Z) = \mathcal{N}(Q) = \mathcal{N}((T^*T)^{\frac{1}{2}}) = \mathcal{N}(T) = \mathcal{N}(W)$. Conclusion: The partial isometries $Z \colon \mathcal{H} \to \mathcal{K}$ and $W \colon \mathcal{H} \to \mathcal{K}$ have the same initial space and they coincide there. That is, $Z = W$. $\qquad\square$

If $T = WQ$ is the polar decomposition of $T$, then $W^*W$ is the orthogonal projection onto $\mathcal{N}(W)^\perp = \mathcal{N}(Q)^\perp = \mathcal{R}(Q)^-$ so that $W^*WQ = Q$. Thus

$$T = WQ \quad \text{implies} \quad W^*T = Q.$$

Here is another corollary of Theorem 5.89.

**Corollary 5.90.** *Let $T = WQ$ be the polar decomposition of a bounded linear transformation $T \in \mathcal{B}[\mathcal{H}, \mathcal{K}]$.*

(a)  $W \in \mathcal{B}[\mathcal{H}, \mathcal{K}]$ *is an isometry if and only if* $\mathcal{N}(T) = \{0\}$.

(b)  $W \in \mathcal{B}[\mathcal{H}, \mathcal{K}]$ *is a coisometry if and only if* $\mathcal{R}(T)^- = \mathcal{K}$.

*Proof.* Recall that $\mathcal{N}(T) = \mathcal{N}(W)$ and $\mathcal{R}(W) = \mathcal{R}(T)^-$ whenever $T = WQ$ is the polar decomposition of $T$.

(a) A partial isometry $W$ is an isometry if and only if $\mathcal{N}(W)^\perp = \mathcal{H}$, which means that $\mathcal{N}(W) = \{0\}$.

(b) $W$ is a coisometry if and only if $W^*$ is an isometry. But $W^*$ is a partial isometry with $\mathcal{N}(W^*)^\perp = \mathcal{R}(W)$ (Proposition 5.88). Hence $W^*$ is an isometry if and only if $\mathcal{R}(W) = \mathcal{K}$.     $\square$

Therefore, if $T = WQ$ is the polar decomposition of $T$, then $W$ is unitary (i.e., a surjective isometry or, equivalently, an isometry and a coisometry) if and only if $T$ is injective and has dense range (i.e.. if and only if $T$ is quasiinvertible). In particular, if $T \in \mathcal{G}[\mathcal{H}, \mathcal{K}]$ (i.e., if $T$ is invertible), then $T = U(T^*T)^{\frac{1}{2}}$, where $U \in \mathcal{G}[\mathcal{H}, \mathcal{K}]$ is unitary.

## *Suggested Reading*


Akhiezer and Glazman [1]

Arveson [1]

Bachman and Narici [1]

Balakrishnan [1]

Beals [1]

Berberian [3]

Berezansky, Sheftel and Us [1]

Conway [1], [2]

Davidson [1]

Dunford and Schwartz [2]

Fillmore [1], [2]

Gohberg and Kreĭn [1]

Halmos [1], [4]

Kato [1]

Kubrusly [1]

Murphy [1]

Naylor and Sell [1]

Pearcy [1], [2]

Putnam [1]

Radjavi and Rosenthal [1]

Reed and Simon [1]

Riesz and Sz.-Nagy [1]

Schatten [1]

Stone [1]

Sz.-Nagy and Foiaş [1]

Weidmann [1]


# Problems

**Problem 5.1.** Let $\| \ \|$ be the norm induced by an inner product $\langle \ ; \ \rangle$ on a linear space $\mathcal{X} \neq \{0\}$. Show that, for every $x \in \mathcal{X}$,

$$\|x\| = \sup_{\|y\|=1} |\langle x ; y \rangle| = \sup_{\|y\| \leq 1} |\langle x ; y \rangle| = \sup_{y \neq 0} \frac{|\langle x ; y \rangle|}{\|y\|}.$$

*Hint*: If $\|x\| \neq 0$, then $\|x\| = \langle \frac{x}{\|x\|} ; x \rangle$. Use the Schwarz inequality.

**Problem 5.2.** Let $(\mathcal{X}, \langle\ ;\ \rangle)$ be an inner product space and let $\|\ \|$ be the norm on $\mathcal{X}$ induced by $\langle\ ;\ \rangle$. Take arbitrary vectors $x$ and $y$ in $\mathcal{X}$ and consider the following assertions.

(a) $\|x + y\| = \|x\| + \|y\|$.

(a′) $\|x - y\| = \big| \|x\| - \|y\| \big|$.

(b) $|\langle x ; y \rangle| = \|x\| \|y\|$.

(b′) $\langle x ; y \rangle = \|x\| \|y\|$.

(c)   $x$ and $y$ are *collinear* (i.e., either one of them is null or $y = \alpha x$ for some nonzero $\alpha$ in $\mathbb{F}$).

(c′)   $x$ and $y$ are *proportional* (i.e., either one of them is null or $y = \alpha x$ for some real $\alpha > 0$).

Show that the diagram below exhibits *all* possible implications among the above assertions.

$$(\text{a}) \iff (\text{a}') \iff (\text{b}') \iff (\text{c}')$$
$$\Downarrow \qquad\qquad\qquad \Downarrow$$
$$(\text{b}) \iff (\text{c})$$

*Hint*: Consider the following auxiliary assertion.

(b″)   $\mathrm{Re}\,\langle x ; y \rangle = \|x\| \|y\|$.

Recall that $\|x \pm y\|^2 = \|x\|^2 \pm 2\mathrm{Re}\,\langle x ; y \rangle + \|y\|^2$ for every $x, y \in \mathcal{X}$. Use this identity to show that (a)$\Leftrightarrow$(b″), (a′)$\Leftrightarrow$(b″) and, by setting $z = \langle x ; y \rangle \|y\|^{-2} y$ and $z = |\langle x ; y \rangle| \|y\|^{-2} y$, that (b)$\Rightarrow$(c) and (b′)$\Rightarrow$(c′). Now use the Schwarz inequality to show that (b″)$\Leftrightarrow$(b′). The remaining implications are trivially verified. Show that (c)$\nRightarrow$(c′) and conclude that there may be no other implication in the above diagram.

**Problem 5.3.** Let $\mathcal{X}$ and $\mathcal{Y}$ be inner product spaces and take arbitrary $T$ and $L$ in $\mathcal{L}[\mathcal{X}, \mathcal{Y}]$. Prove the following identities.

(a) $\langle Tx ; Ly \rangle + \langle Ty ; Lx \rangle = \frac{1}{2}\big(\langle T(x + y) ; L(x + y)\rangle - \langle T(x - y) ; L(x - y)\rangle\big)$

for every $x, y \in \mathcal{X}$. If $\mathcal{X}$ is a *complex* inner product space, then

$$
\begin{aligned}
(\text{b}) \quad \langle Tx ; Ly \rangle \;=\; & \tfrac{1}{4}\big(\langle T(x + y) ; L(x + y)\rangle - \langle T(x - y) ; L(x - y)\rangle \\
& + \; i\,\langle T(x + iy) ; L(x + iy)\rangle - i\langle T(x - iy) ; L(x - iy)\rangle\big)
\end{aligned}
$$

for every $x, y \in \mathcal{X}$. Note that these yield the polarization identities.

*Hint:* $\langle T(x + \alpha y) ; L(x + \alpha y)\rangle = \langle Tx ; Lx\rangle + \bar{\alpha}\langle Tx ; Ly\rangle + \alpha\langle Ty ; Lx\rangle + |\alpha|^2\langle Ty ; Ly\rangle$. Set $\alpha = \pm 1$ to get (a), and also $\alpha = \pm i$ to get (b).

**Problem 5.4.** Let $T \in \mathcal{B}[\mathcal{X}]$ be an operator on an inner product space $\mathcal{X}$. Show that the following assertions are pairwise equivalent.

(a) $T = O$.

(b) $Tx = 0$ for all $x \in \mathcal{X}$.

(c) $\langle Tx ; y\rangle = 0$ for all $x, y \in \mathcal{X}$.

Now consider the following further assertion.

(d) $\langle Tx ; x\rangle = 0$ for all $x \in \mathcal{X}$.

Clearly, (c) implies (d). If $\mathcal{X}$ is a *complex* inner product space, then show that these four assertions are all pairwise equivalent.

*Hint:* Use Problem 5.3(b) to show that (d) implies (c) if $\mathcal{X}$ is complex.

**Problem 5.5.** Let $\{T_n\}$ be a sequence of bounded linear transformations of a Hilbert space $\mathcal{H}$ into a Hilbert space $\mathcal{K}$ (i.e., $T_n \in \mathcal{B}[\mathcal{H}, \mathcal{K}]$ for each $n$). Show that uniform, strong and weak boundedness coincide. In other words, show that the following assertions are pairwise equivalent.

(a) $\sup_n \|T_n\| < \infty$.

(b) $\sup_n \|T_n x\| < \infty$ for every $x \in \mathcal{H}$.

(c) $\sup_n |\langle T_n x ; y\rangle| < \infty$ for every $x \in \mathcal{H}$ and every $y \in \mathcal{K}$.

Now set $\mathcal{K} = \mathcal{H}$ and consider the following further assertion.

(d) $\sup_n |\langle T_n x ; x\rangle| < \infty$ for every $x \in \mathcal{H}$.

Clearly, (c) implies (d). If $\mathcal{K} = \mathcal{H}$ is a *complex* Hilbert space, then show that these four assertions are all pairwise equivalent.

*Hint:* Check that (a)$\Rightarrow$(b)$\Rightarrow$(c). Now take an arbitrary $x \in \mathcal{H}$. For each $n$ consider the functional $x_n : \mathcal{K} \to \mathbb{F}$ given by

$$x_n(y) = \langle y ; T_n x\rangle$$

for every $y \in \mathcal{K}$. Show that each $x_n$ is linear and bounded. If (c) holds (so that $\sup_n |x_n(y)| < \infty$ for every $y \in \mathcal{K}$), then the Banach–Steinhaus Theorem ensures that $\sup_n \|x_n\| < \infty$ because $\mathcal{K}$ is a Banach space. But

$$\|T_n x\| = \sup_{\|y\|=1} |x_n(x)| = \|x_n\|$$

for each $n$ (cf. Problem 5.1). Thus conclude that (c)$\Rightarrow$(b). A straightforward application of the Banach–Steinhaus Theorem ensures that (b)$\Rightarrow$(a) because $\mathcal{H}$ is a Banach space. Finally, if $\mathcal{K} = \mathcal{H}$ is complex, then use Problem 5.3(b) to show that (d)$\Rightarrow$(c).

**Problem 5.6.** Let $(\mathcal{X}, \langle \, ; \, \rangle)$ be an inner product space and equip the field $\mathbb{F}$ with its usual metric. Use the Schwarz inequality and Corollary 3.8 to prove the following assertions.

(a)  $\langle \cdot \, ; y \rangle \colon \mathcal{X} \to \mathbb{F}$ is a continuous function for every $y \in \mathcal{X}$.

(b)  $\langle x \, ; \cdot \rangle \colon \mathcal{X} \to \mathbb{F}$ is a continuous function for every $x \in \mathcal{X}$.

Equip the direct sum $\mathcal{X} \oplus \mathcal{X}$ with the inner product $\langle \, ; \, \rangle_\oplus$ of Example 5E. That is, $\langle x \, ; y \rangle_\oplus = \langle x_1 \, ; y_1 \rangle + \langle x_2 \, ; y_2 \rangle$ for every $x = (x_1, x_2)$ and $y = (y_1, y_2)$ in $\mathcal{X} \oplus \mathcal{X}$. Show that

(c)  $\langle \, ; \, \rangle_\oplus \colon \mathcal{X} \oplus \mathcal{X} \to \mathbb{F}$ is a continuous function.

> *Hint*: Take an arbitrary convergent sequence $\{x(n) = (x_1(n), x_2(n))\}$ in $\mathcal{X} \oplus \mathcal{X}$ and let $x = (x_1, x_2) \in \mathcal{X} \oplus \mathcal{X}$ be its limit. Verify that
>
> $$\begin{aligned}
> \langle x_1(n) - x_1 + x_1 \, ; x_2(n) - x_2 + x_2 \rangle &= \langle x_1(n) - x_1 \, ; x_2(n) - x_2 \rangle \\
> &\quad + \langle x_1(n) - x_1 \, ; x_2 \rangle + \langle x_1 \, ; x_2(n) - x_2 \rangle + \langle x_1 \, ; x_2 \rangle,
> \end{aligned}$$
>
> and use the Schwarz inequality to show that
>
> $$\begin{aligned}
> |\langle x_1(n) \, ; x_2(n) \rangle - \langle x_1 \, ; x_2 \rangle| \; \leq \; & \|x_1(n) - x_1\| \, \|x_2(n) - x_2\| \\
> & + \|x_2\| \, \|x_1(n) - x_1\| + \|x_1\| \, \|x_2(n) - x_2\|.
> \end{aligned}$$
>
> Since $\|x(n) - x\|_\oplus^2 = \|x_1(n) - x_1\|^2 + \|x_1(n) - x_1\|^2$, $x(n) \to x$ in $\mathcal{X} \oplus \mathcal{X}$ if and only if $x_1(n) \to x_1$ and $x_1(n) \to x_1$ in $\mathcal{X}$, which implies $\langle x_1(n) \, ; x_2(n) \rangle \to \langle x_1 \, ; x_2 \rangle$ in $\mathbb{F}$.

**Problem 5.7.** Let $A$ and $B$ be subsets of a Hilbert space $\mathcal{H}$. Prove:

(a)  $A \subseteq B$ and $A^\perp \subseteq B$  implies  $\bigvee B = \mathcal{H}$.

In particular, if $\mathcal{M}$ is a subspace of $\mathcal{H}$, then

(b)  $A \subseteq \mathcal{M}$ and $A^\perp \subseteq \mathcal{M}$  implies  $\mathcal{M} = \mathcal{H}$.

*Hint*: $B^{\perp\perp} = \mathcal{H}$ because $B^\perp \subseteq A^\perp \cap A^{\perp\perp} = \{0\}$. Use Proposition 5.15.

**Problem 5.8.** Let $\mathcal{X}$ be an inner product space and let $\mathcal{H}$ be a Hilbert space. Prove the following propositions. If $\{A_\gamma\}_{\gamma \in \Gamma}$ is a nonempty family of subsets of $\mathcal{X}$, then

(a) $\bigcap_{\gamma\in\Gamma}A_\gamma^\perp = \left(\bigcup_{\gamma\in\Gamma}A_\gamma\right)^\perp$.

If $\{\mathcal{M}_\gamma\}_{\gamma\in\Gamma}$ is a nonempty family of linear manifolds of $\mathcal{X}$, then

(b) $\bigcap_{\gamma\in\Gamma}\mathcal{M}_\gamma^\perp = \left(\sum_{\gamma\in\Gamma}\mathcal{M}_\gamma\right)^\perp$.

> *Hint*: Proposition 5.12. Recall: $\bigvee\left(\bigcup_{\gamma\in\Gamma}\mathcal{M}_\gamma\right) = \left(\sum_{\gamma\in\Gamma}\mathcal{M}_\gamma\right)^-$.

If $\{\mathcal{M}_\gamma\}_{\gamma\in\Gamma}$ is a nonempty family of subspaces of $\mathcal{H}$, then

(c) $\bigcap_{\gamma\in\Gamma}\mathcal{M}_\gamma = \{0\}$ if and only if $\left(\sum_{\gamma\in\Gamma}\mathcal{M}_\gamma^\perp\right)^- = \mathcal{H}$,

(d) $\left(\sum_{\gamma\in\Gamma}\mathcal{M}_\gamma\right)^- = \mathcal{H}$ if and only if $\bigcap_{\gamma\in\Gamma}\mathcal{M}_\gamma^\perp = \{0\}$.

*Hint*: Proposition 5.12 and 5.15 and item (b).

If $\{\mathcal{M}_\gamma\}_{\gamma\in\Gamma}$ is a nonempty family of orthogonal subspaces of $\mathcal{H}$, then

(e) $\mathcal{H} = \left(\sum_{\gamma\in\Gamma}\mathcal{M}_\gamma\right)^-$ implies $\mathcal{M}_\alpha = \bigcap_{\alpha\neq\gamma\in\Gamma}\mathcal{M}_\gamma^\perp$ for every $\alpha\in\Gamma$.

> *Hint*: Take an arbitrary $\alpha\in\Gamma$. Show: $\mathcal{M}_\alpha \perp \sum_{\alpha\neq\gamma\in\Gamma}\mathcal{M}_\gamma$, and hence $\mathcal{M}_\alpha \perp \left(\sum_{\alpha\neq\gamma\in\Gamma}\mathcal{M}_\gamma\right)^-$ (cf. Proposition 5.12). Verify each of the following steps:

$$\left(\sum_{\gamma\in\Gamma}\mathcal{M}_\gamma\right)^- = \bigvee\left(\mathcal{M}_\alpha\cup\bigcup_{\alpha\neq\gamma\in\Gamma}\mathcal{M}_\gamma\right) \subseteq \bigvee\left(\mathcal{M}_\alpha\cup\bigvee\left(\bigcup_{\alpha\neq\gamma\in\Gamma}\mathcal{M}_\gamma\right)\right)$$

$$= \mathcal{M}_\alpha + \left(\sum_{\alpha\neq\gamma\in\Gamma}\mathcal{M}_\gamma\right)^- \subseteq \mathcal{H}$$

> (see Theorem 5.10(b)). Then $\mathcal{H} = \mathcal{M}_\alpha + \left(\sum_{\alpha\neq\gamma\in\Gamma}\mathcal{M}_\gamma\right)^-$. Conclude that $\mathcal{M}_\alpha^\perp = \left(\sum_{\alpha\neq\gamma\in\Gamma}\mathcal{M}_\gamma\right)^-$ by Proposition 5.19, and therefore $\mathcal{M}_\alpha = \left(\sum_{\alpha\neq\gamma\in\Gamma}\mathcal{M}_\gamma\right)^\perp$ (Propositions 5.12 and 5.15). Apply (b).

Remark: Recall from Section 4.3 that the collection $\mathrm{Lat}(\mathcal{H})$ of all subspaces of $\mathcal{H}$ is a complete lattice in the inclusion ordering, where $\mathcal{M}\wedge\mathcal{N} = \mathcal{M}\cap\mathcal{N}$ and $\mathcal{M}\vee\mathcal{N} = (\mathcal{M}+\mathcal{N})^-$. Therefore, according to item (b) and Proposition 5.15, $\mathcal{M}\vee\mathcal{N} = (\mathcal{M}^\perp\cap\mathcal{N}^\perp)^\perp$.

**Problem 5.9.** Let $\mathcal{X}$ and $\mathcal{Y}$ be inner product spaces and consider the setup of Problem 4.42. If there exists a unitary transformation $U\in\mathcal{G}[\mathcal{X},\mathcal{Y}]$ intertwining $T\in\mathcal{B}[\mathcal{X}]$ and $S\in\mathcal{B}[\mathcal{Y}]$; that is, if

$$UT = SU,$$

then we say that $T$ and $S$ are *unitarily equivalent* (notation: $T\cong S$). If this happens, then it is clear that $\mathcal{X}$ and $\mathcal{Y}$ are unitarily equivalent inner product spaces. As in the case of similarity (on linear spaces) or isometric equivalence (on normed spaces), show that unitary equivalence has the defining properties of an equivalence relation. Now take an arbitrary polynomial $p$ in one variable (cf. Problem 2.20) and show that $Up(T) = p(S)U$ whenever $UT = SU$ for every unitary $U \in \mathcal{G}[\mathcal{X}, \mathcal{Y}]$. In particular, $T \cong S$ implies $T^n \cong S^n$ for every positive integer $n$.

**Problem 5.10.** This is the uncountable version of the Orthogonal Structure Theorem. Prove it.

(a) Let $\mathcal{H}$ be a Hilbert space. Suppose $\{\mathcal{M}_\gamma\}_{\gamma \in \Gamma}$ is an uncountable family of pairwise orthogonal subspaces of $\mathcal{H}$. If $x \in \left(\sum_{\gamma \in \Gamma} \mathcal{M}_\gamma\right)^-$, then there exists a unique summable family $\{u_\gamma\}_{\gamma \in \Gamma}$ of vectors in $\mathcal{H}$ with $u_\gamma \in \mathcal{M}_\gamma$ for each $\gamma$ such that $x = \sum_{\gamma \in \Gamma} u_\gamma$. Moreover, $\|x\|^2 = \sum_{\gamma \in \Gamma} \|u_\gamma\|^2$. Conversely, if $\{u_\gamma\}_{\gamma \in \Gamma}$ is a square-summable family of vectors in $\mathcal{H}$ with $u_\gamma \in \mathcal{M}_\gamma$ for each $\gamma \in \Gamma$, then $\{u_\gamma\}_{\gamma \in \Gamma}$ is summable and $\sum_{\gamma \in \Gamma} u_\gamma$ lies in $\left(\sum_{\gamma \in \Gamma} \mathcal{M}_\gamma\right)^-$.

*Hint*: Set $x(n) = \sum_{k \in N_n} x(n)_k$ in $\sum_{k \in N_n} \mathcal{M}_k$, where each $N_n$ is a finite subset of $\Gamma$ such that $N_n \subseteq N_{n+1}$ and $\#N_n \geq n$. Consider the set $N_\infty = \bigcup_{n \geq 1} N_n$ and construct a countable family of orthogonal vectors in $\mathcal{H}$, say $\{u_k\}_{k \in N_\infty}$, as in the proof of Theorem 5.16.

(b) Show that $\mathcal{H}$ must be nonseparable if $\{\mathcal{M}_\gamma\}_{\gamma \in \Gamma}$ has uncountably many nonzero subspaces. (*Hint*: Proposition 5.43 and the Axiom of Choice.) Even in this case, the sum $x = \sum_{\gamma \in \Gamma} u_\gamma$ has only a countable number of nonzero vectors. (Why?)

(c) Restate Corollaries 5.17 and 5.18 in light of item (a). Now rewrite Examples 5I and 5J for an uncountable family $\{\mathcal{M}_\gamma\}_{\gamma \in \Gamma}$ of pairwise orthogonal subspaces of a Hilbert space $(\mathcal{H}, \langle\ ;\ \rangle)$ and conclude that $\bigoplus_{\gamma \in \Gamma} \mathcal{M}_\gamma \cong \left(\sum_{\gamma \in \Gamma} \mathcal{M}_\gamma\right)^-$.

**Problem 5.11.** Let $\{\mathcal{M}_\gamma\}$ be a collection of orthogonal subspaces of a Hilbert space $\mathcal{H}$ that spans $\mathcal{H}$ (i.e., $\bigvee_\gamma \mathcal{M}_\gamma = \left(\sum_\gamma \mathcal{M}_\gamma\right)^- = \mathcal{H}$) and let $B_\gamma$ be an orthonormal basis for each $\mathcal{M}_\gamma$. Show that $\bigcup_\gamma B_\gamma$ is an orthonormal basis for $\mathcal{H}$.

*Hint*: $\bigcup_\gamma B_\gamma$ is an orthonormal set in $\mathcal{H}$ and $\bigcup_\gamma (\bigvee B_\gamma) \subseteq \bigvee(\bigcup_\gamma B_\gamma)$ (cf. Sections 2.3, 3.5 and 4.3). Also verify each of the following steps. $\left(\sum_\gamma \mathcal{M}_\gamma\right)^- = \bigvee(\bigcup_\gamma \mathcal{M}_\gamma) = \bigvee(\bigcup_\gamma \bigvee B_\gamma) \subseteq \bigvee(\bigvee(\bigcup_\gamma B_\gamma)) = \bigvee(\bigcup_\gamma B_\gamma) \subseteq \mathcal{H}$.

**Problem 5.12.** Completeness is necessary in Theorem 5.10(b): *If $\mathcal{M}$ and $\mathcal{N}$ are orthogonal subspaces of an incomplete inner product space $\mathcal{X}$, then it may happen that $\mathcal{M} + \mathcal{N}$ is not closed in $\mathcal{X}$.* Indeed, set

$$\mathcal{X} = \operatorname{span}\{e_k\}_{k=0}^\infty,$$

where $e_0 = \{\frac{1}{i}\}_{i=1}^\infty \in \ell_+^2$ and $\{e_k\}_{k=1}^\infty$ is the canonical orthonormal basis for the Hilbert space $\ell_+^2$ (cf. Example 5L(b)). Recall that $\mathcal{X}$ is the linear manifold of $\ell_+^2$

consisting of all (finite) linear combinations of vectors from $\{e_k\}_{k=0}^{\infty}$. Now consider the following linear manifolds of $\mathcal{X}$.

$$
\begin{aligned}
\mathcal{M} &= \left\{ u \in \{v_k\}_{k=1}^{\infty} \in \mathcal{X}: \ v_{2k-1} = 0 \ \text{ for all } \ k \geq 1 \right\}, \\
\mathcal{N} &= \left\{ v \in \{v_k\}_{k=1}^{\infty} \in \mathcal{X}: \ v_{2k} = 0 \ \text{ for all } \ k \geq 1 \right\}.
\end{aligned}
$$

It is clear that $\mathcal{M} \perp \mathcal{N}$, isn't it? Moreover, they are subspaces of $\mathcal{X}$.

(a)  Show that $\mathcal{M}$ and $\mathcal{N}$ are both closed in the inner product space $\mathcal{X}$.

*Hint*: Take a sequence $\{u_n\}_{n=1}^{\infty}$, with each $u_n = \{v_{k(n)}\}_{k=1}^{\infty}$ in $\mathcal{M}$, such that $u_n \to x = \{\xi_k\}_{k=1}^{\infty}$ in $\mathcal{X}$. Split the series $\|u_n - x\|^2 = \sum_{k=1}^{\infty} |v_{k(n)} - \xi_k|^2$ into the sum of a series running over the odd integers and a series running over the even integers and conclude that $\|u_n - x\|^2 = \sum_{k=1}^{\infty} |\xi_{2k-1}|^2 + \sum_{k=1}^{\infty} |v_{2k(n)} - \xi_{2k}|^2$. This implies that $x \in \mathcal{M}$.

(b)  Exhibit an $\mathcal{M}+\mathcal{N}$-valued sequence that converges in $\mathcal{X}$ to $e_0 \in \mathcal{X}$.

*Hint*: For each $n \geq 1$ define $u_n = \{v_{k(n)}\}_{k=1}^{\infty}$ and $v_n = \{v_{k(n)}\}_{k=1}^{\infty}$ as follows. $v_{k(n)} = \frac{1}{k}$ if $1 \leq k \leq n$ is an even integer, and $v_{k(n)} = 0$ otherwise. $v_{k(n)} = \frac{1}{k}$ if $1 \leq k \leq n$ is an odd integer, and $v_{k(n)} = 0$ otherwise. Show that $u_n, v_n \in \operatorname{span}\{e_k\}_{k=1}^{n} \subset \mathcal{X}, u_n \in \mathcal{M}, v_n \in \mathcal{N}$, and $u_n + v_n = (1, \frac{1}{2}, \ldots, \frac{1}{n}, 0, 0, 0, \cdots)$ for every $n \geq 1$.

(c)  Show (by contradiction) that $e_0 \notin \mathcal{M} + \mathcal{N}$.

Conclude from (b) and (c) that $\mathcal{M} + \mathcal{N}$ is not closed in $\mathcal{X}$.

**Problem 5.13.**  Orthogonality is necessary in Theorem 5.10(b): *If $\mathcal{M}$ and $\mathcal{N}$ are nonorthogonal subspaces of a Hilbert space $\mathcal{H}$, then it may happen that $\mathcal{M} + \mathcal{N}$ is not closed in $\mathcal{H}$ even if $\mathcal{M} \cap \mathcal{N} = \{0\}$.* Set

$$
v_k = \left(1 - \tfrac{1}{k^2}\right)^{\frac{1}{2}} e_{2k-1} + \tfrac{1}{k} e_{2k} \quad \text{in} \quad \ell_+^2
$$

for every $k \geq 1$, where $\{e_k\}_{k=1}^{\infty}$ is the canonical orthonormal basis for the Hilbert space $\ell_+^2$, and consider the following subspaces of $\ell_+^2$.

$$
\mathcal{M} = \bigvee \{e_{2k-1}\}_{k=1}^{\infty} \quad \text{and} \quad \mathcal{N} = \bigvee \{v_k\}_{k=1}^{\infty}.
$$

(a)  Show that $\{v_k\}_{k=1}^{\infty}$ is an orthonormal sequence, and hence an orthonormal basis for the Hilbert space $\mathcal{N}$.

(b)  Apply the Fourier Series Theorem to show that $\mathcal{M} \cap \mathcal{N} = \{0\}$.

*Hint*: Take $x \in \mathcal{M} \cap \mathcal{N}$ and consider its Fourier series expansions: $x = \sum_{k=1}^{\infty} \langle x ; e_{2k-1} \rangle e_{2k-1} = \sum_{k=1}^{\infty} \langle x ; v_k \rangle v_k$. Now verify: $0 = \langle x ; e_{2j} \rangle = \sum_{k=1}^{\infty} \langle x ; v_k \rangle \langle v_k ; e_{2j} \rangle = \frac{1}{j} \langle x ; v_j \rangle$ for all $j \geq 1$. Conclude that $x = 0$.

Show that the series $\sum_{k=1}^{\infty} \frac{1}{k} e_{2k}$ converges in $\ell_+^2$ and set $x = \sum_{k=1}^{\infty} \frac{1}{k} e_{2k}$ in $\ell_+^2$. Note that $x_n = \sum_{k=1}^{n} \frac{1}{k} e_{2k} = \sum_{k=1}^{n} v_k - \sum_{k=1}^{n}(1 - \frac{1}{k^2})^2 e_{2k-1}$ lies in $\mathcal{M} + \mathcal{N}$ for each $n \geq 1$. Moreover, check that $x_n \to x$ in $\ell_+^2$ as $n \to \infty$.

(c) Apply the Fourier Series Theorem to show that $x \notin \mathcal{M} + \mathcal{N}$.

*Hint*: Suppose $x = u + v$ with $u \in \mathcal{M}$ and $v \in \mathcal{N}$. Verify that $\langle x \, ; e_{2n} \rangle = \frac{1}{n}$, $\langle u \, ; e_{2n} \rangle = 0$, and $\langle v \, ; e_{2n} \rangle = \langle v \, ; v_n \rangle \frac{1}{n}$ for each $n \geq 1$. Now conclude that $\frac{1}{n} \langle v \, ; v_n \rangle = \frac{1}{n}$, and hence $\langle v \, ; v_n \rangle = 1$ for every $n \geq 1$, which is a contradiction (since $\|v\|^2 = \sum_{n=1}^{\infty} |\langle v \, ; v_n \rangle|^2$).

Therefore, $\mathcal{M} + \mathcal{N}$ is not closed in $\ell_+^2$ by the Closed Set Theorem.

**Problem 5.14.** Let $\{e_\gamma\}_{\gamma \in \Gamma}$ be any orthonormal basis for a complex Hilbert space $\mathcal{H}$. For each $x \in \mathcal{H}$ consider its Fourier series expansion with respect to $\{e_\gamma\}_{\gamma \in \Gamma}$; that is, $x = \sum_{\gamma \in \Gamma} \langle x \, ; e_\gamma \rangle e_\gamma$. Verify that $\{\mathrm{Re} \, \langle x \, ; e_\gamma \rangle e_\gamma\}_{\gamma \in \Gamma}$ and $\{\mathrm{Im} \, \langle x \, ; e_\gamma \rangle e_\gamma\}_{\gamma \in \Gamma}$ are both summable families of vectors in $\mathcal{H}$. Set

$$\mathrm{Re} \, x = \sum_{\gamma \in \Gamma} \mathrm{Re} \, \langle x \, ; e_\gamma \rangle e_\gamma \quad \text{and} \quad \mathrm{Im} \, x = \sum_{\gamma \in \Gamma} \mathrm{Im} \, \langle x \, ; e_\gamma \rangle e_\gamma$$

in $\mathcal{H}$. Show that $x = \mathrm{Re} \, x + i \, \mathrm{Im} \, x$ and prove the following identities.

(a) $\langle \mathrm{Re} \, x \, ; e_\gamma \rangle = \mathrm{Re} \, \langle x \, ; e_\gamma \rangle$ and $\langle \mathrm{Im} \, x \, ; e_\gamma \rangle = \mathrm{Im} \, \langle x \, ; e_\gamma \rangle$ for every $\gamma \in \Gamma$.

(b) $\langle \mathrm{Re} \, x \, ; \mathrm{Im} \, x \rangle = \langle \mathrm{Im} \, x \, ; \mathrm{Re} \, x \rangle$ and $\|x\|^2 = \|\mathrm{Re} \, x\|^2 + \|\mathrm{Im} \, x\|^2$.

(c) $\mathrm{Re}(\alpha x) = \mathrm{Re} \, \alpha \, \mathrm{Re} \, x - \mathrm{Im} \, \alpha \, \mathrm{Im} \, x$ and $\mathrm{Im}(\alpha x) = \mathrm{Re} \, \alpha \, \mathrm{Im} \, x - \mathrm{Im} \, \alpha \mathrm{Re} \, x$.

(d) $\langle \mathrm{Re}(\alpha x) \, ; \mathrm{Im}(\alpha x) \rangle =$
$\mathrm{Re} \, \alpha \, \mathrm{Im} \, \alpha \big(\|\mathrm{Re} \, x\|^2 - \|\mathrm{Im} \, x\|^2\big) + \langle \mathrm{Re} \, x \, ; \mathrm{Im} \, x \rangle \big((\mathrm{Re} \, \alpha)^2 - (\mathrm{Im} \, \alpha)^2\big)$.

Use the above results (and the polar representation for $\alpha \in \mathbb{C}$) to prove the Orthogonal Normalization Lemma: *For every nonzero vector $x$ in a complex Hilbert space there exists an $\alpha \in \mathbb{C}$ such that $\|\alpha x\| = 1$ and $\mathrm{Re}(\alpha x) \perp \mathrm{Im}(\alpha x)$.*

**Problem 5.15.** Let $\{e_\gamma\}_{\gamma \in \Gamma}$ be an orthonormal family in an infinite-dimensional Hilbert space $\mathcal{H}$. Let $\{e_k\}_{k=1}^{\infty}$ be a countably infinite subset of $\{e_\gamma\}_{\gamma \in \Gamma}$, set $\mathcal{M} = \bigvee \{e_k\}_{k=1}^{\infty}$ and, for each positive integer $n$, set $\mathcal{M}_n = \mathrm{span} \, \{e_k\}_{k=1}^{n}$. Let $P_n \colon \mathcal{H} \to \mathcal{H}$ be the orthogonal projection onto the finite-dimensional subspace $\mathcal{M}_n$. Show that $P_n \xrightarrow{s} P$, where $P \colon \mathcal{H} \to \mathcal{H}$ is the orthogonal projection onto $\mathcal{M}$. (*Hint*: Corollary 5.55.)

**Problem 5.16.** Let $\{P_k\}_{k=0}^{\infty}$ be a resolution of the identity on a Hilbert space $\mathcal{H}$ so that $\sum_{k=0}^{\infty} P_k x = x$ for every $x \in \mathcal{H}$; that is, $\sum_{k=0}^{n} P_k \xrightarrow{s} I$ as $n \to \infty$. Suppose $P_k \neq O$ for every $k \geq 0$ and let $\{\lambda_k\}_{k=0}^{\infty}$ be a bounded sequence of scalars. According to Proposition 5.61 the weighted sum of projections $\sum_{k=0}^{\infty}\lambda_k P_k$ is the operator on $\mathcal{H}$ for which $\sum_{k=0}^{n}\lambda_k P_k \overset{s}{\longrightarrow} \sum_{k=0}^{\infty}\lambda_k P_k$ as $n \to \infty$. Theorem 5.59 says that $\mathcal{H} = \left(\sum_{k=0}^{\infty}\mathcal{R}(P_k)\right)^{-}$, and $\left(\sum_{k=0}^{\infty}\mathcal{R}(P_k)\right)^{-}$ is unitarily equivalent to the orthogonal direct sum $\bigoplus_{k=0}^{\infty}\mathcal{R}(P_k)$ equipped with its usual inner product (see Examples 5I and 5J). Moreover, the natural unitary transformation $\Phi$ of the Hilbert space $\bigoplus_{k=0}^{\infty}\mathcal{R}(P_k)$ onto the Hilbert space $\left(\sum_{k=0}^{\infty}\mathcal{R}(P_k)\right)^{-}$ is given by $\Phi(\{u_k\}_{k=0}^{\infty}) = \sum_{k=0}^{\infty}u_k$ for every $\{u_k\}_{k=0}^{\infty}$ in $\bigoplus_{k=0}^{\infty}\mathcal{R}(P_k)$. Set $I_k = P_k|_{\mathcal{R}(P_k)}$ in $\mathcal{B}[\mathcal{R}(P_k)]$ for each $k$ and consider the operator $\bigoplus_{k=0}^{\infty}\lambda_k I_k$ in $\mathcal{B}[\bigoplus_{k=0}^{\infty}\mathcal{R}(P_k)]$ (cf. Problem 4.16): $\left(\bigoplus_{k=0}^{\infty}\lambda_k I_k\right)\{u_k\}_{k=0}^{\infty} = \{\lambda_k u_k\}_{k=0}^{\infty}$ for every $\{u_k\}_{k=0}^{\infty}$ in $\bigoplus_{k=0}^{\infty}\mathcal{R}(P_k)$. Show that

$$\Phi\left(\bigoplus_{k=0}^{\infty}\lambda_k I_k\right)\Phi^{-1}x = \left(\sum_{k=0}^{\infty}\lambda_k P_k\right)x$$

for all $x$ in $\mathcal{H}$. (*Hint:* $\Phi^{-1}\left(\sum_{k=0}^{\infty}P_k x\right) = \{P_k x\}_{k=0}^{\infty}$.) In other words, as the orthogonal projections $\{P_k\}_{k=0}^{\infty}$ are orthogonal to each other, a weighted sum of projections is identified with an orthogonal sum of scalar operators. In fact, these are unitarily equivalent operators:

$$\sum_{k=0}^{\infty}\lambda_k P_k \cong \bigoplus_{k=0}^{\infty}\lambda_k I_k \quad \text{with} \quad I_k = P_k|_{\mathcal{R}(P_k)}.$$

**Problem 5.17.** Let $\{e_k\}_{k=1}^{\infty}$ be an orthonormal basis for a (separable infinite-dimensional) Hilbert space $\mathcal{H}$ and let $\{\lambda_k\}_{k=1}^{\infty}$ be a sequence of scalars. For each $n \geq 1$ consider the mapping $T_n : \mathcal{H} \to \mathcal{H}$ defined by

$$T_n x = \sum_{k=1}^{n}\lambda_k \langle x \, ; e_k\rangle e_k \quad \text{for every} \quad x \in \mathcal{H}.$$

Verify that $T_n$ lies in $\mathcal{B}[\mathcal{H}]$ for each positive integer $n$ and show that

$$\sup_{k}|\lambda_k| < \infty \quad \text{if and only if} \quad T_n \overset{s}{\longrightarrow} T,$$

where $T \in \mathcal{B}[\mathcal{H}]$ is the weighted sum of projections

$$T x = \sum_{k=1}^{\infty}\lambda_k \langle x \, ; e_k\rangle e_k \quad \text{for every} \quad x = \sum_{k=1}^{\infty}\langle x \, ; e_k\rangle e_k \in \mathcal{H}$$

(cf. Theorem 5.48, Proposition 5.57, Definition 5.60, Proposition 5.61). Also check that and $\|T\| = \sup_{k}|\lambda_k|$. In this case, $T$ is called a *diagonal operator with respect to the basis* $\{e_k\}_{k=1}^{\infty}$. Conversely, take any $T \in \mathcal{B}[\mathcal{H}]$. If there exists an orthogonal basis for $\mathcal{H}$ and a bounded sequence of scalars such that $T$ is the strong limit of $\{T_n\}_{n=1}^{\infty}$, then $T$ is a *diagonalizable operator*. Suppose $\sup_{k}|\lambda_k| < \infty$. Show that

$$\lim_{k}\lambda_k = 0 \quad \text{if and only if} \quad T_n \overset{u}{\longrightarrow} T$$

(cf. Problem 4.53) and prove the following assertion (see Example 4J.)

There exists $T^{-1} \in \mathcal{L}[\mathcal{R}(T), \mathcal{H}]$ if and only if $\lambda_k \neq 0$ for every $k \geq 1$

and, in this case, $\mathcal{R}(T)^- = \mathcal{H}$. (*Hint*: span $\{e_k\}_{k=1}^\infty \subseteq \mathcal{R}(T)$ if $\lambda_k \neq 0$ for all $k$ — see Problem 3.47.) Finally, prove the following proposition.

There exists $T^{-1} \in \mathcal{B}[\mathcal{H}]$ if and only if $\inf_k |\lambda_k| > 0$

(see Example 4J again) and, in this case,

$$T^{-1}x = \sum_{k=1}^\infty \lambda_k^{-1} \langle x ; e_k \rangle e_k \quad \text{for every} \quad x \in \mathcal{H}.$$

**Problem 5.18.** Consider the setup of the previous problem. Use the Fourier expansion of $x \in \mathcal{H}$ to show by induction that

$$T^n x = \sum_{k=1}^\infty \lambda_k^n \langle x ; e_k \rangle e_k \quad \text{for every} \quad x \in \mathcal{H}$$

and every positive integer $n$. Now prove the following propositions.

(a) $T^n \xrightarrow{\;u\;} 0$ if and only if $\sup_k |\lambda_k| < 1$.

(b) $T^n \xrightarrow{\;s\;} 0$ if and only if $|\lambda_k| < 1$ for every $k \geq 1$.

(c) $T^n \xrightarrow{\;w\;} 0$ if and only if $T^n \xrightarrow{\;s\;} 0$.

*Hint*: For (a) and (b) see Example 4H. For (c) note that $T^n e_j = \lambda_j^n e_j$, and hence $|\langle T^n e_j ; e_j \rangle| = |\lambda_j|^n$. If $|\lambda_j| \geq 1$ for some $j$, then $T^n \xrightarrow{\;w\;} \!\!\!\!\!\!/\; 0$.

**Problem 5.19.** Let $\{e_k\}_{k=1}^\infty$ be an orthonormal basis for a Hilbert space $\mathcal{H}$. Show that $\mathcal{M}$ (defined below) is a dense linear manifold of $\mathcal{H}$.

$$\mathcal{M} = \{x \in \mathcal{H} : \; \textstyle\sum_{k=1}^\infty |\langle x ; e_k \rangle| < \infty\}.$$

*Hint*: Let $T$ be a diagonal operator as in Problem 5.17 with $\lambda_k \neq 0$ for all $k$ (so that $\mathcal{R}(T)^- = \mathcal{H}$) and $\sum_{k=1}^\infty |\lambda_k|^2 < \infty$. Use the Schwarz inequality in $\ell_+^2$ to verify that $\sum_{j=1}^\infty |\langle Tx ; e_j \rangle| = \sum_{j=1}^\infty |\lambda_j| |\langle x ; e_j \rangle| \leq (\sum_{j=1}^\infty |\lambda_j|^2)^{\frac{1}{2}} (\sum_{j=1}^\infty |\langle x ; e_j \rangle|^2)^{\frac{1}{2}} < \infty$ for all $x \in \mathcal{H}$. Hence $\mathcal{R}(T) \subseteq \mathcal{M}$.

**Problem 5.20.** Let $\{x_n\}$ be an $\mathcal{M}$-valued sequence and let $x$ be a vector in $\mathcal{M}$, where $\mathcal{M}$ is a subspace of a Hilbert space $\mathcal{H}$. Show that

$$x_n \xrightarrow{\;w\;} x \; \text{ in } \; \mathcal{M} \quad \text{if and only if} \quad x_n \xrightarrow{\;w\;} x \; \text{ in } \; \mathcal{H}.$$

*Hint*: $x_n \xrightarrow{w} x$ in $\mathcal{M}$ if and only if $\langle x_n \, ; u \rangle \to \langle x \, ; u \rangle$ for every $u \in \mathcal{M}$ because $\mathcal{M}$ is a Hilbert space (Theorem 5.62). Recall: $\mathcal{H} = \mathcal{M} + \mathcal{M}^{\perp}$ (Theorem 5.20), and show that $x_n \xrightarrow{w} x$ in $\mathcal{M}$ implies $x_n \xrightarrow{w} x$ in $\mathcal{H}$.

**Problem 5.21.** Let $\{T_n\}$ be a sequence in $\mathcal{B}[\mathcal{X}, \mathcal{Y}]$ where $\mathcal{X}$ and $\mathcal{Y}$ are inner product spaces. Prove the following propositions.

(a) If $T_n \xrightarrow{w} T$ for some $T \in \mathcal{B}[\mathcal{X}, \mathcal{Y}]$, then $\|T\| \leq \liminf_n \|T_n\|$.

    *Hint*: Let $\{T_{n_k}\}$ be a subsequence of $\{T_n\}$ such that $\lim_k \|T_{n_k}\| = \liminf_n \|T_n\|$. If $T_n \xrightarrow{w} T$, then $\langle Tx \, ; y \rangle = \lim_k \langle T_{n_k} x \, ; y \rangle$ (why?) and hence $|\langle Tx \, ; y \rangle| \leq \liminf_n \|T_n\| \, \|x\| \, \|y\|$, for every $x, y \in \mathcal{H}$. Now apply Corollary 5.71.

(b) If $\sup_n \|T_n\| < \infty$ and $\{\langle T_n a \, ; b \rangle\}$ converges in $\mathbb{F}$ for every $a$ in a dense subset $A$ of $\mathcal{X}$ and every $b$ in a dense subset $B$ of $\mathcal{Y}$, then $\{\langle T_n x \, ; y \rangle\}$ converges in $\mathbb{F}$ for every $x \in \mathcal{X}$ and every $y \in \mathcal{Y}$.

    *Hint*: See the hint in Problem 4.45(b) and recall that $\mathbb{F}$ is complete.

(c) Take $T \in \mathcal{B}[\mathcal{X}, \mathcal{Y}]$. If $\sup_n \|T_n\| < \infty$ and $\langle T_n a \, ; b \rangle \to \langle Ta \, ; b \rangle$ for every $a$ in a dense subset $A$ of $\mathcal{X}$ and every $b$ in a dense subset $B$ of $\mathcal{Y}$, then $\langle T_n x \, ; y \rangle \to \langle Tx \, ; y \rangle$ for every $x \in \mathcal{X}$ and every $y \in \mathcal{Y}$.

    *Hint*: $\langle (T_n - T)x \, ; y \rangle = \langle (T_n - T)(x - a_\varepsilon) + (T_n - T)a_\varepsilon \, ; y - b_\varepsilon + b_\varepsilon \rangle$.

(d) If $\mathcal{X}$ and $\mathcal{Y}$ are Hilbert spaces, and if the hypothesis of (b) or (c) holds true, then $T_n \xrightarrow{w} T$.

**Problem 5.22.** Let $\{T_n\}$ be a sequence in $\mathcal{B}[\mathcal{X}, \mathcal{Y}]$ and let $\{S_n\}$ be a sequence in $\mathcal{B}[\mathcal{Y}, \mathcal{Z}]$, where $\mathcal{X}$, $\mathcal{Y}$ and $\mathcal{Z}$ are inner product spaces. Take $T \in \mathcal{B}[\mathcal{X}, \mathcal{Y}]$, $S \in \mathcal{B}[\mathcal{Y}, \mathcal{Z}]$, and prove the following propositions.

(a) If $\sup_n \|S_n\| < \infty$, $S_n \xrightarrow{w} S$ and $T_n \xrightarrow{s} T$, then $S_n T_n \xrightarrow{w} ST$.

    *Hint*: $S_n T_n - ST = S_n(T_n - T) + (S_n - S)T$.

(b) If $\mathcal{Y}$ and $\mathcal{Z}$ are Hilbert spaces, $S_n \xrightarrow{w} S$ and $T_n \xrightarrow{s} T$, then $S_n T_n \xrightarrow{w} ST$.

(c) If $S_n \xrightarrow{u} S$, $\sup_n \|T_n\| < \infty$, and $T_n \xrightarrow{w} T$, then $S_n T_n \xrightarrow{w} ST$.

    *Hint*: $S_n T_n - ST = (S_n - S)T_n + S(T_n - T)$.

(d) If $\mathcal{X}$ and $\mathcal{Y}$ are Hilbert spaces, $S_n \xrightarrow{u} S$ and $T_n \xrightarrow{w} T$, then $S_n T_n \xrightarrow{w} ST$.

Show that addition of weakly convergent sequences of bounded linear transformations is again a weakly convergent sequence of bounded linear transformations whose limit is the sum of the limits of each summand.

Remarks: It is easy to show, even if $\mathcal{X} = \mathcal{Y} = \mathcal{Z}$ is a Hilbert space, that

$$S_n \xrightarrow{w} S \quad \text{and} \quad T_n \xrightarrow{u} T \quad \text{does not imply} \quad S_n T_n \xrightarrow{s} ST.$$

For instance, if $T_n = I$ for all $n$ and $\{S_n\}$ is any operator sequence that converge weakly to zero but does not converges strongly, then $S_n T_n = S_n \xrightarrow{s}\!\!\!\!/\; O = OI$. It is also easy to show that

$$S_n \xrightarrow{s} S \quad \text{and} \quad T_n \xrightarrow{w} T \quad \text{does not imply} \quad S_n T_n \xrightarrow{w} ST.$$

Samples: There exists an isometry $S_+$ (thus $S_+^{*n} S_+^n = I$ and $\|S_+^n x\| = 1$ for all $n$ and $x$, so that $S_+^{*n} S^n \xrightarrow{w}\!\!\!\!/\; O$ and $S_+^n \xrightarrow{s}\!\!\!\!/\; O$) for which $S_+^{*n} \xrightarrow{s} O$ (and hence $S_+^{*n} \xrightarrow{w} O$ and $S_+^n \xrightarrow{w} O$) — see Problem 5.29(b,c) below.

**Problem 5.23.** Let $\{e_k\}_{k=1}^\infty$ be any orthonormal basis for a separable Hilbert space $\mathcal{H}$. Prove the following proposition. A sequence $\{T_n\}$ of operators in $\mathcal{B}[\mathcal{H}]$ converges weakly to $T \in \mathcal{B}[\mathcal{H}]$ if and only if

$$\sup_n \|T^n\| < \infty \quad \text{and} \quad \langle T_n e_j \,;\, e_k \rangle \to \langle T e_j \,;\, e_k \rangle \quad \text{as } n \to \infty \text{ for every } j, k \geq 1.$$

*Hint*: The "only if" part is easy (cf. Problems 5.5 and Proposition 5.67). Conversely, suppose $\sup_n \|T^n\| < \infty$ and $\lim_n \langle (T_n - T)e_j \,;\, e_k \rangle = 0$ for every $j, k \geq 1$. Use Theorem 5.48 and Problem 4.14(b) to show that

$$\limsup_n |\langle (T_n - T)e_j \,;\, y \rangle| \leq \limsup_n \sum_{k=1}^\infty |\langle (T_n - T)e_j \,;\, e_k \rangle| |\langle e_k \,;\, y \rangle| = 0$$

for each $j \geq 1$ and every $y \in \mathcal{M}$, where $\mathcal{M}$ is the linear manifold of Problem 5.19. Repeat the argument to show:

$$\limsup_n |\langle (T_n - T)x \,;\, y \rangle| = \limsup_n |\langle x \,;\, (T_n - T)^* y \rangle|$$

$$< \limsup_n \sum_{k=1}^\infty |\langle x \,;\, e_k \rangle| |\langle (T_n - T)e_k \,;\, y \rangle| = 0,$$

so that $\lim_n |\langle (T_n - T)x \,;\, y \rangle| = 0$, for every $x, y \in \mathcal{M}$. Now use Problems 5.19 and 5.21(d) to conclude that $T_n \xrightarrow{w} T$.

**Problem 5.24.** Let $\mathcal{X}$ be a linear space and take any $L \in \mathcal{L}[\mathcal{X}]$. Recall that the $n$th power of $L$, $L^n$, is the composition of $L$ with itself $n$ times. By setting $L^0 = I$, the power sequence $\{L^n\}_{n \geq 0}$ can be recursively defined by $L^{n+1} = LL^n$ for every $n \geq 0$. Show by induction:

(a)  $L^{n+k} = L^n L^k = L^k L^n$ and $(L^n)^k = (L^k)^n = L^{nk}$ for every $k, n \geq 0$.

Now let $\mathcal{X}$ be an inner product space and take any power bounded operator $T$ in $\mathcal{B}[\mathcal{X}]$ (i.e., $\sup_n \|T^n\| < \infty$). If $T^n \xrightarrow{w} P$ for some operator $P \in \mathcal{B}[\mathcal{X}]$, then use Problem 5.22 to show that

(b)  $PT^k = T^k P = P = P^k$ for every $k \geq 1$, so that $P$ is a projection

(not necessarily an orthogonal projection). Show by induction that

(c)  $(T - P)^n = T^n - P$ for every $n \geq 1$, so that $(T - P)^n \xrightarrow{w} O$.

If $\mathcal{X}$ is a Hilbert space and $T \in \mathcal{B}[\mathcal{X}]$, then show that

(d)  $T^{n*} = T^{*n}$ and $\|T^{*n}\| = \|T^n\|$ for every $n \geq 0$.

**Problem 5.25.** Let $\mathcal{X}$ and $\mathcal{Y}$ be two normed spaces. Recall that $T \in \mathcal{B}[\mathcal{X}]$ and $S \in \mathcal{B}[\mathcal{Y}]$ are similar if there exists $W \in \mathcal{G}[\mathcal{X}, \mathcal{Y}]$ such that $WT = SW$ (i.e., $T = W^{-1}SW$ — see Problem 4.42). Consider a sequence $\{T_n\}$ of operators in $\mathcal{B}[\mathcal{X}]$ and a sequence $\{S_n\}$ of operators in $\mathcal{B}[\mathcal{Y}]$. Suppose there exists $W \in \mathcal{G}[\mathcal{X}, \mathcal{Y}]$ such that $WT_n = S_n W$ for every integer $n$. Show that

$$S_n \xrightarrow{u} S \quad \text{implies} \quad T_n \xrightarrow{u} W^{-1}SW,$$
$$S_n \xrightarrow{s} S \quad \text{implies} \quad T_n \xrightarrow{s} W^{-1}SW.$$

In particular, if $WT = SW$, and if $S^n \xrightarrow{s} P$ (or $S^n \xrightarrow{u} P$) for some $P \in \mathcal{B}[\mathcal{Y}]$, then show that $T^n \xrightarrow{s} W^{-1}PW$ (or $T^n \xrightarrow{u} W^{-1}PW$) and $W^{-1}PW$ is a projection in $\mathcal{B}[\mathcal{X}]$ (cf. Problem 4.55(c)).

*Hint:* $T_n - W^{-1}S_n W = W^{-1}(S_n - S)W$.

Now let $\mathcal{X}$ and $\mathcal{Y}$ be two Hilbert spaces. Recall that $T \in \mathcal{B}[\mathcal{X}]$ and $S \in \mathcal{B}[\mathcal{Y}]$ are unitarily equivalent if there exists a unitary $U \in \mathcal{G}[\mathcal{X}, \mathcal{Y}]$ such that $UT = SU$ (i.e., $T = U^{-1}SU$ — see Problem 5.9). Suppose there exists a unitary $U \in \mathcal{G}[\mathcal{X}, \mathcal{Y}]$ such that $UT_n = S_n U$ for every integer $n$. Show that

$$S_n \xrightarrow{w} S \quad \text{implies} \quad T_n \xrightarrow{w} U^{-1}SU.$$

In particular, if $UT = SU$, and if $S^n \xrightarrow{w} P$ for some $P \in \mathcal{B}[\mathcal{Y}]$, then show that $T^n \xrightarrow{w} U^{-1}PU$ and $U^{-1}PU$ is a projection in $\mathcal{B}[\mathcal{X}]$ (not necessarily orthogonal — cf. Problem 5.24(b)).

*Hint:* $\langle (T_n - U^{-1}S_n U)x \, ; y \rangle = \langle U^{-1}(S_n - S)Ux \, ; y \rangle = \langle (S_n - S)Ux \, ; Uy \rangle$.

**Problem 5.26.** Take a $\mathcal{B}[\mathcal{H}, \mathcal{K}]$-valued sequence $\{T_n\}$, where $\mathcal{H}$ and $\mathcal{K}$ are Hilbert spaces, and take $T \in \mathcal{B}[\mathcal{H}, \mathcal{K}]$. Show that

$$T_n \xrightarrow{u} T \quad \text{if and only if} \quad T_n^* \xrightarrow{u} T^*,$$
$$T_n \xrightarrow{w} T \quad \text{if and only if} \quad T_n^* \xrightarrow{w} T^*.$$

That is, the adjoint operation preserves uniform and weak convergence. But it does not preserve strong convergence. In fact, as we shall see shortly (Problem 5.29(b,c) below),

$$T_n \xrightarrow{s} T \quad \text{does not imply} \quad T_n^* \xrightarrow{s} T^*.$$

However, if $T_n \xrightarrow{w} T$ and $T_n^* \xrightarrow{s} S$, then $S = T^*$. Why?

**Problem 5.27.** Let $\mathcal{H}$ and $\mathcal{K}$ be Hilbert spaces. Show that

(a) $V \in \mathcal{B}[\mathcal{H}, \mathcal{K}]$ is an isometry if and only if $V^{*n} V^n = I$ (where $I$ is the identity on $\mathcal{H}$) for every $n \geq 1$,

(b) $U \in \mathcal{B}[\mathcal{H}, \mathcal{K}]$ is a unitary transformation if and only if $U^{*n} U^n = I$ (identity on $\mathcal{H}$) and $U^n U^{*n} = I$ (identity on $\mathcal{K}$) for every $n \geq 1$.

Let $T \in \mathcal{B}[\mathcal{H}]$ be a diagonalizable operator (Problem 5.17). Show that

(c) $T^* x = \sum_{k=1}^{\infty} \bar{\lambda}_k \langle x \,;\, e_k \rangle e_k$ for every $x \in \mathcal{H}$,

(d) $T$ is unitary if and only if $|\lambda_k| = 1$ for every $k \geq 1$.

**Problem 5.28.** Let $\mathcal{H}_i$ and $\mathcal{K}_i$ be Hilbert spaces for $i = 1, 2$ and consider the Hilbert spaces $\mathcal{H}_1 \oplus \mathcal{H}_2$ and $\mathcal{K}_1 \oplus \mathcal{K}_2$ of Example 5E. Take $T_{ij} \in \mathcal{B}[\mathcal{H}_j, \mathcal{H}_i]$ for $i, j = 1, 2$ and consider transformation

$$T = \begin{pmatrix} T_{11} & T_{12} \\ T_{21} & T_{22} \end{pmatrix} \in \mathcal{B}[\mathcal{H}_1 \oplus \mathcal{H}_2, \mathcal{K}_1 \oplus \mathcal{K}_2]$$

of Problem 4.17. Show that

$$T^* = \begin{pmatrix} T_{11}^* & T_{21}^* \\ T_{12}^* & T_{22}^* \end{pmatrix} \in \mathcal{B}[\mathcal{K}_1 \oplus \mathcal{K}_2, \mathcal{H}_1 \oplus \mathcal{H}_2].$$

In particular, if $\mathcal{K}_i = \mathcal{H}_i$ for $i = 1, 2$ so that $T, T^* \in \mathcal{B}[\mathcal{H}_1 \oplus \mathcal{H}_2]$, then

$$T = T_{11} \oplus T_{22} = \begin{pmatrix} T_{11} & O \\ O & T_{22} \end{pmatrix} \quad \text{implies} \quad T^* = T_{11}^* \oplus T_{22}^* - \begin{pmatrix} T_{11}^* & O \\ O & T_{22}^* \end{pmatrix}.$$

Verify that $T = \bigoplus_k T_k \in \mathcal{B}[\bigoplus_k \mathcal{H}_k]$ implies $T^* = \bigoplus_k T_k^* \in \mathcal{B}[\bigoplus_k \mathcal{H}_k]$, where $\{\mathcal{H}_k\}$ is a sequence of Hilbert spaces, $\bigoplus_k \mathcal{H}_k$ is the Hilbert space $(\lceil\bigoplus_{k=1}^{\infty} \mathcal{H}_k\rceil_2, \langle \,;\, \rangle)$ of Examples 5F and 5G, and $T$ is the direct sum of $\{T_k\}$ as in Problem 4.16. Consider the usual identification $\mathcal{H}_k \cong \bigoplus_{j<k}\{0\} \oplus \mathcal{H}_k \oplus \bigoplus_{j>k}\{0\}$, so that we may interpret each $\mathcal{H}_k$ as a subspace of $\bigoplus_k \mathcal{H}_k$. Show that $\mathcal{H}_k$ reduces $T = \bigoplus_k T_k$, $T|_{\mathcal{H}_k} = T_k$, and $(T|_{\mathcal{H}_k})^* = T^*|_{\mathcal{H}_k}$ for every $k$ (cf. Corollary 5.75).

**Problem 5.29.** An operator $S_+$ acting on a Hilbert space $\mathcal{H}$ is a *unilateral shift* if there exists an infinite sequence $\{\mathcal{H}_k\}_{k=0}^{\infty}$ of nonzero pairwise orthogonal subspaces of $\mathcal{H}$ such that $\mathcal{H} = \bigoplus_{k=0}^{\infty} \mathcal{H}_k$ and $S_+$ maps each $\mathcal{H}_k$ isometrically onto $\mathcal{K}_{k+1}$.

Observe that $S_+|_{\mathcal{H}_k} : \mathcal{H}_k \to \mathcal{H}_{k+1}$ is a unitary transformation (i.e., a surjective isometry), and hence $\dim \mathcal{H}_{k+1} = \dim \mathcal{H}_k$, for every $k \geq 0$ (Theorem 5.49). Such a common dimension is the *multiplicity* of $S_+$. The adjoint of $S_+ \in \mathcal{B}[\mathcal{H}]$, $S_+^* \in \mathcal{B}[\mathcal{H}]$, is referred to as a *backward unilateral shift*. We shall write $\bigoplus_{k=0}^{\infty} x_k$ for $\{x_k\}_{k=0}^{\infty}$ in $\bigoplus_{k=0}^{\infty} \mathcal{H}_k$. Prove the following assertions.

(a) $S_+ : \mathcal{H} \to \mathcal{H}$ and $S_+^* : \mathcal{H} \to \mathcal{H}$ are given by the formulas

$$S_+ x = 0 \oplus \bigoplus_{k=1}^{\infty} U_k\, x_{k-1} \quad \text{and} \quad S_+^* x = \bigoplus_{k=0}^{\infty} U_{k+1}^*\, x_{k+1}$$

for every $x = \bigoplus_{k=0}^{\infty} x_k$ in $\mathcal{H} = \bigoplus_{k=0}^{\infty} \mathcal{H}_k$, with 0 denoting the origin of $\mathcal{H}_0$, where $\{U_{k+1}\}_{k=0}^{\infty}$ is an arbitrary sequence of unitary transformations $U_{k+1}:$ $\mathcal{H}_k \to \mathcal{H}_{k+1}$, so that $S_+|_{\mathcal{H}_k} = U_{k+1}$, for each $k \geq 0$.

$S_+$ and $S_+^*$ are identified with the following infinite matrix of operators.

$$S_+ = \begin{pmatrix} O & & & \\ U_1 & O & & \\ & U_2 & O & \\ & & U_3 & O \\ & & & & \ddots \end{pmatrix} \quad \text{and} \quad S_+^* = \begin{pmatrix} O & U_1^* & & & \\ & O & U_2^* & & \\ & & O & U_3^* & \\ & & & O & \\ & & & & \ddots \end{pmatrix}.$$

(b) $S_+^*$ is a strongly stable coisometry. That is, $S_+$ is an isometry and $S_+^{*n} \xrightarrow{\ s\ } O$ (so that $S_+$ is an isometry that is not a coisometry).

*Hint:* $\|S_+ x\|^2 = \|x\|^2$, and $S_+^{*n} x = \bigoplus_{k=0}^{\infty} U_{k+1}^* \cdots U_{k+n}^* x_{k+n}$ (by induction) so that $\|S_+^{*n} x\|^2 = \sum_{k=n}^{\infty} \|x_k\|^2$.

(c) $S_+^n \xrightarrow{\ w\ } O$ but $\{S_+^n\}$ does not converge strongly.

*Hint:* $S_+^{*n} \xrightarrow{\ w\ } O$ and $\|S_+^n x\| = \|x\|$ (i.e., $S_+^{*n} S_+^n = I$).

Let $\mathcal{K}_0$ be an arbitrary Hilbert space unitarily equivalent to $\mathcal{H}_0$. Let $U_0 : \mathcal{K}_0 \to \mathcal{H}_0$ be a unitary transformation so that $\dim \mathcal{H}_k = \dim \mathcal{K}_0$ for all $k$. Consider the Hilbert space $\ell_+^2(\mathcal{K}_0) = \bigoplus_{k=0}^{\infty} \mathcal{K}_0$ of Example 5F.

(d) $U = \bigoplus_{k=0}^{\infty} U_k \cdots U_0 : \ell_+^2(\mathcal{K}_0) \to \mathcal{H}$ is a unitary transformation, and

$$U^* S_+ U = \begin{pmatrix} O & & & \\ I & O & & \\ & I & O & \\ & & I & O \\ & & & & \ddots \end{pmatrix} \quad \text{in} \quad \mathcal{B}[\ell_+^2(\mathcal{K}_0)].$$

Thus $S_+$ is unitarily equivalent to $U^* S_+ U$, which is a unilateral shift of multiplicity $\dim \mathcal{K}_0$, called the *canonical unilateral shift* on $\ell_+^2(\mathcal{K}_0)$.

**Problem 5.30.** An operator $S$ acting on a Hilbert space $\mathcal{H}$ is a *bilateral shift* if there exists an infinite family $\{\mathcal{H}_k\}_{k=-\infty}^{\infty}$ of nonzero pairwise orthogonal subspaces of $\mathcal{H}$ such that $\mathcal{H} = \bigoplus_{k=-\infty}^{\infty} \mathcal{H}_k$ and $S$ maps each $\mathcal{H}_k$ isometrically onto $\mathcal{K}_{k+1}$.

As it happened in the case of a unilateral shift, the above definition ensures that $S|_{\mathcal{H}_k}\colon \mathcal{H}_k \to \mathcal{H}_{k+1}$ is a surjective isometry. Then the subspaces $\mathcal{H}_k$ of $\mathcal{H}$ are all unitarily equivalent, and their common dimension is the *multiplicity* of $S$. The adjoint $S^* \in \mathcal{B}[\mathcal{H}]$ of $S \in \mathcal{B}[\mathcal{H}]$ is referred to as a *backward bilateral shift*. Prove the following assertions.

(a) $S\colon \mathcal{H} \to \mathcal{H}$ and $S^*\colon \mathcal{H} \to \mathcal{H}$ are given by the formulas

$$Sx = \bigoplus_{k=-\infty}^{\infty} U_k\, x_{k-1} \quad \text{and} \quad S^*x = \bigoplus_{k=-\infty}^{\infty} U^*_{k+1} x_{k+1}$$

for every $x = \bigoplus_{k=-\infty}^{\infty} x_k$ in $\mathcal{H} = \bigoplus_{k=-\infty}^{\infty}\mathcal{H}_k$, where $\{U_k\}_{k=-\infty}^{\infty}$ is an arbitrary family of unitary transformations $U_{k+1}\colon \mathcal{H}_k \to \mathcal{H}_{k+1}$, so that $S|_{\mathcal{H}_k} = U_{k+1}$, for each integer $k$.

$S$ and $S^*$ are identified with the following (doubly) infinite matrix of operators, where the inner parenthesis indicates the zero-zero entry.

$$S = \begin{pmatrix} \ddots & & & & \\ & O & & & \\ & U_{-1} & O & & \\ & & U_0 & (O) & \\ & & & U_1 & O \\ & & & & \ddots \end{pmatrix} \quad \text{and} \quad S^* = \begin{pmatrix} \ddots & & & & \\ & O & U^*_{-1} & & \\ & & O & U^*_0 & \\ & & & (O) & U^*_1 \\ & & & & O \\ & & & & \ddots \end{pmatrix}.$$

(b) $S$ is unitarily equivalent to the *canonical bilateral shift* acting on $\ell^2(\mathcal{K}_0) = \bigoplus_{k=-\infty}^{\infty}\mathcal{K}_0$,

$$\begin{pmatrix} \ddots & & & & \\ & O & & & \\ & I & O & & \\ & & I & (O) & \\ & & & I & O \\ & & & & \ddots \end{pmatrix} \quad \text{in} \quad \mathcal{B}[\ell^2(\mathcal{K}_0)],$$

where $\mathcal{K}_0$ is any Hilbert space with the property that $\dim \mathcal{K}_0$ is the multiplicity of $S$. (*Hint*: Problem 5.29(d).)

(c) $S$ is a weakly stable unitary operator. In other words, $S$ is an isometry and a coisometry (i.e., $S^{*n}S^n = S^n S^{*n} = I$) and $S^n \xrightarrow{w} O$.

*Hint*: Let $S_0$ denote the above canonical unilateral shift on $\ell^2(\mathcal{K}_0)$, which is unitarily equivalent to $S$. Take any $x = \bigoplus_{k=-\infty}^{\infty} x_k$ in $\ell^2(\mathcal{K}_0) = \bigoplus_{k=-\infty}^{\infty}\mathcal{K}_0$. Show by induction that $S_0^n x = \bigoplus_{k=-\infty}^{\infty} x_{k-n}$, and hence

$$\langle S_0^n x\,;x\rangle = \sum_{k=-\infty}^{\infty} \langle x_{k-n}\,;x_k\rangle = \sum_{k=-\infty}^{-1} \langle x_k\,;x_{k+n}\rangle + \sum_{k=0}^{\infty}\langle x_k\,;x_{k+n}\rangle.$$

Now apply the Schwarz inequality in $\mathcal{K}_0$ and in $\ell^2$ to show that

$$\sum_{k=-\infty}^{-1} |\langle x_k \,;\, x_{k+n}\rangle| \leq \sum_{k=-\infty}^{-1} \|x_k\|\,\|x_{k+n}\|$$

$$\leq \left(\sum_{k=-\infty}^{-1} \|x_k\|^2\right)^{\frac{1}{2}} \left(\sum_{k=-\infty}^{-1} \|x_{k+n}\|^2\right)^{\frac{1}{2}} \leq \left(\sum_{k=-\infty}^{-1} \|x_{k+n}\|^2\right)^{\frac{1}{2}} \|x\|.$$

Similarly, $\sum_{k=0}^{\infty} |\langle x_k \,;\, x_{k+n}\rangle| \leq \left(\sum_{k=0}^{\infty}\|x_{k+n}\|^2\right)^{\frac{1}{2}} \|x\|$. Next verify:

$$\lim_n \sum_{k=-\infty}^{-1} \|x_{k+n}\|^2 = \lim_n \sum_{k=0}^{\infty} \|x_{k+n}\|^2 = 0.$$

Hence $S_0^n \xrightarrow{w} O$ (cf. Proposition 5.67). Use Problem 5.25 to conclude that $S^n \xrightarrow{w} O$.

(d) Both $\{S^n\}$ and $\{S^{*n}\}$ do not converge strongly.

   *Hint*: $S^{*n} \xrightarrow{w} O$, and $\|S^n x\| = \|S^{*n} x\| = \|x\|$.


**Problem 5.31.** According to Problem 5.29 unilateral shifts exist only on infinite-dimensional spaces. Let $\mathcal{H}$ be a *separable* Hilbert space.

(a) $S_+ \in \mathcal{B}[\mathcal{H}]$ is a unilateral shift of *multiplicity one* if and only if

$$S_+ e_k = e_{k+1} \quad \text{for every} \quad k \in \mathbb{N}_0$$

for some orthonormal basis $\{e_k\}_{k=0}^{\infty}$ for $\mathcal{H}$. That is, if and only if it shifts some orthonormal basis for $\mathcal{H}$ indexed by $\mathbb{N}_0$ (or by any set that is in a one-to-one order-preserving correspondence with $\mathbb{N}_0$). Moreover, in this case, show also that

$$S_+^* e_0 = 0 \quad \text{and} \quad S_+^* e_k = e_{k-1} \quad \text{for every} \quad k \in \mathbb{N}_0.$$

   *Hint*: Set $\mathcal{H}_k = \operatorname{span}\{e_k\}$, so that $\dim \mathcal{H}_k = 1$, for each $k \in \mathbb{N}_0$.

(b) Verify that the operator $S_+ \in \mathcal{B}[\ell_+^2]$ of Problem 4.39 is, in fact, the canonical unilateral shift of multiplicity one on $\ell_+^2$, which shifts the canonical orthonormal basis for $\ell_+^2$. (*Hint*: $\ell_+^2 = \bigoplus_{k=0}^{\infty}\mathbb{C}$.)

(c) Let $\{e_k\}_{k=-\infty}^{\infty}$ be the canonical orthonormal basis for $\ell^2$. Re-index this basis as follows. Set, for every $n \in \mathbb{N}$,

$$f_n = e_{\frac{1-n}{2}} \quad \text{if } n \text{ is odd} \quad \text{and} \quad f_n = e_{\frac{n}{2}} \quad \text{if } n \text{ is even.}$$

Check that $\{f_n\}_{n=1}^{\infty}$ is an orthonormal basis for $\ell^2$ and that the operator $S_+$ in $\mathcal{B}[\ell^2]$ given by the following (doubly) infinite matrix,

$$
S_+ \;=\;
\begin{pmatrix}
 & & & & & \ddots \\
 & & & & 1 & \\
 & & & 1 & & \\
 & & (0) & & & \\
 & 0 & 1 & & & \\
 0 & 1 & & & & \\
 \ddots & 1 & & & &
\end{pmatrix},
$$

is a unilateral shift on $\ell^2$ that shifts the orthonormal basis $\{f_n\}_{n=1}^{\infty}$.

**Problem 5.32.** According to Problem 5.30 bilateral shifts exist only on infinite-dimensional spaces. Let $\mathcal{H}$ be a *separable* Hilbert space.

(a) $S \in \mathcal{B}[\mathcal{H}]$ is a bilateral shift of *multiplicity* 1 if and only if

$$
Se_k = e_{k+1} \quad \text{for every} \quad k \in \mathbb{Z}
$$

for some orthonormal basis $\{e_k\}_{k=-\infty}^{\infty}$ for $\mathcal{H}$. That is, if and only if it shifts some orthonormal basis for $\mathcal{H}$ indexed by $\mathbb{Z}$ (or by any set that is in a one-to-one order-preserving correspondence with $\mathbb{Z}$). Moreover, in this case, show also that

$$
S^* e_k = e_{k-1} \quad \text{for every} \quad k \in \mathbb{Z}.
$$

*Hint*: Set $\mathcal{H}_k = \operatorname{span}\{e_k\}$, so that $\dim \mathcal{H}_k = 1$, for each $k \in \mathbb{Z}$.

(b) Verify that the operator $S \in \mathcal{B}[\ell^2]$ of Problem 4.40 is, in fact, the canonical bilateral shift of multiplicity one on $\ell^2$, which shifts the canonical orthonormal basis for $\ell^2$. (*Hint*: $\ell^2 = \bigoplus_{k=-\infty}^{\infty}\mathbb{C}$.)

(c) Let $\{e_n\}_{n=1}^{\infty}$ be the canonical orthonormal basis for $\ell_+^2$. Re-index this basis as follows. Set, for every $k \in \mathbb{Z}$,

$$
f_k = e_{1-2k} \;\; \text{if} \;\; k \le 0 \quad \text{and} \quad f_k = e_{2k} \;\; \text{if} \;\; n \;\; k > 0.
$$

Check that $\{f_k\}_{k=\infty}^{\infty}$ is an orthonormal basis for $\ell_+^2$ and that the operator $S$ in $\mathcal{B}[\ell_+^2]$ given by the following infinite matrix,

$$
S =
\begin{pmatrix}
b & A & & & \\
 & B & A & & \\
 & & B & A & \\
 & & & B & \ddots \\
 & & & & \ddots
\end{pmatrix},
\quad \text{with } b = \begin{pmatrix}0\\1\end{pmatrix},\; A = \begin{pmatrix}0 & 1\\0 & 0\end{pmatrix} \text{ and } B = \begin{pmatrix}0 & 0\\1 & 0\end{pmatrix},
$$

is a bilateral shift on $\ell_+^2$ that shifts the orthonormal basis $\{f_k\}_{k=-\infty}^{\infty}$.

**Problem 5.33.** Consider the orthonormal basis $\{e_k\}_{k=-\infty}^{\infty}$ for the Hilbert space $L^2(\Gamma)$ of Example 3L(c), where $\Gamma$ denotes the unit circle about the origin of the complex plane and, for each $k \in \mathbb{Z}$, $e_k(z) = z^k$ for every $z \in \Gamma$. Define a mapping $U : L^2(\Gamma) \to L^2(\Gamma)$ as follows. If $f \in L^2(\Gamma)$, then $Uf$ is given by

$$(Uf)(z) = zf(z) \quad \text{for every} \quad z \in \Gamma.$$

(a) Verify that $Uf$, in fact, lies in $L^2(\Gamma)$ for every $f$ in $L^2(\Gamma)$. Moreover, show that $U \in \mathcal{B}[L^2(\Gamma)]$.

(b) Show that $U$ is a bilateral shift of multiplicity one on $L^2(\Gamma)$ that shifts the orthonormal basis $\{e_k\}_{k=-\infty}^{\infty}$.

(c) Prove the Riemann–Lebesgue Lemma: If $f \in L^2(\Gamma)$, then $\int_\Gamma z^k f(z)\,dz \to 0$ as $k \to \pm\infty$.

*Hint:* $(U^k f)(z) = z^k f(z)$ so that $\langle U^k f\,;\,1 \rangle = \int_\Gamma z^k f(z)\,dz$, where $1(z) = 1$ for all $z \in \Gamma$. Recall that $U^k \xrightarrow{w} O$ (cf. Problem 5.30(c)).

**Problem 5.34.** Let $\mathcal{H}$ be a Hilbert space and take $T, S \in \mathcal{B}[\mathcal{H}]$. Use Problem 4.20 and Corollary 5.75 to prove the following assertion. If $S$ commutes with both $T$ and $T^*$, then $\mathcal{N}(S)$ and $\mathcal{R}(S)^-$ reduce $T$.

**Problem 5.35.** Take $T \in \mathcal{B}[\mathcal{H}, \mathcal{K}]$, where $\mathcal{H}$ and $\mathcal{K}$ are Hilbert spaces.

(a) $T$ is injective $\iff$ $T^*T$ is injective $\iff$ $\mathcal{R}(T^*)^- = \mathcal{H}$.

(a*) $T^*$ is injective $\iff$ $TT^*$ is injective $\iff$ $\mathcal{R}(T)^- = \mathcal{K}$.

Prove (a) and (a*). *Hint:* Propositions 5.15 and 5.76.

(b) $T$ is surjective $\iff$ $T^*$ has a bounded inverse on $\mathcal{R}(T^*)$.

(b*) $T^*$ is surjective $\iff$ $T$ has a bounded inverse on $\mathcal{R}(T)$.

Prove (b) and (b*). *Hint:* Corollary 4.24 and Proposition 5.77.

**Problem 5.36.** Consider the following assertions under the setup of the previous problem.

| | | | |
|---|---|---|---|
| (a) | $T$ is injective. | (a*) | $T^*$ is injective. |
| (b) | $\dim \mathcal{R}(T) = n$. | (b*) | $\dim \mathcal{R}(T^*) = m$. |
| (c) | $\mathcal{R}(T^*) = \mathcal{H}$. | (c*) | $\mathcal{R}(T) = \mathcal{K}$. |
| (d) | $T^*T \in \mathcal{G}[\mathcal{H}]$. | (d*) | $TT^* \in \mathcal{G}[\mathcal{K}]$. |

If $\dim \mathcal{H} = n$, then (a), (b), (c) and (d) are pairwise equivalent. If $\dim \mathcal{K} = m$, then (a*), (b*), (c*) and (d*) are pairwise equivalent. Prove.

**Problem 5.37.** Let $\mathcal{H}$ and $\mathcal{K}$ be Hilbert spaces and take $T \in \mathcal{B}[\mathcal{H}, \mathcal{K}]$. If $y \in \mathcal{R}(T)$, then there exists a solution $x \in \mathcal{H}$ to the equation $y = Tx$. It is clear that this solution is unique whenever $T$ is injective. If, in addition, $\mathcal{R}(T)$ is closed in $\mathcal{K}$, then this unique solution is given by

$$x = (T^*T)^{-1}T^*y.$$

In other words, suppose $\mathcal{N}(T) = \{0\}$ and $\mathcal{R}(T) = \mathcal{R}(T)^-$. According to Corollary 4.24 there exists $T^{-1} \in \mathcal{B}[\mathcal{R}(T), \mathcal{H}]$. Use Propositions 5.76 and 5.77 to show that there exists $(T^*T)^{-1} \in \mathcal{B}[\mathcal{R}(T^*), \mathcal{H}]$ and

$$T^{-1} = (T^*T)^{-1}T^* \quad \text{on} \quad \mathcal{R}(T).$$

**Problem 5.38.** (Least-Squares). Take $T \in \mathcal{B}[\mathcal{H}, \mathcal{K}]$, where $\mathcal{H}$ and $\mathcal{K}$ are Hilbert spaces. If $y \in \mathcal{K}\backslash\mathcal{R}(T)$, then there is no solution $x \in \mathcal{H}$ to the equation $y = Tx$. Question: Is there a vector $x$ in $\mathcal{H}$ that minimizes $\|y - Tx\|$? Use Theorem 5.13, Proposition 5.76 and Problem 5.37 to prove the following proposition. If $\mathcal{R}(T) = \mathcal{R}(T)^-$, then for each $y \in \mathcal{K}$ there exists $x_y \in \mathcal{H}$ such that

$$\|y - Tx_y\| = \inf_{x \in \mathcal{H}} \|y - Tx\| \quad \text{and} \quad T^*Tx_y = T^*y.$$

Moreover, if $T$ is injective, then $x_y$ is unique and given by

$$x_y = (T^*T)^{-1}T^*y.$$

**Problem 5.39.** Let $\mathcal{H}$ and $\mathcal{K}$ be Hilbert spaces and take $T \in \mathcal{B}[\mathcal{H}, \mathcal{K}]$. If $y \in \mathcal{R}(T)$ and $\mathcal{R}(T) = \mathcal{R}(T)^-$, then show that there exists $x_0 \in \mathcal{H}$ such that $y = Tx_0$ and

$$\|x_0\| \le \|x\|$$

for all $x \in \mathcal{H}$ such that $y = Tx$. That is, if $\mathcal{R}(T) = \mathcal{R}(T)^-$, then for each $y \in \mathcal{R}(T)$ there exists a solution $x_0 \in \mathcal{H}$ to the equation $y = Tx$ with *minimum norm*. Moreover, if $T^*$ is injective, then show that $x_0$ is unique and given by

$$x_0 = T^*(TT^*)^{-1}y.$$

*Hint*: If $\mathcal{R}(T) = \mathcal{R}(T)^-$, then $\mathcal{R}(TT^*) = \mathcal{R}(T)$ (Propositions 5.76 and 5.77). Take $y$ in $\mathcal{R}(T)$. Thus $y = TT^*z$ for some $z$ in $\mathcal{K}$. Set $x_0 = T^*z$ in $\mathcal{H}$ so that $y = Tx_0$. If $x \in \mathcal{H}$ is such that $y = Tx$, then verify: $\|x_0\|^2 = \langle T^*z ; x_0 \rangle = \langle z ; Tx \rangle = \langle T^*z ; x \rangle = \langle x_0 ; x \rangle \le \|x_0\| \|x\|$. If $\mathcal{N}(T^*) = \{0\}$, then $\mathcal{N}(TT^*) = \{0\}$ (Proposition 5.76). Since $\mathcal{R}(TT^*) = \mathcal{R}(T) = \mathcal{R}(T)^-$, it follows by Corollary 4.24 that $TT^*$ has a bounded inverse on $\mathcal{R}(T)$. Hence $z = (TT^*)^{-1}y$ is unique, and so is $x_0 = T^*z$.

**Problem 5.40.** Show that $T \in \mathcal{B}_0[\mathcal{H}, \mathcal{K}]$ if and only if $T^* \in \mathcal{B}_0[\mathcal{K}, \mathcal{H}]$, where $\mathcal{H}$ and $\mathcal{K}$ are Hilbert spaces. Moreover, $\dim \mathcal{R}(T) = \dim \mathcal{R}(T^*)$.

*Hint*: $\mathcal{B}_0[\mathcal{H}, \mathcal{K}]$ denotes the set of all finite-rank bounded linear transformations of $\mathcal{H}$ into $\mathcal{K}$. If $T \in \mathcal{B}_0[\mathcal{H}, \mathcal{K}]$, then $\mathcal{R}(T) = \mathcal{R}(T)^-$. (Why?) Use Propositions 5.76 and 5.77 to show that $\mathcal{R}(T^*) = T^*(\mathcal{R}(T))$. Thus conclude: $\dim \mathcal{R}(T^*) \leq \dim \mathcal{R}(T)$ (cf. Problems 2.17 and 2.18).

**Problem 5.41.** Let $T \in \mathcal{B}[\mathcal{H}, \mathcal{Y}]$ be a bounded linear transformation of a Hilbert space $\mathcal{H}$ into a normed space $\mathcal{Y}$. Show that the following assertions are pairwise equivalent.

(a)  $T$ is compact    (i.e., $T \in \mathcal{B}_\infty[\mathcal{H}, \mathcal{Y}]$).

(b)  $Tx_n \to Tx$ in $\mathcal{Y}$ whenever $x_n \xrightarrow{w} x$ in $\mathcal{H}$.

(c)  $Tx_n \to 0$ in $\mathcal{Y}$ whenever $x_n \xrightarrow{w} 0$ in $\mathcal{H}$.

*Hint*: Problem 4.69 for (a)$\Rightarrow$(b). Conversely, let $\{x_n\}$ be a bounded sequence in $\mathcal{H}$. Apply Lemma 5.69 to ensure the existence of a subsequence $\{x_{n_k}\}$ of $\{x_n\}$ such that $\{Tx_{n_k}\}$ converges in $\mathcal{Y}$ whenever (b) holds true. Now conclude that $T$ is compact (Theorem 4.52(d)). Hence (b)$\Rightarrow$(a). Trivially, (b)$\Rightarrow$(c). On the other hand, if $x_n \xrightarrow{w} x$ in $\mathcal{H}$, then verify that $T(x_n - x) \to 0$ in $\mathcal{Y}$ whenever (c) holds; that is, (c)$\Rightarrow$(b).

**Problem 5.42.** If $T \in \mathcal{B}[\mathcal{H}, \mathcal{K}]$, where $\mathcal{H}$ and $\mathcal{K}$ are Hilbert spaces, then show that the following assertions are pairwise equivalent.

(a)  $T$ is compact    (i.e., $T \in \mathcal{B}_\infty[\mathcal{H}, \mathcal{K}]$).

(b)  $T$ is the (uniform) limit in $\mathcal{B}[\mathcal{H}, \mathcal{K}]$ of a sequence of finite-rank bounded linear transformations of $\mathcal{H}$ into $\mathcal{K}$. That is, there exists a $\mathcal{B}_0[\mathcal{H}, \mathcal{K}]$-valued sequence $\{T_n\}$ such that $\|T_n - T\| \to 0$.

(c)  $T^*$ is compact    (i.e., $T^* \in \mathcal{B}_\infty[\mathcal{K}, \mathcal{H}]$).

*Hint: Take any $T \in \mathcal{B}_\infty[\mathcal{H}, \mathcal{K}]$ and let $\{e_k\}_{k=1}^\infty$ be an orthonormal basis for $\mathcal{R}(T)^-$. If $P_n : \mathcal{K} \to \mathcal{K}$ is the orthogonal projection onto $\bigvee\{e_k\}_{k=1}^n$, then $P_n T \xrightarrow{u} T$.* Indeed, $\mathcal{R}(T)^-$ is separable (Proposition 4.57) and the existence of $P_n$ is ensured by Theorem 5.52. Show that $P_n \xrightarrow{s} P$, where $P : \mathcal{K} \to \mathcal{K}$ is the orthogonal projection onto $\mathcal{R}(T)^-$ (cf. Problem 5.15). Now use Problem 4.57(b) to establish that $P_n T \xrightarrow{u} PT = T$. Set $T_n = P_n T$ and verify that each $T_n$ lies in $\mathcal{B}_0[\mathcal{H}, \mathcal{K}]$. Hence (a)$\Rightarrow$(b). For the converse, see Corollary 4.55. Thus (a)$\Leftrightarrow$(b), which implies (a)$\Leftrightarrow$(c) (Proposition 5.65(d) and Problem 5.40).

(d)  Verify that $\mathcal{B}_0[\mathcal{H}, \mathcal{K}]$ is dense in $\mathcal{B}_\infty[\mathcal{H}, \mathcal{K}]$.

**Problem 5.43.** An operator $T \in \mathcal{B}[\mathcal{H}]$ on a Hilbert space $\mathcal{H}$ is an involution if $T^2 = I$ (cf. Problem 1.11). A *symmetry* is a unitary involution. Show that the following assertions are pairwise equivalent.

(a)  $T$ is a unitary involution.

(b)  $T$ is a self-adjoint involution.

(c)  $T$ is self-adjoint and unitary.


**Problem 5.44.**  Let $\mathcal{H}$ be a Hilbert space. Show that the set of all self-adjoint operators from $\mathcal{B}[\mathcal{H}]$ is weakly closed in $\mathcal{B}[\mathcal{H}]$.

*Hint*: $|\langle Tx\,;\,y\rangle - \langle x\,;\,Ty\rangle| = |\langle Tx\,;\,y\rangle - \langle T_n x\,;\,y\rangle + \langle x\,;\,T_n y\rangle - \langle x\,;\,Ty\rangle| \leq |\langle (T_n - T)x\,;\,y\rangle| + |\langle (T_n - T)y\,;\,x\rangle|$ whenever $T_n^* = T_n$.

**Problem 5.45.**  Let $S$ and $T$ be self-adjoint operators in $\mathcal{B}[\mathcal{H}]$, where $\mathcal{H}$ is a Hilbert space. Prove the following results.

(a)  $T + S$ is self-adjoint.

(b)  $\alpha T$ is self-adjoint if and only if $\alpha \in \mathbb{R}$.

Therefore, if $\mathcal{H}$ is a *real* Hilbert space, then the set of all self-adjoint operators from $\mathcal{B}[\mathcal{H}]$ is a subspace of $\mathcal{B}[\mathcal{H}]$.

(c)  $TS$ is self-adjoint if and only if $TS = ST$.

(d)  $p(T) = p(T)^*$ for every polynomial $p$ with real coefficients.

(e)  $T^{2n} \geq O$ and $\|T^{2^n}\| = \|T\|^{2^n}$ for each $n \geq 1$. (*Hint*: Proposition 5.78.)


**Problem 5.46.**  If an operator $T \in \mathcal{B}[\mathcal{H}]$ acting on a complex Hilbert space $\mathcal{H}$ is such that $T = A + iB$, where $A$ and $B$ are self-adjoint operators in $\mathcal{B}[\mathcal{H}]$, then the representation $T = A + iB$ is called the *Cartesian decomposition* of $T$. Prove the following propositions.

(a)  Every operator $T \in \mathcal{B}[\mathcal{H}]$ acting on a complex Hilbert space $\mathcal{H}$ has a unique Cartesian decomposition.

   *Hint*: Set $A = \frac{1}{2}(T^* + T)$ and $B = \frac{i}{2}(T^* - T)$.

(b)  $T^*T = TT^*$ if and only if $AB = BA$. In this case, $T^*T = A^2 + B^2$ and $\max\{\|A\|^2, \|B\|^2\} \leq \|T\|^2 \leq \|A^2\| + \|B^2\|$.


**Problem 5.47.**  If $T \in \mathcal{B}[\mathcal{H}]$ is a *self-adjoint* operator acting on a *real* Hilbert space $\mathcal{H}$, then show that

$$\langle Tx\,;\,y\rangle = \tfrac{1}{4}\big(\langle T(x+y)\,;\,x+y\rangle - \langle T(x-y)\,;\,x-y\rangle\big)$$

for every $x, y \in \mathcal{H}$. (*Hint*: Problem 5.3(a).)

**Problem 5.48.**  Let $\mathcal{H}$ be any (real or complex) Hilbert space.

(a) If $\{T_n\}$ is a sequence of *self-adjoint* operators in $\mathcal{B}[\mathcal{H}]$, then the five assertions of Proposition 5.67 are all pairwise equivalent even in a real Hilbert space.

*Hint*: If $T_n^* = T_n$ and the real sequence $\{\langle T_n x \, ; x \rangle\}$ converges in $\mathbb{R}$ for every $x \in \mathcal{H}$, and if $\mathcal{H}$ is real, then use Problem 5.47 to show that the real sequence $\{\langle T_n x \, ; y \rangle\}$ converges in $\mathbb{R}$ for every $x, y \in \mathcal{H}$. Now apply Proposition 5.67.

(b) If $\{T_n\}$ is a sequence of *self-adjoint* operators in $\mathcal{B}[\mathcal{H}]$, then the four assertions of Problem 5.5 are all pairwise equivalent even in a real Hilbert space. (*Hint*: Problems 5.5 and 5.47.)

**Problem 5.49.** The set $\mathcal{B}^+[\mathcal{H}]$ of all nonnegative operators on a Hilbert space $\mathcal{H}$ is a weakly closed convex cone in $\mathcal{B}[\mathcal{H}]$.

*Hint*: If $Q_n \geq O$ for every positive integer $n$ and $Q_n \xrightarrow{w} Q$, then $Q \geq O$ $\big(0 \leq \langle Q_n x \, ; x \rangle \leq |\langle (Q_n - Q) x \, ; x \rangle| + \langle Q x \, ; x \rangle\big)$. See Problems 2.2 and 2.21.

**Problem 5.50.** Let $\mathcal{H}$ and $\mathcal{K}$ be Hilbert spaces and take $T \in \mathcal{B}[\mathcal{H}, \mathcal{K}]$. Recall: $T^* T \in \mathcal{B}^+[\mathcal{H}]$ and $T T^* \in \mathcal{B}^+[\mathcal{K}]$. Show that

(a)  $T^* T > O$  if and only if  $T$ is injective,

(b)  $T^* T \in \mathcal{G}^+[\mathcal{H}]$  if and only if  $T \in \mathcal{G}[\mathcal{H}, \mathcal{K}]$,

(a*)  $T T^* > O$  if and only if  $T^*$ is injective,

(b*)  $T T^* \in \mathcal{G}^+[\mathcal{K}]$  if and only if  $T^* \in \mathcal{G}[\mathcal{K}, \mathcal{H}]$.

**Problem 5.51.** Let $\mathcal{H}$ be a Hilbert space and take $Q$, $R$ and $T$ in $\mathcal{B}[\mathcal{H}]$. Verify the following implications.

(a)  $Q \geq O$  implies  $T^* Q T \geq O$.

(b)  $Q \geq O$  and  $R \geq O$  imply  $Q + R \geq O$.

(c)  $Q > O$  and  $R \geq O$  imply  $Q + R > O$.

(d)  $Q \succ O$  and  $R \geq O$  imply  $Q + R \succ O$.

**Problem 5.52.** Let $Q$ be an operator acting on a Hilbert space $\mathcal{H}$. Prove the following propositions.

(a)  $Q \geq O$  implies  $Q^n \geq O$  for every integer $n \geq 0$.

(b)  $Q > O$  implies  $Q^n > O$  for every integer $n \geq 0$.

(c)  $Q \succ O$  implies  $Q^n \succ O$  for every integer $n \geq 0$.

(d)  $Q \succ O$ implies $Q^{-1} \succ O$.

(e)  If $p$ is an arbitrary polynomial with positive coefficients, then $p(Q) \geq O$, $p(Q) > O$ or $p(Q) \succ O$ whenever $Q \geq O, Q > O$ or $Q \succ O$, respectively.

*Hints*: (a), (b) and (c) are trivially verified for $n = 0, 1$. Suppose $n \geq 2$.

(a)  Show that $\langle Q^n x ; x \rangle = \| Q^{\frac{n}{2}} x \|^2$ for every $x \in \mathcal{H}$ if $n$ is even, and $\langle Q^n x ; x \rangle = \langle Q Q^{\frac{n-1}{2}} x ; Q^{\frac{n-1}{2}} x \rangle$ for every $x \in \mathcal{H}$ if $n$ is odd.

(b,c)  $Q > O$ if and only if $Q \geq O$ and $\mathcal{N}(Q) \neq \{0\}$; and $Q \succ O$ if and only if $Q \geq O$ and $Q$ is bounded below. In both cases, $Q \neq O$. Note that (i) $\langle Q^{2n} x ; x \rangle = \| Q^n x \|^2$ and (ii) $\langle Q^{2n-1} x ; x \rangle \geq \| Q \|^{-1} \| Q^n x \|^2$, for every $x$ in $\mathcal{H}$ and every $n \geq 1$. The inequality in (ii) is a consequence of Proposition 5.82: $\| Q^n x \|^2 = \| Q Q^{n-1} x \|^2 \leq \| Q \| \langle Q Q^{n-1} x ; Q^{n-1} x \rangle$. Apply (i) to show that (b) and (c) hold for $n = 2$, and hence they hold for $n = 3$ by (ii). Conclude the proofs by induction.

(d)  $\| x \|^2 = \| Q Q^{-1} x \|^2 \leq \| Q \| \langle Q Q^{-1} x ; Q^{-1} x \rangle = \| Q \| \langle Q^{-1} x ; x \rangle$. Why?

**Problem 5.53.**  Let $\mathcal{H}$ be a Hilbert space and take $Q, R \in \mathcal{B}[\mathcal{H}]$. Prove:

(a)  $O \prec Q \prec R$ implies $O \prec R^{-1} \prec Q^{-1}$,

(b)  $O \prec Q \leq R$ implies $O \prec R^{-1} \leq Q^{-1}$,

(c)  $O \prec Q < R$ implies $O \prec R^{-1} < Q^{-1}$.

*Hints*: Consider the result in Problem 5.22(d).

(a)  If $O \prec Q \prec R$, then $Q^{-1} \succ O$, $R^{-1} \succ O$ and $(R - Q)^{-1} \succ O$. But $Q^{-1} - R^{-1} = Q^{-1}(R - Q)R^{-1} = ((R - Q + Q)(R - Q)^{-1}Q)^{-1} = (Q + Q(R - Q)^{-1}Q)^{-1}$ and $Q + Q(R - Q)^{-1}Q \succ O$. This is enough to ensure that $Q^{-1} - R^{-1} \succ O$.

(b)  If $O \prec Q \leq R$, then $Q^{-1} \succ O$, $R^{-1} \succ O$ (there exists $\alpha > 0$ such that $\alpha \| x \|^2 \leq \langle R^{-1} x ; x \rangle$ for all $x \in \mathcal{H}$) and $O \prec Q \leq R \prec \frac{n+1}{n} R$. But $Q^{-1} - R^{-1} = Q^{-1} - (\frac{n+1}{n} R)^{-1} - \frac{1}{n+1} R^{-1}$, and $Q^{-1} - (\frac{n+1}{n} R)^{-1} \succ O$ by item (a). Therefore,

$$
\begin{aligned}
\langle (Q^{-1} - R^{-1}) x ; x \rangle &= \left\langle \left( Q^{-1} - \left( \frac{n+1}{n} R \right)^{-1} \right) x ; x \right\rangle - \frac{1}{n+1} \langle R^{-1} x ; x \rangle \\
&\geq -\frac{\alpha}{n+1} \| x \|^2
\end{aligned}
$$

for every $x \in \mathcal{H}$ and all $n \geq 1$, which implies that $\langle (Q^{-1} - R^{-1}) x ; x \rangle \geq 0$ for all $x \in \mathcal{H}$.

(c)  If $O \prec Q < R$, then $Q^{-1} \succ O$, $R^{-1} \succ O$ and $R - Q > O$. Thus there exists $\alpha > 0$ such that $\alpha \| x \| \leq \| Q^{-1} x \|$ for all $x \in \mathcal{H}$, $R^{-1} \in \mathcal{G}[\mathcal{H}]$ and $\mathcal{N}(R - Q) =$

$\{0\}$. Hence $0 < \alpha \|(R - Q)R^{-1}x\| \leq \|Q^{-1}(R - Q)R^{-1}x\| = \|(Q^{-1} - R^{-1})x\|$ for every nonzero vector $x$ in $\mathcal{H}$, and therefore $\mathcal{N}(Q^{-1} - R^{-1}) = \{0\}$. Recall: $Q^{-1} - R^{-1} \geq O$ by item (b).

**Problem 5.54.** Show that the following equivalences hold for every $T \in \mathcal{B}[\mathcal{H}, \mathcal{K}]$, where $\mathcal{H}$ and $\mathcal{K}$ are Hilbert spaces (apply Corollary 5.83).

$$T^{*n}T^n \xrightarrow{\ s\ } O \quad \Longleftrightarrow \quad T^{*n}T^n \xrightarrow{\ w\ } O \quad \Longleftrightarrow \quad T^n \xrightarrow{\ s\ } O.$$

Now conclude that *for a self-adjoint operator the concepts of strong and weak stabilities coincide* (i.e., if $T^* = T$, then $T^n \xrightarrow{\ s\ } O \Longleftrightarrow T^n \xrightarrow{\ w\ } O$).

**Problem 5.55.** Take $Q, T \in \mathcal{B}[\mathcal{H}]$, where $\mathcal{H}$ is a Hilbert space. Prove:

(a)  $-I \leq T^* = T \leq I$  if and only if  $T^* = T$  and  $\|T\| \leq 1$.

    *Hint*: Use Propositions 5.78 and 5.79 to show the "only if" part. On the other hand, use Proposition 5.79 and recall: $|\langle Tx\,;x\rangle| \leq \|T\|\|x\|^2$.

(b)  $O \leq Q \leq I \iff O \leq Q$ and $\|Q\| \leq 1 \iff Q^* = Q$ and $Q^2 \leq Q$.

**Problem 5.56.** Take $P, Q, T \in \mathcal{B}[\mathcal{H}]$ on a Hilbert space $\mathcal{H}$. Prove:

(a)  If $T^* = T$ and $T^n \xrightarrow{\ w\ } P$, then $P$ is an orthogonal projection.

    *Hint*: Problems 5.24 and 5.44 and Proposition 5.81.

(b)  If $O \leq Q \leq I$, then $Q^{n+1} \leq Q^n$ for every integer $n \geq 0$.

    *Hint*: Let $n$ be a *positive* integer and take any $x \in \mathcal{H}$. If $n$ is even, then use Problem 5.55(b) and Proposition 5.82 to verify: $\langle Q^n x\,;x\rangle = \|Q^{\frac{n}{2}}x\|^2 \leq \|Q\|\langle QQ^{\frac{n-2}{2}}x\,;Q^{\frac{n-2}{2}}x\rangle \leq \langle Q^{n-1}x\,;x\rangle$. If $n$ is odd, then $\langle Q^n x\,;x\rangle = \langle QQ^{\frac{n-1}{2}}x\,;Q^{\frac{n-1}{2}}x\rangle \leq \langle Q^{\frac{n-1}{2}}x\,;Q^{\frac{n-1}{2}}x\rangle = \langle Q^{n-1}x\,;x\rangle$.

(c)  If $O \leq Q \leq I$, then $Q^n \xrightarrow{\ s\ } P$ and $P$ is an orthogonal projection.

    *Hint*: Problems 5.55(b), 4.47(a), 5.24, items (a,b), Proposition 5.84.

**Problem 5.57.** This is our first problem that uses the square root of a nonnegative operator (Theorem 5.85). Take $T \in \mathcal{B}[\mathcal{H}]$ acting on a complex Hilbert space $\mathcal{H}$ and prove the following propositions.

(a)  If $T \neq O$ is self-adjoint, then $U_{\pm}(T) = \|T\|^{-1}\big(T \pm i(\|T\|^2 I - T^2)^{\frac{1}{2}}\big)$ are unitary operators in $\mathcal{B}[\mathcal{H}]$.

*Hint*: $\|T\|^{-2}T^2 \leq I$ so that $O \leq \|T\|^2 I - T^2$ (cf. Problems 5.45 and 5.55). See Proposition 5.73.

(b) Every operator on a complex Hilbert space is a linear combination of four unitary operators.

*Hint*: If $O \neq T = T^*$, then show that $T = \frac{\|T\|}{2}U_+(T) + \frac{\|T\|}{2}U_-(T)$. Apply the Cartesian decomposition (Problem 5.46) if $O \neq T \neq T^*$.

**Problem 5.58.** If $Q \in \mathcal{B}^+[\mathcal{H}]$, where $\mathcal{H}$ is a Hilbert space, then show that (cf. Theorem 5.85 and Proposition 5.86)

(a) $\langle Qx ; x \rangle = \|Q^{\frac{1}{2}}x\|^2 \leq \|Q\|^{\frac{1}{2}} \langle Q^{\frac{1}{2}}x ; x \rangle$ for every $x \in \mathcal{H}$,

(b) $\langle Q^{\frac{1}{2}}x ; x \rangle \leq \langle Qx ; x \rangle^{\frac{1}{2}} \|x\|$ for every $x \in \mathcal{H}$,

(c) $Q^{\frac{1}{2}} > 0$ if and only if $Q > O$,

(d) $Q^{\frac{1}{2}} \succ 0$ if and only if $Q \succ O$.

**Problem 5.59.** Take $Q, R \in \mathcal{B}^+[\mathcal{H}]$ on a Hilbert space $\mathcal{H}$. Prove:

(a) If $Q \leq R$ and $QR = RQ$, then $Q^2 \leq R^2$.

(b) $Q \leq R$ does not imply $Q^2 \leq R^2$.

*Hints*: (a) $\langle R Q^{\frac{1}{2}}x ; Q^{\frac{1}{2}}x \rangle = \langle Q R^{\frac{1}{2}}x ; R^{\frac{1}{2}}x \rangle$. (b) $Q = \begin{pmatrix} 1 & 0 \\ 0 & 0 \end{pmatrix}$ and $R = \begin{pmatrix} 2 & 1 \\ 1 & 1 \end{pmatrix}$.

**Problem 5.60.** Let $Q$ and $R$ be nonnegative operators acting on a Hilbert space. Use Problem 5.52 and Theorem 5.85 to prove that

$$QR = RQ \quad \text{implies} \quad Q^n R^m \geq O \quad \text{for every } m, n \geq 1.$$

Show that $p(Q)q(R) \geq O$ for every pair of polynomials $p$ and $q$ with positive coefficients whenever $Q \geq O$ and $R \geq O$ commute.

**Problem 5.61.** Let $\mathcal{H}$ and $\mathcal{K}$ be Hilbert spaces. Take any $T \in \mathcal{B}[\mathcal{H}, \mathcal{K}]$ and recall that $T^*T$ lies in $\mathcal{B}^+[\mathcal{H}]$. Set

$$|T| = (T^*T)^{\frac{1}{2}}$$

in $\mathcal{B}^+[\mathcal{H}]$ so that $|T|^2 = (T^*T)$. Prove the following assertions.

(a) $\|T\| = \||T|^2\|^{\frac{1}{2}} = \||T|\| = \||T|^{\frac{1}{2}}\|^2$.

(b) $\langle |T|x ; x \rangle = \||T|^{\frac{1}{2}}x\|^2 \leq \||T|x\| \|x\|$ for every $x \in \mathcal{H}$.

(c)  $\|Tx\|^2 = \||T|x\|^2 \le \|T\|\langle|T|x\,;x\rangle$ for every $x \in \mathcal{H}$.

Moreover, if $\mathcal{H} = \mathcal{K}$ (i.e., if $T \in \mathcal{B}[\mathcal{H}]$), then show that

(d)  $T^n \xrightarrow{s} O \iff |T^n| \xrightarrow{s} O \iff |T^n| \xrightarrow{w} O,$

(e)  $\mathcal{B}^+[\mathcal{H}] = \{T \in \mathcal{B}[\mathcal{H}]:\ T = |T|\}$   (i.e., $T \ge O$ if and only if $T = |T|$).

**Problem 5.62.** Let $Q$ be a nonnegative operator on a Hilbert space.

$$Q \text{ is compact if and only if } Q^{\frac{1}{2}} \text{ is compact.}$$

*Hint*: If $Q^{\frac{1}{2}}$ is compact then $Q$ is compact by Proposition 4.54. On the other hand, $\|Q^{\frac{1}{2}}x_n\|^2 = \langle Qx_n\,;x_n\rangle \le \sup_k\|x_k\|\,\|Qx_n\|$ (Problem 5.41).

Take $T \in \mathcal{B}[\mathcal{H}, \mathcal{K}]$, where $\mathcal{H}$ and $\mathcal{K}$ are Hilbert spaces. Also prove that

$$T \in \mathcal{B}_\infty[\mathcal{H}, \mathcal{K}] \iff T^*T \in \mathcal{B}_\infty[\mathcal{H}] \iff |T| \in \mathcal{B}_\infty[\mathcal{H}] \iff |T|^{\frac{1}{2}} \in \mathcal{B}_\infty[\mathcal{H}].$$

**Problem 5.63.** Consider a sequence $\{Q_n\}$ of nonnegative operators on a Hilbert space $\mathcal{H}$ (i.e., $Q_n \ge O$ for all $n$). Show that

(a)  $Q_n \xrightarrow{s} Q$ implies $Q_n^{1/2} \xrightarrow{s} Q^{1/2}$,

(b)  If $Q_n$ is compact for every $n$ and $Q_n \xrightarrow{u} Q$, then $Q_n^{1/2} \xrightarrow{u} Q^{1/2}$.

*Hints*: (a)  $Q \ge O$ by Problem 5.49. Recall: $Q^{1/2}$ is the strong limit of a sequence $\{p_k(Q)\}$ of polynomials in $Q$, where the polynomials $\{p_k\}$ themselves do not depend on $Q$; that is, $p_k(Q) \xrightarrow{s} Q^{1/2}$ for every $Q \ge O$ (cf. proof of Theorem 5.85). Verify that

$$
\begin{aligned}
\|(Q_n^{1/2} - Q^{1/2})x\| \ \le\ & \|(Q_n^{1/2} - p_k(Q_n))x\| \\
& + \|(p_k(Q_n) - p_k(Q))x\| + \|(p_k(Q) - Q^{1/2}))x\|.
\end{aligned}
$$

Take any $\varepsilon \ge 0$ and any $x \in \mathcal{H}$. Show that there exist positive integers $n_\varepsilon$ and $k_\varepsilon$ such that $\|(p_{k_\varepsilon}(Q) - Q^{1/2})x\| < \varepsilon$, $\|(p_{k_\varepsilon}(Q_n) - p_{k_\varepsilon}(Q))x\| < \varepsilon$ for every $n \ge n_\varepsilon$ (because $Q_n^j \xrightarrow{s} Q^j$ for every positive integer $j$ by Problem 4.46), and $\|(Q_{n_\varepsilon}^{1/2} - p_{k_\varepsilon}(Q_{n_\varepsilon}))x\| < \varepsilon$.

(b)  $Q \in \mathcal{B}_\infty[\mathcal{H}]$ (Theorem 4.53). Since $Q_n^{1/2} \xrightarrow{s} Q^{1/2}$ by part (a), $Q_n^{1/2}Q^{1/2} \xrightarrow{u} Q$ (Problems 5.62 and 4.57). So $(Q_n^{1/2} - Q^{1/2})^2 = Q_n + Q - Q_n^{1/2}Q^{1/2} - (Q_n^{1/2}Q^{1/2})^* \xrightarrow{u} O$ (Problem 5.26). $Q_n^{1/2} - Q^{1/2}$ is self-adjoint so that $\|Q_n^{1/2} - Q^{1/2}\|^2 = \|(Q_n^{1/2} - Q^{1/2})^2\|$ (Problem 5.45).

**Problem 5.64.** Let $\{e_\gamma\}_{\gamma\in\Gamma}$ and $\{f_\gamma\}_{\gamma\in\Gamma}$ be orthonormal bases for a Hilbert space $\mathcal{H}$ and take an arbitrary $T \in \mathcal{B}[\mathcal{H}]$. Use the Parseval identity to show that

$$\sum_{\gamma\in\Gamma} \|Te_\gamma\|^2 = \sum_{\gamma\in\Gamma} \|T^*f_\gamma\|^2 = \sum_{\alpha\in\Gamma}\sum_{\beta\in\Gamma} |\langle Te_\alpha; f_\beta\rangle|^2$$

whenever the family of nonnegative numbers $\{\|Te_\gamma\|^2\}_{\gamma\in\Gamma}$ is summable; that is, whenever $\sum_{\gamma\in\Gamma}\|Te_\gamma\|^2 < \infty$ (cf. Proposition 5.31). Apply the above result to $|T|^{\frac{1}{2}}$ and show that

$$\sum_{\gamma\in\Gamma}\langle |T|e_\gamma; e_\gamma\rangle = \sum_{\gamma\in\Gamma}\langle |T|f_\gamma; f_\gamma\rangle$$

whenever $\sum_{\gamma\in\Gamma}\langle |T|e_\gamma; e_\gamma\rangle < \infty$. Outcome: If the sum $\sum_{\gamma\in\Gamma}\langle |T|e_\gamma; e_\gamma\rangle$ exists in $\mathbb{R}$, then it is independent of the choice of the orthonormal basis $\{e_\gamma\}_{\gamma\in\Gamma}$ for $\mathcal{H}$. An operator $T \in \mathcal{B}[\mathcal{H}]$ is *trace-class* (or *nuclear*) if $\sum_{\gamma\in\Gamma}\langle |T|e_\gamma; e_\gamma\rangle < \infty$ (equivalently, if $\sum_{\gamma\in\Gamma}\||T|^{\frac{1}{2}}e_\gamma\|^2 < \infty$) for some orthonormal basis $\{e_\gamma\}_{\gamma\in\Gamma}$ for $\mathcal{H}$. Let $\mathcal{B}_1[\mathcal{H}]$ denote the subset of $\mathcal{B}[\mathcal{H}]$ consisting of all trace-class operators on $\mathcal{H}$. If $T \in \mathcal{B}_1[\mathcal{H}]$, then set

$$\|T\|_1 = \sum_{\gamma\in\Gamma}\langle |T|e_\gamma; e_\gamma\rangle = \sum_{\gamma\in\Gamma}\||T|^{\frac{1}{2}}e_\gamma\|^2.$$

**Problem 5.65.** Let $T \in \mathcal{B}[\mathcal{H}]$ be an operator on a Hilbert space $\mathcal{H}$, and let $\{e_\gamma\}_{\gamma\in\Gamma}$ be an orthonormal basis for $\mathcal{H}$. If $|T|^2$ is trace-class (i.e., if $\sum_{\gamma\in\Gamma}\||T|e_\gamma\|^2 < \infty$ or, equivalently, $\sum_{\gamma\in\Gamma}\|Te_\gamma\|^2 < \infty$ — Problem 5.61(c)), then $T$ is called a *Hilbert–Schmidt* operator. Let $\mathcal{B}_2[\mathcal{H}]$ denote the subset of $\mathcal{B}[\mathcal{H}]$ made up of all Hilbert–Schmidt operators on $\mathcal{H}$. Take $T \in \mathcal{B}_2[\mathcal{H}]$. According to Problems 5.61 and 5.64 set

$$\|T\|_2 = \|T^*T\|_1^{\frac{1}{2}} = \||T|^2\|_1^{\frac{1}{2}} = \Big(\sum_{\gamma\in\Gamma}\||T|e_\gamma\|^2\Big)^{\frac{1}{2}} = \Big(\sum_{\gamma\in\Gamma}\|Te_\gamma\|^2\Big)^{\frac{1}{2}}$$

for any orthonormal basis $\{e_\gamma\}_{\gamma\in\Gamma}$ for $\mathcal{H}$. Prove the following results.

(a) $T \in \mathcal{B}_2[\mathcal{H}]$ if and only if $|T| \in \mathcal{B}_2[\mathcal{H}]$, if and only if $|T|^2 \in \mathcal{B}_1[\mathcal{H}]$, and $\|T\|_2^2 = \||T|\|_2^2 = \||T|^2\|_1$.

(b) $T \in \mathcal{B}_1[\mathcal{H}]$ if and only if $|T| \in \mathcal{B}_1[\mathcal{H}]$, if and only if $|T|^{\frac{1}{2}} \in \mathcal{B}_2[\mathcal{H}]$, and $\|T\|_1 = \||T|\|_1 = \||T|^{\frac{1}{2}}\|_2^2$.

(c) $T^* \in \mathcal{B}_2[\mathcal{H}]$ and $\|T^*\|_2 = \|T\|_2$ if $T \in \mathcal{B}_2[\mathcal{H}]$. (*Hint*: Problem 5.64.)

(d) $\|T\| \le \|T\|_2$ for every $T \in \mathcal{B}_2[\mathcal{H}]$. (*Hint*: $\|Te\| \le \|T\|_2$ if $\|e\| = 1$.)

(e)  $T + S \in \mathcal{B}_2[\mathcal{H}]$ and $\|T + S\|_2 \leq \|T\|_2 + \|S\|_2$ whenever $T, S \in \mathcal{B}_2[\mathcal{H}]$.

*Hint*: Since $\sum_{\gamma \in \Gamma} \|T e_\gamma\| \|S e_\gamma\| \leq \left(\sum_{\gamma \in \Gamma} \|T e_\gamma\|^2\right)^{\frac{1}{2}} \left(\sum_{\gamma \in \Gamma} \|S e_\gamma\|^2\right)^{\frac{1}{2}} = \|T\|_2 \|S\|_2$ (Schwarz inequality in $\ell_\Gamma^2$), $\|T + S\|_2^2 \leq (\|T\|_2 + \|S\|_2)^2$.

(f)  $\mathcal{B}_2[\mathcal{H}]$ is a linear space and $\| \ \|_2$ is a norm on $\mathcal{B}_2[\mathcal{H}]$.

(g)  $ST$ and $TS$ lie in $\mathcal{B}_2[\mathcal{H}]$, and $\max\{\|ST\|_2, \|TS\|_2\} \leq \|S\| \|T\|_2$, for every $S$ in $\mathcal{B}[\mathcal{H}]$ and every $T$ in $\mathcal{B}_2[\mathcal{H}]$.

*Hint*: $\|STe_\gamma\|^2 \leq \|S\|^2 \|Te_\gamma\|^2$ and $\|(TS)^* e_\gamma\|^2 \leq \|S\|^2 \|T^* e_\gamma\|^2$.

(h)  $\mathcal{B}_2[\mathcal{H}]$ is a two-sided ideal of $\mathcal{B}[\mathcal{H}]$.

**Problem 5.66.** Consider the setup of the previous problem and prove:

(a)  $T + S \in \mathcal{B}_1[\mathcal{H}]$ and $\|T + S\|_1 \leq \|T\|_1 + \|S\|_1$ whenever $T, S \in \mathcal{B}_1[\mathcal{H}]$.

*Hint*: Let $T + S = W|T + S|$, $T = W_1|T|$ and $S = W_2|S|$ be the polar decompositions of $T + S$, $T$ and $S$, respectively, so that $|T + S| = W^*(T + S)$, $|T| = W_1^* T$ and $|S| = W_2^* T$. Verify that

$$
\sum_{\gamma \in \Gamma} \langle |T + S| e_\gamma ; e_\gamma \rangle \leq \sum_{\gamma \in \Gamma} |\langle T e_\gamma ; W^* e_\gamma \rangle| + \sum_{\gamma \in \Gamma} |\langle S e_\gamma ; W^* e_\gamma \rangle|
$$

$$
= \sum_{\gamma \in \Gamma} |\langle |T|^{\frac{1}{2}} e_\gamma ; |T|^{\frac{1}{2}} W_1^* W e_\gamma \rangle| + \sum_{\gamma \in \Gamma} |\langle |S|^{\frac{1}{2}} e_\gamma ; |S|^{\frac{1}{2}} W_2^* W e_\gamma \rangle|
$$

$$
\leq \left( \sum_{\gamma \in \Gamma} \| |T|^{\frac{1}{2}} e_\gamma \|^2 \right)^{\frac{1}{2}} \left( \sum_{\gamma \in \Gamma} \| |T|^{\frac{1}{2}} W_1^* W e_\gamma \|^2 \right)^{\frac{1}{2}}
$$

$$
+ \left( \sum_{\gamma \in \Gamma} \| |S|^{\frac{1}{2}} e_\gamma \|^2 \right)^{\frac{1}{2}} \left( \sum_{\gamma \in \Gamma} \| |S|^{\frac{1}{2}} W_2^* W e_\gamma \|^2 \right)^{\frac{1}{2}}
$$

$$
\leq \| |T|^{\frac{1}{2}} \|_2^2 \| W_1^* W \| + \| |S|^{\frac{1}{2}} \|_2^2 \| W_2^* W \| = \|T\|_1 + \|S\|_1 .
$$

(Problem 5.65(b,g); recall: $\|W\| = \|W_1\| = \|W_2\| = 1$.)

(b)  $\mathcal{B}_1[\mathcal{H}]$ is a linear space and $\| \ \|_1$ is a norm on $\mathcal{B}_1[\mathcal{H}]$.

(c)  $\mathcal{B}_1[\mathcal{H}] \subseteq \mathcal{B}_2[\mathcal{H}]$ (i.e., every trace-class operator is Hilbert–Schmidt). If $T$ lies in $\mathcal{B}_1[\mathcal{H}]$, then $\|T\|_2 \leq \|T\|_1$.

*Hint*: Problem 5.65(a,b,g) to prove the inclusion, and Problems 5.61(c) and 5.65(b) to prove the inequality.

(d)  $\mathcal{B}_2[\mathcal{H}] \subseteq \mathcal{B}_\infty[\mathcal{H}]$ (i.e., every Hilbert–Schmidt operator is compact).

*Hint*: Take $T \in \mathcal{B}_2[\mathcal{H}]$ so that $T^* \in \mathcal{B}_2[\mathcal{H}]$ (Proposition 5.65(c)), and hence $\sum_{\gamma \in \Gamma} \|T^* e_\gamma\|^2 < \infty$. Take an arbitrary integer $n \geq 1$. There exists a finite $N_n \subseteq \Gamma$ such that $\sum_{k \in N} \|T^* e_k\|^2 < \frac{1}{n}$ for all finite $N \subseteq \Gamma \backslash N_n$ (cf. Theorem 5.27). Thus $\sum_{\gamma \in \Gamma \backslash N_n} \|T^* e_\gamma\|^2 < \frac{1}{n}$. Recall that $Tx = \sum_{\gamma \in \Gamma} \langle Tx ; e_\gamma \rangle e_\gamma$ (cf. Theorem 5.48) and define $T_n \colon \mathcal{H} \to \mathcal{H}$ by $T_n x = \sum_{k \in N_n} \langle Tx ; e_k \rangle e_k$. Show that $T_n$ lies in $\mathcal{B}_0[\mathcal{H}]$ and $\|(T - T_n)x\|^2 = \sum_{\gamma \in \Gamma \backslash N_n} |\langle Tx ; e_\gamma \rangle|^2 \leq \sum_{\gamma \in \Gamma \backslash N_n} \|T^* e_\gamma\|^2 \|x\|^2$. Therefore, $\|T_n - T\| \to 0$. Conclude: $T \in \mathcal{B}_\infty[\mathcal{H}]$ (Problem 5.42).

(e) $T \in \mathcal{B}_1[\mathcal{H}]$ if and only if $T = AB$ for some $A, B \in \mathcal{B}_2[\mathcal{H}]$.

*Hint*: Let $T = W|T| = W|T|^{\frac{1}{2}}|T|^{\frac{1}{2}}$ be the polar decomposition of $T$. If $T \in \mathcal{B}_1[\mathcal{H}]$, then use Problem 5.65(b,g). Conversely, suppose $T = AB$, where $A, B \in \mathcal{B}_2[\mathcal{H}]$. Since $|T| = W^*T$, $|T| = W^*AB$ with $A^*W \in \mathcal{B}_2[\mathcal{H}]$ (Problem 5.65(c,g)). Verify that $\sum_{\gamma \in \Gamma} \langle |T| e_\gamma ; e_\gamma \rangle \leq \sum_{\gamma \in \Gamma} \|Be_\gamma\| \|A^*We_\gamma\| \leq (\sum_{\gamma \in \Gamma} \|Be_\gamma\|^2)^{\frac{1}{2}} (\sum_{\gamma \in \Gamma} \|A^*We_\gamma\|^2)^{\frac{1}{2}}$, and hence $\|T\|_1 \leq \|B\|_2 \|A^*W\|_2$.

(f) $ST$ and $TS$ lie in $\mathcal{B}_1[\mathcal{H}]$ for every $T$ in $\mathcal{B}_1[\mathcal{H}]$ and every $S$ in $\mathcal{B}[\mathcal{H}]$.

*Hint*: Apply (e). $T = AB$ for some $A, B \in \mathcal{B}_2[\mathcal{H}]$. $SA$ and $BS$ lie in $\mathcal{B}_2[\mathcal{H}]$, and so $ST = SAB$ and $TS = ABS$ lie in $\mathcal{B}_1[\mathcal{H}]$.

(g) $\mathcal{B}_1[\mathcal{H}]$ is a two-sided ideal of $\mathcal{B}[\mathcal{H}]$.

**Problem 5.67.** Let $\{e_\gamma\}_{\gamma \in \Gamma}$ be an arbitrary orthonormal basis for a Hilbert space $\mathcal{H}$ and take any $T \in \mathcal{B}_1[\mathcal{H}]$. Show that

$$\sum_{\gamma \in \Gamma} |\langle Te_\gamma ; e_\gamma \rangle| < \infty.$$

*Hint*: $2|\langle Te_\gamma ; e_\gamma \rangle| = 2|\langle ABe_\gamma ; e_\gamma \rangle| \leq 2\|Be_\gamma\| \|A^*e_\gamma\|$ for $A, B \in \mathcal{B}_2[\mathcal{H}]$ (Problem 5.66(e)), and hence $2|\langle Te_\gamma ; e_\gamma \rangle| \leq \|Be_\gamma\|^2 + \|A^*e_\gamma\|^2$. Then

$$\sum_{\gamma \in \Gamma} |\langle Te_\gamma ; e_\gamma \rangle| \leq \frac{1}{2}(\|A\|_2^2 + \|B\|_2^2)$$

(cf. Problem 5.65(c)).

Therefore, according to Corollary 5.29, $\{\langle Te_\gamma ; e_\gamma \rangle\}$ is a summable family of scalars (since $\mathbb{F}$ is a Banach space). Let $\sum_{\gamma \in \Gamma} \langle Te_\gamma ; e_\gamma \rangle$ in $\mathbb{F}$ be its sum and show that

$$\sum_{\gamma \in \Gamma} \langle Te_\gamma ; e_\gamma \rangle \text{ does not depend on } \{e_\gamma\}_{\gamma \in \Gamma}.$$

*Hint*: $\sum_{\alpha \in \Gamma} \langle Te_\alpha ; e_\alpha \rangle = \sum_{\alpha \in \Gamma} \sum_{\beta \in \Gamma} \langle Te_\alpha ; f_\beta \rangle \langle f_\beta ; e_\alpha \rangle$, where $\{e_\gamma\}_{\gamma \in \Gamma}$ and $\{f_\gamma\}_{\gamma \in \Gamma}$ are any orthonormal bases for $\mathcal{H}$ (cf. Theorem 5.48(c)). Now observe that $\sum_{\beta \in \Gamma} \sum_{\alpha \in \Gamma} \langle Te_\alpha ; f_\beta \rangle \langle f_\beta ; e_\alpha \rangle = \sum_{\beta \in \Gamma} \langle f_\beta ; T^* f_\beta \rangle = \sum_{\beta \in \Gamma} \langle Tf_\beta ; f_\beta \rangle$.

If $T \in \mathcal{B}_1[\mathcal{H}]$ and $\{e_\gamma\}_{\gamma\Gamma}$ is any orthonormal basis for $\mathcal{H}$, then set

$$\mathrm{tr}(T) = \sum_{\gamma\in\Gamma} \langle Te_\gamma ; e_\gamma \rangle \quad \text{so that} \quad \|T\|_1 = \mathrm{tr}(|T|).$$

Hence $\mathcal{B}_1[\mathcal{H}] = \{T \in \mathcal{B}[\mathcal{H}]: \mathrm{tr}(|T|) < \infty\}$. The number $\mathrm{tr}(T)$ is called the *trace* of $T \in \mathcal{B}_1[\mathcal{H}]$ (thus the terminology "trace-class"). Warning: $T \in \mathcal{B}[\mathcal{H}]$ and $\sum_{\gamma\in\Gamma} \langle Te_\gamma ; e_\gamma \rangle < \infty$ for *some* orthonormal basis $\{e_\gamma\}_{\gamma\in\Gamma}$ for $\mathcal{H}$ does not imply $T \in \mathcal{B}_1[\mathcal{H}]$. However, if $\sum_{\gamma\in\Gamma} \langle |T|e_\gamma ; e_\gamma \rangle < \infty$ for some orthonormal basis $\{e_\gamma\}_{\gamma\in\Gamma}$ for $\mathcal{H}$, then $T \in \mathcal{B}_1[\mathcal{H}]$ (Problem 5.64).

**Problem 5.68.** Consider the setup of the previous problem and prove:

(a)  $\mathrm{tr}: \mathcal{B}_1[\mathcal{H}] \to \mathbb{F}$ is a linear functional.

(b)  $|\mathrm{tr}(T)| \leq \|T\|_1$ for every $T \in \mathcal{B}_1[\mathcal{H}]$ (i.e., $\mathrm{tr}: (\mathcal{B}_1[\mathcal{H}], \| \ \|_1) \to \mathbb{F}$ is a contraction, and hence a bounded linear functional).

*Hint*: Let $T = W|T|$ be the polar decomposition of $T$. Recall that $\|W\| = 1$ and verify:

$$|\mathrm{tr}(T)| \leq \sum_{\gamma\in\Gamma} |\langle |T|^{\frac{1}{2}} e_\gamma ; |T|^{\frac{1}{2}} W^* e_\gamma \rangle|$$

$$\leq \left( \sum_{\gamma\in\Gamma} \| |T|^{\frac{1}{2}} e_\gamma \|^2 \right)^{\frac{1}{2}} \left( \sum_{\gamma\in\Gamma} \| |T|^{\frac{1}{2}} W^* e_\gamma \|^2 \right)^{\frac{1}{2}} \leq \|T\|_1$$

(Problem 5.65).

(c)  $\mathrm{tr}(T^*) = \overline{\mathrm{tr}(T)}$ for every $T \in \mathcal{B}_1[\mathcal{H}]$.

(d)  $\mathrm{tr}(TS) = \mathrm{tr}(ST)$ whenever $T \in \mathcal{B}_1[\mathcal{H}]$ and $S \in \mathcal{B}[\mathcal{H}]$.

*Hint*: $\overline{\mathrm{tr}(TS)} = \sum_{\alpha\in\Gamma} \langle T^*e_\alpha; Se_\alpha \rangle = \sum_{\alpha\in\Gamma}\sum_{\beta\in\Gamma} \langle T^*e_\alpha; f_\beta \rangle \langle f_\beta; Se_\alpha \rangle$ and $\overline{\mathrm{tr}(TS)} = \sum_{\beta\in\Gamma} \langle S^* f_\beta; Tf_\beta \rangle = \sum_{\beta\in\Gamma}\sum_{\alpha\in\Gamma} \langle S^* f_\beta ; e_\alpha \rangle \langle e_\alpha ; Tf_\beta \rangle$ (cf. Problem 5.66(f), item (c), and Theorem 5.48(c)).

(e)  $|\mathrm{tr}(S|T|)| = |\mathrm{tr}(|T|S)| \leq \|S\| \|T\|_1$ if $T \in \mathcal{B}_1[\mathcal{H}]$ and $S \in \mathcal{B}[\mathcal{H}]$.

*Hint*: Problems 5.65(b,g) and 5.66(f). Verify: $\left| \sum_{\gamma\in\Gamma} \langle S|T|e_\gamma ; e_\gamma \rangle \right| \leq \sum_{\gamma\in\Gamma} |\langle |T|^{\frac{1}{2}} e_\gamma ; |T|^{\frac{1}{2}} S^* e_\gamma \rangle| \leq \| |T|^{\frac{1}{2}} \|_2 \| |T|^{\frac{1}{2}} S^* \|_2$. See item (d).

(f)  $T^* \in \mathcal{B}_1[\mathcal{H}]$ and $\|T^*\|_1 = \|T\|_1$ for every $T \in \mathcal{B}_1[\mathcal{H}]$.

*Hint*: Let $T = W_1|T|$ and $T^* = W_2|T^*|$ be the polar decompositions of $T$ and $T^*$. Since $|T^*| = W_2^* T^* = W_2^*|T|W_1^*$, it follows by Problems 5.65(b) and 5.66(f) that $T^* \in \mathcal{B}_1[\mathcal{H}]$. Show that $\|T^*\|_1 = \mathrm{tr}(|T^*|) = \mathrm{tr}(W_2^*|T|W_1^*) \leq \|W_1^* W_2^*\| \|T\|_1$ (Problem 5.65(b) and items (d) and (e)). But $\|W_1^* W_2^*\| \leq \|W_1\| \|W_2\| = 1$. Therefore, $\|T^*\|_1 \leq \|T\|_1$. Dually, $\|T\|_1 \leq \|T^*\|_1$.

(g) $\max\{\|ST\|_1, \|TS\|_1\} \leq \|S\|\|T\|_1$ whenever $T \in \mathcal{B}_1[\mathcal{H}]$ and $S \in \mathcal{B}[\mathcal{H}]$.

*Hint*: Let $T = W|T|$, $ST = W_1|ST|$ and $TS = W_2|TS|$ be the polar decompositions of $T$, $ST$ and $TS$. Verify that $\|ST\|_1 = \mathrm{tr}(|ST|) = \mathrm{tr}(W_1^* S W |T|)$ and $\|TS\|_1 = \mathrm{tr}(|TS|) = \mathrm{tr}(W_2^* W |T|S)$. Use items (d) and (e) and recall that $\|W\| = \|W_1\| = \|W_2\| = 1$.

(h) $\mathcal{B}_0[\mathcal{H}] \subseteq \mathcal{B}_1[\mathcal{H}]$ (i.e., every finite-rank operator is trace-class).

*Hint*: If $\dim \mathcal{R}(T)$ is finite, then $\dim \mathcal{N}(T^*)^\perp$ is finite (Proposition 5.76). Let $\{f_\alpha\}$ be an orthonormal basis for $\mathcal{N}(T^*)$ and let $\{g_k\}$ be a *finite* orthonormal basis for $\mathcal{N}(T^*)^\perp$. Since $\mathcal{H} = \mathcal{N}(T^*) + \mathcal{N}(T^*)^\perp$ (Theorem 5.20), $\{e_\gamma\} = \{f_\alpha\} \cup \{g_k\}$ is an orthonormal basis for $\mathcal{H}$ (Problem 5.11). Now, either $T^*e_\gamma = 0$ or $T^*e_\gamma = T^*g_k$. Show that $\sum_\gamma \langle |T^*|e_\gamma; e_\gamma \rangle = \sum_k \langle |T^*|g_k; g_k \rangle < \infty$ (e.g., see Problem 5.61(c)). Thus $T^* \in \mathcal{B}_1[\mathcal{H}]$ (Problem 5.64), and therefore $T \in \mathcal{B}_1[\mathcal{H}]$ by item (f).

**Problem 5.69.** Let $(\mathcal{B}_1[\mathcal{H}], \|\ \|_1)$ and $(\mathcal{B}_2[\mathcal{H}], \|\ \|_2)$ be the normed spaces of Problems 5.65(f) and 5.66(b). Show that

(a) $(\mathcal{B}_1[\mathcal{H}], \|\ \|_1)$ is a Banach space.

*Hint*: Take a $\mathcal{B}_1[\mathcal{H}]$-valued sequence $\{T_n\}$. If $\{T_n\}$ is Cauchy in $(\mathcal{B}_1[\mathcal{H}], \|\|_1)$, then it is Cauchy in the Banach space $(\mathcal{B}[\mathcal{H}], \|\ \|)$ (Problems 5.65(d) and 5.66(c)), and so $T_n \overset{u}{\longrightarrow} T$ for some $T \in \mathcal{B}[\mathcal{H}]$. Use Problems 5.26 and 4.46 to verify that $|T_n|^2 \overset{u}{\longrightarrow} |T|^2$, and hence $|T_n|^{\frac{1}{2}} \overset{u}{\longrightarrow} |T|^{\frac{1}{2}}$ (Problems 5.63(b) and 5.66(c,d)). Show that

$$\sum_{\gamma \in \Gamma} \||T|^{\frac{1}{2}} e_\gamma\|^2 \leq \limsup_n \sum_{\gamma \in \Gamma} \||T_n|^{\frac{1}{2}} e_\gamma\|^2 \leq \sup_n \|T_n\|_1 < \infty$$

(recall: $\{T_n\}$ is Cauchy in $(\mathcal{B}_1[\mathcal{H}], \|\ \|_1)$). Thus $T$ lies in $\mathcal{B}_1[\mathcal{H}]$. Since $T_n - T \overset{u}{\longrightarrow} O$, it follows that $|T_n - T|^{\frac{1}{2}} \overset{u}{\longrightarrow} O$ (Problem 5.61(a)). But

$$\|T_n - T\|_1 = \sum_{\gamma \in \Gamma} \||T_n - T|^{\frac{1}{2}} e_\gamma\|^2$$

$$= \sum_{k \in N_\varepsilon} \||T_n - T|^{\frac{1}{2}} e_k\|^2 + \sup_N \sum_{k \in N} \||T_n - T|^{\frac{1}{2}} e_k\|^2 < \infty$$

for every finite $N_\varepsilon \subseteq \Gamma$, where the supremum is taken over all finite $N \subseteq \Gamma \backslash N_\varepsilon$ (Proposition 5.31). Use Theorem 5.27 to conclude that $\|T_n - T\|_1 \to 0$.

Consider the function $\langle\ ;\ \rangle \colon \mathcal{B}_2[\mathcal{H}] \times \mathcal{B}_2[\mathcal{H}] \to \mathbb{F}$ given by

$$\langle T ; S \rangle = \mathrm{tr}(S^*T)$$

for every $S, T \in \mathcal{B}_2[\mathcal{H}]$. Show that $\langle\ ;\ \rangle$ is an inner product on $\mathcal{B}_2[\mathcal{H}]$ that induces the norm $\|\ \|_2$. (*Hint*: Problem 5.68(a,c)). Moreover,

(b)  $(\mathcal{B}_2[\mathcal{H}], \langle\ ;\ \rangle)$ is a Hilbert space.

Recall that $\mathcal{B}_0[\mathcal{H}] \subseteq \mathcal{B}_1[\mathcal{H}] \subseteq \mathcal{B}_2[\mathcal{H}] \subseteq \mathcal{B}_\infty[\mathcal{H}]$ and $\mathcal{B}_0[\mathcal{H}]$ is dense in the Banach space $(\mathcal{B}_\infty[\mathcal{H}], \|\ \|)$. Now show that

(c)  $\mathcal{B}_0[\mathcal{H}]$ is dense in $(\mathcal{B}_1[\mathcal{H}], \|\ \|_1)$ and in $(\mathcal{B}_2[\mathcal{H}], \|\ \|_2)$.


**Problem 5.70.** Two normed spaces $\mathcal{X}$ and $\mathcal{Y}$ are topologically isomorphic if there exists a topological isomorphism between them (i.e., if there exists $W$ in $\mathcal{G}[\mathcal{X}, \mathcal{Y}]$ — see Section 4.6). Two inner product spaces $\mathcal{X}$ and $\mathcal{Y}$ are unitarily equivalent if there exists a unitary transformation between them (i.e., if there exists $U$ in $\mathcal{G}[\mathcal{X}, \mathcal{Y}]$ unitary — see Section 5.6). *Two Hilbert spaces are topologically isomorphic if and only if they are unitarily equivalent.* That is, if $\mathcal{H}$ and $\mathcal{K}$ are Hilbert spaces, then

$$\mathcal{G}[\mathcal{H}, \mathcal{K}] \neq \varnothing \quad \text{if and only if} \quad \{U \in \mathcal{G}[\mathcal{H}, \mathcal{K}]:\ U \text{ is unitary}\} \neq \varnothing.$$

*Hint*: If $W \in \mathcal{G}[\mathcal{H}, \mathcal{K}]$, then $|W| = (W^*W)^{\frac{1}{2}} \in \mathcal{G}^+[\mathcal{H}]$ (see Problems 5.50(b) and 5.58(d)). Show that $U = W|W|^{-1} \in \mathcal{G}[\mathcal{H}, \mathcal{K}]$ is unitary (Proposition 5.73) and that $U|W|$ is the polar decomposition of $W$ (Corollary 5.90).

# 6
# The Spectral Theorem

The Spectral Theorem is a landmark in the theory of operators on Hilbert space, providing a full statement about the nature and structure of *normal* operators. Normal operators play a central role in operator theory; they will be defined in Section 6.1 below. It is customary to say that the Spectral Theorem can be applied to answer essentially all questions on normal operators. This indeed is the case as far as "essentially all" means "almost all" or "all the principal": there exist open questions on normal operators. First we consider the class of normal operators and its relatives (predecessors and successors). Next, the notion of spectrum of an operator acting on a complex Banach space is introduced. The Spectral Theorem for compact normal operators is fully investigated, yielding the concept of diagonalization. The Spectral Theorem for plain normal operators needs measure theory. We would not dare to relegate measure theory to an appendix just to support a proper proof of the Spectral Theorem for plain normal operators. Instead we assume just once, at the very last section of this book, that the reader has some familiarity with measure theory, just enough to grasp the statement of the Spectral Theorem for plain normal operators after having proved it for compact normal operators.

## 6.1 Normal Operators

Throughout this section $\mathcal{H}$ stands for a Hilbert space. An operator $T \in \mathcal{B}[\mathcal{H}]$ is *normal* if it commutes with its adjoint (i.e., $T$ is normal if $T^*T = TT^*$). Here is another characterization of normal operators.

**Proposition 6.1.** *The following assertions are pairwise equivalent.*

(a) $T$ *is normal.*

(b) $\|T^*x\| = \|Tx\|$ *for every* $x \in \mathcal{H}$.

(c) $T^n$ *is normal for every positive integer* $n$.

(d) $\|T^{*n}x\| = \|T^n x\|$ *for every* $x \in \mathcal{H}$ *and every* $n \geq 1$.

*Proof.* If $T \in \mathcal{B}[\mathcal{H}]$, then $\|T^*x\|^2 - \|Tx\|^2 = \langle (TT^* - T^*T)x \, ; x \rangle$ for every $x \in \mathcal{H}$. Since $TT^* - T^*T$ is self-adjoint, it follows by Corollary 5.80 that $TT^* = T^*T$ if and only if $\|T^*x\| = \|Tx\|$ for every $x \in \mathcal{H}$. This shows that (a)$\Leftrightarrow$(b). Therefore, as $T^{*n} = T^{n*}$ for every $n \geq 1$ (cf. Problem 5.24), (c)$\Leftrightarrow$(d). If $T^*$ commutes with $T$, then it commutes with $T^n$ and, dually, $T^n$ commutes with $T^{*n} = T^{n*}$. Hence (a)$\Rightarrow$(c). Since (d)$\Rightarrow$(b) trivially, the proposition is proved.     $\square$

Clearly, *every self-adjoint operator is normal* ($T^* = T$ implies $T^*T = TT^* = T^2$), and so are the nonnegative operators and, in particular, the orthogonal projections — cf. Proposition 5.81). It is also clear that *every unitary operator is normal* ($U \in \mathcal{B}[\mathcal{H}]$ is unitary if and only if $U^*U = UU^* = I$ — cf. Proposition 5.73). In fact, normality distinguishes the orthogonal projections among the projections, and the unitaries among the isometries.

**Proposition 6.2.** $P \in \mathcal{B}[\mathcal{H}]$ *is an orthogonal projection if and only if it is a normal projection.*

*Proof.* If $P$ is an orthogonal projection, then it is a self-adjoint projection (Proposition 5.81), and hence a normal projection. On the other hand, if $P$ is normal, then $\|P^*x\| = \|Px\|$ for every $x \in \mathcal{H}$ (by the previous proposition) so that $\mathcal{N}(P^*) = \mathcal{N}(P)$. If $P$ is a projection, then $\mathcal{R}(P) = \mathcal{N}(I - P)$ so that $\mathcal{R}(P) = \mathcal{R}(P)^-$ by Proposition 4.13. Therefore, if $P$ is a normal projection, then $\mathcal{N}(P)^\perp = \mathcal{N}(P^*)^\perp = \mathcal{R}(P)^- = \mathcal{R}(P)$ (Proposition 5.76) so that $\mathcal{R}(P) \perp \mathcal{N}(P)$, and hence $P$ is an orthogonal projection.     $\square$

**Proposition 6.3.** *Let* $U$ *be an operator in* $\mathcal{B}[\mathcal{H}]$. *The following assertions are pairwise equivalent.*

(a) $U$ *is unitary*    (i.e., $U^*U = UU^* = I$).

(b) $\|U^{*n}x\| = \|U^n x\| = \|x\|$ *for every* $x \in \mathcal{H}$ *and every* $n \geq 1$.

(c) $\|U^*x\| = \|Ux\| = \|x\|$ *for every* $x \in \mathcal{H}$.

(d) $U$ *is a normal isometry.*

*Proof.* (a)$\Leftrightarrow$(b) by Propositions 4.37 and 5.73, (b)$\Leftrightarrow$(c) by Proposition 4.37, and (c)$\Leftrightarrow$(d) Propositions 4.37 and 5.61.                    $\square$

Given any operator $T \in \mathcal{B}[\mathcal{H}]$ set

$$D = T^*T - TT^*$$

in $\mathcal{B}[\mathcal{H}]$. Recall that $D = D^*$ (i.e., $D$ is always self-adjoint). Moreover,

$$T \text{ is normal if and only if } D = O.$$

An operator $T \in \mathcal{B}[\mathcal{H}]$ is *quasinormal* if it commutes with $T^*T$. That is, if $T^*TT = TT^*T$. Equivalently, if $(T^*T - TT^*)T = O$. Therefore,

$$T \text{ is quasinormal if and only if } DT = O.$$

It is clear that *every normal operator is quasinormal*. Observe that *every isometry is quasinormal*. (Proof: If $V \in \mathcal{B}[\mathcal{H}]$ is an isometry, then $V^*V = I$ so that $V^*VV - VV^*V = O$.)

**Proposition 6.4.** *If $T = WQ$ is the polar decomposition of an operator $T \in \mathcal{B}[\mathcal{H}]$, then* (a)
$$WQ = QW \text{ if and only if } T \text{ is quasinormal.}$$

*In this case we have $WT = TW$ and $QT = TQ$. Moreover,* (b) *if $T$ is normal, then $W|_{\mathcal{N}(W)^\perp}$ is unitary. That is, the partial isometry of the polar decomposition of any normal operator is, in fact, a "partial unitary transformation" in the following sense. $W = UP$, where $P$ is the orthogonal projection onto $\mathcal{N}^\perp$, with $\mathcal{N} = \mathcal{N}(T) = \mathcal{N}(W) = \mathcal{N}(Q)$, and $U \colon \mathcal{N}^\perp \to \mathcal{H}$ is unitary.*

*Proof.* (a) Let $T = WQ$ be the polar decomposition of $T$ so that $Q^2 = T^*T$ (Theorem 5.89). If $WQ = QW$, then $Q^2W = QWQ = WQ^2$, and hence $TT^*T = WQQ^2 = Q^2WQ = T^*TT$ (i.e., $T$ is quasinormal). Conversely, if $TT^*T = T^*TT$, then $TQ^2 = Q^2T$. Thus $TQ = QT$ by Theorem 5.85 (for $Q = (Q^2)^{\frac{1}{2}}$) so that $WQQ = QWQ$; that is, $(WQ - QW)Q = O$. Therefore, $\mathcal{R}(Q)^- \subseteq \mathcal{N}(WQ - QW)$ and so

$$\mathcal{N}(Q)^\perp \subseteq \mathcal{N}(WQ - QW)$$

by Proposition 5.76 (for $Q = Q^*$). Recall that $\mathcal{N}(Q) = \mathcal{N}(W)$ (Theorem 5.89). If $u \in \mathcal{N}(Q)$, then $u \in \mathcal{N}(W)$ so that $(WQ - QW)u = 0$. Hence

$$\mathcal{N}(Q) \subseteq \mathcal{N}(WQ - QW).$$

The above displayed inclusions imply $\mathcal{N}(WQ - QW) = \mathcal{H}$ (Problem 5.7(b)); that is, $WQ = QW$. Since $T = WQ$, it follows at once that $W$ and $Q$ commute with $T$ whenever they commute with each other.

(b) Recall from Theorem 5.89 that the null spaces of $T$, $W$ and $Q$ coincide. Thus (cf. Proposition 5.86) set

$$\mathcal{N} = \mathcal{N}(T) = \mathcal{N}(W) = \mathcal{N}(Q) = \mathcal{N}(Q^2).$$

According to Proposition 5.87, $W = VP$, where $V: \mathcal{N}^\perp \to \mathcal{H}$ is an isometry and $P: \mathcal{H} \to \mathcal{H}$ is the orthogonal projection onto $\mathcal{N}^\perp$. Since $\mathcal{R}(Q)^- = \mathcal{N}(Q)^\perp = \mathcal{N}^\perp = \mathcal{R}(P)$, it follows that $PQ = Q$. Taking the adjoint and recalling that $P = P^*$ (Proposition 5.81) we get

$$PQ = QP = Q.$$

Moreover, since $V \in \mathcal{B}[\mathcal{N}^\perp, \mathcal{H}]$, its adjoint $V^*$ lies in $\mathcal{B}[\mathcal{H}, \mathcal{N}^\perp]$. Then $\mathcal{R}(V^*) \subseteq \mathcal{N}^\perp = \mathcal{R}(P)$, which implies that $PV^* = V^*$. Hence

$$VPV^* = VV^*.$$

These identities hold for the polar decomposition of every $T \in \mathcal{B}[\mathcal{H}]$. Now suppose $T$ is normal so that $T$ is quasinormal. By part (a) we get

$$Q^2 = T^*T = TT^* = WQQW^* = Q^2WW^* = Q^2VPV^* = Q^2VV^*.$$

Therefore, $Q^2(I - VV^*) = O$ and hence

$$\mathcal{R}(I - VV^*) \subseteq \mathcal{N}(Q^2) = \mathcal{N}.$$

However, according to Proposition 5.76 and recalling that $\mathcal{R}(T)^- = \mathcal{R}(W) = \mathcal{R}(V)$, we also get

$$\begin{aligned}
\mathcal{R}(I - VV^*) &= \mathcal{R}(VV^*) = \mathcal{R}(V) = R(T)^- \\
&= \mathcal{R}(TT^*)^- = \mathcal{R}(T^*T)^- = \mathcal{R}(Q^2)^- = \mathcal{N}(Q^2)^\perp = \mathcal{N}^\perp.
\end{aligned}$$

Hence $\mathcal{R}(I - VV^*) = \{0\}$ (for $\mathcal{R}(I - VV^*) \subseteq \mathcal{N} \cap \mathcal{N}^\perp = \{0\}$), which means that $I - VV^* = O$. That is, $VV^* = I$ so that the isometry $V$ also is a coisometry. Outcome: $V$ is unitary (Proposition 5.73). $\qquad\qquad\square$

A *part* of an operator is a restriction of it to an invariant subspace. For instance, *every unilateral shift is a part of some bilateral shift* (of the same multiplicity). This takes a little proving. In this sense, every unilateral shift has an extension that is a bilateral shift. Recall that unilateral shifts are isometries, and bilateral shifts are unitary operators (see Problems 5.29 and 5.30). The above italicized result can be extended as follows. *Every isometry is a part of a unitary operator*. This takes a little proving too. Since every isometry is quasinormal, and since every unitary operator is normal, we might expect that *every quasinormal operator is a part of a normal operator*. This actually is the case. We shall call an operator *subnormal* if it is a part of a normal operator. Equivalently, if it has a normal extension. Precisely, an operator $T$ on a Hilbert space $\mathcal{H}$ is subnormal if there exists a Hilbert space $\mathcal{K}$ including $\mathcal{H}$ and a normal operator $N$ on $\mathcal{K}$ such that $\mathcal{H}$ is $N$-invariant (i.e., $N(\mathcal{H}) \subseteq \mathcal{H}$) and $T$

is the restriction of $N$ to $\mathcal{H}$ (i.e., $T = N|_{\mathcal{H}}$). In other words, $T \in \mathcal{B}[\mathcal{H}]$ is subnormal if $\mathcal{H}$ is a subspace of a larger Hilbert space $\mathcal{K}$, so that $\mathcal{K} = \mathcal{H} \oplus \mathcal{H}^{\perp}$ by Theorem 5.25, and the operator

$$N = \begin{pmatrix} T & X \\ O & Y \end{pmatrix} : \mathcal{H} \oplus \mathcal{H}^{\perp} \to \mathcal{H} \oplus \mathcal{H}^{\perp}$$

in $\mathcal{B}[\mathcal{K}]$ is normal for some $X \in \mathcal{B}[\mathcal{H}^{\perp}, \mathcal{H}]$ and $Y \in \mathcal{B}[\mathcal{H}^{\perp}]$ (see Example 2O). Recall that, writing the orthogonal direct sum decomposition $\mathcal{K} = \mathcal{H} \oplus \mathcal{H}^{\perp}$, we are identifying $\mathcal{H} \subseteq \mathcal{K}$ with $\mathcal{H} \oplus \{0\}$ (which is a subspace of $\mathcal{H} \oplus \mathcal{H}^{\perp}$) and $\mathcal{H}^{\perp} \subseteq \mathcal{K}$ with $\{0\} \oplus \mathcal{H}^{\perp}$ (which also is a subspace of $\mathcal{H} \oplus \mathcal{H}^{\perp}$).

**Proposition 6.5.** *Every quasinormal operator is subnormal.*

*Proof.* Suppose $T \in \mathcal{B}[\mathcal{H}]$ is quasinormal.

*Claim.* $\mathcal{N}(T)$ reduces $T$.

*Proof.* If $T$ is quasinormal, then $T^*T$ commutes with both $T$ and $T^*$ so that $\mathcal{N}(T^*T)$ reduces $T$ (cf. Problem 5.34). But $\mathcal{N}(T^*T) = \mathcal{N}(T)$ by Proposition 5.76. $\square$

Thus $T = O \oplus S$ on $\mathcal{H} = \mathcal{N}(T) \oplus \mathcal{N}(T)^{\perp}$, with $O = T|_{\mathcal{N}(T)} : \mathcal{N}(T) \to \mathcal{N}(T)$ and $S = T|_{\mathcal{N}(T)^{\perp}} : \mathcal{N}(T)^{\perp} \to \mathcal{N}(T)^{\perp}$. Note that $T^*T = O \oplus S^*S$, and so

$$(O \oplus S^*S)(O \oplus S) = T^*TT = TT^*T = (O \oplus S)(O \oplus S^*S).$$

Then $O \oplus S^*SS = O \oplus SS^*S$, and hence $S^*SS = SS^*S$. That is, $S$ is quasinormal. Since $\mathcal{N}(S) = \mathcal{N}(T|_{\mathcal{N}(T)^{\perp}}) = \{0\}$, it follows by Corollary 5.90 that the partial isometry of the polar decomposition of $S \in \mathcal{B}[\mathcal{N}(T)^{\perp}]$ is an isometry. Therefore $S = VQ$, where $V \in \mathcal{B}[\mathcal{N}(T)^{\perp}]$ is an isometry (so that $V^*V = I$) and $Q \in \mathcal{B}[\mathcal{N}(T)^{\perp}]$ is nonnegative. But $S = VQ = QV$ by Proposition 6.4, and hence $S^* = QV^* = V^*Q$. Set

$$U = \begin{pmatrix} V & I - VV^* \\ O & V^* \end{pmatrix} \quad \text{and} \quad R = \begin{pmatrix} Q & O \\ O & Q \end{pmatrix}$$

in $\mathcal{B}[\mathcal{N}(T)^{\perp} \oplus \mathcal{N}(T)^{\perp}]$. Observe that $U$ is unitary. In fact,

$$U^*U = \begin{pmatrix} V^* & O \\ I - VV^* & V \end{pmatrix} \begin{pmatrix} V & I - VV^* \\ O & V^* \end{pmatrix}$$

$$= \begin{pmatrix} I & O \\ O & I \end{pmatrix} = \begin{pmatrix} V & I - VV^* \\ O & V^* \end{pmatrix} \begin{pmatrix} V^* & O \\ I - VV^* & V \end{pmatrix} = UU^*.$$

Also note that the nonnegative operator $R$ commutes with U:

$$UR = \begin{pmatrix} VQ & (I - VV^*)Q \\ O & V^*Q \end{pmatrix} = \begin{pmatrix} S & Q(I - VV^*) \\ O & S^* \end{pmatrix}$$

$$= \begin{pmatrix} QV & Q(I - VV^*) \\ O & QV^* \end{pmatrix} = RU.$$

Now set $N = UR$ in $\mathcal{B}[\mathcal{N}(T)^\perp \oplus \mathcal{N}(T)^\perp]$. The middle operator matrix says that $S$ is a part of $N$ (i.e., $\mathcal{N}(T)^\perp$ is $N$-invariant and $S = N|_{\mathcal{N}(T)^\perp}$). Finally, note that

$$N^*N \; = \; RU^*UR \; = \; R^2 \; = \; R^2U^*U \; = \; UR^2U^* \; = \; NN^*$$

and, therefore, $N$ is normal. Conclusion: $S$ is subnormal, and so is $T = O \oplus S$ because $T$ is a part of the normal operator $O \oplus N$ on $\mathcal{N}(T) \oplus \mathcal{N}(T)^\perp \oplus \mathcal{N}(T)^\perp$.
$\square$

An operator $T \in \mathcal{B}[\mathcal{H}]$ is *hyponormal* if $TT^* \leq T^*T$. In other words,

$$T \text{ is hyponormal if and only if } D \geq O.$$

Recall that $T^*T$ and $TT^*$ are nonnegative, and $D = (T^*T - TT^*)$ is self-adjoint, for every $T \in \mathcal{B}[\mathcal{H}]$.

**Proposition 6.6.** *An operator $T \in \mathcal{B}[\mathcal{H}]$ is hyponormal if and only if $\|T^*x\| \leq \|Tx\|$ for every $x \in \mathcal{H}$.*

*Proof.* $TT^* \leq T^*T$ if and only if $\langle TT^*x\,;x \rangle \leq \langle T^*Tx\,;x \rangle$ or, equivalently, $\|T^*x\| \leq \|Tx\|$ for every $x \in \mathcal{H}$.
$\square$

An operator $T \in \mathcal{B}[\mathcal{H}]$ is *cohyponormal* if its adjoint $T^* \in \mathcal{B}[\mathcal{H}]$ is hyponormal (i.e., if $T^*T \leq TT^*$ or, equivalently, if $D \leq O$, which means by the above proposition that $\|Tx\| \leq \|T^*x\|$ for every $x \in \mathcal{H}$). Hence $T$ *is normal if and only if it is both hyponormal and cohyponormal* (cf. Propositions 6.1 and 6.6). If an operator is either hyponormal or cohyponormal, then it is called *seminormal*. Every normal operator is trivially hyponormal. The next proposition goes beyond that.

**Proposition 6.7.** *Every subnormal operator is hyponormal.*

*Proof.* If $T \in \mathcal{B}[\mathcal{H}]$ is subnormal, then $\mathcal{H}$ is a subspace of a larger Hilbert space $\mathcal{K}$ so that $\mathcal{K} = \mathcal{H} \oplus \mathcal{H}^\perp$, and the operator

$$N = \begin{pmatrix} T & X \\ O & Y \end{pmatrix} : \mathcal{H} \oplus \mathcal{H}^\perp \to \mathcal{H} \oplus \mathcal{H}^\perp$$

in $\mathcal{B}[\mathcal{K}]$ is normal for some $X \in \mathcal{B}[\mathcal{H}^\perp, \mathcal{H}]$ and $Y \in \mathcal{B}[\mathcal{H}^\perp]$. Then

$$\begin{pmatrix} T^*T & T^*X \\ X^*T & X^*X + Y^*Y \end{pmatrix} = \begin{pmatrix} T^* & O \\ X^* & Y^* \end{pmatrix} \begin{pmatrix} T & X \\ O & Y \end{pmatrix} = N^*N$$

$$= NN^* = \begin{pmatrix} T & X \\ O & Y \end{pmatrix} \begin{pmatrix} T^* & O \\ X^* & Y^* \end{pmatrix} = \begin{pmatrix} TT^* + XX^* & XY^* \\ YX^* & YY^* \end{pmatrix}.$$

Therefore $T^*T = TT^* + XX^*$, and hence $T^*T - TT^* = XX^* \geq O$.
$\square$

Let $\mathcal{X}$ be a normed space and take any $T \in \mathcal{B}[\mathcal{X}]$. A trivial induction (Problem 4.47(a)) shows that $\|T^n\| \leq \|T\|^n$ for every $n \geq 0$.

**Lemma 6.8.** *If* $\mathcal{X}$ *is a normed space and* $T \in \mathcal{B}[\mathcal{X}]$, *then the real-valued sequence* $\{\|T^n\|^{\frac{1}{n}}\}$ *converges in* $\mathbb{R}$.

*Proof.* The proof uses the following bit of elementary number theory. Take an arbitrary $m \in \mathbb{N}$. Every $n \in \mathbb{N}$ can be written as $n = m p_n + q_n$ for $p_n, q_n \in \mathbb{N}$, where $q_n < m$. Hence

$$\|T^n\| = \|T^{mp_n + q_n}\| = \|T^{mp_n} T^{q_n}\| \leq \|T^{mp_n}\| \|T^{q_n}\| \leq \|T^m\|^{p_n} \|T^{q_n}\|.$$

Set $\mu = \max_{0 \leq k \leq m-1} \{\|T^k\|\}$ and recall that $q_n \leq m - 1$. Then

$$\|T^n\|^{\frac{1}{n}} \leq \|T^m\|^{\frac{p_n}{n}} \mu^{\frac{1}{n}} = \mu^{\frac{1}{n}} \|T^m\|^{\frac{1}{m} - \frac{q_n}{mn}}.$$

Since $\mu^{\frac{1}{n}} \to 1$ and $\|T^m\|^{\frac{1}{m} - \frac{q_n}{mn}} \to \|T^m\|^{\frac{1}{m}}$ as $n \to \infty$, it follows that

$$\limsup_n \|T^n\|^{\frac{1}{n}} \leq \|T^m\|^{\frac{1}{m}}$$

for every $m \in \mathbb{N}$. Therefore, $\limsup_n \|T^n\|^{\frac{1}{n}} \leq \liminf_n \|T^n\|^{\frac{1}{n}}$ and so (cf. Problem 3.13) $\{\|T^n\|^{\frac{1}{n}}\}$ converges in $\mathbb{R}$. $\quad\square$

We shall denote the limit of $\{\|T^n\|^{\frac{1}{n}}\}$ by $r(T)$:

$$r(T) = \lim_n \|T^n\|^{\frac{1}{n}}.$$

According to the above proof we get $r(T) \leq \|T^n\|^{\frac{1}{n}}$ for every $n \geq 1$. In particular, $r(T) \leq \|T\|$. Also note that $r(T^k)^{\frac{1}{k}} = \lim_n \|T^{kn}\|^{\frac{1}{kn}} = r(T)$ for each $k \geq 1$, because $\{\|T^{kn}\|^{\frac{1}{kn}}\}$ is a subsequence of the convergent sequence $\{\|T^n\|^{\frac{1}{n}}\}$. Thus $r(T^k) = r(T)^k$ for every positive integer $k$. Therefore, if $T \in \mathcal{B}[\mathcal{X}]$ is an operator on a normed space $\mathcal{X}$, then

$$r(T)^n = r(T^n) \leq \|T^n\| \leq \|T\|^n \quad \text{for each integer } n \geq 0.$$

Definition: If $r(T) = \|T\|$, then we say that $T \in \mathcal{B}[\mathcal{X}]$ is *normaloid*.

**Proposition 6.9.** *An operator* $T \in \mathcal{B}[\mathcal{X}]$ *on a normed space* $\mathcal{X}$ *is normaloid if and only if* $\|T^n\| = \|T\|^n$ *for every integer* $n \geq 0$.

*Proof.* If $r(T) = \|T\|$, then $\|T^n\| = \|T\|^n$ every $n \geq 0$ by the above displayed inequalities. Conversely, if $\|T^n\| = \|T\|^n$ every $n \geq 0$, then $r(T) = \lim_n \|T^n\|^{\frac{1}{n}} = \|T\|$. $\quad\square$

**Proposition 6.10.** *Every hyponormal operator is normaloid.*

*Proof.* Let $T \in \mathcal{B}[\mathcal{H}]$ be a hyponormal operator.

*Claim 1.* $\|T^n\|^2 \le \|T^{n+1}\|\,\|T^{n-1}\|$ for every positive integer $n$.

*Proof.* Take any $T \in \mathcal{B}[\mathcal{H}]$ and note that

$$\|T^n x\|^2 = \langle T^n x\,;\,T^n x\rangle = \langle T^* T^n x\,;\,T^{n-1} x\rangle \le \|T^* T^n x\|\,\|T^{n-1} x\|$$

for each integer $n \ge 1$ and every $x \in \mathcal{H}$. If $T$ is hyponormal, then

$$\|T^* T^n x\|\,\|T^{n-1} x\| \le \|T^{n+1} x\|\,\|T^{n-1} x\| \le \|T^{n+1}\|\,\|T^{n-1}\|\,\|x\|^2$$

by Proposition 6.6, and hence

$$\|T^n x\|^2 \le \|T^{n+1}\|\,\|T^{n-1}\|\,\|x\|^2$$

for each $n \ge 1$ and every $x \in \mathcal{H}$, which ensures the claimed result. $\square$

*Claim 2.* $\|T^k\| = \|T\|^k$ for every positive integer $k \le n$, for all $n \ge 1$.

*Proof.* The above result holds trivially if $T = O$ and it also holds trivially for $n = 1$ (for all $T \in \mathcal{B}[\mathcal{H}]$). Let $T \neq O$ and suppose the above result holds for some integer $n \ge 1$. By Claim 1 we get

$$\|T\|^{2n} = (\|T\|^n)^2 = \|T^n\|^2 \le \|T^{n+1}\|\,\|T^{n-1}\| = \|T^{n+1}\|\,\|T\|^{n-1}.$$

Therefore, as $\|T^n\| \le \|T\|^n$ for every $n \ge 0$, and since $T \neq O$,

$$\|T\|^{n+1} = \|T\|^{2n}(\|T\|^{n-1})^{-1} \le \|T^{n+1}\| \le \|T\|^{n+1}.$$

Hence $\|T^{n+1}\| = \|T\|^{n+1}$. Then the claimed result holds for $n + 1$ whenever it holds for $n$, which concludes the proof by induction. $\square$

Outcome: $\|T^n\| = \|T\|^n$ for every integer $n \ge 0$ so that $T$ is normaloid by Proposition 6.9. $\qquad\square$

Since $\|T^{*n}\| = \|T^n\|$ for each $n \ge 0$ (cf. Problem 5.24), it follows that $r(T^*) = r(T)$. Thus $T$ is normaloid if and only if $T^*$ is normaloid, and hence every seminormal operator is normaloid. Summing up: An operator $T$ is normal if it commutes with its adjoint, quasinormal if it commutes with $T^*T$, subnormal if it is a restriction of a normal operator to an invariant subspace, hyponormal if $T T^* \le T^* T$, and normaloid if $r(T) = \|T\|$. These classes are related by proper inclusion as follows.

$$\text{Normal} \subset \text{Quasinormal} \subset \text{Subnormal} \subset \text{Hyponormal} \subset \text{Normaloid}.$$

**Example 6A.** We shall verify that the above inclusions are, in fact, proper. The unilateral shift will do the whole job. First recall that a unilateral shift $S_+$ is an isometry but not a coisometry, and hence $S_+$ is a nonnormal quasinormal operator. Since $S_+$ is subnormal, $A = I + S_+$ is subnormal (i.e., if $N$ is a normal extension of $S_+$, then $I + N$ is a normal extension of $A$). However, since $S_+$ is a nonnormal isometry, $A^*A\,A - A\,A^*A = A^*A\,S_+ - S_+ A^*A = S_+^* S_+ - S_+ S_+^* \neq O$, and therefore $A$ is not quasinormal. Check that $B = S_+^* + 2S_+$ is hyponormal but $B^2$ is not hyponormal. Since the square of every subnormal operator is again a subnormal operator, it follows that $B$ is not subnormal. Finally, $S_+^*$ is normaloid (cf. Proposition 6.9) but not hyponormal.

# 6.2   The Spectrum of an Operator

Let $T \in \mathcal{L}[\mathcal{D}(T), \mathcal{X}]$ be a linear transformation, where $\mathcal{X} \neq \{0\}$ is a normed space and $\mathcal{D}(T)$, the domain of $T$, is a linear manifold of $\mathcal{X}$. Let $I$ be the identity on $\mathcal{X}$. The *resolvent set* $\rho(T)$ of $T$ is the set of all scalars $\lambda \in \mathbb{F}$ for which $(\lambda I - T) \in \mathcal{L}[\mathcal{D}(T), \mathcal{X}]$ has a densely defined continuous inverse. That is,

$$\rho(T) = \big\{\lambda \in \mathbb{F}: \ (\lambda I - T)^{-1} \in \mathcal{B}[\mathcal{R}(\lambda I - T), \mathcal{D}(T)] \text{ and } \mathcal{R}(\lambda I - T)^{-} = \mathcal{X}\big\}.$$

Henceforward, all linear transformations are operators on a complex Banach space. In other words, $T \in \mathcal{B}[\mathcal{X}]$, where $\mathcal{D}(T) = \mathcal{X} \neq \{0\}$ is a *complex Banach space*, and the linear $T : \mathcal{X} \to \mathcal{X}$ is *bounded*. In such a case (i.e., in the Banach algebra $\mathcal{B}[\mathcal{X}]$), Corollary 4.24 ensures that the resolvent set $\rho(T)$, defined as above, is precisely the set of all those complex numbers $\lambda$ for which $(\lambda I - T) \in \mathcal{B}[\mathcal{X}]$ is invertible (i.e., has a bounded inverse on $\mathcal{X}$). Equivalently (cf. Theorem 4.22),

$$\begin{aligned}
\rho(T) &= \big\{\lambda \in \mathbb{C}: \ (\lambda I - T) \in \mathcal{G}[\mathcal{X}]\big\} \\
&= \big\{\lambda \in \mathbb{C}: \ \mathcal{N}(\lambda I - T) = \{0\} \text{ and } \mathcal{R}(\lambda I - T) = \mathcal{X}\big\}.
\end{aligned}$$

The complement of $\rho(T)$, denoted by $\sigma(T)$, is the *spectrum* of $T$. Thus

$$\sigma(T) = \mathbb{C}\backslash\rho(T) = \big\{\lambda \in \mathbb{C}: \ \mathcal{N}(\lambda I - T) \neq \{0\} \text{ or } \mathcal{R}(\lambda I - T) \neq \mathcal{X}\big\}.$$

**Proposition 6.11.** *If* $\lambda \in \rho(T)$, *then* $\delta = \|(\lambda I - T)^{-1}\|^{-1}$ *is a positive number. The open ball* $B_\delta(\lambda)$ *with center at* $\lambda$ *and radius* $\delta$ *is included in* $\rho(T)$, *and hence* $\delta \leq d(\lambda, \sigma(T))$.

*Proof.* If $\lambda \in \rho(T)$, then $(\lambda I - T) \in \mathcal{G}[\mathcal{X}]$ so that $(\lambda I - T)^{-1}$ is nonzero and bounded. Hence $\delta = \|(\lambda I - T)^{-1}\|^{-1} > 0$. Let $B_\delta(0)$ be the nonempty open ball of radius $\delta$ about the origin of the complex plane $\mathbb{C}$ and take an arbitrary $\mu$ in $B_\delta(0)$. Since $|\mu| < \|(\lambda I - T)^{-1}\|^{-1}$, it follows that $\|\mu(\lambda I - T)^{-1}\| < 1$. Then $I - \mu(\lambda I - T)^{-1} \in \mathcal{G}[\mathcal{X}]$ (see Problem 4.48(a)), and so $(\lambda - \mu)I - T = (\lambda I - T)[I - \mu(\lambda I - T)^{-1}]$ also lies in $\mathcal{G}[\mathcal{X}]$ by Corollary 4.23. Outcome: $\lambda - \mu \in \rho(T)$ so that

$$B_\delta(\lambda) = B_\delta(0) + \lambda = \big\{\nu \in \mathbb{C}: \ \nu = \mu + \lambda \text{ for some } \mu \in B_\delta(0)\big\} \subseteq \rho(T).$$

Therefore, $d(\lambda - \varsigma) = |\lambda - \varsigma| > \delta$ for every $\varsigma \in \sigma(T) = \mathbb{C}\backslash\rho(T)$, and hence $d(\lambda, \sigma(T)) = \inf_{\varsigma \in \sigma(T)} |\lambda - \varsigma| \geq \delta$.   $\square$

**Corollary 6.12.** *The resolvent set* $\rho(T)$ *is nonempty and open, and the spectrum* $\sigma(T)$ *is compact.*

*Proof.* If $T \in \mathcal{B}[\mathcal{X}]$ is an operator on a Banach space $\mathcal{X}$, then the von Neumann expansion (Problem 4.47) ensures that $\lambda \in \rho(T)$ whenever $\|T\| < |\lambda|$. Since $\sigma(T) = \mathbb{C}\backslash\rho(T)$, this is equivalent to

$$|\lambda| \leq \|T\| \quad \text{for every} \quad \lambda \in \sigma(T).$$

Thus $\sigma(T)$ is bounded, and therefore $\rho(T) \neq \varnothing$. Proposition 6.11 says that $\rho(T)$ includes a nonempty open ball centered at each one of its points. That is, $\rho(T)$ is open and, consequently, $\sigma(T)$ is closed. In $\mathbb{C}$, closed and bounded means compact (Theorem 3.83). $\qquad\square$

The mapping $R \colon \rho(T) \to \mathcal{G}[\mathcal{X}]$ such that $R(\lambda) = (\lambda I - T)^{-1}$ for every $\lambda$ in $\rho(T)$ is called the *resolvent function* of $T$. Since

$$R(\lambda) - R(\mu) = R(\lambda)\big(\mathcal{R}(\mu)^{-1} - R(\lambda)^{-1}\big)R(\mu),$$

we get

$$R(\lambda) - R(\mu) = (\mu - \lambda)R(\lambda)R(\mu)$$

for every $\lambda, \mu \in \rho(T)$. This is referred to as the *resolvent identity*. Swapping $\lambda$ and $\mu$ in the resolvent identity, it follows that $R(\lambda)$ and $R(\mu)$ commute for every $\lambda, \mu \in \rho(T)$. Also note that $TR(\lambda) = R(\lambda)T$ for every $\lambda \in \rho(T)$ (trivially, $R(\lambda)^{-1}R(\lambda) = R(\lambda)R(\lambda)^{-1}$). To prove the next proposition we need a piece of elementary complex analysis. Let $\Lambda$ be a nonempty and open subset of the complex plane $\mathbb{C}$. Take a function $f \colon \Lambda \to \mathbb{C}$ and a point $\mu \in \Lambda$. Suppose $f'(\mu)$ is a complex number with the following property. For every $\varepsilon > 0$ there exists $\delta > 0$ such that $\left|\frac{f(\lambda)-f(\mu)}{\lambda-\mu} - f'(\mu)\right| < \varepsilon$ for all $\lambda$ in $\Lambda$ for which $0 < |\lambda - \mu| < \delta$. If there exists such an $f'(\mu) \in \mathbb{C}$, then it is called the *derivative* of $f$ at $\mu$. If $f'(\mu)$ exists for every $\mu$ in $\Lambda$, then $f \colon \Lambda \to \mathbb{C}$ is *analytic* on $\Lambda$. A function $f \colon \mathbb{C} \to \mathbb{C}$ is *entire* if it is analytic on the whole complex plane $\mathbb{C}$. The *Liouville Theorem* is the result we need. It says that *every bounded entire function is constant*.

**Proposition 6.13.** *The spectrum $\sigma(T)$ is nonempty.*

*Proof.* Let $T \in \mathcal{B}[\mathcal{X}]$ be an operator on a complex Banach space $\mathcal{X}$. Take an arbitrary nonzero element $\varphi$ in the dual $\mathcal{B}[\mathcal{X}]^*$ of $\mathcal{B}[\mathcal{X}]$ (i.e., an arbitrary nonzero bounded linear functional $\varphi \colon \mathcal{B}[\mathcal{X}] \to \mathbb{C}$ — note: $\mathcal{B}[\mathcal{X}] \neq \{O\}$ because $\mathcal{X} \neq \{0\}$, and so $\mathcal{B}[\mathcal{X}]^* \neq \{0\}$ by Corollary 4.64). Recall that $\rho(T)$ is nonempty and open in $\mathbb{C}$.

*Claim 1.* $\varphi \circ R \colon \rho(T) \to \mathbb{C}$ is bounded.

*Proof.* The resolvent function $R \colon \rho(T) \to \mathcal{G}[\mathcal{X}]$ is continuous (reason: scalar multiplication and addition are continuous mappings, and also is inversion by Problem 4.48(c)) so that $\|R(\cdot)\| \colon \rho(T) \to \mathbb{R}$ is continuous. Then $\sup_{|\lambda| \leq \|T\|} \|R(\lambda)\| < \infty$ by Theorem 3.86. On the other hand, if $|\lambda| > \|T\|$, then $\|R(\lambda)\| = \|(\lambda I - T)^{-1}\| \leq (|\lambda| - \|T\|)^{-1}$ (Problem 4.47(h)) so that $\|R(\lambda)\| \to 0$ as $|\lambda| \to \infty$. Thus, as the function $\|R(\cdot)\| \colon \rho(T) \to \mathbb{R}$ is continuous, $\sup_{|\lambda| > \|T\|} \|R(\lambda)\| < \infty$. Hence $\sup_{\lambda \in \rho(T)} \|R(\lambda)\| < \infty$, and therefore

$$\sup_{\lambda \in \rho(T)} \|(\varphi \circ R)(\lambda)\| \leq \|\varphi\| \sup_{\lambda \in \rho(T)} \|R(\lambda)\| < \infty. \quad\square$$

*Claim 2.* $\varphi \circ R : \rho(T) \to \mathbb{C}$ is analytic.

*Proof.* If $\lambda$ and $\mu$ are distinct points in $\rho(T)$, then

$$\frac{R(\lambda) - R(\mu)}{\lambda - \mu} + R(\mu)^2 = \big(R(\mu) - R(\lambda)\big) R(\mu)$$

by the resolvent identity. Set $f = \varphi \circ R : \rho(T) \to \mathbb{C}$, and let $f' : \rho(T) \to \mathbb{C}$ be defined by $f'(\lambda) = -\varphi(R(\lambda)^2)$ for each $\lambda \in \rho(T)$. Therefore,

$$\left| \frac{f(\lambda) - f(\mu)}{\lambda - \mu} - f'(\mu) \right| = \left| \varphi\big((R(\mu) - R(\lambda)) R(\mu)\big) \right|$$

$$\leq \|\varphi\| \, \|R(\mu)\| \, \|R(\mu) - R(\lambda)\|$$

so that $f : \rho(T) \to \mathbb{C}$ is analytic because $R : \rho(T) \to \mathcal{G}[\mathcal{X}]$ is continuous. $\square$

If $\sigma(T) = \varnothing$ (i.e., if $\rho(T) = \mathbb{C}$), then $\varphi \circ R : \mathbb{C} \to \mathbb{C}$ is a bounded entire function, and so a constant function by the Liouville Theorem. But we have just seen (proof of Claim 1) that $\|R(\lambda)\| \to 0$ as $|\lambda| \to \infty$. Hence $\varphi(R(\lambda)) \to 0$ as $|\lambda| \to \infty$ (since $\varphi$ is continuous). Then $\varphi \circ R = 0$ for all $\varphi \in \mathcal{B}[\mathcal{H}]^*$ so that $R = O$ (Corollary 4.64). That is, $(\lambda I - T)^{-1} = O$ for every $\lambda \in \mathbb{C}$, which is a contradiction ($O \notin \mathcal{G}[\mathcal{X}]$). Thus $\sigma(T) \neq \varnothing$. $\square$

Remark: $\sigma(T)$ is compact and nonempty, and so is its boundary $\partial\sigma(T)$. Therefore, $\partial\sigma(T) = \partial\rho(T) \neq \varnothing$ (see Problem 3.41).

The spectrum $\sigma(T)$ is the set of all $\lambda$ in $\mathbb{C}$ such that $(\lambda I - T)$ fails to be invertible (i.e., fails to have a bounded inverse on $\mathcal{R}(\lambda I - T) = \mathcal{X}$). According to the origin of such a failure, $\sigma(T)$ can be split into many disjoint parts. A classical partition comprises three parts. The set of those $\lambda$ such that $(\lambda I - T)$ has no inverse is the *point spectrum*:

$$\sigma_P(T) = \big\{ \lambda \in \mathbb{C} : \ \mathcal{N}(\lambda I - T) \neq \{0\} \big\}.$$

A scalar $\lambda \in \mathbb{C}$ is called an *eigenvalue* of $T$ if there exists a nonzero vector $x$ in $\mathcal{X}$ such that $Tx = \lambda x$. Equivalently, if $\mathcal{N}(\lambda I - T) \neq \{0\}$. If $\lambda \in \mathbb{C}$ is an eigenvalue of $T$, then the nonzero vectors in $\mathcal{N}(\lambda I - T)$ are the *eigenvectors* of $T$, and $\mathcal{N}(\lambda I - T)$ is the *eigenspace* (which, in fact, is a subspace of $\mathcal{X}$ — cf. Proposition 4.13), associated with $\lambda$. The *multiplicity* of an eigenvalue is the dimension of the respective eigenspace. After this quick digression on eigenvalues and eigenvectors, note that *the point spectrum of $T$ is precisely the set of all eigenvalues of $T$*. The set of those $\lambda$ for which $(\lambda I - T)$ has a densely defined but unbounded inverse on its range is the *continuous spectrum*:

$$\sigma_C(T) = \big\{ \lambda \in \mathbb{C} : \ \mathcal{N}(\lambda I - T) = \{0\}, \ \mathcal{R}(\lambda I - T)^- = \mathcal{X} \ \text{and} \ \mathcal{R}(\lambda I - T) \neq \mathcal{X} \big\}$$

(see Corollary 4.24 again). If $(\lambda I - T)$ has an inverse on its range that is not densely defined, then $\lambda$ belongs to the *residual spectrum*:

$$\sigma_R(T) = \big\{ \lambda \in \mathbb{C} : \ \mathcal{N}(\lambda I - T) = \{0\} \ \text{and} \ \mathcal{R}(\lambda I - T)^- \neq \mathcal{X} \big\}.$$

452     6. *The Spectral Theorem*

The collection $\{\sigma_P(T), \sigma_C(T), \sigma_R(T)\}$ of subsets of $\sigma(T)$ is a disjoint covering of it. That is, they are pairwise disjoint and

$$\sigma(T) = \sigma_P(T) \cup \sigma_C(T) \cup \sigma_R(T).$$

The diagram below summarizes such a partition of the spectrum, where $\sigma_R(T) = \sigma_{R_1}(T) \cup \sigma_{R_2}(T)$ and $\sigma_P(T) = \bigcup_{i=1}^{4} \sigma_{P_i}(T)$. Here we adopt the following notation: $T_\lambda = (\lambda I - T)$, $\mathcal{N}_\lambda = \mathcal{N}(T_\lambda)$, and $\mathcal{R}_\lambda = \mathcal{R}(T_\lambda)$.

| | | $\mathcal{R}_\lambda^- = \mathcal{X}$ | | $\mathcal{R}_\lambda^- \neq \mathcal{X}$ | |
| --- | --- | --- | --- | --- | --- |
| | | $\mathcal{R}_\lambda^- = \mathcal{R}_\lambda$ | $\mathcal{R}_\lambda^- \neq \mathcal{R}_\lambda$ | $\mathcal{R}_\lambda^- \neq \mathcal{R}_\lambda$ | $\mathcal{R}_\lambda^- = \mathcal{R}_\lambda$ |
| $\mathcal{N}_\lambda = \{0\}$ | $T_\lambda^{-1} \in \mathcal{B}[\mathcal{R}_\lambda, \mathcal{X}]$ | $\rho(T)$ | $\varnothing$ | $\varnothing$ | $\sigma_{R_1}(T)$ |
| | $T_\lambda^{-1} \notin \mathcal{B}[\mathcal{R}_\lambda, \mathcal{X}]$ | $\varnothing$ | $\sigma_C(T)$ | $\sigma_{R_2}(T)$ | $\varnothing$ |
| $\mathcal{N}_\lambda \neq \{0\}$ | | $\sigma_{P_1}(T)$ | $\sigma_{P_2}(T)$ | $\sigma_{P_3}(T)$ | $\sigma_{P_4}(T)$ |

The brace at the right of the last three value columns of the first two rows is labeled $\sigma_{AP}(T)$, and the brace under the last two columns is labeled $\sigma_{CP}(T)$.

Recall that $\sigma(T) \neq \varnothing$ but any of the above disjoint parts of the spectrum may be empty (as we shall see in Section 6.5). However, if $\sigma_P(T)$ is nonempty, then every set of eigenvectors associated with *distinct* eigenvalues is linearly independent.

**Proposition 6.14.** *Let $\{\lambda_\gamma\}_{\gamma \in \Gamma}$ be any family of distinct eigenvalues of $T$. For each $\gamma \in \Gamma$ let $x_\gamma$ be an eigenvector associated with $\lambda_\gamma$ (i.e., $0 \neq x_\gamma \in \mathcal{N}(\lambda_\gamma I - T) \neq \{0\}$). The set $\{x_\gamma\}_{\gamma \in \Gamma}$ is linearly independent.*

*Proof.* Consider the set $\{x_\gamma\}_{\gamma \in \Gamma}$ (whose existence is ensured by the Axiom of Choice). Let $\{x_i\}_{i=1}^{n}$ be an arbitrary *finite* subset of $\{x_\gamma\}_{\gamma \in \Gamma}$.

*Claim.* $\{x_i\}_{i=1}^{n}$ is linearly independent.

*Proof.* If $n = 1$, then linear independence is trivial. Suppose $\{x_i\}_{i=1}^{n}$ is linearly independent for some $n \geq 1$. If $\{x_i\}_{i=1}^{n+1}$ is not linearly independent, then $x_{n+1} = \sum_{i=1}^{n} \alpha_i x_i$, where $\{\alpha_k\}_{i=1}^{n}$ is a family of complex numbers with at least one nonzero number among them. Thus $\lambda_{n+1} x_{n+1} = T x_{n+1} = \sum_{i=1}^{n} \alpha_i T x_i = \sum_{i=1}^{n} \alpha_i \lambda_i x_i$. If $\lambda_{n+1} = 0$, then $\lambda_i \neq 0$ for every $i \neq n + 1$ and $\sum_{i=1}^{n} \alpha_i \lambda_i x_i = 0$ so that $\{x_i\}_{i=1}^{n}$ is not linearly independent, which is a contradiction. If $\lambda_{n+1} \neq 0$, then $x_{n+1} = \sum_{i=1}^{n} \alpha_i \lambda_{n+1}^{-1} \lambda_i x_i$, and therefore $\sum_{i=1}^{n} \alpha_i (1 - \lambda_{n+1}^{-1} \lambda_i) x_i = 0$ so that $\{x_i\}_{i=1}^{n}$ is not linearly independent (because $\lambda_i \neq \lambda_{n+1}$ for every $i \neq n + 1$), which is again a contradiction. This completes the proof by induction. $\square$

Outcome: $\{x_\gamma\}_{\gamma \in \Gamma}$ is linearly independent by Proposition 2.3. $\square$

There are some overlapping parts of the spectrum which are commonly used too. For instance, the *compression spectrum* $\sigma_{CP}(T)$ and the *approximate point spectrum* (or *approximation spectrum*) $\sigma_{AP}(T)$ are defined by

$$
\begin{aligned}
\sigma_{CP}(T) &= \{\lambda \in \mathbb{C}: \ \mathcal{R}(\lambda I - T) \text{ is not dense in } \mathcal{X}\} \\
&= \sigma_{P_3}(T) \cup \sigma_{P_4}(T) \cup \sigma_R(T), \\
\sigma_{AP}(T) &= \{\lambda \in \mathbb{C}: \ (\lambda I - T) \text{ is not bounded below}\} \\
&= \sigma_P(T) \cup \sigma_C(T) \cup \sigma_{R_2}(T) = \sigma(T) \backslash \sigma_{R_1}(T).
\end{aligned}
$$

Next we give an alternative definition of $\sigma_{AP}(T)$ which may come as a motivation for the term "approximate point spectrum". The elements of it are sometimes referred to as the *approximate eigenvalues* of $T$.

**Proposition 6.15.** *The following assertions are pairwise equivalent.*

(a) $\lambda \in \sigma_{AP}(T)$.

(b) *There exists an $\mathcal{X}$-valued sequence $\{x_n\}$ of unit vectors such that* $\|(\lambda I - T)x_n\| \to 0$.

(c) *For every $\varepsilon > 0$ there exists a unit vector $x_\varepsilon \in \mathcal{X}$ such that $\|(\lambda I - T)x_\varepsilon\| < \varepsilon$.*

*Proof.* It is clear that (c) implies (b). If (b) holds true, then there is no constant $\alpha > 0$ such that $\alpha = \alpha\|x_n\| \le \|(\lambda I - T)x_n\|$ for all $n$, and hence $(\lambda I - T)$ is not bounded below. Hence (b) implies (a). If $(\lambda I - T)$ is not bounded below, then there is no constant $\alpha > 0$ such that $\alpha\|x\| \le \|(\lambda I - T)x\|$ for all $x \in \mathcal{X}$ or, equivalently, for every $\varepsilon > 0$ there exists a nonzero $y_\varepsilon$ in $\mathcal{X}$ such that $\|(\lambda I - T)y_\varepsilon\| < \varepsilon\|y_\varepsilon\|$. By setting $x_\varepsilon = \|y_\varepsilon\|^{-1} y_\varepsilon$ it follows that (a) implies (c). $\qquad\square$

**Proposition 6.16.** *The approximate point spectrum is nonempty, closed in $\mathbb{C}$, and includes the boundary $\partial\sigma(T)$ of the spectrum.*

*Proof.* Take an arbitrary $\lambda \in \partial\sigma(T)$. Recall that $\rho(T) \ne \varnothing$ (Corollary 6.12) and $\partial\sigma(T) = \partial\rho(T) \subset \rho(T)^-$ (Problem 3.41). Hence there exists a sequence $\{\lambda_n\}$ in $\rho(T)$ such that $\lambda_n \to \lambda$ (Proposition 3.32). Since

$$
(\lambda_n I - T) - (\lambda I - T) = (\lambda_n - \lambda)I
$$

for every $n$, it follows that $(\lambda_n I - T) \to (\lambda I - T)$ in $\mathcal{B}[\mathcal{X}]$; that is $\{(\lambda_n I - T)\}$ in $\mathcal{G}[\mathcal{X}]$ converges in $\mathcal{B}[\mathcal{X}]$ to $(\lambda I - T) \in \mathcal{B}[\mathcal{X}] \backslash \mathcal{G}[\mathcal{X}]$ (each $\lambda_n$ lies in $\rho(T)$ and $\lambda \in \partial\sigma(T) \subseteq \sigma(T)$ because $\sigma(T)$ is closed). If $\sup_n \|(\lambda_n I - T)^{-1}\| < \infty$, then $(\lambda I - T) \in \mathcal{G}[\mathcal{X}]$ (cf. hint to Problem 4.48(c)), which is a contradiction. Therefore,

$$
\sup_n \|(\lambda_n I - T)^{-1}\| = \infty.
$$

For each integer $n$ take $y_n$ in $\mathcal{X}$ with $\|y_n\| = 1$ such that $\|(\lambda_n I - T)^{-1}\| - \frac{1}{n} \le \|(\lambda_n I - T)^{-1}y_n\| \le \|(\lambda_n I - T)^{-1}\|$. Thus $\sup_n \|(\lambda_n I - T)^{-1}y_n\| = \infty$, and hence

$\inf_n \|(\lambda_n I - T)^{-1} y_n\|^{-1} = 0$, so that there exist subsequences of $\{\lambda_n\}$ and $\{y_n\}$, say $\{\lambda_k\}$ and $\{y_k\}$, for which

$$\|(\lambda_k I - T)^{-1} y_k\|^{-1} \to 0.$$

Set $x_k = \|(\lambda_k I - T)^{-1} y_k\|^{-1} (\lambda_k I - T)^{-1} y_k$ and get a sequence $\{x_k\}$ of unit vectors in $\mathcal{X}$ such that $\|(\lambda_k I - T) x_k\| = \|(\lambda_k I - T)^{-1} y_k\|^{-1}$. Since

$$\|(\lambda I - T) x_k\| = \|(\lambda_k I - T) x_k - (\lambda_k - \lambda) x_k\| \leq \|(\lambda_k I - T)^{-1} y_k\|^{-1} + |\lambda_k - \lambda|$$

and $\lambda_k \to \lambda$, it follows that $\|(\lambda I - T) x_k\| \to 0$. Hence $\lambda \in \sigma_{AP}(T)$ according to Proposition 6.15. Therefore,

$$\partial \sigma(T) \subseteq \sigma_{AP}(T).$$

This inclusion clearly implies that $\sigma_{AP}(T) \neq \varnothing$ (for $\sigma(T)$ is closed and nonempty). Moreover, as $\sigma_{AP}(T) \subseteq \sigma(T)$, we also get $\partial \sigma_{AP}(T) \subseteq \partial \sigma(T) \subseteq \sigma_{AP}(T)$ so that $\sigma_{AP}(T)$ is closed in $\mathbb{C}$ (see Problem 3.41). $\qquad\square$

Remark: $\sigma_{R_1}(T)$ *is open in* $\mathbb{C}$. Proof: $\sigma_{AP}(T)$ is closed in $\mathbb{C}$ and includes $\partial \sigma(T)$. Then $\mathbb{C} \backslash \sigma_{R_1}(T) = \rho(T) \cup \sigma_{AP}(T) = \rho(T) \cup \partial \sigma(T) \cup \sigma_{AP}(T) = \rho(T) \cup \partial \rho(T) \cup \sigma_{AP}(T) = \rho(T)^- \cup \sigma_{AP}(T)$, which is closed in $\mathbb{C}$.

For the next proposition we assume that $T \in \mathcal{B}[\mathcal{H}]$, where $\mathcal{H} \neq \{0\}$ is a *complex Hilbert space*. If $\Lambda$ is any subset of $\mathbb{C}$, then set

$$\Lambda^* = \{\bar{\lambda} \in \mathbb{C}: \ \lambda \in \Lambda\}$$

so that $\Lambda^{**} = \Lambda$, $(\mathbb{C} \backslash \Lambda)^* = \mathbb{C} \backslash \Lambda^*$ and $(\Lambda_1 \cup \Lambda_2)^* = \Lambda_1^* \cup \Lambda_2^*$.

**Proposition 6.17.** *If* $T^* \in \mathcal{B}[\mathcal{H}]$ *is the adjoint of* $T \in \mathcal{B}[\mathcal{H}]$, *then*

$$\rho(T) = \rho(T^*)^*, \quad \sigma(T) = \sigma(T^*)^*, \quad \sigma_C(T) = \sigma_C(T^*)^*,$$

*and the residual spectrum of* $T$ *is given by the formula*

$$\sigma_R(T) = \sigma_P(T^*)^* \backslash \sigma_P(T).$$

*As for the subparts of the point and residual spectrum,*

$$\sigma_{P_1}(T) = \sigma_{R_1}(T^*)^*, \quad \sigma_{P_2}(T) = \sigma_{R_2}(T^*)^*,$$

$$\sigma_{P_3}(T) = \sigma_{P_3}(T^*)^*, \quad \sigma_{P_4}(T) = \sigma_{P_4}(T^*)^*.$$

*For the compression and approximate point spectrum we get*

$$\sigma_{CP}(T) = \sigma_P(T^*)^*,$$

$$\partial\sigma(T) \subseteq \sigma_{AP}(T) \cap \sigma_{AP}(T^*)^* = \sigma(T)\backslash(\sigma_{P_1}(T) \cup \sigma_{R_1}(T)).$$

*Proof.* Since $S \in \mathcal{G}[\mathcal{H}]$ if and only if $S^* \in \mathcal{G}[\mathcal{H}]$, it follows that $\rho(T) = \rho(T^*)^*$. Hence $\sigma(T)^* = (\mathbb{C}\backslash\rho(T))^* = \mathbb{C}\backslash\rho(T^*) = \sigma(T^*)$. Recall that $\mathcal{R}(S)^- = \mathcal{R}(S)$ if and only if $\mathcal{R}(S^*)^- = \mathcal{R}(S^*)$, and $\mathcal{N}(S) = \{0\}$ if and only if $\mathcal{R}(S^*)^- = \mathcal{H}$ (cf. Proposition 5.77 and Problem 5.35). Thus $\sigma_{P_1}(T) = \sigma_{R_1}(T^*)^*$, $\sigma_{P_2}(T) = \sigma_{R_2}(T^*)^*$, $\sigma_{P_3}(T) = \sigma_{P_3}(T^*)^*$, and also $\sigma_{P_4}(T) = \sigma_{P_4}(T^*)^*$. Applying the same argument, $\sigma_C(T) = \sigma_C(T^*)^*$ and $\sigma_{CP}(T) = \sigma_P(T^*)^*$. Therefore

$$\sigma_R(T) = \sigma_{CP}(T)\backslash\sigma_P(T) \quad \text{implies} \quad \sigma_R(T) = \sigma_P(T^*)^*\backslash\sigma_P(T).$$

Moreover, by using the above properties and the definition of

$$\sigma_{AP}(T^*) = \sigma_P(T^*) \cup \sigma_C(T^*) \cup \sigma_{R_2}(T^*) = \sigma_{CP}(T)^* \cup \sigma_C(T)^* \cup \sigma_{P_2}(T)^*,$$

we get

$$\sigma_{AP}(T^*)^* = \sigma_{CP}(T) \cup \sigma_C(T) \cup \sigma_{P_2}(T).$$

So $\sigma_{AP}(T^*) \cap \sigma_{AP}(T) = \sigma(T)\backslash(\sigma_{P_1}(T) \cup \sigma_{R_1}(T))$. But $\sigma(T)$ is closed and $\sigma_{R_1}(T)$ is open (and so is $\sigma_{P_1}(T) = \sigma_{R_1}(T^*)^*$) in $\mathbb{C}$. This implies that $\sigma_{P_1}(T) \cup \sigma_{R_1}(T) \subseteq \sigma(T)^\circ$ and $\partial\sigma(T) \subseteq \sigma(T)\backslash(\sigma_{P_1}(T) \cup \sigma_{R_1}(T))$ (cf. Problem 3.41(b,d)).  $\square$

Remark: We have just seen that $\sigma_{P_1}(T)$ *is open in* $\mathbb{C}$.

**Corollary 6.18.** *Let $\mathcal{H} \neq \{0\}$ be a complex Hilbert space, and let $\Gamma$ denote the unit circle about the origin of the complex plane.*

(a) *If $H \in \mathcal{B}[\mathcal{H}]$ is hyponormal, then $\sigma_P(H)^* \subseteq \sigma_P(H^*)$ and $\sigma_R(H^*) = \varnothing$.*

(b) *If $N \in \mathcal{B}[\mathcal{H}]$ is normal, then $\sigma_P(N^*) = \sigma_P(N)^*$ and $\sigma_R(N) = \varnothing$.*

(c) *If $U \in \mathcal{B}[\mathcal{H}]$ is unitary, then $\sigma(U) \subset \Gamma$.*

(d) *If $A \in \mathcal{B}[\mathcal{H}]$ is self-adjoint, then $\sigma(A) \subseteq \mathbb{R}$.*

(e) *If $Q \in \mathcal{B}[\mathcal{H}]$ is nonnegative, then $\sigma(Q) \subseteq [0, \infty)$.*

(f) *If $R \in \mathcal{B}[\mathcal{H}]$ is strictly positive, then $\sigma(R) \subseteq [\alpha, \infty)$ for some $\alpha > 0$.*

(g) *If $P \in \mathcal{B}[\mathcal{H}]$ is a nontrivial projection, then $\sigma(P) = \sigma_P(P) = \{0, 1\}$.*

(h) *If $J \in \mathcal{B}[\mathcal{H}]$ is a nontrivial involution, then $\sigma(J) = \sigma_P(J) = \{-1, 1\}$.*

*Proof.* Take any $T \in \mathcal{B}[\mathcal{H}]$ and any $\lambda \in \mathbb{C}$. It is readily verified that

$$(\lambda I - T)^*(\lambda I - T) - (\lambda I - T)(\lambda I - T)^* = T^*T - TT^*.$$

Hence $(\lambda I - T)$ is hyponormal if and only if $T$ is hyponormal. If $H$ is hyponormal, then $(\lambda I - H)$ is hyponormal and so (cf. Proposition 6.6)

$$\|(\bar{\lambda} I - H^*)x\| \le \|(\lambda I - H)x\| \quad \text{for every } x \in \mathcal{H} \text{ and every } \lambda \in \mathbb{C}.$$

Suppose $\lambda \in \sigma_P(H)$. Then $\mathcal{N}(\lambda I - H) \ne \{0\}$ and the above inequality ensures that $\mathcal{N}(\bar{\lambda} I - H^*) \ne \{0\}$. Thus $\bar{\lambda} \in \sigma_P(H^*)$. Conclusion: if $H$ is hyponormal, then $\sigma_P(H) \subseteq \sigma_P(H^*)^*$. Equivalently, $\sigma_P(H)^* \subseteq \sigma_P(H^*)$, and therefore (Proposition 6.17) $\sigma_R(H^*) = \sigma_P(H)^* \backslash \sigma_P(H^*) = \varnothing$. This proves (a). Since $N$ is normal if and only if it is both hyponormal and cohyponormal, this also proves (b). Now let $U$ be unitary (i.e., a normal isometry). If $|\lambda| > 1$, then $\lambda \in \rho(U)$ because $\|U\| = 1$ (cf. Problem 4.47). If $|\lambda| < 1$, then $\|\lambda U^*\| < 1$ because $\|U^*\| = \|U\| = 1$, and hence $(I - \lambda U^*) \in \mathcal{G}[\mathcal{H}]$ (cf. Problem 4.48). Since $U^* = U^{-1} \in \mathcal{G}[\mathcal{H}]$, this implies by Corollary 4.23 that $(\lambda I - U) = -U(I - \lambda U^*) \in \mathcal{G}[\mathcal{H}]$, and therefore $\lambda \in \rho(U)$. Outcome: if $\lambda \in \sigma(U)$, then $|\lambda| = 1$, which proves (c). Next let $A$ be self-adjoint (i.e., $A^* = A$). For every $\alpha, \beta \in \mathbb{R}$

$$\langle i\beta x ; (\alpha I - A)x \rangle = -\langle (\alpha I - A)x ; i\beta x \rangle = -\overline{\langle i\beta x ; (\alpha I - A)x \rangle}$$

so that $2\,\mathrm{Re}\,\langle i\beta x ; (\alpha I - A)x \rangle = 0$. Hence

$$\begin{aligned}
\|(\lambda I - A)x\|^2 &= \|i\,\mathrm{Im}\,\lambda x + (\mathrm{Re}\,\lambda I - A)x\|^2 \\
&= |\mathrm{Im}\,\lambda|^2 \|x\|^2 + \|(\mathrm{Re}\,\lambda I - A)x\|^2 \ge |\mathrm{Im}\,\lambda|^2 \|x\|^2
\end{aligned}$$

for every $x \in \mathcal{H}$ and every $\lambda \in \mathbb{C}$. If $\lambda$ is not real, then $(\lambda I - A)$ is bounded below, which means that $\lambda \in \rho(A) \cup \sigma_R(A) = \rho(A)$ because $\sigma_R(A) = \varnothing$ according to (b). This shows that (d) holds true. If $Q$ is nonnegative (i.e., $\langle Qx ; x \rangle \ge 0$ for every $x \in \mathcal{H}$) and $\lambda \in \sigma(Q)$, then $\lambda \in \mathbb{R}$ by item (d) and $\|(\lambda I - Q)x\|^2 = |\lambda|^2 \|x\|^2 - 2\lambda \langle Qx ; x \rangle + \|Qx\|^2$ for every $x \in \mathcal{H}$. If $\lambda < 0$, then $\|(\lambda I - Q)x\|^2 \ge |\lambda|^2 \|x\|^2$ for every $x \in \mathcal{H}$, and hence $(\lambda I - Q)$ is bounded below. Using the same argument of the previous item we get the result in (e). If $O \prec R$, then $O \le R \in \mathcal{G}[\mathcal{H}]$, and so $\sigma(R) \in [0, 1]$ by item (e) and $0 \in \rho(R)$. Since $\rho(R)$ is open, $\sigma(R)$ must be bounded away from zero, which proves (f). Now let $P$ be a projection so that $I - P$ is a projection as well, $\mathcal{R}(I - P) = \mathcal{N}(P)$, and $\mathcal{R}(P) = \mathcal{N}(I - P)$ (see Section 2.9). If $0 \notin \sigma_P(T)$, then $\mathcal{N}(P) = \{0\}$ so that $\mathcal{R}(I - P) = \{0\}$, and hence $P = I$. Similarly, if $1 \notin \sigma_P(T)$, then $\mathcal{N}(I - P) = \{0\}$ so that $\mathcal{R}(P) = \{0\}$, and hence $P = O$. Therefore, if $P$ is a nontrivial projection (i.e., $O \ne P = P^2 \ne I$), then $0, 1 \in \sigma_P(T)$. Moreover, if $\lambda$ is any complex number such that $0 \ne \lambda \ne 1$, then

$$(\lambda I - P)\left(\tfrac{1}{\lambda} I + \tfrac{1}{\lambda(\lambda-1)} P\right) = I = \left(\tfrac{1}{\lambda} I + \tfrac{1}{\lambda(\lambda-1)} P\right)(\lambda I - P),$$

which means that $(\lambda I - P) \in \mathcal{G}[\mathcal{H}]$ (Theorem 4.22), and so $\lambda \in \rho(T)$. This concludes the proof of (g). Finally, let $J$ be an involution; that is, $J^2 = I$. In this case, $(I - J)(-I - J) = 0 = (-I - J)(I - J)$ so that $\mathcal{R}(-I - J) \subseteq \mathcal{N}(I - J)$ and $\mathcal{R}(I - J) \subseteq \mathcal{N}(-I - J)$. If $1 \notin \sigma_P(J)$ or $-1 \notin \sigma_P(J)$, then $\mathcal{N}(I - J) = \{0\}$

or $\mathcal{N}(-I - J) = \{0\}$, which implies $\mathcal{R}(I + J) = \{0\}$ or $\mathcal{R}(I - J) = \{0\}$, and hence $J = I$ or $J = -I$. Thus, if $J$ is a nontrivial involution (i.e., $J \neq \pm I$), then $\pm 1 \in \sigma_P(T)$. Moreover, if $\lambda$ in $\mathbb{C}$ is such that $\lambda^2 \neq 1$ (i.e., $\lambda \neq \pm 1$), then $\lambda \in \rho(T)$. Indeed,

$$(\lambda I - J)\left(\tfrac{-\lambda}{1-\lambda^2} I - \tfrac{1}{1-\lambda^2} J\right) = I = \left(\tfrac{-\lambda}{1-\lambda^2} I - \tfrac{1}{1-\lambda^2} J\right)(\lambda I - J),$$

so that $(\lambda I - J) \in \mathcal{G}[\mathcal{H}]$, which concludes the proof of (h). $\qquad\square$

## 6.3  Spectral Radius

We open this section with the Spectral Mapping Theorem for polynomials. Let us just mention that there are versions of it that hold for functions other than polynomials. If $\Lambda$ is any subset of $\mathbb{C}$, and $p\colon \mathbb{C} \to \mathbb{C}$ is any polynomial with complex coefficients, then set

$$p(\Lambda) = \{p(\lambda) \in \mathbb{C}\colon \ \lambda \in \Lambda\}.$$

**Theorem 6.19.** (The Spectral Mapping Theorem). *If $T \in \mathcal{B}[\mathcal{X}]$, where $\mathcal{X}$ is a complex Banach space, then*

$$\sigma(p(T)) = p(\sigma(T))$$

*for every polynomial $p$ with complex coefficients.*

*Proof.* If $p$ is a constant polynomial (i.e., if $p(T) = \alpha I$ for some $\alpha \in \mathbb{C}$), then the result is trivially verified (and has nothing to do with $T$: $\sigma(\alpha I) = \alpha\sigma(I) = \{\alpha\}$, since $\rho(\alpha I) = \mathbb{C}\backslash\{\alpha\}$ for every $\alpha \in \mathbb{C}$). Let $p\colon \mathbb{C} \to \mathbb{C}$ be an arbitrary nonconstant polynomial with complex coefficients,

$$p(z) = \sum_{i=0}^{n} \alpha_i z^i, \quad \text{with } n \geq 1 \text{ and } \alpha_n \neq 0,$$

for every $z \in \mathbb{C}$. Take an arbitrary $\mu \in \mathbb{C}$ and consider the factorization

$$\mu - p(z) = \alpha_n^{-1} \prod_{i=1}^{n}(z_i - z),$$

where $\{z_i\}_{i=1}^n$ are the roots of $\mu - p(z)$, so that

$$\mu I - p(T) = \alpha_n^{-1} \prod_{i=1}^{n}(z_i I - T).$$

If $\mu \in \sigma(p(T))$, then there exists $z_j \in \sigma(T)$ for some $j = 1, \dots, n$. Indeed, if $z_i \in \rho(T)$ for every $i = 1, \dots, n$, then $\alpha_n^{-1} \prod_{i=1}^{n}(z_i I - T) \in \mathcal{G}[\mathcal{X}]$ so that $\mu \in \rho(T)$. However,

$$\mu - p(z_j) = \alpha_n^{-1} \prod_{i=1}^{n}(z_i - z_j) = 0,$$

and so $p(z_j) = \mu$. Hence $\mu = p(z_j) \in \{p(\lambda) \in \mathbb{C}\colon \lambda \in \sigma(T)\} = p(\sigma(T))$ because $z_j \in \sigma(T)$. Therefore,

$$\sigma(p(T)) \subseteq p(\sigma(T)).$$

Conversely, if $\mu \in p(\sigma(T)) = \{p(\lambda) \in \mathbb{C}\colon \lambda \in \sigma(T)\}$, then $\mu = p(\lambda)$ for some $\lambda \in \sigma(T)$. Thus $\mu - p(\lambda) = 0$ so that $\lambda = z_j$ for some $j = 1, \dots, n$, and hence

$$\begin{aligned}
\mu I - p(T) &= \alpha_n^{-1} \prod_{i=1}^{n}(z_i I - T) \\
&= (z_j I - T)\alpha_n^{-1} \prod_{j \neq i=1}^{n}(z_i I - T) = \alpha_n^{-1} \prod_{j \neq i=1}^{n}(z_i I - T)(z_j I - T)
\end{aligned}$$

for $(z_j I - T)$ commutes with $(z_i I - T)$ for every $i$. If $\mu \in \rho(p(T))$, then $(\mu I - p(T)) \in \mathcal{G}[\mathcal{X}]$ so that

$$\begin{aligned}
(z_j I - T)\Big(\alpha_n^{-1} &\prod_{j \neq i=1}^{n}(z_i I - T)(\mu I - p(T))^{-1}\Big) \\
&= (\mu I - p(T))(\mu I - p(T))^{-1} = I = (\mu I - p(T))^{-1}(\mu I - p(T)) \\
&= \Big((\mu I - p(T))^{-1}\alpha_n^{-1} \prod_{j \neq i=1}^{n}(z_i I - T)\Big)(z_j I - T).
\end{aligned}$$

This means that $(z_j I - T)$ has a right and a left inverse, and so it is injective and surjective (cf. Problems 1.5 and 1.6). The Inverse Mapping Theorem (Theorem 4.22) ensures that $(z_j I - T) \in \mathcal{G}[\mathcal{X}]$, and therefore $\lambda = z_j \in \rho(T)$. But this contradicts the fact that $\lambda \in \sigma(T)$. Conclusion: $\mu \notin \rho(p(T))$; that is, $\mu \in \sigma(p(T))$. Hence

$$p(\sigma(T)) \subseteq \sigma(p(T)). \qquad \square$$

In particular, $\mu \in \sigma(T)^n = \{\lambda^n \in \mathbb{C}\colon \lambda \in \sigma(T)\}$ if and only if $\mu \in \sigma(T^n)$:

$$\sigma(T^n) = \sigma(T)^n \quad \text{for every} \quad n \geq 0;$$

and $\mu \in \alpha\sigma(T) = \{\alpha\lambda \in \mathbb{C}\colon \lambda \in \sigma(T)\}$ if and only if $\mu \in \sigma(\alpha T)$:

$$\sigma(\alpha T) = \alpha\sigma(T) \quad \text{for every} \quad \alpha \in \mathbb{C}.$$

It is also worth noticing (even though this is not a particular case of the Spectral Mapping Theorem for polynomials) that if $T \in \mathcal{G}[\mathcal{X}]$, then

$$\sigma(T^{-1}) = \sigma(T)^{-1}.$$

That is, $\mu \in \sigma(T)^{-1} = \{\lambda^{-1} \in \mathbb{C}: 0 \neq \lambda \in \sigma(T)\}$ if and only if $\mu \in \sigma(T^{-1})$. Indeed, if $T \in \mathcal{G}[\mathcal{X}]$ (so that $0 \in \rho(T)$) and $\mu \neq 0$, then $-\mu T^{-1}(\mu^{-1}I - T) = \mu I - T^{-1}$, and hence $\mu^{-1} \in \rho(T)$ if and only if $\mu \in \rho(T^{-1})$. Also note that if $T \in \mathcal{B}[\mathcal{H}]$, then

$$\sigma(T^*) = \sigma(T)^*,$$

where $\mathcal{H}$ is a complex Hilbert space (cf. Proposition 6.17).

Let $T \in \mathcal{B}[\mathcal{X}]$ be an operator on a complex Banach space $\mathcal{X}$. The *spectral radius* of $T$ is the number

$$r_\sigma(T) = \sup_{\lambda \in \sigma(T)} |\lambda| = \max_{\lambda \in \sigma(T)} |\lambda|.$$

The first identity defines the spectral radius $r_\sigma(T)$; the second one is a consequence of Theorem 3.86. (Reason: $\sigma(T) \neq \varnothing$ is compact in $\mathbb{C}$ and the function $| \ |: \mathbb{C} \to \mathbb{R}$ is continuous.)

**Corollary 6.20.** $r_\sigma(T^n) = r_\sigma(T)^n$ *for every* $n \geq 0$.

*Proof.* Take an arbitrary integer $n \geq 0$. Since $\sigma(T^n) = \sigma(T)^n$, it follows that $\mu \in \sigma(T^n)$ if and only if $\mu = \lambda^n$ for some $\lambda \in \sigma(T)$. Hence $\sup_{\mu \in \sigma(T^n)} |\mu| = \sup_{\lambda \in \sigma(T)} |\lambda^n| = \sup_{\lambda \in \sigma(T)} |\lambda|^n = \left(\sup_{\lambda \in \sigma(T)} |\lambda|\right)^n.$ $\qquad\square$

Remarks: Recall that $\lambda \in \sigma(T)$ only if $|\lambda| \leq \|T\|$ (cf. proof of Corollary 6.12), and so $r_\sigma(T) \leq \|T\|$. Therefore, according to Corollary 6.20,

$$r_\sigma(T^n) = r_\sigma(T)^n \leq \|T^n\| \leq \|T\|^n \quad \text{for every} \quad n \geq 0.$$

Thus $r_\sigma(T) \leq 1$ *whenever $T$ is power bounded*. Indeed, if $\sup_n \|T^n\| < \infty$, then $r_\sigma(T)^n = r_\sigma(T^n) \leq \sup_k \|T^k\|$ and $\lim_n (\sup_k \|T^k\|)^{\frac{1}{n}} = 1$, so that

$$\sup_n \|T^n\| < \infty \quad \text{implies} \quad r_\sigma(T) \leq 1.$$

Also note that the spectral radius of a nonzero operator may be null. Sample: $r_\sigma(T) = 0$ for every nilpotent operator $T$ (i.e., whenever $T^n = O$ for some positive integer $n$). An operator $T \in \mathcal{B}[\mathcal{X}]$ is *quasinilpotent* if $r_\sigma(T) = 0$, so that *every nilpotent operator is quasinilpotent*. Observe that $\sigma(T) = \sigma_P(T) = \{0\}$ if $T$ is nilpotent. Indeed, if $T^{n-1} \neq O$ and $T^n = O$, then $T(T^{n-1}x) = 0$ for every $x \in \mathcal{X}$, so that $\{0\} \neq \mathcal{R}(T^{n-1}) \subseteq \mathcal{N}(T)$, and hence $\lambda = 0$ is an eigenvalue of $T$. Since $\sigma_P(T)$ may be empty for a quasinilpotent operator $T$ (as we shall see in Examples 6F and 6G of Section 6.5), it follows that the inclusion below is proper:

$$\text{Nilpotent} \quad \subset \quad \text{Quasinilpotent}.$$

The next proposition is the so-called *Gelfand–Beurling formula* for the spectral radius. The proof of it requires another piece of elementary complex analysis, namely, *every analytic function has a power series representation*. Precisely, if $f: \Lambda \to \mathbb{C}$ is analytic, and if the annulus $B_{\alpha,\beta}(\mu) = \{\lambda \in \mathbb{C}: \ 0 \le \alpha < |\lambda - \mu| < \beta\}$ lies in the open set $\Lambda \subseteq \mathbb{C}$, then $f$ has a *unique Laurent expansion* about the point $\mu$, viz., $f(\lambda) = \sum_{k=-\infty}^{\infty} \gamma_k (\lambda - \mu)^k$ for every $\lambda \in B_{\alpha,\beta}(\mu)$.

**Proposition 6.21.** $r_\sigma(T) = \lim_n \|T^n\|^{\frac{1}{n}}$.

*Proof.* Since $r_\sigma(T)^n \le \|T^n\|$ for every positive integer $n$,

$$r_\sigma(T) \le \lim_n \|T^n\|^{\frac{1}{n}}.$$

(Reason: the limit of the sequence $\{\|T^n\|^{\frac{1}{n}}\}$ exists for every $T \in \mathcal{B}[\mathcal{X}]$, according to Lemma 6.8.). Now recall the von Neumann expansion for the resolvent function $R: \rho(T) \to \mathcal{G}[\mathcal{X}]$:

$$R(\lambda) = (\lambda I - T)^{-1} = \lambda^{-1} \sum_{i=0}^{\infty} T^i \lambda^{-i}$$

for every $\lambda \in \rho(T)$ such that $|\lambda| > \|T\|$, where the above series converges in the (uniform) topology of $\mathcal{B}[\mathcal{X}]$ (cf. Problem 4.47). Take an arbitrary bounded linear functional $\varphi: \mathcal{B}[\mathcal{X}] \to \mathbb{C}$ in $\mathcal{B}[\mathcal{X}]^*$. Since $\varphi$ is continuous,

$$\varphi(R(\lambda)) = \lambda^{-1} \sum_{i=0}^{\infty} \varphi(T^i) \lambda^{-i}$$

for every $\lambda \in \rho(T)$ such that $|\lambda| > \|T\|$.

*Claim.* The displayed identity holds whenever $|\lambda| > r_\sigma(T)$.

*Proof.* $\lambda^{-1} \sum_{i=0}^{\infty} \varphi(T^i) \lambda^{-i}$ is a Laurent expansion of $\varphi(R(\lambda))$ about the origin for every $\lambda \in \rho(T)$ such that $|\lambda| > \|T\|$. But $\varphi \circ \mathcal{R}$ is analytic on $\rho(T)$ (cf. Claim 2 in Proposition 6.13) so that $\varphi(R(\lambda))$ has a unique Laurent expansion about the origin for every $\lambda \in \rho(T)$, and hence for every $\lambda \in \mathbb{C}$ such that $|\lambda| > r_\sigma(T)$. Then $\varphi(R(\lambda)) = \lambda^{-1} \sum_{i=0}^{\infty} \varphi(T^i) \lambda^{-i}$, which holds for every $|\lambda| > \|T\| \ge r_\sigma(T)$, must be the Laurent expansion about the origin for every $\lambda \in \mathbb{C}$ such that $|\lambda| > r_\sigma(T)$. $\square$

Therefore, if $|\lambda| > r_\sigma(T)$, then $\varphi((\lambda^{-1}T)^i) = \varphi(T^i)\lambda^{-i} \to 0$ (cf. Problem 4.7(c)) for every $\varphi \in \mathcal{B}[\mathcal{X}]^*$. But this implies that $\{(\lambda^{-1}T)^i\}$ is bounded in the (uniform) topology of $\mathcal{B}[\mathcal{X}]$ (cf. Problem 4.67(d)). That is, $\lambda^{-1}T$ is power bounded. Hence $|\lambda|^{-n}\|T^n\| \le \sup_i \|(\lambda^{-1}T)^i\| < \infty$, so that

$$|\lambda|^{-1}\|T^n\|^{\frac{1}{n}} \le \left(\sup_i \|(\lambda^{-1}T)^i\|\right)^{\frac{1}{n}},$$

for every positive integer $n$ whenever $|\lambda| > r_\sigma(T)$. Then $|\lambda|^{-1} \lim_n \|T^n\|^{\frac{1}{n}} \leq 1$ so that $\lim_n \|T^n\|^{\frac{1}{n}} \leq |\lambda|$ whenever $|\lambda| > r_\sigma(T)$. In other words, $\lim_n \|T^n\|^{\frac{1}{n}} \leq r_\sigma(T) + \varepsilon$ for every $\varepsilon > 0$. Outcome:

$$\lim_n \|T^n\|^{\frac{1}{n}} \leq r_\sigma(T). \qquad \square$$

Observe the following immediate consequences of Proposition 6.21.

$$r_\sigma(\alpha T) = |\alpha| r_\sigma(T) \quad \text{for every} \quad \alpha \in \mathbb{C}$$

and, if $\mathcal{H}$ is a complex Hilbert space and $T \in \mathcal{B}[\mathcal{H}]$, then

$$r_\sigma(T^*) = r_\sigma(T).$$

An important application of the Gelfand–Beurling formula reads as follows: $T$ *is uniformly stable if and only if* $r_\sigma(T) < 1$. In fact, there exists in the current literature a large collection of equivalent conditions for uniform stability. We shall consider below just a few of them.

**Proposition 6.22.** *Let $T \in \mathcal{B}[\mathcal{X}]$ be an operator on a complex Banach space $\mathcal{X}$. The following assertions are pairwise equivalent.*

(a) $T^n \xrightarrow{u} O.$

(b) $r_\sigma(T) < 1.$

(c) $\|T^n\| \leq \beta \alpha^n$ *for every $n \geq 0$, for some $\beta \geq 1$ and some $\alpha \in (0, 1)$.*

(d) $\sum_{n=0}^{\infty} \|T^n\|^p < \infty$ *for an arbitrary $p > 0$.*

(e) $\sum_{n=0}^{\infty} \|T^n x\|^p < \infty$ *for all $x \in \mathcal{X}$, for an arbitrary $p > 0$.*

*Proof.* Since $r_\sigma(T)^n = r_\sigma(T^n) \leq \|T^n\|$ for every $n \geq 0$, it follows that (a) $\Rightarrow$ (b). Suppose $r_\sigma(T) < 1$ and take any $\alpha \in (r_\sigma(T), 1)$. The Gelfand–Beurling formula says that $\lim_n \|T^n\|^{\frac{1}{n}} = r_\sigma(T)$. Therefore, there exists an integer $n_\alpha \geq 1$ such that $\|T^n\| \leq \alpha^n$ for every $n \geq n_\alpha$, and hence (b)$\Rightarrow$(c) with $\beta = \max_{0 \leq n \leq n_\alpha} \|T^n\| \alpha^{-n_\alpha}$. It is trivially verified that (c)$\Rightarrow$(d)$\Rightarrow$(e). If assertion (e) holds true, then $\sup_n \|T^n x\| < \infty$ for every $x \in \mathcal{X}$, and so $\sup_n \|T^n\| < \infty$ by the Banach–Steinhaus Theorem (Theorem 4.43). Moreover, for $m \geq 1$ and $p > 0$ arbitrary,

$$\|m^{\frac{1}{p}} T^m x\|^p = \sum_{n=0}^{m-1} \|T^{m-n} T^n x\|^p \leq \Big(\sup_n \|T^n\|\Big)^p \sum_{n=0}^{\infty} \|T^n x\|^p.$$

Thus $\sup_m \|m^{\frac{1}{p}} T^m x\| < \infty$ for every $x \in \mathcal{X}$ whenever (e) holds true. Since $m^{\frac{1}{p}} T^m$ lies in $\mathcal{B}[\mathcal{X}]$ for each $m \geq 1$, it follows that $\sup_m \|m^{\frac{1}{p}} T^m\| < \infty$ by using the Banach–Steinhaus Theorem again. Hence

$$0 \leq \|T^n\| \leq n^{-\frac{1}{p}} \sup_m \|m^{\frac{1}{p}} T^m\|$$

for every $n \geq 1$ so that $\|T^n\| \to 0$ as $n \to \infty$. Therefore, (e)$\Rightarrow$(a). $\qquad\square$

The next result extends the von Neumann expansion of Problem 4.47.

**Corollary 6.23.** *Let $\mathcal{X}$ be a complex Banach space. Take any operator $T$ in $\mathcal{B}[\mathcal{X}]$ and any nonzero complex number $\lambda$.*

(a) *$r_\sigma(T) < |\lambda|$ if and only if $\{\sum_{i=0}^{n}(\frac{T}{\lambda})^i\}$ converges uniformly. In this case, $\lambda$ lies in $\rho(T)$, $(\lambda I - T)^{-1} = \frac{1}{\lambda}\sum_{i=0}^{\infty}(\frac{T}{\lambda})^i$ where $\sum_{i=0}^{\infty}(\frac{T}{\lambda})^i$ denotes the uniform limit of $\{\sum_{i=0}^{n}(\frac{T}{\lambda})^i\}$, and $\|(\lambda I - T)^{-1}\| \leq (|\lambda| - \|T\|)^{-1}$.*

(b) *If $r_\sigma(T) = |\lambda|$ and $\{\sum_{i=0}^{n}(\frac{T}{\lambda})^i\}$ converges strongly, then $\lambda$ lies in $\rho(T)$ and $(\lambda I - T)^{-1} = \frac{1}{\lambda}\sum_{i=0}^{\infty}(\frac{T}{\lambda})^i$ where $\sum_{i=0}^{\infty}(\frac{T}{\lambda})^i$ denotes the strong limit of $\{\sum_{i=0}^{n}(\frac{T}{\lambda})^i\}$.*

(c) *If $|\lambda| < r_\sigma(T)$, then $\{\sum_{i=0}^{n}(\frac{T}{\lambda})^i\}$ does not converge strongly.*

*Proof.* If $\{\sum_{i=0}^{n}(\frac{T}{\lambda})^i\}$ converges uniformly, then $(\frac{T}{\lambda})^n \xrightarrow{u} O$ (cf. Problem 4.7), and hence $|\lambda|^{-1} r_\sigma(T) = r_\sigma(\frac{T}{\lambda}) < 1$ by Proposition 6.22. Conversely, if $r_\sigma(T) < |\lambda|$, then $\lambda \in \rho(T)$ so that $(\lambda I - T) \in \mathcal{G}[\mathcal{X}]$, and $r_\sigma(\frac{T}{\lambda}) = |\lambda|^{-1} r_\sigma(T) < 1$. Hence $\{(\frac{T}{\lambda})^n\}$ is an absolutely summable sequence in $\mathcal{B}[\mathcal{X}]$ by Proposition 6.22. Now follow the steps of Problem 4.47 to conclude all the properties of item (a). If $\{\sum_{i=0}^{n}(\frac{T}{\lambda})^i\}$ converges strongly, then $(\frac{T}{\lambda})^n x \to 0$ in $\mathcal{X}$ for every $x \in \mathcal{X}$ (cf. Problem 4.7 again) so that $\sup_n \|(\frac{T}{\lambda})^n x\| < \infty$ for every $x \in \mathcal{X}$. Then $\sup_n \|(\frac{T}{\lambda})^n\| < \infty$ by the Banach–Steinhaus Theorem (i.e., $\frac{T}{\lambda}$ is power bounded), and hence $|\lambda|^{-1} r_\sigma(T) = r_\sigma(\frac{T}{\lambda}) \leq 1$. This proves assertion (c). Moreover,

$$(\lambda I - T)\frac{1}{\lambda}\sum_{i=0}^{n}\left(\tfrac{T}{\lambda}\right)^i = \frac{1}{\lambda}\sum_{i=0}^{n}\left(\tfrac{T}{\lambda}\right)^i(\lambda I - T) = I - \left(\tfrac{T}{\lambda}\right)^{n+1} \xrightarrow{s} I.$$

Therefore, $(\lambda I - T)^{-1} = \frac{1}{\lambda}\sum_{i=0}^{\infty}(\frac{T}{\lambda})^i$, where $\sum_{i=0}^{\infty}(\frac{T}{\lambda})^i \in \mathcal{B}[\mathcal{X}]$ is the strong limit of $\{\sum_{i=0}^{n}(\frac{T}{\lambda})^i\}$, which concludes the proof of (b). $\qquad\square$

## 6.4   Numerical Radius

What Proposition 6.21 says is that $r_\sigma(T) = r(T)$, where $r(T)$ is the limit of the numerical sequence $\{\|T^n\|^{\frac{1}{n}}\}$ (whose existence was proved in Lemma 6.8). We shall then adopt one and the same notation (the simplest, of course) for both of them: the limit of $\{\|T^n\|^{\frac{1}{n}}\}$ and the spectral radius. Thus, from now on, we write

$$r(T) \;=\; \sup_{\lambda \in \sigma(T)} |\lambda| \;=\; \max_{\lambda \in \sigma(T)} |\lambda| \;=\; \lim_n \|T^n\|^{\frac{1}{n}}.$$

Therefore, *a normaloid operator acting on a complex Banach space is precisely an operator whose norm coincides with the spectral radius.* Recall that, in a complex Hilbert space $\mathcal{H}$, every normal operator is normaloid, and so is every nonnegative operator. Since $T^*T$ is always nonnegative, it follows that (cf. Proposition 5.65)

$$r(T^*T) \;=\; r(TT^*) \;=\; \|T^*T\| \;=\; \|TT^*\| \;=\; \|T\|^2 \;=\; \|T^*\|^2$$

for every $T \in \mathcal{B}[\mathcal{H}]$. Also note that $T$ *is normaloid if and only if there exists $\lambda$ in $\sigma(T)$ such that $|\lambda| = \|T\|$.* However, such a $\lambda$ can never be in the residual spectrum. In fact, for every $T \in \mathcal{B}[\mathcal{H}]$,

$$\sigma_R(T) \;\subseteq\; \{\lambda \in \mathbb{C}: \ |\lambda| < \|T\|\}.$$

(If $\lambda \in \sigma_R(T) = \sigma_P(T^*)^* \backslash \sigma_P(T)$, then there exists $0 \neq x \in \mathcal{H}$ such that $0 < \|Tx - \lambda x\|^2 = \|Tx\|^2 - 2\mathrm{Re}\,\langle \bar{\lambda} x\,; T^*x \rangle + |\lambda|^2\|x\|^2 = \|Tx\|^2 - |\lambda|^2\|x\|^2$, and hence $|\lambda| < \|T\|$.)

The *numerical range* of an operator $T$ acting on a complex Hilbert space $\mathcal{H} \neq \{0\}$ is the (nonempty) set

$$W(T) \;=\; \{\lambda \in \mathbb{C}: \ \lambda = \langle Tx\,; x \rangle \text{ for some } \|x\| = 1\}.$$

It can be shown that $W(T)$ is always convex in $\mathbb{C}$ and, clearly,

$$W(T^*) \;=\; W(T)^*.$$

**Proposition 6.24.** $\sigma_P(T) \cup \sigma_R(T) \subseteq W(T)$ *and* $\sigma(T) \subseteq W(T)^-$.

*Proof.* Take $T \in \mathcal{B}[\mathcal{H}]$, where $\mathcal{H} \neq \{0\}$ is a complex Hilbert space.

(a) If $\lambda \in \sigma_P(T)$, then there exists a unit vector $x \in \mathcal{H}$ such that $Tx = \lambda x$. Hence $\langle Tx\,; x \rangle = \lambda\|x\|^2 = \lambda$; that is $\lambda \in W(T)$. If $\lambda \in \sigma_R(T)$, then $\bar{\lambda} \in \sigma_P(T^*)$ (Proposition 6.17). Thus $\bar{\lambda} \in W(T^*)$ so that $\lambda \in W(T)$.

(b) If $\lambda \in \sigma_{AP}(T)$, then there exists a sequence $\{x_n\}$ of unit vectors in $\mathcal{H}$ such that $\|(\lambda I - T)x_n\| \to 0$ (Proposition 6.15). Therefore,

$$0 \leq |\lambda - \langle Tx_n\,; x_n \rangle| \;=\; |\langle (\lambda I - T)x_n\,; x_n \rangle| \;\leq\; \|(\lambda I - T)x_n\| \to 0$$

so that $\langle Tx_n\,x_n \rangle \to \lambda$. Since each $\langle Tx_n\,; x_n \rangle$ lies in $W(T)$, it follows by the Closed Set Theorem that $\lambda \in W(T)^-$. Hence

$$\sigma_{AP}(T) \;\subseteq\; W(T)^-,$$

and so $\sigma(T) = \sigma_R(T) \cup \sigma_{AP}(T) \subseteq W(T)^-$ according to item (a).    $\square$

The *numerical radius* of $T \in \mathcal{B}[\mathcal{H}]$ is the number

$$w(T) = \sup_{\lambda \in W(T)} |\lambda| = \sup_{\|x\|=1} |\langle Tx \,;\, x\rangle|.$$

It is readily verified that

$$w(T^*) = w(T) \quad \text{and} \quad w(T^*T) = \|T\|^2.$$

Unlike the spectral radius, the numerical radius is a norm on $\mathcal{B}[\mathcal{H}]$. That is, $0 \leq w(T)$ for every $T \in \mathcal{B}[\mathcal{H}]$ and $0 < w(T)$ whenever $T \neq O$, $w(\alpha T) = |\alpha| w(T)$ and $w(T + S) \leq w(T) + w(S)$ for every $\alpha \in \mathbb{C}$ and every $S, T \in \mathcal{B}[\mathcal{H}]$. Warning: the numerical radius does not have the "operator norm property" in the sense that the inequality $w(ST) \leq w(S)w(T)$ is *not* true for all operators $S, T \in \mathcal{B}[\mathcal{H}]$, but the power inequality holds (i.e., $w(T^n) \leq w(T)^n$ for all $T \in \mathcal{B}[\mathcal{H}]$ and every positive integer $n$ — the proof is tricky). Nevertheless, the numerical radius is a norm equivalent to the (induced uniform) operator norm of $\mathcal{B}[\mathcal{H}]$ and dominates the spectral radius, as in the following proposition.

**Proposition 6.25.**  $0 \leq r(T) \leq w(T) \leq \|T\| \leq 2w(T)$.

*Proof.* Since $\sigma(T) \subseteq W(T)$, we get $r(T) \leq w(T)$. Moreover,

$$w(T) = \sup_{\|x\|=1} |\langle Tx \,;\, x\rangle| \leq \sup_{\|x\|=1} \|Tx\| = \|T\|.$$

From Problem 5.3, recalling that $|\langle Tz \,;\, z\rangle| \leq \sup_{\|u\|=1} |\langle Tu \,;\, u\rangle| \|z\|^2 = w(T)\|z\|^2$ for every $z \in \mathcal{H}$ (cf. proof of Proposition 5.78), and by the parallelogram law,

$$\begin{aligned}
|\langle Tx \,;\, y\rangle| &\leq \tfrac{1}{4}\big(|\langle T(x+y) \,;\, (x+y)\rangle| + |\langle T(x-y) \,;\, (x-y)\rangle| \\
&\quad + |\langle T(x+iy) \,;\, (x+iy)\rangle| + |\langle T(x-iy) \,;\, (x-iy)\rangle|\big) \\
&\leq \tfrac{1}{4} w(T)\big(\|x+y\|^2 + \|x-y\|^2 + \|x+iy\|^2 + \|x-iy\|^2\big) \\
&= w(T)\big(\|x\|^2 + \|y\|^2\big) \leq 2w(T)
\end{aligned}$$

whenever $\|x\| = \|y\| = 1$. Therefore, according to Corollary 5.71,

$$\|T\| = \sup_{\|x\|=\|y\|=1} |\langle Tx \,;\, y\rangle| \leq 2w(T).$$    $\square$

An operator $T \in \mathcal{B}[\mathcal{H}]$ is *spectraloid* if $r(T) = w(T)$. The next result is a straightforward application of the previous proposition.

**Corollary 6.26.** *Every normaloid operator is spectraloid.*

Indeed, $r(T) = \|T\|$ implies $r(T) = w(T)$ by Proposition 6.25. However, Proposition 6.25 also ensures that $r(T) = \|T\|$ implies $w(T) = \|T\|$, so that $w(T) = \|T\|$

is a property of every normaloid operator on $\mathcal{H}$. What comes out as a nice surprise is that this property can be viewed as a third definition of a normaloid operator on a complex Hilbert space.

**Proposition 6.27.** $T \in \mathcal{B}[\mathcal{H}]$ *is normaloid if and only if* $w(T) = \|T\|$.

*Proof.* The easy half of the proof was presented above. Now suppose $w(T) = \|T\|$ (and $T \neq O$ — otherwise the result is trivially verified). Recall that $W(T)^-$ is compact in $\mathbb{C}$ (for $W(T)$ is clearly bounded). Thus $\max_{\lambda \in W(T)^-} |\lambda| = \sup_{\lambda \in W(T)^-} |\lambda| = \sup_{\lambda \in W(T)} |\lambda| = w(T) = \|T\|$, and hence there exists $\lambda \in W(T)^-$ such that $\lambda = \|T\|$. Since $W(T)$ is always nonempty, it follows by Proposition 3.32 that there exists a sequence $\{\lambda_n\}$ in $W(T)$ that converges to $\lambda$. In other words, there exists a sequence $\{x_n\}$ of unit vectors in $\mathcal{H}$ ($\|x_n\| = 1$ for each $n$) such that $\lambda_n = \langle Tx_n ; x_n \rangle \to \lambda$, where $|\lambda| = \|T\| \neq 0$. If $S = \lambda^{-1}T \in \mathcal{B}[\mathcal{H}]$, then

$$\langle Sx_n ; x_n \rangle \to 1.$$

*Claim.*   $\|Sx_n\| \to 1$  and  $\mathrm{Re}\,\langle Sx_n ; x_n \rangle \to 1$.

*Proof.* $|\langle Sx_n ; x_n \rangle| \leq \|Sx_n\| \leq \|S\| = 1$ for each $n$. But $\langle Sx_n ; x_n \rangle \to 1$ implies that $|\langle Sx_n ; x_n \rangle| \to 1$ (and hence $\|Sx_n\| \to 1$) and also that $\mathrm{Re}\,\langle Sx_n ; x_n \rangle \to 1$. Both arguments follow by continuity. $\square$

Then $\|(I - S)x_n\|^2 = \|Sx_n - x_n\|^2 = \|Sx_n\|^2 - 2\mathrm{Re}\,\langle Sx_n; x_n \rangle + \|x_n\|^2 \to 0$ so that $1 \in \sigma_{AP}(S) \subseteq \sigma(S)$ (cf. Proposition 6.15). Hence $r(S) \geq 1$ and $r(T) = r(\lambda S) = |\lambda| r(S) \geq |\lambda| = \|T\|$, which implies $r(T) = \|T\|$ (since $r(T) \leq \|T\|$ for every operator $T$).                    $\square$

Therefore, the class of all normaloid operators on $\mathcal{H}$ coincides with the class of all operators $T \in \mathcal{B}[\mathcal{H}]$ for which

$$\|T\| = \sup_{\|x\|=1} |\langle Tx ; x \rangle|.$$

This includes the normal operators and, in particular, the self-adjoint operators (see Proposition 5.78). This includes the isometries too. In fact, every isometry is quasinormal, and hence normaloid. Thus

$$r(V) = w(V) = \|V\| = 1 \quad \text{whenever} \quad V \in \mathcal{B}[\mathcal{H}] \text{ is an isometry.}$$

(The above identity can be directly verified by Propositions 6.21 and 6.25, once $\|V^n\| = 1$ for every positive integer $n$ — cf. Proposition 4.37.)

Remark: *If* $T \in \mathcal{B}[\mathcal{H}]$ *is spectraloid and quasinilpotent, then* $T = O$. Proof: If $w(T) = r(T) = 0$, then $T = O$ by Proposition 6.25. Particular cases: *The unique normal (or hyponormal, or normaloid) quasinilpotent operator is the null operator.*

In other words, if $T \in \mathcal{B}[\mathcal{H}]$ is normal (or hyponormal, or normaloid) and $r(T) = 0$ (i.e., $\sigma(T) = \{0\}$), then $T = O$.

**Corollary 6.28.** *If there exists $\lambda \in W(T)$ such that $|\lambda| = \|T\|$, then $T$ is normaloid and $\lambda \in \sigma_P(T)$. In other words, if there exists a unit vector $x$ such that $\|T\| = \langle Tx ; x \rangle$, then $r(T) = w(T) = \|T\|$ and $\langle Tx ; x \rangle \in \sigma_P(T)$.*

*Proof.* If $\lambda \in W(T)$ is such that $|\lambda| = \|T\|$, then $w(T) = \|T\|$ (see Proposition 6.25) so that $T$ is normaloid by Proposition 6.27. Moreover, since $\lambda = \langle Tx ; x \rangle$ for some unit vector $x$, it follows that $\|T\| = |\lambda| = |\langle Tx ; x \rangle| \leq \|Tx\|\|x\| \leq \|T\|$, and hence $|\langle Tx ; x \rangle| = \|Tx\|\|x\|$. Then $Tx = \alpha x$ for some $\alpha \in \mathbb{C}$ (cf. Problem 5.2) so that $\alpha \in \sigma_P(T)$. But $\alpha = \alpha\|x\|^2 = \langle \alpha x ; x \rangle = \langle Tx ; x \rangle = \lambda$.    $\square$

Remark: Using the inequality $\|T^n\| \leq \|T\|^n$, which holds for every operator $T$, we have shown in Proposition 6.9 that $T$ is normaloid if and only if $\|T^n\| = \|T\|^n$ for every $n \geq 0$. Now, using the inequality $w(T^n) \leq w(T)^n$, which also holds for every operator $T$, we can show that *$T$ is spectraloid if and only if $w(T^n) = w(T)^n$ for every $n \geq 0$.* Indeed, according to Propositions 6.20 and 6.25,

$$r(T)^n = r(T^n) \leq w(T^n) \leq w(T)^n \quad \text{for every} \quad n \geq 0.$$

Hence $r(T) = w(T)$ implies $w(T^n) = w(T)^n$. Conversely, since

$$w(T^n)^{\frac{1}{n}} \leq \|T^n\|^{\frac{1}{n}} \to r(T) \leq w(T),$$

if follows that $w(T^n) = w(T)^n$ implies $r(T) = w(T)$.

## 6.5   Examples of Spectra

Every closed and bounded subset of the complex plane (i.e., every compact subset of $\mathbb{C}$) is the spectrum of some operator.

**Example 6B.** Take any $T \in \mathcal{B}[\mathcal{H}]$, where $\mathcal{X}$ is a *finite-dimensional complex normed space*. Then $\mathcal{X}$ and its manifolds are all Banach spaces (Corollaries 4.28 and 4.29). Moreover, $\mathcal{N}(\lambda I - T) = \{0\}$ if and only if $(\lambda I - T) \in \mathcal{G}[\mathcal{X}]$ (cf. Problem 4.38(c)). That is, $\mathcal{N}(\lambda I - T) = \{0\}$ if and only if $\lambda \in \rho(T)$, and hence

$$\sigma_C(T) = \sigma_R(T) = \varnothing.$$

Furthermore, since $\mathcal{R}(\lambda I - T)$ is a subspace of $\mathcal{X}$ for every $\lambda \in \mathbb{C}$, it also follows that $\sigma_{P_2}(T) = \sigma_{P_3}(T) = \varnothing$ (see diagram of Section 6.2). Finally, if $\mathcal{N}(\lambda I - T) \neq \{0\}$, then $\mathcal{R}(\lambda I - T) \neq \mathcal{X}$ whenever $\mathcal{X}$ is finite-dimensional (cf. Problems 2.6 and 2.17), and so $\sigma_{P_1}(T) = \varnothing$. Therefore,

$$\sigma(T) = \sigma_P(T) = \sigma_{P_4}(T).$$

**Example 6C.** Let $T \in \mathcal{B}[\mathcal{H}]$ be a *diagonalizable operator* on a complex (separable infinite-dimensional) Hilbert space $\mathcal{H}$. That is, according to Problem 5.17 there exists an orthonormal basis $\{e_k\}_{k=1}^{\infty}$ for $\mathcal{H}$ and a sequence $\{\lambda_k\}_{k=1}^{\infty}$ in $\ell_+^{\infty}$ such that, for every $x \in \mathcal{H}$,

$$Tx = \sum_{k=1}^{\infty} \lambda_k \langle x ; e_k \rangle e_k.$$

Take an arbitrary $\lambda \in \mathbb{C}$ and note that $(\lambda I - T) \in \mathcal{B}[\mathcal{H}]$ is again a diagonalizable operator. Indeed, $(\lambda I - T)x = \sum_{k=1}^{\infty}(\lambda - \lambda_k)\langle x ; e_k \rangle e_k$ for every $x \in \mathcal{H}$. Since $\mathcal{N}(\lambda I - T) = \{0\}$ if and only if $\lambda \neq \lambda_k$ for every $k \geq 1$ (that is, there exists $(\lambda I - T)^{-1} \in \mathcal{L}[\mathcal{R}(\lambda I - T), \mathcal{H}]$ if and only if $\lambda - \lambda_k \neq 0$ for every $k \geq 1$ — cf. Problem 5.17), it follows that

$$\sigma_P(T) = \left\{\lambda \in \mathbb{C}: \ \lambda = \lambda_k \text{ for some } k \geq 1\right\}.$$

Similarly, since $T^* \in \mathcal{B}[\mathcal{H}]$ also is a diagonalizable operator, given by $T^*x = \sum_{k=1}^{\infty} \overline{\lambda}_k \langle x ; e_k \rangle e_k$ for every $x \in \mathcal{H}$ (e.g., Problem 5.27(c)), we get

$$\sigma_P(T^*) = \left\{\lambda \in \mathbb{C}: \ \lambda = \overline{\lambda}_k \text{ for some } k \geq 1\right\}.$$

Therefore, $\sigma_R(T) = \sigma_R(T^*)^* \backslash \sigma_P(T) = \varnothing$. Moreover, $\lambda$ lies in $\rho(T)$ if and only if $(\lambda I - T) \in \mathcal{G}[\mathcal{H}]$ or, equivalently, if and only if $\inf_k |\lambda - \lambda_k| > 0$ (cf. Problem 5.17). Then

$$\sigma(T) = \sigma_P(T) \cup \sigma_C(T) = \left\{\lambda \in \mathbb{C}: \ \inf_k |\lambda - \lambda_k| = 0\right\},$$

and hence $\sigma(T)\backslash\sigma_P(T)$ is the set of all cluster points of the sequence $\{\lambda_k\}_{k=1}^{\infty}$ (i.e., the set of all accumulation points of the set $\{\lambda_k\}_{k=1}^{\infty}$):

$$\sigma_C(T) = \left\{\lambda \in \mathbb{C}: \ \inf_k |\lambda - \lambda_k| = 0 \ \text{ and } \ \lambda \neq \lambda_k \text{ for every } k \geq 1\right\}.$$

Note that $\sigma_{P_1}(T) = \sigma_{P_2}(T) = \varnothing$ (reason: $T^*$ is a diagonalizable operator so that $\sigma_R(T^*) = \varnothing$ — see Proposition 6.17). If $\lambda_j \in \sigma_P(T)$ also is an accumulation point of $\sigma_P(T)$, then it lies in $\sigma_{P_3}(T)$; otherwise (i.e., if it is an isolated point of $\sigma_P(T)$), it lies in $\sigma_{P_4}(T)$. Indeed, consider a new set $\{\lambda_k\}'$ without this point $\lambda_j$ and the associated diagonalizable operator $T'$ so that $\lambda_j \in \sigma_C(T')$, and hence $\mathcal{R}(\lambda_j I - T')$ is not closed, which means that $\mathcal{R}(\lambda_j I - T)$ is not closed. If $\{\lambda_k\}$ is a constant sequence, say $\lambda_k = \mu$ for all $k$, then $T = \mu I$ is a *scalar operator* and, in this case,

$$\sigma(\mu I) = \sigma_P(\mu I) = \sigma_{P_4}(\mu I) = \{\mu\}.$$

Now recall that $\mathbb{C}$ (equipped with its usual metric) is a separable metric space (Example 3P) so that it includes a countable dense subset, and so does every compact subset $\Sigma$ of $\mathbb{C}$. Let $\Lambda$ be any countable dense subset of $\Sigma$, and let $\{\lambda_k\}_{k=1}^{\infty}$ be an enumeration of it (if $\Sigma$ is finite, say $\#\Sigma = n$, then set $\lambda_k = 0$ for all $k > n$). Observe that $\sup_k |\lambda_k| < \infty$ because $\Sigma$ is bounded. Thus consider a diagonalizable operator $T \in \mathcal{B}[\mathcal{H}]$ such that $Tx = \sum_{k=1}^{\infty} \lambda_k \langle x \,;\, e_k \rangle e_k$ for every $x \in \mathcal{H}$. As we have just seen,

$$\sigma(T) = \Lambda^- = \Sigma;$$

that is, $\sigma(T)$ is the set of all points of adherence of $\Lambda = \{\lambda_k\}_{k=1}^{\infty}$, which means the closure of $\Lambda$. This confirms the statement that introduced this section. Precisely, *every closed and bounded subset of the complex plane is the spectrum of some diagonalizable operator on $\mathcal{H}$.*

**Example 6D.** Let $\Delta$ and $\Gamma$ denote the open unit disc and the unit circle in the complex plane centered at the origin, respectively. In this example we shall characterize each part of the spectrum of a *unilateral shift* of arbitrary multiplicity. Let $S_+$ be a unilateral shift acting on a (complex) Hilbert space $\mathcal{H}$, and let $\{\mathcal{H}_k\}_{k=0}^{\infty}$ be the underlying sequence of orthogonal subspaces of $\mathcal{H} = \bigoplus_{k=0}^{\infty} \mathcal{H}_k$ (Problem 5.29). Recall that

$$S_+ x = 0 \oplus \bigoplus_{k=1}^{\infty} U_k \, x_{k-1} \quad \text{and} \quad S_+^* x = \bigoplus_{k=0}^{\infty} U_{k+1}^* x_{k+1}$$

for every $x = \bigoplus_{k=0}^{\infty} x_k$ in $\mathcal{H} = \bigoplus_{k=0}^{\infty} \mathcal{H}_k$, with $0$ denoting the origin of $\mathcal{H}_0$, where $\{U_{k+1}\}_{k=0}^{\infty}$ is an arbitrary sequence of unitary transformations $U_{k+1} \colon \mathcal{H}_k \to \mathcal{H}_{k+1}$. Since a unilateral shift is an isometry, we get

$$r(S_+) = 1.$$

Take $x = \bigoplus_{k=0}^{\infty} x_k \in \mathcal{H}$ and $\lambda \in \mathbb{C}$. If $x \in \mathcal{N}(\lambda I - S_+)$, then $\lambda x_0 \oplus \bigoplus_{k=1}^{\infty} \lambda x_k = 0 \oplus \bigoplus_{k=1}^{\infty} U_k x_{k-1}$. Hence $\lambda x_0 = 0$ and, for every $k \geq 0$, $\lambda x_{k+1} = U_{k+1} x_k$. If $\lambda = 0$, then $x = 0$. If $\lambda \neq 0$, then $x_0 = 0$ and $x_{k+1} = \lambda^{-1} U_{k+1} x_k$, so that $\|x_0\| = 0$ and $\|x_{k+1}\| = |\lambda|^{-1} \|x_k\|$, for each $k \geq 0$. Thus $\|x_k\| = |\lambda|^{-k} \|x_0\| = 0$ for every $k \geq 0$, and therefore $x = 0$. Conclusion: $\mathcal{N}(\lambda I - S_+) = \{0\}$ for all $\lambda \in \mathbb{C}$. Equivalently,

$$\sigma_P(S_+) = \varnothing.$$

Now take any $x_0 \neq 0$ in $\mathcal{H}_0$ and any $\lambda \in \Delta$. Consider the sequence $\{x_k\}_{k=0}^{\infty}$, with each $x_k$ in $\mathcal{H}_k$, recursively defined by $x_{k+1} = \lambda U_{k+1} x_k$ so that $\|x_{k+1}\| = |\lambda| \|x_k\|$ for every $k \geq 0$. Then $\|x_k\| = |\lambda|^k \|x_0\|$ for every $k \geq 1$, and hence $\sum_{k=0}^{\infty} \|x_k\|^2 = \|x_0\|^2 \big(1 + \sum_{k=1}^{\infty} |\lambda|^{2k}\big) < \infty$, which implies that the nonzero $x = \bigoplus_{k=0}^{\infty} x_k$ lies in $\bigoplus_{k=0}^{\infty} \mathcal{H}_k = \mathcal{H}$. Moreover, since $\lambda x_k = U_{k+1}^* x_{k+1}$ for each $k \geq 0$, $\lambda x = S_+^* x$. Therefore, $0 \neq x \in \mathcal{N}(\lambda I - S_+^*)$. Conclusion: $\mathcal{N}(\lambda I - S_+^*) \neq \{0\}$ for all $\lambda \in \Delta$. Equivalently, $\Delta \subseteq \sigma_P(S_+^*)$. On the other hand, if $\lambda \in \sigma_P(S_+^*)$, then there exists $0 \neq x = \bigoplus_{k=0}^{\infty} x_k \in \bigoplus_{k=0}^{\infty} \mathcal{H}_k = \mathcal{H}$ such that $S_+^* x = \lambda x$. Thus $U_{k+1}^* x_{k+1} = \lambda x_k$ so that $\|x_{k+1}\| = |\lambda| \|x_k\|$ for each $k \geq 0$, and hence $\|x_k\| = |\lambda|^k \|x_0\|$ for every $k \geq 1$. Therefore, $x_0 \neq 0$ (because $x \neq 0$) and $\big(1 + \sum_{k=1}^{\infty} |\lambda|^{2k}\big) \|x_0\|^2 = \sum_{k=0}^{\infty} \|x_k\|^2 = \|x\|^2 < \infty$, which implies that $|\lambda| < 1$ (i.e., $\lambda \in \Delta$). Conclusion: $\sigma_P(S_+^*) \subseteq \Delta$. Then

$$\sigma_P(S_+^*) = \Delta.$$

But the spectrum of any operator $T$ on $\mathcal{H}$ is a closed set included in the closed disc $\{\lambda \in \mathbb{C}\colon |\lambda| \leq r(T)\}$, which is the disjoint union of $\sigma_P(T), \sigma_R(T)$ and $\sigma_C(T)$, where $\sigma_R(T) = \sigma_P(T^*)^* \backslash \sigma_P(T)$ (Proposition 6.17). Hence

$$\sigma_P(S_+) = \sigma_R(S_+^*) = \varnothing, \quad \sigma_R(S_+) = \sigma_P(S_+^*) = \Delta, \quad \sigma_C(S_+) = \sigma_C(S_+^*) = \Gamma.$$

**Example 6E.** The spectrum of a *bilateral shift* is simpler than that of a unilateral shift, for bilateral shifts are unitary (i.e., besides being isometries they are normal too). Let $S$ be a bilateral shift of arbitrary multiplicity acting on a (complex) Hilbert space $\mathcal{H}$, and let $\{\mathcal{H}_k\}_{k=-\infty}^{\infty}$ be the underlying family of orthogonal subspaces of $\mathcal{H} = \bigoplus_{k=-\infty}^{\infty} \mathcal{H}_k$ (Problem 5.30). Recall that

$$Sx = \bigoplus_{k=-\infty}^{\infty} U_k \, x_{k-1} \quad \text{and} \quad S^*x = \bigoplus_{k=\infty}^{\infty} U_{k+1}^* x_{k+1}$$

for every $x = \bigoplus_{k=-\infty}^{\infty} x_k$ in $\mathcal{H} = \bigoplus_{k=-\infty}^{\infty} \mathcal{H}_k$, where $\{U_k\}_{k=-\infty}^{\infty}$ is an arbitrary family of unitary transformations $U_{k+1}\colon \mathcal{H}_k \to \mathcal{H}_{k+1}$. Suppose there exists $\lambda \in \Gamma \cap \rho(S)$ so that $\mathcal{R}(\lambda I - S) = \mathcal{H}$ and $|\lambda| = 1$. Take any $y_0 \neq 0$ in $\mathcal{H}_0$ and set $y_k = 0 \in \mathcal{H}_k$ for each $k \neq 0$. Consider the vector $y = \bigoplus_{k=-\infty}^{\infty} y_k \in \mathcal{H} = \mathcal{R}(\lambda I - S)$ and let $x = \bigoplus_{k=-\infty}^{\infty} x_k \in \mathcal{H}$ be any inverse image of $y$ under $(\lambda I - S)$; that is, $(\lambda I - S)x = y$. Since $y_0 \neq 0$ it follows that $y \neq 0$, and hence $x \neq 0$. On the other hand, since $y_k = 0$ for every $k \neq 0$, it also follows that $\lambda x_k = U_k x_{k-1} + y_k = U_k x_{k-1}$ so that $\|x_k\| = \|x_{k-1}\|$ for every $k \neq 0$. Therefore, $\|x_j\| = \|x_{-1}\|$ for every $j \leq -1$ and $\|x_j\| = \|x_0\|$ for every $j \geq 0$, and hence $x = 0$. (Reason: $\|x\|^2 = \sum_{k=-\infty}^{\infty} \|x_k\|^2 = \sum_{j=-\infty}^{-1} \|x_j\|^2 + \sum_{j=0}^{\infty} \|x_j\|^2 < \infty$.) Thus the existence of a complex number $\lambda$ in $\Gamma \cap \rho(S)$ leads to a contradiction. Conclusion: $\Gamma \cap \rho(S) = \varnothing$. Equivalently, $\Gamma \subseteq \sigma(S)$. Finally, recall that $S$ is unitary and so $\sigma(S) \subseteq \Gamma$ (Corollary 6.18(c)). Outcome:

$$\sigma(S) = \Gamma.$$

Now take an arbitrary pair $\{\lambda, x\}$ with $\lambda$ in $\sigma(S)$ and $x = \bigoplus_{k=-\infty}^{\infty} x_k$ in $\mathcal{H}$. If $x \in \mathcal{N}(\lambda I - S)$, then $\bigoplus_{k=-\infty}^{\infty} \lambda x_k - \bigoplus_{k=-\infty}^{\infty} U_k x_{k-1}$ and so $\lambda x_k = U_k x_{k-1}$ for every $k$. Since $|\lambda| = 1$ (because $\sigma(S) = \Gamma$), $\|x_k\| = \|x_{k-1}\|$ for every $k$. Hence $x = 0$ ($\|x\|^2 = \sum_{k=-\infty}^{\infty} \|x_k\|^2$ is finite). Conclusion: $\mathcal{N}(\lambda I - S) = \{0\}$ for all $\lambda \in \sigma(S)$. Equivalently,

$$\sigma_P(S) = \varnothing.$$

But $S$ is normal so that $\sigma_R(S) = \varnothing$ (Corollary 6.18(b)). Recalling that $\sigma(S^*) = \sigma(S)^*$ and $\sigma_C(S^*) = \sigma_C(S)^*$ (Proposition 6.17) we get

$$\sigma(S) = \sigma(S^*) = \sigma_C(S^*) = \sigma_C(S) = \Gamma.$$

Consider a weighted sum of projections $D = \sum_k \alpha_k P_k$ on $\ell_+^2(\mathcal{H})$ or on $\ell^2(\mathcal{H})$, where $\{\alpha_k\}$ is a bounded family of scalars and $\mathcal{R}(P_k) \cong \mathcal{H}$ for all $k$. This is identified with an orthogonal direct sum of scalar operators $D = \bigoplus_k \alpha_k I$ (Problem 5.16), and is referred to as a *diagonal operator* on $\ell^2_+(\mathcal{H})$ or on $\ell^2(\mathcal{H})$, respectively. A *weighted shift* is the product of a shift and a diagonal operator. Such a definition implicitly assumes that the shift (unilateral or bilateral, of any multiplicity) acts on the direct sum of countably infinite copies of a single Hilbert space $\mathcal{H}$. Explicitly, a *weighted unilateral shift* on $\ell^2_+(\mathcal{H})$ is the product of a unilateral shift on $\ell^2_+(\mathcal{H})$ and a diagonal operator on $\ell^2_+(\mathcal{H})$. Similarly, a *weighted bilateral shift* on $\ell^2(\mathcal{H})$ is the product of a bilateral shift on $\ell^2(\mathcal{H})$ and a diagonal operator on $\ell^2(\mathcal{H})$. Diagonal operators acting on $\ell^2_+(\mathcal{H})$ and on $\ell^2(\mathcal{H})$, $D_+ = \bigoplus_{k=0}^{\infty}\alpha_k I$ and $D = \bigoplus_{k=-\infty}^{\infty}\alpha_k I$, with $I$ standing for the identity on $\mathcal{H}$, are denoted by $D_+ = \mathrm{diag}(\{\alpha_k\}_{k=0}^{\infty})$ and $D = \mathrm{diag}(\{\alpha_k\}_{k=-\infty}^{\infty})$, respectively. Likewise, weighted shifts acting on $\ell^2_+(\mathcal{H})$ and on $\ell^2(\mathcal{H})$, $T_+ = S_+ D_+$ and $T = SD$, will be denoted by $T_+ = \mathrm{shift}(\{\alpha_k\}_{k=0}^{\infty})$ and $T = (\{\alpha_k\}_{k=-\infty}^{\infty})$, respectively, whenever $S_+$ is the *canonical unilateral shift* on $\ell^2_+(\mathcal{H})$ and $S$ is the *canonical bilateral shift* on $\ell^2(\mathcal{H})$ (see Problems 5.29 and 5.30).

**Example 6F.** Let $\{\alpha_k\}_{k=0}^{\infty}$ be a bounded sequence in $\mathbb{C}$ such that

$$\alpha_k \neq 0 \ \text{ for every } \ k \geq 0 \quad \text{and} \quad \alpha_k \to 0 \ \text{ as } \ k \to \infty,$$

and consider the *weighted unilateral shift* $T_+ = \mathrm{shift}(\{\alpha_k\}_{k=0}^{\infty})$ on $\ell^2_+(\mathcal{H})$, where $\mathcal{H} \neq \{0\}$ is a complex Hilbert space. $T_+$ and $T_+^*$ are given by

$$T_+ x = S_+ D_+ x = 0 \oplus \bigoplus_{k=1}^{\infty}\alpha_{k-1}x_{k-1} \ \text{ and } \ T_+^* x = D_+^* S_+^* x = \bigoplus_{k=0}^{\infty}\overline{\alpha}_k x_{k+1}$$

for every $x = \bigoplus_{k=0}^{\infty}x_k$ in $\ell^2_+(\mathcal{H}) = \bigoplus_{k=0}^{\infty}\mathcal{H}$, with 0 denoting the origin of $\mathcal{H}$. Applying the same argument used in Example 6D to show that $\sigma(S_+) = \varnothing$, we get $\mathcal{N}(\lambda I - T_+) = \{0\}$ for all $\lambda \in \mathbb{C}$. Indeed, if $x = \bigoplus_{k=0}^{\infty}x_k \in \mathcal{N}(\lambda I - T_+)$, then $\lambda x_0 \oplus \bigoplus_{k=1}^{\infty}\lambda x_k = 0 \oplus \bigoplus_{k=1}^{\infty}\alpha_{k-1}x_{k-1}$ so that $\lambda x_0 = 0$ and $\lambda x_{k+1} = \alpha_k x_k$ for every $k \geq 0$. Thus $x = 0$ if $\lambda = 0$ (for $\alpha_k \neq 0$) and, if $\lambda \neq 0$, then $x_0 = 0$ and $\|x_{k+1}\| \leq \sup_k |\alpha_k| |\lambda|^{-1} \|x_k\|$ for every $k \geq 0$, which also implies that $x = 0$. Outcome:

$$\sigma_P(T_+) = \varnothing.$$

Now note that the vector $x = \bigoplus_{k=0}^{\infty}x_k$ with $0 \neq x_0 \in \mathcal{H}$ and $x_k = 0 \in \mathcal{H}$ for every $k \geq 1$ lies in $\ell^2_+(\mathcal{H})$ but not in $\mathcal{R}(T_+)^- \subseteq \{0\} \oplus \bigoplus_{k=1}^{\infty}\mathcal{H}$. Hence $\mathcal{R}(T_+)^- \neq \ell^2_+(\mathcal{H})$ and so $0 \in \sigma_{CP}(T_+)$. Since $\sigma_P(T_+) = \varnothing$,

$$0 \in \sigma_R(T_+).$$

However, if $\lambda \neq 0$, then $\mathcal{R}(\lambda I - T_+) = \ell^2_+(\mathcal{H})$. In fact, suppose $\lambda \neq 0$ and take any $y = \bigoplus_{k=0}^{\infty}y_k$ in $\ell^2_+(\mathcal{H})$. Set $x_0 = \lambda^{-1} y_0$ and, for each $k \geq 0$, $x_{k+1} = \lambda^{-1}(\alpha_k x_k + y_{k+1})$. Since $\alpha_k \to 0$, there exists a positive $k_\lambda$ such that $\alpha := |\lambda|^{-1}\sup_{k \geq k_\lambda}|\alpha_k| \leq \frac{1}{2}$. Then $\|\alpha_{k+1}x_{k+1}\| \leq \alpha(\|\alpha_k x_k\| + \|y_{k+1}\|)$, so that $\|\alpha_{k+1}x_{k+1}\|^2 \leq \alpha^2(\|\alpha_k x_k\| + \|y_{k+1}\|)^2 \leq 2\alpha^2(\|\alpha_k x_k\|^2 + \|y_k\|^2)$, for every $k \geq k_\lambda$. Hence $\sum_{k=k_\lambda}^{\infty}\|\alpha_{k+1}x_{k+1}\|^2 \leq$

$\frac{1}{2}\sum_{k=k_\lambda}^{\infty}\|\alpha_k x_k\|^2 + \frac{1}{2}\|y\|^2$, which implies that $\sum_{k=0}^{\infty}\|\alpha_k x_k\|^2 < \infty$, and therefore

$$|\lambda|^2\left(\sum_{k=0}^{\infty}\|x_{k+1}\|^2\right) \le \sum_{k=0}^{\infty}(\|\alpha_k x_k\| + \|y_{k+1}\|)^2$$

$$\le 2\left(\sum_{k=0}^{\infty}\|\alpha_k x_k\|^2 + \|y\|^2\right) < \infty$$

so that $x = \bigoplus_{k=0}^{\infty}x_k$ lies in $\ell_+^2(\mathcal{H})$. But $(\lambda I - T_+)x = \lambda x_0 \oplus \bigoplus_{k=1}^{\infty}\lambda x_k - \alpha_{k-1}x_{k-1} = y$ and so $y \in \mathcal{R}(\lambda I - T_+)$. Outcome: $\mathcal{R}(\lambda I - T_+) = \ell_+^2(\mathcal{H})$. Since $\mathcal{N}(\lambda I - T_+) = \{0\}$ for all $\lambda \in \mathbb{C}$, $\lambda \in \rho(T_+)$ for every nonzero $\lambda \in \mathbb{C}$. Conclusion: $\sigma(T_+) = \sigma_R(T_+) = \{0\}$. Moreover, as $\sigma_{R_1}(T)$ is an open set for every operator $T$, we get

$$\sigma(T_+) = \sigma_R(T_+) = \sigma_{R_2}(T_+) = \{0\},$$

and hence

$$\sigma(T_+^*) = \sigma_P(T_+^*) = \sigma_{P_2}(T_+^*) = \{0\}.$$

This was our first sample of a quasinilpotent operator $(r(T_+) = 0)$ that is not nilpotent $(\sigma_P(T_+) = \varnothing)$. The next example exhibits another one. It is worth noticing that $\sigma(\mu I - T_+) = \{\mu - \lambda \in \mathbb{C}: \lambda \in \sigma(T_+)\} = \{\mu\}$ by the Spectral Mapping Theorem, and therefore $\sigma(\overline{\mu}I - T_+^*) = \{\overline{\mu}\}$. Moreover, if $x$ is an eigenvector of $T_+^*$, then $T_+^* x = 0$ so that $(\overline{\mu}I - T_+^*)x = \overline{\mu}x$; that is, $\overline{\mu} \in \sigma_P(\overline{\mu}I - T_+^*)$. Thus

$$\sigma(\mu I - T_+) = \sigma_R(\mu I - T_+) = \{\mu\} \quad \text{and} \quad \sigma(\overline{\mu}I - T_+^*) = \sigma_P(\overline{\mu}I - T_+^*) = \{\overline{\mu}\}.$$

**Example 6G.** Let $\{\alpha_k\}_{k=-\infty}^{\infty}$ be a bounded family in $\mathbb{C}$ such that

$$\alpha_k \ne 0 \quad \text{for every} \quad k \in \mathbb{Z} \quad \text{and} \quad \alpha_k \to 0 \quad \text{as} \quad |k| \to \infty,$$

and consider the *weighted bilateral shift* $T = \text{shift}(\{\alpha_k\}_{k=-\infty}^{\infty})$ on $\ell^2(\mathcal{H})$, where $\mathcal{H} \ne \{0\}$ is a complex Hilbert space. $T$ and $T^*$ are given by

$$T = SDx = \bigoplus_{k=-\infty}^{\infty}\alpha_{k-1}x_{k-1} \quad \text{and} \quad T^* = D^*S^*x = \bigoplus_{k=-\infty}^{\infty}\overline{\alpha}_k x_{k+1}$$

for $x = \bigoplus_{k=-\infty}^{\infty}x_k$ in $\ell^2(\mathcal{H}) = \bigoplus_{k=-\infty}^{\infty}\mathcal{H}$. Take any $\lambda \in \mathbb{C}$. If $x = \bigoplus_{k=-\infty}^{\infty}x_k \subseteq \mathcal{N}(\lambda I - T)$, then $\bigoplus_{k=-\infty}^{\infty}\lambda x_k - \alpha_{k-1}x_{k-1} = 0$ so that $\lambda x_{k+1} = \alpha_k x_k$ for every $k \in \mathbb{Z}$. If $\lambda = 0$, then $x = 0$. If $\lambda \ne 0$, then $\|x_{k+1}\| \le |\lambda|^{-1}\sup_k|\alpha_k|\|x_k\|$ for every $k \in \mathbb{Z}$. But $\lim_{k \to -\infty}\|x_k\| = 0$ $(\|x\|^2 = \sum_{k=-\infty}^{\infty}\|x_k\|^2 < \infty)$ so that $x = 0$. Hence $\mathcal{N}(\lambda I - T) = \{0\}$ for all $\lambda \in \mathbb{C}$, and so

$$\sigma_P(T) = \varnothing.$$

Take any vector $y = \bigoplus_{k=-\infty}^{\infty} y_k$ in $\ell^2(\mathcal{H})$ and any $\lambda \neq 0$ in $\mathbb{C}$. Since $\alpha_k \to 0$ as $|k| \to \infty$, there exists a positive integer $k_\lambda$ and a finite set $K_\lambda = \{k \in \mathbb{Z}: -k_\lambda \leq k \leq k_\lambda\}$ such that $\alpha := \sup_{k \in \mathbb{Z} \setminus K_\lambda} |\frac{\alpha_k}{\lambda}| \leq \frac{1}{2}$. Then

$$\sum_{j=-\infty}^{k} \left|\frac{\alpha_k}{\lambda}\right| \cdots \left|\frac{\alpha_j}{\lambda}\right| \left\|\frac{y_j}{\lambda}\right\| \leq \sup_k \left\|\frac{y_k}{\lambda}\right\| \left(2 \sum_{j=0}^{\infty} \alpha^j + \#K_\lambda \sup_k \left|\frac{\alpha_k}{\lambda}\right|\right) < \infty$$

for all $k \in \mathbb{Z}$, which implies that the infinite series $\sum_{j=-\infty}^{k} \frac{\alpha_k}{\lambda} \cdots \frac{\alpha_j}{\lambda} \frac{y_j}{\lambda}$ is absolutely convergent (thus convergent in $\mathcal{H}$) for every $k \in \mathbb{Z}$. Set

$$x_k = \sum_{j=-\infty}^{k-1} \frac{\alpha_{k-1}}{\lambda} \cdots \frac{\alpha_j}{\lambda} \frac{y_j}{\lambda} + \frac{y_k}{\lambda}$$

in $\mathcal{H}$ so that $x_{k+1} = \frac{\alpha_k}{\lambda} x_k + \frac{y_{k+1}}{\lambda}$ for each $k \in \mathbb{Z}$. If $k \in \mathbb{Z} \setminus K_\lambda$, then $\|\alpha_k x_k\| \leq \alpha(\|\alpha_{k-1} x_{k-1}\| + \|y_k\|)$ and so $\|\alpha_k x_k\|^2 \leq 2\alpha^2(\|\alpha_{k-1} x_{k-1}\|^2 + \|y_k\|^2)$. Hence $\sum_{k \in \mathbb{Z} \setminus K_\lambda} \|\alpha_k x_k\|^2 \leq \frac{1}{2}(\sum_{k \in \mathbb{Z} \setminus K_\lambda} \|\alpha_{k-1} x_{k-1}\|^2 + \|y\|^2)$. Then $\sum_{k=-\infty}^{\infty} \|\alpha_k x_k\|^2 < \infty$, and therefore

$$|\lambda|^2 \left(\sum_{k=-\infty}^{\infty} \|x_{k+1}\|^2\right) \leq \sum_{k=-\infty}^{\infty} (\|\alpha_k x_k\| + \|y_{k+1}\|)^2$$

$$\leq 2 \left(\sum_{k=-\infty}^{\infty} \|\alpha_k x_k\|^2 + \|y\|^2\right) < \infty.$$

Thus $x = \bigoplus_{k=-\infty}^{\infty} x_k$ lies in $\ell^2(\mathcal{H})$. But $(\lambda I - T)x = \bigoplus_{k=-\infty}^{\infty} \lambda x_k - \alpha_{k-1} x_{k-1} = y$ and so $y \in \mathcal{R}(\lambda I - T)$. Outcome: $\mathcal{R}(\lambda I - T) = \ell^2(\mathcal{H})$. Since $\mathcal{N}(\lambda I - T) = \{0\}$ for all $\lambda \in \mathbb{C}$, it follows that every nonzero $\lambda \in \mathbb{C}$ lies in $\rho(T)$. Conclusion:

$$\sigma(T) = \{0\}.$$

However, if $x \in \mathcal{N}(T^*)$, then $\overline{\alpha}_k x_{k+1} = 0$ so that $x_{k+1} = 0$ (since $\alpha_k \neq 0$) for every $k \in \mathbb{Z}$ and hence $x = 0$. That is, $\mathcal{N}(T^*) = \{0\}$ or, equivalently, (cf. Problem 5.35) $\mathcal{R}(T)^- = \ell^2(\mathcal{H})$. This implies that $0 \neq \sigma_R(T)$ and, as $\sigma_P(T) = \varnothing$, we finally get

$$\sigma(T) = \sigma_C(T) = \sigma_C(T^*) = \sigma(T^*) = \{0\}.$$

Note: Using the Spectral Mapping Theorem (as we did in the previous example) it can be shown that

$$\sigma(\mu I - T) = \sigma_C(\mu I - T) = \{\mu\} \quad \text{and} \quad \sigma(\overline{\mu} I - T^*) = \sigma_C(\overline{\mu} I - T^*) = \{\overline{\mu}\}.$$

**Example 6H.** Let $F \in \mathcal{B}[\mathcal{H}]$ be an operator on a complex Hilbert space $\mathcal{H} \neq \{0\}$. Consider the operator $T \in \mathcal{B}[\ell_+^2(\mathcal{H})]$ defined by

$$Tx = 0 \oplus \bigoplus_{k=1}^{\infty} F x_{k-1} \quad \text{so that} \quad T^*x = \bigoplus_{k=0}^{\infty} F^* x_{k+1}$$

for every $x = \bigoplus_{k=0}^{\infty} x_k$ in $\ell_+^2(\mathcal{H}) = \bigoplus_{k=0}^{\infty} \mathcal{H}$, where $0$ is the origin of $\mathcal{H}$. These can be identified with the following infinite matrix of operators.

$$T = \begin{pmatrix} O \\ F & O \\ & F & O \\ & & F & O \\ & & & & \ddots \end{pmatrix} \quad \text{and} \quad T^* = \begin{pmatrix} O & F^* \\ & O & F^* \\ & & O & F^* \\ & & & O \\ & & & & \ddots \end{pmatrix}.$$

It is readily verified by induction that $T^n x = \bigoplus_{k=0}^{n-1} 0 \oplus \bigoplus_{k=n}^{\infty} F^n x_{k-n}$, and hence $\|T^n x\|^2 = \sum_{k=0}^{\infty} \|F^n x_k\|^2$ so that $\|T^n x\| \leq \|F^n\| \|x\|$ for all $x \in \ell_+^2(\mathcal{H})$, which implies that $\|T^n\| \leq \|F^n\|$ for each $n \geq 1$. On the other hand, take any nonzero vector $y_0$ in $\mathcal{H}$, set $y_k = 0 \in \mathcal{H}$ for all $k \geq 1$, and consider the vector $y = \bigoplus_{k=0}^{\infty} y_k$ in $\ell_+^2(\mathcal{H})$ such that $\|y\| = \|y_0\| \neq 0$. Thus $\|T^n\| = \sup_{\|x\|=1} \|T^n x\| \geq \sup_{\|y_0\|=1} \|T^n y\| = \sup_{\|y_0\|=1} \|F^n y_0\| = \|F^n\|$ for each $n \geq 1$. Therefore, $\|T^n\| = \|F^n\|$ for every $n \geq 1$. Then the Gelfand–Beurling formula for the spectral radius ensures that

$$r(T) = r(F).$$

Moreover, $y \neq 0$ and $T^* y = 0$ so that

$$0 \in \sigma_P(T^*),$$

and hence $0 \in \sigma(T)$. Take an arbitrary $\lambda \in \rho(T)$ so that $\lambda \neq 0$ and $\mathcal{R}(\lambda I - T) = \ell_+^2(\mathcal{H})$. Since $y = y_0 \oplus \bigoplus_{k=1}^{\infty} 0$ lies in $\ell_+^2(\mathcal{H})$ for every $y_0 \in \mathcal{H}$, it follows that $y \in \mathcal{R}(\lambda I - T)$. That is, $y = (\lambda I - T)x$ for some $x = \bigoplus_{k=0}^{\infty} x_k$ in $\ell_+^2(\mathcal{H})$ and so $y_0 \oplus \bigoplus_{k=1}^{\infty} 0 = \lambda x_0 \oplus \bigoplus_{k=1}^{\infty} \lambda x_k - F x_{k-1}$. Therefore, $x_0 = \lambda^{-1} y_0$ and $x_{k+1} = \lambda^{-1} F x_k$ for every $k \geq 0$, and hence $x_k = (\lambda^{-1} F)^k x_0 = \lambda^{-1} (\lambda^{-1} F)^k y_0$. Since $x \in \ell_+^2(\mathcal{H})$ for every $y_0 \in \mathcal{H}$, $\|x\|^2 = \sum_{k=0}^{\infty} \|x_k\|^2 = |\lambda|^{-2} \sum_{k=0}^{\infty} \|(\lambda^{-1} F)^k y_0\|^2 < \infty$ for every $y_0 \in \mathcal{H}$, so that $r(\lambda^{-1} F) < 1$ by Proposition 6.22. Conclusion: If $\lambda \in \rho(T)$, then $r(F) < |\lambda|$. Equivalently, if $|\lambda| \leq r(F)$, then $\lambda \in \sigma(T)$; that is, $\{\lambda \in \mathbb{C}: |\lambda| \leq r(F)\} \subseteq \sigma(T)$. But $\sigma(T) \subseteq \{\lambda \in \mathbb{C}: |\lambda| \leq r(F)\}$ because $r(F) = r(T)$. Therefore (since $\sigma(T^*) = \sigma(T)^*$ for every operator $T$)

$$\sigma(T) = \{\lambda \in \mathbb{C}: \ |\lambda| \leq r(F)\} = \sigma(T^*).$$

Now recall that $\lambda \in \sigma_P(T)$ if and only if $Tx = \lambda x$ (i.e., $\lambda x_0 = 0$ and $\lambda x_{k+1} = F x_k$ for every $k \geq 0$) for some nonzero $x = \bigoplus_{k=0}^{\infty} x_k$ in $\ell_+^2(\mathcal{H})$. If $0 \in \sigma_P(T)$, then $F x_k = 0$ for all $k \geq 0$ for some nonzero $x = \bigoplus_{k=0}^{\infty} x_k$ in $\ell_+^2(\mathcal{H})$ so that $0 \in \sigma_P(F)$. Conversely, if $0 \in \sigma_P(F)$, then there exists $x_0 \neq 0$ in $\mathcal{H}$ such that $F x_0 = 0$. Set $x = \bigoplus_{k=0}^{\infty} (k+1)^{-1} x_0$, a nonzero vector in $\ell_+^2(\mathcal{H})$, so that $Tx = 0 \oplus \bigoplus_{k=1}^{\infty} k^{-1} F x_0 = 0$. Hence $0 \in \sigma_P(T)$. Outcome: $0 \in \sigma_P(T)$ if and only if $0 \in \sigma_P(F)$. If $\lambda \neq 0$ lies in $\sigma_P(T)$, then $x_0 = 0$ and $x_{k+1} = \lambda^{-1} F x_k$ for every $k \geq 0$ so that $x = 0$, which is a contradiction. Therefore, if $\lambda \neq 0$, then $\lambda \neq \sigma_P(T)$. Conclusion:

$$\sigma_P(T) = \begin{cases} \{0\}, & 0 \in \sigma_P(F), \\ \varnothing, & 0 \notin \sigma_P(F). \end{cases}$$

Since $\sigma_R(T^*) = \sigma_P(T)^* \backslash \sigma_P(T^*)$, $\sigma_P(T)^* \subseteq \{0\}$ and $0 \in \sigma_P(T^*)$,

$$\sigma_R(T^*) = \varnothing, \quad \text{and hence} \quad \sigma_C(T^*) = \big\{\lambda \in \mathbb{C}: \ |\lambda| \le r(F)\big\} \backslash \sigma_P(T^*).$$

If $\sigma_P(T^*) \ne \{0\}$, then there exists $0 \ne \lambda \in \sigma_P(T^*)$, which means that $T^*x = \lambda x$ for some nonzero $x = \bigoplus_{k=0}^{\infty} x_k$ in $\ell_+^2(\mathcal{H})$. Hence there exists $0 \ne x_j \in \mathcal{H}$ such that $F^* x_{k+1} = \lambda x_k$ for every $k \ge j$. A trivial induction shows that $F^{*k} x_{j+k} = \lambda^k x_j$ for every $k \ge 0$. Then $x_j \in \bigcap_{k=0}^{\infty} \mathcal{R}(F^{*k})$ because $\lambda \ne 0$ so that $\bigcap_{k=0}^{\infty} \mathcal{R}(F^{*k}) \ne \{0\}$. Conclusion:

$$\bigcap_{k=0}^{\infty} \mathcal{R}(F^{*k}) = \{0\} \quad \text{implies} \quad \sigma_P(T^*) = \{0\},$$

and, in this case,

$$\sigma_C(T^*) = \big\{\lambda \in \mathbb{C}: \ |\lambda| \le r(F)\big\} \backslash \{0\}.$$

Sample: Set $F = S_+^*$ on $\mathcal{H} = \ell_+^2(\mathcal{K})$ for any complex Hilbert space $\mathcal{K} \ne \{0\}$. Therefore, $r(F) = r(S_+^*) = 1$ (cf. Example 6E) and $\mathcal{R}(F^{*k}) = \mathcal{R}(S_+^k) = \bigoplus_{j=0}^{k}\{0\} \oplus \bigoplus_{j=k+1}^{\infty} \mathcal{K} \subseteq \ell_+^2(\mathcal{K})$ so that $\bigcap_{k=0}^{\infty} \mathcal{R}(F^{*k}) = \{0\}$. With $\Delta^-$ denoting the closed unit disc about the origin we get

$$\sigma_P(T^*) = \sigma_R(T) = \{0\}, \ \ \sigma_R(T^*) = \sigma_P(T) = \varnothing, \ \ \sigma_C(T^*) = \sigma_C(T) = \Delta^- \backslash \{0\}.$$

Summing up: A backward unilateral shift of unilateral shifts (i.e., $T^*$ with $F^* = S_+$) and a unilateral shift of backward unilateral shifts (i.e., $T$ with $F = S_+^*$) have a continuous spectrum equal to the punctured disc $\Delta^- \backslash \{0\}$. This was our first example of operators for which the continuous spectrum is not included in the boundary of the spectrum.

## 6.6   The Spectrum of a Compact Operator

The spectral theory of compact operators is an essential feature for the Spectral Theorem for compact normal operators of the next section. Normal operators were defined on a Hilbert space, and therefore we assume throughout this section that the compact operators act on a complex Hilbert space $\mathcal{H} \ne \{0\}$, although the spectral theory for compact operators can be developed on a Banach space as well. Recall that $\mathcal{B}_\infty[\mathcal{H}]$ denotes the class of all compact operators on $\mathcal{H}$.

**Proposition 6.29.** *If $T \in \mathcal{B}_\infty[\mathcal{H}]$ and $\lambda$ is any nonzero complex number, then $\mathcal{R}(\lambda I - T)$ is a subspace of $\mathcal{H}$.*

*Proof.* Take any compact operator $K \in \mathcal{B}_\infty[\mathcal{M}, \mathcal{X}]$, where $\mathcal{X} \ne \{0\}$ is a complex Banach space and $\mathcal{M}$ is a subspace of $\mathcal{X}$. Let $I$ be the identity on $\mathcal{M}$, let $\lambda$ be any nonzero complex number, and consider the operator $(\lambda I - K) \in \mathcal{B}[\mathcal{M}, \mathcal{X}]$.

*Claim.* If $\mathcal{N}(\lambda I - K) = \{0\}$, then $\mathcal{R}(\lambda I - K)$ is closed in $\mathcal{X}$.

*Proof.* If $\mathcal{N}(\lambda I - K) = \{0\}$ and $\mathcal{R}(\lambda I - K)$ is not closed in $\mathcal{X} \neq \{0\}$, then $\lambda I - K$ is not bounded below (Corollary 4.24). This means that for every $\varepsilon > 0$ there exists $0 \neq x_\varepsilon \in \mathcal{M}$ such that $\|(\lambda I - K)x_\varepsilon\| < \varepsilon\|x_\varepsilon\|$. Thus $\inf_{\|x\|=1}\|(\lambda I - K)x\| = 0$ and there exists a sequence $\{x_n\}$ of unit vectors in $\mathcal{M}$ for which $\|(\lambda I - K)x_n\| \to 0$. Since $K$ is compact and $\{x_n\}$ is bounded, it follows by Theorem 4.52 that $\{Kx_n\}$ has a convergent subsequence, say $\{Kx_k\}$, so that $Kx_k \to y \in \mathcal{X}$. However, $\|\lambda x_k - y\| = \|\lambda x_k - Kx_k + Kx_k - y\| \leq \|(\lambda I - K)x_k\| + \|Kx_k - y\| \to 0$. Then $\{\lambda x_k\}$ also converges in $\mathcal{X}$ to $y$, and hence $y \in \mathcal{M}$ (for $\mathcal{M}$ is closed in $\mathcal{X}$ — Theorem 3.30). Moreover, $y \neq 0$ (since $0 \neq |\lambda| = \|\lambda x_k\| \to \|y\|$) and, as $K$ is continuous, $Ky = K \lim_k \lambda x_k = \lambda \lim_k Kx_k = \lambda y$ so that $y \in \mathcal{N}(\lambda I - K)$. Therefore we get $\mathcal{N}(\lambda I - K) \neq \{0\}$, which is a contradiction. $\square$

Take any $T \in \mathcal{B}[\mathcal{H}]$. Recall that $(\lambda I - T)|_{\mathcal{N}(\lambda I - T)^\perp} \in \mathcal{B}[\mathcal{N}(\lambda I - T)^\perp, \mathcal{H}]$ is injective (i.e., $\mathcal{N}((\lambda I - T)|_{\mathcal{N}(\lambda I - T)^\perp}) = \{0\}$ — cf. Remark that follows Proposition 5.12) and coincides with $\lambda I - T|_{\mathcal{N}(\lambda I - T)^\perp}$ on $\mathcal{N}(\lambda I - T)^\perp$. If $T$ is compact, then so is $T|_{\mathcal{N}(\lambda I - T)^\perp} \in \mathcal{B}[\mathcal{N}(\lambda I - T)^\perp, \mathcal{H}]$. (Reason: $\mathcal{N}(\lambda I - T)^\perp$ is a subspace of $\mathcal{H}$, and the restriction of a compact linear transformation to a linear manifold is a compact linear transformation — see Section 4.9.) Since $\mathcal{H} \neq 0$, it follows by the above claim that $(\lambda I - T)|_{\mathcal{N}(\lambda I - T)^\perp} = \lambda I - T|_{\mathcal{N}(\lambda I - T)^\perp}$ has a closed range for all $\lambda \neq 0$. But it is readily verified that $\mathcal{R}((\lambda I - T)|_{\mathcal{N}(\lambda I - T)^\perp}) = \mathcal{R}(\lambda I - T)$.   $\square$

**Proposition 6.30.** *If $T \in \mathcal{B}_\infty[\mathcal{H}]$ and $\lambda$ is any nonzero complex number, then $\mathcal{R}(\lambda I - T) = \mathcal{H}$ whenever $\mathcal{N}(\lambda I - T) = \{0\}$.*

*Proof.* Take any $\lambda \neq 0$ in $\mathbb{C}$ and any $T \in \mathcal{B}[\mathcal{H}]$. Suppose $\mathcal{N}(\lambda I - T) = \{0\}$ and $\mathcal{R}(\lambda I - T) \neq \mathcal{H}$ (recall: $\mathcal{H} \neq \{0\}$), and consider the sequence $\{\mathcal{M}_n\}_{n=0}^\infty$ of linear manifolds of $\mathcal{H}$ recursively defined by

$$\mathcal{M}_{n+1} = (\lambda I - T)(\mathcal{M}_n) \text{ for every } n \geq 0, \quad \text{with} \quad \mathcal{M}_0 = \mathcal{H}.$$

It can be verified by induction that

$$\mathcal{M}_{n+1} \subseteq \mathcal{M}_n \quad \text{for every} \quad n \geq 0.$$

Indeed, $\mathcal{M}_1 = \mathcal{R}(\lambda I - T) \subseteq \mathcal{H} = \mathcal{M}_0$ and, if the above inclusion holds for some $n \geq 0$, then $\mathcal{M}_{n+2} = (\lambda I - T)(\mathcal{M}_{n+1}) \subseteq (\lambda I - T)(\mathcal{M}_n) = \mathcal{M}_{n+1}$, which concludes the induction. The previous proposition ensures that $\mathcal{R}(\lambda I - T)$ is a subspace of $\mathcal{H}$, and so $(\lambda I - T) \in \mathcal{G}[\mathcal{H}, \mathcal{R}(\lambda I - T)]$ by Corollary 4.24. Hence (another induction plus Theorem 3.24),

$$\{\mathcal{M}_n\}_{n=0}^\infty \text{ is a decreasing sequence of subspaces of } \mathcal{H}.$$

Moreover, if $\mathcal{M}_{n+1} = \mathcal{M}_n$ for some $n$, then there exists an integer $k \geq 1$ such that $\mathcal{M}_{k+1} = \mathcal{M}_k \neq \mathcal{M}_{k-1}$ (for $\mathcal{M}_0 = \mathcal{H} \neq \mathcal{R}(\lambda I - T) = \mathcal{M}_1$). But this leads to a contradiction: if $\mathcal{M}_{k+1} = \mathcal{M}_k$, then $(\lambda I - T)(\mathcal{M}_k) = \mathcal{M}_k$ so that

$\mathcal{M}_k = (\lambda I - T)^{-1}(\mathcal{M}_k) = \mathcal{M}_{k-1}$. Outcome: $\mathcal{M}_{n+1}$ is properly included in $\mathcal{M}_n$ for each $n$; that is,

$$\mathcal{M}_{n+1} \subset \mathcal{M}_n \quad \text{for every} \quad n \geq 0.$$

Hence $\mathcal{M}_{n+1}$ is a proper subspace of $\mathcal{M}_n$ (see Problem 3.38). By Lemma 4.33, for each $n \geq 0$ there exists $x_n \in \mathcal{M}_n$ with $\|x_n\| = 1$ such that $\frac{1}{2} < d(x_n, \mathcal{M}_{n+1})$. Recall that $\lambda \neq 0$, take any pair of integers $0 \leq m < n$, and set

$$x = x_n + \lambda^{-1}\big((\lambda I - T)x_m - (\lambda I - T)x_n\big)$$

so that $Tx_n - Tx_m = \lambda(x - x_m)$. Since $x$ lies in $\mathcal{M}_{m+1}$,

$$\|Tx_n - Tx_m\| = |\lambda|\|x - x_m\| > \tfrac{1}{2}|\lambda|,$$

which implies that the sequence $\{Tx_n\}$ has no convergent subsequence (every sub-sequence of $\{Tx_n\}$ is not a Cauchy sequence). Since $\{x_n\}$ is bounded, this ensures that $T$ is not compact (cf. Theorem 4.52). Conclusion: If $\lambda \neq 0$ and $T \in \mathcal{B}[\mathcal{H}]$ is such that $\mathcal{N}(\lambda I - T) = \{0\}$ and $\mathcal{R}(\lambda I - T) \neq \mathcal{H}$, then $T \notin \mathcal{B}_\infty[\mathcal{H}]$. Equivalently, if $T \in \mathcal{B}_\infty[\mathcal{H}]$ and $\mathcal{N}(\lambda I - T) = \{0\}$ for $\lambda \neq 0$, then $\mathcal{R}(\lambda I - T) = \mathcal{H}$. $\square$

**Corollary 6.31.** *If* $T \in \mathcal{B}_\infty[\mathcal{H}]$, *then* $0 \neq \lambda \in \rho(T) \cup \sigma_{P_4}(T)$ *so that*

$$\sigma(T)\backslash\{0\} = \sigma_P(T)\backslash\{0\} \subseteq \sigma_{P_4}(T).$$

*Proof.* Take $0 \neq \lambda \in \mathbb{C}$. Since $\mathcal{H} \neq \{0\}$, Propositions 6.29 and 6.30 say that $\lambda \in \rho(T) \cup \sigma_{P_1}(T) \cup \sigma_{P_4}(T) \cup \sigma_{R_1}(T)$ and $\lambda \in \rho(T) \cup \sigma_P(T)$ (see the diagram of Section 6.2). Then $\lambda \in \rho(T) \cup \sigma_{P_1}(T) \cup \sigma_{P_4}(T)$ and hence $\bar{\lambda} \in \rho(T)^* \cup \sigma_{P_1}(T)^* \cup \sigma_{P_4}(T)^* = \rho(T^*) \cup \sigma_{R_1}(T^*) \cup \sigma_{P_4}(T^*)$ (by Proposition 6.17). But $T^* \in \mathcal{B}_\infty[\mathcal{H}]$ whenever $T \in \mathcal{B}_\infty[\mathcal{H}]$ (cf. Problem 5.42) so that $\bar{\lambda} \in \rho(T^*) \cup \sigma_{P_1}(T^*) \cup \sigma_{P_4}(T^*)$, and therefore $\bar{\lambda} \in \rho(T^*) \cup \sigma_{P_4}(T^*)$. That is, $\lambda \in \rho(T) \cup \sigma_{P_4}(T)$ whenever $\lambda \neq 0$. $\square$

**Example 6I.** If $T \in \mathcal{B}_0[\mathcal{H}]$ (i.e., $T$ is a finite-rank operator on $\mathcal{H}$), then

$$\sigma(T) = \sigma_P(T) = \sigma_{P_4}(T) \text{ is finite.}$$

Indeed, if $\dim \mathcal{H} < \infty$, then $\sigma(T) = \sigma_P(T) = \sigma_{P_4}(T)$ (Example 6B). Suppose $\dim \mathcal{H} = \infty$. Since $\mathcal{B}_0[\mathcal{H}] \subseteq \mathcal{B}_\infty[\mathcal{H}]$, it follows by Corollary 6.31 that $0 \neq \lambda \in \rho(T) \cup \sigma_{P_4}(T)$. Moreover, since $\dim \mathcal{R}(T) < \infty$ and $\dim \mathcal{H} = \infty$, it also fol-lows that $\mathcal{R}(T)^- = \mathcal{R}(T) \neq \mathcal{H}$ and $\mathcal{N}(T) \neq \{0\}$ (recall from Problem 2.17 that $\dim \mathcal{N}(T) + \dim \mathcal{R}(T) = \dim \mathcal{H}$). Then $0 \in \sigma_{P_4}(T)$ (cf. diagram of Section 6.2). Hence $\sigma(T) = \sigma_P(T) = \sigma_{P_4}(T)$. If $\sigma_P(T)$ is infinite, then there exists an infinite set of linearly independent eigenvectors of $T$ (Proposition 6.14). Since every eigenvec-tor of $T$ lies in $\mathcal{R}(T)$ this implies that $\dim \mathcal{R}(T) = \infty$ (see Theorem 2.5), which is a contradiction. Conclusion: $\sigma_P(T)$ must be finite. In particular, this shows that the spectrum in Example 6B is, clearly, finite.

**Example 6J.** A glance at the spectra of some compact operators:

(a)  The operator $A = \begin{pmatrix} 0 & 0 \\ 0 & 1 \end{pmatrix}$ on $\mathbb{C}^2$ is obviously compact. Its spectrum is given by (cf. Example 6B and 6I)

$$\sigma(A) = \sigma_P(A) = \sigma_{P_4}(A) = \{0, 1\}.$$

(b)  The diagonal operator $D = \mathrm{diag}\{\lambda_k\}_{k=0}^{\infty} \in \mathcal{B}[\ell_+^2]$ with $\lambda_k \to 0$ is compact (Example 4N). By Example 6C, $\sigma_{P_4}(D) = \{\lambda_k\}_{k=0}^{\infty} \backslash \{0\}$ and

$$\sigma(D) = \sigma_{P_4}(D) \cup \begin{cases} \sigma_C(D), & \lambda_k \neq 0 \text{ for all } k \geq 0 \quad (\text{with } \sigma_C(D) = \{0\}), \\ \sigma_{P_3}(D), & \lambda_k = 0 \text{ for some } k \geq 0 \quad (\text{with } \sigma_{P_3}(D) = \{0\}). \end{cases}$$

(c)  The unilateral weighted shift $T_+ = \mathrm{shift}(\{\alpha_k\}_{k=0}^{\infty})$ acting on $\ell_+^2$, as introduced in Example 6F, is compact (reason: $T_+ = S_+ D_+$ and $D_+$ is compact). We saw there (Example 6F) that

$$\sigma(T_+) = \sigma_R(T_+) = \sigma_{R_2}(T_+) = \{0\},$$

Moreover, $T_+^*$ also is compact (Problem 5.42) and (Example 6F)

$$\sigma(T_+^*) = \sigma_P(T_+^*) = \sigma_{P_2}(T_+^*) = \{0\}.$$

(d)  Finally, consider the bilateral weighted shift $T = \mathrm{shift}(\{\alpha_k\}_{k=-\infty}^{\infty})$ of Example 6G acting on $\ell^2$. The same argument as above shows that $T$ is compact and (cf. Example 6G)

$$\sigma(T) = \sigma_C(T) = \{0\}.$$

**Corollary 6.32.** *If an operator $T$ on $\mathcal{H}$ is compact and normaloid, then $\sigma_P(T) \neq \varnothing$ and there exists $\lambda \in \sigma_P(T)$ such that $|\lambda| = \|T\|$.*

*Proof.* Recall that $\mathcal{H} \neq \{0\}$. If $T$ is normaloid (i.e., $r(T) = \|T\|$), then $\sigma(T) = \{0\}$ only if $T = O$. If $T = O$ and $\mathcal{H} \neq \{0\}$, then $0 \in \sigma_P(T)$ and $\|T\| = 0$. If $T \neq O$, then $\sigma(T) \neq \{0\}$ and $\|T\| = r(T) = \max_{\lambda \in \sigma(T)}|\lambda|$ so that there exists $\lambda$ in $\sigma(T)$ such that $|\lambda| = \|T\|$. Moreover, if $T$ is compact and $\sigma(T) \neq \{0\}$, then $\varnothing \neq \sigma(T) \backslash \{0\} \subseteq \sigma_P(T)$ by Theorem 3.30, and hence $r(T) = \max_{\lambda \in \sigma(T)}|\lambda| = \max_{\lambda \in \sigma_P(T)}|\lambda| = \|T\|$. Thus there exists $\lambda \in \sigma_P(T)$ such that $|\lambda| = \|T\|$. $\square$

**Proposition 6.33.** *If $T \in \mathcal{B}_\infty[\mathcal{H}]$ and $\{\lambda_n\}$ is an infinite sequence of distinct elements in $\sigma(T)$, then $\lambda_n \to 0$.*

*Proof.* Take any $T \in \mathcal{B}[\mathcal{H}]$ and let $\{\lambda_n\}$ be an infinite sequence of distinct elements in $\sigma(T)$. If $\lambda_{n'} = 0$ for some $n'$, then the subsequence $\{\lambda_k\}$ of $\{\lambda_n\}$ consisting of all points of $\{\lambda_n\}$ except $\lambda_{n'}$ is a sequence of distinct nonzero elements in $\sigma(T)$. Since $\lambda_k \to 0$ implies $\lambda_n \to 0$, there is no loss of generality in assuming that $\{\lambda_n\}$ is a sequence of distinct *nonzero* elements in $\sigma(T)$ indexed by $\mathbb{N}$. Moreover, if

$T$ is compact and $0 \neq \lambda_n \in \sigma(T)$, then Corollary 6.31 says that $\lambda_n \in \sigma_P(T)$ for every $n \geq 1$. Let $\{x_n\}_{n=1}^{\infty}$ be a sequence of eigenvectors associated with $\{\lambda_n\}_{n=1}^{\infty}$ (i.e., $Tx_n = \lambda_n x_n$ with $x_n \neq 0$ for every $n \geq 1$), which is a sequence of linearly independent vectors by Proposition 6.14. Set

$$\mathcal{M}_n = \mathrm{span}\,\{x_i\}_{i=1}^{n} \quad \text{for each} \quad n \geq 1$$

so that each $\mathcal{M}_n$ is a subspace of $\mathcal{H}$ with $\dim \mathcal{M}_n = n$, and

$$\mathcal{M}_n \subset \mathcal{M}_{n+1} \quad \text{for every} \quad n \geq 1.$$

Actually, each $\mathcal{M}_n$ is properly included in $\mathcal{M}_{n+1}$ because $\{x_i\}_{i=1}^{n+1}$ is linearly independent and hence $x_{n+1} \in \mathcal{M}_{n+1} \backslash \mathcal{M}_n$. From now on the proof is similar to that of Proposition 6.30. Since each $\mathcal{M}_n$ is a proper subspace of $\mathcal{M}_{n+1}$, it follows by Lemma 4.33 that for every $n \geq 1$ there exists $y_{n+1} \in \mathcal{M}_{n+1}$ with $\|y_{n+1}\| = 1$ such that $\frac{1}{2} < d(y_{n+1}, \mathcal{M}_n)$. Write $y_{n+1} = \sum_{i=1}^{n+1} \alpha_i x_i$ in $\mathcal{M}_{n+1}$ so that

$$(\lambda_{n+1} I - T) y_{n+1} = \sum_{i=1}^{n+1} \alpha_i (\lambda_{n+1} - \lambda_i) x_i = \sum_{i=1}^{n} \alpha_i (\lambda_{n+1} - \lambda_i) x_i \in \mathcal{M}_n.$$

Recall that $\lambda_n \neq 0$ for all $n$, take any pair of integers $1 \leq m < n$, and set

$$y = y_m - \lambda_m^{-1}(\lambda_m I - T) y_m + \lambda_n^{-1}(\lambda_n I - T) y_n$$

so that $T(\lambda_m^{-1} y_m) - T(\lambda_n^{-1} y_n) = y - y_n$. Since $y$ lies in $\mathcal{M}_{n-1}$,

$$\|T(\lambda_m^{-1} y_m) - T(\lambda_n^{-1} y_n)\| = \|y - y_n\| > \tfrac{1}{2},$$

which implies that the sequence $\{T(\lambda_n^{-1} y_n)\}$ has no convergent subsequence. If $T$ is compact, then Theorem 4.52 ensures that $\{\lambda_n^{-1} y_n\}$ is an unbounded sequence. That is, $\sup_n |\lambda_n|^{-1} = \sup_n \|\lambda_n^{-1} y_n\| = \infty$, and hence $\inf_n |\lambda_n| = 0$. As $\lambda_n \neq 0$ for all $n$, this means that $\lambda_n \to 0$. $\qquad\qquad\qquad\square$

**Corollary 6.34.** *Take any compact operator* $T \in \mathcal{B}_{\infty}[\mathcal{H}]$.

(a)  $0$ *is the only possible accumulation point of* $\sigma(T)$.

(b)  *If* $\lambda \in \sigma(T) \backslash \{0\}$, *then* $\lambda$ *is an isolated point of* $\sigma(T)$.

(c)  $\sigma(T) \backslash \{0\}$ *is a discrete subset of* $\mathbb{C}$.

(d)  $\sigma(T)$ *is countable.*

*Proof.* If $\lambda \neq 0$, then the previous proposition says that there is no sequence of distinct points in $\sigma(T)$ that converges to $\lambda$. Thus $\lambda \neq 0$ is not an accumulation point of $\sigma(T)$ by Proposition 3.28. Therefore, if $\lambda \in \sigma(T) \backslash \{0\}$, then it is not an accumulation point of $\sigma(T)$, which means (by definition) that it is an isolated point of $\sigma(T)$. Hence $\sigma(T)\backslash\{0\}$ consists entirely of isolated points, which means (by definition again) that it is a discrete subset of $\mathbb{C}$. But $\mathbb{C}$ is separable, and every discrete subset of a separable metric space is countable (this is a consequence of Theorem 3.35 and Corollary 3.36 — see the observations that follow Proposition 3.37). Then $\sigma(T)\backslash\{0\}$ is countable and so is $\sigma(T)$. $\qquad\square$

The point $\lambda = 0$ may be anywhere. That is, if $T \in \mathcal{B}_\infty[\mathcal{H}]$, then $\lambda = 0$ may lie in $\sigma_P(T)$, $\sigma_R(T)$, $\sigma_C(T)$ or $\rho(T)$ (see Example 6J). However, if $0 \in \rho(T)$, then $\mathcal{H}$ must be finite-dimensional. Indeed, if $0 \in \rho(T)$, then $T^{-1} \in \mathcal{B}[\mathcal{H}]$ so that $I = T^{-1}T$ is compact by Proposition 4.54, which implies that $\mathcal{H}$ is finite-dimensional (see Corollary 4.34). Moreover, the eigenspaces associated with nonzero eigenvalues of a compact operator also are finite-dimensional, as in the next proposition.

**Proposition 6.35.** *If $T \in \mathcal{B}_\infty[\mathcal{H}]$ and $\lambda$ is a nonzero complex number, then* $\dim \mathcal{N}(\lambda I - T) = \dim \mathcal{N}(\bar\lambda I - T^*) < \infty$.

*Proof.* Take any $\lambda \neq 0$ in $\mathbb{C}$ and any $T \in \mathcal{B}_\infty[\mathcal{H}]$. If $\dim \mathcal{N}(\lambda I - T) = 0$, then $\mathcal{N}(\lambda I - T) = \{0\}$ so that $\lambda \in \rho(T)$ by Corollary 6.31 and hence $\bar\lambda \in \rho(T^*)$ by Proposition 6.17. Therefore $\mathcal{N}(\bar\lambda I - T^*) = \{0\}$, which means that $\dim \mathcal{N}(\bar\lambda I - T^*) = 0$. Dually, since $T \in \mathcal{B}_\infty[\mathcal{H}]$ if and only if $T^* \in \mathcal{B}_\infty[\mathcal{H}]$ (cf. Problem 5.42), $\dim \mathcal{N}(\bar\lambda I - T^*) = 0$ implies $\dim \mathcal{N}(\lambda I - T) = 0$. That is,

$$\dim \mathcal{N}(\lambda I - T) = 0 \quad \text{if and only if} \quad \dim \mathcal{N}(\bar\lambda I - T^*) = 0.$$

Suppose $\dim \mathcal{N}(\lambda I - T) \neq 0$, and so $\dim \mathcal{N}(\bar\lambda I - T^*) \neq 0$. Note that $\mathcal{N}(\lambda I - T) \neq \{0\}$ is an invariant subspace for $T$ (if $Tx = \lambda x$, then $T(Tx) = \lambda(Tx)$), and also that $T|_{\mathcal{N}(\lambda I - T)} = \lambda I$ of $\mathcal{N}(\lambda I - T)$ into itself. If $T$ is compact, then $T|_{\mathcal{N}(\lambda I - T)}$ is compact (Section 4.9) and so is $\lambda I \neq O$ on $\mathcal{N}(\lambda I - T) \neq \{0\}$. But $\lambda I \neq O$ is not compact in an infinite-dimensional normed space (by Corollary 4.34) so that $\dim \mathcal{N}(\lambda I - T) < \infty$. Dually, as $T^*$ is compact, $\dim \mathcal{N}(\bar\lambda I - T^*) < \infty$. Therefore, there exists positive integers $m$ and $n$ such that

$$\dim \mathcal{N}(\lambda I - T) = m \quad \text{and} \quad \dim \mathcal{N}(\bar\lambda I - T^*) = n.$$

Let $\{e_i\}_{i=1}^m$ and $\{f_i\}_{i=1}^n$ be orthonormal bases for the Hilbert spaces $\mathcal{N}(\lambda I - T)$ and $\mathcal{N}(\bar\lambda I - T^*)$, respectively. Set $k = \min\{m, n\} \geq 1$ and consider the mappings $S \colon \mathcal{H} \to \mathcal{H}$ and $S^* \colon \mathcal{H} \to \mathcal{H}$ defined by

$$Sx = \sum_{i=1}^{k} \langle x \, ; e_i \rangle f_i \quad \text{and} \quad S^*x = \sum_{i=1}^{k} \overline{\langle x \, ; f_i \rangle} e_i$$

for every $x \in \mathcal{H}$. It is clear that $S$ and $S^*$ lie in $\mathcal{B}[\mathcal{H}]$, and also that $S^*$ is the adjoint of $S$: $\langle Sx \, ; y \rangle = \langle x \, ; S^*y \rangle$ for every $x, y \in \mathcal{H}$. Actually,

$$\mathcal{R}(S) \subseteq \bigvee\{f_i\}_{i=1}^k \subseteq \mathcal{N}(\bar\lambda I - T^*) \quad \text{and} \quad \mathcal{R}(S^*) \subseteq \bigvee\{e_i\}_{i=1}^k \subseteq \mathcal{N}(\lambda I - T)$$

so that $S, S^* \in \mathcal{B}_0[\mathcal{H}]$, and hence $T + S$ and $T^* + S^*$ lie in $\mathcal{B}_\infty[\mathcal{H}]$ by Theorem 4.53 (for $\mathcal{B}_0[\mathcal{H}] \subseteq \mathcal{B}_\infty[\mathcal{H}]$). First suppose that $m \leq n$ (and so $k = m$). If $x$ is a vector in $\mathcal{N}(\lambda I - (T + S))$, then $(\lambda I - T)x = Sx$. But $\mathcal{R}(S) \subseteq \mathcal{N}(\bar{\lambda} I - T^*) = \mathcal{R}(\lambda I - T)^\perp$ (Proposition 5.76), and hence $(\lambda I - T)x = Sx = 0$. Then $x \in \mathcal{N}(\lambda I - T) = \mathrm{span}\, \{e_i\}_{i=1}^m$ so that $x = \sum_{i=1}^m \alpha_i e_i$ (for some family of scalars $\{\alpha_i\}_{i=1}^m$), and therefore $0 = Sx = \sum_{j=1}^m \alpha_j S e_j = \sum_{j=1}^m \alpha_j \sum_{i=1}^m \langle e_j \, ; e_i \rangle f_i = \sum_{i=1}^m \alpha_i f_i$, which implies that $\alpha_i = 0$ for every $i = 1, \dots, m$ (reason: $\{f_i\}_{i=1}^m$ is an orthonormal set, thus linearly independent — Proposition 5.34.) That is, $x = 0$. Outcome: $\mathcal{N}(\lambda I - (T + S)) = \{0\}$. Hence $\lambda \in \rho(T + S)$ according to Corollary 6.31 (once $T + S \in \mathcal{B}_\infty[\mathcal{H}]$ and $\lambda \neq 0$). Conclusion:

$$m \leq n \quad \text{implies} \quad \mathcal{R}(\lambda I - (T + S)) = \mathcal{H}.$$

Dually, using exactly the same argument,

$$n \leq m \quad \text{implies} \quad \mathcal{R}(\bar{\lambda} I - (T^* + S^*)) = \mathcal{H}.$$

If $m < n$, then $k = m < m + 1 \leq n$ and $f_{m+1} \in \mathcal{R}(\lambda I - (T + S)) = \mathcal{H}$ so that there exists $v \in \mathcal{H}$ for which $(\lambda I - (T + S))v = f_{m+1}$. Hence

$$\begin{aligned}
1 = \langle f_{m+1} \, ; f_{m+1} \rangle &= \langle (\lambda I - (T + S))v \, ; f_{m+1} \rangle \\
&= \langle (\lambda I - T)v \, ; f_{m+1} \rangle - \langle Sv \, ; f_{m+1} \rangle = 0,
\end{aligned}$$

which is a contradiction. Indeed, $\langle (\lambda I - T)v \, ; f_{m+1} \rangle = \langle Sv \, ; f_{m+1} \rangle = 0$ for $f_{m+1} \in \mathcal{N}(\bar{\lambda} I - T^*) = \mathcal{R}(\lambda I - T)^\perp$ and $Sv \in \mathcal{R}(S) \subseteq \mathrm{span}\{f_i\}_{i=1}^m$. If $n < m$, then $k = n < n + 1 \leq m$, and $e_{n+1} \in \mathcal{R}(\bar{\lambda} I - (T^* + S^*)) = \mathcal{H}$ so that there exists $u \in \mathcal{H}$ for which $(\bar{\lambda} I - (T^* + S^*))u = e_{n+1}$. Hence

$$\begin{aligned}
1 = \langle e_{m+1} \, ; e_{m+1} \rangle &= \langle (\bar{\lambda} I - (T^* + S^*))u \, ; e_{n+1} \rangle \\
&= \langle (\bar{\lambda} I - T^*)u \, ; e_{n+1} \rangle - \langle S^* u \, ; e_{n+1} \rangle = 0,
\end{aligned}$$

which is a contradiction too (for $e_{n+1} \in \mathcal{N}(\lambda I - T) = \mathcal{R}(\bar{\lambda} I - T^*)^\perp$ and $S^* u \in \mathcal{R}(S^*) \subseteq \mathrm{span}\{e_i\}_{i=1}^n$). Therefore, $m = n$. $\qquad\square$

Together, the statements of Propositions 6.29 and 6.35 are sometimes referred to as the *Fredholm Alternative*.

## 6.7    The Spectral Theorem for Compact Normal Operators

Throughout this section $\mathcal{H} \neq \{0\}$ is a complex Hilbert space. Let $\{\lambda_\gamma\}_{\gamma \in \Gamma}$ be a bounded family of complex numbers, let $\{P_\gamma\}_{\gamma \in \Gamma}$ be a resolution of the identity on

$\mathcal{H}$, and let $T \in \mathcal{B}[\mathcal{H}]$ be a (bounded) weighted sum of projections (cf. Definition 5.60 and Proposition 5.61):

$$Tx = \sum_{\gamma \in \Gamma} \lambda_\gamma P_\gamma x \quad \text{for every} \quad x \in \mathcal{H}.$$

**Proposition 6.36.** *Every weighted sum of projections is normal.*

*Proof.* Note that $\{\overline{\lambda}_\gamma\}_{\gamma \in \Gamma}$ is a bounded family of complex numbers, and consider the weighted sum of projections $T^* \in \mathcal{B}[\mathcal{H}]$ given by

$$T^*x = \sum_{\gamma \in \Gamma} \overline{\lambda}_\gamma P_\gamma x \quad \text{for every} \quad x \in \mathcal{H}.$$

Since each $P_\gamma$ is self-adjoint (Proposition 5.81), it is readily verified that $T^*$ is, in fact, the adjoint of $T \in \mathcal{B}[\mathcal{H}]$. Indeed, take $x = \sum_{\gamma \in \Gamma} P_\gamma x$ and $y = \sum_{\gamma \in \Gamma} P_\gamma y$ arbitrary in $\mathcal{H}$ (recall: $\{P_\gamma\}_{\gamma \in \Gamma}$ is a resolution of the identity on $\mathcal{H}$). Therefore, as $\mathcal{R}(P_\alpha) \perp \mathcal{R}(P_\beta)$ whenever $\alpha \neq \beta$,

$$\begin{aligned}
\langle Tx \,;\, y \rangle &= \Big\langle \textstyle\sum_{\alpha \in \Gamma} \lambda_\alpha P_\alpha x \,;\, \sum_{\beta \in \Gamma} P_\beta y \Big\rangle = \sum_{\alpha \in \Gamma} \sum_{\beta \in \Gamma} \lambda_\alpha \langle P_\alpha x \,;\, P_\beta y \rangle \\
&= \sum_{\gamma \in \Gamma} \lambda_\gamma \langle P_\gamma x \,;\, P_\gamma x \rangle = \sum_{\beta \in \Gamma} \sum_{\alpha \in \Gamma} \lambda_\alpha \langle P_\beta x \,;\, P_\alpha y \rangle \\
&= \Big\langle \textstyle\sum_{\beta \in \Gamma} P_\beta x \,;\, \sum_{\alpha \in \Gamma} \overline{\lambda}_\alpha P_\alpha y \Big\rangle = \langle x \,;\, T^*y \rangle.
\end{aligned}$$

Moreover, since $P_\gamma^2 = P_\gamma$ for all $\gamma$ and $P_\alpha P_\beta = P_\beta P_\alpha = O$ if $\alpha \neq \beta$,

$$\begin{aligned}
T^*Tx &= \textstyle\sum_{\alpha \in \Gamma} \overline{\lambda}_\alpha P_\alpha \sum_{\beta \in \Gamma} \lambda_\beta P_\beta x = \sum_{\alpha \in \Gamma} \sum_{\beta \in \Gamma} \overline{\lambda}_\alpha \lambda_\beta P_\alpha P_\beta x \\
&= \textstyle\sum_{\gamma \in \Gamma} |\lambda_\gamma|^2 P_\gamma x = \sum_{\alpha \in \Gamma} \sum_{\beta \in \Gamma} \lambda_\alpha \overline{\lambda}_\beta P_\alpha P_\beta x \\
&= \textstyle\sum_{\alpha \in \Gamma} \lambda_\alpha P_\alpha \sum_{\beta \in \Gamma} \overline{\lambda}_\beta P_\beta x = T T^*x
\end{aligned}$$

for every $x \in \mathcal{H}$. That is, $T$ is normal. $\qquad\square$

Particular case: Diagonal operators and, more generally, diagonalizable operators on a separable Hilbert space (defined in Problem 5.17), are normal operators. In fact, the concept of weighted sum of projections on any Hilbert space can be thought of as a generalization of the concept of diagonalizable operators on a separable Hilbert space. The next proposition shows that such a generalization preserves the spectral properties (compare with Example 6C).

**Proposition 6.37.** *If* $T \in \mathcal{B}[\mathcal{H}]$ *is a weighted sum of projections, then*

$$\sigma_P(T) = \big\{\lambda \in \mathbb{C}\colon\ \lambda = \lambda_\gamma \text{ for some } \lambda \in \Gamma\big\}, \quad \sigma_R(T) \neq \varnothing, \quad \text{and}$$

$$\sigma_C(T) = \big\{\lambda \in \mathbb{C}\colon\ \lambda \neq \lambda_\gamma \text{ for all } \lambda \in \Gamma \text{ and } \inf_{\gamma \in \Gamma} |\lambda - \lambda_\gamma| = 0\big\}.$$

*Proof.* Take any $x = \sum_{\gamma\Gamma} P_\gamma x$ in $\mathcal{H}$ (recall that $\{P_\gamma\}_{\gamma\in\Gamma}$ is a resolution of the identity on $\mathcal{H}$) so that $\|x\|^2 = \sum_{\gamma\in\Gamma}\|P_\gamma x\|^2$ by Theorem 5.32. Thus, for any $\lambda \in \mathbb{C}$, $(\lambda I - T)x = \sum_{\gamma\in\Gamma}(\lambda - \lambda_\gamma)P_\gamma x$ so that $\|(\lambda I - T)x\|^2 = \sum_{\gamma\Gamma}|\lambda - \lambda_\gamma|^2\|P_\gamma x\|^2$ (cf. Theorem 5.32 again). If $\mathcal{N}(\lambda I - T) \neq \{0\}$, then there exists $x \neq 0$ in $\mathcal{H}$ such that $(\lambda I - T)x = 0$. Hence $\sum_{\gamma\in\Gamma}\|P_\gamma x\|^2 \neq 0$ and $\sum_{\gamma\in\Gamma}|\lambda - \lambda_\gamma|^2\|P_\gamma x\|^2 = 0$, which implies that $\|P_\alpha x\| \neq 0$ for some $\alpha \in \Gamma$ and $|\lambda - \lambda_\alpha|\|P_\alpha x\| = 0$. Therefore, $\lambda = \lambda_\alpha$. Conversely, take any $\alpha \in \Gamma$ and an arbitrary nonzero vector $x$ in $\mathcal{R}(P_\alpha)$ (recall: $P_\gamma \neq O$ and so $\mathcal{R}(P_\gamma) \neq \{0\}$ for every $\gamma \in \Gamma$). But $\mathcal{R}(P_\alpha) \perp \mathcal{R}(P_\gamma)$ for $\alpha \neq \gamma$ so that $\mathcal{R}(P_\alpha) \perp \bigcup_{\alpha\neq\gamma\in\Gamma}\mathcal{R}(P_\gamma)$. Hence $\mathcal{R}(P_\alpha) \subseteq (\bigcup_{\alpha\neq\gamma\in\Gamma}\mathcal{R}(P_\gamma))^\perp = \bigcap_{\alpha\neq\gamma\in\Gamma}\mathcal{R}(P_\gamma)^\perp = \bigcap_{\alpha\neq\gamma\in\Gamma}\mathcal{N}(P_\gamma)$ (cf. Problem 5.8(a) and Propositions 5.76(a) and 5.81(b)). Therefore $x \in \mathcal{N}(P_\gamma)$ for every $\alpha \neq \gamma \in \Gamma$, which implies that $\|(\lambda_\alpha I - T)x\|^2 = \sum_{\gamma\in\Gamma}|\lambda_\alpha - \lambda_\gamma|^2\|P_\gamma x\|^2 = 0$, and so $\mathcal{N}(\lambda_\alpha I - T) \neq \{0\}$. Conclusion: $\mathcal{N}(\lambda I - T) \neq \{0\}$ if and only if $\lambda = \lambda_\alpha$ for some $\alpha \in \Gamma$. In other words,

$$\sigma_P(T) = \big\{\lambda \in \mathbb{C}: \ \lambda = \lambda_\gamma \text{ for some } \gamma \in \Gamma\big\}.$$

We have just seen that $\mathcal{N}(\lambda I - T) = \{0\}$ if and only if $\lambda \neq \lambda_\gamma$ for all $\gamma \in \Gamma$. In this case (i.e., if $(\lambda I - T)$ is injective), there exists $(\lambda I - T)^{-1}$ in $\mathcal{L}[\mathcal{R}(\lambda I - T), \mathcal{H}]$, a weighted sum of projections on $\mathcal{R}(\lambda I - T)$:

$$(\lambda I - T)^{-1}x = \sum_{\gamma\in\Gamma}(\lambda - \lambda_\gamma)^{-1}P_\gamma x \quad \text{for every} \quad x \in \mathcal{R}(\lambda I - T).$$

Indeed, if $\lambda \neq \lambda_\gamma$ for all $\gamma \in \Gamma$, $\sum_{\alpha\in\Gamma}(\lambda - \lambda_\alpha)^{-1}P_\alpha\sum_{\beta\in\Gamma}(\lambda - \lambda_\beta)P_\beta x = \sum_{\alpha\in\Gamma}\sum_{\beta\in\Gamma}(\lambda - \lambda_\alpha)^{-1}(\lambda - \lambda_\beta)P_\alpha P_\beta = \sum_{\gamma\in\Gamma}P_\gamma = x$ for every $x$ in $\mathcal{H}$. According to Proposition 5.61 $(\lambda I - T)^{-1} \in \mathcal{B}[\mathcal{H}]$ if and only if $\lambda \neq \lambda_\gamma$ for all $\gamma \in \Gamma$ and $\sup_{\gamma\in\Gamma}|\lambda - \lambda_\gamma|^{-1} < \infty$. Equivalently, if and only if $\inf_{\gamma\in\Gamma}|\lambda - \lambda_\gamma| > 0$. In other words,

$$\rho(T) = \big\{\lambda \in \mathbb{C}: \ \inf_{\gamma\in\Gamma}|\lambda - \lambda_\gamma| > 0\big\}.$$

But $T$ is normal by Proposition 6.36 so that $\sigma_R(T) = \varnothing$ (Corollary 6.18), and hence $\sigma_C(T) = \rho(T)\backslash\sigma_P(T)$. $\qquad\square$

**Proposition 6.38.** *A weighted sum of projections $T \in \mathcal{B}[\mathcal{H}]$ is compact if and only if the following triple condition holds. $\sigma(T)$ is countable, $0$ is the only possible accumulation point of $\sigma(T)$, and $\dim \mathcal{R}(P_\gamma) < \infty$ for every $\gamma$ such that $\lambda_\gamma \neq 0$.*

*Proof.* Let $T \in \mathcal{B}[\mathcal{H}]$ be a weighted sum of projections.

*Claim.* $\mathcal{R}(P_\gamma) \subseteq \mathcal{N}(\lambda_\gamma I - T)$ for every $\gamma$.

*Proof.* Take an arbitrary $\gamma$. If $x \in \mathcal{R}(P_\gamma)$, then $x = P_\gamma x$ (Problem 1.4) so that $Tx = TP_\gamma x = \sum_\alpha \lambda_\alpha P_\alpha P_\gamma x = \lambda_\gamma P_\gamma x = \lambda_\gamma x$ (for $P_\alpha \perp P_\gamma$ whenever $\gamma \neq \alpha$), and hence $x \in \mathcal{N}(\lambda_\gamma I - T)$. $\square$

If $T$ is compact, then $\sigma(T)$ is countable and 0 is the only possible accumulation point of $\sigma(T)$ (Corollary 6.34), and $\dim \mathcal{N}(\lambda I - T) < \infty$ whenever $\lambda \neq 0$ (Proposition 6.35) so that $\dim \mathcal{R}(P_\gamma) < \infty$ for every $\gamma$ such that $\lambda_\gamma \neq 0$ by the above claim. Conversely, if $T = O$, then $T$ is trivially compact. Thus suppose $T \neq O$. Since $T$ is normal (Proposition 6.36), $r(T) > 0$ (reason: the unique normal operator with a null spectral radius is the null operator — see the remark that precedes Corollary 6.28) so that there exists $\lambda \neq 0$ in $\sigma_P(T)$ by Corollary 6.31. If $\sigma(T)$ is countable, then let $\{\lambda_k\}$ be any enumeration of the countable set $\sigma_P(T)\backslash\{0\} = \sigma(T)\backslash\{0\}$. Hence

$$Tx = \sum_k \lambda_k P_k x \quad \text{for every} \quad x \in \mathcal{H}$$

(Proposition 6.37), where $\{P_k\}$ is included in a resolution of the identity on $\mathcal{H}$ (which is itself a resolution of the identity on $\mathcal{H}$ if $0 \notin \sigma_P(T)$). If $\{\lambda_k\}$ is finite, say $\{\lambda_k\} = \{\lambda_k\}_{k=1}^n$, then $\mathcal{R}(T) = \sum_{k=1}^n \mathcal{R}(P_k)$. If $\dim \mathcal{R}(P_k) < \infty$ for every $k$, then $\dim\left(\sum_{k=1}^n \mathcal{R}(P_k)\right)^- < \infty$ (according to Problem 5.11), and so $T \in \mathcal{B}_0[\mathcal{H}] \subseteq \mathcal{B}_\infty[\mathcal{H}]$. Now suppose $\{\lambda_k\}$ is countably infinite. Since $\sigma(T)$ is compact (Corollary 6.12), it follows by Theorem 3.80 and Proposition 3.77 that $\{\lambda_k\}$ has an accumulation point in $\sigma(T)$. If 0 is the only possible accumulation point of $\sigma(T)$, then 0 is the unique accumulation point of $\{\lambda_k\}$. Therefore, for each integer $n \geq 1$ consider the partition $\{\lambda_k\} = \{\lambda_{k'}\} \cup \{\lambda_{k''}\}$, where $|\lambda_{k'}| \geq \frac{1}{n}$ and $|\lambda_{k''}| < \frac{1}{n}$. Note that $\{\lambda_{k'}\}$ is a finite subset of $\sigma(T)$ (it has no accumulation point), and hence $\{\lambda_{k''}\}$ is an infinite subset of $\sigma(T)$. Set

$$T_n = \sum_{k'} \lambda_{k'} P_{k'} \in \mathcal{B}[\mathcal{H}] \quad \text{for each} \quad n \geq 1.$$

We have just seen that $\dim \mathcal{R}(T_n) < \infty$. That is, $T_n \in \mathcal{B}_0[\mathcal{H}]$ for every $n \geq 1$. However, as $P_j \perp P_k$ whenever $j \neq k$, we get (cf. Corollary 5.9)

$$\|(T - T_n)x\|^2 = \left\|\sum_{k''} \lambda_{k''} P_{k''} x\right\|^2 \leq \sup_{k''} |\lambda_{k''}|^2 \sum_{k''} \|P_{k''}x\|^2 \leq \tfrac{1}{n^2}\|x\|^2$$

for all $x \in \mathcal{H}$, so that $T_n \xrightarrow{u} T$. Hence $T \in \mathcal{B}_\infty[\mathcal{H}]$ by Definition 4.45.     $\square$

Before considering the Spectral Theorem for compact normal operators we need a few spectral properties of normal operators.

**Proposition 6.39.** *If* $T \in \mathcal{B}[\mathcal{H}]$ *is normal, then*

$$\mathcal{N}(\lambda I - T) = \mathcal{N}(\bar{\lambda} I - T^*) \quad \text{for every} \quad \lambda \in \mathbb{C}.$$

*Proof.* Take an arbitrary $\lambda \in \mathbb{C}$. If $T$ is normal, then $(\lambda I - T)$ is normal (cf. proof of Corollary 6.18) so that $\|(\bar{\lambda} I - T^*)x\| = \|(\lambda I - T)x\|$ for every $x \in \mathcal{H}$ by Proposition 6.1(b).     $\square$

**Proposition 6.40.** *Take* $\lambda, \mu \in \mathbb{C}$. *If* $T \in \mathcal{B}[\mathcal{H}]$ *is normal, then*

$$\mathcal{N}(\lambda I - T) \perp \mathcal{N}(\mu I - T) \quad \text{whenever} \quad \lambda \neq \mu.$$

*Proof.* Suppose $x \in \mathcal{N}(\lambda I - T)$ and $y \in \mathcal{N}(\mu I - T)$ so that $\lambda x = Tx$ and $\mu y = Ty$. Since $\mathcal{N}(\lambda I - T) = \mathcal{N}(\bar{\lambda} I - T^*)$ by the previous proposition, it follows that $\bar{\lambda} x = T^* x$. Thus $\langle \mu y ; x \rangle = \langle Ty ; x \rangle = \langle y ; T^* x \rangle = \langle y ; \bar{\lambda} x \rangle = \langle \lambda y ; x \rangle$, and hence $(\mu - \lambda) \langle y ; x \rangle = 0$, which implies that $\langle y ; x \rangle = 0$ whenever $\mu \neq \lambda$. $\qquad\square$

**Proposition 6.41.** *If* $T \in \mathcal{B}[\mathcal{H}]$ *is normal, then* $\mathcal{N}(\lambda I - T)$ *reduces* $T$ *for every* $\lambda \in \mathbb{C}$.

*Proof.* Take an arbitrary $\lambda \in \mathbb{C}$ and an arbitrary $T \in \mathcal{B}[\mathcal{H}]$. Recall that $\mathcal{N}(\lambda I - T)$ is a subspace of $\mathcal{H}$ (Proposition 4.13). Moreover, it is clear that $\mathcal{N}(\lambda I - T)$ is $T$-invariant (if $Tx = \lambda x$, then $T(Tx) = \lambda(Tx)$). Similarly, $\mathcal{N}(\bar{\lambda} I - T^*)$ is $T^*$-invariant. Now suppose $T \in \mathcal{B}[\mathcal{H}]$ is a normal operator. Proposition 6.39 says that $\mathcal{N}(\lambda I - T) = \mathcal{N}(\bar{\lambda} I - T^*)$, and hence $\mathcal{N}(\lambda I - T)$ also is $T^*$-invariant. Then $\mathcal{N}(\lambda I - T)$ reduces $T$ according to Corollary 5.75. $\qquad\square$

**Corollary 6.42.** *If* $\{\lambda_\gamma\}_{\gamma \in \Gamma}$ *is any (nonempty) family of distinct scalars, and if* $T \in \mathcal{B}[\mathcal{H}]$ *is a normal operator, then the (topological) sum* $\left(\sum_{\gamma \in \Gamma} \mathcal{N}(\lambda_\gamma I - T)\right)^-$ *reduces* $T$.

*Proof.* For each $\gamma \in \Gamma \neq \varnothing$ write $\mathcal{N}_\gamma = \mathcal{N}(\lambda_\gamma I - T)$, which is a subspace of $\mathcal{H}$ (Proposition 4.13). According to Proposition 6.40 $\{\mathcal{N}_\gamma\}_{\gamma \in \Gamma}$ is a family of pairwise orthogonal subspaces of $\mathcal{H}$. Take an arbitrary $x \in \left(\sum_{\gamma \in \Gamma} \mathcal{N}_\gamma\right)^-$. If $\Gamma$ is finite, then $\left(\sum_{\gamma \in \Gamma} \mathcal{N}_\gamma\right)^- = \sum_{\gamma \in \Gamma} \mathcal{N}_\gamma$ (Corollary 5.11); otherwise apply the Orthogonal Structure Theorem (i.e., Theorem 5.16 if $\Gamma$ is countably infinite or Problem 5.10 if $\Gamma$ is uncountable). In any case (finite, countably infinite or uncountable $\Gamma$), $x = \sum_{\gamma \in \Gamma} u_\gamma$ with each $u_\gamma$ in $\mathcal{N}_\gamma$. Moreover, for each $\gamma \in \Gamma$, $Tu_\gamma$ and $T^* u_\gamma$ lie in $\mathcal{N}_\gamma$ because $\mathcal{N}_\gamma$ reduces $T$ by Proposition 6.41 (cf. Corollary 5.75). Thus, since $T$ and $T^*$ are linear and continuous, it follows that $Tx = \sum_{\gamma \in \Gamma} Tu_\gamma \in \left(\sum_{\gamma \in \Gamma} \mathcal{N}_\gamma\right)^-$ and $T^* x = \sum_{\gamma \in \Gamma} T^* u_\gamma \in \left(\sum_{\gamma \in \Gamma} \mathcal{N}_\gamma\right)^-$. Therefore, $\left(\sum_{\gamma \in \Gamma} \mathcal{N}_\gamma\right)^-$ reduces $T$ (cf. Corollary 5.75 again). $\qquad\square$

Every (bounded) weighted sum of projections is normal (Proposition 6.36), and every compact weighted sum of projections has a countable set of distinct eigenvalues (Propositions 6.37 and 6.38). The Spectral Theorem for compact normal operators ensures the converse.

**Theorem 6.43.** (The Spectral Theorem). *If* $T \in \mathcal{B}[\mathcal{H}]$ *is compact and normal, then there exists a countable resolution of the identity* $\{P_k\}$ *on* $\mathcal{H}$ *and a (similarly indexed) bounded set of scalars* $\{\lambda_k\}$ *such that*

$$T = \sum_k \lambda_k P_k,$$

*where $\{\lambda_k\} = \sigma_P(T)$, the set of all (distinct) eigenvalues of $T$, and each $P_k$ is the orthogonal projection onto the eigenspace $\mathcal{N}(\lambda_k I - T)$. Moreover, if the above countable weighted sum of projections is infinite, then it converges in the (uniform) topology of $\mathcal{B}[\mathcal{H}]$.*

*Proof.* If $T$ is compact and normal, then it has a nonempty point spectrum (Corollary 6.32) and its eigenspaces span $\mathcal{H}$. In other words,

*Claim.* $\left(\sum_{\lambda \in \sigma_P(T)} \mathcal{N}(\lambda I - T)\right)^- = \mathcal{H}$.

*Proof.* Set $\mathcal{M} = \left(\sum_{\lambda \in \sigma_P(T)} \mathcal{N}(\lambda I - T)\right)^-$, which is a subspace of $\mathcal{H}$. Suppose $\mathcal{M} \neq \mathcal{H}$ so that $\mathcal{M}^\perp \neq \{0\}$ (Proposition 5.15). Consider the restriction $T|_{\mathcal{M}^\perp}$ of $T$ to $\mathcal{M}^\perp$. If $T$ is normal, then $\mathcal{M}$ reduces $T$ (Corollary 6.42) so that $\mathcal{M}^\perp$ is $T$-invariant, and hence $T|_{\mathcal{M}^\perp} \in \mathcal{B}[\mathcal{M}^\perp]$ is normal (cf. Problem 6.17). If $T$ is compact, then $T|_{\mathcal{M}^\perp}$ is compact (see Section 4.9). Thus $T|_{\mathcal{M}^\perp}$ is a compact normal operator on the Hilbert space $\mathcal{M}^\perp \neq \{0\}$, and therefore $\sigma_P(T|_{\mathcal{M}^\perp}) \neq \varnothing$ by Corollary 6.32. That is, there exist $\lambda \in \mathbb{C}$ and $0 \neq x \in \mathcal{M}^\perp$ such that $T|_{\mathcal{M}^\perp} x = \lambda x$ and so $Tx = \lambda x$. Hence $\lambda \in \sigma_P(T)$ and $x \in \mathcal{N}(\lambda I - T) \subseteq \mathcal{M}$. But this leads to a contradiction, viz., $0 \neq x \in \mathcal{M} \cap \mathcal{M}^\perp = \{0\}$. Outcome: $\mathcal{M} = \mathcal{H}$. $\square$

Since $T$ is compact, the nonempty set $\sigma_P(T)$ is countable (Corollary 6.34) and bounded (for $T \in \mathcal{B}[\mathcal{H}]$). Then write

$$\sigma_P(T) = \{\lambda_k\}_{k \in N},$$

where $\{\lambda_k\}_{k \in N}$ is a finite or infinite sequence of distinct elements in $\mathbb{C}$ consisting of all eigenvalues of $T$. Here, either $N = \{1, \dots, m\}$ for some $m \in \mathbb{N}$ if $\sigma_P(T)$ is finite, or $N = \mathbb{N}$ if $\sigma_P(T)$ is (countably) infinite. Recall that each $\mathcal{N}(\lambda_k I - T)$ is a subspace of $\mathcal{H}$ (Proposition 4.13). Moreover, since $T$ is normal, Proposition 6.40 says that $\mathcal{N}(\lambda_k I - T) \perp \mathcal{N}(\lambda_j I - T)$ whenever $k \neq j$. Therefore, $\{\mathcal{N}(\lambda_k I - T)\}_{k \in N}$ is a sequence of orthogonal subspaces of $\mathcal{H}$ such that $\mathcal{H} = \left(\sum_{k \in N} \mathcal{N}(\lambda_k I - T)\right)^-$ by the above claim. Then the sequence $\{P_k\}_{k \in N}$ consisting of the orthogonal projections onto each $\mathcal{N}(\lambda_k I - T)$ is a resolution of the identity on $\mathcal{H}$ (see Theorem 5.59). This implies that $x = \sum_{k \in N} P_k x$ and, since $T$ is linear and continuous, $Tx = \sum_{k \in N} T P_k x$ for every $x \in \mathcal{H}$. But $P_k x \in \mathcal{R}(P_k) = \mathcal{N}(\lambda_k I - T)$, and so $T P_k x = \lambda_k P_k x$, for each $k \in N$ and every $x \in \mathcal{H}$. Hence

$$Tx = \sum_{k \in N} \lambda_k P_k x \quad \text{for every} \quad x \in \mathcal{H}.$$

Conclusion: $T$ is a countable weighted sum of projections. If $N$ is finite, then the theorem is proved. Now suppose $N$ is infinite (i.e., $N = \mathbb{N}$). In this case, the above identity says that $\sum_{k=1}^{n} \lambda_k P_k \overset{s}{\longrightarrow} T$ (see the observation that follows the proof of Proposition 5.61). We show next that the above convergence actually is uniform.

Indeed, for any $n \in \mathbb{N}$,

$$\left\| \left(T - \sum_{k=1}^{n} \lambda_k P_k\right) x \right\|^2 = \left\| \sum_{k=n+1}^{\infty} \lambda_k P_k x \right\|^2 = \sum_{k=n+1}^{\infty} |\lambda_k|^2 \| P_k x \|^2$$

$$\leq \sup_{k \geq n+1} |\lambda_k|^2 \sum_{k=n+1}^{\infty} \| P_k x \|^2 \leq \sup_{k \geq n} |\lambda_k|^2 \|x\|^2.$$

(Reason: $\mathcal{R}(P_j) \perp \mathcal{R}(P_k)$ whenever $j \neq k$ and $x = \sum_{k=1}^{\infty} P_k x$ so that $\|x\|^2 = \sum_{k=1}^{\infty} \| P_k x \|^2$ — see Corollary 5.9.) Hence

$$0 \leq \left\| T - \sum_{k=1}^{n} \lambda_k P_k \right\| = \sup_{\|x\|=1} \left\| \left(T - \sum_{k=1}^{n} \lambda_k P_k\right) x \right\| \leq \sup_{k \geq n} |\lambda_k|$$

for all $n \in \mathbb{N}$. Since $T$ is compact and $\{\lambda_n\}_{n=1}^{\infty}$ is a sequence of distinct elements in $\sigma(T)$, it follows by Proposition 6.33 that $\lambda_n \to 0$. Therefore $\lim_n \sup_{k \geq n} |\lambda_k| = \lim \sup_n |\lambda_n| = 0$, and so $\sum_{k=1}^{n} \lambda_k P_k \xrightarrow{u} T$.    $\square$

In other words, if $T \in \mathcal{B}[\mathcal{H}]$ is compact and normal, then the family of orthogonal projections $\{P_\lambda\}_{\lambda \in \sigma_P(T)}$ onto each eigenspace $\mathcal{N}(\lambda I - T)$ is a resolution of the identity on $\mathcal{H}$, and $T$ is a weighted sum of projections:

$$T = \sum_{\lambda \in \sigma_P(T)} \lambda P_\lambda.$$

This was naturally identified in Problem 5.16 with an orthogonal sum of scalar operators $\bigoplus_{\lambda \in \sigma_P(T)} \lambda I_\lambda$, where $I_\lambda = P_\lambda|_{\mathcal{R}(P_\lambda)}$. Here $\mathcal{R}(P_\lambda) = \mathcal{N}(\lambda I - T)$. Under such a natural identification we also write

$$T = \bigoplus_{\lambda \in \sigma_P(T)} \lambda P_\lambda.$$

These representations are referred to as the *spectral decomposition* of a compact normal operator $T$. The next result states the Spectral Theorem for compact normal operators in terms of an orthonormal basis for $\mathcal{N}(T)^\perp$ consisting of eigenvectors of $T$.

**Corollary 6.44.** *Let $T \in \mathcal{B}[\mathcal{H}]$ be compact and normal.*

(a) *For each $\lambda \in \sigma_P(T) \backslash \{0\}$ there exists a finite orthonormal basis $\{e_k(\lambda)\}_{k=1}^{n_\lambda}$ for $\mathcal{N}(\lambda I - T)$ consisting entirely of eigenvectors of $T$,*

(b) *$\{e_k\} = \bigcup_{\lambda \in \sigma_P(T) \backslash \{0\}} \{e_k(\lambda)\}_{k=1}^{n_\lambda}$ is a countable orthonormal basis for $\mathcal{N}(T)^\perp$ made up of eigenvectors of $T$, and*

(c) *$T x = \sum_{\lambda \in \sigma_P(T) \backslash \{0\}} \lambda \sum_{k=1}^{n_\lambda} \langle x \, ; e_k(\lambda) \rangle e_k(\lambda)$ for every $x \in \mathcal{H}$, so that*

(d)  $Tx = \sum_k \mu_k \langle x ; e_k \rangle e_k$ *for every* $x \in \mathcal{H}$, *where* $\{\mu_k\}$ *is a sequence containing all nonzero eigenvalues of* $T$, *finitely repeated according to the multiplicity of the respective eigenspace.*

*Proof.* We have already seen that $\sigma_P(T)$ is nonempty and countable (cf. proof of the previous theorem). Recall that $\sigma_P(T) = \{0\}$ if and only if $T = O$ (Corollary 6.32) or, equivalently, if and only if $\mathcal{N}(T)^\perp = \{0\}$ (i.e., $\mathcal{N}(T) = \mathcal{H}$). If $T = O$ (i.e., $T = OI$), then the above assertions hold trivially ($(\sigma_P(T)\backslash\{0\} = \varnothing$, $\{e_k\} = \varnothing$, $\mathcal{N}(T)^\perp = \{0\}$ and $Tx = 0x = 0$ for every $x \in \mathcal{H}$, because the empty sum is null). Thus suppose $T \neq O$ (so that $\mathcal{N}(T)^\perp \neq \{0\}$), and take an arbitrary $\lambda \neq 0$ in $\sigma_P(T)$. According to Proposition 6.35, dim $\mathcal{N}(\lambda I - T)$ is finite, say, dim $\mathcal{N}(\lambda I - T) = n_\lambda$ for some *positive* integer $n_\lambda$. This implies the existence of a finite orthonormal basis $\{e_k(\lambda)\}_{k=1}^{n_\lambda}$ for the Hilbert space $\mathcal{N}(\lambda I - T) \neq \{0\}$ (cf. Proposition 5.39). Observe that $e_k(\lambda)$ is an eigenvector of $T$ for each $k = 1, \ldots , n_\lambda$ (because $0 \neq e_k(\lambda) \in \mathcal{N}(\lambda I - T)$).

*Claim.* $\bigcup_{\lambda \in \sigma_P(T)\backslash\{0\}} \{e_k(\lambda)\}_{k=1}^{n_\lambda}$ is an orthonormal basis for $\mathcal{N}(T)^\perp$.

*Proof.* We know (cf. Claim in the proof of Theorem 6.43) that

$$\mathcal{H} = \left( \sum_{\lambda \in \sigma_P(T)} \mathcal{N}(\lambda I - T) \right)^{-}.$$

Therefore, according to Problem 5.8(b,d,e),

$$\mathcal{N}(T) = \bigcap_{\lambda \in \sigma_P(T)\backslash\{0\}} \mathcal{N}(\lambda I - T)^\perp = \left( \sum_{\lambda \in \sigma_P(T)\backslash\{0\}} \mathcal{N}(\lambda I - T) \right)^\perp$$

(because $\{\mathcal{N}(\lambda I - T)\}_{\lambda \in \sigma_P(T)}$ is a nonempty family of orthogonal subspaces of $\mathcal{H}$ — Proposition 6.40). Hence

$$\mathcal{N}(T)^\perp = \left( \sum_{\lambda \in \sigma_P(T)\backslash\{0\}} \mathcal{N}(\lambda I - T) \right)^{-}$$

(Proposition 5.15), and the claimed result follows by part (a), Corollary 6.40, and Problem 5.11. $\square$

Note that $\{e_k\} = \bigcup_{\lambda \in \sigma_P(T)\backslash\{0\}} \{e_k(\lambda)\}_{k=1}^{n_\lambda}$ is countable by Corollary 1.11. Finally, consider the decomposition $\mathcal{H} = \mathcal{N}(T) + \mathcal{N}(T)^\perp$ of Theorem 5.20, and take an arbitrary $x \in \mathcal{H}$ so that $x = u + v$ with $u \in \mathcal{N}(T)$ and $v \in \mathcal{N}(T)^\perp$. Consider the Fourier series expansion

$$v = \sum_k \langle v ; e_k \rangle e_k = \sum_{\lambda \in \sigma_P(T)\backslash\{0\}} \sum_{k=1}^{n_\lambda} \langle v ; e_k(\lambda) \rangle e_k(\lambda)$$

(cf. Theorem 5.48) of $v$ in terms of the orthonormal basis

$$\{e_k\} = \bigcup_{\lambda \in \sigma_P(T) \setminus \{0\}} \{e_k(\lambda)\}_{k=1}^{n_\lambda}$$

for the Hilbert space $\mathcal{N}(T)^\perp \neq \{0\}$. Since $T$ is linear and continuous, and since $Te_k(\lambda) = \lambda e_k(\lambda)$ for each $k = 1, \cdots, n_\lambda$ and each $\lambda \in \sigma_P(T) \setminus \{0\}$, it follows that

$$Tx = Tu + Tv = Tv = \sum_{\lambda \in \sigma_P(T) \setminus \{0\}} \sum_{k=1}^{n_\lambda} \langle v \, ; \, e_k(\lambda) \rangle \, T e_k(\lambda)$$

$$= \sum_{\lambda \in \sigma_P(T) \setminus \{0\}} \lambda \sum_{k=1}^{n_\lambda} \langle v \, ; \, e_k(\lambda) \rangle \, e_k(\lambda).$$

However, $\langle x \, ; \, e_k(\lambda) \rangle = \langle u \, ; \, e_k(\lambda) \rangle + \langle v \, ; \, e_k(\lambda) \rangle = \langle v \, ; \, e_k(\lambda) \rangle$ because $u \in \mathcal{N}(T)$ and $e_k(\lambda) \in \mathcal{N}(T)^\perp$.    $\square$

Remark: *If $T \in \mathcal{B}[\mathcal{H}]$ is compact and normal, and if $\mathcal{H}$ is nonseparable, then $0 \in \sigma_P(T)$ and $\mathcal{N}(T)$ is nonseparable.* Indeed, for $T = O$ the italicized result is trivial ($T = O$ implies $0 \in \sigma_P(T)$ and $\mathcal{N}(T) = \mathcal{H}$). On the other hand, if $T \neq O$, then $\mathcal{N}(T)^\perp \neq \{0\}$ is separable because it has a countable orthonormal basis $\{e_k\}$ (Theorem 5.44 and Corollary 6.44). If $\mathcal{N}(T)$ is separable, then it also has a countable orthonormal basis, say $\{f_k\}$, and hence $\{e_k\} \cup \{f_k\}$ is a countable orthonormal basis for $\mathcal{H} = \mathcal{N}(T) + \mathcal{N}(T)^\perp$ (Problem 5.11) so that $\mathcal{H}$ is separable. Moreover, if $0 \notin \sigma_P(T)$, then $\mathcal{N}(T) = \{0\}$, and therefore $\mathcal{H} = \mathcal{N}(T)^\perp$ is separable.

$\mathcal{N}(T)$ reduces $T$ (Proposition 6.41), and hence $T = T|_{\mathcal{N}(T)^\perp} \oplus O$. By Problem 5.17 and Corollary 6.44(d), *if $T \in \mathcal{B}[\mathcal{H}]$ is compact and normal, then $T|_{\mathcal{N}(T)^\perp} \in \mathcal{B}[\mathcal{N}(T)^\perp]$ is diagonalizable.* Precisely, $T|_{\mathcal{N}(T)^\perp}$ is a diagonal operator with respect to the orthonormal basis $\{e_k\}$ for the separable Hilbert space $\mathcal{N}(T)^\perp$. Generalizing: An operator $T \in \mathcal{B}[\mathcal{H}]$ (not necessarily compact) acting on any Hilbert space $\mathcal{H}$ (not necessarily separable) is *diagonalizable* if there exist a resolution of the identity $\{P_\gamma\}_{\gamma \in \Gamma}$ on $\mathcal{H}$ and a bounded family of scalars $\{\lambda_\gamma\}_{\gamma \in \Gamma}$ such that

$$Tu = \lambda_\gamma u \quad \text{whenever} \quad u \in \mathcal{R}(P_\gamma).$$

Take an arbitrary $x = \sum_{\gamma \in \Gamma} P_\gamma x$ in $\mathcal{H}$. Since $T$ is linear and continuous, $Tx = \sum_{\gamma \in \Gamma} T P_\gamma x = \sum_{\gamma \in \Gamma} \lambda_\gamma P_\gamma x$ so that $T$ is a weighted sum of projections (which is normal by Proposition 6.36). Thus we write (cf. Problem 5.16)

$$T = \sum_{\gamma \in \Gamma} \lambda_\gamma P_\gamma \quad \text{or} \quad T = \bigoplus_{\gamma \in \Gamma} \lambda_\gamma P_\gamma.$$

Conversely, if $T$ is a weighted sum of projections ($Tx = \sum_{\gamma \in \Gamma} \lambda_\gamma P_\gamma x$ for every $x \in \mathcal{H}$), then $Tu = \sum_{\gamma \in \Gamma} \lambda_\gamma P_\gamma u = \sum_{\gamma \in \Gamma} \lambda_\gamma P_\gamma P_\alpha u = \lambda_\alpha u$ for every $u \in \mathcal{R}(P_\alpha)$

(since $P_\gamma P_\alpha = O$ whenever $\gamma \neq \alpha$ and $u = P_\alpha u$ whenever $u \in \mathcal{R}(P_\alpha)$), and hence $T$ is diagonalizable. Outcome: *An operator $T$ on $\mathcal{H}$ is diagonalizable if and only if it is a weighted sum of projections for some bounded sequence of scalars $\{\lambda_\gamma\}_{\gamma \in \Gamma}$ and some resolution of the identity $\{P_\gamma\}_{\gamma \in \Gamma}$ on $\mathcal{H}$. In this case, $\{P_\gamma\}_{\gamma \in \Gamma}$ is said to diagonalize $T$.*

**Corollary 6.45.** *If $T \in \mathcal{B}[\mathcal{H}]$ is compact, then $T$ is normal if and only if $T$ is diagonalizable. Let $\{P_k\}$ be a resolution of the identity on $\mathcal{H}$ that diagonalizes a compact and normal operator $T \in \mathcal{B}[\mathcal{H}]$ into its spectral decomposition, and take any operator $S \in \mathcal{B}[\mathcal{H}]$. The following assertions are pairwise equivalent.*

(a)   *$S$ commutes with $T$ and with $T^*$.*

(b)   *$\mathcal{R}(P_k)$ reduces $S$ for every $k$.*

(c)   *$S$ commutes with every $P_k$.*

*Proof.* Let $T \in \mathcal{B}_\infty[\mathcal{H}]$ be any compact operator. If $T$ is normal, then the Spectral Theorem ensures that it is diagonalizable. The converse is trivial since every diagonalizable operator is normal. Now suppose $T$ is compact and normal so that

$$T = \sum_k \lambda_k P_k,$$

where $\{P_k\}$ is a resolution of the identity on $\mathcal{H}$ and $\{\lambda_k\} = \sigma_P(T)$ is the set of all (distinct) eigenvalues of $T$ (Theorem 6.43). Recall from the proof of Proposition 6.36 that

$$T^* = \sum_k \bar{\lambda}_k P_k.$$

Take any $\lambda \in \mathbb{C}$. If $S$ commutes with $T$ and with $T^*$, then $(\lambda I - T)$ commutes with $S$ and with $S^*$, so that $\mathcal{N}(\lambda I - T)$ is an invariant subspace for both $S$ and $S^*$ (Problem 4.20(c)). Hence $\mathcal{N}(\lambda I - T)$ reduces $S$ (Corollary 5.75), which means that $S$ commutes with the orthogonal projection onto $\mathcal{N}(\lambda I - T)$ (cf. observation that precedes Proposition 5.74). In particular, since $\mathcal{R}(P_k) = \mathcal{N}(\lambda_k I - T)$ for each $k$ (Theorem 6.43), $\mathcal{R}(P_k)$ reduces $S$ for every $k$, which means that $S$ commutes with every $P_k$. Then (a)$\Rightarrow$(b)$\Leftrightarrow$(c). It is readily verified that (c)$\Rightarrow$(a). Indeed, if $SP_k = P_k S$ for every $k$, then $ST = \sum_k \lambda_k SP_k = \sum_k \lambda_k P_k S = TS$ and $ST^* = \sum_k \bar{\lambda}_k SP_k = \sum_k \bar{\lambda}_k P_k S = T^* S$ (recall that $S$ is linear and continuous).   $\square$

## 6.8   A Glimpse at the Spectral Theorem for Normal Operators

What is the role played by compact operators in the Spectral Theorem? First note that, if $T$ is compact, then its spectrum (and so its point spectrum) is countable. But this is not crucial once we know how to deal with uncountable sums. In particular, we know how to deal with an uncountable weighted sum of projections $Tx = \sum_{\gamma \in \Gamma} \lambda_\gamma P_\gamma x$ (recall that, even in this case, the above sum has only a countable number of nonzero vectors for each $x$). What really brings a compact operator into play is that a compact normal operator has a nonempty point spectrum and, more than that, it has enough eigenspaces to span $\mathcal{H}$ (see the claim in the proof of Theorem 6.43). That makes the difference, for a normal (noncompact) operator may have an empty point spectrum (witness: a bilateral shift) or it may have eigenspaces but not enough to span the whole space $\mathcal{H}$ (sample: an orthogonal direct sum of a bilateral shift with an identity).

However, the Spectral Theorem survives the lack of compactness if the point spectrum is replaced with the full spectrum (which is never empty). But this has a price: a suitable statement of the Spectral Theorem for plain normal operators requires some knowledge of measure theory, and a proper proof requires a sound knowledge of it. We shall not prove the two fundamental theorems of this final section. Instead, we just state them, and verify some of their basic consequences. Thus we assume here (and only here) that the reader has, at least, some familiarity with measure theory in order to grasp the definition of *spectral measure* and, therefore, the statement of the Spectral Theorem. Operators will be acting on complex Hilbert spaces $\mathcal{H} \neq \{0\}$ or $\mathcal{K} \neq \{0\}$.

**Definition 6.46.** Let $\Omega$ be a set in the complex plane $\mathbb{C}$ and let $\Sigma_\Omega$ be the $\sigma$-algebra of Borel sets in $\Omega$. A (complex) *spectral measure* in a (complex) Hilbert space $\mathcal{H}$ is a mapping $P: \Sigma_\Omega \to \mathcal{B}[\mathcal{H}]$ such that

(a)   $P(\Lambda)$ is an orthogonal projection for every $\Lambda \in \Sigma_\Omega$,

(b)   $P(\varnothing) = O$  and  $P(\Omega) = I$,

(c)   $P(\Lambda_1 \cap \Lambda_2) = P(\Lambda_1)P(\Lambda_2)$ for every $\Lambda_1, \Lambda_2 \in \Sigma_\Omega$,

(d)   if $\{\Lambda_k\}$ is a countable collection of pairwise disjoint sets in $\Sigma_\Omega$, then

$$P\left(\bigcup_k \Lambda_k\right) = \sum_k P(\Lambda_k).$$

If $\{\Lambda_k\}_{k\in\mathbb{N}}$ is a countably infinite collection of pairwise disjoint sets in $\Sigma_\Omega$, then the above identity means convergence in the strong topology:

$$\sum_{k=1}^{n} P(\Lambda_k) \xrightarrow{\ s\ } P\left(\bigcup_{k\in\mathbb{N}} \Lambda_k\right).$$

Indeed, Since $\Lambda_j \cap \Lambda_k = \varnothing$ whenever $j \neq k$, it follows by properties (b) and (c) that $P(\Lambda_j)P(\Lambda_k) = P(\Lambda_j \cap \Lambda_k) = P(\varnothing) = O$ for $j \neq k$, so that $\{P(\Lambda_k)\}_{k\in\mathbb{N}}$ is an orthogonal sequence of orthogonal projections in $\mathcal{B}[\mathcal{H}]$. Then, according to Proposition 5.58, $\{\sum_{k=1}^{n} P(\Lambda_k)\}_{n\in\mathbb{N}}$ converges strongly to the orthogonal projection in

$\mathcal{B}[\mathcal{H}]$ onto $\left(\sum_{k\in\mathbb{N}}\mathcal{R}(P(\Lambda_k))\right)^- = \bigvee\left(\bigcup_{k\in\mathbb{N}}\mathcal{R}(P(\Lambda_k))\right)$. Therefore, what property (d) says (in the case of a countably infinite collection of pairwise disjoint Borel sets $\{\Lambda_k\}_{k\in\mathbb{N}}$) is that $P\left(\bigcup_{k\in\mathbb{N}}\Lambda_k\right)$ coincides with the orthogonal projection in $\mathcal{B}[\mathcal{H}]$ onto $\bigvee\left(\bigcup_{k\in\mathbb{N}}\mathcal{R}(P(\Lambda_k))\right)$. This generalizes the concept of a resolution of the identity on $\mathcal{H}$. In fact, if $\{\Lambda_k\}_{k\in\mathbb{N}}$ is a partition of $\Omega$, then the orthogonal sequence of orthogonal projections $\{P(\Lambda_k)\}_{k\in\mathbb{N}}$ is such that

$$\sum_{k=1}^{n} P(\Lambda_k) \xrightarrow{\ s\ } P\left(\bigcup_{k\in\mathbb{N}}\Lambda_k\right) = P(\Omega) = I.$$

Now take $x, y \in \mathcal{H}$ and consider the mapping $p_{x,y} : \Sigma_\Omega \to \mathbb{C}$ defined by

$$p_{x,y}(\Lambda) = \langle P(\Lambda)x \,;\, y \rangle$$

for every $\Lambda \in \Sigma_\Omega$. $p_{x,y}$ is an ordinary complex-valued countably additive measure on $\Sigma_\Omega$. Let $\varphi : \Omega \to \mathbb{C}$ be any bounded $\Sigma_\Omega$-measurable function. The integral of $\varphi$ with respect to $p_{x,y}$, $\int \varphi(\lambda)\,d\,p_{x,y}$, will be denoted by $\int \varphi(\lambda)\,d\langle P_\lambda x \,;\, y\rangle$. Moreover, there exists a unique $F \in \mathcal{B}[\mathcal{H}]$ such that

$$\langle Fx \,;\, y \rangle = \int \varphi(\lambda)\,d\langle P_\lambda x \,;\, y \rangle \quad \text{for every} \quad x, y \in \mathcal{H}.$$

Indeed, let $f : \mathcal{H} \times \mathcal{H} \to \mathbb{C}$ be defined by $f(x, y) = \int \varphi(\lambda)\,d\langle P_\lambda x \,;\, y\rangle$, and note that $|f(x, y)| \le \int |\varphi(\lambda)|\,d\langle P_\lambda x \,;\, y\rangle \le \|\varphi\|_\infty \int d\langle P_\lambda x \,;\, y\rangle = \|\varphi\|_\infty \langle P(\Omega)x \,;\, y\rangle = \|\varphi\|_\infty \langle x \,;\, y\rangle = \|\varphi\|_\infty \|x\|\,\|y\|$, for every $x, y \in \mathcal{H}$. Outcome: for each $y \in \mathcal{H}$ the functional $f(\cdot, y) : \mathcal{H} \to \mathbb{C}$, which is clearly linear, is bounded too. Then, by the Riesz Representation Theorem, for each $y \in \mathcal{H}$ there exists a unique $z_y \in \mathcal{H}$ such that $f(x, y) = \langle x \,;\, z_y \rangle$ for every $x \in \mathcal{H}$. This establishes a mapping $\Phi : \mathcal{H} \to \mathcal{H}$ that assigns to each $y \in \mathcal{H}$ this unique $z_y \in \mathcal{H}$ (i.e., $\Phi y = z_y$), and $f(x, y) = \langle x \,;\, \Phi y \rangle$ for every $x, y \in \mathcal{H}$. It is easy to show that $\Phi$ is unique and lies in $\mathcal{B}[\mathcal{H}]$ (cf. proof of Proposition 5.65(a,b)). Thus there exists a unique $F \in \mathcal{B}[\mathcal{H}]$, viz., $F = \Phi^*$, such that $\langle Fx \,;\, y \rangle = f(x, y)$ for every $x, y \in \mathcal{H}$. The notation

$$F = \int \varphi(\lambda)\,dP_\lambda$$

is just short for the identity $\langle Fx \,;\, y \rangle = \int \varphi(\lambda)\,d\langle P_\lambda x \,;\, y \rangle$ for every $x, y \in \mathcal{H}$. Note that $\langle F^*x \,;\, y \rangle = \langle \Phi x \,;\, y \rangle = \overline{\langle y \,;\, \Phi x \rangle} = \overline{f(y, x)} = \overline{\int \varphi(\lambda)\,d\langle P_\lambda y \,;\, x \rangle} = \int \overline{\varphi(\lambda)}\,d\langle P_\lambda x \,;\, y \rangle)$ for every $x, y \in \mathcal{H}$, and hence

$$F^* = \int \overline{\varphi(\lambda)}\,dP_\lambda.$$

If $\psi : \Omega \to \mathbb{C}$ is a bounded $\Sigma_\Omega$-measurable function and $G = \int \psi(\lambda)\,dP_\lambda$, then it can be shown (by using the *Radon–Nikodým Theorem*) that $FG = \int \varphi(\lambda)\psi(\lambda)\,dP_\lambda$.

In particular $F^*F = \int |\varphi(\lambda)|^2 \, dP_\lambda = FF^*$ so that $F$ is normal. The Spectral Theorem states the converse.

**Theorem 6.47.** (The Spectral Theorem). *If $N \in \mathcal{B}[\mathcal{H}]$ is normal, then there exists a unique spectral measure $P$ on $\Sigma_{\sigma(N)}$ such that*

$$N = \int \lambda \, dP_\lambda,$$

*If $\Lambda$ is a nonempty relatively open subset of $\sigma(N)$, then $P(\Lambda) \neq O$.*

The representation $N = \int \lambda \, dP_\lambda$ is usually referred to as the *spectral decomposition* of $N$. Note that $N^*N = \int |\lambda|^2 \, dP_\lambda = NN^*$.

**Theorem 6.48.** (Fuglede). *Let $N = \int \lambda \, dP_\lambda$ be the spectral decomposition of a normal operator $N \in \mathcal{B}[\mathcal{H}]$. If $S \in \mathcal{B}[\mathcal{H}]$ commutes with $N$, then $S$ commutes with $P(\Lambda)$ for every $\Lambda \in \Sigma_{\sigma(N)}$.*

In other words, if $SN = NS$, then $SP(\Lambda) = P(\Lambda)S$, and hence each subspace $\mathcal{R}(P(\Lambda))$ reduces $S$, which means that $\{\mathcal{R}(P(\Lambda))\}_{\Lambda \in \Sigma_{\sigma(N)}}$ is a family of reducing subspaces for every operator that commutes with a normal operator $N = \int \lambda \, dP_\lambda$. If $\sigma(N)$ has a single point, say $\sigma(N) = \{\lambda\}$, then $N = \lambda I$ (by uniqueness of the spectral measure); that is, $N$ is a scalar operator so that every subspace of $\mathcal{H}$ reduces $N$. Hence, if $N$ is nonscalar, then $\sigma(N)$ has more than one point (and $\dim \mathcal{H} > 1$). If $\lambda, \mu \in \sigma(N)$ and $\lambda \neq \mu$, then let $\Delta_\lambda$ denote the open disc of radius $\frac{1}{2}|\lambda - \mu|$ centered at $\lambda$. Put $\Lambda_\lambda = \sigma(N) \cap \Delta_\lambda$ and $\Lambda'_\lambda = \sigma(N)\backslash\Delta_\lambda$ in $\Sigma_{\sigma(N)}$, so that $\sigma(N)$ is the disjoint union of $\Lambda_\lambda$ and $\Lambda'_\lambda$. Note that $P(\Lambda_\lambda) \neq O$ and $P(\Lambda'_\lambda) \neq O$ (for $\Lambda_\lambda$ and $\sigma(N)\backslash\Delta_\lambda^-$ are nonempty relatively open subsets of $\sigma(N)$, and $\sigma(N)\backslash\Delta_\lambda^- \subseteq \Lambda'_\lambda$). Then $I = P(\sigma(N)) = P(\Lambda_\lambda \cup \Lambda'_\lambda) = P(\Lambda_\lambda) + P(\Lambda'_\lambda)$, and therefore $P(\Lambda_\lambda) = I - P(\Lambda'_\lambda) \neq I$. Thus

$$\{0\} \neq \mathcal{R}(P(\Lambda_\lambda)) \neq \mathcal{H}.$$

Conclusions: If $\dim \mathcal{H} > 1$, then *every normal operator has a nontrivial reducing subspace*. Actually, *every nonscalar normal operator has a nontrivial hyperinvariant subspace which reduces every operator that commutes with it.* In fact, *every operator that commutes with a nonscalar normal operator is reducible.*

**Corollary 6.49.** (Fuglede–Putnam). *If $N_1 \in \mathcal{B}[\mathcal{H}]$ and $N_2 \in \mathcal{B}[\mathcal{K}]$ are normal operators, and if $X \in \mathcal{B}[\mathcal{H}, \mathcal{K}]$ intertwines $N_1$ to $N_2$, then $X$ intertwines $N_1^*$ to $N_2^*$ (i.e., if $XN_1 = N_2X$, then $XN_1^* = N_2^*X$).*

*Proof.* Let $N = \int \lambda \, dP_\lambda$ be a normal operator in $\mathcal{B}[\mathcal{H}]$, let $\Lambda$ be an arbitrary set in $\Sigma_{\sigma(N)}$, and take $S \in \mathcal{B}[\mathcal{H}]$.

*Claim.* $SN = NS \iff SP(\Lambda) = P(\Lambda)S \iff SN^* = N^*S$.

*Proof.* If $SN = NS$, then $SP(\Lambda) = P(\Lambda)S$ for every $\Lambda \in \Sigma_{\sigma(N)}$ by Theorem 6.48. Thus $\langle SN^*x ; y \rangle = \langle N^*x ; S^*y \rangle = \int \bar{\lambda} \, d \langle P_\lambda x ; S^*y \rangle = \int \bar{\lambda} \, d \langle SP_\lambda x ; y \rangle = \int \bar{\lambda} \, d \langle P_\lambda Sx ; y \rangle = \langle N^*Sx ; y \rangle$ for every $x, y \in \mathcal{H}$, and hence $SN^* = N^*S$ so that $NS^* = S^*N$. This implies that $P(\Lambda)S^* = S^*P(\Lambda)$, so that $SP(\Lambda) = P(\Lambda)S$, for every $\Lambda \in \Sigma_{\sigma(N)}$ (cf. Theorem 6.48 again). Therefore, $\langle SNx ; y \rangle = \langle Nx ; S^*y \rangle = \int \lambda \, d \langle P_\lambda x ; S^*y \rangle = \int \lambda \, d \langle SP_\lambda x ; y \rangle = \int \lambda \, d \langle P_\lambda Sx ; y \rangle = \langle NSx ; y \rangle$ for every $x, y \in \mathcal{H}$, and hence $SN = NS$. $\square$

Take $N_1 \in \mathcal{B}[\mathcal{H}]$, $N_2 \in \mathcal{B}[\mathcal{K}]$, and $X \in \mathcal{B}[\mathcal{H}, \mathcal{K}]$. Set $N = N_1 \oplus N_2 = \left( \begin{smallmatrix} N_1 & O \\ O & N_2 \end{smallmatrix} \right)$ and $S = \left( \begin{smallmatrix} O & O \\ X & O \end{smallmatrix} \right)$ in $\mathcal{B}[\mathcal{H} \oplus \mathcal{K}]$. If $N_1$ and $N_2$ are normal operators, then $N$ is clearly normal. If $XN_1 = N_2X$, then $SN = NS$ and so $SN^* = N^*S$ by the above claim. Hence $XN_1^* = N_2^*X$. $\square$

The claim in the above proof ensures that $S \in \mathcal{B}[\mathcal{H}]$ *commutes with $N$ and with $N^*$ if and only if $S$ commutes with $P(\Lambda)$ or, equivalently, $\mathcal{R}(P(\Lambda))$ reduces $S$, for every $\Lambda \in \Sigma_{\sigma(N)}$* (compare with Corollary 6.45).

**Corollary 6.50.** *Take $N_1 \in \mathcal{B}[\mathcal{H}]$, $N_2 \in \mathcal{B}[\mathcal{K}]$, and $X \in \mathcal{B}[\mathcal{H}, \mathcal{K}]$. If $N_1$ and $N_2$ are normal operators and $XN_1 = N_2X$, then*

(a)  $\mathcal{N}(X)$ *reduces $N_1$  and  $\mathcal{R}(X)^-$ reduces $N_2$*

*so that $N_1|_{\mathcal{N}(X)^\perp} \in \mathcal{B}[\mathcal{N}(X)^\perp]$ and $N_2|_{\mathcal{R}(X)^-} \in \mathcal{B}[\mathcal{R}(X)^-]$. Moreover,*

(b)  $N_1|_{\mathcal{N}(X)^\perp}$ *and $N_2|_{\mathcal{R}(X)^-}$ are unitarily equivalent.*

*Proof.* (a) Since $XN_1 = N_2X$, it follows that $\mathcal{N}(X)$ is $N_1$-invariant and $\mathcal{R}(X)$ is $N_2$-invariant (and so $\mathcal{R}(X)^-$ is $N_2$-invariant — cf. Problem 4.18). Indeed, if $Xx = 0$, then $X(N_1x) = N_2(Xx) = 0$; and $N_2Xx = XN_1x \in \mathcal{R}(X)$ for every $x \in \mathcal{H}$. Corollary 6.49 ensures that $XN_1^* = N_2^*X$, and hence $\mathcal{N}(X)$ is $N_1^*$-invariant and $\mathcal{R}(X)^-$ is $N_2^*$-invariant. Therefore (Corollary 5.75), $\mathcal{N}(X)$ reduces $N_1$ and $\mathcal{R}(X)^-$ reduces $N_2$.

(b) Let $X = WQ$ be the polar decomposition of $X$, where $Q = (X^*X)^{\frac{1}{2}}$ (Theorem 5.89). Observe that $XN_1 = N_2X$ implies $X^*N_2^* = N_1^*X^*$, which in turn implies $X^*N_2 = N_1X^*$ by Corollary 6.49. Then

$$Q^2N_1 = X^*XN_1 = X^*N_2X = N_1X^*X = N_1Q^2$$

so that $QN_1 = N_1Q$ (Theorem 5.85). Therefore, $WN_1Q = WQN_1 = XN_1 = N_2X = N_2WQ$. That is,

$$(WN_1 - N_2W)Q = O.$$

Thus $\mathcal{R}(Q)^- \subseteq \mathcal{N}(WN_1 - N_2W)$, and so

$$\mathcal{N}(Q)^\perp \subseteq \mathcal{N}(WN_1 - N_2W)$$

(for $Q = Q^*$ so that $\mathcal{R}(Q)^- = \mathcal{N}(Q)^\perp$ by Proposition 5.76). Recall that

$$\mathcal{N}(W) = \mathcal{N}(Q) = \mathcal{N}(Q^2) = \mathcal{N}(X^*X) = \mathcal{N}(X)$$

(cf. Propositions 5.76 and 5.86, and Theorem 5.89). If $u \in \mathcal{N}(Q)$ then $N_2 W u = 0$, and $N_1 u = N_1|_{\mathcal{N}(X)} u \in \mathcal{N}(X) = \mathcal{N}(W)$ (because $\mathcal{N}(X)$ is $N_1$-invariant) so that $W N_1 u = 0$. Hence

$$\mathcal{N}(Q) \subseteq \mathcal{N}(W N_1 - N_2 W).$$

The above displayed inclusions imply that $\mathcal{N}(W N_1 - N_2 W) = \mathcal{K}$ (cf. Problem 5.7(b)). Equivalently,

$$W N_1 = N_2 W.$$

Now $W = VP$, where $V : \mathcal{N}(W)^\perp \to \mathcal{K}$ is an isometry and $P : \mathcal{H} \to \mathcal{H}$ is the orthogonal projection onto $\mathcal{N}(W)^\perp$ (Proposition 5.87). Then

$$V P N_1 = N_2 V P \quad \text{so that} \quad V P N_1|_{\mathcal{N}(X)^\perp} = N_2 V P|_{\mathcal{N}(X)^\perp}.$$

Since $\mathcal{R}(P) = \mathcal{N}(W)^\perp = \mathcal{N}(X)^\perp$ is $N_1$-invariant (recall: $\mathcal{N}(X)$ reduces $N_1$), it follows that $N_1(\mathcal{N}(X)^\perp) \subseteq \mathcal{N}(X)^\perp = \mathcal{R}(P)$, and hence

$$V P N_1|_{\mathcal{N}(X)^\perp} = V N_1|_{\mathcal{N}(X)^\perp}.$$

Since $\mathcal{R}(V) = \mathcal{R}(W) = \mathcal{R}(X)^-$ (cf. Theorem 5.89 and the observation that precedes Proposition 5.88), it also follows that

$$N_2 V P|_{\mathcal{N}(X)^\perp} = N_2 V P|_{\mathcal{R}(P)} = N_2 V = N_2|_{\mathcal{R}(X)^-} V.$$

But $V : \mathcal{N}(W)^\perp \to \mathcal{R}(V)$ is a unitary transformation (i.e., a surjective isometry) of the Hilbert space $\mathcal{N}(X)^\perp = \mathcal{N}(W)^\perp \subseteq \mathcal{H}$ onto the Hilbert space $\mathcal{R}(X)^- = \mathcal{R}(V) \subseteq \mathcal{K}$. Conclusion:

$$V N_1|_{\mathcal{N}(X)^\perp} = N_2|_{\mathcal{R}(X)^-} V$$

so that the operators $N_1|_{\mathcal{N}(X)^\perp} \in \mathcal{B}[\mathcal{N}(X)^\perp]$ and $N_2|_{\mathcal{R}(X)^-} \in \mathcal{B}[\mathcal{R}(X)^-]$ are unitarily equivalent. $\qquad\qquad\square$

An immediate consequence of Corollary 6.50: *If a quasiinvertible linear transformation intertwines two normal operators, then these normal operators are unitarily equivalent.* That is, if $N_1 \in \mathcal{B}[\mathcal{H}]$ and $N_2 \in \mathcal{B}[\mathcal{K}]$ are normal operators, and if $X N_1 = N_2 X$, where $X \in \mathcal{B}[\mathcal{H}, \mathcal{K}]$ is such that $\mathcal{N}(X) = \{0\}$ (equivalently, $\mathcal{N}(X)^\perp = \mathcal{H}$) and $\mathcal{R}(X)^- = \mathcal{K}$, then $U N_1 = N_2 U$ for a unitary transformation $U \in \mathcal{B}[\mathcal{H}, \mathcal{K}]$. This happens, in particular, when $X$ is invertible (i.e., if $X \in \mathcal{G}[\mathcal{H}, \mathcal{K}]$). Outcome: *Two similar normal operators are unitarily equivalent.*

Applying Theorems 6.47 and 6.48, we saw that normal operators (on a complex Hilbert space of dimension greater than one) have a nontrivial invariant subspace. This also is the case for compact operators (on a complex Banach space of dimension greater than one). The definitive result in this line was presented by Lomonosov in 1973: *An operator has a nontrivial invariant subspace if it commutes with a non-scalar operator that commutes with a nonzero compact operator*. In fact, *every nonscalar operator that commutes with a nonscalar compact operator* (itself, in particular) *has a nontrivial hyperinvariant subspace*. Recall that, on an infinite-dimensional normed space, the only scalar compact operator is the null operator. On a finite-dimensional normed space every operator is compact, and hence every operator on a complex finite-dimensional normed space of dimension greater than one has a nontrivial invariant subspace and, if it is nonscalar, a nontrivial hyperinvariant subspace as well. This prompts the most celebrated open question in operator theory, namely, the *invariant subspace problem*: *Does every operator* (on an infinite-dimensional complex separable Hilbert space) *have a nontrivial invariant subspace*? All the qualifications are crucial here. Note that the operator $\left(\begin{smallmatrix} 0 & 1 \\ -1 & 0 \end{smallmatrix}\right)$ on $\mathbb{R}^2$ has no nontrivial invariant subspace (when acting on the Euclidean *real* space but, of course, it has a nontrivial invariant subspace when acting on the unitary *complex* space $\mathbb{C}^2$). Thus the above question actually refers to complex spaces and, henceforward, we assume that all spaces are complex. The problem has a negative answer if we replace Hilbert space with Banach space. This (the invariant subspace problem in a Banach space) remained as an open question for a long period up to the mid-1980s, when it was solved by Read (1984) and Enflo (1987) who constructed a Banach-space operator without a nontrivial invariant subspace. As we have just seen, the problem has an affirmative answer in a finite-dimensional space (of dimension greater than one). It has an affirmative answer in a nonseparable Hilbert space too. Indeed, let $T$ be any operator on a nonseparable Hilbert space $\mathcal{H}$, and let $x$ be any nonzero vector in $\mathcal{H}$. Consider the orbit of $x$ under $T$, $\{T^n x\}_{n \geq 0}$, so that $\bigvee \{T^n x\}_{n \geq 0} \neq \{0\}$ is an invariant subspace for $T$ (cf. Problem 4.23). Since $\{T^n x\}_{n \geq 0}$ is a countable set, it follows by Proposition 4.9(b) that $\bigvee \{T^n x\}_{n \geq 0} \neq \mathcal{H}$. Hence $\bigvee \{T^n x\}_{n \geq 0}$ is a nontrivial invariant subspace for $T$. Completeness and boundedness are also crucial here. In fact, it can be shown that (1) there exists an operator on an infinite-dimensional complex separable (incomplete) inner product space which has no nontrivial invariant subspace, and that (2) there exists a (not necessarily bounded) linear transformation of a complex separable Hilbert space into itself without nontrivial invariant subspaces. However, for (bounded linear) operators on an infinite-dimensional complex separable Hilbert space, the invariant subspace problem remains a recalcitrant open question.

## *Suggested Reading*

Akhiezer and Glazman [1], [2]  
Bachman and Narici [1]  
Beals [1]  
Beauzamy [1]  
Berberian [1], [3]

Helmberg [1]  
Kubrusly [1]  
Martin and Putinar [1]  
Naylor and Sell [1]  
Pearcy [1], [2]

Berezansky, Sheftel and Us [1], [2]

Clancey [1]

Colojoară and Foiaş [1]

Conway [1], [2], [3]

Douglas [1]

Dowson [1]

Dunford and Schwartz [3]

Fillmore [1], [2]

Halmos [1], [4]

Putnam [1]

Radjavi and Rosenthal [1]

Riesz and Sz.-Nagy [1]

Sunder [1]

Sz.-Nagy and Foiaş [1]

Taylor and Lay [1]

Weidmann [1]

Xia [1]

Yoshino [1]

# Problems

**Problem 6.1.** Let $\mathcal{H}$ be a Hilbert space. Show that the set of all normal operators from $\mathcal{B}[\mathcal{H}]$ is closed in $\mathcal{B}[\mathcal{H}]$.

*Hint*: $(T^* - S^*)(T - S) + (T^* - S^*)S + S^*(T - S) = T^*T - S^*S$, and hence $\|T^*T - S^*S\| \leq \|T - S\|^2 + 2\|S\|\|T - S\|$ for every $T, S \in \mathcal{B}[\mathcal{H}]$. Verify the above inequality. Now let $\{N_n\}_{n=1}^\infty$ be a sequence of normal operators in $\mathcal{B}[\mathcal{H}]$ that converges in $\mathcal{B}[\mathcal{H}]$ to $N \in \mathcal{B}[\mathcal{H}]$. Check that

$$
\begin{aligned}
\|N^*N - NN^*\| &= \|N^*N - N_n^*N_n + N_nN_n^* - NN^*\| \\
&\leq \|N_n^*N_n - N^*N\| + \|N_nN_n^* - NN^*\| \\
&\leq 2\big(\|N_n - N\|^2 + 2\|N\|\|N_n - N\|\big).
\end{aligned}
$$

Conclude: *The (uniform) limit of a uniformly convergent sequence of normal operators is normal*. Finally, apply the Closed Set Theorem.

**Problem 6.2.** Let $S$ and $T$ be normal operators acting on the same Hilbert space. Prove the following assertions.

(a)  $\alpha T$ is normal for every scalar $\alpha$.

(b)  If $S^*T = TS^*$, then $S + T$, $TS$ and $ST$ are normal operators.

(c)  $T^{*n}T^n = T^nT^{*n} = (T^*T)^n = (TT^*)^n$ for every integer $n \geq 0$.

   *Hint*: Problem 5.24 and Proposition 6.1.

**Problem 6.3.** Let $T$ be a contraction on a Hilbert space $\mathcal{H}$. (i.e., $T \in \mathcal{B}[\mathcal{H}]$ and $\|T\| \leq 1$). Show that

(a)  $T^{*n}T^n \xrightarrow{s} A$,

(b)  $O \leq A \leq I$    (i.e. $A \in \mathcal{B}^+[\mathcal{H}]$ and $\|A\| \leq 1$),

(c) $T^{*n}A T^n = A$ for every integer $n \geq 0$.

*Hint*: Proposition 5.84, Problem 5.49 and Proposition 5.68, Problems 4.45(a), 5.55 and 5.24(a).

Note that, according to Problem 5.54, *a contraction $T$ is strongly stable if and only if $A = O$*. Since $A \geq O$, it follows by Proposition 5.81 that $A$ is an orthogonal projection if and only if it is idempotent (i.e., if and only if $A = A^2$). In general, $A$ is not a projection.

(d) Set $T = \text{shift}(\alpha, 1, 1, 1, \dots)$ in $\mathcal{B}[\ell_+^2]$, a weighted shift on $\mathcal{H} = \ell_+^2$ with $|\alpha| \in (0, 1)$, and show that this is a contraction for which $A = \text{diag}(|\alpha|^2, 1, 1, 1, \dots)$ in $\mathcal{B}[\ell_+^2]$ is not a projection.

(e) Show that $A = A^2$ if and only if $A T = T A$.

    *Hint*: Use part (c) to verify that $\langle A T x \,;\, T A x \rangle = \|Ax\|^2$. Since $\|T\| \leq 1$, check that $\|A T x - T A x\|^2 \leq \|A T x\|^2 - \|Ax\|^2$. Thus, recalling that $\|T\| \leq 1$, $\|A\| \leq 1$, and applying part (c), verify that $\|Ax\|^2 \leq \|A T x\|^2 \leq \|A^{\frac{1}{2}} T x\|^2 = \langle T^*A T x \,;\, x \rangle = \langle Ax \,;\, x \rangle = \|A^{\frac{1}{2}}x\|^2$. Conclude: $A = A^2$ implies $A T = T A$. For the converse, use parts (a) and (c).

(f) Show that $A = A^2$ whenever $T$ is a normal contraction.

    *Hint*: Problems 6.2(c) and 5.24.

It can be shown that $A = A^2$ whenever $T$ is a cohyponormal contraction.

**Problem 6.4.** Consider the Hilbert space $L^2(\Gamma)$ of Example 5L(c), where $\Gamma$ denotes the unit circle about the origin of complex plane. Recall that, in this context, the terms "bounded function", "equality", "inequality", "belongs" and "for all", are interpreted in the sense of equivalence classes. Let $\varphi \colon \Gamma \to \mathbb{C}$ be a bounded function. Show that

(a) $\varphi f$ lies in $L^2(\Gamma)$ for every $f \in L^2(\Gamma)$.

Thus consider the mapping $M_\varphi \colon L^2(\Gamma) \to L^2(\Gamma)$ defined by

$$M_\varphi f = \varphi f \quad \text{for every} \quad f \in L^2(\Gamma).$$

That is, $(M_\varphi f)(z) = \varphi(z) f(z)$ for all $z \in \Gamma$. This mapping is called the *multiplication operator* on $L^2(\Gamma)$. It is easy to show that $M_\varphi$ is linear and bounded (i.e., $M_\varphi \in \mathcal{B}[L^2(\Gamma)]$). Prove the following propositions.

(b) $\|M_\varphi\| = \|\varphi\|_\infty$.

*Hint*: First verify that $\|M_\varphi f\| \le \|\varphi\|_\infty \|f\|$ for every $f \in L^2(\Gamma)$. Take $\varepsilon > 0$ and consider the set $\Gamma_\varepsilon = \{z \in \Gamma: \ \|\varphi\|_\infty - \varepsilon < |\varphi(z)|\}$. Let $f_\varepsilon$ be the characteristic function of $\Gamma_\varepsilon$. Check that $f_\varepsilon \in L^2(\Gamma)$, and also that $\|M_\varphi f_\varepsilon\| \ge (\|\varphi\|_\infty - \varepsilon)\|f - \varepsilon\|$.

(c)  $M_\varphi^* g = \overline{\varphi} g$ for every $g \in L^2(\Gamma)$.

(d)  $M_\varphi$ is a normal operator.

(e)  $M_\varphi$ is unitary if and only if $\varphi(z) \in \Gamma$ for all $z \in \Gamma$.

(f)  $M_\varphi$ is self-adjoint if and only if $\varphi(z) \in \mathbb{R}$ for all $z \in \Gamma$.

(g)  $M_\varphi$ is nonnegative if and only if $\varphi(z) \ge 0$ for all $z \in \Gamma$.

(h)  $M_\varphi$ is positive if and only if $\varphi(z) > 0$ for all $z \in \Gamma$.

(i)  $M_\varphi$ is strictly positive if and only if $\varphi(z) \ge \alpha > 0$ for all $z \in \Gamma$.

**Problem 6.5.** If $T$ is a quasinormal operator, then

(a)  $(T^*T)^n T = T(T^*T)^n$ for every $n \ge 0$,

(b)  $|T^n| = |T|^n$ for every $n \ge 0$,    and

(c)  $T^n \xrightarrow{s} O \iff |T|^n \xrightarrow{s} O \iff |T|^n \xrightarrow{w} O$.

*Hint*: Prove (a) by induction. Assertion (b) holds trivially for $n = 0, 1$, for every operator $T$. Let $T$ be a quasinormal operator (so that (a) holds true) and suppose $|T^n| = |T|^n$ for some $n \ge 1$. Then verify that $|T^{n+1}|^2 = T^*T^{*n}T^nT = T^*|T^n|^2T = T^*|T|^{2n}T = T^*(T^*T)^nT = T^*T(T^*T)^n = (T^*T)^{n+1} = |T|^{2(n+1)} = (|T|^{n+1})^2$. Now conclude the induction by recalling that the square root is unique. Use Problem 5.61(d) and part (b) to prove (c).

**Problem 6.6.** Every quasinormal operator is hyponormal. Give a direct proof.

*Hint*: Let $T$ be an operator on a Hilbert space $\mathcal{H}$. Take an arbitrary $x = u + v \in \mathcal{H} = \mathcal{N}(T^*) + \mathcal{N}(T^*)^\perp = \mathcal{N}(T^*) + \mathcal{R}(T)^-$, with $u \in \mathcal{N}(T^*)$ and $v \in \mathcal{R}(T)^-$, so that $v = \lim_n v_n$ where $\{v_n\}$ is an $\mathcal{R}(T)$-valued sequence (cf. Proposition 4.13, Theorem 5.20, Propositions 5.76 and 3.27). Set $D = T^*T - TT^*$ and check that $\langle Du\,;u \rangle = \|Tu\|^2$. If $T$ is quasinormal (i.e., if $DT = O$), then verify that $\langle Du\,;v \rangle = \lim_n \langle u\,;Dv_n \rangle = 0$, $\langle Dv\,;u \rangle = \lim_n \langle Dv_n\,;u \rangle = 0$, and $\langle Dv\,;v \rangle = \lim_n \langle Dv_n\,;v \rangle = 0$. Finally, show that $\langle Dx\,;x \rangle \ge 0$.

**Problem 6.7.** If $T \in \mathcal{G}[\mathcal{H}]$ is hyponormal, then $T^{-1}$ is hyponormal.

*Hint*: $O \le D = T^*T - TT^*$. Then (Problem 5.51(a)) $O \le T^{-1}DT^{-1*}$. Now show that $I \le T^{-1}T^*TT^{*-1}$, and hence $T^*T^{-1}T^{*-1}T \le I$ (see Problems 1.10 and

5.53(b)). Verify: $O \leq T^{-1*}(I - T^*T^{-1}T^{*-1}T)T^{-1}$, and conclude that $T^{-1}$ is hyponormal.

**Problem 6.8.** If $T \in \mathcal{G}[\mathcal{H}]$ is hyponormal and both $T$ and $T^{-1}$ are contractions, then $T$ is normal.

*Hint*: $\|Tx\| = \|TT^{*-1}T^*x\| \leq \|T^*x\|$. Thus $T$ is cohyponormal.

**Problem 6.9.** Let $\mathcal{H}$ be a Hilbert space. Take any $T \in \mathcal{B}[\mathcal{H}]$ and any $x \in \mathcal{H}$. Show that

(a) $T^*Tx = \|T\|^2 x$ if and only if $\|Tx\| = \|T\|\|x\|$.

*Hint*: If $\|Tx\| = \|T\|\|x\|$, then $\langle T^*Tx ; \|T\|^2 x \rangle = \|T\|^4\|x\|^2$. Therefore, by using the above identity, verify that $\|T^*Tx - \|T\|^2 x\|^2 = \|T^*Tx\|^2 - \|T\|^4\|x\|^2 \leq (\|T^*T\|^2 - \|T\|^4)\|x\|^2 = 0$.

Set $\mathcal{M} = \{x \in \mathcal{H}: \|Tx\| = \|T\|\|x\|\}$. Prove the following assertions.

(b) $\mathcal{M}$ is a subspace of $\mathcal{H}$.

*Hint*: $\mathcal{M} = \mathcal{N}(\|T\|^2 I - T^*T)$.

(c) If $T$ is hyponormal, then $\mathcal{M}$ is $T$-invariant.

*Hint*: If $x \in \mathcal{M}$ and if $T$ is a hyponormal operator, then show that $\|T(Tx)\| \leq \|T\|\|Tx\| = \|\|T\|^2 x\| = \|T^*Tx\| \leq \|T(Tx)\|$.

(d) If $T$ is normal, then $\mathcal{M}$ reduces $T$.

*Hint*: $\mathcal{M}$ is invariant for both $T$ and $T^*$ whenever $T$ is normal.

Note: $\mathcal{M}$ may be trivial (samples: $T = I$ and $T = \mathrm{diag}(\{\frac{k}{k+1}\}_{k=1}^{\infty})$).

**Problem 6.10.** Let $\mathcal{H} \neq \{0\}$ and $\mathcal{K} \neq \{0\}$ be complex Hilbert spaces. Take $T \in \mathcal{B}[\mathcal{H}]$ and $W \in \mathcal{G}[\mathcal{H}, \mathcal{K}]$ arbitrary. Recall that $\mathcal{H}$ and $\mathcal{K}$ are unitarily equivalent, according to Problem 5.70. Show that

$$\sigma_P(T) = \sigma_P(WTW^{-1}) \quad \text{and} \quad \rho(T) = \rho(WTW^{-1}).$$

Thus conclude that (see Proposition 6.17),

$$\sigma_R(T) = \sigma_R(WTW^{-1}) \quad \text{and} \quad \sigma(T) = \sigma(WTW^{-1}).$$

Finally, verify that

$$\sigma_C(T) = \sigma_C(WTW^{-1}).$$

Outcome: *Similarity preserves each part of the spectrum*, and hence *similarity preserves the spectral radius*: $r(T) = r(WTW^{-1})$. In other words, if $T \in \mathcal{B}[\mathcal{H}]$

and $S \in \mathcal{B}[\mathcal{K}]$ are similar (i.e., if $T \approx S$), then $\sigma_P(T) = \sigma_P(S)$, $\sigma_R(T) = \sigma_R(S)$, $\sigma_C(T) = \sigma_C(S)$, and so $r(T) = r(S)$. Use Problem 4.41 to show that *unitary equivalence also preserves the norm* (i.e., if $T \cong S$, then $\|T\| = \|S\|$).

Note: Similarity preserves nontrivial invariant subspaces (Problem 4.29).

**Problem 6.11.** Let $A \in \mathcal{B}[\mathcal{H}]$ be a self-adjoint operator on a complex Hilbert space $\mathcal{H} \neq \{0\}$. Use Corollary 6.18(d) to check that $\pm i \in \rho(A)$ so that $(A + iI)$ and $(A - iI)$ both lie in $\mathcal{G}[\mathcal{H}]$. Consider the operator

$$U = (A - iI)(A + iI)^{-1} = (A + iI)^{-1}(A - iI)$$

in $\mathcal{G}[\mathcal{H}]$, where $U^{-1} = (A + iI)(A - iI)^{-1} = (A - iI)^{-1}(A + iI)$ (cf. Corollary 4.23). It is clear that $A$ commutes with $(A + iI)^{-1}$ and with $(A - iI)^{-1}$ because every operator trivially commutes with its resolvent function. Show that

(a)  $U$ is unitary,

(b)  $U = I - 2i(A + iI)^{-1}$,

(c)  $1 \in \rho(U)$ and $A = i(I + U)(I - U)^{-1} = i(I - U)^{-1}(I + U)$.

*Hint*: (a) Verify that $\|(A \pm iI)x\|^2 = \|Ax\|^2 + \|x\|^2$ (reason: $A = A^*$, and hence $2\,\mathrm{Re}\,\langle Ax \,;\, ix \rangle = 0$) for every $x \in \mathcal{H}$. Take any $y \in \mathcal{H}$ so that $y = (A + iI)x$ for some $x \in \mathcal{H}$ (recall: $\mathcal{R}(A + iI) = \mathcal{H}$). Then $Uy = (A - iI)x$ and $\|Uy\|^2 = \|(A - iI)x\|^2 = \|(A + iI)x\|^2 = \|y\|^2$ so that $U$ is an (invertible) isometry. (b) $(A - iI) = -2iI + (A + iI)$. (c) $(I - U)^{-1} = \frac{1}{2i}(A + iI)$ and $(I + U) = I + (A - iI)(A + iI)^{-1}$.

Conversely, let $U \in \mathcal{G}[\mathcal{H}]$ be a unitary operator with $1 \in \rho(U)$ (so that $(I - U) \in \mathcal{G}[\mathcal{H}]$) and consider the operator

$$A = i(I + U)(I - U)^{-1} = i(I - U)^{-1}(I + U)$$

in $\mathcal{B}[\mathcal{H}]$. Recall again: $U$ commutes with $(I - U)^{-1}$. Show that

(d)  $A = iI + 2iU(I - U)^{-1} = -iI + 2i(I - U)^{-1}$,

(e)  $A$ is self-adjoint,

(f)  $\pm i \in \rho(A)$ and $U = (A - iI)(A + iI)^{-1} = (A + iI)^{-1}(A - iI)$.

*Hint*: (d) Verify that $i(I - U) + 2iU = i(I + U) = -i(I - U) + 2iI$. (e) $(I - U)^{-1*}U^* = (I - U^*)^{-1}U^{-1} = (U(I - U^*))^{-1} = -(I - U)^{-1}$ so that $A^* = -iI - 2i(I - U)^{-1*}U^* = -iI + 2i(I - U)^{-1} = A$. (f) Using assertion (d) we get $(A - iI) = 2iU(I - U)^{-1}$ and $(A + iI) = 2i(I - U)^{-1}$ so that $(A + iI)^{-1} = \frac{1}{2i}(I - U)$.

Summing up: Set $U = (A - iI)(A + iI)^{-1}$ for an arbitrary self-adjoint operator $A$. $U$ is unitary with $1 \in \rho(U)$ and $i(I + U)(I - U)^{-1} = A$. Conversely, set $A = i(I + U)(I - U)^{-1}$ for any unitary operator $U$ with $1 \in \rho(U)$. $A$ is self-adjoint and $(A - iI)(A + iI)^{-1} = U$.

Outcome: There exists a one-to-one correspondence between the class of all self-adjoint operators and the class of all unitary operators for which 1 belongs to the resolvent set: $A \mapsto (A - iI)(A + iI)^{-1}$ with inverse $U \mapsto i(I + U)(I - U)^{-1}$. If $A$ is self-adjoint, then the unitary $U = (A - iI)(A + iI)^{-1}$ is called the *Cayley transform* of $A$. What is behind such a one-to-one correspondence is the *Möbius transformation* $z \mapsto \frac{z-i}{z+i}$ that maps the open upper half plane onto the open unit disc, and the extended real line onto the unit circle.

**Problem 6.12.** Let $T \in \mathcal{B}[\mathcal{H}]$ be any operator on a complex Hilbert space $\mathcal{H} \neq \{0\}$. Prove the following assertions.

(a) $\mathcal{N}(\lambda I - T)$ is a subspace of $\mathcal{H}$, which is $T$-invariant for every $\lambda \in \mathbb{C}$. Moreover, $T|_{\mathcal{N}(\lambda I - T)} = \lambda I \colon \mathcal{N}(\lambda I - T) \to \mathcal{N}(\lambda I - T)$. That is, if restricted to the invariant subspace $\mathcal{N}(\lambda I - T)$, then $T$ acts as a scalar operator on $\mathcal{N}(\lambda I - T)$, and hence $T|_{\mathcal{N}(\lambda I - T)}$ is normal.

Remark: Obviously, if $\mathcal{N}(\lambda I - T) = \{0\}$ (i.e., if $\lambda \notin \sigma_P(T)$), then $T|_{\mathcal{N}(\lambda I - T)} = \lambda I \colon \{0\} \to \{0\}$ coincides with the null operator: on the null space every operator is null or, equivalently, the only operator on the null space is the null operator.

(b) Every eigenspace of a nonscalar operator $T$ is a nontrivial invariant subspace for $T$ (i.e., if $\lambda \in \sigma_P(T)$, then $\{0\} \neq \mathcal{N}(\lambda I - T) \neq \mathcal{H}$ and $\mathcal{N}(\lambda I - T)$ is $T$-invariant).

(c) If $\sigma_P(T) \cup \sigma_P(T^*) \neq \varnothing$, then $T$ has a nontrivial invariant subspace.

(d) If $T$ has no nontrivial invariant subspace, then $\sigma(T) = \sigma_C(T)$.

*Hint*: For parts (c) and (d) use Propositions 5.74 and 6.17, respectively.

**Problem 6.13.** We have already seen in Section 6.3 that $\sigma(T^{-1}) = \sigma(T)^{-1} = \{\lambda^{-1} \in \mathbb{C} \colon \lambda \in \sigma(T)\}$ for every $T \in \mathcal{G}[\mathcal{X}]$, where $\mathcal{X} \neq \{0\}$ is a complex Banach space. Exhibit a diagonal operator $T$ in $\mathcal{G}[\mathbb{C}^2]$ for which $r(T^{-1}) \neq r(T)^{-1}$.

**Problem 6.14.** Let $T$ be an arbitrary operator on a complex Banach space $\mathcal{X} \neq \{0\}$, take any $\lambda \in \rho(T)$ (so that $(\lambda I - T) \in \mathcal{G}[\mathcal{X}]$), and set

$$d = d(\lambda, \sigma(T)),$$

the distance of $\lambda$ to $\sigma(T)$. Since $\sigma(T)$ is nonempty (bounded) and closed, if follows that $d$ is a positive real number (cf. Problem 3.43(b)). Show that the spectral radius of the inverse of $(\lambda I - T)$ coincides with the inverse of the distance of $\lambda \in \rho(T)$ to the spectrum of $T$. That is,

(a) $$r((\lambda I - T)^{-1}) = d^{-1}.$$

*Hint*: $d = \inf_{\mu \in \sigma(T)} |\lambda - \mu|$ so that $d^{-1} = \sup_{\mu \in \sigma(T)} |\lambda - \mu|^{-1}$ and, from the Spectral Mapping Theorem (Theorem 6.19), $\sigma(\lambda I - T) = \{\lambda - \mu \in \mathbb{C} \colon \mu \in \sigma(T)\}$. However, since $\sigma((\lambda I - T)^{-1}) = \sigma(\lambda I - T)^{-1} = \{\mu^{-1} \in \mathbb{C} \colon \mu \in \sigma(\lambda I - T)\}$, show that $\sigma((\lambda I - T)^{-1}) = \{(\lambda - \mu)^{-1} \in \mathbb{C} \colon \mu \in \sigma(T)\}$.

Now let $\mathcal{X}$ be a Hilbert space and prove the following implication.

(b)    If $T$ is hyponormal, then $\|(\lambda I - T)^{-1}\| = d^{-1}$.

*Hint*: If $T$ is hyponormal, then $(\lambda I - T)$ is hyponormal (cf. proof of Corollary 6.18) and so is $(\lambda I - T)^{-1}$ (Problem 7.7). Hence $(\lambda I - T)^{-1}$ is normaloid by Proposition 6.10. Apply (a).

**Problem 6.15.** Let $\mathcal{M}$ be a subspace of a Hilbert space $\mathcal{H}$ and take $T \in \mathcal{B}[\mathcal{H}]$. If $\mathcal{M}$ is $T$-invariant, then $(T|_{\mathcal{M}})^* = PT^*|_{\mathcal{M}}$ in $\mathcal{B}[\mathcal{M}]$, where $P \colon \mathcal{H} \to \mathcal{H}$ is the orthogonal projection onto $\mathcal{M}$.

*Hint*: Use Proposition 5.81 to verify that $\langle (T|_{\mathcal{M}})^* u \, ; v \rangle = \langle u \, ; T|_{\mathcal{M}} v \rangle = \langle u \, ; T v \rangle = \langle u \, ; T P v \rangle = \langle P T^* u \, ; v \rangle = \langle P T^*|_{\mathcal{M}} u \, ; v \rangle$ for every $u, v \in \mathcal{M}$.

In other words, if $\mathcal{M}$ is $T$-invariant, then $T(\mathcal{M}) \subseteq \mathcal{M}$ (so that $T|_{\mathcal{M}}$ lies in $\mathcal{B}[\mathcal{M}]$) but $T^*(\mathcal{M})$ may not be included in $\mathcal{M}$; it has to be projected there: $PT^*(\mathcal{M}) \subseteq \mathcal{M}$ (so that $PT^*|_{\mathcal{M}}$ lies in $\mathcal{B}[\mathcal{M}]$ and coincides with $(T|_{\mathcal{M}})^*$). If $\mathcal{M}$ reduces $T$ (i.e., if $\mathcal{M}$ also is $T^*$-invariant), then $T^*(\mathcal{M})$ does not need to be projected on $\mathcal{M}$; it is already there (i.e., if $\mathcal{M}$ reduces $T$, then $T^*(\mathcal{M}) \subseteq \mathcal{M}$ and $(T|_{\mathcal{M}})^* = T^*|_{\mathcal{M}}$ — see Corollary 5.75).

**Problem 6.16.** Let $\mathcal{M}$ be an invariant subspace for $T \in \mathcal{B}[\mathcal{H}]$.

(a) If $T$ is hyponormal, then $T|_{\mathcal{M}}$ is hyponormal.

(b) If $T$ is hyponormal and $T|_{\mathcal{M}}$ is normal, then $\mathcal{M}$ reduces $T$.

*Hint*: Since $\mathcal{M}$ is $T$-invariant, $T|_{\mathcal{M}} \in \mathcal{B}[\mathcal{M}]$. Use Problem 6.15 (and Propositions 5.81 and 6.6) to show that $\|(T|_{\mathcal{M}})^* u\| \leq \|T^* u\| \leq \|T|_{\mathcal{M}} u\|$ for every $u \in \mathcal{M}$. Moreover, if $T|_{\mathcal{M}}$ is normal, say $T|_{\mathcal{M}} = N$, then $T = \left( \begin{smallmatrix} N & X \\ O & Y \end{smallmatrix} \right)$ in $\mathcal{B}[\mathcal{M} \oplus \mathcal{M}^\perp]$ (cf. Example 2O and Proposition 5.51). Since $N$ is normal and $T$ is hyponormal, verify that

$$O \leq D = T^* T - T T^* = \begin{pmatrix} -XX^* & N^*X - XY^* \\ X^*N - YX^* & XX^* \end{pmatrix} \in \mathcal{B}[\mathcal{M} \oplus \mathcal{M}^\perp].$$

Take any $u$ in $\mathcal{M}$, set $x = (u, 0)$ in $\mathcal{M} \oplus \mathcal{M}^\perp$, and show that $\langle Dx \, ; x \rangle = -\|X^* x\|^2$. Conclude: $T = \left( \begin{smallmatrix} N & O \\ O & Y \end{smallmatrix} \right) = N \oplus Y$, and hence $\mathcal{M}$ reduces $T$.

**Problem 6.17.** This is a rather important result. Let $\mathcal{M}$ be an invariant subspace for a normal operator $T \in \mathcal{B}[\mathcal{H}]$. Show that $T|_{\mathcal{M}}$ is normal if and only if $\mathcal{M}$ reduces $T$.

*Hint*: If $T$ is normal, then it is hyponormal. Apply Problem 6.16 to verify that $\mathcal{M}$ reduces $T$ whenever $T|_{\mathcal{M}}$ is normal. Conversely, if $\mathcal{M}$ reduces $T$, then write $T = N_1 \oplus N_2$ on $\mathcal{M} \oplus \mathcal{M}^{\perp}$, where $N_1 = T|_{\mathcal{M}}$ in $\mathcal{B}[\mathcal{M}]$ and $N_2 = T|_{\mathcal{M}^{\perp}}$ in $\mathcal{B}[\mathcal{M}^{\perp}]$. Now verify that both $N_1$ and $N_2$ are normal operators whenever $T$ is normal.

**Problem 6.18.** Let $T$ be a compact operator on a complex Hilbert space $\mathcal{H}$ and let $\Delta$ denote the open unit disc about the origin of the complex plane.

(a) Show that $\sigma_P(T) \subseteq \Delta$ implies $T^n \xrightarrow{u} O$.

 *Hint*: Corollary 6.31 and Proposition 6.22.

(b) Show that $T^n \xrightarrow{w} O$ implies $\sigma_P(T) \subseteq \Delta$.

 *Hint*: If $\lambda \in \sigma_P(T)$, then verify that there exists a unit vector $x$ in $\mathcal{H}$ such that $T^n x = \lambda^n x$ for every positive integer $n$. Thus $|\lambda|^n \to 0$, and hence $|\lambda| < 1$, whenever $T^n \xrightarrow{w} O$ (cf. Proposition 5.67).

Conclude: *The concepts of weak, strong, and uniform stabilities coincide for a compact operator on a complex Hilbert space.*

**Problem 6.19.** If $T \in \mathcal{B}[\mathcal{H}]$ is hyponormal, then

$$\mathcal{N}(\lambda I - T) \subseteq \mathcal{N}(\bar{\lambda} I - T^*) \quad \text{for every} \quad \lambda \in \mathbb{C}.$$

*Hint*: Adapt the proof of Proposition 6.39.

**Problem 6.20.** Take $\lambda, \mu \in \mathbb{C}$. If $T \in \mathcal{B}[\mathcal{H}]$ is hyponormal, then

$$\mathcal{N}(\lambda I - T) \perp \mathcal{N}(\mu I - T) \quad \text{whenever} \quad \lambda \neq \mu.$$

*Hint*: Adapt the proof of Proposition 6.40 by using Problem 6.19.

**Problem 6.21.** If $T \in \mathcal{B}[\mathcal{H}]$ is hyponormal, then $\mathcal{N}(\lambda I - T)$ reduces $T$ for every $\lambda \in \mathbb{C}$.

*Hint*: Adapt the proof of Proposition 6.41. First observe that, if $\lambda x = Tx$, then $T^* x = \bar{\lambda} x$ (by Problem 6.19). Next verify that $\lambda T^* x = \lambda \bar{\lambda} x = \bar{\lambda} \lambda x = \bar{\lambda} T x = T \bar{\lambda} x = T T^* x$. Then conclude: $\mathcal{N}(\lambda I - T)$ is $T^*$-invariant. Note: $T|_{\mathcal{N}(\lambda I - T)}$ is a scalar operator on $\mathcal{N}(\lambda I - T)$, and hence a normal operator (cf. Problem 6.12(a)).

A *pure hyponormal* operator is a hyponormal operator that has no normal direct summand (i.e., it has no reducing subspace on which it acts as a normal operator). Use Problem 6.17 to show that a pure hyponormal operator has an empty point spectrum.

**Problem 6.22.** Let $T \in \mathcal{B}[\mathcal{H}]$ be a hyponormal operator and set

$$\mathcal{M} = \left( \sum_{\lambda \in \sigma_P(T)} \mathcal{N}(\lambda I - T) \right)^{-}.$$

Show that

$$\mathcal{M} \text{ reduces } T \text{ and } T|_{\mathcal{M}} \text{ is normal.}$$

*Hint*: If $\sigma_{P(T)} = \varnothing$, then the result is trivial (for the empty sum is null). Thus suppose $\sigma_P(T) \neq \varnothing$. First note that $\{\mathcal{N}(\lambda I - T)\}_{\lambda \in \sigma_P(T)}$ is an orthogonal family of nonzero subspaces of the Hilbert space $\mathcal{M}$ (Problem 6.20). Now choose one of the following ways.

(1) Adapt the proof of Corollary 6.42, with the help of Problems 6.20 and 6.21, to verify that $\mathcal{M}$ reduces $T$. Use Theorem 5.59 and Problem 5.10 to check that the family $\{P_\lambda\}_{\lambda \in \sigma_P(T)}$ consisting of the nonzero orthogonal projections $P_\lambda \in \mathcal{B}[\mathcal{M}]$ onto each $\mathcal{N}(\lambda I - T)$ is a resolution of the identity on $\mathcal{M}$. Take any $u \in \mathcal{M}$. Verify that $u = \sum_\lambda P_\lambda u$, and hence $T|_{\mathcal{M}} u = Tu = \sum_\lambda T P_\lambda u = \sum_\lambda \lambda P_\lambda u$ (reason: $P_\lambda u \in \mathcal{N}(\lambda I - T)$, where the sums run over $\sigma_P(T)$). Conclude that $T|_{\mathcal{M}} \in \mathcal{B}[\mathcal{M}]$ is a weighted sum of projections. Apply Proposition 6.36.

(2) Use Example 5J and Problem 5.10 to identify the topological sum $\mathcal{M}$ with the orthogonal direct sum $\bigoplus_{\lambda \in \sigma_P(T)} \mathcal{N}(\lambda I - T)$. Since each $\mathcal{N}(\lambda I - T)$ reduces $T$ (Problem 6.21), it follows that $\mathcal{M}$ reduces $T$, and also that each $\mathcal{N}(\lambda I - T)$ reduces $T|_{\mathcal{M}} \in \mathcal{B}[\mathcal{M}]$. Therefore, $T|_{\mathcal{M}} = \bigoplus_{\lambda \in \sigma_P(T)} T|_{\mathcal{N}(\lambda I - T)}$. But each $T|_{\mathcal{N}(\lambda I - T)}$ is normal (in fact, a scalar operator — Problem 6.12), which implies that $T|_{\mathcal{M}}$ is normal (actually, a weighted sum of projections).

**Problem 6.23.** Every compact hyponormal operator is normal.

*Hint*: Let $T \in \mathcal{B}[\mathcal{H}]$ be a compact hyponormal operator on a Hilbert space $\mathcal{H}$ and consider the subspace $\mathcal{M}$ of Problem 6.22. Show that $\sigma_P(T|_{\mathcal{M}^\perp}) = \varnothing$: if $\lambda \in \sigma_P(T|_{\mathcal{M}^\perp})$, then there exists a *nonzero* vector $v \in \mathcal{M}^\perp$ such that $\lambda v = T|_{\mathcal{M}^\perp} v = Tv$, and hence $v \in \mathcal{N}(\lambda I - T) \subseteq \mathcal{M}$, which is a contradiction. Now recall that $T|_{\mathcal{M}^\perp}$ is compact (Section 4.9) and hyponormal (Problem 6.16). Use Corollary 6.32 to conclude that $\mathcal{M}^\perp = \varnothing$. Apply Problem 6.22 to show that $T$ is normal.

Remark: According to the above result, on a finite-dimensional Hilbert space, quasi-normality, subnormality and hyponormality all collapse to normality (and so isometries become unitaries — see Problem 4.38(d)).

**Problem 6.24.** Let $T \in \mathcal{B}[\mathcal{H}]$ be a weighted sum of projections on a complex Hilbert space $\mathcal{H} \neq \{0\}$. That is,

$$Tx = \sum_\gamma \lambda_\gamma P_\gamma x \quad \text{for every} \quad x \in \mathcal{H},$$

where $\{P_\gamma\}$ is a resolution of the identity on $\mathcal{H}$ (with $P_\gamma \neq O$ for all $\gamma$) and $\{\lambda_\gamma\}$ is a (similarly indexed) bounded family of scalars. Recall from Proposition 6.36 that $T$ is normal. Now prove the following equivalences.

(a) $T$ is unitary $\iff \lambda_\gamma \in \Gamma$ for all $\gamma \iff \sigma(T) \subseteq \Gamma$.

(b) $T$ is self-adjoint $\iff \lambda_\gamma \in \mathbb{R}$ for all $\gamma \iff \sigma(T) \subseteq \mathbb{R}$.

(c) $T$ is nonnegative $\iff \lambda_\gamma \subseteq [0,\infty)$ for all $\gamma \iff \sigma(T) \subseteq [0,\infty)$.

(d) $T$ is positive $\iff \lambda_\gamma \subseteq (0,\infty)$ for all $\gamma$.

(e) $T$ is strictly positive $\iff \lambda_\gamma \subseteq [\alpha,\infty)$ for all $\gamma \iff \sigma(T) \subseteq [\alpha,\infty)$.

(f) $T$ is a projection $\iff \lambda_\gamma \subseteq \{0,1\}$ for all $\gamma \iff \sigma(T) = \sigma_P(T) \subseteq \{0,1\}$.

Note: In part (a), $\Gamma$ denotes the unit circle about the origin of the complex plane. In part (e), $\alpha$ is some positive real number. In part (f), projection means orthogonal projection (Proposition 6.2).

**Problem 6.25.** Let $T$ be an operator on a complex Hilbert space $\mathcal{H} \neq \{0\}$. Show that

(a) $T$ is diagonalizable if and only if $\mathcal{H}$ has an orthogonal basis made up of eigenvectors of $T$.

*Hint*: If $\{e_\gamma\}$ is an orthonormal basis for $\mathcal{H}$, where each $e_\gamma$ is an eigenvector of $T$, then use the Fourier Series Theorem to show that the resolution of the identity on $\mathcal{H}$ of Proposition 5.57 diagonalizes $T$. Conversely if $T$ is diagonalizable, then every nonzero vector in each $\mathcal{R}(P_\gamma)$ is an eigenvector of $T$. Let $B_\gamma$ be an orthonormal basis for the Hilbert space $\mathcal{R}(P_\gamma)$. Since $\left(\sum_\gamma \mathcal{R}(P_\gamma)\right)^- - \mathcal{H}$ (see Theorem 5.59 and Problem 5.10), use Problem 5.11 to verify that $\bigcup_\gamma B_\gamma$ is an orthonormal basis for $\mathcal{H}$ consisting of eigenvectors of $T$.

If there exists an orthonormal basis $\{e_\gamma\}$ for $\mathcal{H}$ and a (similarly indexed) bounded family of scalars $\{\lambda_\gamma\}$ such that $Tx = \sum_\gamma \lambda_\gamma \langle x ; e_\gamma \rangle e_\gamma$ for every $x \in \mathcal{H}$, then we say that $T$ is a *diagonal operator with respect to the basis* $\{e_\gamma\}$ (cf. Problem 5.17). Use part (a) to show that

(b) $T$ is diagonalizable if and only if it is a diagonal operator with respect to some orthonormal basis for $\mathcal{H}$.

Now let $\{e_\gamma\}_{\gamma \in \Gamma}$ be an orthonormal basis for $\mathcal{H}$ and consider the Hilbert space $\ell_\Gamma^2$ of Example 5K. Let $\{\lambda_\gamma\}_{\gamma \in \Gamma}$ be a bounded family of scalars and consider the mapping $D : \ell_\Gamma^2 \to \ell_\Gamma^2$ defined by

$$Dx = \{\lambda_\gamma \xi_\gamma\}_{\gamma \in \Gamma} \quad \text{for every} \quad x = \{\xi_\gamma\}_{\gamma \in \Gamma} \in \ell_\Gamma^2.$$

In fact, $Dx \in \ell_\Gamma^2$ for all $x \in \ell_\Gamma^2$, $D \in \mathcal{B}[\ell_\Gamma^2]$, and $\|D\| = \sup_{\gamma \in \Gamma} |\lambda_\gamma|$ (*hint*: Example 4H). This is called a *diagonal operator* on $\ell_\Gamma^2$. Show that

(c)  $T$ is diagonalizable if and only if it is unitarily equivalent to a diagonal operator.

*Hint*: Let $\{e_\gamma\}_{\gamma\in\Gamma}$ be an orthonormal basis for $\mathcal{H}$ and consider the natural mapping (cf. Theorem 5.48) $U : \mathcal{H} \to \ell_\Gamma^2$ given by

$$U x = \{\langle x \, ; e_\gamma \rangle\}_{\gamma\in\Gamma} \quad \text{for every} \quad x = \sum_{\gamma\in\Gamma} \langle x \, ; e_\gamma \rangle e_\gamma.$$

Verify that $U$ is unitary (i.e., a linear surjective isometry — see the proof of Theorem 5.49), and use part (b) to show that the diagram

$$
\begin{array}{ccc}
\mathcal{H} & \xrightarrow{\ T\ } & \mathcal{H} \\[2pt]
U \downarrow & & \uparrow U^* \\[2pt]
\ell_\Gamma^2 & \xrightarrow{\ D\ } & \ell_\Gamma^2
\end{array}
$$

commutes if and only if $T$ is diagonalizable, where $D$ is a diagonal operator on $\ell_\Gamma^2$.

**Problem 6.26.**  If $T$ is a normal operator, then

(a)  $T$ is unitary if and only if $\sigma(T) \subseteq \Gamma$,

(b)  $T$ is self-adjoint if and only if $\sigma(T) \subseteq \mathbb{R}$,

(c)  $T$ is nonnegative if and only if $\sigma(T) \subseteq [0, \infty)$,

(d)  $T$ is strictly positive if and only if $\sigma(T) \subseteq [\alpha, \infty)$ for some $\alpha > 0$,

(e)  $T$ is an orthogonal projection if and only if $\sigma(T) \subseteq \{0, 1\}$.

*Hint*: Recall that $\Gamma$ denotes the unit circle about the origin of complex plane. Half of this problem was proved in Corollary 6.18. To prove the other half use the Spectral Theorem: $T = \int_{\sigma(T)} \lambda \, d P_\lambda$, $T^* = \int_{\sigma(T)} \overline{\lambda} \, d P_\lambda$, $T^* T = \int_{\sigma(T)} |\lambda|^2 \, d P_\lambda = T T^*$, $\langle T x \, ; x \rangle = \int_{\sigma(T)} \lambda \, d \langle P_\lambda x \, ; x \rangle$ for all $x \in \mathcal{H}$.

**Problem 6.27.**  Let $T \in \mathcal{B}[\mathcal{H}]$ be a hyponormal operator. Prove the following implications.

(a)  If $\sigma(T) \subseteq \Gamma$, then $T$ is unitary.

*Hint*: If the spectrum of $T$ is included in the unit circle $\Gamma$ about the origin, then $0 \in \rho(T)$ so that $T \in \mathcal{G}[\mathcal{H}]$. Moreover, since $T$ is hyponormal, it follows that $\|T\| = r(T) = 1$. Now use Problem 6.7 to check that $T^{-1}$ is hyponormal. Verify that $\|T^{-1}\| = 1$ (recall: $\sigma(T^{-1}) = \sigma(T)^{-1} = \Gamma$) and conclude from Problem 6.8 that $T$ is normal. Finally, apply Problem 6.26(a) to show that $T$ is unitary.

(b) If $\sigma(T) \subseteq \mathbb{R}$, then $T$ is self-adjoint.

> *Hint*: Show that scaling and translation by the identity of any hyponormal operator are again hyponormal (i.e., if $T$ is hyponormal, then $\alpha T$ and $(I - T)$ are hyponormal for every $\alpha \in \mathbb{C}$, and hence $(\lambda I - T)$ is hyponormal for every $\lambda \in \mathbb{C}$ as we saw in the proof of Corollary 6.18). Let $T$ be a hyponormal operator such that $\sigma(T) \subseteq \mathbb{R}$. Suppose $T \neq O$ (otherwise the result is trivial) and set $T' = (2\|T\|)^{-1}T$, which is hyponormal. Verify that $\sigma(T') = (2\|T\|)^{-1}\sigma(T) \subseteq \mathbb{R}$ and $\|T'\| = \frac{1}{2}$. Also verify that $(\frac{i}{2}I - T')$ is a hyponormal operator in $\mathcal{G}[\mathcal{H}]$ and use Problem 6.14 to show that $\|(\frac{i}{2}I - T')^{-1}\| = \frac{1}{2}$. Moreover, $\|\frac{i}{2}I - T'\| \leq \frac{1}{2} + \|T'\| = 1$. Outcome: Both $(\frac{i}{2}I - T')$ and $(\frac{i}{2}I - T')^{-1}$ are contractions. Use Problem 6.8 to conclude that $(\frac{i}{2}I - T')$ is normal. Now undo scaling and translation: for instance, apply Problem 6.2 (with $S = -\frac{i}{2}I$) to verify that $-T'$ is normal, and so is $T$. Finally, show that $T$ is self-adjoint by using Problem 6.26(b).

(c) If $\sigma(T) \subseteq [0, \infty)$, then $T$ is nonnegative.

(d) If $\sigma(T) \subseteq [\alpha, \infty)$ for some $\alpha > 0$, then $T$ is strictly positive.

(e) If $\sigma(T) \subseteq \{0, 1\}$, then $T$ is an orthogonal projection.

*Hint*: Use part (b) and Problem 6.26 to prove (c), (d) and (e).

**Problem 6.28.** (a) An isolated point of the spectrum of a normal operator is an eigenvalue.

*Hint*: Consider the spectral representation $N = \int \lambda \, dP_\lambda$ of a normal operator on a Hilbert space $\mathcal{H}$ and let $\lambda_0$ be an isolated point of $\sigma(N)$. Apply Theorems 6.47 and 6.48 to show that:

(1) $P(\{\lambda_0\}) \neq O$,

(2) $\mathcal{R}(P(\{\lambda_0\})) \neq \{0\}$ reduces $N$,

(3) $N|_{\mathcal{R}(P(\{\lambda_0\}))} = NP(\{\lambda_0\}) = \int \lambda \, \chi_{\{\lambda_0\}} \, dP_\lambda = \lambda_0 P(\{\lambda_0\})$, where $\chi_{\{\lambda_0\}}$ is the characteristic function of $\{\lambda_0\}$, and

(4) $(\lambda_0 I - N)u = \lambda_0 P(\{\lambda_0\})u - N|_{\mathcal{R}(P(\{\lambda_0\}))}u = 0$ for every $u \in \mathcal{R}(P(\{\lambda_0\}))$.

An important result in operator theory is the Riesz Decomposition Theorem, which reads as follows. *If $T$ is an operator on a complex Hilbert space, and if $\sigma(T) = \sigma_1 \cup \sigma_2$, where $\sigma_1$ and $\sigma_2$ are disjoint nonempty closed sets in $\mathbb{C}$, then $T$ has a complementary (not necessarily orthogonal) pair of nontrivial invariant subspaces $\{\mathcal{M}_1, \mathcal{M}_2\}$ such that $\sigma(T|_{\mathcal{M}_1}) = \sigma_1$ and $\sigma(T|_{\mathcal{M}_2}) = \sigma_2$.* Now prove the following assertion.

(b) An isolated point of the spectrum of a hyponormal operator is an eigenvalue.

*Hint*: Let $\lambda_1$ be an isolated point of the spectrum $\sigma(T)$ of a hyponormal operator $T \in \mathcal{B}[\mathcal{H}]$. Verify that $\sigma(T) = \{\lambda_1\} \cup \sigma_2$ for some nonempty closed set $\sigma_2$ that does not contain $\lambda_1$. Apply the Riesz Decomposition Theorem to ensure that $T$ has a nontrivial invariant subspace $\mathcal{M}$ such that $\sigma(T|_\mathcal{M}) = \{\lambda_1\}$. Set $H = T|_\mathcal{M}$ on $\mathcal{M} \neq \{0\}$. Show that $(\lambda_1 I - H)$ is a hyponormal (thus normaloid) operator for which $\sigma(\lambda_1 I - H) = \{0\}$, and conclude that $T|_\mathcal{M} = H = \lambda_1 I$ in $\mathcal{B}[\mathcal{M}]$.

(c)  A pure hyponormal operator has no isolated point in its spectrum.

*Hint*: Problem 6.21.

**Problem 6.29.**  Let $S$ and $T$ be normal operators acting on the same Hilbert space. Prove the following assertion.

$$\text{If } ST = TS, \text{ then } S + T, \ TS \text{ and } ST \text{ are normal operators.}$$

*Hint*: Corollary 6.49 and Problem 6.2.

**Problem 6.30.**  The operators in this problem act on a complex Hilbert space of dimension greater than one. Recall from Problem 4.22:

(a)  Every nilpotent operator has a nontrivial invariant subspace.

It is an open question whether every quasinilpotent operator has a nontrivial invariant subspace. Shift from nilpotent to normal operators, and recall from the Spectral Theorem:

(b)  Every normal operator has a nontrivial invariant subspace.

Now prove the following propositions.

(c)  Every quasinormal operator has a nontrivial invariant subspace.

   *Hint*: $(T^*T - TT^*)T = O$. Use Problem 4.21.

(d)  Every isometry has a nontrivial invariant subspace.

*Every subnormal operator has a nontrivial invariant subspace*. This is a deep result proved by S. Brown in 1978. However, it is still unknown whether every hyponormal operator has a nontrivial invariant subspace.

# References


N.I. AKHIEZER AND I.M. GLAZMAN
- [1] *Theory of Linear Operators in Hilbert Space – Volume I* (Pitman, London, 1981).
- [2] *Theory of Linear Operators in Hilbert Space – Volume II* (Pitman, London, 1981).

W. ARVESON
- [1] *An Invitation to $C^*$-Algebras* (Springer, New York, 1976).

G. BACHMAN AND L. NARICI
- [1] *Functional Analysis* (Academic Press, New York, 1966).

A.V. BALAKRISHNAN
- [1] *Applied Functional Analysis* 2nd edn. (Springer, New York, 1980).

S. BANACH
- [1] *Theory of Linear Operations* (North-Holland, Amsterdam, 1987).

R. BEALS
- [1] *Topics in Operator Theory* (The University of Chicago Press, Chicago, 1971).

B. BEAUZAMY
- [1] *Introduction to Operator Theory and Invariant Subspaces* (North-Holland, Amsterdam, 1988).

S.K. BERBERIAN
- [1] *Notes on Spectral Theory* (Van Nostrand, New York, 1966).

[2] *Lectures in Functional Analysis and Operator Theory* (Springer, New York, 1974).

[3] *Introduction to Hilbert Space* 2nd edn. (Chelsea, New York, 1976).

Y.M. BEREZANSKY, Z.G. SHEFTEL AND G.F. US

[1] *Functional Analysis – Volume I* (Birkhäuser, Basel, 1996).

[2] *Functional Analysis – Volume II* (Birkhäuser, Basel, 1996).

K.G. BINMORE

[1] *The Foundations of Analysis – A Straightforward Introduction – Book I: Logic, Sets and Numbers* (Cambridge University Press, Cambridge, 1980).

A. BROWN AND C. PEARCY

[1] *Introduction to Operator Theory I – Elements of Functional Analysis* (Springer, New York, 1977).

[2] *An Introduction to Analysis* (Springer, New York, 1995).

S.W. BROWN

[1] Some invariant subspaces for subnormal operators, *Integral Equations Operator Theory* **1** (1978) 310–333.

G. CANTOR

[1] Ein Beitrag zur Mannigfaltigkeitslehre, *J. für Math.* **84** (1878) 242–258.

K. CLANCEY

[1] *Seminormal Operators* (Springer, Berlin, 1979).

I. COLOJOARĂ AND C. FOIAŞ

[1] *Theory of Generalized Spectral Operators* (Gordon and Breach, New York, 1968).

J.B. CONWAY

[1] *A Course in Functional Analysis* 2nd edn. (Springer, New York, 1990).

[2] *The Theory of Subnormal Operators* (Mathematical Surveys and Monographs Vol.36, Amer. Math. Soc., Providence, 1991).

[3] *A Course in Operator Theory* (Graduate Studies in Mathematics Vol.21, Amer. Math. Soc., Providence, 2000).

J.N. CROSSLEY et al.

[1] *What is Mathematical Logic* (Oxford University Press, Oxford, 1972).

P. COHEN

[1] The independence of the continuum hypothesis, *Proc. Nat. Acad. Sci.* **50** (1963) 1143–1148.

K.R. DAVIDSON

[1] *C*-Algebras by Example* (Fields Institute Monographs Vol.6, Amer. Math. Soc., Providence, 1996).

J. Dieudonné

[1] *Foundations of Modern Analysis* (Academic Press, New York, 1969).

R.G. Douglas

[1] *Banach Algebra Techniques in Operator Theory* (Academic Press, New York, 1972; 2nd edn. Springer, New York, 1998).

H.R. Dowson

[1] *Spectral Theory of Linear Operators* (Academic Press, New York, 1978).

J. Dugundji

[1] *Topology* (Allyn & Bacon, Boston, 1960).

N. Dunford and J.T. Schwartz

[1] *Linear Operators – Part I: General Theory* (Interscience, New York, 1958).

[2] *Linear Operators – Part II: Spectral Theory – Self Adjoint Operators in Hilbert Space* (Interscience, New York, 1963).

[3] *Linear Operators – Part III: Spectral Operators* (Interscience, New York, 1971).

A. Dvoretzky and C.A. Rogers

[1] Absolute and unconditional convergence in normed linear spaces, *Proc. Nat. Acad. Sci.* **36** (1950) 192–197.

P. Enflo

[1] A counterexample to the approximation problem in Banach spaces, *Acta Math.* **130** (1973) 309–317.

[2] On the invariant subspace problem for Banach spaces, *Acta Math.* **158** (1987) 213–313.

S. Feferman

[1] Some applications of the notion of forcing and generic sets, *Fund. Math.* **56** (1965) 325–345.

P.A. Fillmore

[1] *Notes in Operator Theory* (Van Nostrand, New York, 1970).

[2] *A User's Guide to Operator Algebras* (Wiley, New York, 1996).

A.A. Fraenkel, Y. Bar-Hillel and A. Levy

[1] *Foundations of Set Theory* 2nd edn. (North-Holland, Amsterdam, 1973).

K. Gödel

[1] Consistence-proof for the generalized continuum-hypothesis, *Proc. Nat. Acad. Sci.* **25** (1939) 220–224.

C. Goffman and G. Pedrick

[1] *A First Course in Functional Analysis* 2nd edn. (Chelsea, New York, 1983).

I.C. Gohberg and M.G. Kreĭn

[1] *Introduction to Nonselfadjoint Operators* (Translations of Mathematical Monographs Vol.18, Amer. Math. Soc., Providence, 1969).

S. GOLDBERG

[1] *Unbounded Linear Operators* (Dover, New York, 1985).

P.R. HALMOS

[1] *Introduction to Hilbert Space and the Theory of Spectral Multiplicity* 2nd edn. (Chelsea, New York, 1957; reprinted: AMS Chelsea, Providence, 1998).

[2] *Finite-Dimensional Vector Spaces* (Van Nostrand, New York, 1958; reprinted: Springer, New York, 1974).

[3] *Naive Set Theory* (Van Nostrand, New York, 1960; reprinted: Springer, New York, 1974).

[4] *A Hilbert Space Problem Book* (Van Nostrand, New York, 1967; 2nd edn. Springer, New York, 1982).

G. HELMBERG

[1] *Introduction to Spectral Theory in Hilbert Space* (North-Holland, Amsterdam, 1969).

I.N. HERSTEIN

[1] *Topics in Algebra* (Xerox, Lexington, 1964).

E. HILLE AND R.S. PHILLIPS

[1] *Functional Analysis and Semi-Groups* (Colloquium Publications Vol.31, Amer. Math. Soc., Providence, 1957).

V.I. ISTRĂŢESCU

[1] *Introduction to Linear Operator Theory* (Marcel Dekker, New York, 1981).

T. KATO

[1] *Perturbation Theory for Linear Operators* 2nd edn. (Springer, Berlin, 1980).

L.V. KANTOROVICH AND G.P. AKILOV

[1] *Functional Analysis* 2nd edn. (Pergamon Press, Oxford, 1982).

J.L. KELLEY

[1] *General Topology* (Van Nostrand, New York, 1955; reprinted: Springer, New York, 1975).

A.N. KOLMOGOROV AND S.V. FOMIN

[1] *Introductory Real Analysis* (Prentice-Hall, Englewood Cliffs, 1970).

E. KREYSZIG

[1] *Introduction to Functional Analysis with Applications* (Wiley, New York, 1978).

C.S. KUBRUSLY

[1] *An Introduction to Models and Decompositions in Operator Theory* (Birkhäuser, Boston, 1997).

V.I. LOMONOSOV

[1] Invariant subspaces for the family of operators which commute with a completely continuous operator, *Functional Anal. Appl.* **7** (1973) 213–214.

S. MacLane and G. Birkhoff
  [1]  *Algebra* (Macmillan, New York, 1967).

M. Martin and M. Putinar
  [1]  *Lectures on Hyponormal Operators* (Birkhäuser, Basel, 1989).

I.J. Maddox
  [1]  *Elements of Functional Analysis* 2nd edn. (Cambridge University Press, Cambridge, 1988).

G.H. Moore
  [1]  *Zermelo's Axiom of Choice* (Springer, New York, 1982).

G. Murphy
  [1]  *C*-Algebras and Operator Theory* (Academic Press, San Diego, 1990).

A.W. Naylor and G.R. Sell
  [1]  *Linear Operator Theory in Engineering and Science* (Rolt, Rinehart & Winston, New York, 1971; reprinted: Springer, New York, 1982).

C.M. Pearcy
  [1]  *Some Recent Developments in Operator Theory* (CBMS Regional Conference Series in Mathematics No.36, Amer. Math. Soc., Providence, 1978).
  [2]  *Topics in Operator Theory* (Mathematical Surveys No.13, Amer. Math. Soc., Providence, 2nd pr. 1979).

C.R. Putnam
  [1]  *Commutation Properties of Hilbert Space Operators and Related Topics* (Springer, Berlin, 1967).

H. Radjavi and P. Rosenthal
  [1]  *Invariant Subspaces* (Springer, New York, 1973).

C.J. Read
  [1]  A solution to the invariant subspace problem, *Bull. London Math. Soc.* **16** (1984) 337–401.

M. Reed and B. Simon
  [1]  *Methods of Modern Mathematical Physics I: Functional Analysis* 2nd edn. (Academic Press, New York, 1980).

F. Riesz and B. Sz-Nagy
  [1]  *Functional Analysis* (Frederick Ungar, New York, 1955).

A.P. Robertson and W.I. Robertson
  [1]  *Topological Vector Spaces* 2nd edn. (Cambridge University Press, Cambridge, 1973).

S. Roman
  [1]  *Advanced Linear Algebra* (Springer, New York, 1992).

H.L. ROYDEN
  [1] *Real Analysis* 3rd edn. (Macmillan, New York, 1988).

W. RUDIN
  [1] *Functional Analysis* 2nd edn. (McGraw-Hill, New York, 1991).

R. SCHATTEN
  [1] *Norm Ideals of Completely Continuous Operators* (Springer, Berlin, 1970).

L. SCHWARTZ
  [1] *Analyse – Topologie Générale et Analyse Fonctionnelle* 2ème édn. (Hermann, Paris, 1970).

W. SIERPINSKI
  [1] L'hypothese généralisée du continu et l'axiome du choix, *Fund. Math.* **34** (1947) 1–5.

G.F. SIMMONS
  [1] *Introduction to Topology and Modern Analysis* (McGraw-Hill, New York, 1963).

D.R. SMART
  [1] *Fixed Point Theorems* (Cambridge University Press, Cambridge, 1974).

M.H. STONE
  [1] *Linear Transformations in Hilbert Space* (Colloquium Publications Vol.15, Amer. Math. Soc., Providence, 1932).

V.S. SUNDER
  [1] *Functional Analysis – Spectral Theory* (Birkhäuser, Basel, 1998).

P. SUPPES
  [1] *Axiomatic Set Theory* (Dover, New York, 1963).

W.A. SUTHERLAND
  [1] *Introduction to Metric and Topological Spaces* (Oxford University Press, Oxford, 1975).

B. SZ.-NAGY AND C. FOIAŞ
  [1] *Harmonic Analysis of Operators on Hilbert Space* (North-Holland, Amsterdam, 1970).

A.E. TAYLOR AND D.C. LAY
  [1] *Introduction to Functional Analysis* 2nd edn. (Wiley, New York, 1980; enlarged edn. of A.E. TAYLOR, 1958).

R.L. VAUGHT
  [1] *Set Theory – An Introduction* 2nd edn. (Birkhäuser, Boston, 1995).

J. WEIDMANN
  [1] *Linear Operators in Hilbert Spaces* (Springer, New York, 1980).

R.L. WILDER

[1] *Introduction to the Foundation of Mathematics* 2nd edn. (Wiley, New York, 1965; reprinted: Krieger, Malabar, 1983).

D. XIA

[1] *Spectral Theory of Hyponormal Operators* (Birkhäuser, Basel, 1983).

T. YOSHINO

[1] *Introduction to Operator Theory* (Longman, Harlow, 1993).

K. YOSIDA

[1] *Functional Analysis* 6nd edn. (Springer, Berlin, 1981).

# Index